FLUID POWER SYSTEMS

Second Edition

atp AMERICAN TECHNICAL PUBLISHERS
Orland Park, Illinois 60467-5756

Patrick J. Klette

American Technical Publishers Editorial Staff

Editor in Chief:
Jonathan F. Gosse
Vice President—Production:
Peter A. Zurlis
Digital Media Manager:
Carl R. Hansen
Art Manager:
Jennifer M. Hines
Technical Editor:
James T. Gresens
Copy Editors:
Talia J. Lambarki
Dane K. Hamann

Cover Design:
Robert M. McCarthy
Illustration/Layout:
Robert M. McCarthy
Mark S. Maxwell
Nick W. Basham
Digital Media Development:
Robert E. Stickley
Nicole S. Polak
Daniel Kundrat
Hannah A. Swidergal
Kathleen A. Moster

2 3 4 5 6 7 8 9 – 14 – 9 8 7 6 5 4

Printed in the United States of America

ISBN 978-0-8269-3634-9

 This book is printed on recycled paper.

Acknowledgments

Atlas Copco
Atlas Technologies, Inc.
Briggs & Stratton Corporation
Chicago Pneumatic
Clippard Instrument Laboratory, Inc.
CNH America
Curtis-Toledo, Inc.
Enerpac
Flow Ezy Filters, Inc.
Fluke Corporation
The Gates Rubber Company
GOMACO Corporation
Hurst Jaws of Life
Klein Tools, Inc.
Lifting Gear Hire Corporation
The Lincoln Electric Company
Parker Motion & Control
Power Team, Division of SPX Corporation
Prince Manufacturing Corporation
Ridge Tool Company
Rothenberger USA, Inc.
Salisbury
Saylor-Beall Manufacturing Company
Schroeder Industries
SKF Condition Monitoring
The Snell Group
Snorkel
Village of Homewood Fire Department, Homewood, IL
Village of Mokena Fire Department, Mokena, IL
W.W. Grainger, Inc.

Contents

chapter 1 **Fluid Power Systems in Industry** 1

Fluid Power • History of Fluid Power • Future of Fluid Power • Fluid Power System Applications • Advantages and Disadvantages of Fluid Power Applications • Fluid Power Safety • Personal Protective Equipment • Lockouts and Tagouts • Fluid Power Hazards • Fluid Power Safety Training • Fluid Power Associations and Certifications

Chapter Review 15
Activities 21

chapter 2 **Fluid Power System Principles** 25

Energy • Forms of Energy • Fluid Power Energy Transmission • Fluid Power System Efficiency • Work • Power • Fluid Power System Variables • Force • Area • Pressure • Fluid Power Formulas • Pressure Types • Pressure Measurement • Vacuum

Chapter Review 45
Activities 49

chapter 3 **Hydraulic System Fundamentals** 59

Hydraulic Principles • Properties of Liquids • Principles of Fluid Flow • Pressure and Resistance Relationship • Pressure Control • Maximum Resistance • Hydraulic System Pressure Supplements • Accumulators • Force Multiplication Systems • Intensifiers • Hydraulic Power Units • Fluid Power System Diagrams • Pictorial Diagrams • Cutaway Diagrams • Schematic Diagrams

Chapter Review 82
Activities 87

chapter 4 **Fluid Conductors and Connectors** 93

Velocity and Conductors • Velocity • Conductors and Connectors • Piping • Tubing • Hoses • Connector Thread Identification • Thread Types • Cleaning Conductors before Installation

Chapter Review 111
Activities 114

chapter 5 **Hydraulic Pumps** 117

Hydraulic Pumps • Hydraulic Pump Ratings • Positive-Displacement Pumps • Positive-Displacement Pump Operation • Hydraulic Pump Types • Gear Pumps • Gear Pump Assembly • Vane Pumps • Vane Pump Assembly • Piston Pumps • Double and Triple Pumps • Hydraulic Pump Schematic Symbols

Chapter Review 143
Activities 149

chapter 6 **Directional Control** 155

Check Valves • Direct-Acting Check Valves • Pilot-Operated Check Valves • Directional Control Valves • Directional Control Valve Schematic Diagrams • Two-Position, Two-Way Directional Control Valves • Two-Position, Three-Way Directional Control Valves • Two-Position, Four-Way Directional Control Valves • Three-Position, Four-Way Directional Control Valves • Directional Control Valve Mounting Methods • Directional Control Valve Actuators • Spring Actuators • Pilot Pressure Actuators • Solenoid Actuators • Detent Actuators • Proportional Actuators • Logic Valves • AND Logic Valves • Shuttle Valves

Chapter Review 183
Activities 189

chapter 7 **Flow Control** 195

Hydraulic Flow Control • Pressure Differential • Orifices • Flow Control Valves • Piston Speed • Metering Fluid Flow • Metering Single-Acting Cylinders • Metering Double-Acting Cylinders • Bleed-Off Systems • Flow Dividers • Spool-Type Flow Dividers • Rotary-Type Flow Dividers

Chapter Review 217
Activities 223

chapter 8 **Hydraulic Actuators** 229

Hydraulic Cylinders • Ram Cylinders • Single-Acting Cylinders • Double-Acting Cylinders • Seals • Cylinder Mounting Methods • Hydraulic Motors • Motor Torque • Hydraulic Motor Types

Chapter Review 251
Activities 257

chapter 9 **Pressure Control** 265

Direct-Acting Pressure Control Valves • Direct-Acting Relief Valves • Sequence Valves • Counterbalance Valves • Pressure-Reducing Valves • Unloading Valves • Remote-Controlled Sequence Valves • Remote-Controlled Counterbalance Valves • Brake Valves • Pilot-Operated Pressure Control Valves • Pilot-Operated Relief Valves • Pilot-Operated Pressure Reducing Valves • Other Pilot-Operated Pressure Control Valves

Chapter Review 285
Activities 291

chapter 10 **Hydraulic Fluid Maintenance** 297

Contaminants • Sizing Contaminants • Contaminant Sources • Contaminant Effects • Contaminant Protection • Filter and Strainer Ratings • Hydraulic System Filtration • Hydraulic Fluid Cleanliness Ratings • Hydraulic System Heat • Hydraulic System Heat Generation • Heat Exchangers • Reservoirs • Hydraulic Fluid Storage • Breather/Filler Cap Assemblies • Hydraulic Fluid Cooling and Cleaning • Receiving Hydraulic Fluid from the System • Sending Hydraulic Fluid to the Pump • Other Reservoir Components

Chapter Review 321
Activities 329

chapter 11 **Pneumatic System Fundamentals** 335

Basic Gas Physics • Volume • Absolute Pressure • Absolute Temperature • Boyle's Law • Charles's Law • Gay-Lussac's Law • Combined Gas Law • Pneumatic Systems • Compressing Air • Airflow • Cfm Ratings

Chapter Review 349
Activities 355

chapter 12 **Pneumatic System Compression and Control** 359

Pneumatic Systems • Prime Movers • Air Compressor Types • Receivers • Control Devices • Compressor Control • Directional Control Valves • Flow Control Valves • Work Devices • Pneumatic Cylinders • Air Motors • Vacuum Cups

Chapter Review 383
Activities 389

chapter 13 **Pneumatic System Conditioning** 393

Pneumatic System Contaminants • Solid Contaminants • Liquid Contaminants • Pneumatic Filters • Intake Filters • Inline Filters • Pneumatic Conditioning Devices • Intercoolers • Moisture Separators • Air Dryers • Filter-Regulator-Lubricators (FRLs) • Compressed Air Preparation Procedure

Chapter Review 409
Activities 413

chapter 14 **Fluid Power System Electrical Control** 417

Electrical Quantities • Current • Voltage • Resistance • Basic Electrical Circuits • Parallel Circuits • Series Circuits • Series-Parallel Circuits • Electrical Control Circuits • Electrical Control Circuit Components • Electric Diagrams and Applications • Single-Acting Cylinder Control Applications • Double-Acting Cylinder Control Applications

Chapter Review 439
Activities 445

chapter 15 **Fluid Power System Maintenance and Troubleshooting** 453

Fluid Power System Maintenance • Hydraulic System Maintenance • Pneumatic System Maintenance • Fluid Power System Troubleshooting • Troubleshooting Hydraulic Systems • Troubleshooting Pneumatic Systems • Fluid Power System Troubleshooting Methods • Fluid Power System Troubleshooting Procedures

Chapter Review 471
Activities 475

Appendix ______ 483

Glossary ______ 517

Index ______ 529

Interactive DVD Features

- *Using This Interactive DVD*
- *Quick Quizzes™*
- *Illustrated Glossary*
- *Flash Cards*
- *Interactive Schematics*
- *Review Questions*
- *Media Library*
- *ATPeResources.com*

FluidSIM® Hydraulics (Student Version) CD-ROM

Introduction

Fluid Power Systems is a comprehensive text/workbook that covers topics specific to the design, application, and maintenance of hydraulic and pneumatic systems. The text/workbook includes fluid power systems, components, and devices that are related to industrial and commercial applications such as pumps, valves, actuators, electrical controls, and troubleshooting techniques. Each component, device, or system is introduced with descriptions, operational procedures, common applications, system examples, and operating characteristics. Schematic symbols are introduced throughout the text/workbook to assist the learner with schematic diagram comprehension. Detailed illustrations provide examples of each component, device, or system, and its related operation. Chapter objectives at the beginning of each chapter provide learning goals for the topics introduced. Review questions and activities at the end of each chapter provide a variety of assessment opportunities. Relevant chapters include activities related to the FluidSIM® Hydraulics (student version) software included in the Interactive DVD and CD-ROM package. This new edition includes expanded content on hydraulic pumps, fluid conductors, connectors, means of transmission, and troubleshooting procedures.

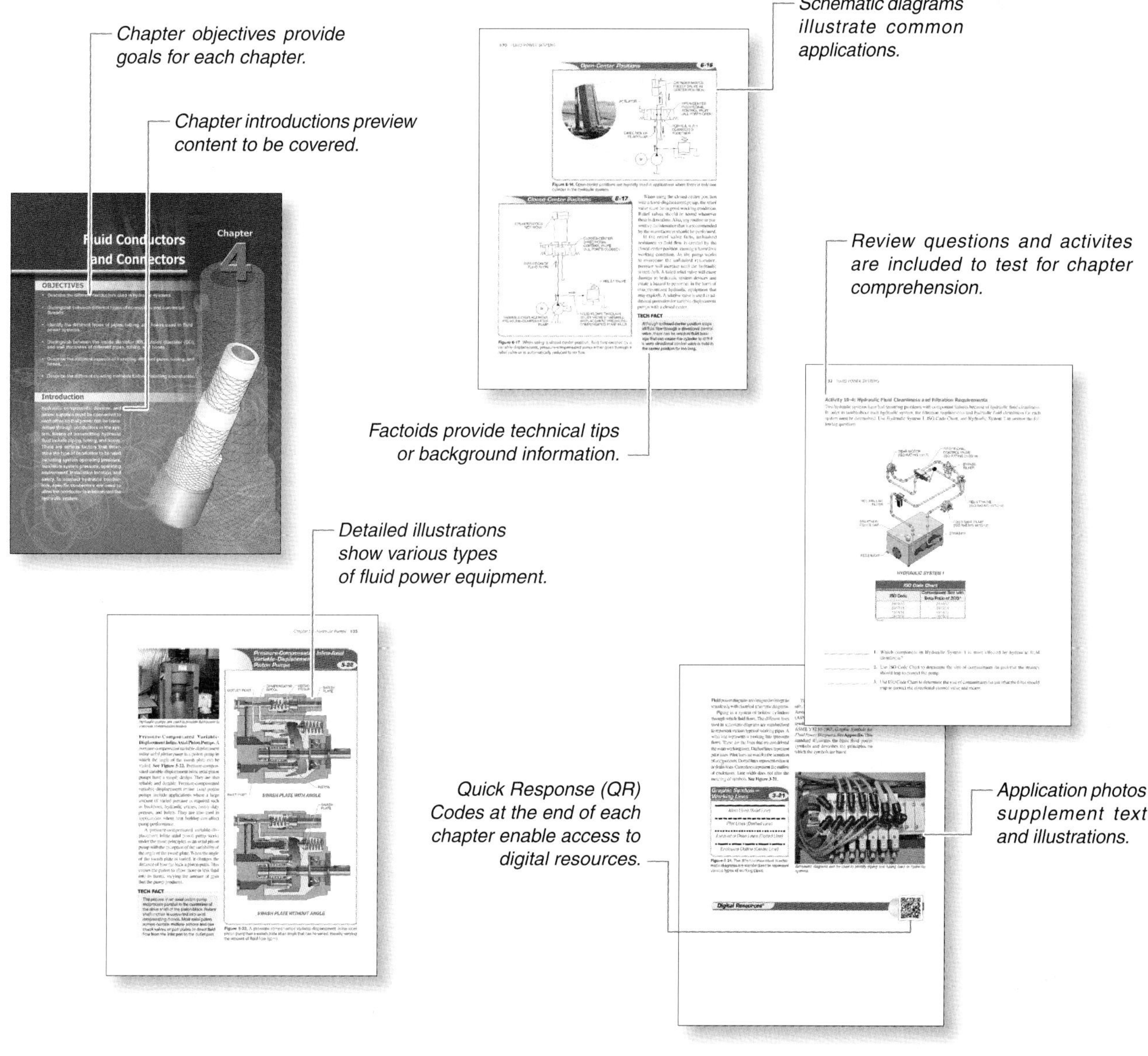

Interactive DVD Features

The Interactive DVD included with this textbook provides an array of learning tools that reinforce and enhance the information detailed in the book. Information about using the *Fluid Power Systems* Interactive DVD is included on the last page.

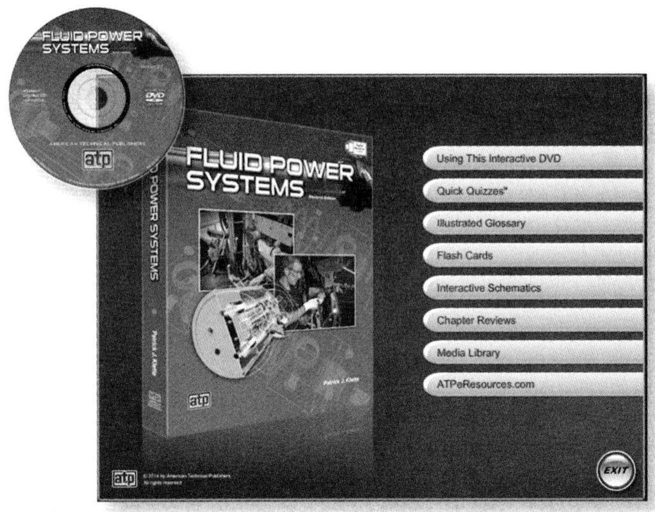

The Interactive DVD is a self-study aid that includes the following:

- Quick Quizzes™ reinforce fundamental concepts, with 10 questions per chapter
- An Illustrated Glossary of industry terms include links to illustrations, video clips, and animated graphics
- Flash Cards enable a review of fluid power terms, definitions, and schematic symbols
- Interactive Schematics provide PDF files that link to photographs and information from parts of specific schematic diagrams
- Chapter Reviews from the end of each chapter are in Microsoft Word format
- Media Library consists of animated illustrations or video clips that expand upon text/workbook content
- ATPeResources.com provides a comprehensive array of instructional resources

To obtain information on related training products, visit the American Technical Publishers website at www.atplearning.com.

The Publisher

Chapter 1

Fluid Power Systems in Industry

OBJECTIVES

- Describe how fluid power is used in modern applications.
- List and describe common safety hazards when working around and on fluid power systems.
- List and describe common fluid power safety practices and related training programs.
- List the different fluid power industry associations and certifications.

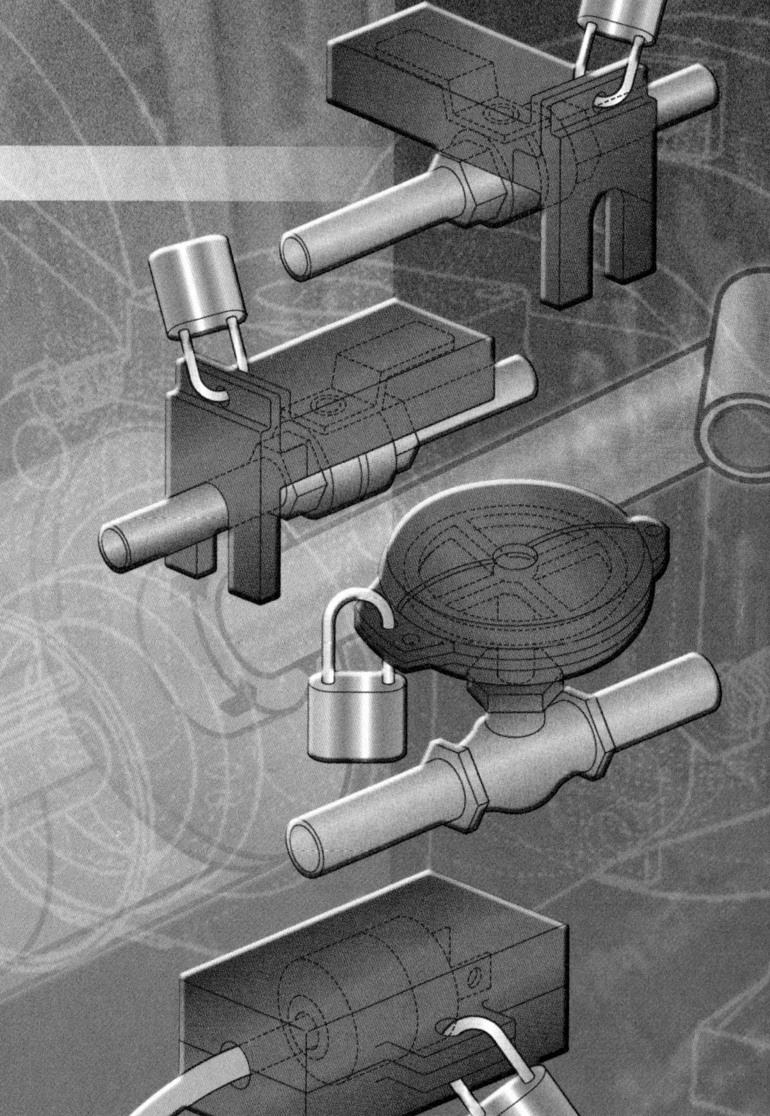

INTRODUCTION

Fluid power systems are used to control and provide power for various industrial and commercial processes and equipment. The fluid used in fluid power systems can be either liquid or gas. For a fluid power system to function, there has to be a power source, a load, and a means of transmission.

Proper safety rules apply when working with any type of fluid power system. Fluid power trade associations certify fluid power mechanics, technicians, and other specialists.

TERMS

Fluid power is the technology of using a fluid to transmit energy from one location to another.

A **fluid** is a liquid or a gas that takes the shape of its container.

A **hydraulic system** is a system that uses liquid under pressure to create movement.

A **pneumatic system** is a system that uses gas under pressure to create movement.

FLUID POWER

Fluid power is the technology of using a fluid to transmit power from one location to another. A *fluid* is a liquid or a gas that takes the shape of its container. Fluids that are used in the fluid power systems are categorized as hydraulic and pneumatic. A *hydraulic system* is a system that uses liquid under pressure to create movement. Typically, the fluid used is petroleum-based oil, but historically water was most commonly used. A *pneumatic system* is a system that uses gas under pressure to create movement. The gas most often used is air. Fluid in a closed fluid power system exerts pressure equally in all directions. **See Figure 1-1.**

History of Fluid Power

Fluid power systems have been used as an energy source for thousands of years. China and Egypt used fluid power as early as 4000 BC to control the flow of water to help irrigate farmland. The Romans and the Greeks built aqueducts to supply water to cities in large quantity. The Egyptians, Babylonians, Indians, and Chinese all used clepsydras, or water clocks, which were devices that could keep time without the use of the sun. The Romans also used fluid power to create the first decorative water fountains.

Much of the science behind fluid power technology was not understood until the 17th century when the French mathematician and physicist Blaise Pascal made the discovery that fluid in a closed system takes the shape of the container that it occupies and exerts pressure equally in all directions. This discovery became known as Pascal's law. Despite this discovery, fluid power technology was not widely used before the Industrial Revolution. For example, one of the first practical applications of Pascal's law was the hydraulic press invented by the English inventor, Joseph Bramah and patented in 1795. The hydraulic press, also known as the Bramah press, was used to shape and form large pieces of metal and is still used in many modern hydraulic applications.

The invention of the steam engine allowed water to be put under pressure and made to produce large amounts of work that humans and other technology could not accomplish efficiently. Fluid power was mainly used to do heavy lifting. During the 1870s, many textile mills in England used water placed under pressure to run textile production machines. However, inventions and innovations in the fields of electrical and mechanical power quickly overtook fluid power as the main technology used in the late 19th century.

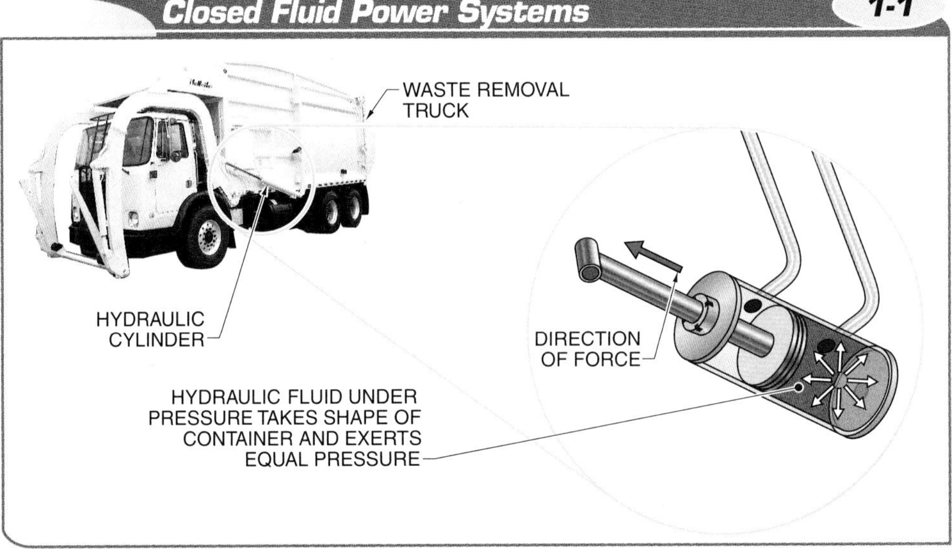

Figure 1-1. Fluid in a closed system takes the shape of the container that it occupies and exerts pressure equally in all directions.

As electricity was increasingly used to perform work that fluid power used to perform, invention and innovation in the fluid power field slowed dramatically. However, the American inventor George Westinghouse, later known for his work with electrical technology, invented automotive air brakes. Automotive air brakes, which helped to build Westinghouse's fortune, were soon being used on trains, buses, automobiles, and trucks.

It was eventually determined that it was easier to use fluid power for different types of movement, rather than electrical or mechanical power. In 1906, the United States military replaced electricity with hydraulics to lift and move the guns on the battleship USS Virginia. By that time, petroleum-based oil had replaced water as the main liquid used in hydraulic systems because it operates at higher pressure, leaks less, and has a much wider range of operating temperatures. Interest in fluid power was renewed.

The first self-contained fluid power system used in aircraft, similar to modern systems, was developed by the United States government in 1926. Soon, many of the control systems on aircraft were hydraulic systems. For example, hydraulic systems were integrated into aircraft produced during the Second World War. Modern fluid power systems have been incorporated into every industry. As with most technologies, as the demand for fluid power increases, fluid power technology becomes more modernized.

Future of Fluid Power

No foreseeable future technology will be able to produce the power, control, or speed that fluid power systems are capable of producing. Instead, technological innovations and advancements for fluid power systems will be made. Although it is impossible to predict the exact future of a technology, there are signs that make it possible to accurately foresee these technological advances.

For example, fluid power systems will become more integrated with electronic control. The integration of fluid power technology with electronics will allow fluid power systems to become more energy efficient and allow more precise control of individual systems. It will also allow fluid power systems to be integrated into more applications and make them more cost effective.

Another advancement will be for more environmentally friendly fluid power equipment, devices, and systems. Currently, some hydraulic equipment uses water-based hydraulic fluid rather than oil-based hydraulic fluid. There are advantages and disadvantages to using water-based hydraulic fluid rather than oil-based hydraulic fluid. Advantages of using water-based hydraulic fluid include good heat dissipation, good product availability, lower viscosity than oil, and that it is less hazardous to the environment. The lower viscosity makes water-based fluid flow easier than oil-based fluid. This gives it many advantages in a fluid power system. The main disadvantage of using water-based hydraulic fluid is that the internal sealing methods for water-based hydraulic equipment have not been developed for pressure over 2500 psi. Also, water-based hydraulic fluid does not lubricate internal moving parts as well as oil-based hydraulic fluid, which can cause internal heat buildup in the equipment and parts to wear faster.

Equipment with these types of capabilities will be used more often. Research also continues on innovations that will decrease fluid leakage problems for both hydraulic and pneumatic systems.

Additional innovations include the development of bio-hydraulic fluid to help minimize the effects of hydraulic systems on the environment. *Bio-hydraulic fluid* is environmentally nonhazardous hydraulic fluid that is composed of synthetic chemicals and vegetable-based oil to lower the hazardous effects from leaks and spills. The use of bio-hydraulic fluid can also increase the efficiency of fluid power systems by requiring less energy to produce the same amount of power.

TERMS

Bio-hydraulic fluid is environmentally nonhazardous hydraulic fluid that is composed of synthetic chemicals and vegetable-based oil to lower the hazardous effects from leaks and spills.

Fluid power systems will also become more powerful and decrease in size. For example, hydraulic systems are being designed to produce 30,000 psi, compared to a current maximum of 12,000 psi. Fluid power equipment will also be increasingly integrated with renewable energy systems. For example, fluid power systems are currently used with many renewable systems, such as large wind generators, hydroelectric plants, and hybrid systems on commercial vehicles, such as waste-hauling trucks and road construction equipment. The demand for fluid power equipment in renewable energy will increase as new types of renewable energy sources are developed, such as power-generating ocean buoys and large-scale solar power plants.

FLUID POWER SYSTEM APPLICATIONS

Fluid power is a system used to transmit power. Fluid power systems are used to move large loads and must be able to withstand high pressure because of the large amounts of resistance. The two types of fluid power systems used in industrial applications are hydraulic systems and pneumatic systems. Hydraulic system applications consist of equipment used to move medium to heavy loads, while pneumatic system applications consist of equipment used to move light to medium loads.

Hydraulic systems are commonly used in amusement park equipment.

Modern fluid power systems are present in many different applications. For example, elevators, hydraulic cranes, aerial lift systems, industrial robotic systems, construction equipment, and forestry equipment all use hydraulic fluid power systems to transmit power. **See Figure 1-2.**

Fluid power systems are a vital part of every industry and many types of equipment, including agriculture, mining, food processing, recycling and waste management, aerospace, robotics, construction, power generation, steel and metals processing, paper mills, heavy duty off-road vehicles, plastics processing, wind turbines, solar energy plants, chemical refineries, ship building, die casting, forging, medical instrumentation, and transportation. Most manufacturing facilities use fluid power systems to produce, package, and/or ship their product.

Fluid power is used to perform work that cannot be physically performed by humans. For example, a building can be demolished and rebuilt in a matter of a several months or bridges can be built that connect distant landmasses. Lifting and suspending hundreds of pounds is easily accomplished with fluid power. For example, a car can be lifted and held in place with a hydraulic jack so that the lower portion of the car can be accessed. Also, by using fluid power systems, a large amount of work can be accomplished in the most productive manner with a small amount of energy. For example, two tons of dirt from the back of a truck bed can be moved in a matter of minutes. It would require several hours of manual labor to move the same amount of material.

TECH FACT

The Joint Industry Conference (JIC) developed the first hydraulic standard symbols in 1948 and the first pneumatic standard symbols in 1950. These symbols were based on patterns from electrical symbols.

Figure 1-2. Modern fluid power systems are present in applications such as elevators, hydraulic cranes, aerial lifts, industrial robots, construction equipment, and forestry equipment.

TERMS

Personal protective equipment (PPE) is clothing and/or equipment worn by a worker to reduce the possibility of an injury.

Advantages and Disadvantages of Fluid Power Applications

There are several advantages and disadvantages to using fluid power technology. For some applications, fluid power technology is the only technology capable of completing the work that needs to be performed. Other times, a choice must be made about which technology would be best suited for a specific application. For example, a conveyor line that requires rotational force can be powered by electric, hydraulic, or pneumatic sources. If there is a choice between using different power sources, the advantages and disadvantages of each must be considered.

The advantages of hydraulic and pneumatic systems include the following:

- Large amounts of force can be created to move extremely heavy loads with hydraulic power.
- Large amounts of torque can be developed using hydraulic power (even at low speeds), while extremely fast speeds can be developed using pneumatic power.
- Excellent speed and accuracy control for linear movements is possible.
- They can be used in a wide range of environments, including extreme cold, extreme heat, and dirty or hazardous locations.
- Manual, electrical, or automatic control is possible.
- They have constant speed, force, and torque.
- They are composed of simple mechanical systems.
- Standardized schematic symbols are used for system components.

The disadvantages of hydraulic and pneumatic systems include the following:

- The systems can operate at extremely high pressures, which can cause many safety hazards.
- Fluid power systems have high costs associated with components, hydraulic fluid (purchase and disposal), generated energy, and maintenance requirements.
- They have high noise levels.
- Hydraulic fluid or pneumatic lubricating oil spills can create environmental and slipping hazards.

FLUID POWER SAFETY

Many safety concerns arise during the operation of fluid power systems. The high pressure used to accomplish work in fluid power systems creates conditions for accidents that can cause injury or death. Technicians that work near or with fluid power systems and related equipment must be trained in common fluid power safety practices and personal protection. To work safely, technicians must be aware of fluid power hazards associated with each piece of equipment and lock out and tag out equipment as required.

Personal Protective Equipment

Personal protective equipment (PPE) is clothing and/or equipment worn by a worker to reduce the possibility of an injury. Fluid power technicians must wear appropriate PPE when working on a fluid power system. All PPE must meet the Occupational Safety and Health Administration (OSHA) 29 Code of Federal Regulations (CFR) 1910 Subpart I – *Personal Protective Equipment* standards. PPE includes head protection, eye protection, arm and hand protection, protective clothing, hearing protection, and foot protection. Special working conditions may require additional forms of protection.

Head Protection. Protective helmets, or hard hats, are used in the workplace to prevent injury from impact, falling and flying objects, and electrical shock. Protective helmets protect workers by resisting penetration and absorbing the blow of an impact. The shell of the protective helmet is made of a durable, lightweight material. A shock-absorbing lining consisting of crown straps and an adjustable headband keeps the shell of the protective helmet away from the head and allows ventilation. **See Figure 1-3.**

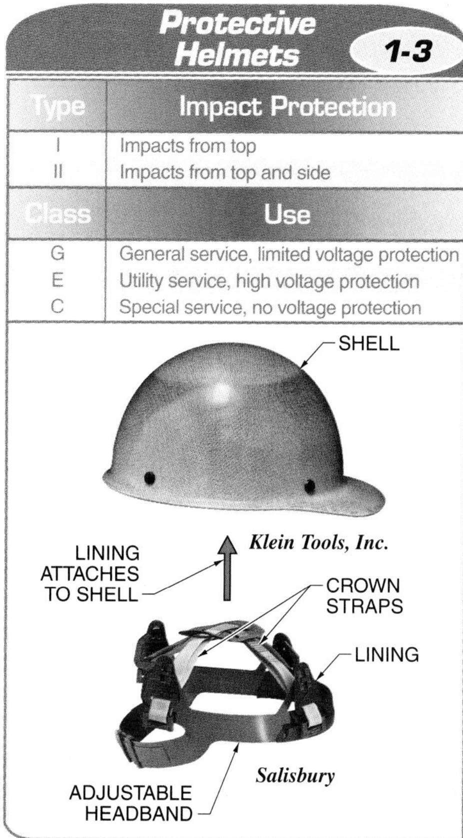

Protective Helmets 1-3

Type	Impact Protection
I	Impacts from top
II	Impacts from top and side
Class	**Use**
G	General service, limited voltage protection
E	Utility service, high voltage protection
C	Special service, no voltage protection

Figure 1-3. Protective helmets are used in the workplace to prevent injury from impact, falling and flying objects, and electrical shock.

Standards for protective helmets are specified in ANSI Z89.1, *Industrial Head Protection.* Protective helmets are identified by type and class for protection against specific hazardous conditions. Type I helmets protect against impacts at the top of the head. Type II helmets protect against top and side impacts. Type II helmet designs often include wide brims all around the helmet to help deflect side impacts.

Electrical protection for helmets is classified as Class G, Class E, and Class C. Class G (general) protective helmets protect against impact hazards and contact with voltages up to 2200 V. Class G helmets are the most common helmets used in manufacturing, mining, and construction. Class E (electrical) protective helmets protect against impact hazards and contact with high voltages up to 20,000 V. Class E protective helmets are used by electrical workers and maintenance technicians subject to electrical hazards. Class C (conductive) protective helmets are designed for impact protection only and should not be used when there is danger from electrical hazards.

Hard hats must be inspected daily for any damage. When damage such as scrapes, cracks, dents, and gouges are found, the hard hat must be replaced. Although there is no OSHA regulation on hard hat expiration dates, a hard hat should only be used up to five years from the manufactured date stamped on the inside of the hard hat. If the hard hat is used outdoors or in environments where high temperatures or chemicals are present, the hard hat should be replaced every two years. It is also recommended that the suspension inside of the hard hat be replaced every 12 months.

Eye and Face Protection. Proper eye and face protection must always be worn when working on fluid power systems. Eye and face protection prevents injury from hydraulic fluid and other debris that may spray from a broken hose. OSHA requires eye and face protection when there is a reasonable probability of preventing injury to the eyes or face from flying particles, molten metal, liquid chemicals, chemical gases, radiant energy, or any combination of these hazards. Eye and face protection includes safety glasses, face shields, and goggles. **See Figure 1-4.**

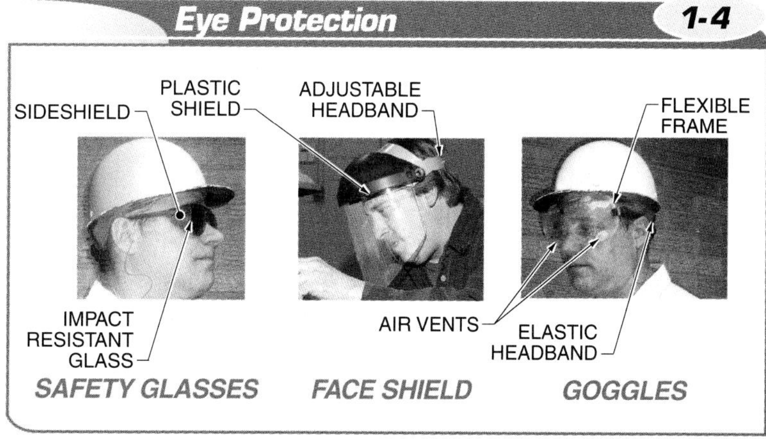

Figure 1-4. Eye protection must be worn to prevent injury to the eyes or face from flying particles, molten metal, liquid chemicals, chemical gases, radiant energy, or any combination of these hazards.

TERMS

Safety glasses are an eye protection device with special impact-resistant glass or plastic lenses, reinforced frames, and side shields.

A **face shield** is an eye and face protection device that covers the entire face with a plastic shield and is used for protection from flying objects.

Safety glasses are an eye protection device with special impact-resistant glass or plastic lenses, reinforced frames, and side shields. Plastic frames are designed to keep the lenses secured in the frame if an impact occurs and minimize the shock hazard when working with electrical equipment. Side shields provide additional protection from flying objects. Tinted-lens safety glasses protect against low-voltage arc hazards. Safety glasses standards are set by ANSI Z87.1-2010, *Certified Safety Glasses*. They are rated as either nonimpact or impact. Impact-rated safety glasses must pass a high impact test. High-impact safety glasses are marked "Z87+". Safety glasses that are rated as nonimpact are not tested and are marked "Z87".

A *face shield* is an eye and face protection device that covers the entire face with a plastic shield and is used for protection from flying objects. Goggles fit snugly against the face to seal the areas around the eyes and can be used over prescription glasses. Goggles with clear lenses protect against small flying particles or splashing liquids.

Safety glasses, face shields, and goggle lenses must be properly maintained to provide protection and clear visibility. Lens cleaners are available that clean without the risk of lens damage. Pitted or scratched lenses reduce vision and can cause lenses to fail on impact. Side shields must be provided on all eye protection. The eye protection required varies with the task performed. Eye protection used for extended periods should be cleaned and disinfected regularly.

Persons wearing corrective glasses must wear face shields or goggles, or have glasses made from impact-resistance lens material with side shields. Eye protection must comply with OSHA 29 CFR 1910.133, *Eye and Face Protection*. Standards for eye protection are specified in ANSI Z87.1-2010, *American National Standard for Occupational and Education Eye and Face Protection.*

Eyewash stations must be provided to allow the quick flushing from eyes of contaminants from accidental exposure. Per ANSI Z358.1-2004, *Emergency Eyewash and Shower Equipment,* specifies that an emergency eyewash station must be located within 10 sec or about 55′ of hazards. **See Figure 1-5.**

Arm and Hand Protection. Arm and hand protection is required to prevent injuries to the arms and hands from burns, cuts, electrical shock, amputation, and absorption of chemicals. A wide assortment of gloves, hand pads, sleeves, and wristlets are available to provide protection against various hazards. The work activities of fluid power technicians determine the degree of dexterity required and the duration, frequency, and degree of exposure to potential hazards.

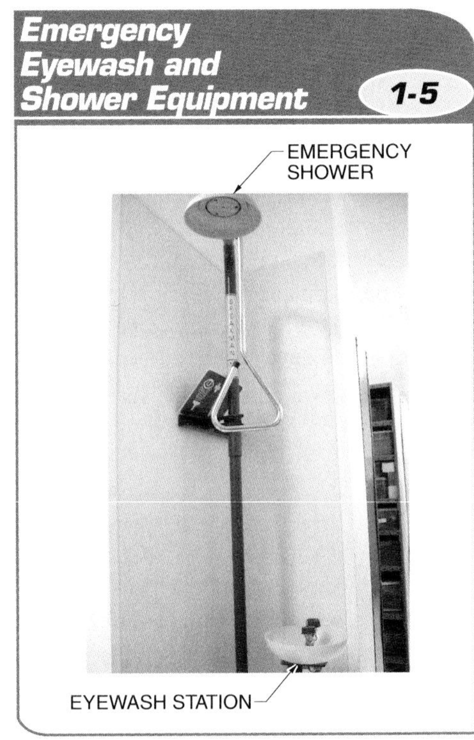

Figure 1-5. Eyewash stations allow the quick flushing of contaminants from eyes after accidental exposure.

Selection of gloves is based on required protection and glove test data. Documentation from the manufacturer verifies that gloves meet standards for safety and protection, such as OSHA 29 CFR 1910.138, *Hand Protection*. ANSI and the International Safety Equipment Association (ISEA) collaborated to develop another standard, ANSI/ISEA 105-2005, *Hand Protection Selection Criteria*. For example, protection against chemicals measures the ability of the gloves to restrict the passing of chemicals through to the skin. Other gloves made from wire mesh, leather, and canvas provide insulation from cuts and burns.

Gloves are generally sized and can be loose-fitting or snug-fitting. Loose-fitting gloves are easier to put on and take off. However, gloves that are too large may pose a safety hazard when working around machinery. Snug-fitting gloves provide better sensitivity and control than loose-fitting gloves.

Certain tasks require the use of specific types of gloves. For example, neoprene gloves provide hand protection from hydraulic oil and a variety of chemicals and acids. Disposable gloves can be discarded after use. **See Figure 1-6.**

Protective Clothing. Protective clothing is clothing that provides protection from contact with sharp objects, hot equipment, and harmful materials such as hydraulic fluid. Durable materials such as denim should be worn and the fit should be snug, yet allow ample movement. Clothing made of synthetic materials such as nylon, polyester, or rayon should not be worn because these materials are flammable and can melt to the skin.

Hearing Protection. Exposure to high noise levels from fluid power equipment can cause hearing loss or impairment. A hearing conservation program is outlined in OSHA 29 CFR 1910.95, *Occupational Noise Exposure*. The hearing conservation program includes monitoring the facility, the proper fitting of hearing protection devices, training, and worker notification. The program is required whenever workers are exposed to noise equal to or exceeding an 8 hr time-weighted average (TWA) of 85 decibels (dB). This level is equivalent to the sound level of busy street traffic. However, since the requirement evaluates the effects of sound over time, workers may be permissibly exposed to higher average sound levels when it is for shorter periods of time.

Hearing protection devices include earplugs and earmuffs. An *earplug* is a moldable device inserted into the ear canal to reduce the level of noise reaching the eardrum. An *earmuff* is an ear protection device worn over the ears to reduce the level of noise reaching the eardrum. A tight seal around the earmuff is required for proper protection. Earmuffs are often attached to a protective helmet for complete head protection near fluid power equipment. **See Figure 1-7.**

TERMS

An **earplug** is a moldable device inserted into the ear canal to reduce the level of noise reaching the eardrum.

An **earmuff** is an ear protection device worn over the ears to reduce the level of noise reaching the eardrum.

Hand Protection 1-6

Figure 1-6. Neoprene gloves can provide hand protection from hydraulic oil and a variety of chemicals and acids.

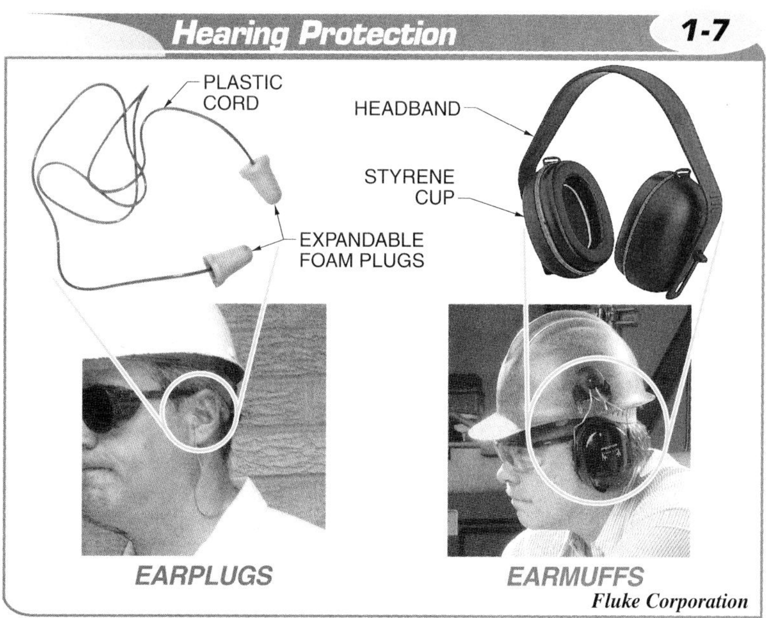

Figure 1-7. Ear protection devices reduce the level of noise reaching the eardrum.

TERMS

A **noise reduction rating (NRR) number** is a number that indicates how many decibels the noise level is reduced by.

Lockout is the use of locks, chains, or other lockout devices to prevent the startup and operation of specific equipment.

Tagout is the process of attaching a danger tag to the source of power to indicate that the equipment may not be operated until the tag is removed.

Hearing protection devices are rated for noise reduction. A *noise reduction rating (NRR) number* is a number that indicates how many decibels the noise level is reduced by. For example, an NRR of 27 means that the noise level is reduced by 27 dB when tested at the factory. For field use, 7 is subtracted from the NRR. Thus, the effective reduction in the field is approximately 20 dB.

Foot Protection. According to the U.S. Bureau of Labor Statistics, a typical foot injury is caused by objects with an average weight of approximately 65 lb falling less than 4′. Fluid power technicians perform many tasks that require handling similar objects. Safety shoes with steel-reinforced toes provide protection against injuries caused by compression and impact. **See Figure 1-8.**

Foot Protection 1-8

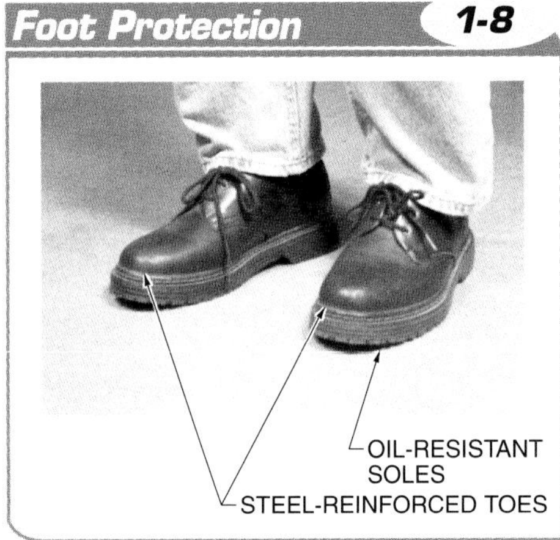

Figure 1-8. Safety shoes protect against falling objects and have soles that are not affected by hydraulic oil.

Some safety shoes have protective metal insoles and metatarsal or toe guards for additional protection. Oil-resistant soles and heels are not affected by hydraulic oil and provide improved traction. Protective footwear must comply with ANSI Z41-1991, *Personal Protection – Protective Footwear.* For the most protection, safety shoes should be replaced on an annual schedule.

Lockouts and Tagouts

Lockouts and tagouts are applied to power sources to prevent equipment operation during inspection, maintenance, and repair. **See Figure 1-9.** Power sources include all electrical, pneumatic, and hydraulic power to the equipment. *Lockout* is the use of locks, chains, or other lockout devices to prevent the startup and operation of specific equipment. *Tagout* is the process of attaching a danger tag to the source of power to indicate that the equipment may not be operated until the tag is removed.

Lockout devices are lightweight enclosures that allow the lockout of standard control devices such as valves, switches, and plugs. Colors are used to match standard hazard-level color codes. A multiple lockout hasp is used when more than one worker must lock out a power source. Locks used to lock out a device may be color-coded and individually keyed.

A tagout does not prevent the startup of equipment, but serves as a warning to operating and service personnel. A tagout is used when a lockout is not possible. Danger tags may include warnings such as "Do Not Operate" and provide space to enter information on the worker, date, and tagout reason. Tagouts are attached by hand and must be easy to read, durable, and resistant to accidental removal.

Written lockout/tagout procedures must be established for each piece of equipment in the facility, and personnel must be trained in the lockout/tagout procedures. If the lockout extends beyond a shift change, incoming workers must follow the established procedures of the lockout.

Warning: Lockout/tagout procedures must conform to OSHA 29 CFR 1910.147, *The Control of Hazardous Energy (Lockout/Tagout),* and company rules and procedures. A lockout/tagout shall not be removed by any person other than the authorized person who installed it, except in an emergency. In an emergency, only supervisory personnel may remove a lockout/tagout, and only upon notification of the authorized person who installed it.

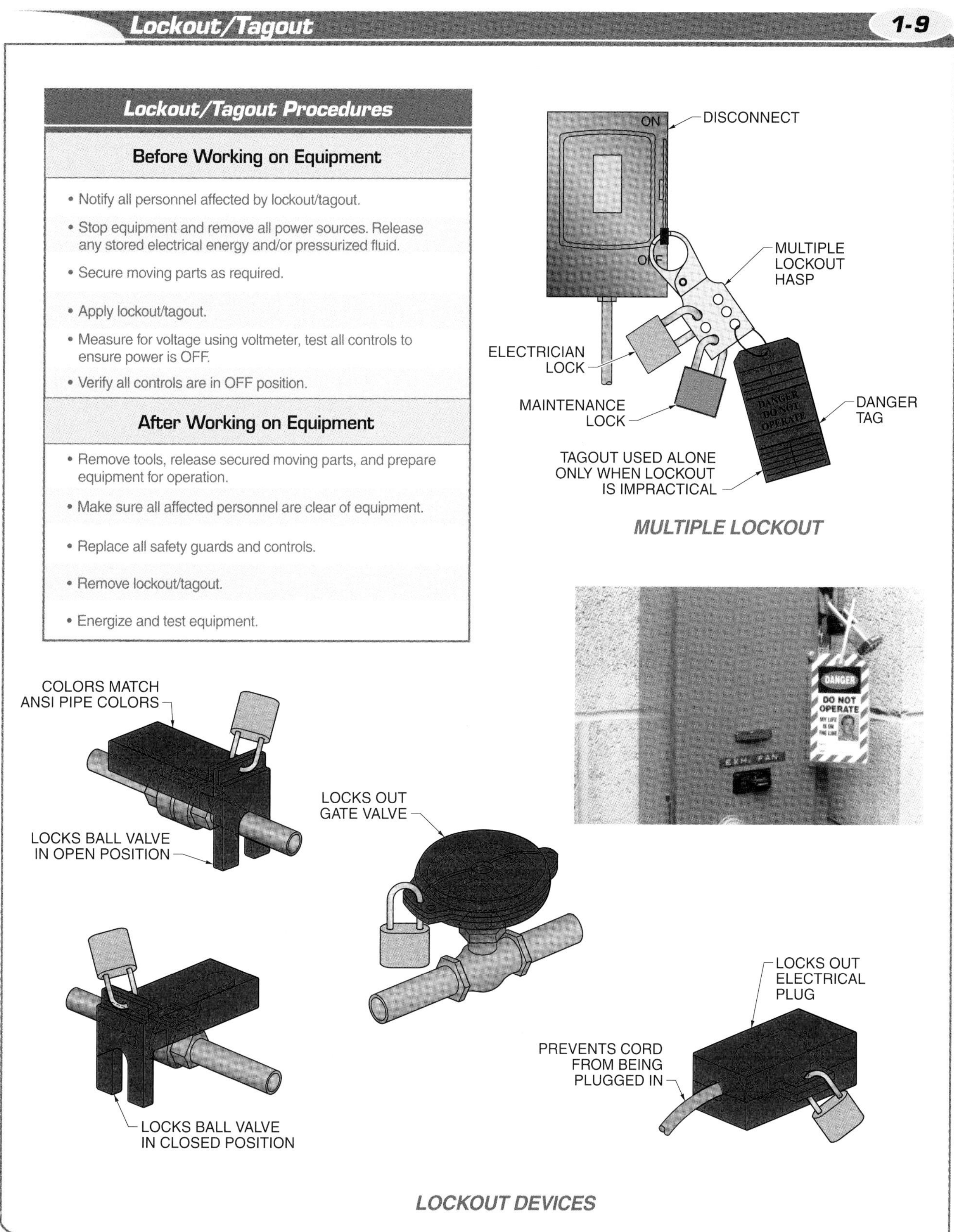

Figure 1-9. Lockouts and tagouts are applied to power sources to prevent equipment operation during inspection, maintenance, and repair.

Fluid Power Hazards

There are many hazards that must be taken into consideration when working with fluid power systems and related equipment. The foremost hazard is working on a system that is pressurized or could have residual pressure in the lines. A technician must always verify that there is no pressure present in the system and that none can build up.

A technician must always follow internal lockout and tagout procedures prior to working on a system. The technician must verify that there is no pressure stored in the system that could be released during the bleed-off, or release, of any residual pressure that may be in the system. If a system is pressurized while work is performed on it, damage to the machine could occur, possibly causing injury and/or death to those working nearby.

Hydraulic System Safety. Leaks and spills from hydraulic equipment such as hoses, pipes, and reservoirs (tanks) can cause additional hazards. These hazards include the possibility of injury caused by slips and falls when technicians are working near fluid power systems. Because of this, all excess fluid should be immediately removed. Commercial and industrial hydraulic spill kits are typically used to absorb and remove any excess oil or water.

It is hazardous to attempt to locate leaks in pressurized hydraulic systems. When a puddle of hydraulic fluid is discovered, a technician should never attempt to locate the source of the leak by hand. Pressurized hydraulic fluid can spray out of small holes in hoses or pipes and be injected underneath the skin, causing severe injury.

Pneumatic System Safety. When working around or on pneumatic equipment, safety glasses are always mandatory. There is always a possibility of leaks dispersing into the atmosphere or a hose becoming loose, endangering a technician's eyes. Pneumatic systems also tend to be loud and disperse air at high decibels, which can cause permanent hearing loss. Earplugs or earmuffs should always be worn if there is the possibility that decibels can reach high enough levels to damage the ears.

Pneumatic systems can operate at high speeds, and equipment can extend or retract quickly. Technicians should be aware of the operation of a pneumatic system and know when the equipment extends or retracts. Failure to understand the control and operation of pneumatic equipment can cause safety hazards that may result in serious injury or even death.

FLUID POWER SAFETY TRAINING

Along with PPE and lockout/tagout procedures, fluid power safety is an essential component of safety training programs in a facility. Training that is specific to fluid power equipment and systems includes multimedia presentations (videos), seminars, or multiple day training courses. Technicians responsible for fluid power systems should consider training that is specific to the types of fluid power systems that they might work with.

Fluid power system accidents can be reduced by following basic safety rules. Fluid power system safety rules include the following:

- Remove and lock out all potential power sources in a system, including mechanical, hydraulic, and electrical energy sources.
- Lower all elevated components. Secure elevated components that cannot be lowered.
- Bleed off all hydraulic oil pressure from the system, including the hydraulic fluid in the accumulators.
- Use caution when disassembling components that may contain springs.
- Charge pressurized fluid power containers with nitrogen only. Never use compressed air, which can be explosive around oil, and never overcharge a pressurized fluid power container.
- Do not perform work on hydraulic elevators. Although maintenance technicians maintain most hydraulic systems, elevators are maintained and tested by certified personnel from companies specializing in elevator installation and repair.

FLUID POWER ASSOCIATIONS AND CERTIFICATIONS

Fluid power technicians must be certified because of the nature of the tasks they perform. Working with direction from system designers and engineers, fluid power technicians must review project instructions and systems to determine test specifications, procedures, objectives, and technical problems and their possible solutions.

Technicians must also prepare and revise fluid power systems. Technicians must obtain data for development, standardization, and quality control by setting up and testing fluid power systems and components under operating conditions. Technicians may recommend modifications to existing systems and components to improve performance. Fluid power technicians are responsible for writing technical reports and preparing graphs and diagrams to describe the operation and performance of developmental or operational fluid power systems.

There are a number of fluid power industry trade associations that certify mechanics, technicians, and other specialists in the fluid power industry. These associations help individuals in fluid power industries to earn certification, keep up-to-date with current trends in the field, and locate manufacturers of fluid power components.

Fluid power industry trade associations also supply general knowledge on topics related to fluid power and provide training. For example, the Fluid Power Safety Institute (FPSI) is an organization created to help promote the concept of specific fluid power safety practices and procedures. The FPSI is a good resource for fluid power safety. There are also a number of industry web sites available that can provide technical information on fluid power safety. **See Figure 1-10.**

Fluid Power Industry Trade Associations **1-10**

The FPDA Motion and Control Network www.fpda.org	Comprised of over 300 distributor and manufacturer members. The FPDA represents motion solution providers who offer fluid power, automation, and electromechanical technologies and distribution services to enhance customer performance and profitability by sharing resources, business opportunities, knowledge, and training.
The Fluid Power Educational Foundation (FPEF) www.fpef.org	Created to stimulate, advance, and support the sciences and technologies of hydraulics and pneumatics in North America. Supports education and research initiatives that provide meaningful impact to the advancement of the fluid power industry and is wholly supported by fluid power industry firms and trade associations, which enable the FPEF to bring fluid power to students of all grade levels.
The Fluid Power Safety Institute™ (FPSI) www.fluidpowersafety.org	Networks and advocates hydraulic safety. Works to prevent occupational hazards in fluid power, and to attain a firm commitment from both hydraulic-related organizations and educational facilities to implement safety in the workplace and with their learning materials. Establishes on-going discussions of safety via their web site, and posts the latest information on hydraulic safety products, web sites, and news.
The International Fluid Power Society (IFPS) www.ifps.org	Facilitates and promotes the advancement of technology and professionalism of the fluid power and motion control industry through awareness, education, and certification.
The National Fluid Power Association (NFPA) www.nfpa.com	Provides a forum for manufacturers, distributors, suppliers, customers, and educators to work on the advancement of fluid power technology for the purpose of strengthening the industry and creating success for member organizations.

Figure 1-10. A number of fluid power industry trade associations offer certification to mechanics, technicians, and other specialists in the fluid power industry.

There are various professional fluid power industry certifications available for fluid power technicians. Certification can be earned at every level of fluid power. All certifications are available through the International Fluid Power Society (IFPS). Most require a written test and practical test to receive the certification. Most certifications are valid for up to 5 years, at which time they must be renewed. **See Figure 1-11.**

Professional Fluid Power Industry Certifications **1-11**

Mechanic Certifications	Technician Certifications	Specialist Certifications
• Mobile Hydraulic Mechanic (MHM)	• Mobile Hydraulic Technician (MHT)	• Hydraulic Specialist (HS)
• Industrial Hydraulic Mechanic (IHM)	• Industrial Hydraulic Technician	• Pneumatic Specialist (PS)
• Pneumatic Mechanic (PM)	• Pneumatic Technician (PT)	• Electronic Controls Specialist (ECS)
• Master Mechanic	• Master Technician (MT)	• Fluid Power Specialist (FPS)
		• Fluid Power Engineer (FPE)

Figure 1-11. Various professional fluid power industry certifications are available for fluid power technicians.

Chapter Review

Name: ______________________________ Date: ______________

MULTIPLE CHOICE

________ **1.** Most professional certifications for fluid power technicians are valid for up to ___ year(s), at which time they must be renewed.
A. 1
B. 2
C. 3
D. 5

________ **2.** ___ protective helmets protect against impacts at the top of the head only.
A. Type I
B. Type II
C. Type III
D. Type IV

________ **3.** An OSHA-approved hearing conservation program is required whenever workers are exposed to noise equal to or exceeding an 8 hr time-weighted average (TWA) of ___ decibels (dB).
A. 55
B. 85
C. 90
D. 100

________ **4.** ___ is clothing and/or equipment worn by a worker to reduce the possibility of an injury.
A. Fluid power protective equipment (FPPE)
B. Hydraulic protective equipment (HPE)
C. Personal protective equipment (PPE)
D. Pneumatic protective equipment (PPE)

________ **5.** ANSI Z358.1-2004, *Emergency Eyewash and Shower Equipment,* specifies that an emergency eyewash station must be located within ___ sec or about ___′ of hazards.
A. 5; 25
B. 10; 55
C. 10; 50
D. 15; 40

________ **6.** All professional certifications for fluid power technicians are available through the ___.
A. Fluid Power Educational Foundation (FPEF)
B. Fluid Power Safety Institute (FPSI)
C. International Fluid Power Society (IFPS)
D. National Fluid Power Association (NFPA)

______ 7. An ear protection device with an NRR of 33 means that in an area where the noise level is at 101 dB, the level of noise entering the ear would be at ___ dB.
A. 68
B. 87
C. 92
D. 95

______ 8. ___ law states that fluid in a closed system takes the shape of the container that it occupies and exerts pressure equally in all directions.
A. Blaise's
B. Boyle's
C. Gay-Lussac's
D. Pascal's

______ 9. ___ protective helmets protect against top and side impacts.
A. Type I
B. Type II
C. Type III
D. Type IV

______ 10. ___ are an eye and face protection device that covers the entire face with a plastic shield and is used for protection from flying objects.
A. Face shields
B. Goggles
C. Safety glasses
D. All of the above

COMPLETION

______ 1. Fluid in a closed fluid power system exerts ___ equally in all directions.

______ 2. ___ system applications consist of equipment used to move light loads that are typically less than 100 lb.

______ 3. A(n) ___ is an ear protection device worn over the ears to reduce the level of noise reaching the eardrum.

______ 4. ___ are an eye protection device with a flexible frame that is secured on the face with an elastic headband.

______ 5. ___ are used to protect a worker's head by resisting penetration and absorbing the blow of an impact.

______ 6. A(n) ___ system uses gas under pressure to create movement.

______ 7. ___ is the process of attaching a danger tag to the source of power to indicate that the equipment may not be operated until the tag is removed.

______ 8. ___ protective helmets protect against impact hazards and contact with high voltages up to 20,000 V.

______ 9. ___ is environmentally nonhazardous hydraulic fluid that is composed of synthetic chemicals and vegetable-based oil.

______________________ **10.** In an emergency, only ___ may remove a lockout/tagout, and only upon notification of the authorized person who installed it.

______________________ **11.** A(n) ___ system uses liquid under pressure to create movement.

______________________ **12.** Class ___ protective helmets are the most common class of helmets used in manufacturing, mining, and construction industries.

______________________ **13.** ___ are an eye protection device with special impact-resistant glass or plastic lenses, reinforced frames, and side shields.

______________________ **14.** A(n) ___ number indicates how many decibels a noise level is reduced by.

______________________ **15.** ___ is the technology of using a fluid to transmit power from one location to another.

TRUE/FALSE

T F **1.** Before performing work on a fluid power system, a technician must always verify that there is no pressure present in the system and that none can build up.

T F **2.** Certification for fluid power technicians can only be earned after completing fluid power training for mobile hydraulic equipment.

T F **3.** A tagout is used when a lockout is not possible.

T F **4.** Persons who normally wear corrective glasses do not need to wear OSHA-approved safety glasses.

T F **5.** Class C (conductive) protective helmets are designed for impact protection only and should not be used when there is danger from electrical hazards.

T F **6.** A lockout/tagout should not be removed by any person other than the authorized person who installed it, except in an emergency.

T F **7.** An earmuff is a moldable device inserted into the ear canal to reduce the level of noise reaching the eardrum.

T F **8.** According to the U.S. Bureau of Labor Statistics, a typical foot injury is caused by objects with an average weight of approximately 35 lb falling less than 3′.

T F **9.** Water, rather than petroleum-based oil, was originally used as the fluid in hydraulic systems.

T F **10.** Professional fluid power technician certifications are valid indefinitely and do not need to be renewed.

T F **11.** Pneumatic system applications consist of equipment used to move heavy loads greater than 100 lb.

T F **12.** Although maintenance technicians maintain most hydraulic systems, elevators are maintained and tested by certified personnel from companies specializing in elevator installation and repair.

T F **13.** Clothing made of synthetic materials such as nylon, polyester, or rayon should not be worn when working in industrial environments because these materials are flammable and can melt to the skin.

T F **14.** Much of the science behind fluid power can be attributed to 17th century French mathematician Archimedes.

T F **15.** Tagout is the use of locks, chains, or other lockout devices to prevent the startup and operation of specific equipment.

SHORT ANSWER

1. List at least five operational advantages of using hydraulic and pneumatic systems.

2. List two advantages of using bio-hydraulic fluid.

3. List at least three operational disadvantages of using hydraulic and pneumatic systems.

4. Briefly describe the characteristics of a tagout device.

5. Briefly explain why it can be hazardous to manually locate leaks within a pressurized hydraulic system.

Activity 1-1: Protective Helmet (PPE) Specification

A consulting firm that does work in a number of different industries has a supply of protective helmets (hard hats) that are used by their staff. They need to have a supply of different types of protective helmets for the different industrial environments where they are required to work.

1. List the class of protective helmet that would be the minimum requirement for each industry.

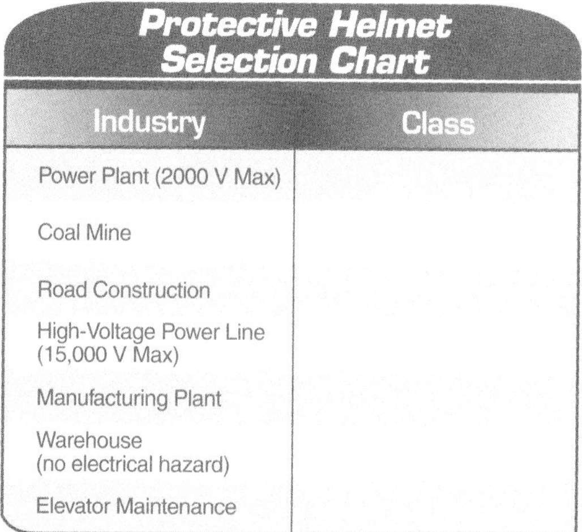

Protective Helmet Selection Chart

Industry	Class
Power Plant (2000 V Max)	
Coal Mine	
Road Construction	
High-Voltage Power Line (15,000 V Max)	
Manufacturing Plant	
Warehouse (no electrical hazard)	
Elevator Maintenance	

Activity 1-2: Personal Protective Equipment (PPE) Requirements

List the minimum PPE requirements that should be worn in each of these working environments.

1. Commercial building construction site:

2. Steel mill:

3. Mobile hydraulic equipment maintenance shop:

4. Road construction site:

5. Shipping and receiving dock:

Activity 1-3: Lockout/Tagout Locations

1. Circle the lockout/tagout locations on the hydraulic system shown.

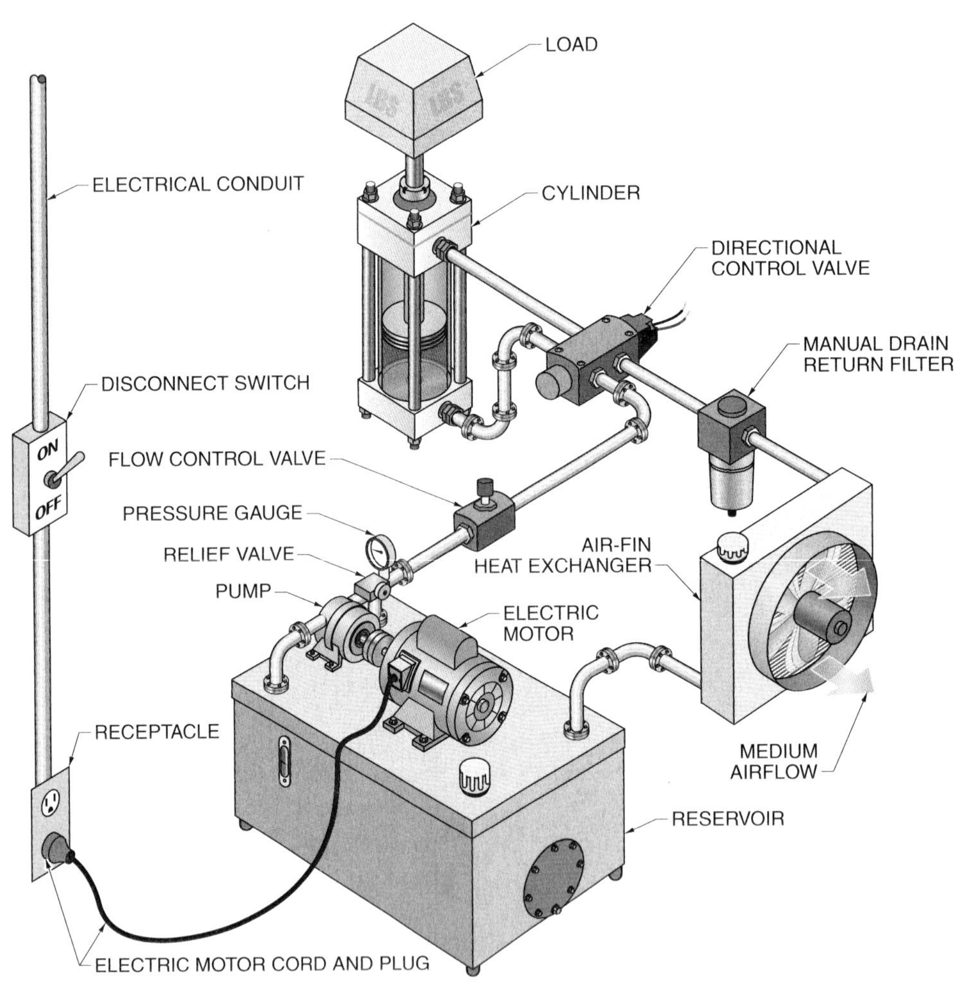

HYDRAULIC SYSTEM

Chapter 2

Fluid Power System Principles

OBJECTIVES

- Define the different states and forms of energy.
- Explain how work, exerted force, foot-pounds, power, and horsepower relate to fluid power.
- Define Pascal's law and describe its importance in fluid power systems.
- Calculate force, area, and pressure.
- Distinguish and describe the difference between pounds per square inch gauge, pounds per square inch absolute, bars, and kilopascals.
- Identify the different types of pressure gauges used in fluid power.

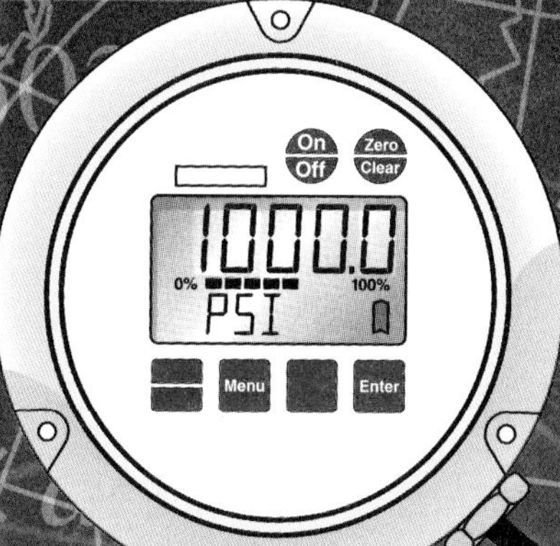

INTRODUCTION

The purpose of a fluid power system is to accomplish work that cannot be performed manually or to accomplish work more efficiently. Fluid power systems convert different types of energy into fluid power energy to accomplish work. Fluid power energy allows mechanical devices to operate. The operation of mechanical devices by fluid power energy must be controlled and measured at all times to ensure safe and efficient operation.

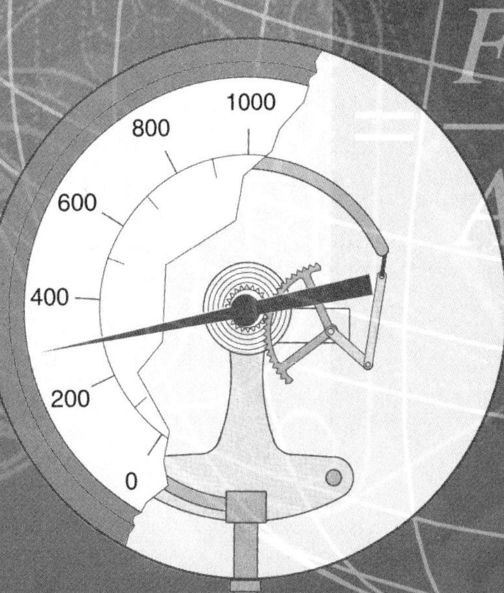

TERMS

The **law of conservation of energy** states that energy can neither be created or destroyed.

Energy is the capacity to do work.

Total energy is the combined forces of different forms of energy.

Static energy, or potential energy, is stored energy ready to be used.

Kinetic energy is the energy of motion.

Thermal energy is the addition of heat to make a particle or molecule move faster or the dissipation of heat that makes a particle or molecule move slower.

ENERGY

Fluid mechanics is based on the law of conservation of energy. The *law of conservation of energy* states that energy can neither be created nor destroyed. Fluid mechanics is comprised of two different types of energy, static energy and kinetic energy. *Energy* is the capacity to do work. For example, a battery in a flashlight takes chemical energy and converts it into electrical energy. Electrical energy is then converted into light energy through a light bulb in the flashlight. In fluid power systems, energy is used to move an object. For example, fluid power energy is used to move the bed of a dump truck. All forms of energy have the ability to do work.

Forms of Energy

Energy can be found in many forms. The law of conservation of energy states that energy can neither be destroyed nor created, although it can be transformed from one form of energy to another. There are several forms of energy in fluid power systems.

Total energy is the combined forces of different forms of energy. In fluid power systems, total energy is the sum of static energy, kinetic energy, and thermal energy. **See Figure 2-1.** The main forms these energies are transformed into are thermal, electrical, mechanical, and sound energy.

Static Energy. *Static energy* (potential energy) is stored energy ready to be used. Static energy is transformed into kinetic energy when a valve is opened, allowing fluid to flow. Fluid flow creates the ability to do work. In a fluid power system, the static energy of a liquid or gas is referred to as hydrostatic fluid power.

In fluid power there are two categories of static energy: pressure and elevation. Pressure static energy is the pressure created then stored for later use within a hydraulic system such as a pneumatic compressor tank or a hydraulic accumulator. Elevation static energy is energy stored due to a mass being located above a given point. For example, the raised boom of a hydraulic crane that is motionless is an example of elevation static energy. The weight of the crane creates static energy due to the higher elevation and weight of the boom (mass).

Kinetic Energy. *Kinetic energy* is the energy of motion. Any moving object, such as fluid in a fluid power system, has kinetic energy. In a fluid power system, the kinetic energy of moving gas or liquid is referred to as hydrodynamic energy. A piston extending in a cylinder is an example of kinetic energy in the form of fluid dynamic energy. In a fluid power system, kinetic energy can be changed into thermal energy because of friction between the fluid and its container.

Total energy in a fluid power system has the ability to change back and forth between static energy and kinetic energy as many times as necessary. For example, in a hydraulic system, hydraulic fluid is in a static state while it is in the reservoir, but changes into a kinetic state when it is pumped to extend a piston. It returns to a static state when it is returned to the reservoir. This pattern repeats itself as long as work is performed by the hydraulic system.

Hydraulic pressure is created when the fluid flow created by a hydraulic pump meets resistance. The transmission of energy throughout a hydraulic system begins when electrical energy is converted into rotating mechanical energy by an electric motor. The rotating mechanical energy of the motor is transferred to the hydraulic pump, which supplies hydraulic energy to the system in the form of moving fluid (kinetic energy). **See Figure 2-2.** Ultimately, the kinetic energy is converted back into mechanical energy by hydraulic cylinders and motors.

Thermal Energy. *Thermal energy* is the addition of heat to make a particle or molecule move faster or the dissipation of heat that makes a particle or molecule move slower. Thermal energy in a fluid power system is usually destructive and therefore cannot be harnessed or used. Once a portion of kinetic energy is converted to thermal energy, the energy is lost to the fluid power system. Thermal energy is what other forms of energy turn into when the system is inefficient. It is considered the energy that is wasted by the system. Any inefficiency in a fluid power system can be analyzed by measuring thermal energy.

Total Energy 2-1

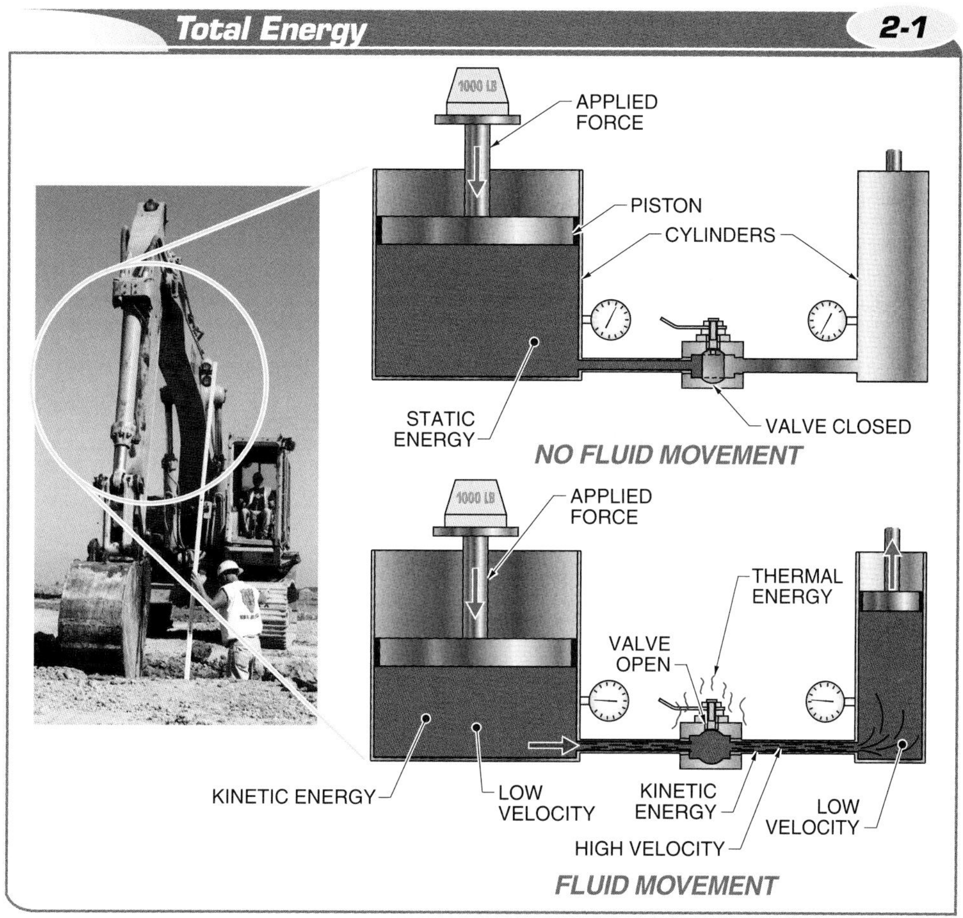

Figure 2-1. In fluid power systems, total energy is the sum of static energy, kinetic energy, and thermal energy.

Hydraulic System Energy Transmission 2-2

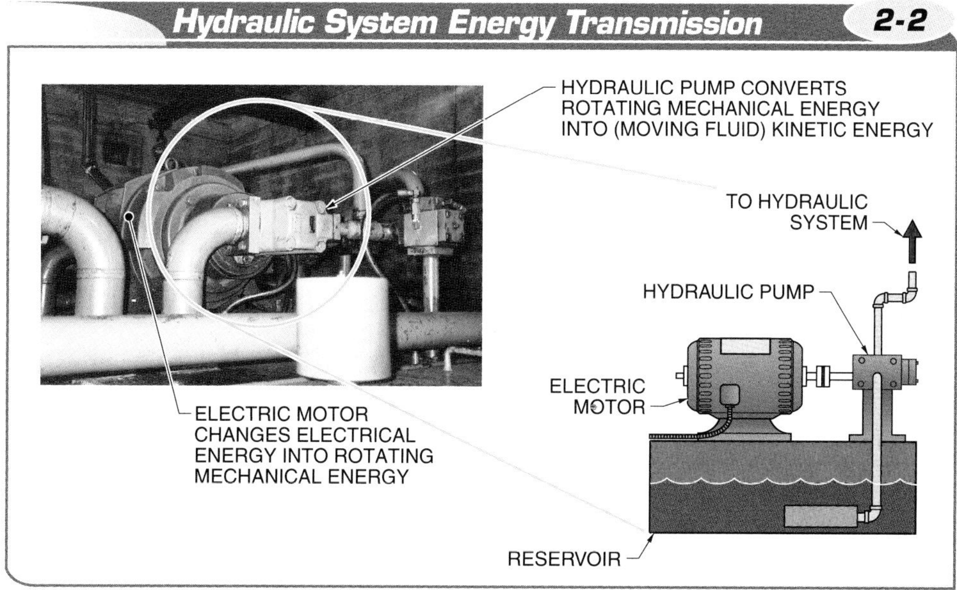

Figure 2-2. Rotating mechanical energy is transferred to a hydraulic pump, which supplies hydraulic energy to the system in the form of kinetic energy (fluid flow).

TERMS

Electrical energy is the flow of electrons (subatomic elements) from atom to atom.

An **electron** is a negatively charged particle that orbits the nucleus of an atom.

An **atom** is the smallest particle that an element can be reduced to while keeping the properties of that element.

Mechanical energy is machine energy.

Sound energy is transmitted vibration that can be sensed by the human ear.

An **actuator** is a mechanical device used for moving or controlling movement of a load.

Electrical Energy. *Electrical energy* is the flow of electrons (subatomic elements) from atom to atom. An *electron* is a negatively charged particle that orbits the nucleus of an atom. An *atom* is the smallest particle that an element can be reduced to while keeping the properties of that element. Electrical energy is the typical energy source used to power the pump in a fluid power system. An electric motor provides rotating mechanical energy to power a pump. The pump provides fluid flow. By controlling electrical energy, many different fluid power system applications are possible. For example, thermal energy in a fluid power system results from the heat that radiates from a pneumatic system compressor tank.

Mechanical Energy. *Mechanical energy* is machine energy. Mechanical energy is energy produced and transferred using gears, pulleys, or belts. Mechanical energy is used to transmit energy from one solid object to another solid object. For example, a conveyer line uses mechanical energy because an electric motor rotates a belt-drive system that is connected to a conveyor belt. The rotating energy from the motor transforms into mechanical energy, causing the conveyor belt to move. **See Figure 2-3.**

Sound Energy. *Sound energy* is transmitted vibration that can be sensed by the human ear. In fluid power systems, sound energy is mainly used for troubleshooting purposes. For example, by listening to an operating fluid power system, an experienced technician can often determine if the system is operating properly based on any unusual sound or noise from the system. If an unusual sound or noise is present, regular troubleshooting methods can be applied to determine the cause of any problems.

TECH FACT

When in operation, fluid power systems can be extremely loud. In certain situations, fluid power equipment can operate above 90 dB. This is loud enough that permanent hearing damage can occur. Proper hearing protection must always be used when working with or near fluid power equipment.

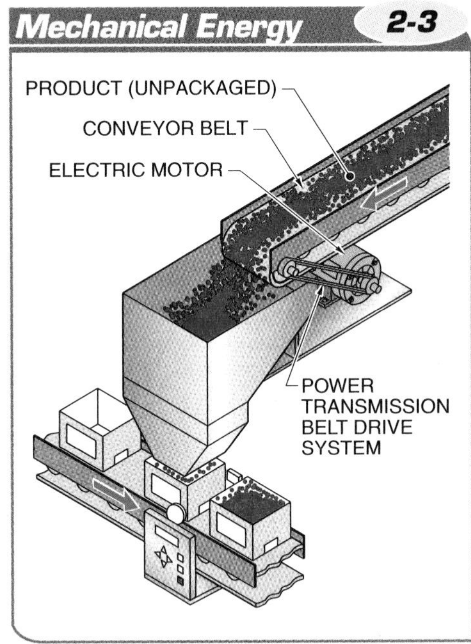

Figure 2-3. Electric energy is changed into rotating mechanical energy by the electric motor and is transmitted to the conveyor belt by a belt drive system.

FLUID POWER ENERGY TRANSMISSION

The transmission of energy using fluid power is accomplished by either hydraulic or pneumatic systems. Hydraulic systems use liquid to transmit energy. Pneumatic systems use gas to transmit energy. Air is the most common fluid used with pneumatics. Although water was the first liquid to be used with hydraulic systems, modern hydraulic equipment uses petroleum-based oil as the main method of hydraulic energy transmission. When comparing hydraulic systems and pneumatic systems, differences in system variables must be taken into consideration. **See Figure 2-4.**

There are also similarities between hydraulic and pneumatic systems. For example, they use similar control valves and actuators. An *actuator* is a mechanical device used for moving or controlling movement of a load. Actuators can be cylinders for linear movement or motors for rotary movement. Hydraulic and pneumatic systems are controlled by electrical and electronic devices. To prevent hazardous situations and loss of energy, hydraulic and pneumatic systems must be under control at all times. Fluid power system energy transmission is based on efficiency (energy loss), work (amount of energy used), and power (the amount of work that needs to be accomplished over a certain amount of time).

Fluid power systems are used in many modern industrial and commercial applications. Another form of system inefficiencies are leaks and resistance to fluid flow in hoses and pipes and the resistance to the movement of mechanical fluid power devices.

TERMS

Fluid power system efficiency is the useful amount of output energy from a system compared to the amount of input energy.

Fluid Power System Efficiency

Fluid power system efficiency is the useful amount of output energy from a fluid power system compared to the amount of input energy. Inefficiencies and energy loss are inherent in every fluid power system. It is impossible to have a hydraulic or a pneumatic system that is 100% efficient. Some of the reasons for inefficiencies are the installed components that the system needs to operate. Certain types of piping, pumps, valves, and meters are more efficient than other types.

Whenever there are inefficiencies in a system, the energy that is wasted is converted into heat and pressure drops. In a hydraulic system, heat can be harmful because it has negative effects on the hydraulic fluid. Premature breakdown of the properties of hydraulic fluid can cause system component breakdown and failure.

Enerpac

Since hydraulic systems move a large amount of weight, hydraulic equipment must be systematically maintained to avoid premature breakdown due to system device failure.

Fluid Power System Differences **2-4**

System Variable	Hydraulic Systems	Pneumatic Systems
Fluid type	Petroleum-based oil	Gas (air)
Fluid cost	Moderate	Conditioning cost only
Fluid flow	Slow; requires pump	Fast, uses air expansion
Work capacity	Heavy	Light to moderate
Compression	Slightly compresses with applied force	Will compress with applied force
Position control	Good	Less control than hydraulics
Mechanical device movement	Consistent and smooth	Fast and inconsistent
Fluid return system	Must return to reservoir	Exhausted into atmosphere
Pressure control	Very precise	Precise
Lubrication	Self-lubricating	Requires external lubrication device
Cleanliness	Not clean	Clean

Figure 2-4. When making comparisons between hydraulics and pneumatics, there are differences in system variables that must be taken into consideration.

TERMS

Resistance is the force that stops, slows, or restricts the movement of fluid or devices in a fluid power system.

Friction is the resistance to movement between two mating surfaces.

Work is the movement of an object (in lb) through a distance (in ft).

Force is anything that changes or tends to change the state of rest or motion of a body.

Resistance. *Resistance* is the force that stops, slows, or restricts the movement of fluid or devices in a fluid power system. The most common type of resistance in a fluid power system is friction. *Friction* is the resistance to movement between two mating surfaces. Friction between two mating surfaces turns the energy that is being used to move the surfaces into heat. Heat from friction is generated by different actions in a fluid power system, such as fluid flowing through a pipe or a piston moving against a cylinder body. **See Figure 2-5.**

Typical points of resistance in fluid power systems include bends in pipes and hoses, flow meters, valves, and pipes and hoses that are narrow in diameter. All of these resistance points can change the energy that a fluid power system transmits and turn it into heat. Because of this, no fluid power system will ever be 100% efficient.

TECH FACT

To have the most efficient fluid power system available, preventive maintenance programs and system maintenance recommendations from the manufacturer must be strictly followed.

Heat from Friction **2-5**

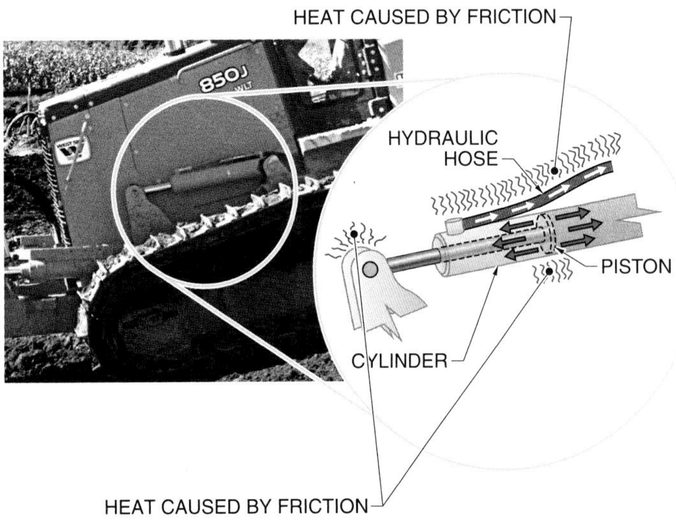

HEAT CAUSED BY FRICTION
HYDRAULIC HOSE
PISTON
CYLINDER
HEAT CAUSED BY FRICTION

Figure 2-5. Heat from friction is generated through different actions in a fluid power system, such as fluid flowing through a hose, or a piston moving against a cylinder body.

Work

Work is the movement of an object (in lb) through a distance (in ft). *Force* is anything that changes or tends to change the state of rest or motion of a body. Work is accomplished when a force overcomes a resistance. All fluid power systems are meant to accomplish some type of work. In most applications, fluid power systems accomplish work by moving something, such as large amounts of backfill material, concrete, or cartons of product. Also, fluid power systems accomplish mechanical work by using actuators such as cylinders, motors, and oscillators. **See Figure 2-6.**

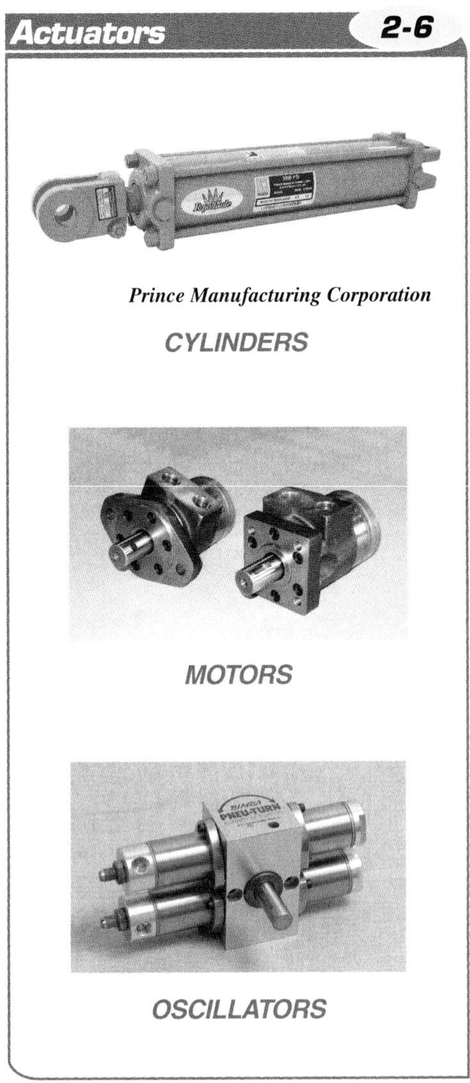

Figure 2-6. Fluid power systems complete work using actuators such as cylinders, motors, and oscillators.

Exerted force is the amount of weight in pounds a fluid power system must produce to move an object. For example, to move a crate of product that weighs 50 lb, the fluid power system must have a minimum of 50 lb of exerted force to move the crate. *Distance* is the extent of advance from one point to another. In fluid power systems, distance is typically measured in feet, although meters are sometimes used for specific situations.

The amount of work (W) produced is calculated by multiplying the force (F) that must be overcome (in lb) by the distance (d) (in ft) over which it acts. **See Figure 2-7.** *Foot-pounds (ft-lb)* is a measure of work. The amount of work produced is calculated by applying the following formula:

$$W = F \times d$$

where

W = work (in ft-lb)

F = force (in lb)

d = distance (in ft)

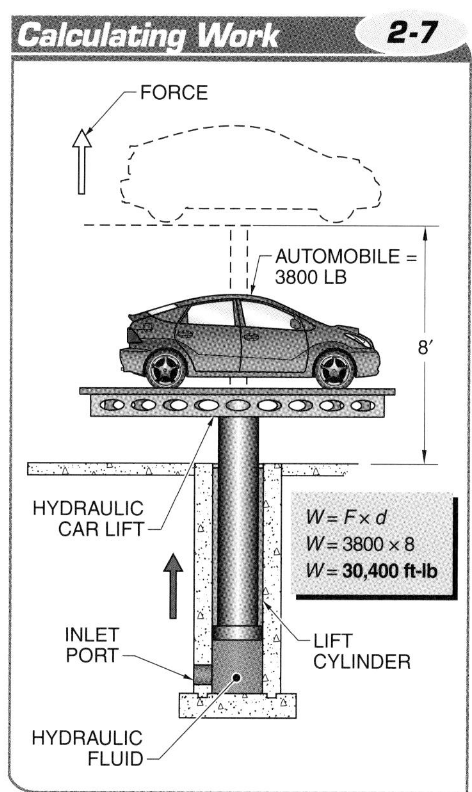

Figure 2-7. The amount of work produced is calculated by multiplying the force that must be overcome by the distance over which it acts.

Example: How much work is accomplished when a hydraulic car lift must elevate a 3800 lb automobile 8′ above the floor of an automotive service center?

$$W = F \times d$$

$$W = 3800 \times 8$$

$$W = \mathbf{30{,}400\ ft\text{-}lb}$$

Resistance must be overcome to perform work. More work would be required if the automobile were heavier, the distance were longer, or a combination of the two. For example, 38,000 ft-lb of work is required to lift a 3800 lb automobile 10′ above the floor (3800 × 10 = 38,000 ft-lb). Less work would be required if the automobile were lighter, the distance were shorter, or a combination of the two. For example, 21,000 ft-lb of work is required to lift a 3500 lb automobile 6′ above the floor (3500 × 6 = 21,000 ft-lb).

Power

Power is the amount of work accomplished over a specific period of time. While work indicates how much energy is needed to move an object, power indicates the rate that work is done. **See Figure 2-8.** Power (P) is measured in foot-pounds per second (ft-lb/sec). Power is calculated by applying the following formula:

$$P = \frac{W}{t}$$

where

P = power (in ft-lb/sec)

W = work (in ft-lb)

t = time (in sec)

Example: What is the power required for a hydraulic car lift to lift a 3800 lb automobile 2′ in 6 sec?

$$P = \frac{W}{t}$$

$$P = \frac{3800 \times 2}{6}$$

$$P = \frac{7600}{6}$$

$$P = \mathbf{1266.67\ ft\text{-}lb/sec}$$

TERMS

Exerted force is the amount of weight in pounds a fluid power system must produce to move an object.

Distance is the extent of advance from one point to another.

Foot-pounds (ft-lb) are a measure of work.

Power is the amount of work accomplished over a specific period of time.

TERMS

Horsepower (HP) is a mechanical unit of measure equal to the force required to move 550 lb, 1 ft in 1 sec.

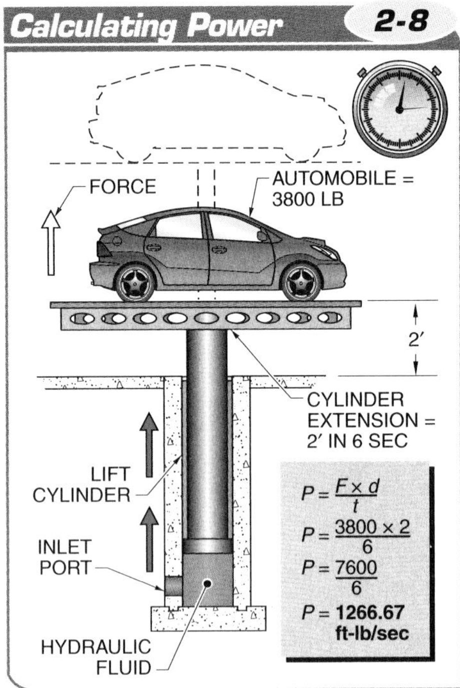

Figure 2-8. Power indicates the rate that work is done.

In a fluid power system, power is commonly expressed in units of horsepower. *Horsepower (HP)* is a mechanical unit of measure equal to the force required to move 550 lb, 1 ft in 1 sec. **See Figure 2-9.** Horsepower is calculated by applying the following formula:

$$HP = \frac{P}{550}$$

where

HP = horsepower

P = power (in ft-lb/sec)

550 = constant

Example: How much horsepower is required for a hydraulic automobile lift to lift a 3800 lb automobile 10′ at a rate of 5 sec?

$$HP = \frac{P}{550}$$

$$HP = \frac{3800 \times (10 \div 5)}{550}$$

$$HP = \frac{3800 \times 2}{550}$$

$$HP = \frac{7600}{550}$$

$$HP = \mathbf{13.82}$$

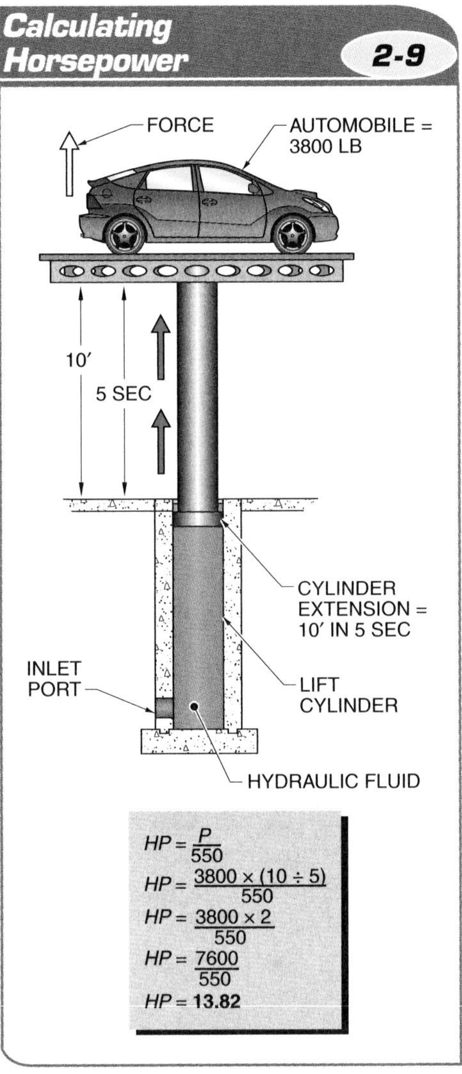

Figure 2-9. Horsepower is a mechanical unit of measure equal to the force required to move 550 lb, 1 ft in 1 sec.

TECH FACT

In 1782, the Scottish inventor of the Watt steam engine, James Watt, originated the constant of 550 used in the horsepower formula. At that time, horses were the main method used to supply power. James Watt wanted to compare the amount of power accomplished from his steam engine to that of a horse and prove that his steam engine was more powerful. In an experiment, Watt discovered that an average horse could move 550 lb a distance of 1′ in 1 sec.

FLUID POWER SYSTEM VARIABLES

Fluid power systems use either a liquid or a gas for the transmission of energy. Hydraulic systems use a liquid such as pressurized oil. A hydraulic system is a fluid power system that transmits energy by using a liquid that is confined. A *liquid* is a relatively noncompressible fluid that can readily flow and assume the shape of a confined space. Pneumatic systems use a gas such as compressed air. A *pneumatic system* is a fluid power system that transmits energy in a confined space using a gas under pressure.

The operation of fluid power systems is based on different interrelated variables. These variables are force, area, and pressure and are based on basic principles derived from Pascal's law. *Pascal's law* is a fluid power law that states that when a force is applied to a confined fluid, the force is felt throughout the fluid undiminished. This means that if fluid trapped in a cylinder has a pressing force on it, that force is distributed equally in all directions within that cylinder. **See Figure 2-10.** The fluid power circle is a visual representation of the force, area, and pressure relationship.

Force

The force produced by a hydraulic system is determined by the area of the surface receiving pressure and the amount of pressure applied. In fluid power systems, force is expressed in units of weight. In the United States, weight is given in pounds, however it is occasionally necessary to convert between English and metric units. Conversions between English and metric units can be determined with an equivalency chart. **See Appendix.**

Pounds are converted to kilograms using 0.45 as the conversion factor. For example, 10 lb of hydraulic oil equals 4.5 kg of hydraulic oil ($10 \times 0.45 = 4.5$). Kilograms are converted to pounds using 2.2 as the conversion factor. For example, 90 kg of force exerted on a cylinder equals 198 lb of force ($90 \times 2.2 = 198$).

Fluids are well suited for transmitting energy through pipes, hoses, and passages because of the characteristics of force. Force is energy, which can produce movement, work, or leverage when applied to a hydraulic application.

Power Team, Division of SPX Corporation

Figure 2-10. Pascal's law states that an applied force placed on a fluid will transmit undiminished in all directions.

TECH FACT

In the 17th century, the French mathematician and physicist, Blaise Pascal provided major contributions to the field of hydraulic principles. The law in fluid power principles that Pascal is most known for is his law on pressure in a confined space. Pascal's law is the most fundamental scientific law of all fluid power principles.

Enerpac

High-tonnage hydraulic cylinders are used during maintenance and construction tasks to provide leverage against heavy workpieces such as bridge beams.

TERMS

A **liquid** is a noncompressible fluid that can readily flow and assume the shape of a confined space.

A **pneumatic system** is a fluid power system that transmits energy in a confined space using a gas under pressure.

Pascal's law is a law that states that when a force is applied to a confined fluid, the force is felt throughout the fluid undiminished.

Area is the number of unit squares equal to the surface of an object.

Pressure is the resistance to flow.

Working pressure is the measure of the force applied to a given area.

Area

Area is the number of unit squares equal to the surface of an object. Area is expressed as square units, such as square feet (sq ft) or square inches (sq in.). Area, as it relates to pressure, is expressed in square inches, or the amount of surface area that the fluid is going to be working on.

In hydraulic systems, area typically refers to the piston face, which is circular in shape. The piston face is the surface that contacts pressure in the fluid. The opposite piston face is ring-shaped because it has a rod attached to it. **See Figure 2-11.** The area of a piston can be determined by using diameter or radius, so it is important to be able to determine the area of a circle by using either measurement.

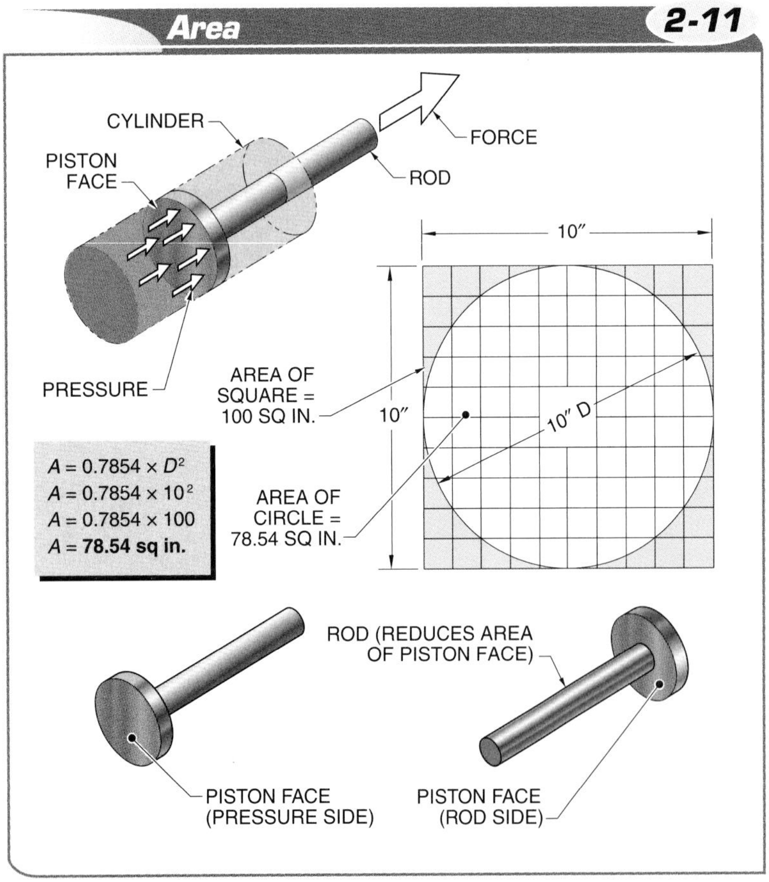

Figure 2-11. In hydraulic systems, area typically refers to diameter of the piston face.

A circle with a diameter equal to a side of a square has 78.54% of the surface area of the square. For example, a circle with a 10″ diameter has an area of 78.54 sq in. while a 10″ × 10″ square has an area of 100 sq in. When the diameter is known, the surface area of a circle is calculated by applying the following formula:

$$A = 0.7854 \times D^2$$

where

A = surface area (in sq in.)

0.7854 = constant

D = diameter (in in.)

Example: What is the area of a 3″ diameter cylinder face?

$$A = 0.7854 \times D^2$$
$$A = 0.7854 \times 3^2$$
$$A = 0.7854 \times 9$$
$$A = \mathbf{7.07\ sq\ in.}$$

Pressure

Pressure is the resistance to flow. *Working pressure* is the measure of the force applied to a given area. Pressure in a hydraulic system is the resistance to fluid flow. Pressure is determined by how much area is available and how much load (in lb) is present on the actuator. These two variables determine the amount of pressure a system can produce. Pressure is measured in pounds per square inch (psi), which translates to how many pounds are applied per a given area in square inches. **See Figure 2-12.** Working pressure is calculated by applying the following formula:

$$p = \frac{F}{A}$$

where

p = working pressure (in psi)

F = force (in lb)

A = area (in sq in.)

Example: How much working pressure is created when a 500 lb load is applied to a cylinder with an area of 3.1416 sq in.?

$$p = \frac{F}{A}$$
$$p = \frac{500}{3.1416}$$
$$p = \mathbf{159.15\ psi}$$

Since pressure, force, and area are interrelated in terms of the movement of an object, it is important to understand the relationship of each variable. A simple method used to understand how the three variables are interrelated is through application of the fluid power circle.

Fluid Power Formulas

The fluid power (Pascal's law) circle is a visual representation of how the formulas for pressure, force, and area are interrelated in a fluid power system. **See Figure 2-13.** To use the fluid power circle, two of the three variables represented must be known. When two of the three variables are known, the unknown variable can be calculated.

When force is known, it is always divided by the other known variable to calculate the unknown variable. For example, if an automotive hydraulic lift needs to lift a truck that weighs 4000 lb and the piston in the cylinder has surface area of 8 sq in., then the fluid power system must produce a system pressure of 500 psi (4000 ÷ 8 = 500). When force is unknown, pressure and area are multiplied.

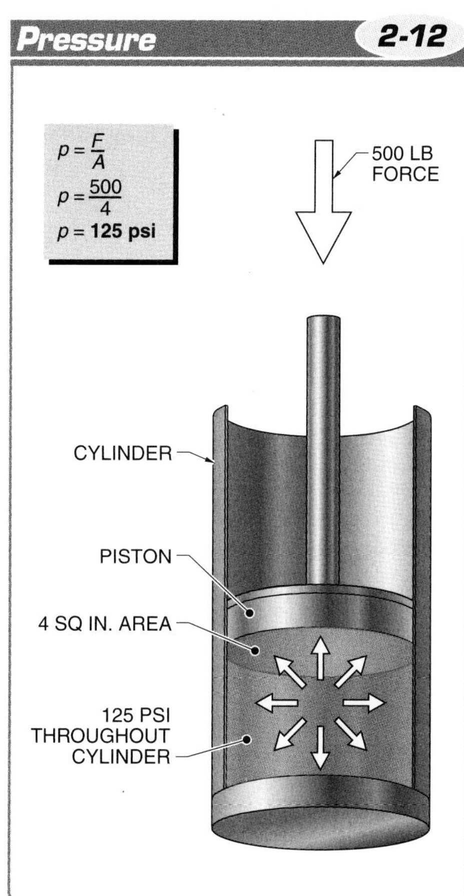

Figure 2-12. Pressure is determined by the area and force present.

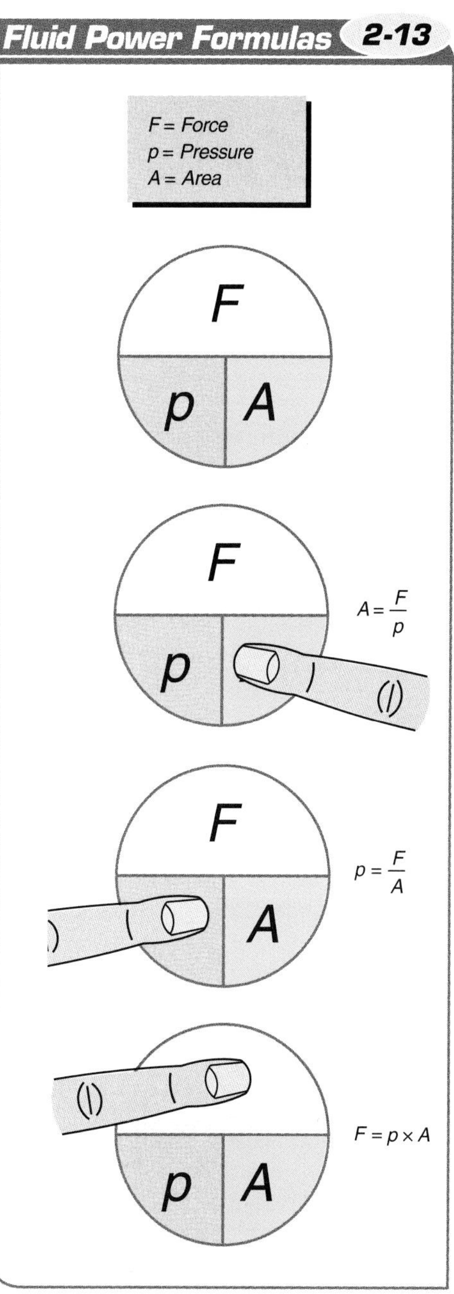

Figure 2-13. The fluid power (Pascal's law) circle is a visual representation of how the formulas for pressure, force, and area are interrelated in a fluid power system.

TERMS

A **fastener** is a mechanical device used to attach two or more members in position, or join two or more members.

All hydraulic cylinders have a piston with a rod attached to one side. In a hydraulic cylinder, the surface area acted on during extension includes the entire surface of the piston and fastener. A *fastener* is a mechanical device used to attach two or more members in position, or join two or more members. The surface area during retraction is reduced by the area of the rod. This means that the amount of pressure required to move a hydraulic cylinder with an attached load is different for extension than retraction. For example, the amount of pressure required for a hydraulic cylinder to retract a splitting wedge in a log splitter is different from the pressure required to extend a splitting wedge in a log splitter. **See Figure 2-14.** When force is known, extension system pressure is calculated by applying the following procedure:

1. Calculate the surface area of the piston face by applying the following formula:

   $A_{ext} = 0.7854 \times D^2$

   where

   A_{ext} = surface area of extension piston face (in sq in.)

   0.7854 = constant

   D = diameter (in in.)

2. Calculate the extension system pressure by applying the following formula:

   $$p = \frac{F}{A_{ext}}$$

   where

   p = system pressure (in psi)

   F = force (in lb)

   A_{ext} = surface area of extension piston face (in sq in.)

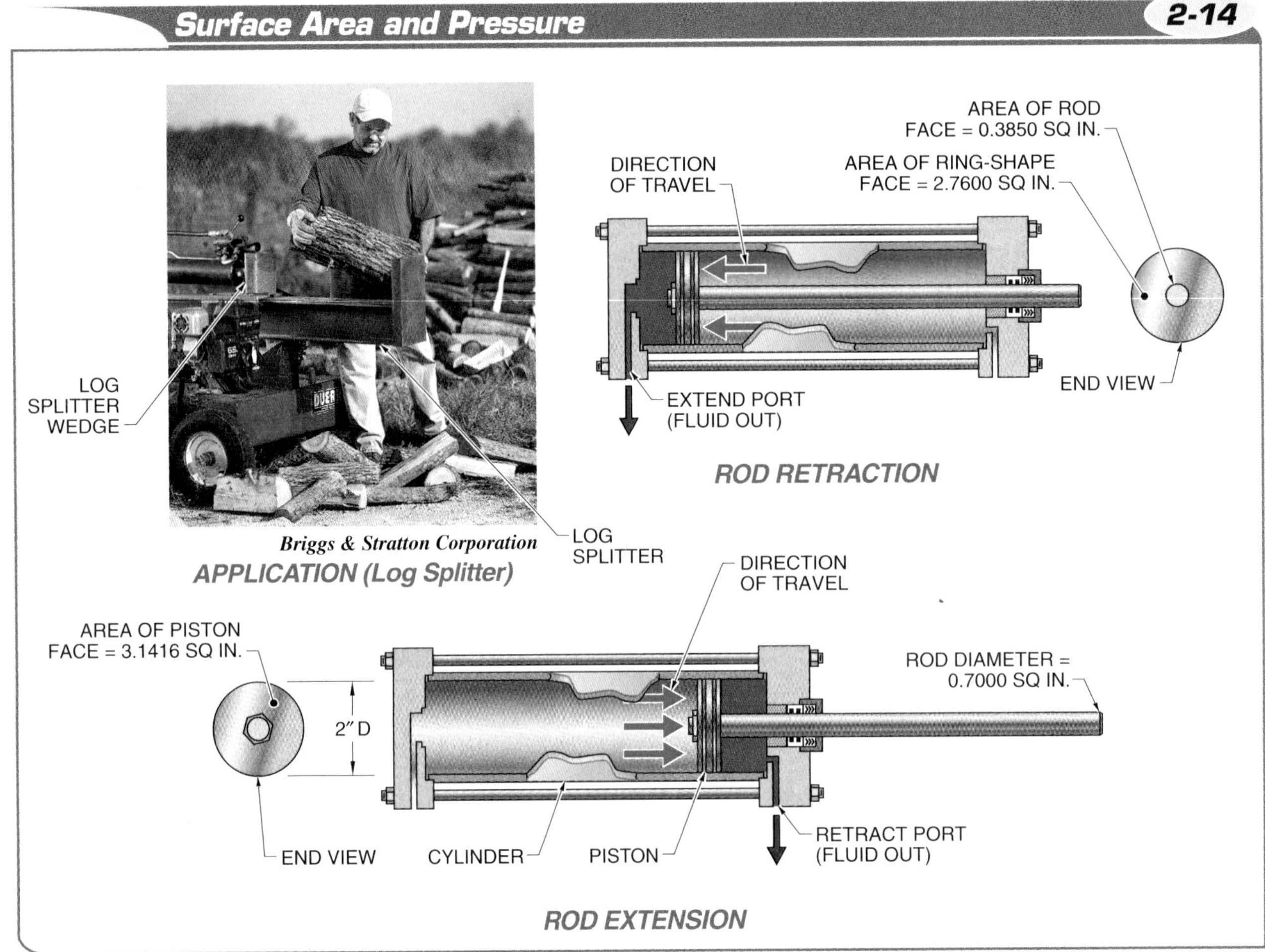

Figure 2-14. Less pressure is required to extend a rod than to retract one due to more surface area.

Example: How much pressure operates in a hydraulic log splitter with a 2″ diameter piston when extending a splitting wedge that weighs 25 lb?

1. Calculate the surface area of the piston face.

$A_{ext} = 0.7854 \times D^2$
$A_{ext} = 0.7854 \times 2^2$
$A_{ext} = 0.7854 \times 4$
$A_{ext} = 3.1416$ sq in.

2. Calculate the extension system pressure.

$$p = \frac{F}{A_{ext}}$$

$$p = \frac{25}{3.1416}$$

$p = \mathbf{7.9578}$ **psi**

The hydraulic cylinder needs to operate with a system pressure of 7.9578 psi when extending the splitting wedge. To calculate how much pressure it takes to retract the splitting wedge, it is necessary to calculate the surface area of the rod face and subtract it from the surface area of the piston face. These two steps provide the surface area of a ring-shaped piston face. When force is known, retraction system pressure is calculated by applying the following procedure:

1. Calculate the surface area of ring-shaped piston face by applying the following formula:

$A = A_{ext} - A_{ret}$
where
A = surface area of ring-shaped piston face (in sq in.)
A_{ext} = surface area of extension piston face (in sq in.)
A_{ret} = surface area of retraction rod face (in sq in.)

2. Calculate the retraction system pressure by applying the following formula:

$$p = \frac{F}{A}$$

where
p = retraction system pressure (in psi)
F = force (in lb)
A = surface area of ring-shaped piston face (in sq in.)

Example: How much pressure operates in a hydraulic log splitter with a 2″ diameter piston that has a rod with a diameter of 0.7″ when retracting a splitting wedge that weighs 25 lb?

1. Calculate the surface area of ring-shaped piston face.

$A = A_{ext} - A_{ret}$
$A = 3.1416 - 0.3850$
$A = 2.7600$ sq in.

2. Calculate retraction system pressure.

$$p_r = \frac{F}{A}$$

$$p_r = \frac{25}{2.7600}$$

$p_r = \mathbf{9.0600}$ **psi**

The hydraulic cylinder needs to operate with a system pressure of 9.06 psi when retracting the splitting wedge. The difference in pressure between extending the splitting wedge and retracting the splitting wedge is 1.10 psi (9.06 – 7.96 = 1.10). The pressure difference is a result of the difference in surface area of the two piston faces.

Enerpac

Hydraulic rams are available in many sizes for different applications.

TERMS

Atmospheric pressure is the pressure created by the weight of the atmosphere at sea level under standard air conditions.

A **vacuum** is any pressure less than atmospheric pressure.

Pressure Types

Pressure is present in all operating fluid power systems. Pressure is affected by variables such as friction due to fluid flow through pipes and hoses, changes in the diameter of pipe and hose sizes, system devices such as meters and gauges, and the viscosity of the fluid used in the system. By analyzing pressure correctly, a technician can more efficiently troubleshoot a fluid power system.

Types of pressures encountered in a fluid power system include atmospheric pressure, gauge pressure, and absolute pressure. Pressure is measured with instruments such as Bourdon tube pressure gauges, Schrader pressure gauges, and digital pressure gauges. A *pressure gauge* is an instrument used to measure pounds per square inch (psi) in a closed system. A *vacuum* is any pressure less than atmospheric pressure.

In the United States, pressure is typically measured in psi. Other units of measure include bars and kilopascals (kPa), which are both derived from an expanded version of the metric system called the International System of Units (SI). Pascals (Pa) are small and precise units of measure. For example, 30 psi is equal to more than 200,000 Pa, and 1 bar is equal to 100,000 Pa. In industry, pascals are typically converted to kilopascals to make the reading more understandable. For example, 200,000 Pa equals 200 kPa. Because equipment rated in pascals is so precise, it is only used in laboratory applications.

Atmospheric Pressure. *Atmospheric pressure* is the pressure created by the weight of the atmosphere at sea level under standard air conditions. A 1 sq in. column of air extending from sea level to the top of the atmosphere weighs 14.7 lb (1.01 bar, 101 kPa). Atmospheric pressure is placing pressure on everything all the time. For example, a vented hydraulic reservoir (oil tank) is considered to have only atmospheric pressure applied to it. **See Figure 2-15.**

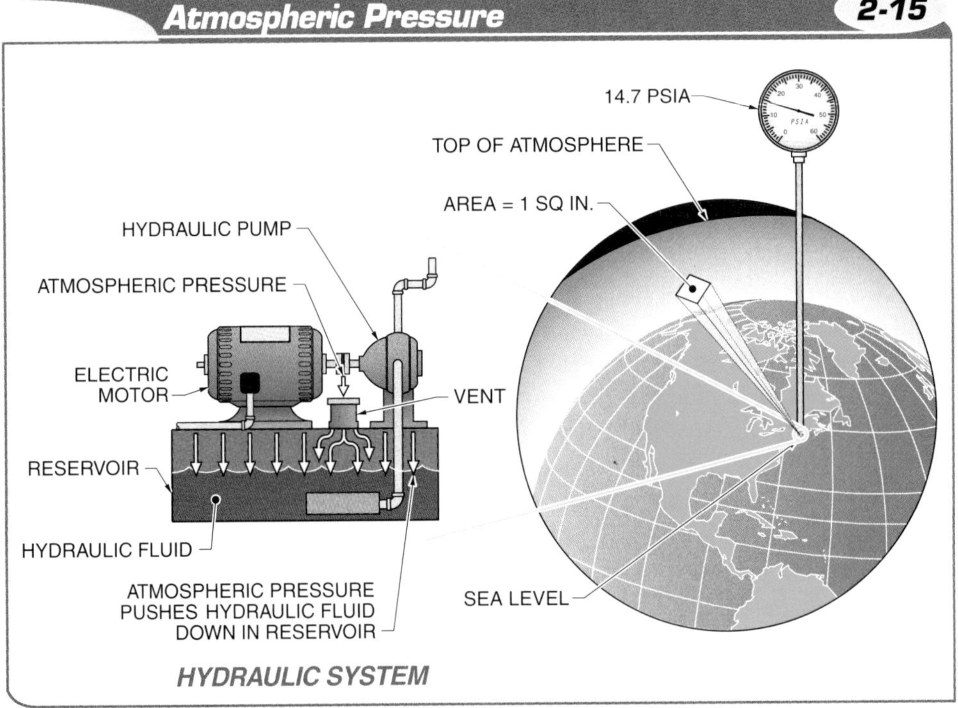

Figure 2-15. Atmospheric pressure is the pressure created by the weight of the atmosphere at sea level under standard conditions.

TECH FACT

Atmospheric pressure changes with altitude. At altitudes higher than sea level, the pressure is less than 14.7 psia. Below sea level, the pressure is greater than 14.7 psia. The change in atmospheric pressure due to altitude also affects absolute pressure. Absolute pressure decreases about 0.4 psia to 0.5 psia for every 1000′ above sea level.

Gauge Pressure. *Gauge pressure* is the amount of pressure above the existing atmospheric pressure and is used to measure pressure inside a closed fluid power system. Gauge pressure is expressed in pounds per square inch gauge (psig). Gauge pressure in a fluid power system does not account for the atmospheric pressure of 14.7 psi.

Most pressure gauges take readings in psig (psi), but some pressure gauges can take readings in bars, kpa, or tons for large industrial equipment. Pounds per square inch gauge and pounds per square inch are the same thing. Psi is simply pounds per square inch gauge without saying "gauge."

Absolute Pressure. *Absolute pressure* is the sum of gauge pressure and atmospheric pressure. **See Figure 2-16.** Absolute pressure is expressed in pounds per square inch absolute (psia). Absolute pressure can be used in systems that have vacuums to avoid using a negative number. For example, a vacuum reading in psi would be –10 psig, but in psia, it would be 4.7 psia. In theory, a perfect vacuum is 0 psia, although it is not possible to get to 0 psia because there will always be a small amount of air inside a container.

Gauge pressure and absolute pressure can be easily converted from one value to another. Gauge pressure is always 14.7 psi less than absolute pressure and is calculated by applying the following formula:

$$psig = psia - 14.7$$

where

psig = pounds per square inch gauge
psia = pounds per square inch absolute
14.7 = atmospheric pressure

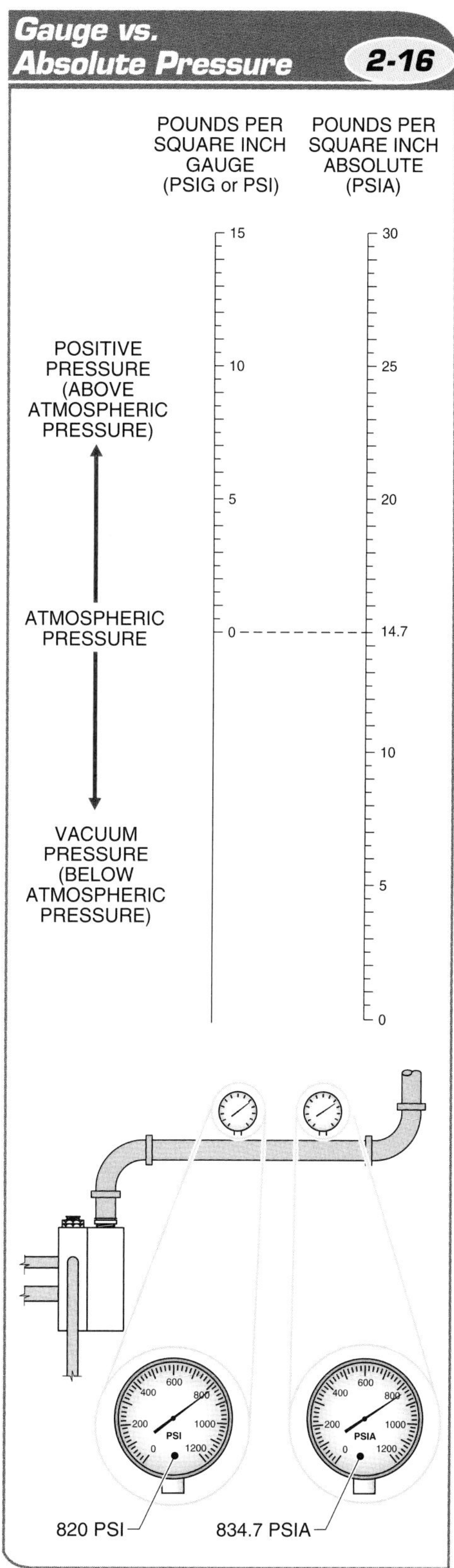

Figure 2-16. The difference between gauge pressure and absolute pressure is 14.7 psi, or atmospheric pressure.

TERMS

Gauge pressure is the amount of pressure above the existing atmospheric pressure and is used to measure pressure inside a closed fluid power system.

A **pressure gauge** is an instrument used to measure pressure in a closed system.

Absolute pressure is the sum of gauge pressure and atmospheric pressure.

TERMS

A **mercury barometer** is an instrument used to measure atmospheric pressure using a column of mercury (Hg).

A **Bourdon tube pressure gauge** is a measurement device that is used to register and measure pressure in fluid power systems.

Example: What is gauge pressure when absolute pressure is equal to 40 psia?

psig = *psia* – 14.7
psig = 40 – 14.7
psig = **25.3 psi**

Absolute pressure is always 14.7 psi greater than gauge pressure and is calculated by applying the following formula:

psia = *psig* + 14.7
where
psia = pounds per square inch absolute
psig = pounds per square inch gauge
14.7 = atmospheric pressure

Example: What is absolute pressure when gauge pressure is equal to 120 psig?

psia = *psig* + 14.7
psia = 120 + 14.7
psia = **134.7 psia**

Pressure Measurement

Pressure measurement is performed with a measurement gauge. Atmospheric pressure is measured with a mercury barometer. Psia, psig, and SI measurements are taken with either a Bourdon tube pressure gauge or a Schrader gauge. New equipment is often fitted with digital pressure gauges.

A *mercury barometer* is an instrument used to measure atmospheric pressure using a column of mercury (Hg). A mercury barometer consists of a glass tube that is completely filled with mercury and closed on one end. **See Figure 2-17.**

The glass tube is inverted with the open end submerged in a dish of mercury. A vacuum is created at the top of the tube as the mercury flows out of the tube. Vacuum is air pressure that is less than atmospheric pressure

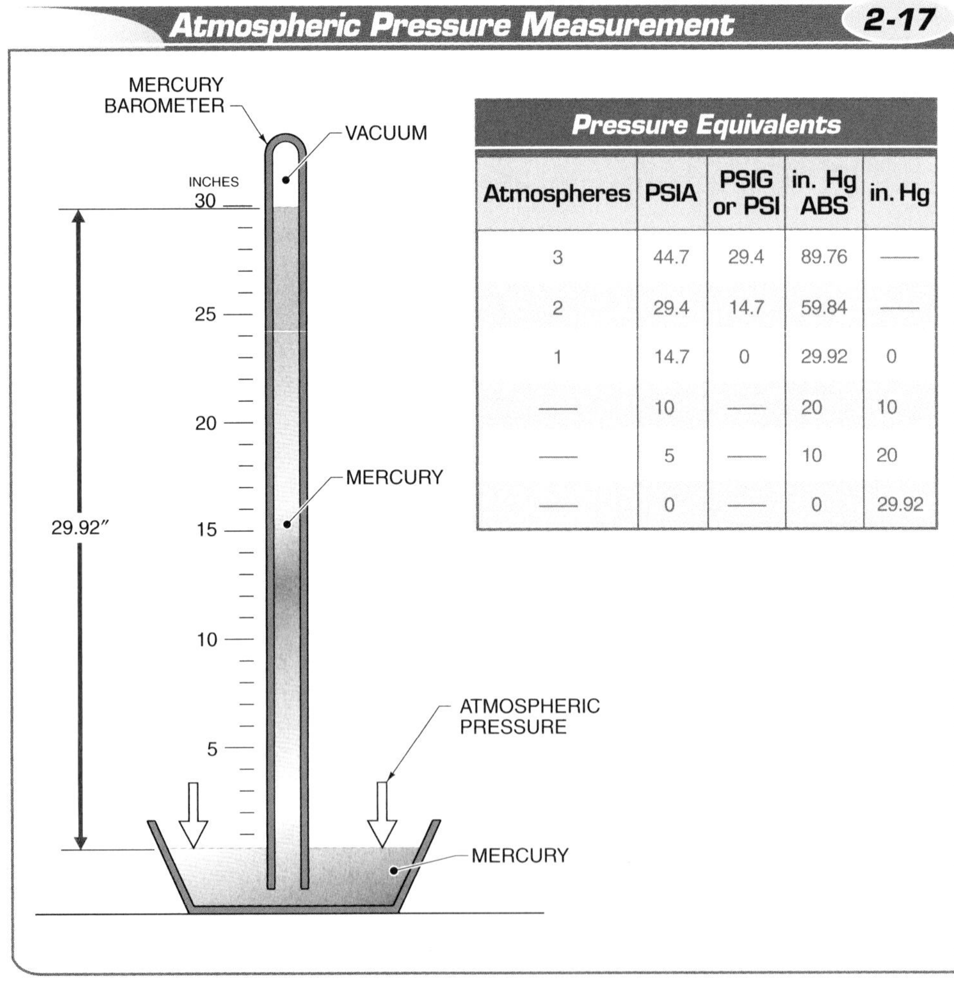

Pressure Equivalents

Atmospheres	PSIA	PSIG or PSI	in. Hg ABS	in. Hg
3	44.7	29.4	89.76	—
2	29.4	14.7	59.84	—
1	14.7	0	29.92	0
—	10	—	20	10
—	5	—	10	20
—	0	—	0	29.92

Figure 2-17. A mercury barometer indicates atmospheric pressure with a column of mercury.

(less than 14.7 psia). The pressure of the atmosphere prevents the mercury in the glass tube from completely flowing out. Thus, the height of the mercury in the glass tube corresponds to the pressure of the atmosphere on the mercury in the open dish.

A mercury barometer is calibrated in inches of mercury (in. Hg). At sea level, the atmosphere can support 29.92 in. Hg in the tube. A barometric pressure of 29.92 in. Hg equals one atmosphere, or 14.7 psia. Pressures above one atmosphere are generally expressed in psi, and pressures below one atmosphere are generally expressed in inches of mercury (in. Hg) representing vacuum. Minor pressure changes are expressed in inches of water column (in. WC). Because water is 13.6 times lighter than mercury, atmospheric pressure at sea level can support a 407.37″ (33.9′) water column (13.6 × 29.92″ = 407.37″).

Bourdon Tube Pressure Gauges. A *Bourdon tube pressure gauge* is a measurement device that is used to register and measure pressure in fluid power systems. A Bourdon tube is a curved device that straightens as pressure increases and bends as pressure decreases. **See Figure 2-18.** Bourdon tube pressure gauges can take readings in psi, bars, and/or kilopascals.

A *Bourdon tube* is a hollow metal tube made of brass or similar material that is bent in the shape of the letter C. One end of the tube is fixed and attached to the inlet of the pressure gauge. The other end is closed and free to move. The tube becomes straighter when more pressure is applied. The closed end of the tube is attached to a linkage arm that actuates a pointer gear. The pointer gear moves the pointer to indicate the pressure on an analog scale. The Bourdon tube pressure gauge is the most commonly used fluid power gauge in industry. The accuracy of the Bourdon tube pressure gauge can range from ±0.1% to 3% of full face value.

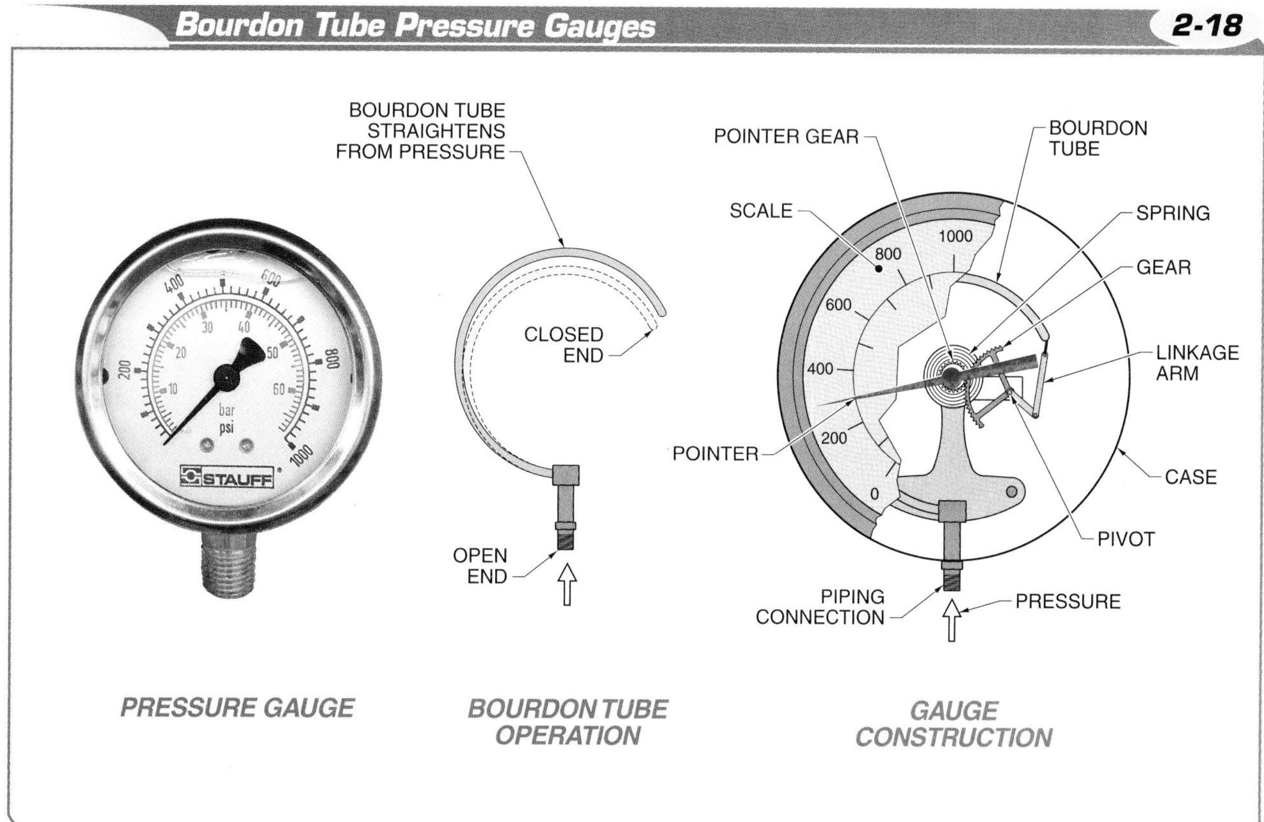

Figure 2-18. A Bourdon tube pressure gauge indicates pressure using the movement of a Bourdon tube.

TERMS

A **Bourdon tube** is a hollow metal tube made of brass or similar material that is bent in the shape of the letter C.

A **Schrader gauge** is a pressure gauge that uses fluid pressure to push a piston against a compression spring that is attached to a pointer.

Some pressure gauges take measurements in both psi and bars. Most equipment is fitted with pressure gauges that measure in two scales. For example, equipment manufactured in Europe is fitted with pressure gauges that are rated in bars and kilopascals. Because of possible use in international markets, equipment manufactured in the United States is sometimes fitted with gauges that give measurements in both psi and bars.

Advantages of using Bourdon tubes include that they have quick response times, can provide readings on two different pressure scales, are inexpensive, and are easy to replace when they fail.

Disadvantages of Bourdon tube pressure gauges are that they do not provide precise readings, that their accuracy can be subject to their environment (such as temperature), that they can be easily damaged if special precautions are not taken, and that they must be at eye level for accurate reading.

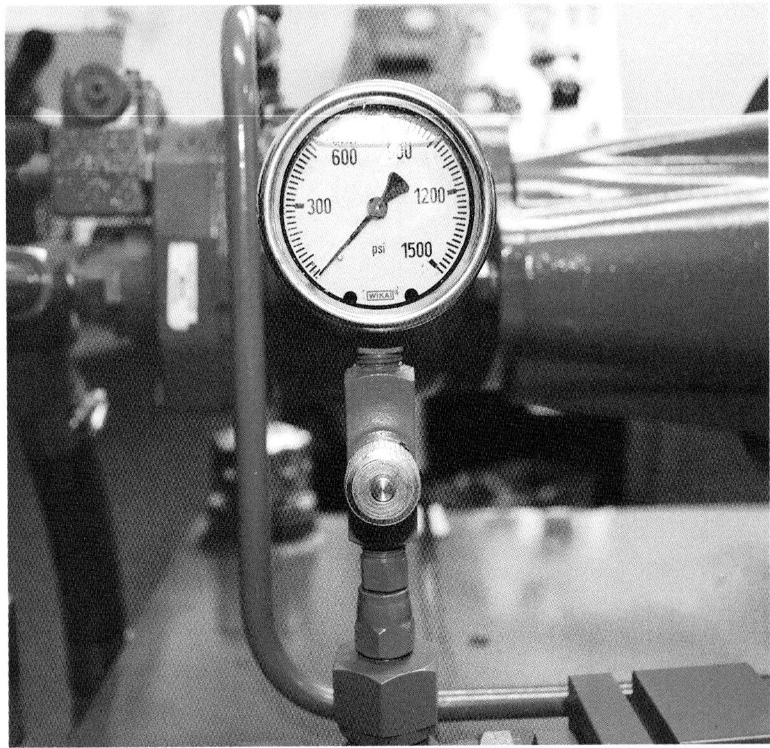

An analog pressure gauge indicates system pressure with a pointer and numerical scale.

When equipment is not rated in both psi and bars, conversions between the two must be made. Converting a pressure reading from bars to psi requires the following formula:

$$psi = bar \times 14.50377$$

where

psi = pressure (in psi)

bar = pressure (in bar)

Example: How many psi are in a system that reads 8 bar on its pressure gauge?

$$psi = bar \times 14.50377$$

$$psi = 8 \times 14.50377$$

$$psi = \mathbf{116.03}$$

Converting a pressure reading from psi to bars requires the following formula:

$$bar = \frac{psi}{14.50377}$$

where

bar = pressure (in bar)

psi = pressure (in psi)

Example: How many bars are in a system that reads 200 psi on its pressure gauge?

$$bar = \frac{psi}{14.50377}$$

$$bar = \frac{200}{14.50377}$$

$$bar = \mathbf{13.79}$$

Schrader Gauges. A *Schrader gauge* is a pressure gauge that uses fluid pressure to push a piston against a compression spring that is attached to a pointer. **See Figure 2-19.** Schrader gauges are also known as spring-loaded piston gauges. The higher the pressure in the system, the more the spring is compressed. Schrader gauges are mostly used in hydraulic systems as an economical method to read hydraulic pressure. Schrader gauges are accurate to about ±10%. Therefore, they should not be used for taking precision measurements.

TECH FACT

A Bourdon tube gauge housing used in a hydraulic system is usually filled with oil to prevent damage to the tube, pointer, or linkage from pressure spikes. Oil-filled Bourdon tube gauges are also more accurate than those that are not oil-filled.

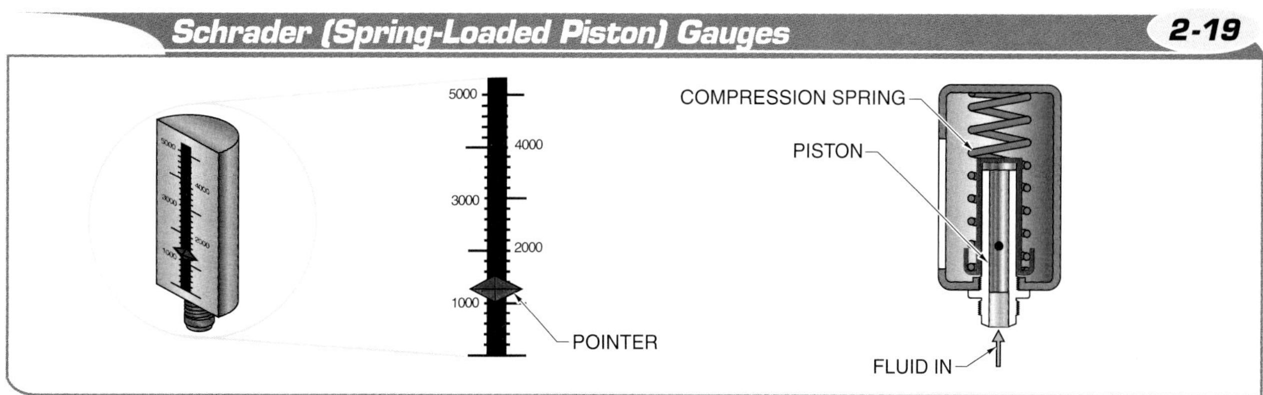

Figure 2-19. A spring-loaded piston gauge uses fluid pressure to push a piston against a compression spring that is attached to a pointer.

Compared to Bourdon tube gauges, the advantages of Schrader gauges are that they have quick response times, are inexpensive, are more durable, and are easily replaced within a hydraulic system. Disadvantages of Schrader gauges are that they are the least accurate of the different types of pressure gauges and usually only take readings on one pressure scale.

Digital Pressure Gauges. A *digital pressure gauge* is a pressure gauge that converts fluid pressure into an electrical signal. In fluid power systems, fluid pressure is converted to electrical current corresponding to the amount of pressure applied. The electrical current is interpreted by an electronic circuit board, which indicates the amount of pressure in the system on a digital display. A computer interface can be used to transmit data to a computer screen. **See Figure 2-20.**

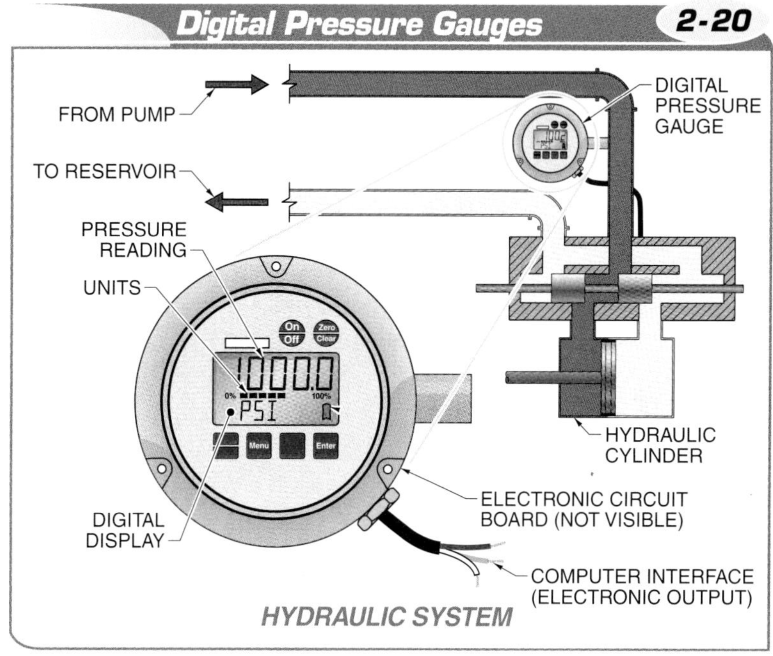

Figure 2-20. A digital pressure gauge converts fluid pressure into an electrical signal and has an easy-to-read digital display.

Most models are powered only by a DC power source (battery) that needs to be replaced within 1000 to 3600 hours of continuous use. The advantages of digital pressure gauges include the following:

- Easy to read digital display—Pressure is shown in decimal numbers that indicate an exact pressure. Also, most have a backlit display that makes it easy to read in poorly lit areas.
- Accuracy from ± 0.1% to 0.5%.
- Readings are provided in pounds per square inch, bars, kilopascals, millibars per hectopascal (mbar/hPa), and millipascals (mPa)
- Good data logging capabilities—Some digital pressure gauges can record and can track minimum and maximum pressure readings via an interface and software.
- Electrical analog output capabilities—Some digital pressure gauges have 4 mA to 20 mA outputs that can be interfaced with industrial computers to provide greater system control.
- Greater safety than with analog pressure gauges—It is safer to use small electrical signals to control the system than to use system pressure.

TERMS

A **digital pressure gauge** is a pressure gauge that converts fluid pressure into an electrical signal.

Disadvantages of digital pressure gauges include the following:

- They are more expensive than analog pressure gauges.
- DC powered digital pressure gauges require batteries, which can be expensive and time consuming to replace.
- AC powered digital pressure gauges require electric power routed throughout the system.

Vacuum

A vacuum is a confined space with a pressure that is less than 14.7 psia. When pressure in a confined space drops to any pressure below 14.7 psia, a vacuum has been created. Vacuum occurs because 14.7 psia surrounds the object that is under 14.7 psia.

Vacuum is used in hydraulic systems to allow fluid from the reservoir to travel into the pump. The strength of a vacuum depends on how far below 14.7 psia it is. The amount of vacuum present typically ranges from 3.0 in. Hg to 29.92 in. Hg. Bourdon tubes are commonly used for vacuum gauges and can take measurements in inches of mercury, bars, psig (or psi) with an accuracy of ±1.5%. Common vacuum gauge readings range from 0 in. Hg to 30 in. Hg, 0 bar to –1 bar, and 0 psig to –14.7 psig. **See Figure 2-21.** The pumps inlet rating depends on the design of the pump and the rating specification assigned by the pump manufacturer.

Pressure vs. Vacuum **2-21**

Vacuum*	Inches of Mercury†	Gauge‡	Pressure§	Atmospheric Pressure‖
10	3.0	– 1.47	– 0.10	13.23
15	4.5	– 2.21	– 0.15	12.50
20	6.0	– 2.94	– 0.20	11.76
25	7.5	– 3.68	– 0.25	11.03
30	9.5	– 4.41	– 0.30	10.29
35	10.5	– 5.15	– 0.35	9.56
40	12.0	– 5.88	– 0.40	8.82
45	13.5	– 6.62	– 0.45	8.09
50	15.0	– 7.35	– 0.50	7.35
55	16.5	– 8.09	– 0.55	6.62
60	18.0	– 8.82	– 0.60	5.88
65	19.5	– 9.56	– 0.65	5.15
70	21.0	– 10.29	– 0.70	4.41
75	22.5	– 11.03	– 0.75	3.68
80	24.0	– 11.76	– 0.80	2.94
85	25.5	– 12.50	– 0.85	2.21
90	27.0	– 13.23	– 0.90	1.47
95	28.5	– 13.97	– 0.95	0.89
100	29.92	– 14.70	– 1.01	0.0

* in %
† in in. Hg
‡ in psi
§ in bar
ll in psia

Figure 2-21. Common vacuum gauge readings range from –1.47 in. Hg to –14.7 in. Hg, and –0.10 bar to –1.01 bar.

Digital Resources

Name: ______________________ Date: ______________

MULTIPLE CHOICE

_______ **1.** ___ indicates the rate that work is done.
- A. Energy
- B. Force
- C. Power
- D. Revolutions per minute (rpm)

_______ **2.** If 225 kg of force is exerted on a hydraulic cylinder, it is equal to ___ lb of force.
- A. 227.2
- B. 450
- C. 454.4
- D. 495

_______ **3.** Pressure is typically measured in psi, bars, and ___.
- A. flow (gpm)
- B. kilopascals (kPa)
- C. pounds (lb)
- D. tons (t)

_______ **4.** If a hydraulic cylinder needs to lift a bridge beam that weighs 60,000 lb and the piston in the cylinder has a surface area of 113.10 sq in., then the hydraulic system must produce system pressure of ___ psi.
- A. 500.35
- B. 530.50
- C. 535
- D. 5300.50

_______ **5.** Most pump manufacturers recommend a vacuum rating less than or equal to ___ in. Hg.
- A. 1
- B. 5
- C. 10
- D. vacuum rating depends on pump design and manufacturer specifications.

_______ **6.** A ___ consists of a glass tube that is completely filled with mercury and closed on one end.
- A. Bourdon tube
- B. mercury barometer
- C. mercury column gauge
- D. Schrader gauge

_______ **7.** ___ energy is the energy of motion.
- A. Kinetic
- B. Static
- C. Thermal
- D. Total

_______________ **8.** A common form of inefficiency in a fluid power system is ___.
A. leaks
B. resistance to movement of mechanical devices
C. resistance in pipes and hoses
D. all of the above

_______________ **9.** ___ is anything that changes or tends to change the state of rest or motion of a body.
A. Energy
B. Force
C. Power
D. Work

_______________ **10.** Fluid mechanics is comprised of static energy and ___ energy.
A. kinetic
B. electrical
C. fluid
D. total

_______________ **11.** Horsepower is a mechanical unit of measure equal to the force required to move 550 lb, ___ ft in ___ sec.
A. 1; 1
B. 1; 10
C. 10; 1
D. 10; 100

_______________ **12.** Energy from gears, belts, or pulleys is an example of ___ energy.
A. electrical
B. mechanical
C. total
D. static

_______________ **13.** The ___ pressure gauge is the most commonly used fluid power pressure gauge in industry.
A. Bourdon tube
B. mercury barometer
C. mercury column gauge
D. Schrader gauge

_______________ **14.** ___ pressure is the amount of pressure above the existing atmospheric pressure and is used to measure pressure inside a closed fluid power system.
A. Absolute
B. Atmospheric
C. Fluid
D. Gauge

_______________ **15.** ___ is the useful amount of output energy from a fluid power system compared to the amount of input energy.
A. Fluid power circle
B. Fluid power system efficiency
C. Kinetic energy
D. Total energy

COMPLETION

______ **1.** ___ pressure is equal to the sum of gauge pressure and atmospheric pressure.

______ **2.** Pressure in a hydraulic system is the resistance to ___.

______ **3.** A(n) ___ pressure gauge has a tube that becomes straighter as pressure is increased.

______ **4.** ___ is used in a hydraulic system to allow fluid from the reservoir to travel into the pump.

______ **5.** Schrader pressure gauges take measurements that are accurate to about ± ___%.

______ **6.** Hydraulic pressure is created when fluid flow from a hydraulic pump meets ___.

______ **7.** ___ pressure is created by the weight of the atmosphere at sea level under standard air conditions.

______ **8.** Cylinders and motors in a fluid power system are considered ___.

______ **9.** ___ is the amount of weight in pounds that a fluid power system must produce to move an object.

______ **10.** A(n) ___ pressure gauge converts fluid pressure into an electrical signal.

______ **11.** ___ must be overcome to perform work.

______ **12.** ___ is the number of unit squares equal to the surface of an object.

______ **13.** Digital pressure gauges are accurate from ± ___% to ___%.

______ **14.** Fluid power system energy transmission is based on efficiency, power, and ___.

______ **15.** ___ is the force that stops, slows, or restricts the movement of fluid or devices in a fluid power system.

______ **16.** ___ is any pressure less than atmospheric pressure.

______ **17.** ___ is the capacity to do work.

______ **18.** A(n) ___ is used to measure atmospheric pressure using a column of mercury.

______ **19.** ___ energy involves the addition of heat to make a particle or molecule move faster.

______ **20.** ___ states that when a force is applied to a confined fluid, the force is felt throughout the fluid undiminished.

TRUE/FALSE

T F **1.** The amount of pressure required to move a cylinder with an attached load is the same for extension and retraction.

T F **2.** Some pressure gauges take measurements in two scales.

T F **3.** In fluid power systems, distance is typically measured in meters (m), although feet (ft) is sometimes used for specific situations.

T F **4.** When pressure is above 14.7 psia, a vacuum is created.

T F **5.** Sound energy is used in a fluid power system for troubleshooting purposes.

T F 6. In a pneumatic system, air is considered to be a gas and not a fluid.

T F 7. Vacuum in a confined space is created when pressure is below 14.7 psia.

T F 8. Pressure can be measured with a Schrader pressure gauge.

T F 9. Thermal energy is created in an efficient fluid power system.

T F 10. A Bourdon tube in its normal position is bent in the shape of the letter C.

T F 11. In a fluid power system, wasted energy is converted into heat and pressure drops.

T F 12. Digital pressure gauges can operate on AC or DC power.

T F 13. A liquid is noncompressible and can assume the shape of a confined space.

T F 14. A circle with a diameter of 10 sq in. has an area of 100 sq in.

T F 15. The strength of a vacuum depends on how far above 14.7 psia it is.

T F 16. Absolute pressure can be used in systems that have vacuums to avoid using a negative number.

T F 17. A properly installed fluid power system is 100% efficient.

T F 18. Force is the movement of an object (in lb) over a distance (in ft).

T F 19. All hydraulic cylinders have a piston with a rod attached to one side.

T F 20. Bourdon tube pressure gauge housings are typically filled with oil to prevent damage to the tube, pointer, or linkage.

T F 21. A measurement on a vacuum gauge that reads of 3.0 inches of mercury (in Hg) is also equal to 29.92 psi.

T F 22. Total energy in a fluid power system has the ability to change back and forth between static energy and kinetic energy as many times as necessary.

T F 23. The piston is the surface that makes contact with the pressure in a cylinder.

T F 24. It is impossible to have a fluid power system that is 100% efficient.

T F 25. Energy can easily be created or destroyed.

SHORT ANSWER

1. How much horsepower is required for the boom of a hydraulic grapple crane to lift a steel billet that weighs 1200 lb a distance of 12′ into a gondola car in 6 sec?

2. What is the surface area of the cap end of a piston with a diameter of 8″?

3. What is the surface area of the cap end of a piston with a diameter of 6″?

4. List four variables that affect pressure in a fluid power system.

5. How much pressure is created when a 22,000 lb load is applied to a 3″ diameter cylinder?

Activity 2-1: Car/Truck Lift System

Hydraulic lifts of varying sizes are used in automotive service centers to lift cars and trucks of various weights in order for service technicians to perform work on the undercarriage of the vehicles. Use the Car/Truck Lift System and information from the chapter to answer the following questions.

1. If Load A weighs 3250 lb and it is lifted a distance of 5′, how much work has been completed?

2. If it takes 7.5 sec to lift Load A a distance of 5′, how much power is required?

3. If it takes 7.5 sec to lift Load A a distance of 7′, how much horsepower is required?

4. If the piston in the lift cylinder has a diameter 18″, what is the surface area of the piston face?

5. If the piston in the lift cylinder has a radius of 9″, what is the surface area of the piston face?

6. If the directional control valve is activated, Load A weighs 6000 lb, and the piston in the lift cylinder has a surface area of 24 sq in., what reading is showing on the pressure gauge (in psi)?

7. What pressure reading is showing if the pressure gauge has its units in psia?

8. What pressure reading is showing if the pressure gauge has its units in bars?

9. If the directional control valve is activated, the pressure gauge is showing 2500 psi, and the piston in the lift cylinder has an area of 11 sq in., how much does Load A weigh?

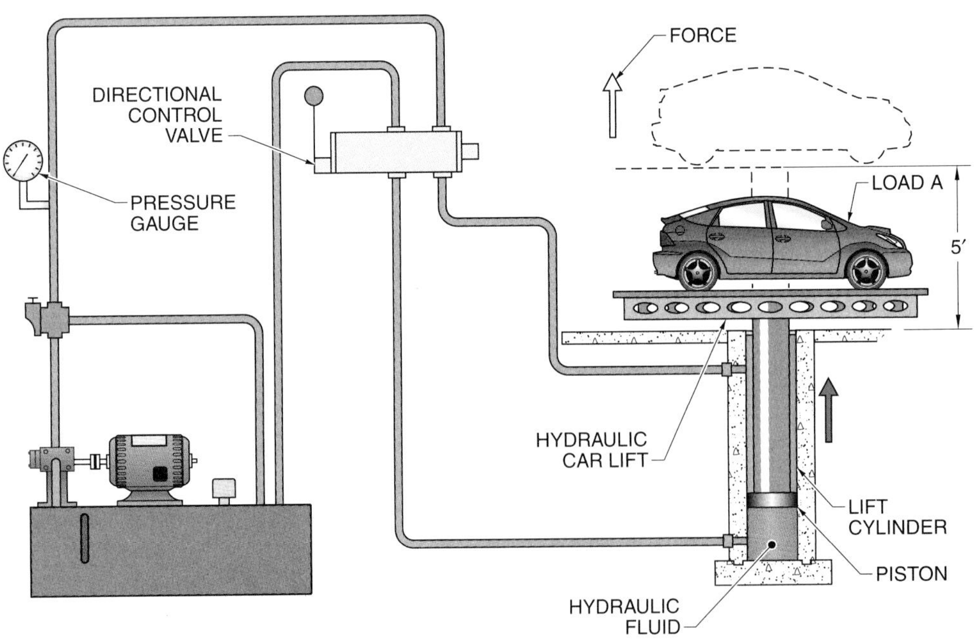

CAR/TRUCK LIFT SYSTEM

Activity 2-2: Material Handling System

Hydraulic systems are often used as part of a material handling system to move crates of product with varying weights on a conveyor line. Cylinder size can also vary depending on the hydraulic system used. Use Material Handling System and information from the chapter to answer the following questions.

1. If Crate A weighs 2250 lb, and the piston in the crate alignment cylinder has a diameter of 5″, what reading does the pressure gauge show (in psi)?

2. If Crate B weighs 2425 lb, and the piston in the crate alignment cylinder has a diameter of 6.5″, what reading does the pressure gauge show (in psia)?

3. If Crate C weighs 830 lb, and the piston in the crate alignment cylinder has a diameter of 8″, what reading does the pressure gauge show (in psi)?

4. If the pressure gauge is showing a reading of 275 psi, and the piston in the crate alignment cylinder has a diameter of 2.5″, what is the weight of Crate D?

5. What would be the pressure reading shown in question 4 if the pressure gauge had its units in psia?

6. What would be the pressure reading shown in question 4 if the pressure gauge had its units in bars?

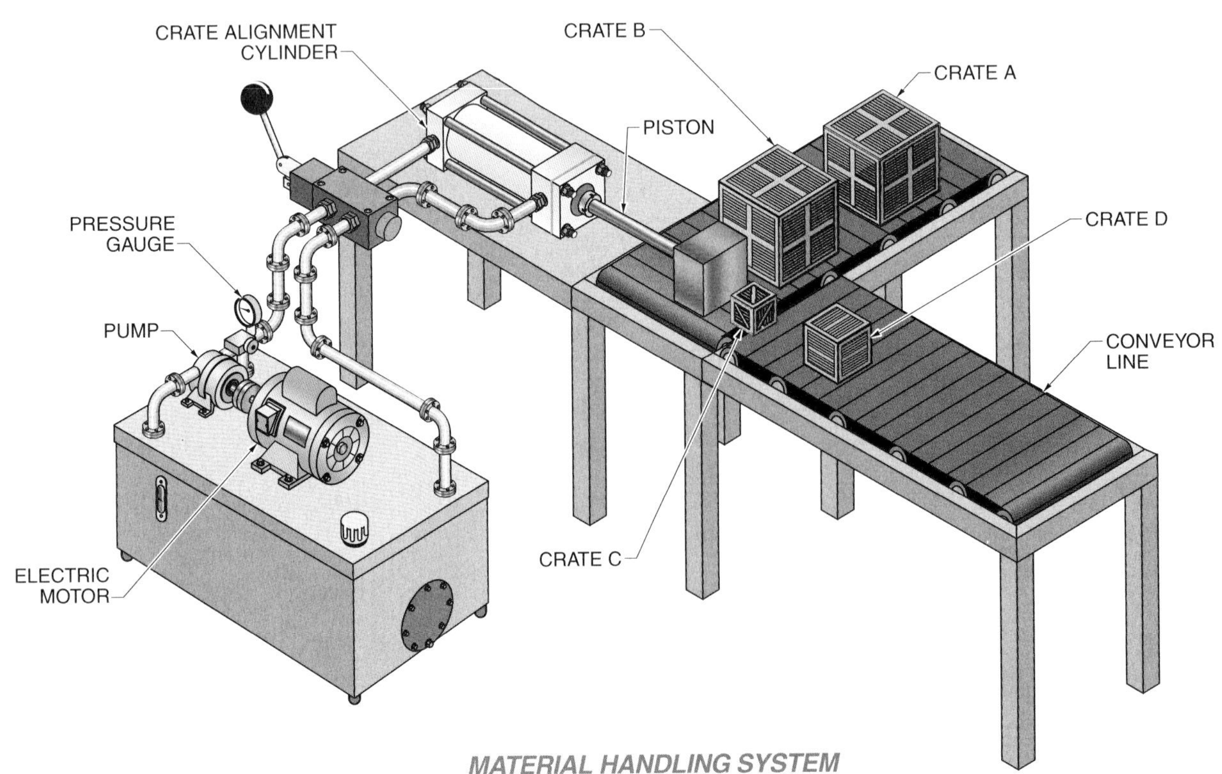

MATERIAL HANDLING SYSTEM

Activity 2-3: Hydraulic Production Press

Hydraulic production presses are typically used to form and bend different types of sheet metal into various shapes. The capabilities of a production press must be taken into consideration when bending the sheet metal. Use Hydraulic Production Press and information from the chapter to answer the following questions.

1. To form a sheet metal workpiece, a hydraulic cylinder on a production press must produce 20,000 lb of force. The die has a surface diameter of 4.5″. How much pressure is required to apply the required force of 20,000 lb?

2. If the production press was manufactured in Europe and the pressure gauge has its units in bars, what is the pressure reading if the required force of 20,000 lb is applied?

3. The rod of the piston in the hydraulic cylinder has a diameter of 2″, and the die on the production press weighs 2500 lb. How much pressure is required on the rod end of the piston in the cylinder to retract the rod and the die?

4. The manufacturer would like to switch over and produce a new product made from a different type of sheet metal workpiece than what is currently being formed in the production press. The sheet metal workpiece for the new product must be formed with 35,000 lb of force. The connection fittings on the hydraulic cylinder for the existing production press have a maximum allowable pressure of 1650 psi. The cap end of the piston in the cylinder has a surface diameter of 4.5″. Will the connection fittings on the cylinder be able to withstand the pressure required to supply 35,000 lb of force?

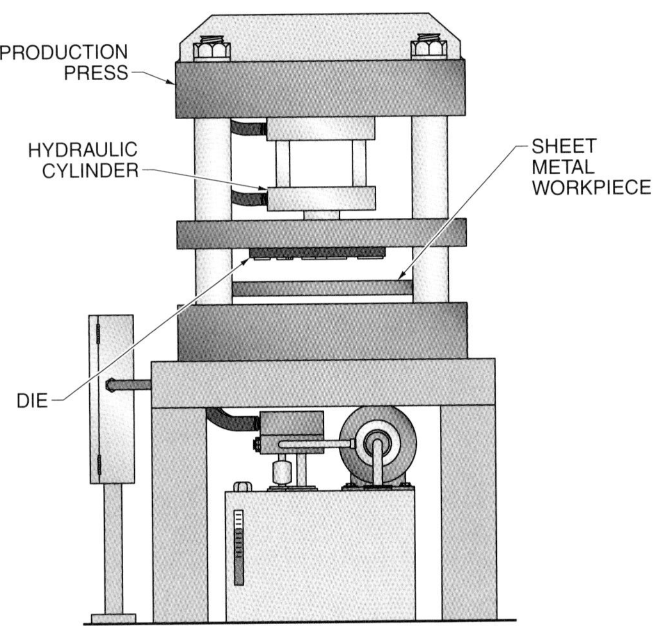

HYDRAULIC PRODUCTION PRESS

Activity 2-4: Tank Dumping System

A factory employee is learning how a tank dumping process works. The employee decides to perform the basic calculations needed to find what type of power and pressure is required for the tank dumping process. Use Tank Dumping System on pg. 56 and information from the chapter to answer the following questions. *Note:* It requires 120,000 lb of force to dump a full production tank and 20,000 lb of force to dump an empty production tank.

1. How much work is accomplished when the rod in the cylinder extends to dump the production tank?

2. If it takes 17 sec for the rod to extend, how much power is exerted?

3. How much horsepower is exerted?

4. How much pressure is required to extend the rod when the liquid from the production tank is dumped into the holding tank?

5. What would the display on Pressure Gauge 1 show if its units were only in bars?

6. How much pressure is required to retract the rod after it has dumped the production tank?

7. What would the display on Pressure Gauge 2 show if its units were only in bars?

8. The cylinder, hoses, connectors, and gauges for the tank dumping process need to be rated for ___ psi or above.

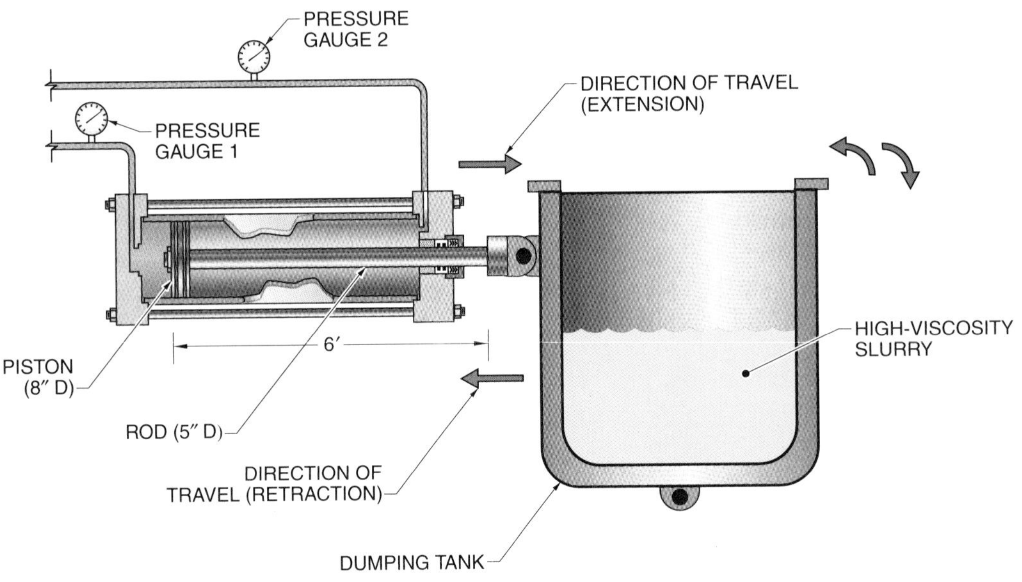

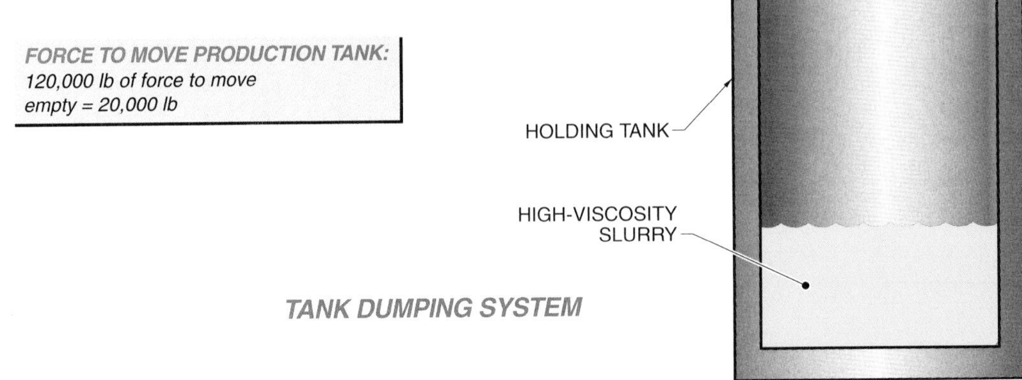

TANK DUMPING SYSTEM

Activity 2-5: Clamping and Bending System

A fabrication shop is planning to purchase a used pneumatic clamping and bending machine to help streamline production. Before the fabrication shop can purchase the clamping and bending machine, they must verify that they will have enough operating pressure within their facility and an inventory of replacement cylinders. Use Clamping and Bending System and information from the table below to answer the following questions.

	Operating Pressure*	Force†	Area‡	Diameter§
CYLINDER A	?	55	?	1.25
CYLINDER B	110	?	0.451	?
CYLINDER C	95	275	?	?

* in psig
† in lb
‡ in sq in
§ in in.

__________ **1.** The surface area of the piston in Cylinder A is ___ sq in.

__________ **2.** The pressure required to operate Cylinder A is ___ psi.

__________ **3.** The force required to operate Cylinder B is ___ lb.

__________ **4.** The diameter of the piston in Cylinder B is ___″.

__________ **5.** The surface area of the piston in Cylinder C is ___ sq in.

__________ **6.** The diameter of the piston in Cylinder C is ___″.

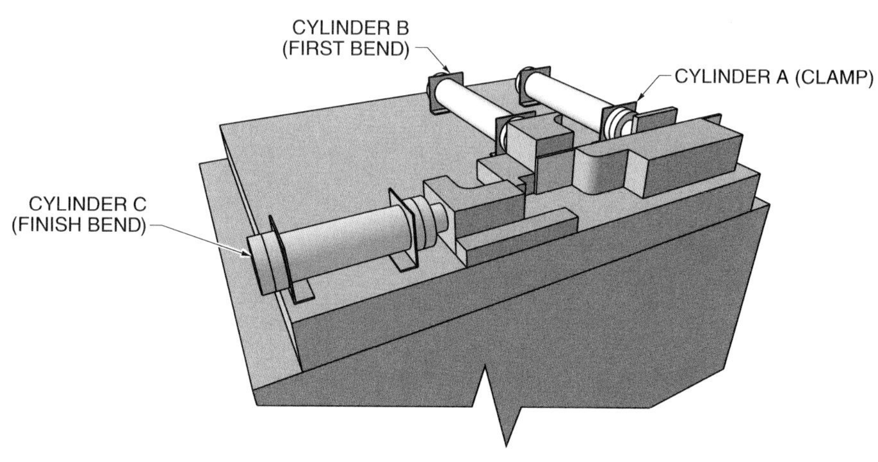

CLAMPING AND BENDING SYSTEM

Chapter 3

Hydraulic System Fundamentals

OBJECTIVES

- Describe the basic characteristics of hydraulic fluid in a hydraulic system.
- Describe the relationship between resistance, and pressure in a hydraulic system.
- Define viscosity of a liquid.
- Describe the different types of accumulators.
- List the differences between cutaway, pictorial, and schematic diagrams used in fluid power systems.

INTRODUCTION

In industrial and mobile fluid power systems, the main sources of energy are hydraulic systems. Hydraulic systems transmit energy by using the energy contained in flowing fluid and converting that energy into mechanical energy. Hydraulic systems can have different components and controls. A fluid power system schematic diagram is used to understand and properly troubleshoot a fluid power system.

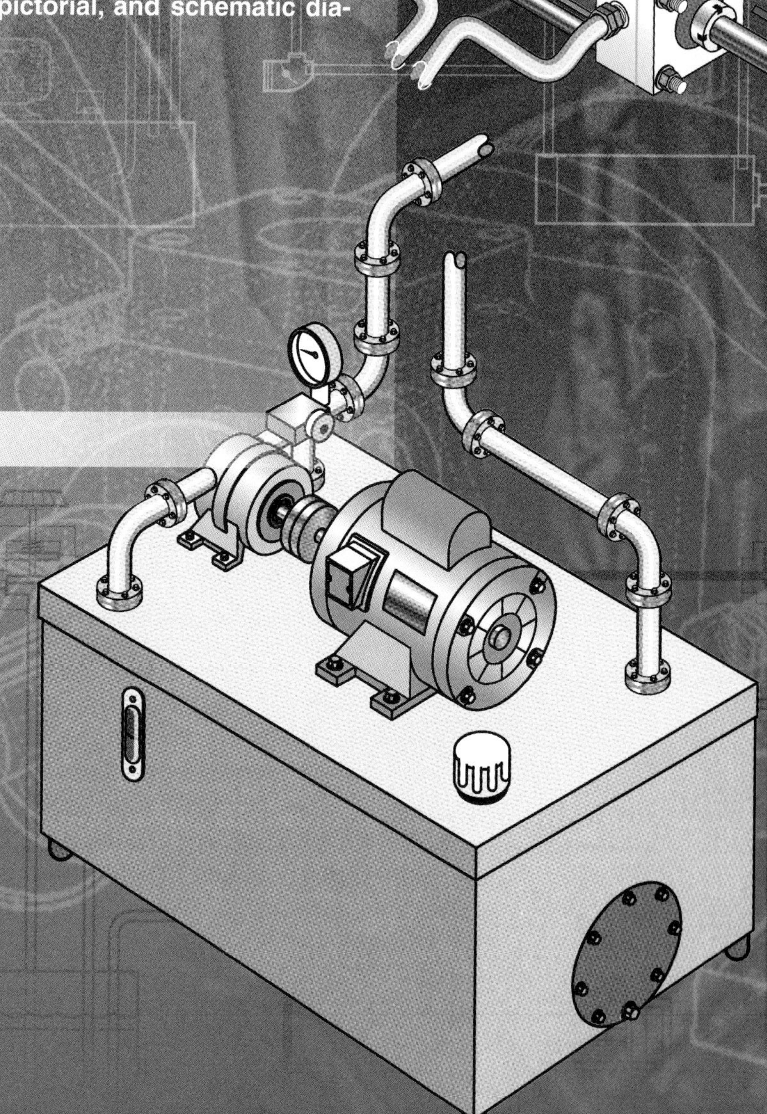

TERMS

Hydraulic fluid is the liquid used in hydraulic systems to transfer energy.

A **molecule** is matter that is composed of atoms and is the smallest particle that a compound can be reduced to while still possessing the chemical properties of that compound.

HYDRAULIC PRINCIPLES

Hydraulic principles are based on fluid flow principals and the properties of liquids, which are used in the transmission of energy. By controlling fluid flow and the resulting pressure, hydraulic systems can produce a large amount of work. Hydraulic systems transmit energy so its specific work can be accomplished.

Properties of Liquids

In fluid power systems, liquid usually refers to hydraulic fluid. *Hydraulic fluid* is the liquid used in hydraulic systems to transfer energy. Most hydraulic fluid is oil-based.

A liquid can take the shape of any container it is in. For example, hydraulic fluid takes the shape of a cylinder in a hydraulic system. However, if the cylinder is not completely filled, it will not be able to transmit power. **See Figure 3-1.**

In a hydraulic system, liquid can transmit energy efficiently because, for all practical purposes, liquids are incompressible. The properties of a liquid are determined by the molecules of the liquid. A *molecule* is matter that is composed of atoms and is the smallest particle that a compound can be reduced to while still possessing the chemical properties of that compound. Molecules are always moving. The molecules of a liquid have a semi-strong attraction to each other that allows them to be very close but still move freely. This allows the liquid to take the shape of any container it is in. The characteristics of liquid molecules allow liquids to be relatively incompressible and still flow inside of a fluid power system. This accounts for the smooth, even movement of a hydraulic actuator, such as a cylinder.

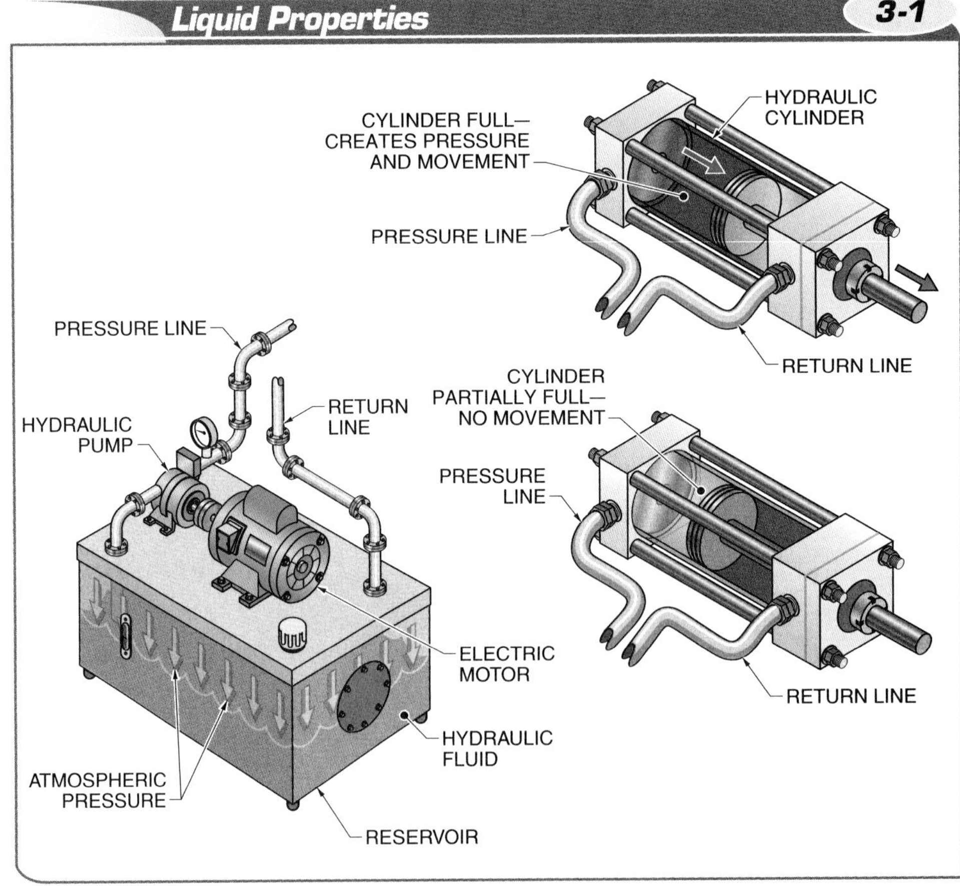

Figure 3-1. Liquid takes the shape of any container it is in, but only as much volume as it can fill.

Unlike air, a liquid is difficult to compress, and unlike a solid, it can be compressed a small amount without destroying its integrity. When a double-acting cylinder is completely filled with hydraulic fluid and pressure is applied to its cap end, the piston in the cylinder will not move if the hydraulic fluid does not have a place to flow. For hydraulic fluid to flow into a double-acting cylinder, it must also flow out of the cylinder and into the reservoir. **See Figure 3-2.** The cylinder needs to be able to release the pressure created in the cylinder (hoses or other weak points) or that part of the system could rupture.

Transmission of Energy. Since hydraulic fluid will only compress approximately 0.5% for every 1000 psi applied to it, the transmission of energy through fluids is almost as efficient as through solids. Pascal's law, stating that pressure is distributed undiminished throughout a closed container, can be applied during the transmission of energy in a hydraulic system. For example, two cylinders both with a piston face area of 10 sq in., are attached together by a pipe that is filled with liquid. An input force of 1000 lb is applied on the first 10 in. area piston, creating an equivalent amount of pressure (100 psi) that will move the second 10 in. area piston with an output force of 1000 lb. **See Figure 3-3.**

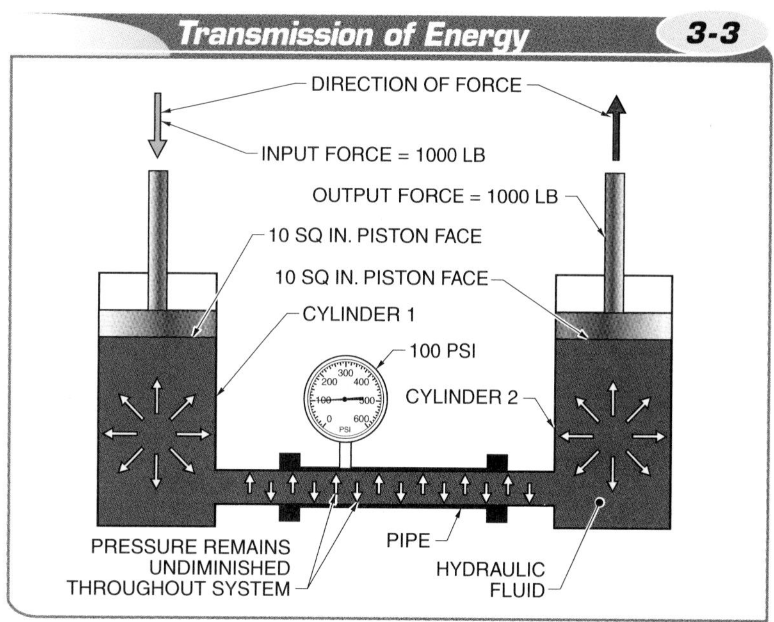

Figure 3-3. Pascal's law can be applied during the transmission of energy in a hydraulic system.

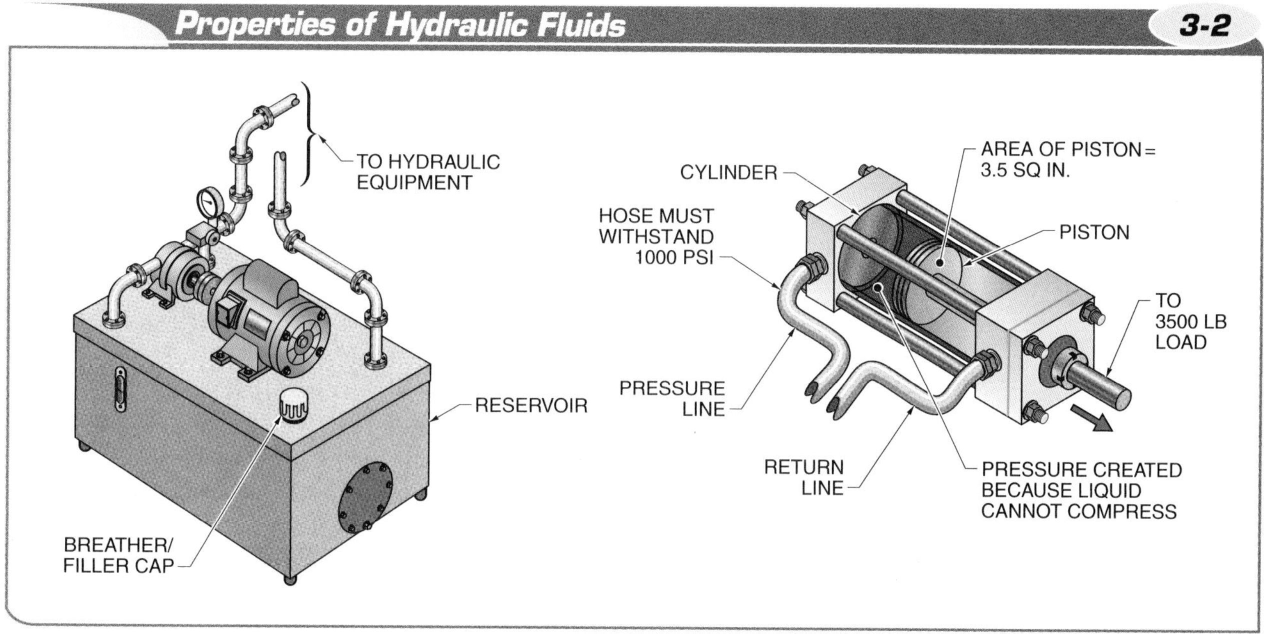

Figure 3-2. In a hydraulic system, liquid can transmit energy because it is almost completely incompressible.

TERMS

Flow is the movement of fluid through piping in gallons per minute (gpm) for hydraulic systems and in standard cubic feet per minute (scfm) in pneumatic systems.

Laminar flow is fluid flow in a hydraulic system passage that is characterized by a slow, smooth movement of fluid in a straight path along the centerline of the passage.

Turbulent flow is fluid flow in a hydraulic system passage that is characterized by a rapid movement of fluid in an erratic, nonlayered, and random pattern.

Fluid flow is the movement of fluid caused by a difference in pressure between two points.

Principles of Fluid Flow

Flow is the movement of fluid through piping in gallons per minute (gpm) for hydraulic systems and in standard cubic feet per minute (scfm) in pneumatic systems.

Fluid flows from high-pressure areas to low-pressure areas. In a pipe, fluid develops a parallel-layered pattern that starts with the hydraulic fluid that is in contact with the pipe's wall. The fluid layer that is in contact with the pipe's wall flows at lowest velocity. The fluid's velocity increases with every layer that is farther away from the wall. The highest velocity is in the middle of the pipe.

When the pipe is properly sized and the bends have the proper radius, the flow has laminar flow. When the pipe is not properly sized and the bends are too tight or there are too many bends located near each other, the fluid can develop turbulent flow. **See Figure 3-4.**

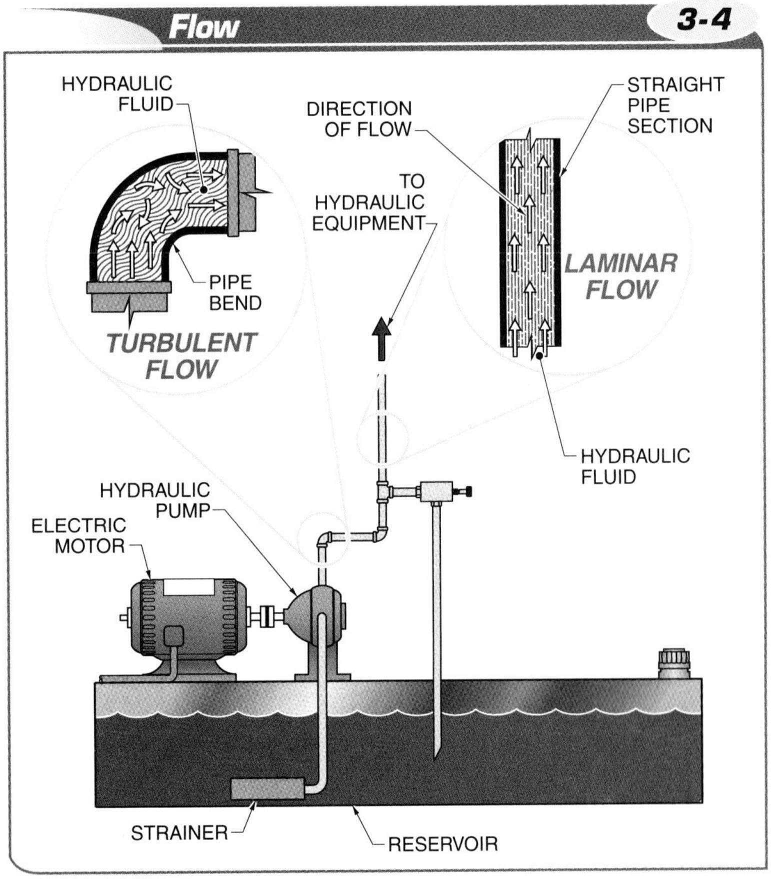

Flow 3-4

Figure 3-4. Laminar flow is preferred in fluid power systems over turbulent flow because it is more efficient and loses less energy.

Laminar flow is fluid flow in a hydraulic system passage that is characterized by a slow, smooth movement of fluid in a straight path along the centerline of the passage. If the pipe is properly sized and the bends have the proper radius, the velocity of the hydraulic fluid should maintain laminar flow. Laminar flow is preferred in fluid power systems because it does not allow unnecessary friction to occur, which achieves minimal energy loss.

Turbulent flow is fluid flow in a hydraulic system passage that is characterized by a rapid movement of fluid in an erratic, nonlayered, and random pattern. Turbulent flow is present in systems where pipes are too small for the system, bends are too sharp, or there are too many bends. This causes the velocity of the hydraulic fluid to increase, which results in the layered patterns distorting and bouncing around inside the pipe, tube, or hose. Turbulent flow in a hydraulic system causes inefficiency because it takes more energy to transmit the fluid. Energy is lost through excessive heat buildup because of unnecessary friction.

Fluid flow is the movement of fluid caused by a difference in pressure between two points. Fluids follow the path of least resistance. In a hydraulic system, fluid flow is produced by a pump. The pump also determines how many gallons of fluid flow per minute. The amount of gallons of fluid flow per minute determines the speed at which the hydraulic cylinders operate.

Fluid flow can have series or parallel flow paths. A *series flow path* is a flow path configuration in which there is only one path for hydraulic fluid to flow. Many components are connected into a hydraulic system in series. For example, the inlet and outlet ports on opposite sides of a filter have separate pipes connected to them. The pipe connected to the filter outlet port is routed to the inlet port of a valve that changes the amount of fluid flow. The outlet port of the valve has another pipe attached to it, which allows fluid flow to continue on through the hydraulic system. Because all of the components and pipes only have one path for fluid to flow, they are attached in series.

A *parallel flow path* is a flow path configuration in which there are two or more paths for hydraulic fluid to flow. Parallel flow paths are used to supply different parts of equipment that receive their power from the same power source (hydraulic pump). Parallel flow paths are also used when the same valve controls two cylinders. For example, a backhoe sometimes has two cylinders attached to its shovel. The same valve controls both cylinders so that when fluid flow is sent to extend the cylinders, it has two paths. One path leads to one cylinder and the other path leads to the other cylinder. *Note*: Almost all hydraulics systems have both series and parallel flow paths.

Volume is the size of a space or chamber measured in cubic units. Regardless of the shape of the space, volume is expressed in cubic inches (cu in.) or cubic feet (cu ft). **See Figure 3-5.**

The two cylinders used to control the boom of an excavator are an example of parallel flow paths.

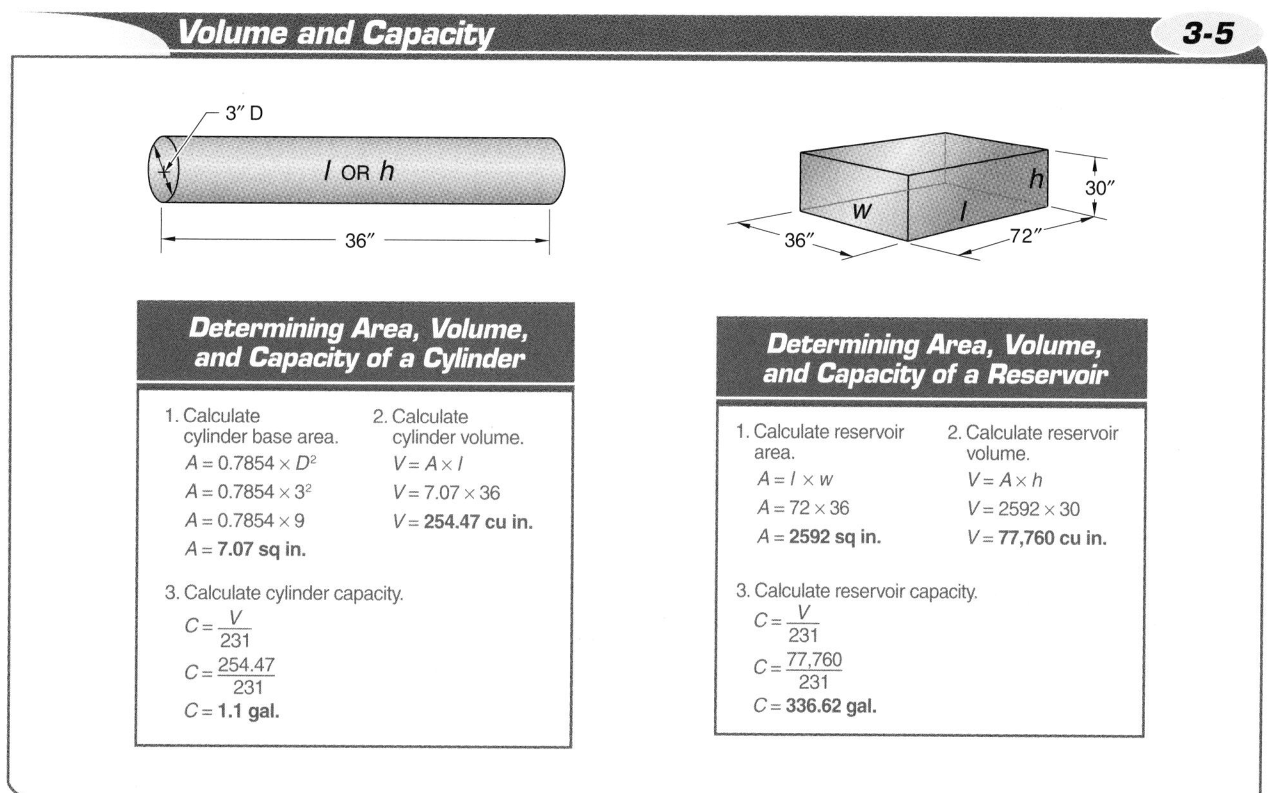

Figure 3-5. Volume is the three-dimensional space of an object measured in cubic units.

TERMS

A **series flow path** is a flow path configuration in which there is only one path for hydraulic fluid to flow.

A **parallel flow path** is a flow path configuration in which there are two or more paths for hydraulic fluid to flow.

Volume is the size of a space or chamber measured in cubic units.

Capacity is the ability to hold or contain something.

Capacity is the ability to hold or contain something. Capacity is based on the volume of an object. Because volume is usually measured in cubic inches, to calculate the capacity of an object in gallons, it is necessary to divide the volume by 231 (there are 231 cu in. in one gallon). The volume and capacity of a cylinder is calculated by applying the following procedure:

1. Calculate cylinder base area.

$$A = 0.7854 \times D^2$$

where
A = area
0.7854 = constant
D^2 = diameter squared

2. Calculate the cylinder volume.

$$V = A \times l$$

where
V = volume
A = area
l = length

3. Calculate cylinder capacity.

$$C = \frac{V}{231}$$

where
C = capacity (in gal.)
V = volume (in cu in.)
231 = constant

Although reservoir capacity can be greater, the correct amount of hydraulic fluid for use in a particular system is indicated with a fluid level sight gauge located on the reservoir.

Example: What is the volume and capacity of a hydraulic cylinder that has a 3″ diameter and is 36″ long?

1. Calculate cylinder base area.

$$A = 0.7854 \times D^2$$
$$A = 0.7854 \times 3^2$$
$$A = 0.7854 \times 9$$
$$A = 7.07 \text{ sq in.}$$

2. Calculate cylinder volume.

$$V = A \times l$$
$$V = 7.07 \times 36$$
$$V = 254.47 \text{ cu in.}$$

3. Calculate cylinder capacity.

$$C = \frac{V}{231}$$
$$C = \frac{254.47}{231}$$
$$C = \mathbf{1.10 \text{ gal.}}$$

Note: The formula for finding the volume of a cylinder may also be expressed as $V = 0.7854 \times D^2 \times l$.

The volume of a rectangular object, such as a reservoir, is calculated by finding the area and multiplying by the height. The volume and capacity of a reservoir is calculated by applying the following procedure:

1. Calculate reservoir area.

$$A = l \times w$$

where
A = area
l = length
w = width

2. Calculate the reservoir volume.

$$V = A \times h$$

where
V = volume
A = area
h = height

3. Calculate reservoir capacity.

$$C = \frac{V}{231}$$

where
C = capacity (in gal.)
V = volume (in cu in.)
231 = cu in. in one gallon

Example: What is the capacity of a rectangular hydraulic fluid tank that is 72″ long, 36″ wide, and 30″ high?

1. Calculate reservoir area.

$$A = l \times w$$
$$A = 72'' \times 36''$$
$$A = 2592 \text{ sq in.}$$

2. Calculate the reservoir volume.

$$V = A \times h$$
$$V = 2592 \times 30$$
$$V = 77{,}760 \text{ cu in.}$$

3. Calculate reservoir capacity.

$$C = \frac{V}{231}$$
$$C = \frac{77{,}760}{231}$$
$$C = \mathbf{336.62\ gal.}$$

Viscosity. *Viscosity* is a measurement of the resistance to flow against an established standard. Fluids that are thick and flow with difficulty have high viscosity. Fluids that are thin and flow easily have low viscosity. For example, liquid plastic resin has high viscosity and water has low viscosity. **See Figure 3-6.** Temperature can affect the viscosity of fluids. For example, when hydraulic fluid is heated, viscosity decreases, causing it to flow more easily. When hydraulic fluid is cooled, viscosity increases, causing it to flow slowly.

Viscosity 3-6

PIPING
THICK FLUID (PLASTIC RESIN)
HIGH VISCOSITY
PIPING
THIN FLUID (WATER)
LOW VISCOSITY

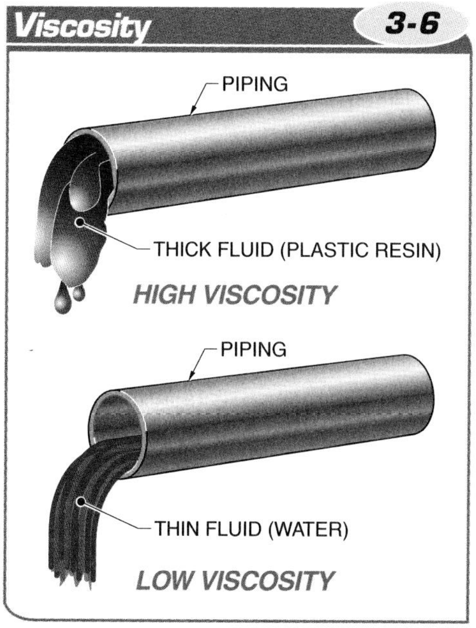

Figure 3-6. Fluids that are thick and flow with difficulty have high viscosity, while liquids that are thin and flow easily have low viscosity.

Viscosity for a specific fluid is determined under laboratory conditions by measuring the time required for a specific amount of a fluid at a set temperature to flow through a precisely sized orifice. Viscosity is measured in Saybolt Universal Seconds (SUS) using a Saybolt viscometer. A *Saybolt viscometer* is a test instrument used to measure fluid viscosity. **See Figure 3-7.** A heated tank or reservoir of test fluid is placed into a heated oil bath. A control valve is opened once the test fluid reaches the required temperature, allowing 60 mL to flow through a precision orifice while being timed with a stopwatch. The measured time is the SUS.

In the United States, a Saybolt viscometer is the most common method used to test hydraulic fluid viscosity. For industrial applications, hydraulic fluid viscosities are typically about 150 SUS at 104°F. The Society of Automotive Engineers (SAE) has established standard numbers for hydraulic fluid viscosity readings. **See Appendix.**

TERMS

Viscosity is a measurement of the resistance to flow against an established standard.

A **Saybolt viscometer** is a test instrument used to measure fluid viscosity.

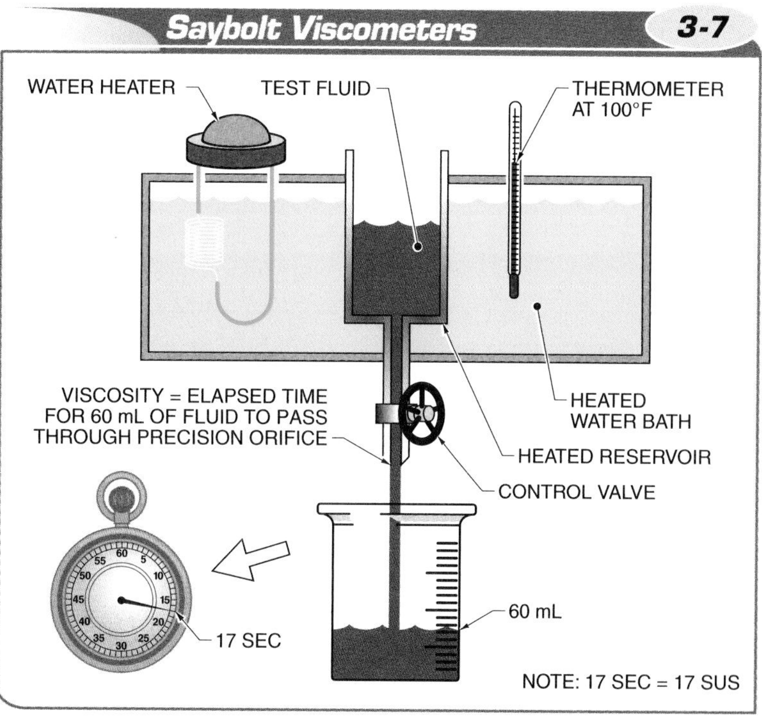

Figure 3-7. A Saybolt viscometer is a test instrument used to measure fluid viscosity.

TERMS

The **viscosity index (VI)** is a number that represents how much the viscosity of a fluid changes in respect to a specific change in temperature.

Flow rate is the amount of fluid that passes a given point in one minute.

A **flow meter** is a meter that measures the flow of hydraulic fluid within a system.

A **digital flow meter** is a flow meter that uses electrical signals to indicate the flow rate of a fluid on a digital display.

Pressure drop is the pressure differential between any two points in a hydraulic system or component.

Viscosity Index. The *viscosity index (VI)* is a number that represents how much the viscosity of a fluid changes in respect to a specific change in temperature.The VI number indicates the relative change in SUS. Hydraulic fluids with a low VI number register a large change in SUS as the temperature changes. Hydraulic fluids with a high VI number change only slightly as the temperature changes. The higher the VI number, the less change there is in viscosity of fluid over a given temperature range. Desirable hydraulic fluids are those that have a high VI number, or a relatively low SUS change.

Flow Rate. *Flow rate* is the amount of fluid that passes a given point in one minute. Flow rate is measured in gallons per minute (gpm) with a flow meter. A *flow meter* is a meter that measures the flow of hydraulic fluid within a system. **See Figure 3-8.** As the hydraulic fluid flows through the meter, a calibrated weight moves an indicator to display the amount of flow. A weighted flow meter has high internal resistance to fluid flow. Because of the calibrated weight in the flow meter, there is usually a drop in pressure across the meter. Flow meters are mounted vertically in the piping with the fluid flow entering at the bottom of the meter. When installing a flow meter, it must be verified that the flow meter is installed in the proper direction. Some flow meters allow flow in both directions, while others allow flow in only one direction. Flow meters are versatile and useful troubleshooting devices when understood and used properly. The equipment manufacturer's manual and specification must always be consulted prior to installation and use.

A *digital flow meter* is a flow meter that uses electrical signals to indicate the flow rate of a fluid on a digital display. Digital flow meters are typically turbine driven and send an analog (4 mA to 20 mA; 1 VDC to 5 VDC) signal to a receiver/display. Digital flow meters have several advantages over mechanical flow meters including:

- less resistance to fluid flow
- greater accuracy
- ability to send analog signals to industrial controls
- an alarm directly attached to the meter
- auxiliary contacts that help control industrial circuit logic
- conversion allowed between gallons per minute (gpm) and liters per minute (Lpm)

Friction is created through a resistance to fluid flow and generates energy that is converted into heat. Friction is generated within a hydraulic system between the hydraulic fluid and the walls of piping, tubing, or hoses, as well as from the viscosity of the hydraulic fluid. The faster a fluid flows, the greater the friction. The resistance to fluid flow is greatest at bends in the piping, extremely long pipe runs, undersized piping that can cause turbulent flow, clogged filters and strainers, worn parts such as damaged cylinder pistons, hose connections, damaged hoses, check valves, and valves with fluid flow. To overcome internal friction, system pressure increases. **See Figure 3-9.**

Pressure and flow rate are independent of each other, but both assist in the movement of a cylinder and a load. Pressure provides the force to move a piston in a cylinder, and therefore the load. The resistance to fluid flow determines pressure. Flow rate is used to affect the piston speed. Flow rate is typically determined by the pump. The pressure of a moving fluid is always higher closer to the pump. *Pressure drop* is the pressure differential between any two points in a hydraulic system or component.

Fluids follow the path of least resistance. For example, when three in-line cylinders are connected to the same pump, the cylinder with the least amount of load on its piston extends first. If cylinder 1 has a load of 200 lb, cylinder 2 has a load of 500 lb, and cylinder 3 has a load of 1000 lb, the piston from cylinder 1 extends first. Once the piston from cylinder 1 extends completely, then the piston from cylinder 2 extends. Once the piston from cylinder 2 extends completely, then the piston from cylinder 3 extends. This occurs because of the relationship between fluid flow and pressure. Fluid will flow only if the pump can supply enough pressure to overcome the load attached to the actuators.

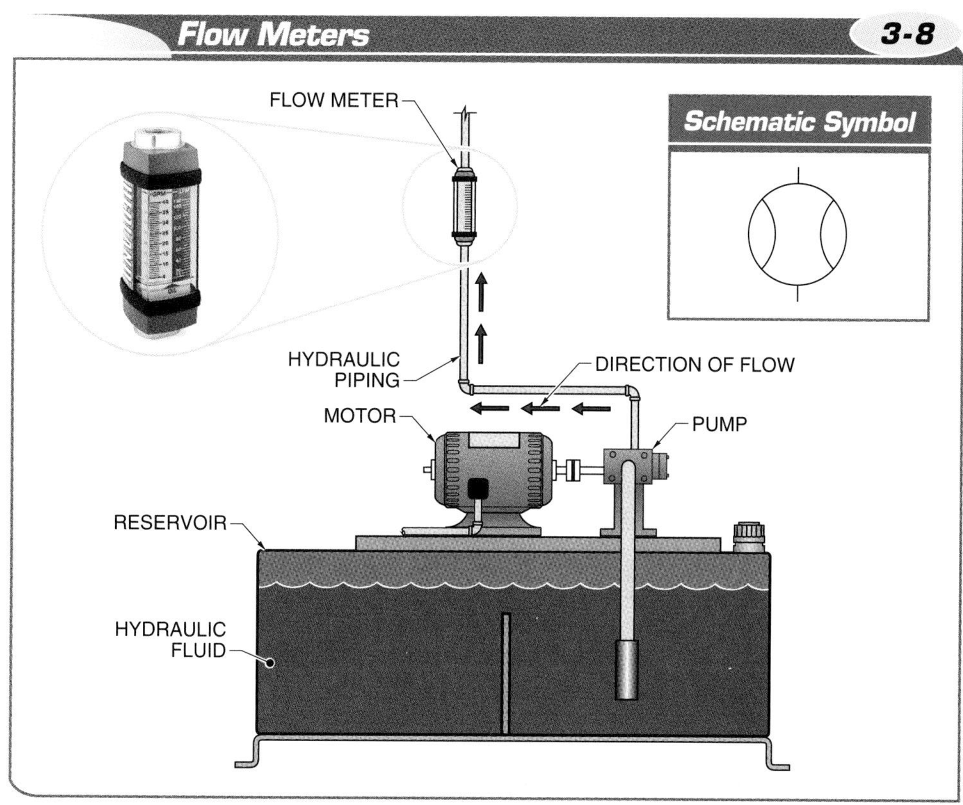

Figure 3-8. A flow meter is a meter that is used to measure the flow of fluid (in gpm) within a system.

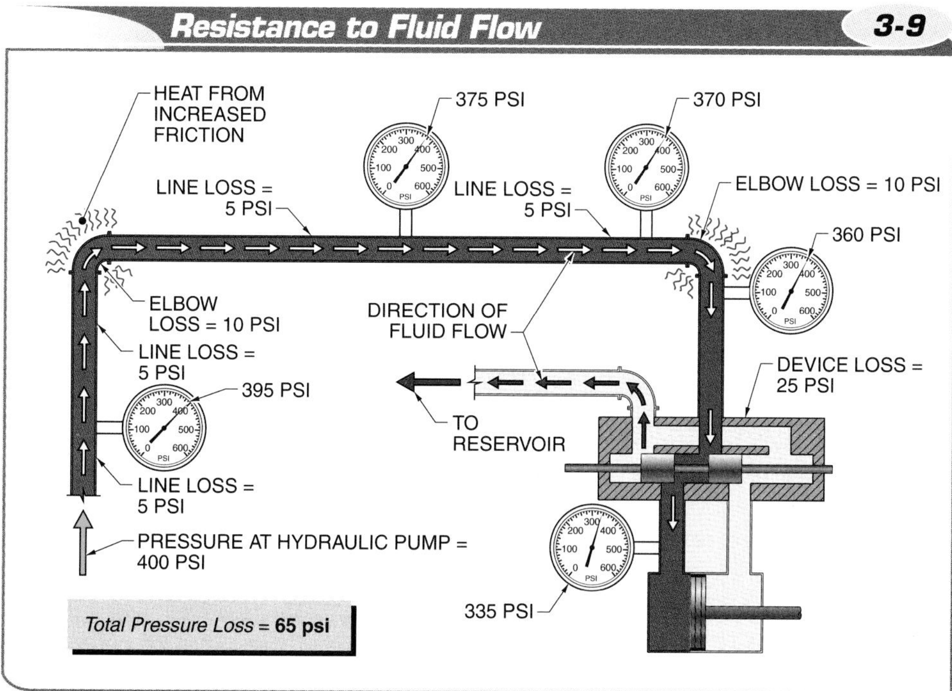

Figure 3-9. Resistance to fluid flow generates friction and the resulting energy is converted into heat.

When the pump starts, it begins to build pressure because there is not yet enough pressure to overcome any of the loads attached to the cylinders. When pressure reaches the proper level, cylinder 1 starts to move. The hydraulic pump only supplies enough pressure to maintain fluid flow. After cylinder 1 is fully extended, flow stops and the pump begins to build pressure again. When the pressure is high enough to move cylinder 2, the pump will only supply enough pressure to maintain flow. After cylinder 2 is fully extended, flow stops and the pump begins to build pressure again. When the pressure is high enough to move cylinder 3, the pump will only supply enough pressure again to maintain flow. **See Figure 3-10.**

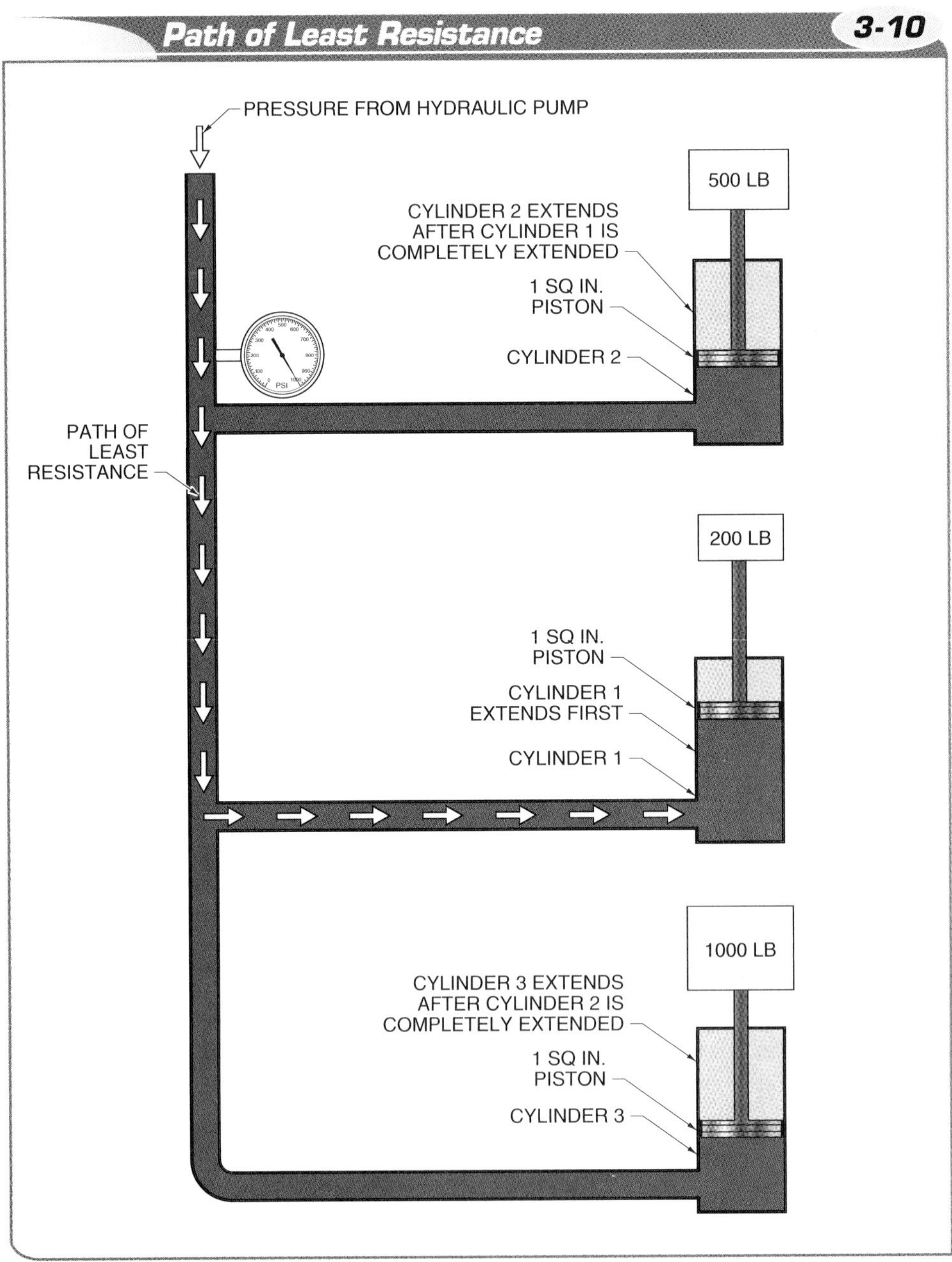

Figure 3-10. Fluids follow the path of least resistance, thus the cylinder with the least amount of load on it extends first.

Pressure and Resistance Relationship

The resistance to fluid flow determines the pressure in a hydraulic system. A hydraulic system is based on the amount of pressure that is required. Resistance to fluid flow in a hydraulic system is determined by two variables, the load on the cylinder and the internal resistance of the flow of fluid to the cylinder.

Pressure Control

Pressure control is an important safety consideration in hydraulic systems. If system pressure is not controlled, a hydraulic pump will continue to create flow, which can cause pressure to keep rising. If pressure continues to rise without any control, the system will burst at its weakest point. A burst pipe or valve is a safety hazard for personnel and equipment in the immediate operating area. To avoid this hazard, valves are used to control pressure. A *valve* is a device that controls the pressure, direction, and/or rate of fluid flow.

Relief Valves. Hydraulic system pressure is maintained by the use of a relief valve. A *relief valve* is a valve that protects a fluid power system from overpressure by setting a maximum operating pressure. Relief valves are normally closed (NC) valves that require a higher-than-spring force to open. **See Figure 3-11.**

A relief valve is composed of a valve body, bias spring, poppet, and bias spring adjustment screw. A *valve body* is a housing for valve components. A *bias spring* is a spring that has an adjustable force. Bias springs are used in many relief valves and can be considered high bias (2000 lb) or low bias (25 lb). A *poppet* is a movable part within the valve body that separates the hydraulic fluid and the bias spring. When pressure overcomes the bias spring, the poppet moves to allow hydraulic fluid to flow through it. A *bias spring adjustment screw* is a screw that is used to adjust the amount of force in a bias spring.

The inlet (primary) port is connected to the system pressure and the discharge (secondary) port is connected to the reservoir. Poppet movement is controlled by a predetermined spring force. The pressure level, or spring force, is usually varied by adjusting a screw. When pressure returns to safe levels, the relief valve returns to a closed position. A relief valve prevents the pressure in a system from overcoming the breaking point and damaging the system by bursting a pipe, tube, or hose. A relief valve operates through the following procedure:

1. Set the required system pressure with a bias spring adjustment screw.
2. Verify maximum pressure by operating the hydraulic system at maximum resistance. The method for getting maximum system pressure depends on the system design. One common method is to allow the cylinder to extend completely out and continue to apply pressure. Take system readings. *Note:* When hydraulic fluid flows through the system to accomplish work, there should not be enough system pressure to overcome the bias spring.

When there is no place for fluid to flow and resistance and pressure quickly increase, the pressure at the relief valve's poppet compresses the bias spring and allows fluid to flow into the reservoir. When the system changes and fluid begins to flow, the pressure lowers and the biasing spring forces the poppet closed. This stops hydraulic fluid from flowing back into the reservoir by forcing the main poppet down against the pressure port of the relief valve.

A bias spring adjustment screw is used to adjust the amount of bias spring force in a relief valve.

TERMS

A **valve body** is a housing for valve components.

A **bias spring** is a spring that has an adjustable force.

A **poppet** is a movable part within the valve body that separates the hydraulic fluid and the bias spring.

A **bias spring adjustment screw** is a screw that is used to adjust the amount of force in a bias spring.

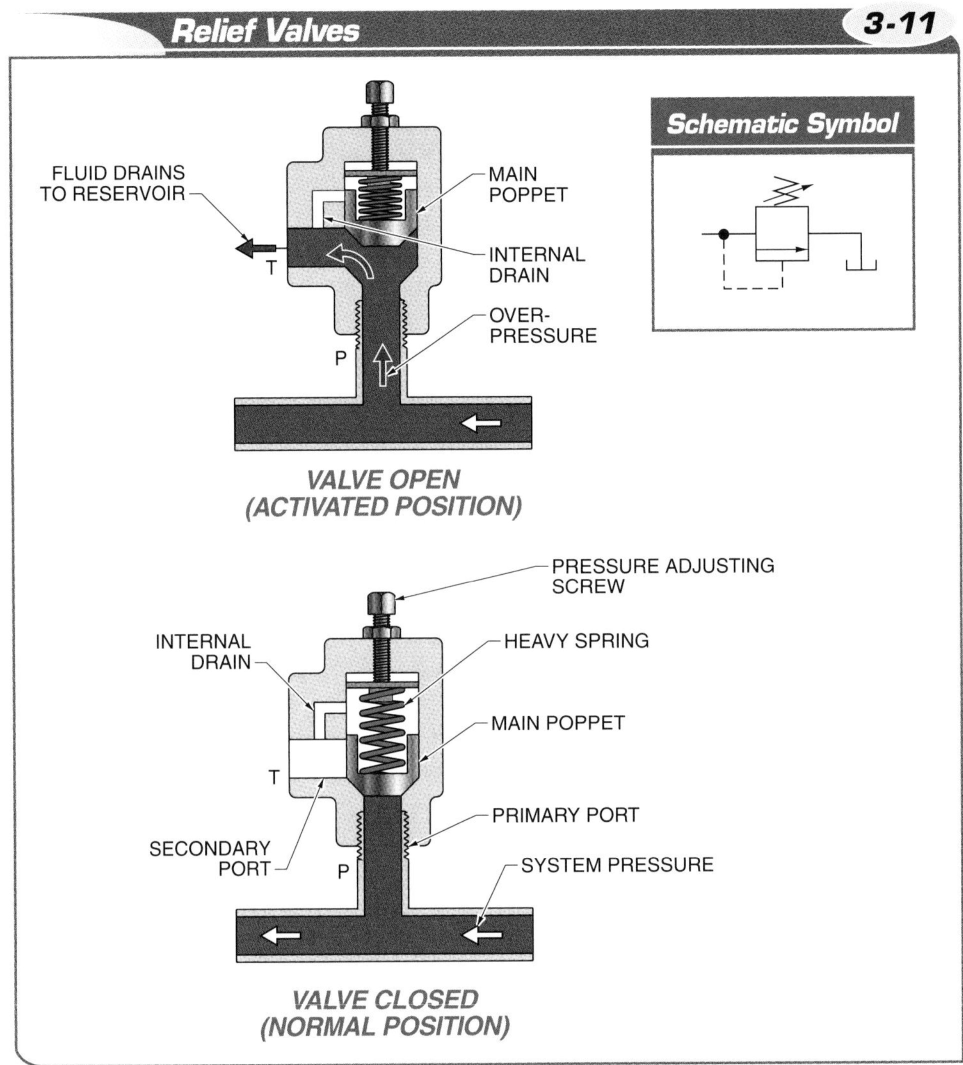

Figure 3-11. A relief valve sets the maximum operating pressure in a hydraulic system.

Maximum Resistance

Maximum resistance occurs when there is no fluid flow in a hydraulic system. For example, there is no fluid flow when a piston reaches the end of a cylinder. When this occurs, resistance rapidly increases to maximum resistance. The system pressure then suddenly increases, or spikes, to its maximum value.

Other forms of resistance in a hydraulic system include valves, bends in the piping, friction between fluid and the piping, and any reduction in the size of the piping the fluid is traveling through. If there is too much resistance, pressure will increase until the weakest point in the system fails and bursts open, or a relief valve opens to allow all of the fluid to flow back into the reservoir, thus relieving all of the extra pressure in the system. **See Figure 3-12.**

TECH FACT

Lubricants on the mating surfaces of hydraulic equipment reduce friction because they form molecular bonds of reduced strength. The best materials for lubricants are typically liquids, such as oil, or soft solids, such as grease, because they form weak molecular bonds.

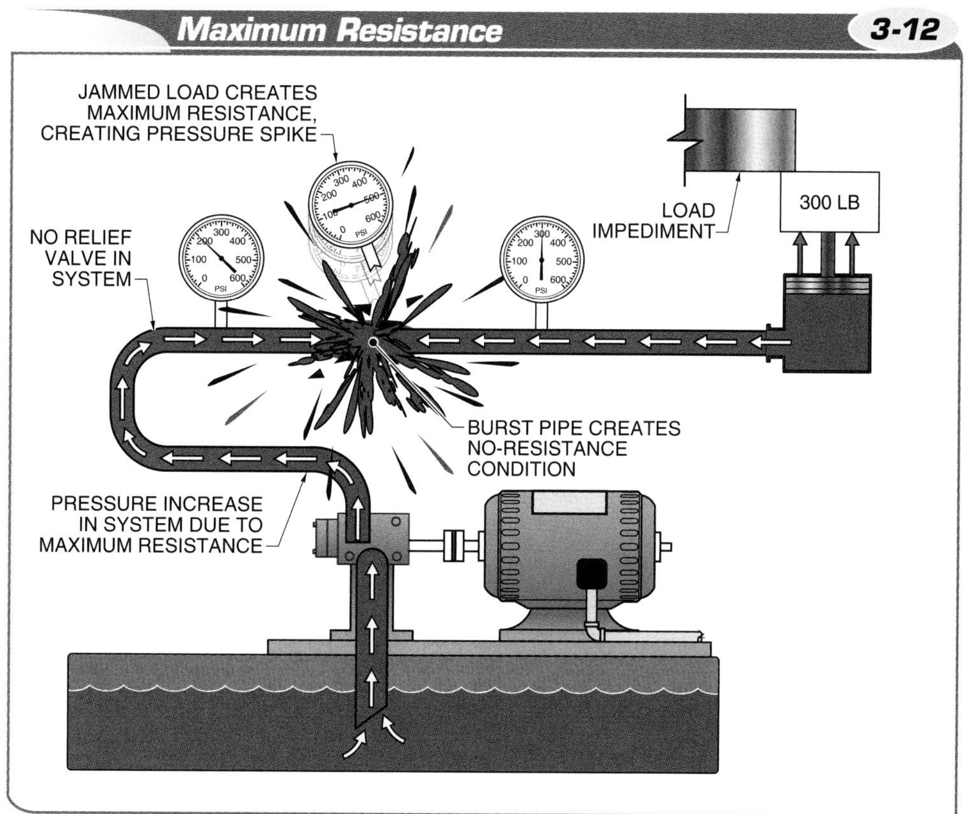

Figure 3-12. When there is too much resistance, pressure increases until the weakest point in the system fails and bursts.

TERMS

An **accumulator** is a vessel in which fluid is stored under pressure for future release of energy.

HYDRAULIC SYSTEM PRESSURE SUPPLEMENTS

Sometimes hydraulic systems must be supplemented for increased efficiency. Hydraulic systems use add-on components to help the system maintain or increase pressure and maintain or increase force. Add-on components are installed in hydraulic systems for specific reasons. For example, pressure supplements are often used so pump pressure rating does not have to be increased to manage increased pressure.

In a hydraulic system, the pressure in the system drops for a short period when several components activate at the same time.

Accumulators

An *accumulator* is a vessel in which fluid is stored under pressure for future release of energy. Accumulators maintain system pressure, develop fluid flow, compensate for thermal expansion and contraction of piping, control noise, and absorb shocks to the system. For example, in an emergency, accumulators can maintain system pressure if a pump fails. This allows a system to complete cycling before shutdown occurs. Sometimes, hydraulic components can produce a short high-pressure burst. An accumulator placed within the hydraulic system can absorb an unwanted pressure surge and help prevent damage to system components.

An accumulator placed close to a vital component provides the fluid flow and pressure needed to maintain the safe operation of the system until it can equalize. For example, in a metal forming press, the first cylinder is used to clamp the metal workpiece in place while the other cylinders are used to form the workpiece into the desired shape. When the cylinders that form the workpiece begin to operate, system pressure will momentarily drop. Pressure must be maintained to the clamping cylinder. If not maintained, the metal workpiece can become dislodged.

TERMS

A **spring-loaded accumulator** is an accumulator that applies force to a fluid by means of a spring.

A **weight-loaded accumulator** is an accumulator that applies force to a fluid by means of heavy weights.

A **gas-charged accumulator** is an accumulator that applies force to a hydraulic fluid by using compressed gas.

A **piston gas-charged accumulator** is an accumulator with a floating piston acting as a barrier between the gas and the hydraulic fluid.

The two categories of accumulators that can be used in hydraulic systems are mechanical and gas charged. The two designs of mechanical accumulators are spring loaded and weight loaded. The three designs of gas-charged accumulators are piston, diaphragm, and bladder. The bladder design is the most commonly used gas-charged accumulator. **See Figure 3-13.**

Warning: Pressurized accumulators create a hazardous situation because their energy is stored, waiting for release. For this reason, all accumulator energy must be released or blocked before attempting to disconnect any hydraulic lines.

Mechanical Accumulators. Mechanical accumulators include spring-loaded and weight-loaded designs. A *spring-loaded accumulator* is an accumulator that applies force to a fluid by means of a spring. Spring-loaded accumulators consist of a cylinder, piston, and spring. The spring presses the piston against the hydraulic fluid inside the cylinder. System pressure and the compression rate of the spring determine the amount of stored energy. In many cases, a mechanical stop in the spring area is used to prevent excessive pressure from overcompressing and damaging the spring. Spring-loaded accumulators discharge fluid at a diminishing rate ranging from the maximum spring compression until the spring is completely relaxed.

An advantage of spring-loaded accumulators is that they can be installed in hydraulic systems in any position and do not have to be vertically mounted. The main use of a spring-loaded accumulator is for surge protection. Spring-loaded accumulators are seldom used in industrial hydraulics due to their size compared to their pressure output ratio. However, they can be found in mobile hydraulic systems that are used in low temperatures (below 32°F) because they are not affected by cold temperature as much as gas-charged accumulators.

A *weight-loaded accumulator* is an accumulator that applies force to a fluid by means of heavy weights. The weights are generally iron (or concrete in large systems) and offer a constant pressure throughout the piston stroke. Because of the constant pressure of the heavy weights, a weight-loaded accumulator may cause excessive pressure surges in a hydraulic system when the accumulator is quickly discharged or suddenly stopped.

Weight-loaded accumulators are used when large volumes of hydraulic fluid are required as backup in a system. Some of these accumulators use a small pump to charge the weight-loaded accumulator and provide energy to the hydraulic system. Weight-loaded accumulators are seldom used due to the need for a large weight to be attached to them. However, they can be found in some older (30 years or more) large stationary presses and forming machines to help maintain pressure at the power unit.

Gas-Charged Accumulators. A *gas-charged accumulator* is an accumulator that applies force to a hydraulic fluid by using compressed gas. Dry nitrogen gas is used in gas-charged accumulators instead of air because the mixture of air and hydraulic fluid vapors can be explosive. Gas-charged accumulators are divided into categories according to the method used to separate the gas from the hydraulic fluid. The methods include using a piston, a diaphragm, or a bladder to separate the gas from the hydraulic fluid.

A *piston gas-charged accumulator* is an accumulator with a floating piston acting as a barrier between the gas and the hydraulic fluid. Gas occupies the volume of space above the piston and is pressurized as hydraulic fluid enters and occupies the space below the piston. When the accumulator is pressurized, the gas pressure is equal to the system pressure.

Applications for piston gas-charged accumulators include working in suspension and load stabilizing systems for construction equipment, reducing noise and smoothing pump pulsations for industrial automation systems, and controlling pressure spikes for fork lifts.

TECH FACT

Weight-loaded accumulators are the only type of accumulator where the applied pressure is constant, whether the accumulator's chamber is nearly filled with hydraulic fluid or nearly empty, and are used in applications that require constant pressure or large volumes.

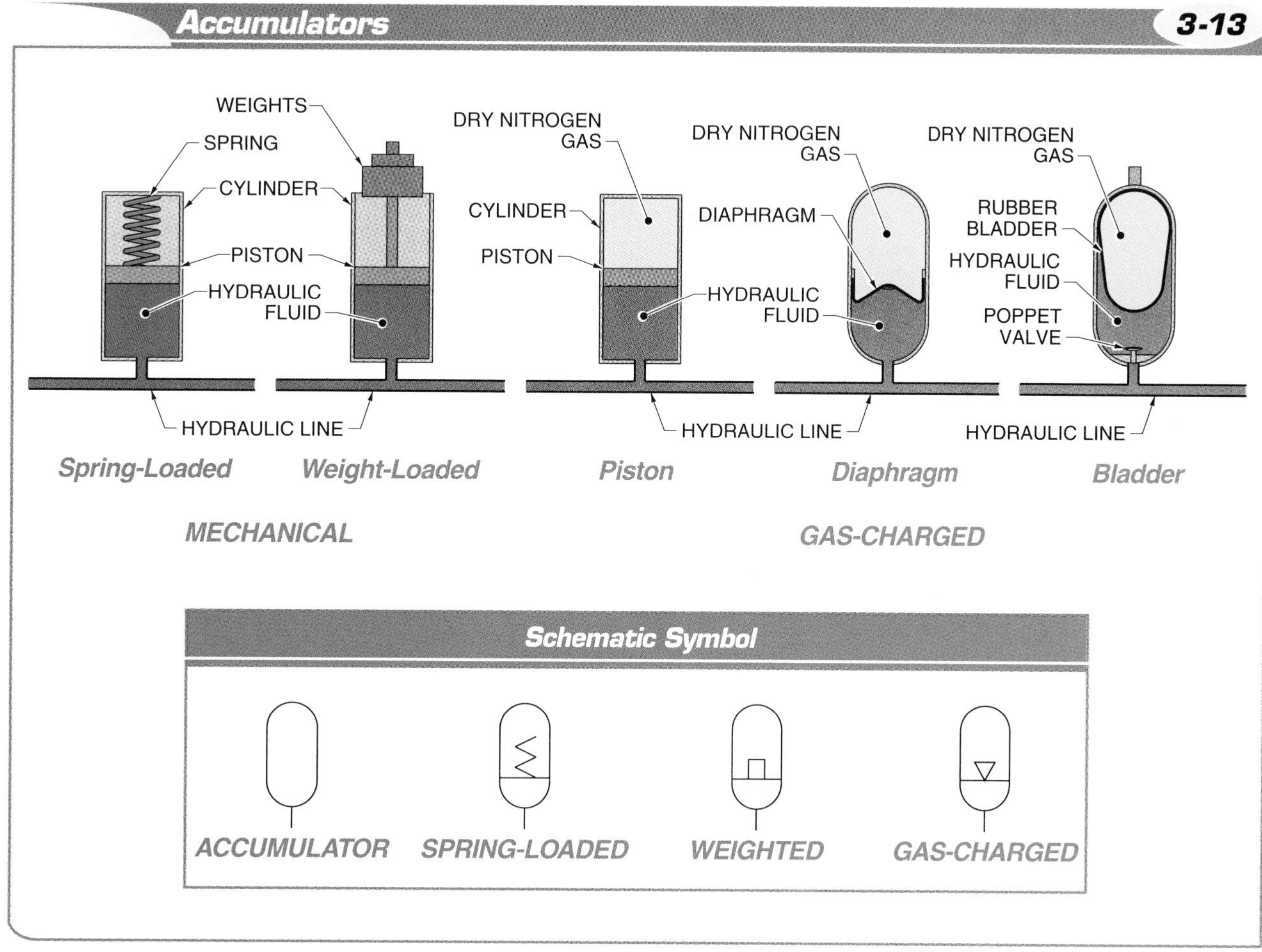

Figure 3-13. Accumulators are categorized as mechanical or gas-charged.

A *diaphragm gas-charged accumulator* is an accumulator with a flexible diaphragm separating the gas and the hydraulic fluid. Diaphragm gas-charged accumulators are generally small and lightweight with capacities up to 1 gal. They are constructed of two steel hemispheres bolted together with a flexible, dish-shaped rubber diaphragm clamped between them. The top hemisphere is pressurized, or precharged, with gas. Hydraulic fluid supplied by system pressure is applied to the bottom hemisphere, compressing the gas. The gas acts as a spring against the diaphragm as system pressure equalizes the pressure in each hemisphere.

Applications for diaphragm gas-charged accumulators include working in steam engines, maintaining hydraulic fluid pressure in agricultural machinery and equipment, and absorbing shock for hydrostatic drives in forestry and construction equipment.

A *bladder gas-charged accumulator* is an accumulator consisting of a seamless steel shell, rubber bladder with a gas valve, and poppet valve. The steel shell is cylindrical in shape and rounded at both ends. A large opening at the bottom of the shell is used to insert the bladder. A small opening at the top of the shell is used for the gas valve of the bladder. The one-piece, pear-shaped bladder is molded of synthetic rubber and includes a molded gas valve fastened to the inside upper end of the shell by a locknut. The bottom of the shell is equipped with a poppet valve at the discharge port. This valve closes off the port when the accumulator fully discharges, preventing the bladder from squeezing out of the discharge port opening.

TERMS

A **diaphragm gas-charged accumulator** is an accumulator with a flexible diaphragm separating the gas and the hydraulic fluid.

A **bladder gas-charged accumulator** is an accumulator consisting of a seamless steel shell, rubber bladder with a gas valve, and poppet valve.

TERMS

Precharge pressure is the pressure of the compressed gas in an accumulator prior to the admission of hydraulic fluid.

A **force multiplication system** is a system that allows the force applied on one cylinder to be increased on another cylinder.

An **intensifier** is a device that converts low-pressure, high-flow-rate fluid to high-pressure, low-flow-rate fluid.

Applications for bladder gas-charged accumulators include supporting suspension and braking systems in mobile hydraulic systems and maintaining pressure and compensating for fluid leakage in industrial hydraulic systems. Also, bladder gas-charged accumulators are used in hybrid garbage trucks to reduce fuel consumption.

Gas-charged accumulators are precharged while empty of hydraulic fluid. *Precharge pressure* is the pressure of the compressed gas in an accumulator prior to the admission of hydraulic fluid. The higher the precharge pressure, the less hydraulic fluid the accumulator can hold. Precharge pressures vary with each application. Typically, an application depends on the system pressure range and the volume of hydraulic fluid required in that range. Precharge pressures should be about 90% of the minimum operating pressure and must not be less than 25% of the maximum system pressure. Following these standards helps to eliminate unnecessary wear on the accumulator. *Note:* Bladder accumulators are typically precharged to 50 psi while diaphragm and piston accumulators are typically not precharged. All three types are charged to the required amount of pressure for machine operation.

Force Multiplication Systems

A *force multiplication system* is a system that allows the force applied on one cylinder to be increased on another cylinder. For example, when two cylinders are connected with a hose, any pressure created by one cylinder will be placed on the opposite cylinder. Force multiplication can be demonstrated when interconnected hydraulic cylinder pistons of different diameters apply force from one cylinder to the other cylinder. **See Figure 3-14.** Force multiplication is calculated by applying the following procedure:

1. Calculate the pressure created by the first cylinder.

$$p_1 = \frac{F_1}{A_1}$$

where

p_1 = pressure created by the first cylinder (in psi)

F_1 = force on the first cylinder (in lb)

A_1 = area of the first cylinder (in sq in.)

2. Calculate the force of the second cylinder.

$$F_2 = p_1 \times A_2$$

where

F_2 = force on the second cylinder (in lb)

p_1 = pressure of the first cylinder (in psi)

A_2 = area of the second cylinder (in sq in.)

Example: If there are two cylinders connected in a hydraulic system and the first cylinder has an area of 10 sq in. while the second cylinder has an area of 20 sq in., how much force can the second cylinder create if 2000 lb of force is placed on the first cylinder?

1. Calculate the pressure created by the first cylinder.

$$p_1 = \frac{F_1}{A_1}$$

$$p_1 = \frac{2000}{10}$$

$$p_1 = 200 \text{ psi}$$

2. Calculate the force of the second cylinder.

$$F_2 = p_1 \times A_2$$

$$F_2 = 200 \times 20$$

$$F_2 = \mathbf{4000\ lb}$$

Note: The system converts 200 psi into 4000 lb, but it has to sacrifice movement. The output piston does not move or travel as far as the input piston.

Intensifiers

An increase in pressure is sometimes necessary to accomplish certain tasks, such as punching holes through a thick metal workpiece with a punch press. To accomplish this, an intensifier is used. An *intensifier* is a device that converts low-pressure, high-flow-rate fluid to high-pressure, low-flow-rate fluid. For example, 1000 psi can be converted to 10,000 psi with a 10:1 hydraulic intensifier. **See Figure 3-15.**

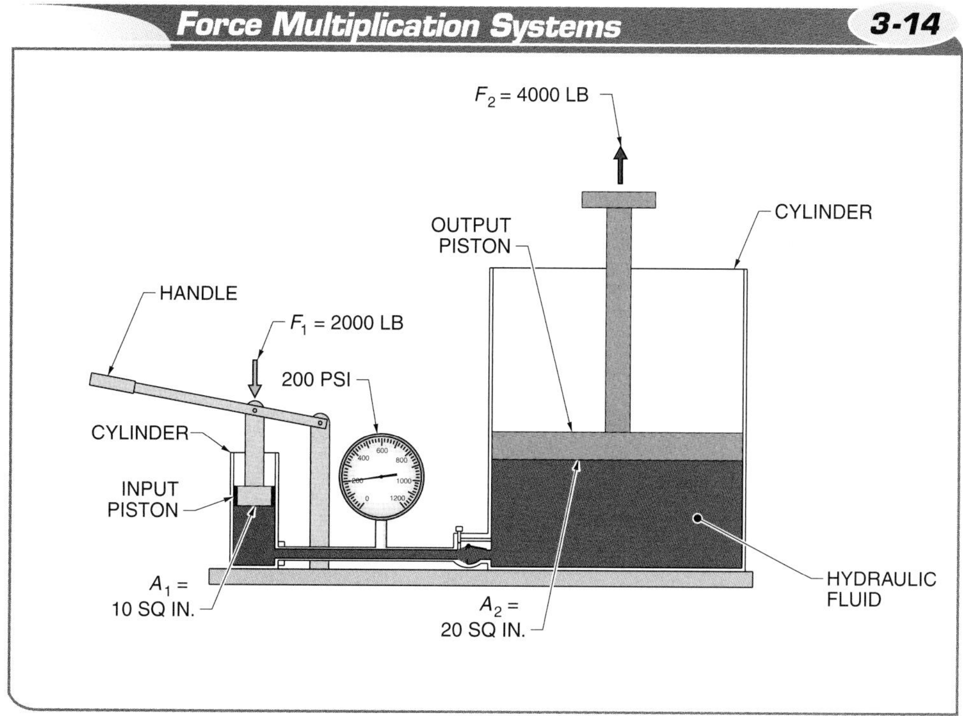

Figure 3-14. A force multiplication system allows the force applied on one cylinder to be increased on another cylinder.

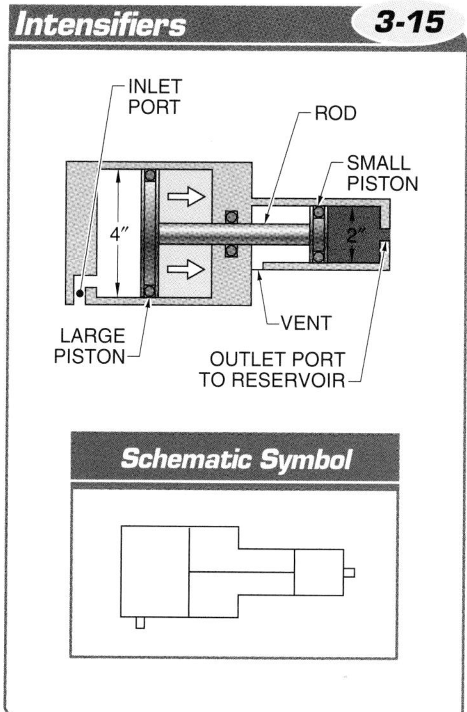

Figure 3-15. An intensifier converts low-pressure, high-flow hydraulic fluid to high-pressure, low-flow hydraulic fluid.

An intensifier consists of a cylinder, a large piston and a small piston connected by a rod, an inlet port, and an outlet port. The two main types of intensifiers are single stroke intensifiers and reciprocating intensifiers. The most common type of intensifier is a single stroke intensifier (boosters), which is typically used in the following applications:

- hydraulic clamping tools
- railway maintenance equipment
- injection molding machines (molding parts lock and core holders)
- hydraulic tools of various kinds (cutting, spreading, and clamping tools)
- torque wrenches, bolt drawers, etc.
- rotating couplings for lathes
- test equipment the can operate up to 40,000 psi
- fork turners on trucks
- concrete crushers
- power packs
- filter presses

TERMS

A **hydraulic power unit** is a self-contained unit that includes all the components required to create flow, regulate pressure, and filter hydraulic fluid.

Single stroke intensifiers operate by applying system pressure to a large piston and transferring that power to a smaller piston to increase the available pressure.

Reciprocating intensifiers are used to provide continuous flow at high pressures. A reciprocating intensifier acts similar to two single stroke intensifiers connected back-to-back. This connection allows the intensifier to produce flow on extension and retraction. Reciprocating intensifiers are used to provide low flow and high pressure without the need for an expensive pump. They are used in the following applications:

- machine tools
- aerial lifts
- hydro forming processes
- die-casting machines
- demolition equipment

Low-pressure fluid enters the inlet port and moves the large piston. The movement of the large piston causes the small piston to move, increasing the pressure of the hydraulic fluid at the outlet port. Intensifiers are often used in production-press clamping systems because a small amount of high pressure is required for the final step in the production process. The pressure increase when using an intensifier is calculated by applying the following procedure:

1. Calculate the amount of force at the first piston.

   $$F_1 = p_i \times A_1$$

   where
   F_1 = force generated by the first piston (in lb)
   p_i = inlet port pressure (in psi)
   A_1 = area of the first piston face (in sq in.)

2. Calculate the pressure at the second piston.

   $$p_o = \frac{F_1}{A_2}$$

   where
   p_o = outlet port pressure (in psi)
   F_1 = force generated by the first piston (in lb)
   A_2 = area of the second piston face (in sq in.)

Example: If 1000 psi is placed at the inlet port and the area of the first piston is 20 sq in. while the area of the second piston is 5 sq in., how much pressure is created at the outlet port?

1. Calculate the amount of force at the first piston.

   $$F_1 = p_i \times A_1$$
   $$F_1 = 1000 \times 20$$
   $$F_1 = 20{,}000 \text{ lb}$$

2. Calculate the pressure at the second piston.

   $$p_o = \frac{F_1}{A_2}$$
   $$p_o = \frac{20{,}000}{5}$$
   $$p_o = \mathbf{4000\ psi}$$

Hydraulic Power Units

Hydraulic power units are often used to transfer hydraulic energy throughout a system. A *hydraulic power unit* is a self-contained unit that includes all the components required to create flow, regulate pressure, and filter hydraulic fluid. **See Figure 3-16.** However, depending on the type of pump used, hydraulic power units are available with various maximum pressure ratings for different hydraulic applications. A hydraulic power unit, sometimes referred to as a pumping unit, regulates maximum pressure only.

TECH FACT

Many modern flow meters are made from polysulfone because of its good heat and chemical resistance properties as well as its rigidity.

Hydraulic power units convert static hydraulic fluid into kinetic energy in the form of fluid flow. Hydraulic power units help protect the fluid power system from contaminants and provide additional safety for the equipment and operator. Most hydraulic power units have the same basic components:

- an electric motor (prime mover) to power the pump
- a pump to create fluid flow

- a reservoir to store hydraulic fluid, supply the pump with hydraulic fluid, and receive returned hydraulic fluid
- a strainer to protect the pump from reservoir dirt
- a pressure gauge to indicate the amount of pressure in the system
- a pressure relief valve to set the maximum pressure in the hydraulic system
- piping, tubing, and hoses to transmit hydraulic fluid

Hydraulic power units are available as self-contained, stand-alone units or as custom-designed units with individual components that meet the specific needs of the end user. Self-contained power units are typically small, sometimes portable, and can be easily moved within a facility. Large individual components are typically installed as part of a commercial or industrial hydraulic system in a separate room or structure.

Stand-Alone Power Units. A *stand-alone power unit* is a self-contained hydraulic power unit. A stand-alone power unit is typically preconnected and only needs electric power to operate. The two types of stand-alone units are stationary units and portable units. Stationary stand-alone power units are built into specific structures. Portable stand-alone power units can be moved wherever they are needed and are less expensive than custom-designed power units.

A disadvantage of portable power units is that they have a small reservoir capacity (usually 5 gal. to 60 gal.). Small reservoirs can only be used in specific commercial and industrial hydraulic systems. Some hydraulic systems require a stand-alone unit built into them. For example, a hydraulic lathe can have a stand-alone hydraulic power unit that is only used for the operation of that particular lathe.

TERMS

A **stand-alone power unit** is a self-contained hydraulic power unit.

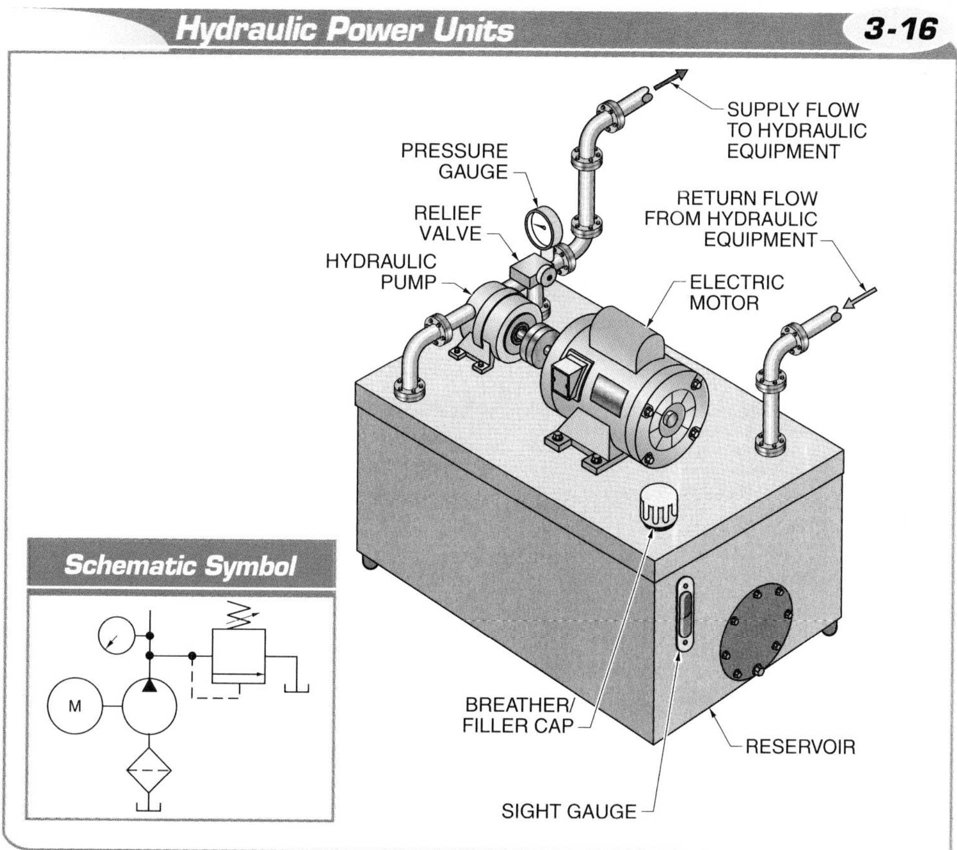

Figure 3-16. A hydraulic power unit is a self-contained unit that contains all the equipment required to create fluid flow.

TERMS

A **custom-designed power unit** is a hydraulic power unit that is designed for a specific use.

A **pictorial diagram** is a graphic representation that shows how devices interconnect in a fluid power system.

A **cutaway diagram** is a diagram showing the internal details of components and the path of fluid flow.

Custom-Designed Power Units. A *custom-designed power unit* is a hydraulic power unit that is designed for a specific use. Custom-designed power units can be sized for almost any power demand. They are typically used in commercial and industrial facilities that are equipped with a large amount of hydraulic equipment.

Custom-designed power units are often housed in a separate facility known as a pumproom or pumphouse, and as such, are easier to access and repair. **See Figure 3-17.** A disadvantage is that, due to their large size and custom design, they are considerably more expensive than stand-alone power units.

Figure 3-17. A custom-designed power unit is often housed in a separate facility known as a pumproom or pumphouse.

TECH FACT

Portable hydraulic power units are used with applications such as the single-acting ram cylinders in lifting jacks and industrial, marine, or construction hydraulic power tools.

FLUID POWER SYSTEM DIAGRAMS

Fluid power systems can be illustrated using a diagram. Common fluid power diagrams are pictorial, cutaway, and schematic diagrams. Each type of diagram has its own advantages and disadvantages. A technician may only have access to one type of diagram.

Pictorial Diagrams

A *pictorial diagram* is a graphic representation that shows how devices interconnect in a fluid power system. **See Figure 3-18.** Pictorial diagrams show where components are roughly located. Technicians use pictorial diagrams to view the general layout of a fluid power system. To make it easier to understand the system, pictorial diagrams show the elements of a system in a graphic format.

Pictorial diagrams can be used to depict individual components, in addition to complete systems. Pictorial diagrams have some disadvantages. They provide only a general idea of the operation of a system and do not illustrate the internal workings of the different components. It may be difficult for technicians to understand the operation of the system and its related components. It is difficult to represent any expansion of the system on a pictorial diagram. If changes are made, the diagram can be difficult to read.

Another disadvantage is that pictorial diagrams are not standardized. Each manufacturer may represent fluid power components differently, which can cause confusion for technicians. Because the disadvantages outnumber the advantages, pictorial diagrams are mainly used during the installation of the equipment.

Cutaway Diagrams

A *cutaway diagram* is a diagram showing the internal details of components and the path of fluid flow. Cutaway diagrams provide more detail than pictorial diagrams by showing the internal parts of a component. This detail helps the technician understand how the system and individual components work internally. For example, a cutaway diagram can show the strainer inside a hydraulic reservoir or a spool inside a directional control valve.

Pictorial Diagrams **3-18**

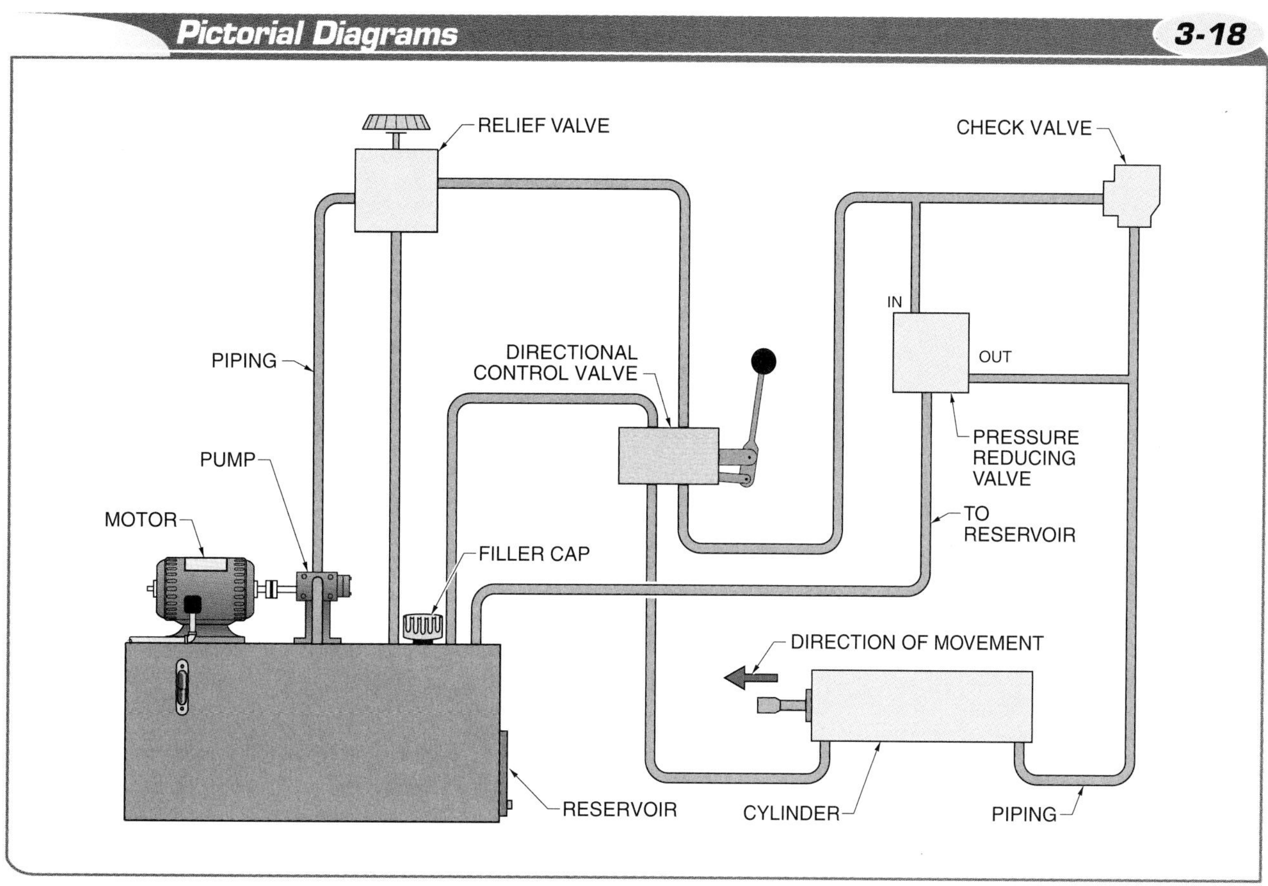

Figure 3-18. A pictorial diagram is a graphic representation that shows how devices interconnect in a fluid power system.

Cutaway diagrams are sometimes provided by the manufacturer of the individual parts when they are ordered. Cutaway diagrams are drawn in double line form, which provides better visualization of the entire system or of an individual component.

Cutaway diagrams have several disadvantages. Because cutaway diagrams are complex, they are difficult to draw. And even simple systems, if not drawn properly, can be extremely difficult to read. It is also difficult to expand or change a cutaway diagram if the system is changed. Cutaway diagrams can be large, so they are limited in what they can show on one piece of paper. Also, because cutaway diagrams are two dimensional, it can be difficult to interpret how the internal working components of a valve should operate. Color coding is sometimes used to help the technician understand the system. **See Figure 3-19.**

Cutaway diagram color coding is as follows:

- Red indicates that fluid is flowing at system operating pressure or at highest working pressure.
- Yellow indicates that flow is controlled by a metering device or is at lowest working pressure.
- Orange indicates an intermediate pressure that is lower than system operating pressure.
- Green indicates intake flow to pump drain line. *Intake flow* is the fluid flow from the reservoir through the filters to the pump.
- Blue indicates exhaust or return flow to the reservoir. *Exhaust flow* is the fluid flow from the cylinder back through the directional control valve to the reservoir.
- White indicates inactive fluid (reservoir fluid).

TERMS

Intake flow is the fluid flow from the reservoir through the filters to the pump.

Exhaust flow is the fluid flow from the cylinder back through the directional control valve to the reservoir.

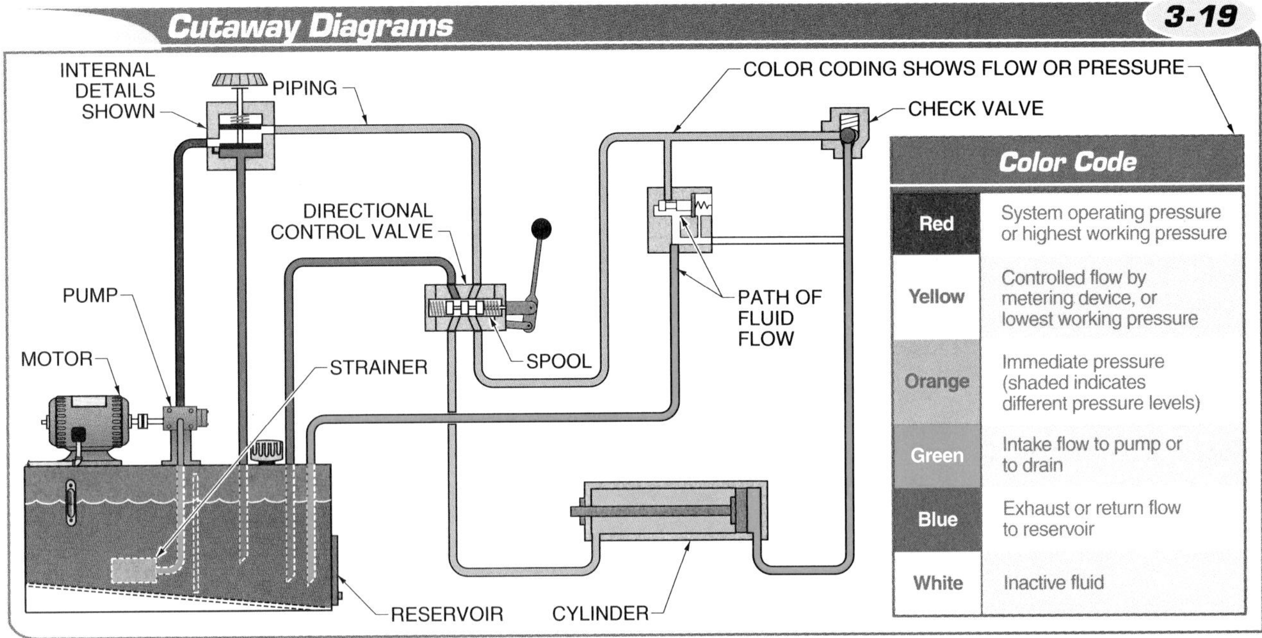

Figure 3-19. A cutaway diagram shows the internal details of components and the path of fluid flow.

TERMS

A **schematic diagram**, also called a graphic diagram, is a diagram that uses standardized lines, shapes, and symbols.

Schematic Diagrams

A *schematic diagram,* also called a graphic diagram, is a diagram that uses standardized lines, shapes, and symbols. Interconnecting lines represent the function of each component in a system. Typically, the symbol resembles the shape of the actual component. However, schematic diagrams do not indicate the exact location of where the component is located within the system. They are used to represent the flow of fluid within a system and how a component functions. For this reason, most schematic diagrams show fluid flowing from the bottom to the top of the diagram and to the right. The power source and pump are typically shown near the bottom while the fluid flows upward toward the components. **See Figure 3-20.**

Schematic diagrams are typically used for design and troubleshooting. By showing the component symbol and internal functions, they provide information about individual components. Schematic diagrams for fluid power systems indicate the following:

- fluid flow path, directions of fluid flow, and connections
- general information on the internal function of each component
- connections between components
- large fluid power systems with small diagrams

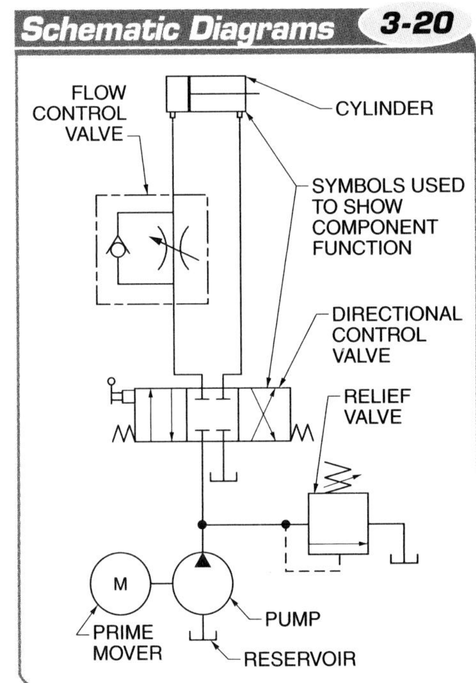

Figure 3-20. A schematic diagram uses standardized lines and shapes with interconnecting lines to represent the function of each component in a system.

Fluid power diagrams are designed to integrate seamlessly with electrical schematic diagrams.

Piping is a system of hollow cylinders through which fluid flows. The different lines used in schematic diagrams are standardized to represent various types of working pipes. A solid line represents a working line (pressure flow). These are the lines that are considered the main working lines. Dashed lines represent pilot lines. Pilot lines are used for the actuation of components. Dotted lines represent exhaust or drain lines. Centerlines represent the outline of enclosures. Line width does not alter the meaning of symbols. **See Figure 3-21.**

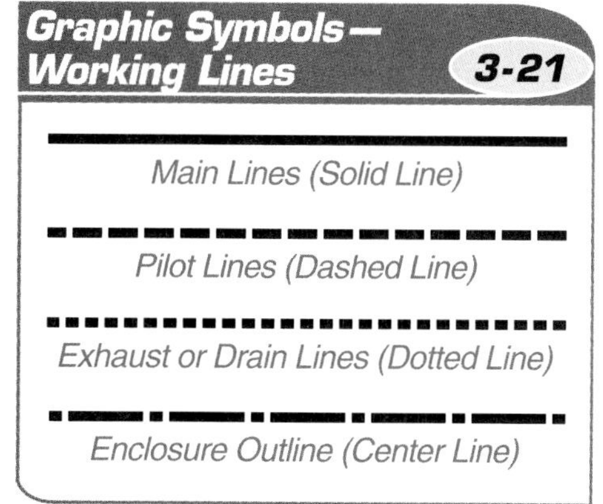

Figure 3-21. The different lines used in schematic diagrams are standardized to represent various types of working pipes.

The American National Standards Institute (ANSI) in collaboration with the American Society of Mechanical Engineers (ASME) has developed a standard regarding symbols for fluid power diagrams, ANSI/ASME Y32.10-1967, *Graphic Symbols for Fluid Power Diagrams*. **See Appendix.** This standard illustrates the basic fluid power symbols and describes the principles on which the symbols are based.

Schematic diagrams can be used to identify piping and tubing lines in hydraulic systems.

Digital Resources

Chapter Review

Hydraulic System Fundamentals — Chapter 3

Name: ______________________________ Date: ______________

MULTIPLE CHOICE

_______________ **1.** The color ___ in a color-coded cutaway diagram indicates intake flow from the reservoir through the filters to the pump.

A. green
B. orange
C. red
D. yellow

_______________ **2.** A ___ valve protects a fluid power system from overpressure by setting a maximum operating pressure.

A. directional control
B. flow control
C. pressure
D. relief

_______________ **3.** The two types of mechanical accumulators include weight- and ___-loaded.

A. bladder
B. diaphragm
C. piston
D. spring

_______________ **4.** In a hydraulic system, fluid flow is produced by a ___.

A. cylinder
B. motor
C. pressurized reservoir
D. pump

_______________ **5.** ___ is the amount of fluid that passes a given point in one minute.

A. Flow rate
B. Fluid flow
C. Viscosity
D. Viscosity index

_______________ **6.** A(n) ___ is a vessel in which fluid is stored under pressure for future release of energy.

A. accumulator
B. cylinder
C. receiver
D. reservoir

_______________ **7.** ___ is the ability to hold or contain something.

A. Area
B. Capacity
C. Cylinder size
D. Volume

_______________ **8.** By controlling ___ and pressure, hydraulic systems can produce a large amount of work.

A. electrical energy
B. fluid flow
C. friction
D. mechanical energy

_______________ **9.** For industrial applications, hydraulic fluid viscosities are typically about ___ SUS at ___ °C.

A. 40; 100
B. 40; 150
C. 100; 40
D. 150; 40

_______________ **10.** Pressure ___ is the pressure difference between two points in a fluid power system.

A. accumulation
B. atmosphere
C. drop
D. relief

_______________ **11.** ___ fluid flow is present in systems where pipes, tubes, and hoses may be too small for the system.

A. Laminar
B. Low-pressure
C. High-pressure
D. Turbulent

_______________ **12.** A ___ flow meter has high internal resistance to fluid flow.

A. digital
B. gas
C. liquid
D. weighted

_______________ **13.** ___ accumulators are typically small and lightweight, with capacities up to 1 gal.

A. Bladder gas-charged
B. Diaphragm gas-charged
C. Piston gas-charged
D. Spring-loaded

_______________ **14.** A(n) ___ diagram is a graphic representation of how devices interconnect within a fluid power system.

A. cutaway
B. engineering
C. pictorial
D. schematic

_______________ **15.** On a schematic diagram, ___ lines are used represent pilot lines.

A. center
B. dashed
C. dotted
D. solid

COMPLETION

______ 1. A(n) ___ is a device that controls the pressure, direction, and/or rate of fluid flow.

______ 2. ___ flow has an erratic nonlayered pattern.

______ 3. ___ pressure is the pressure of compressed gas in an accumulator prior to the admission of hydraulic fluid.

______ 4. A ___ power unit is a self-contained hydraulic power unit.

______ 5. A(n) ___ diagram is also referred to as a graphic diagram.

______ 6. ___ flow has a smooth and continuous layered pattern.

______ 7. A(n) ___ is a device that converts low-pressure, high-flow-rate fluid to high-pressure, low-flow-rate fluid.

______ 8. A(n) ___ allows the force applied on one cylinder to be increased on another cylinder.

______ 9. A(n) ___ measures the flow of hydraulic fluid within a system.

______ 10. ___ is the size of a space or chamber measured in cubic units.

______ 11. A(n) ___ is a self-contained unit that includes all the components required to create flow, regulate pressure, and filter hydraulic fluid.

______ 12. On a schematic diagram, ___ lines are used represent exhaust or drain lines.

______ 13. A(n) ___ is matter composed of atoms and is the smallest particle that a compound can be reduced to while still possessing the chemical properties of that compound.

______ 14. A(n) ___ is a test instrument used to measure fluid viscosity.

______ 15. The two categories of accumulators that can be used in hydraulic systems are gas-charged and ___.

TRUE/FALSE

T F 1. The temperature of a fluid will have no effect on its viscosity.

T F 2. Maximum resistance occurs when there is no fluid flow in a hydraulic system.

T F 3. Air is typically used in gas-charged accumulators.

T F 4. Fluids follow the path of least resistance.

T F 5. Friction generates energy that is converted into heat.

T F 6. Using hydraulic system pressure valves, the inlet port is considered the secondary port and the outlet port is considered the primary port.

T F 7. Most hydraulic fluid is oil-based.

T F 8. The lower the precharge pressure, the less hydraulic fluid an accumulator can hold.

T F 9. Custom-designed hydraulic power units can be sized for almost any power demand.

T F **10.** Hydraulic fluids with a high VI number change drastically as the temperature changes.

T F **11.** Hydraulic fluid is typically pure without any extra additives.

T F **12.** In a hydraulic system, the pressure in the system drops for a short period when several components activate at the same time.

T F **13.** A fluid with high viscosity is thin and flows easily.

T F **14.** A schematic diagram is also called a graphic diagram.

T F **15.** The viscosity of a specific fluid is always determined under conditions in the field.

T F **16.** Hydraulic power units are sometimes referred to as pumping units.

T F **17.** Liquids are easily compressed.

T F **18.** Fluids flow because they move from low-pressure areas to high-pressure areas.

T F **19.** In a hydraulic system, the bladder design is the most commonly used gas-charged accumulator.

T F **20.** There are 128 cu in. in one gallon.

T F **21.** Self-contained hydraulic power units are typically large and stationary.

T F **22.** Resistance to fluid flow in a hydraulic system is greatest at bends in the piping and when fluid flows through valves.

T F **23.** Cutaway diagrams are only drawn in single-line form.

T F **24.** Pictorial diagrams are not standardized.

T F **25.** The shorter the length of a hose, the more internal resistance it will have.

SHORT ANSWER

1. List five functions of accumulators.

2. List at least seven main components of hydraulic power units.

3. List six advantages that digital flow meters have over mechanical flow meters.

4. Briefly explain why energy from a pressurized accumulator must be blocked or released before attempting to disconnect any hydraulic lines.

Activity 3-1: Reservoir Capacity

1. Use Hydraulic System Pictorial Diagram to determine the capacity of the reservoir (in gal.).

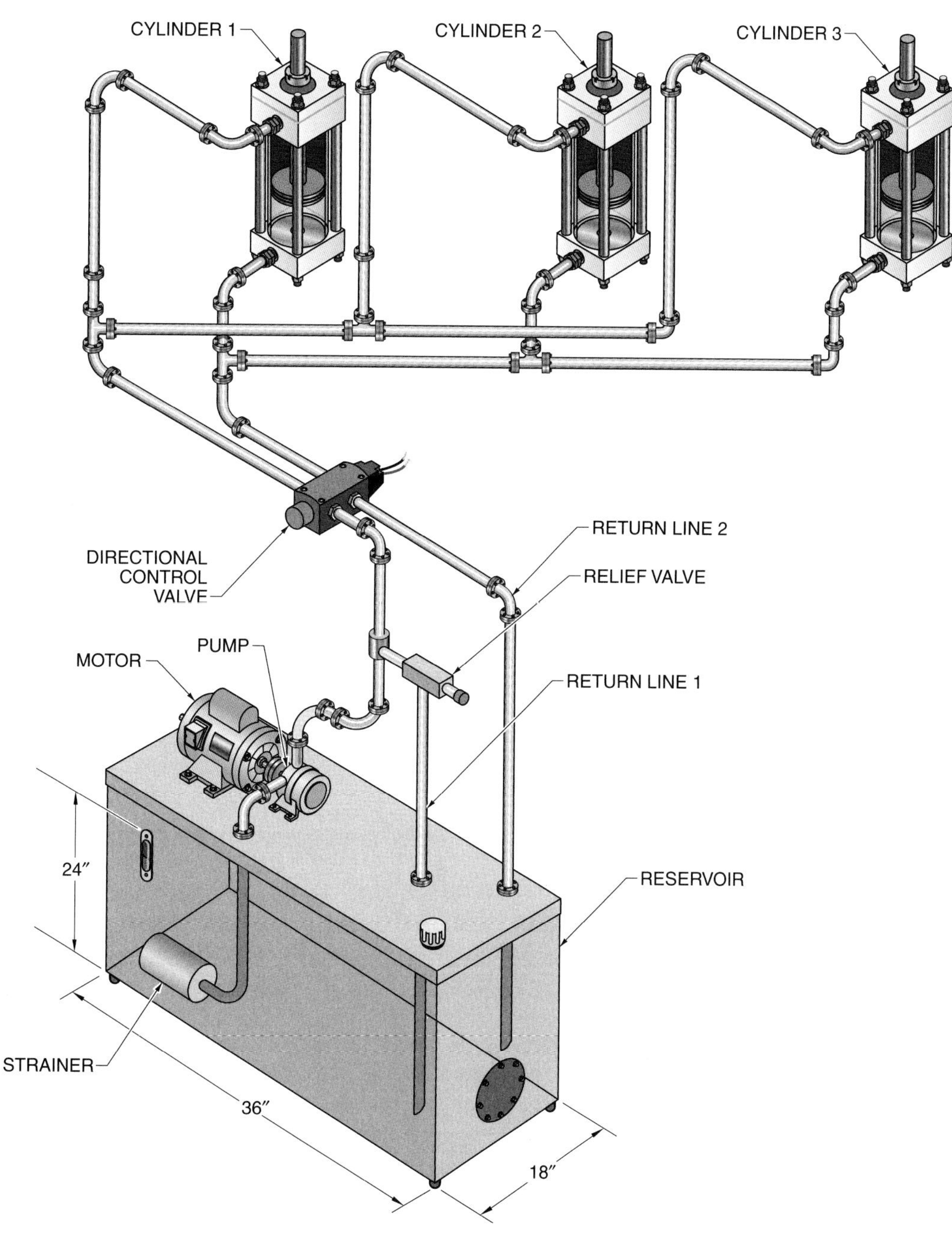

HYDRAULIC SYSTEM PICTORIAL DIAGRAM

Activity 3-2: Hydraulic System Force and Pressure Calculation

Use Hydraulic System to answer the following questions.

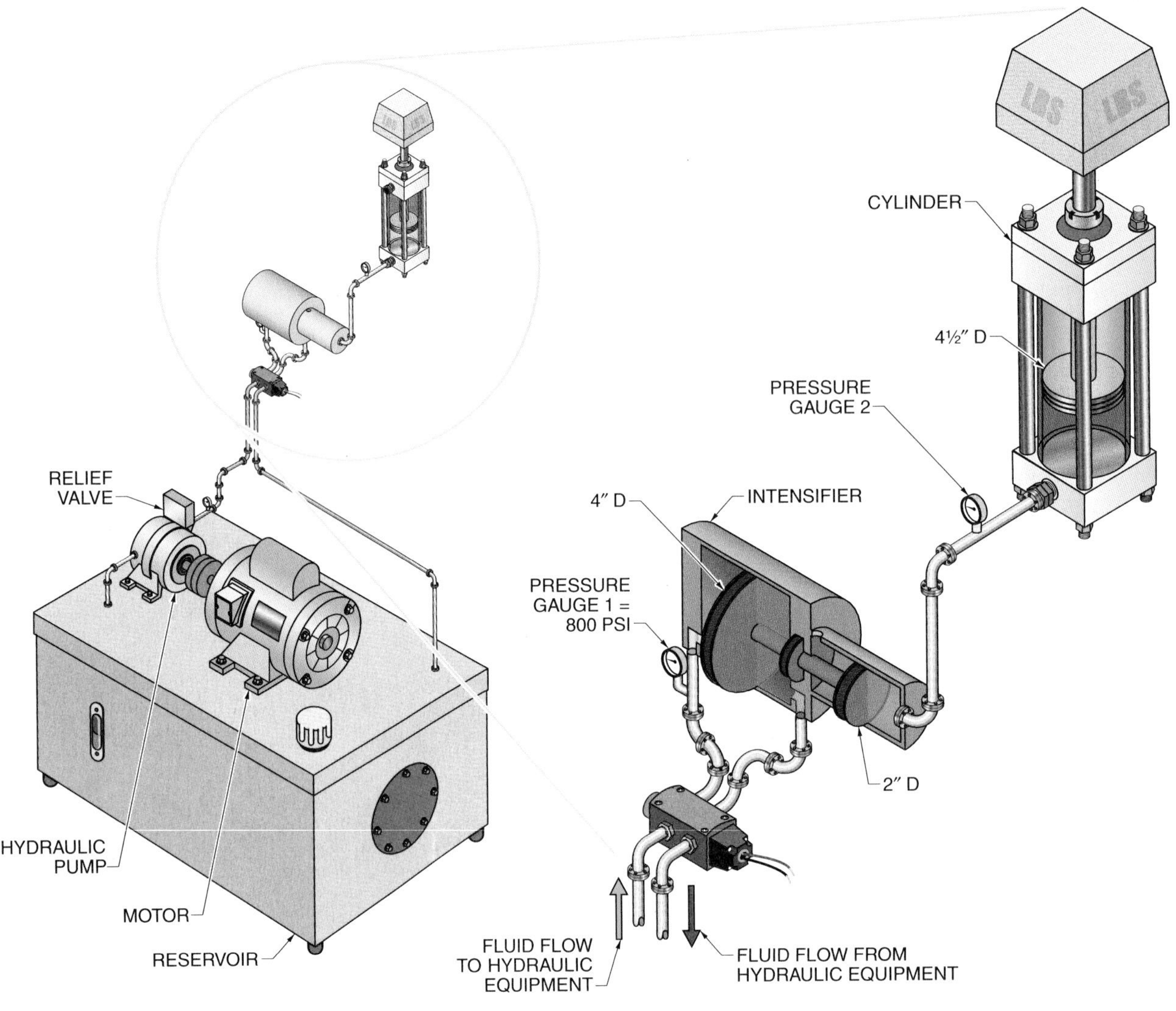

HYDRAULIC SYSTEM

1. How much force is being transmitted through the intensifier?

2. What is the reading (in psi) on Pressure Gauge 2?

3. Under these conditions, what is the greatest load that the cylinder can lift?

Activity 3-3: Conveyor Line Adjustment Mechanism

A production manufacturing facility changes from producing one product to producing two different products every other week. In order to use the same equipment to produce the products, the height of the conveyor line must be adjusted. The plant engineer has designed the hydraulic system so that the height of the conveyor line can be easily adjusted.

1. Using the Conveyor Line Schematic Diagram, draw the means of transmission in Conveyor Line Pictorial Diagram from the directional control valve to Cylinder 1 and Cylinder 2 as they would be connected in the actual hydraulic system.

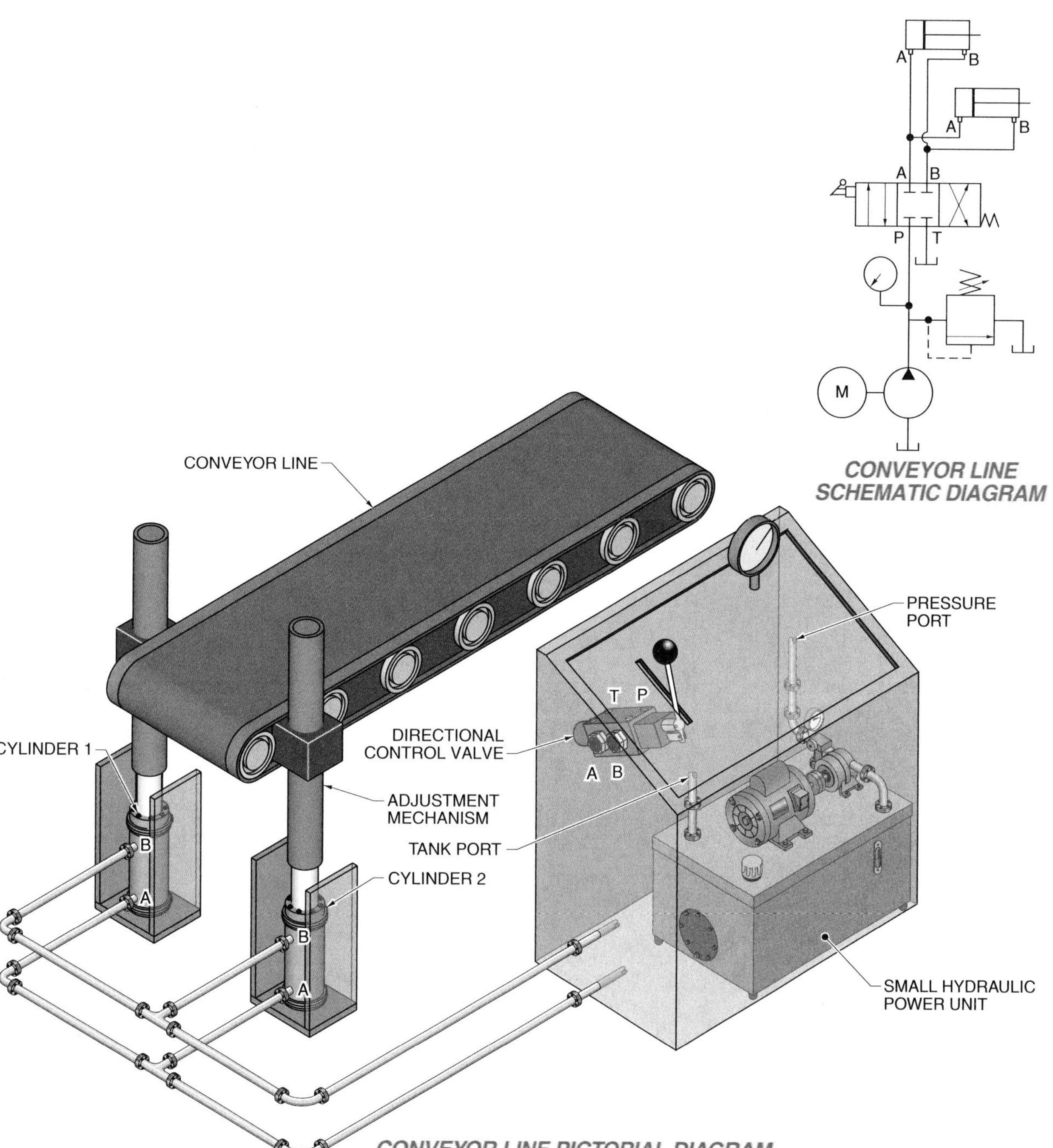

CONVEYOR LINE PICTORIAL DIAGRAM

Activity 3-4: Hydraulic Stamping Press

The production engineering department of a manufacturing facility has designed a custom hydraulic stamping press and needs a schematic diagram drawn of the system.

1. Using the Hydraulic Stamping Press Pictorial Diagram and Hydraulic Schematic Symbols, draw a schematic diagram of the system.

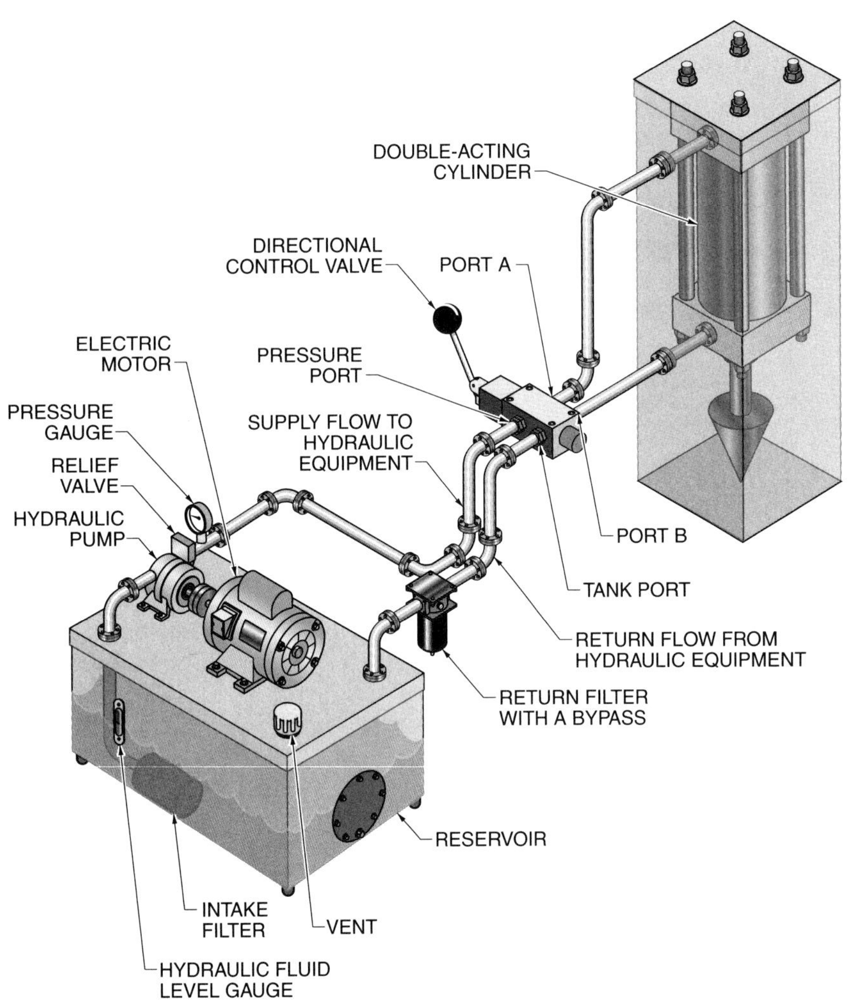

HYDRAULIC STAMPING PRESS PICTORIAL DIAGRAM

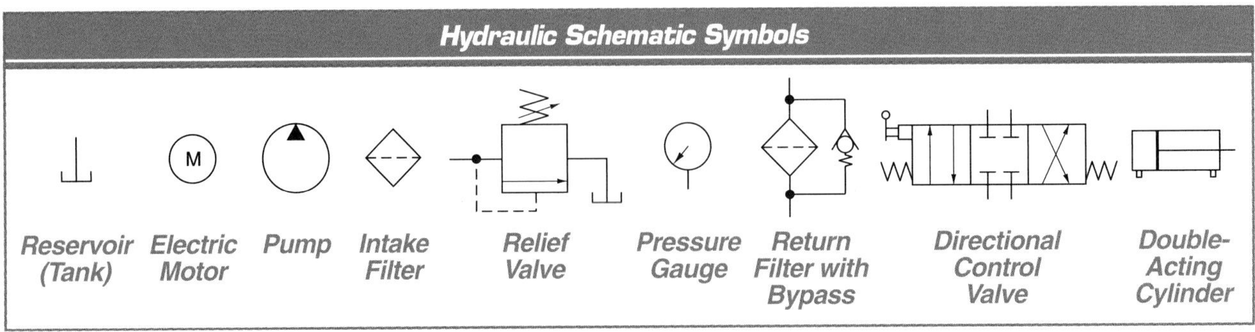

Activity 3-5: FluidSIM® 4.2 Hydraulics Student Version Simulation Software

After installing the FluidSIM® 4.2 Hydraulics Student Version simulation software on the CD-ROM located in the back of the book, click "File" in the menu tool bar. Select "Circuit Preview." Double click "demo_01.ct." to open the file. Select the "Start" button (▶) located in the tool bar and observe the simulation.

1. The first cylinder to extend is cylinder ___.

Click on Switch "S."

2. What happens with the system when the switch is closed?

3. Which cylinder retracts first?

Click the "Help" button (denoted as "?") in the tool bar. Select "Contents." Select "Introduction to Simulating and Creating Circuits" and read through the page. Next, select "The Component Library," and then select "Hydraulic Components" and complete the following section.

Sketch the following symbols as indicated in FluidSIM®.

4. Tank
5. Filter
6. Main line
7. Control line
8. Pump unit (simplified)
9. Check valve
10. T-junction (hydraulic)
11. Double-acting cylinder
12. Single-acting cylinder
13. Pressure relief valve

Fluid Conductors and Connectors

Chapter

OBJECTIVES

- Describe the different conductors used in hydraulic systems.
- Identify the different types of pipes, tubing, and hoses used in fluid power systems.
- Distinguish between different types of connectors and connector threads.
- Describe the different aspects of installing different pipes, tubing, and hoses.
- Distinguish between the inside diameter (ID), outside diameter (OD), and wall thickness of different pipes, tubing, and hoses.
- Describe the different cleaning methods before installing a conductor.

Introduction

Hydraulic components, devices, and power supplies must be connected to each other so that power can be transmitted through conductors in the system. Means of transmitting hydraulic fluid include piping, tubing, and hoses. There are various factors that determine the type of conductor to be used including system operating pressure, maximum system pressure, operating environment, installation location, and safety. To connect hydraulic conductors, specific connectors are used to allow the conductor to interconnect the hydraulic system.

TERMS

A **conductor** is a device used to transmit hydraulic fluid between various components in a hydraulic system.

Velocity is the speed of fluid flow through a hydraulic line.

VELOCITY AND CONDUCTORS

A *conductor* is a device used to transmit hydraulic fluid between various components in a hydraulic system. Conductors include hoses, hard pipes, and tubing. Conductors installed in a hydraulic system are designed to be compatible with the type of hydraulic fluid to be used and be able to withstand the velocity required for the application.

Velocity

Velocity is the speed of fluid flow through a hydraulic line. **See Figure 4-1.** Velocity of a fluid is measured in feet per second (fps). To keep the flow rate constant in differently sized components, the velocity of a fluid can be increased or decreased as needed. For example, if a cylinder needs to be filled with hydraulic fluid within a specific time through small piping, the hydraulic fluid needs to travel at a higher velocity. However, if large piping is used, then the hydraulic fluid travels at a lower velocity. The flow rate (gpm) is the same in both situations.

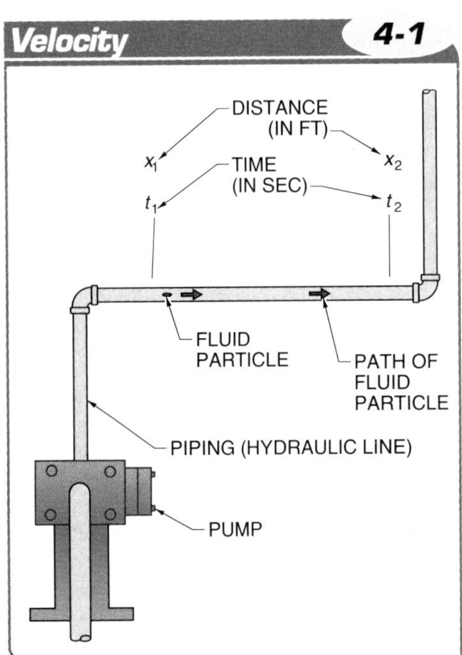

Figure 4-1. In a hydraulic system, liquid can transmit energy because it is almost completely incompressible.

The velocity of the hydraulic fluid in a hydraulic system should not exceed recommended values because it can cause turbulent conditions and result in a loss of pressure and excessive heat. This can cause damage to equipment within the hydraulic system.

The velocity of a fluid particle is calculated by subtracting its initial position from its final position and dividing the result by the value of the initial time subtracted from the final time. Velocity is calculated by applying the following formula:

$$v = \frac{x_2 - x_1}{t_2 - t_1}$$

where

v = velocity (in fps)

x_2 = final position (in ft)

x_1 = initial position (in ft)

t_2 = final time (in sec)

t_1 = initial time (in sec)

Example: What is the velocity of a fluid particle in a hydraulic system that is at point x_1 at 9:33:54 AM and reaches point x_2 at 9:34:30 AM after traveling 50′?

$$v = \frac{x_2 - x_1}{t_2 - t_1}$$

$$v = \frac{50 - 0}{9:34:30 - 9:33:54}$$

$$v = \frac{50}{36}$$

$$v = \mathbf{1.389\ fps}$$

Acceleration. The velocity of a fluid varies from one moment to another as its speed or direction of flow changes. *Acceleration* is an increase in speed and is measured in feet per second squared (ft/sec^2). Acceleration of a fluid is calculated as its change in velocity per unit of time.

TECH FACT

The viscosity index (VI) is used to express the viscosity and temperature behavior of an oil (hydraulic fluid). The calculated viscosity is a factual variable. The greater the viscosity index value, the less of an effect temperature has on the viscosity of oil or hydraulic fluid.

The symbol delta (Δ) is generally used to indicate a change. Acceleration, like velocity, is constantly changing within a hydraulic system. Piping of various diameters, elbows, valves, and other components all affect the velocity and acceleration of fluid within a hydraulic system. Acceleration of a fluid is calculated by applying the formula:

$$a = \frac{\Delta v}{\Delta t}$$

where

a = acceleration (in ft/sec^2)

Δv = average velocity during Δt (in fps)

Δt = time interval elapsed in traveled distance (in sec)

Example: What is the acceleration between measuring points when the fluid flow within a hydraulic system has an initial velocity of 15 fps and changes to 30 fps in 8 sec?

$$a = \frac{\Delta v}{\Delta t}$$

$$a = \frac{30 - 15}{8}$$

$$a = \frac{15}{8}$$

$$a = \mathbf{1.875\ ft/sec^2}$$

Note: If the change in velocity is a negative number, the fluid is decelerating (decreasing in speed).

Velocity of Fluid Flow through Conductor. The difference between fluid flow (gpm) and velocity (fps) is that fluid flow is volume per unit of time, while velocity is distance per unit of time. A fluid in motion is always flowing, but its speed of flow may change. Velocity of fluid flow through conductor depends on the rate of flow in gallons per minute and the cross-sectional area of the piping.

The velocity of fluid flow increases at any restriction in the conductor if the flow rate remains the same in the system. Common restrictions in conductors include valves, elbows, pipes, reducers, etc. The velocity of fluid flow decreases as the cross-sectional area of the conductor or component increases.

Per the law of the conservation of mass, the volumetric flow rate of an incompressible fluid through a conductor is constant at every point in the conductor. If there are no leaks in the system, the velocity of fluid flow must increase at any restriction to maintain a constant flow rate. The velocity increases four times to maintain a constant rate of fluid flow if the diameter of a conductor is halved. **See Figure 4-2.** Velocity of fluid flow through hydraulic conductor is calculated by applying the formula:

$$v = \frac{l_2}{\frac{A \times l_1}{231} \times \frac{60}{Q}}$$

where

v = velocity (in fps)

l_2 = length of conductor (in ft)

A = cross-sectional area of conductor (in sq in.)

l_1 = length of conductor (in in.)

231 = cu in. in one gallon

Q = flow rate (in gpm)

60 = sec in one minute

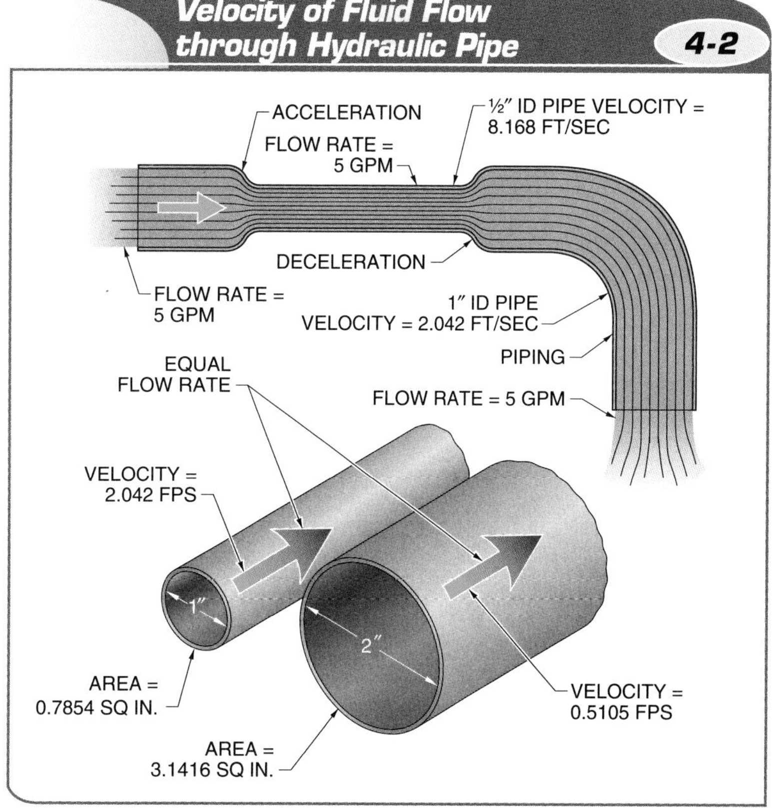

Figure 4-2. Pascal's law, stating that pressure is distributed undiminished throughout a closed container, can be applied during the transmission of energy in a hydraulic system.

Example: What is the velocity of a fluid having a flow rate of 5 gpm through a 12″ section of a conductor with a 1″ diameter?

$$v = \frac{l_2}{\frac{A \times l_1}{231} \times \frac{60}{Q}}$$

$$v = \frac{1}{\frac{0.7854 \times 12}{231} \times \frac{60}{5}}$$

$$v = \frac{1}{\frac{9.4248}{231} \times 12}$$

$$v = \frac{1}{0.0408 \times 12}$$

$$v = \frac{1}{0.4896}$$

$v = \mathbf{2.042\ fps}$

Determining Conductor Size. Determining the proper size conductor for a new system or for the replacement of a worn or undersized conductor is important when working with a hydraulic system. If a conductor is oversized, it will be more expensive than necessary. If a conductor is undersized, the velocity of the fluid will increase to overcome the smaller conductor size. A conductor diameter that is too small will cause turbulent flow, overheat hydraulic fluid through the friction of the fluid against the inside conductor wall, and decrease system efficiency.

Piping in a hydraulic system must be properly sized in order for the system to operate as intended.

Flow velocity requirements can be different for each system, depending on the type of line and hydraulic fluid used. **See Appendix.** Flow velocity in a conductor is related to the area of its inside diameter (ID). For lower velocity, a conductor with a larger area should be used and vice versa. The inside area of a conductor can be found using a few methods. One method is by calculating the inside area of a conductor. The inside area of a conductor can be calculated by applying the following formula:

$$A = \frac{Q \times 0.3208}{V}$$

where
A = pipe area (in sq in.)
Q = fluid flow from pump (in gpm)
0.3208 = constant
V = flow velocity (in fps)

Example: What size pipe is required if the pump needs to supply a maximum flow rate of 30 gpm and the maximum velocity is 18 fps?

$$A = \frac{Q \times 0.3208}{V}$$

$$A = \frac{Q \times 0.3208}{18}$$

$$A = \frac{9.624}{18}$$

$A = \mathbf{0.535\ sq\ in.}$

Once the needed conductor area is known, the area needs to be converted to diameter. This will give the required ID. The area converted to ID is calculated by applying the following formula:

$$ID = \sqrt{\frac{A}{0.7854}}$$

where
ID = ID (in in.)
A = area (in sq in.)
0.7854 = constant

Example: What is the ID of a conductor with an area of 0.75 sq. in?

$$ID = \sqrt{\frac{0.75}{0.7854}}$$

$ID = \sqrt{0.955}$

$ID = \mathbf{0.977\ in.}$

Note: This measurement would typically be rounded up to 1″.

Flow velocity can be calculated by switching the variables in the formula for area and flow velocity. Flow velocity in a pipe is calculated by applying the following formula:

$$V = \frac{(Q \times 0.3208)}{A}$$

Example: What flow velocity is required if the pump has maximum flow rate of 20 gpm and the inside area of a pipe is 0.825 sq in.?

$$V = \frac{(Q \times 0.3208)}{A}$$

$$V = \frac{(20 \times 0.3208)}{0.825}$$

$$V = \frac{6.416}{0.825}$$

$$V = \mathbf{7.777\ fps}$$

Another method that can be used to find the inside area of a conductor is through the use of a nomographic chart. A *nomographic chart* is a chart used to determine the ID of a conductor, flow velocity in fps, and fluid flow in gpm when two variables are known. To use a nomographic chart, find the two known variables and with a straight edge, draw a line that connects the two variables. Draw the line through all of the scales until all variables are found. **See Figure 4-3.**

Example: Use a nomographic chart for the flow capacity of hose assemblies at recommended flow velocities to determine the ID of a conductor with a fluid flow of 30 gpm and velocity of 20 fps.

CONDUCTORS AND CONNECTORS

Hydraulic components, devices, and power supplies must be connected to each other so that power can be transmitted through conductors and connectors in the system. Conductors include piping, tubing, and hoses with related connectors. There are various factors that determine which type of conductor should be used as well as system operating pressure, maximum system pressure, and location.

Piping

Piping is a passage or series of passages in a fluid power system that is constructed of metal, plastic, or plasticized rubber and conforms to ANSI standards. Piping is the most common way of transmitting fluid in both hydraulic and pneumatic systems. Piping is typically connected together with various types of fittings or welds. Piping has threaded male ends that allow them to connect to fittings in order to follow whatever path is required.

The two types of piping used in hydraulic system applications are black and stainless steel. **See Figure 4-4.**

Note: Galvanized piping, which can be used only with pneumatic systems, will not corrode as much as black pipe. Galvanized pipe should not be used with hydraulic systems because it can degrade the hydraulic fluid. Stainless steel pipe is corrosion-resistant and is typically used in specific applications, such as in clean rooms or outdoor locations.

TERMS

A **nomographic chart** is a chart used to determine the ID of a conductor, flow velocity in fps, and fluid flow in gpm when two variables are known.

Piping is a passage or series of passages in a fluid power system that is constructed of metal, plastic, or plasticized rubber and conforms to ANSI standards.

Hydraulic conductors and fittings must be strong enough to withstand the high fluid pressures associated with the operation of hydraulic equipment.

Nomographic Charts 4-3

Flow Capacity of Hose Assemblies at Recommended Flow Velocities

Based on Formula:

$$\text{Area (sq in.)} = \frac{0.3208 \times (\text{GPM})}{\text{Velocity (ft/sec)}}$$

Example: To determine the I.D. needed to transport 20 Gallons Per Minute (GPM) fluid volume.

Draw a straight line from 20 GPM on the left to maximum recommended velocity for pressure lines. The line intersects with the middle vertical column indicating a ¾″ I.D. (–12) hose. This is the smallest hose that should be used.

Recommendations are for oils having a maximum viscosity of 315 S.S.U. at 100°F, operating at temperatures between 65°F and 155°F.

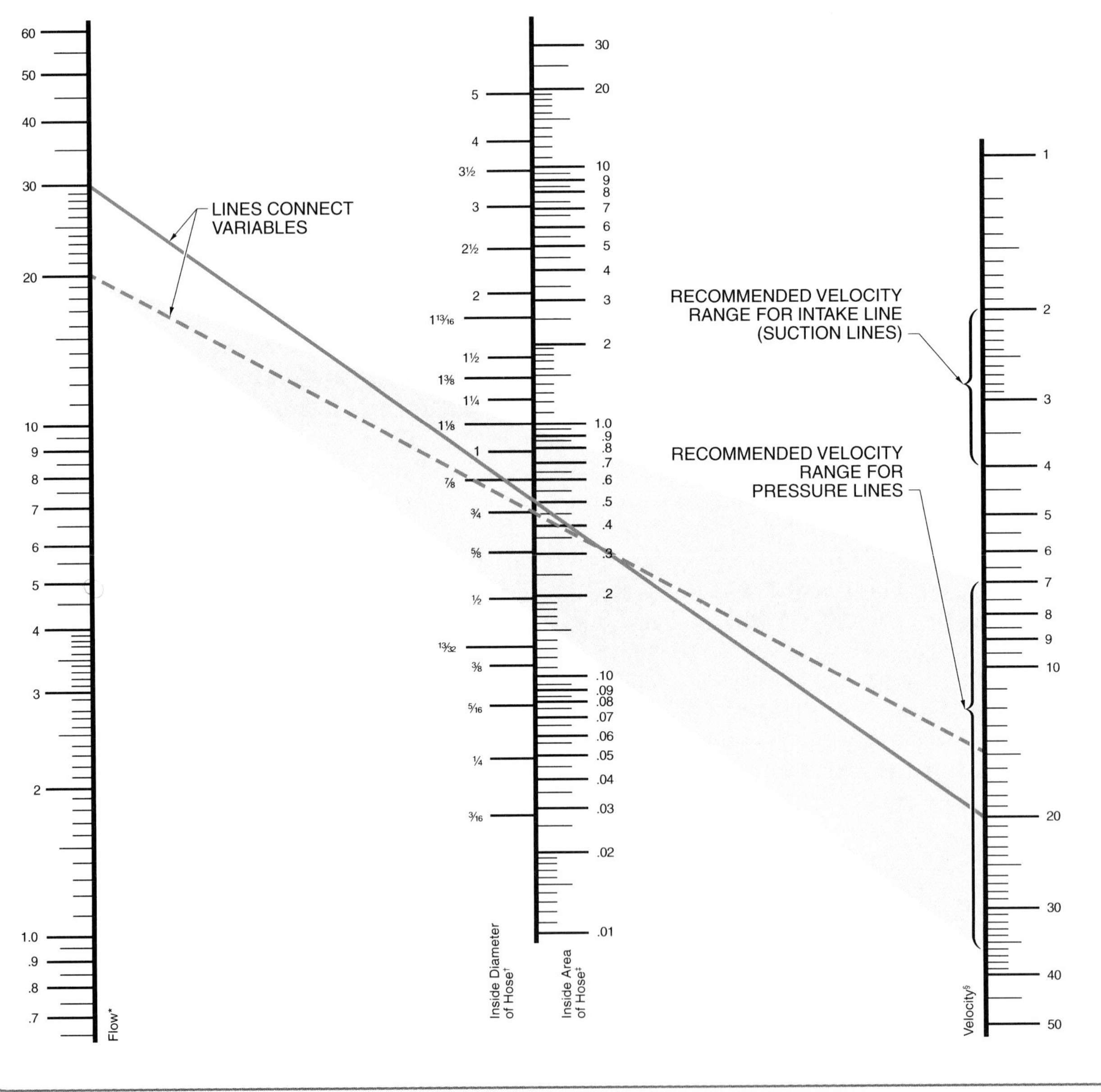

* in GPM
† in in.
‡ in sq in.
§ in FPS

Figure 4-3. A nomographic chart is used to determine the inside diameter of a conductor, flow velocity in fps, and fluid flow in gpm, when two variables are known.

Figure 4-4. The three types of pipes that are used in fluid power applications are black, galvanized, and stainless steel.

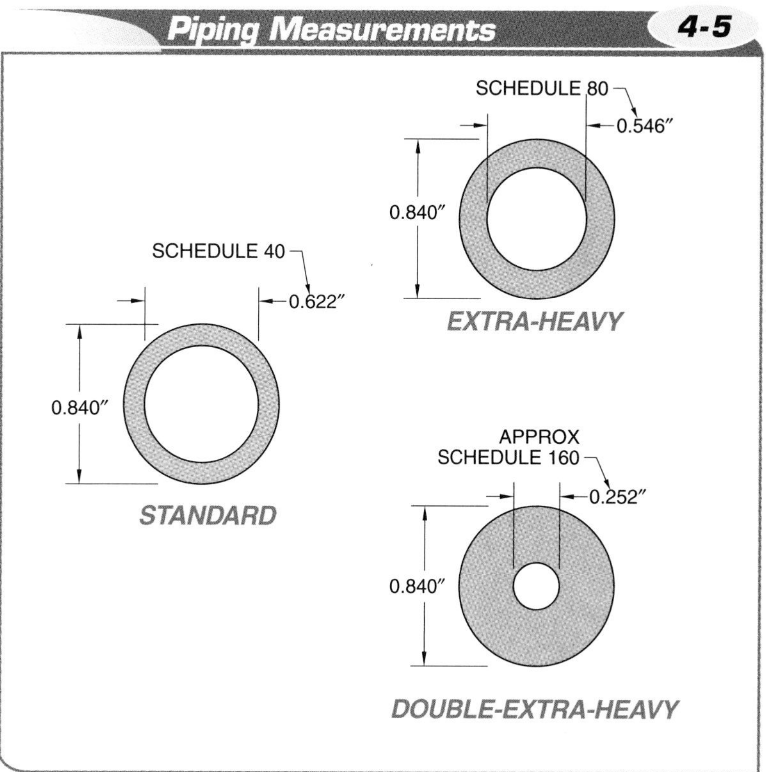

Figure 4-5. The three common pipe schedules that are used in fluid power applications are schedule 40, schedule 80, and schedule 160.

Piping measurements are specified by inside diameter (ID) and outside diameter (OD). *Inside diameter* is the diameter of a pipe measuring between the inside walls. *Outside diameter* is the diameter of a pipe measuring across the outside walls. *Wall thickness* is the difference between the inside diameter and outside diameter divided by two. Wall thickness determines the maximum pressure a pipe can withstand. Pipes are rated by a schedule that ranges from schedule 10 to schedule 160. For example, the three common pipe schedules that are used in fluid power applications are schedule 40, schedule 80, and schedule 160. Each schedule has the same OD, but the higher the schedule, the smaller the ID of the pipe. Thus, the wall thickness will be greater and the pipe can withstand higher pressures. **See Figure 4-5.**

Piping is available in standard OD sizes, ranging from ⅛″ to 12″. For example, when a schematic diagram requires a 1¼″ schedule 80 pipe, it refers to the outside diameter and wall thickness of the piping. **See Appendix.**

Advantages of piping include low cost, ease of installation, and ability to withstand high fluid pressures because of greater wall thicknesses. A disadvantage of piping includes weak points at threaded sections due to reduced wall thickness. Straight threads are preferred over tapered threads because of the reduced possibility of leakage.

Pressure surges that can occur from changes of fluid flow direction at high pressure are greatest at threaded connection points. Pressure spikes may be as great as 50% over operating pressure and can rupture sections with reduced wall thickness through fatigue caused by regular use. Welded connections must be made carefully to ensure that the heat from the welding process does not damage the mating parts of the piping. Pipes are typically welded to flanges, with the flanges bolted together. **See Figure 4-6.** It is impractical to weld piping that is 1″ diameter and smaller.

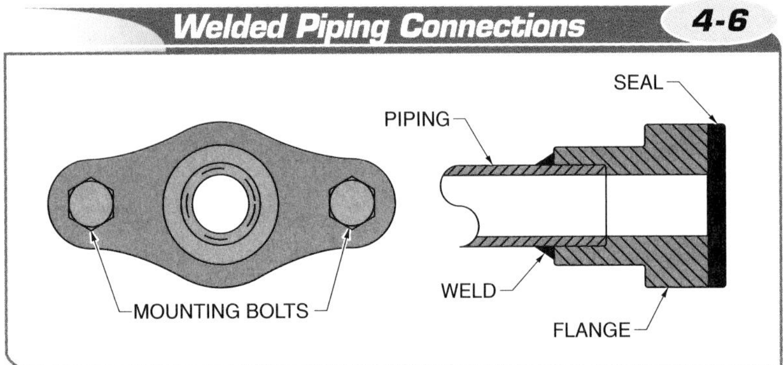

Figure 4-6. The two types of pipe threaders are handheld pipe threaders and pipe-threading machines.

TERMS

Inside diameter is the diameter of a pipe measuring between the inside walls.

Outside diameter is the diameter of a pipe measuring across the outside walls.

Wall thickness is the difference between the inside diameter and outside diameter divided by two.

A **pipe fitting** is a piece of pipe that is used to interconnect pipes and allows them to change direction.

A **pipe threader** is a tool used to cut threads in steel pipe.

Thread sealant is a material applied to the male pipe threads to ensure an airtight connection with the female pipe threads.

Pipe Threading. Pipes must be threaded to be interconnected and redirected. Threading a pipe makes it possible to connect the pipe to a pipe fitting. A *pipe fitting* is a piece of pipe that is used to interconnect pipes and allows them to change direction. A *pipe threader* is a tool used to cut threads in steel pipe. A pipe threader consists of dies, a die head, and one or more handles.

When a pipe is threaded, a pipe threader is used to cut threads that allow the pipe to interconnect with a pipe fitting. The pipe is considered the male end and the pipe fitting is considered the female end. In order for a pipe and a pipe fitting to connect, the OD of the pipe and the ID of the pipe fitting must be the same size and have the same type of threads.

The two types of pipe threaders are handheld pipe threaders and pipe-threading machines. **See Figure 4-7.** A handheld pipe threader is manually attached to the pipe and then threaded by hand. With a pipe-threading machine, the pipe is placed into the machine, all of the settings are configured, and the machine automatically threads the pipe. Regardless of the type of pipe threader used, the technician must ensure that the pipe is threaded only as much as recommended. If pipe threads are cut too deep, the threads will not seal properly and there will always be the possibility of a leak at that thread.

Thread sealant is a material applied to the male pipe threads to ensure an airtight connection with the female pipe threads. Thread sealant must be applied once the pipe is threaded. Thread sealant ensures that the connection between the pipe and the pipe fitting is waterproof and corrosion-resistant. Vibration, pipe size, and hardening time should be considered when selecting the proper sealant. **See Figure 4-8.** Pipes are threaded and joined by applying the following procedure:

1. Secure the pipe properly in a pipe vise or a pipe machine and cut it to the required length using a pipe cutter.
2. Ream the ends of the pipe to remove any burrs on the inside using a pipe reamer or a half-round file, depending on the size of the pipe.
3. Thread the pipe to the proper length using a pipe threader while adequately lubricating the die.
4. Remove the pipe from the pipe machine or remove the pipe threader from the pipe and wipe the threads clean using a cloth.
5. Apply thread sealant to the male threads that will be engaged with the pipe fitting.
6. Start the pipe fitting onto the pipe thread by hand.
7. Tighten the pipe fitting with a pipe wrench approximately two or more rotations. Do not overtighten the pipe fitting.

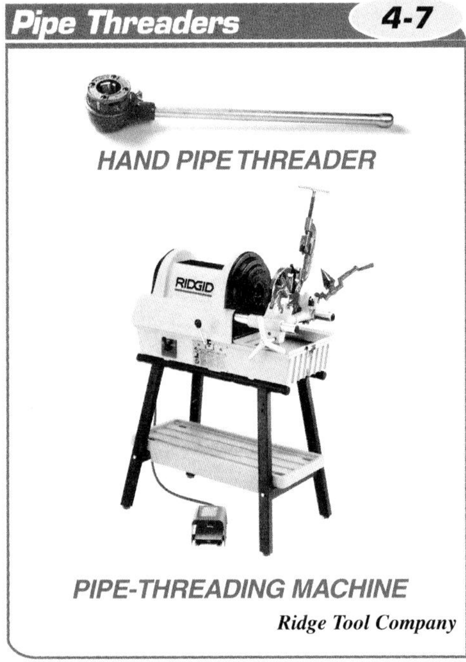

Figure 4-7. Male pipe threads are cut at the ends of a pipe to ensure a proper connection with the pipe fitting.

Figure 4-8. Leave a couple of the end pipe threads free from pipe thread sealant or Teflon® tape to ensure that it does not contaminate the system.

TERMS

Tubing is an extruded tubular material used to convey fluids.

A **union** is a fitting used to connect or disconnect two tubes or two pipes that would otherwise require the removal of other pipes for repair.

When applying pipe thread sealant, it should be applied against pipe threads in a clockwise manner to ensure that it does not get loose when the pipe is screwed into the pipe fitting. A few threads at the end of the pipe should be left free from pipe thread sealant. It will also ensure that none of the pipe thread sealant is pulled away by hydraulic fluid and contaminates the system. **See Figure 4-9.**

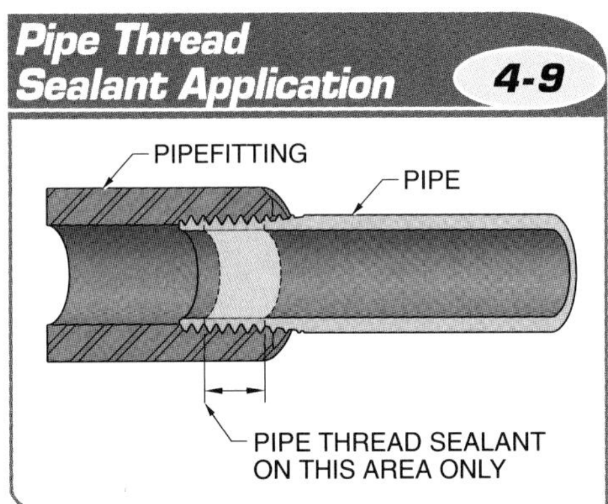

Figure 4-9. A flared tube fitting is connected to tubing in which the end is spread outward.

Piping Installation. When piping is installed, it must be properly supported because of vibrations in the hydraulic system. There are different types of pipe mounts and hangers that give support for any application. The purpose of these supports is to prevent strain or unnecessary pressure on the pipe joints, which can lead to leaks and premature system failure. There are different considerations for hydraulic systems than there are for pneumatic systems. Because of vibrations, and the weight of pipe and hydraulic fluid, the piping in hydraulic systems must be supported more than the piping installed in pneumatic systems.

Tubing

Tubing is an extruded tubular material used to convey fluids. Tubing is more commonly used in hydraulic systems than in pneumatic systems. Tube measurements are stated by the size of the OD. Sizing starts at ⅛″ and increases by $\frac{1}{16}$″ increments until it reaches 1″. After 1″, sizing increases by ¼″ increments up to 4″. Depending on the schedule number, the ID of a tube can be different for the same OD.

Tubing schedule numbers are 5, 10, 15, and 20, with schedule numbers 10 and 20 being the most commonly used. Tubing used for hydraulic systems must always be seamless. As a general rule, tubing used in hydraulic systems may be readily bent to a radius equal to five or six times the nominal (outside) diameter of the tubing.

The greater the difference is between the ID and OD of a tube, the greater its wall thickness. The wall thickness determines the amount of pressure the tubing can withstand. Tubing can be soldered, welded, or formed for compression. However, its wall thickness is always too thin for threading. Carbon steel tubing is commonly used with hydraulic systems. Tubing offers the following distinct advantages over piping when used in hydraulic systems:

- Tubing allows for more laminar fluid flow. Tubing is lighter in weight than piping and has a smooth inner surface that produces less friction, thereby losing less pressure.
- Tubing requires fewer connections than piping because it can be bent.
- Tubing absorbs vibrations better than piping because it has more flexibility.
- Tubing connections make every joint a union, permitting faster assembly and disassembly without the need for thread sealants. A *union* is a fitting used to connect or disconnect two tubes or two pipes that would otherwise require the removal of other pipes for repair.

To properly attach tubing to fluid power components, a tube fitting must be used. Although tubing is more expensive than piping, it is more desirable to use due to its ease of installation, maintenance, workability, and ability to produce laminar flow. The two types of tubing fittings are flared fittings and compression fittings.

Flared Fittings. A *flared fitting* is a fitting that is connected to tubing in which the end is spread outward. With a flared fitting, the nut of the fitting is first placed on the tubing, then the end of the tubing is mushroomed with a flaring tool, and then the nut is brought over the flared tubing to connect the remainder of the flared fitting. The body of the fitting is pressed tightly against the flared end of the tubing. Proper flaring provides a firm, leakproof connection.

Flared fittings generally consist of a body, a sleeve, and a nut made to match the flared tubing. **See Figure 4-10.** A seal is made when the tubing flare is pressed against the angular seat of the fitting. The angles of the body and sleeve of the flared fitting ensure a good seal when the tubing, which is the softest of the pieces, is pressed into the body. The standard flare angle for hydraulic tubing fittings is 37° from the centerline. The flare extends to cover the total angular surface of the sleeve, but not beyond the outside diameter of the sleeve. When tubing is properly flared and tubing nuts are securely tightened, the flare is seated firmly between the sleeve and the body.

Flares that are too short do not provide enough mating area to prevent leaks. Flares that are too long do not allow the nut to thread properly, making installation difficult. Clean, square tube cuts are achieved with a tube cutter. Hacksaws produce rough cuts that are generally not square and should not be used.

Flared fittings may also have a flare angle of 45°. A 45° flare angle is used for low-pressure applications such as pneumatic, refrigeration, or automotive applications. A 45° flare angle should not be used for high-pressure hydraulic systems. Incorrect flares may appear to assemble satisfactorily and may even pass initial pressure tests. However, they are not reliable for continuous service. All tubing flares must conform to the sleeve and the body used to join the tubing sections.

Flared fittings are typically used in hydraulic equipment when attaching tubing to a hose. Flared fitting parts cannot be mixed together, even if they are from the same manufacturer. When a part of the fitting fails, the entire fitting, including the nut, must be replaced.

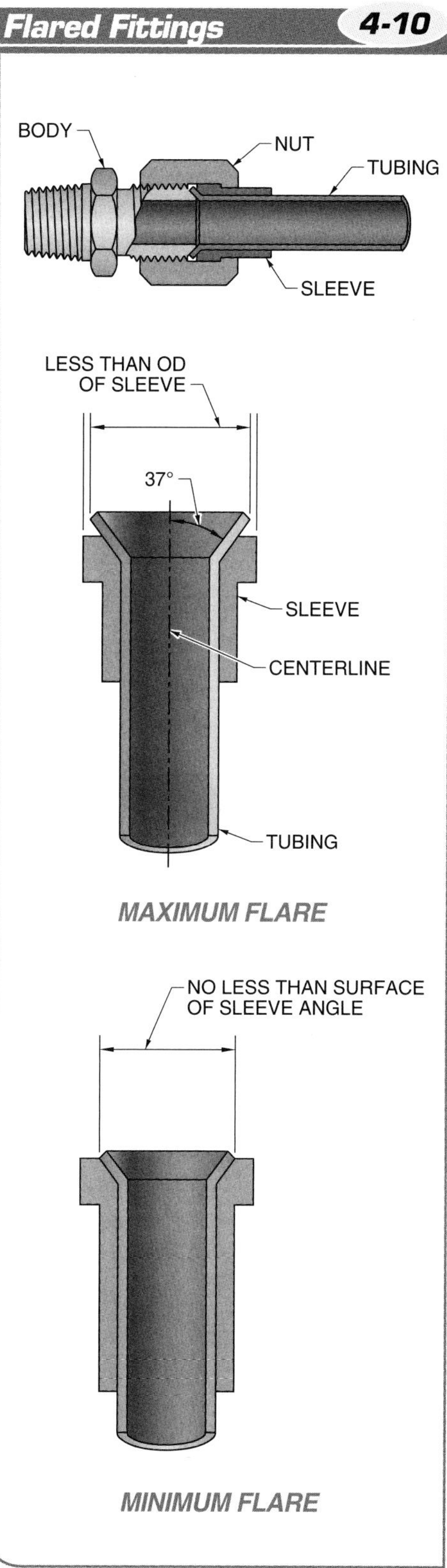

Figure 4-10. A flared tube fitting is tightened by using a torque wrench or by turning the fitting nut while observing witness marks.

TERMS

A **flared fitting** is a fitting that is connected to tubing in which the end is spread outward.

TERMS

A **witness mark** is a mark that is used as a guide to join two parts together.

A **compression fitting** is a type of fitting where the seal is created by a fitting component that is deformed to fit the tubing without threads.

A **ferrule** is a metal sleeve used for joining one piece of tubing to another.

A positive seal on a flared joint is vital to prevent fluid loss, keep out contamination, and maintain hydraulic system pressure. A positive seal is a type of seal that does not allow any amount of fluid leakage. For example, a positive seal is the gasket between the cylinder barrel and the end cap. A nonpositive seal is a seal that allows a certain amount of fluid leakage, which provides a lubricating film between surfaces. For example, a nonpositive seal is a piston that moves within a cylinder barrel.

To create a positive seal using a flared fitting, the proper amount of torque must be applied to the nut of the fitting. A flared fitting nut should be tightened by using a wrench or by manually turning the fitting nut while checking witness marks applied to the sealing nut and body. A *witness mark* is a mark that is used as a guide to join two parts together. Witness marks are typically used on flat sections such as the side of a nut. **See Figure 4-11.** This will help avoid undertightening or overtightening the nut.

Witness marks may be used when a torque wrench is not available. Witness marks are used by applying the following procedure:

1. Assemble the joint with the nut bottomed out and tighten by hand.
2. Mark a line with a permanent felt marker lengthwise on a flat section of the body, as well as the corresponding flat section of the nut.
3. Rotate the nut with a wrench until a determined number of flat sections on the body have been passed.

The number of flat sections passed is based on the size of the fitting. For example, a size 8½ flared fitting should be rotated through two flat sections after being hand tightened. This fitting may also be tightened to 200 in.-lb to 300 in.-lb using a torque wrench.

Compression Fittings. A *compression fitting* is a type of fitting where the seal is created by a fitting component that is deformed to fit the tubing without threads. **See Figure 4-12.** Compression fittings create a seal with a ferrule. A *ferrule* is a metal sleeve used for joining one piece of tubing to another. The ferrule cuts into the tubing by compressing the nut and seat of the fitting. After the completed assembly is hand tightened, the nut is tightened another full turn.

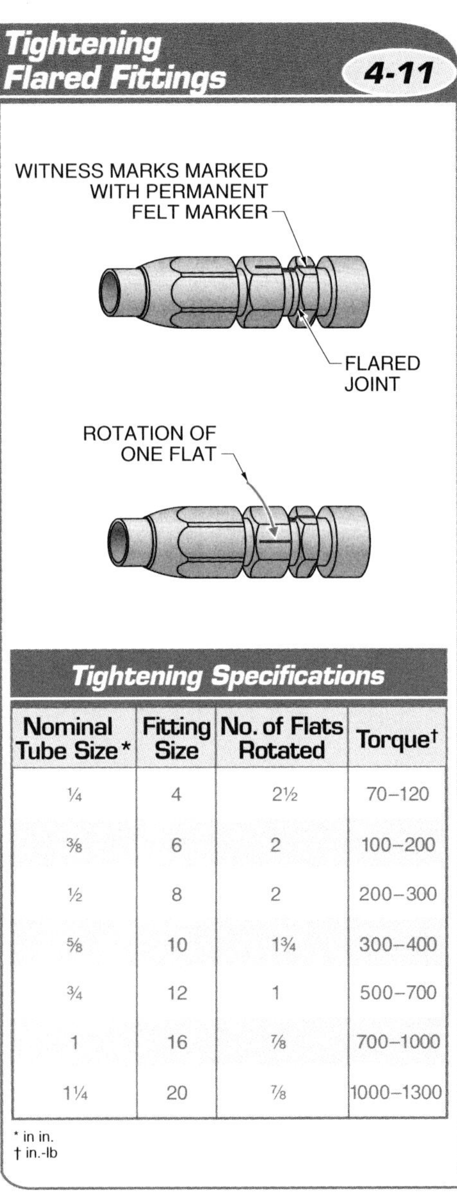

Tightening Specifications

Nominal Tube Size*	Fitting Size	No. of Flats Rotated	Torque†
¼	4	2½	70–120
⅜	6	2	100–200
½	8	2	200–300
⅝	10	1¾	300–400
¾	12	1	500–700
1	16	⅞	700–1000
1¼	20	⅞	1000–1300

* in in.
† in.-lb

Figure 4-11. The impact flaring method uses a flaring tool that is inserted into the tubing end and hammered until the tubing end is flared as required.

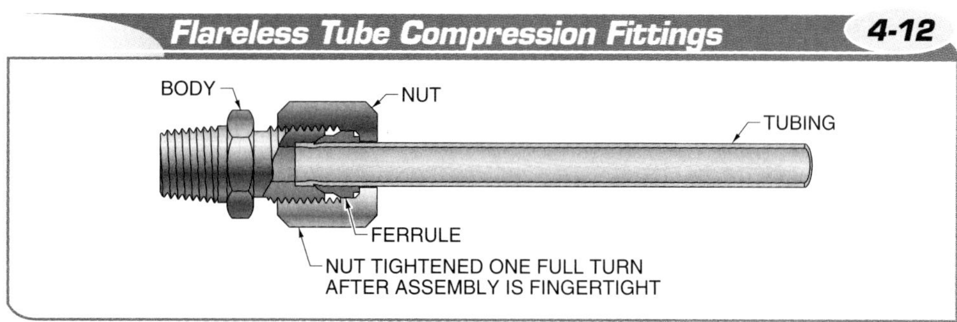

Figure 4-12. Flareless tube fittings create a seal with a ferrule.

TERMS

A **hose** is a flexible tube used for carrying fluids under pressure in both hydraulic and pneumatic systems and allows for movement of components.

Tubing Installation. Tubing can be assembled in a straight line or can be bent to reduce strain from vibration and to compensate for thermal expansion caused by heated liquid. **See Figure 4-13.** A gradual bend is preferred over elbow fittings because elbow fittings have sharp turns with high resistance to flow, which can increase heat in the system. Tubing must be bent with the correct bend radius and without kinks, wrinkles, or flattened bends. Bending tubing is accomplished with either a hand bender for small diameter tubing or a hydraulic bender for large diameter tubing. The method for bending tubing depends on the bender OEM. OEM instructions must always be used when bending tubing. The bending radius should be greater than four times the tubing ID. Tubing must be properly supported to minimize the stresses of vibration.

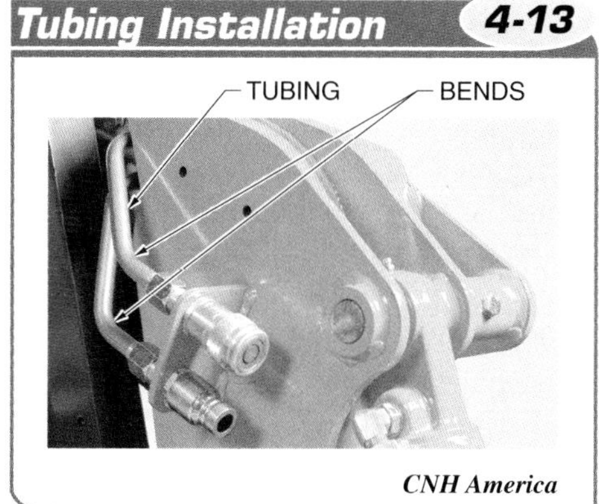

Figure 4-13. Tubing should never be assembled in a straight line, but instead should be bent to reduce vibration strains and to compensate for thermal expansion.

Hoses

A *hose* is a flexible tube used for carrying fluids under pressure in both hydraulic and pneumatic systems and allows for movement of components. For example, if the cylinders in a hydraulic system are moving (pivoting), their means of fluid transmission must also move. OSHA standards also require that hydraulic tools used on or near energized electrical conductors or equipment be supplied with nonconducting hose that can withstand normal operating pressures. Hoses are the only means of transmission capable of this. Hoses are also used to attach pipes or tubes to equipment for protection from vibration and premature wear on joints.

The strength of the hose is dependent on its construction. All hydraulic hoses are constructed with a minimum of three layers. **See Figure 4-14.** The outer layer (cover) is typically made of an oil or weather-resistant material such as synthetic rubber or thermoplastic. This layer helps to reinforce the hose and protects it from outside influences. The middle layer (reinforcement layer) consists of one or two sublayers made out of wire braids wrapped in alternate directions and covered with synthetic rubber or textile.

The more sublayers there are, the stronger the hose will be. The inner layer is the layer that is in direct contact with the fluid. The material used in the inner layer depends on the type of hydraulic fluid being used, but it is typically an oil-resistant synthetic rubber or thermoplastic.

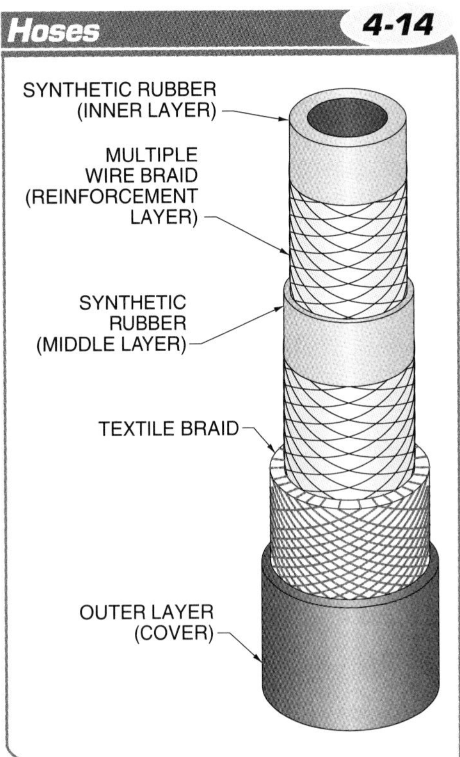

Figure 4-14. Hoses are fabricated in layers for use in high-pressure hydraulic systems.

Hose Selection. When creating a new hose or replacing a hose whose identification markings have been worn off, a common practice is to use the acronym "STAMPED" to determine the proper hose type. "STAMPED" is used to identify the following variables:

- **S**—size of the ID of the hose—If this is a replacement hose, information on the layline should be used to figure out the ID. The standard increments of hose sizes are ¹⁄₁₆″. The ID is represented by a dash number. The dash number is the number of 16th inches. For example, if the number in the layline is -16, then the ID is 1″. If the number in the layline is -5, the ID is ⁵⁄₁₆″. If the hose OEM information is not visible, the hose should be cut and the ID of the hose measured. Before cutting the hose, the total length to the hose should be measured. If this is a new installation, the calculation or the nomographic chart should be used to determine the correct hose ID. Also, it should be verified that the hose is cut to the proper length.
- **T**—temperature of both the hydraulic fluid inside the hose and the ambient temperature around the hose—OEM specifications should be consulted to verify that the hose can withstand the temperature range of both variables.
- **A**—application where the hose is going to be used—If the hose is for a new application, or replacing a hose that prematurely failed, the following items must be considered:
  - minimum bend radius of hose
  - twisting of hose
  - correct operating temperature
  - system pressure
  - hose maximum pressure rating
  - hose and connector OEM
  - application (return line, suction line, or pressure line)
  - type of thread in use
  - unnecessary contact/friction between hose and other equipment
- **M**—material (hydraulic fluid) used—The hydraulic fluid must be compatible with the hose, cover, O-rings, connectors and fittings. Failure to verify that these components are capable of conveying the hydraulic fluid can be catastrophic to the hose assembly.
- **P**—pressure of system where hose is to be installed—One of the most important safety factors when installing a hose is to verify that the system pressure will not be higher than the stated working pressure of the hose. Also, if the system has pressure spikes, it must be verified that the hose is rated to handle them. If not, the hose life will be dramatically shorter. If replacing a hose, it must be verified that any dramatic pressure drop that existed before the replacement of the hose is not a factor. If a pressure drop remains, the system must be inspected for the following conditions:
  - wrong viscosity fluid
  - temperature of the fluid
  - too many 45° and 90° fittings
  - incorrect hose ID

- **E**—ends on couplings—The OEM connector model number should be used to verify that the proper fittings and couplings are being used. If the OEM is unknown or the model number cannot be read, a hose identification kit can be used to determine all of the needed information to find the correct replacement.
- **D**—delivery of fluid, product, and documentation—If a new hose is replacing an old hose, it must be replaced with the same size. If a hose is to be installed in a new or reconfigured system, the required ID or nomographic chart should be used. Also the availability of all needed parts for the entire hose assembly such as delivery time of any needed parts should be considered. Lastly, a final inspection should be conducted and any needed documentation should be completed.

Hose Fittings. The two types of connection fittings available for hoses are threaded (screw-on) fittings and quick-disconnect (push-to-connect) fittings. **See Figure 4-15.** Threaded fittings are attached with a wrench and are used when the hose is installed as a permanent connection. For example, threaded fittings are typically used on permanently installed industrial equipment such as scrap metal compressors.

Quick-disconnect fittings are used when a hose needs to be occasionally disconnected or reconnected. For example, quick-disconnect fittings are commonly used on snowplow attachments for tractors or trucks because the plow systems are installed and uninstalled annually. A *check valve* is a valve that allows fluid flow in one direction, but stops it in the opposite direction. Quick-disconnect fittings often have an interior check valve that allows the hose to be pulled off without losing hydraulic fluid. The check valve ensures that the hoses are completely connected because the male portion unseats the check valve, which allows fluid flow.

Hose Installation. When hoses are installed, they must be installed with the largest bend radius possible to avoid twists and sharp bends. **See Figure 4-16. See Appendix.** The bending radius of flexible hose must be greater than six times the ID. Protective sleeves must encase any hose that is subject to rubbing. Hoses must not be excessively long or excessively short. Hoses that are excessively long have more internal resistance. Hoses tend to decrease in length when pressurized. If a hose is excessively short without any bend or flex, it will fail prematurely. Hose fittings are typically crimped to the ends of hoses. Hydraulic hoses also have a layline. A *layline* is a line mark or print on a hydraulic hose that indicates if an installed hose has been improperly twisted. If a hose is twisted approximately 5°, its service life can be lessened by 70%. **See Figure 4-17.** Hose fittings are typically crimped to the ends of hoses. The same brand of hoses and ends must always be used. Failure to use the same brands cause over 90% of hydraulic hose failures.

TERMS

A **layline** is a line mark or print on a hydraulic hose that indicates if an installed hose has been improperly twisted.

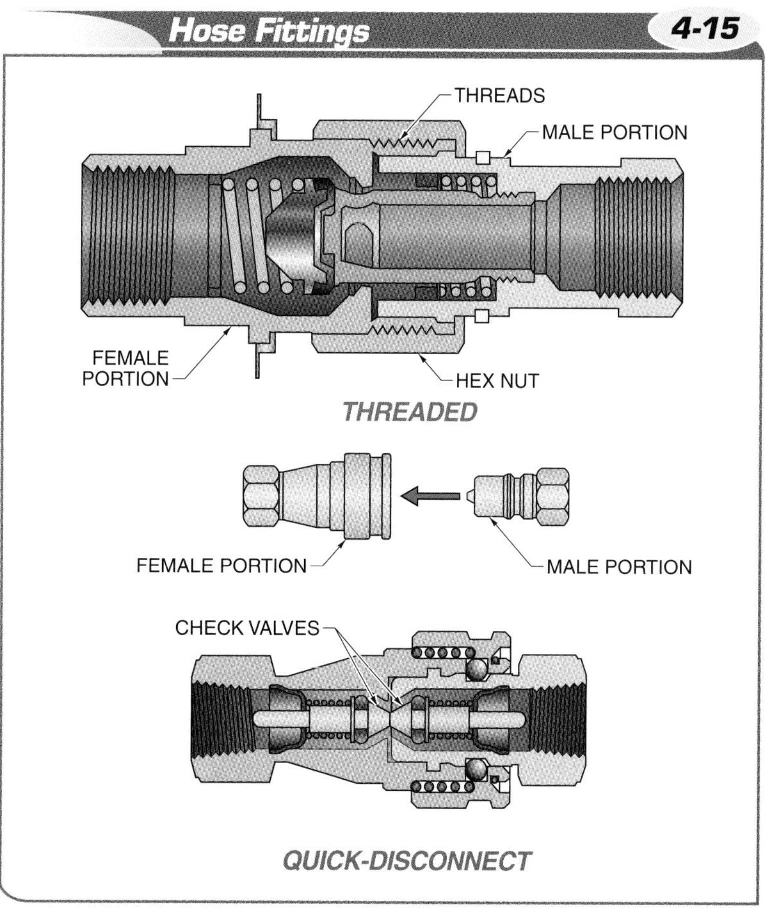

Figure 4-15. Two types of connection fittings for hoses are threaded fittings and quick-disconnect fittings.

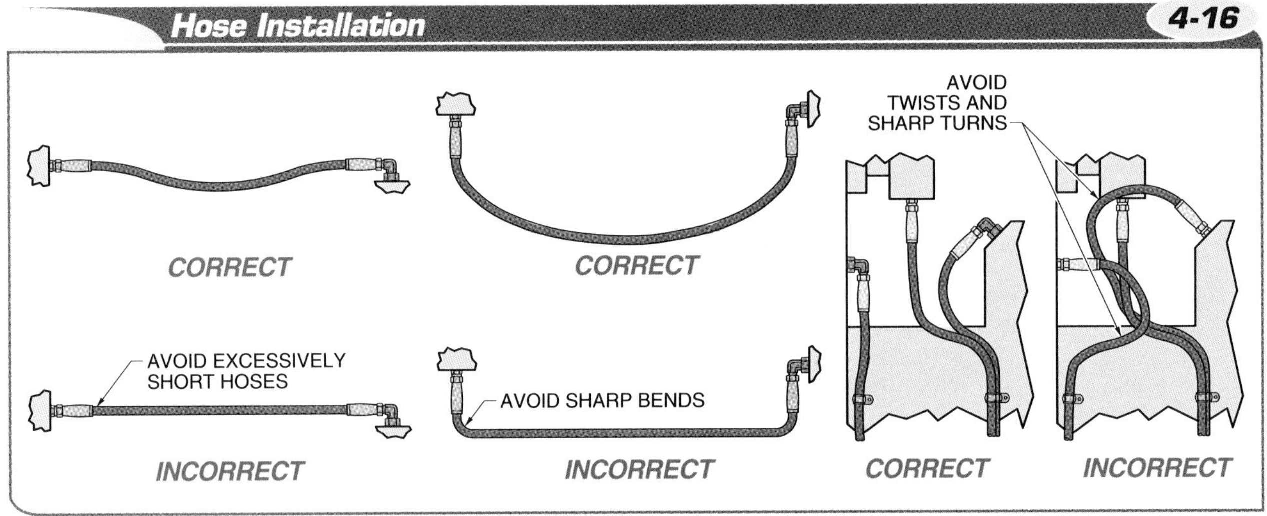

Figure 4-16. Hoses are installed to avoid excessively short lengths, sharp bends, and twists.

TERMS

A **thread screw gauge** is a hand tool used to determine the dimensions and pitch of round and cut threads.

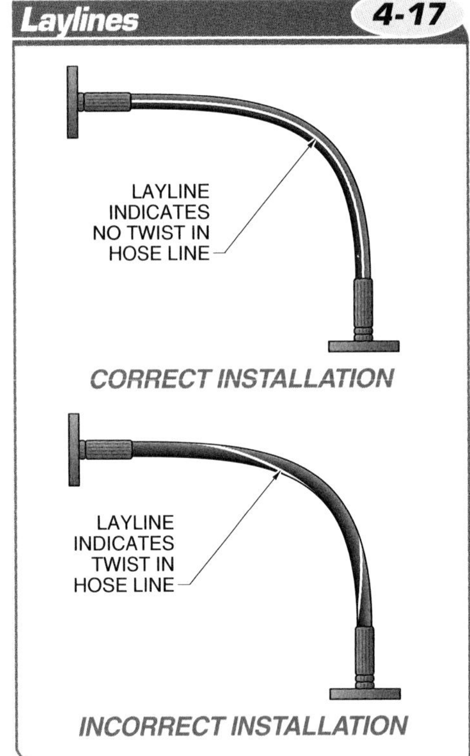

Figure 4-17. A layline is a line mark or print on a hose that indicates if the hose has been improperly twisted.

Warning: Hoses must be crimped per manufacturer's instructions by personnel that have received hands-on training in proper crimping procedures. An improperly crimped hose can rupture under pressure and cause serious injury, property damage, or death.

Connector Thread Identification

There are many different types of connectors used to connect hydraulic systems. It is important to create a safe and leak-free connection by establishing the type of threads being used to verify that the proper connections are being made. To help identify which type of thread is being used or needs to be used, a thread screw gauge can be used. A *thread screw gauge* is a hand tool used to determine the dimensions and pitch of round and cut threads. **See Figure 4-18.**

Connectors are available with two designs. One connector design ties directly to the hose or tube (through compression, screwing, or crimping) and ties into either another connector or the port of a cylinder, manifold, or valve. This type of connector has one female connection side and one male connection side. The other type of connector has two male ends and is used to connect one connector to either another connector or a port. A connector that has two male ends typically has one end that attaches to a hydraulic hose or tube, while the other end attaches to a threaded port. **See Figure 4-19.**

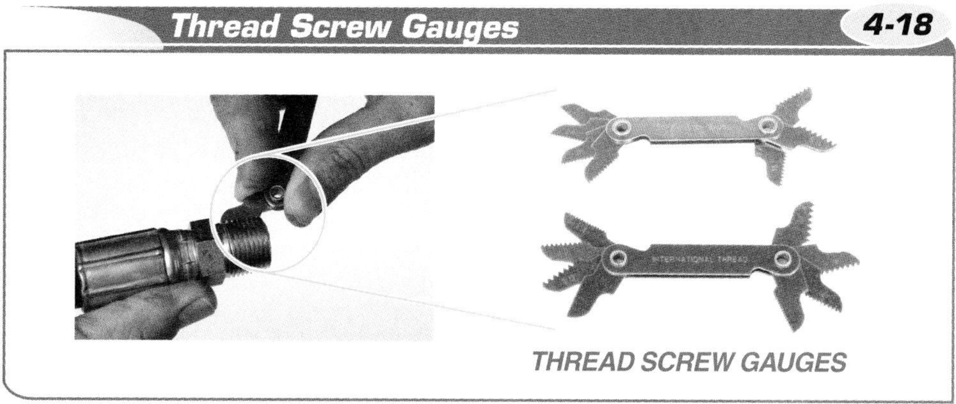

Figure 4-18. Thread screw gauges can be used to verify the type and pitch of unidentified screw threads that are located on certain types of fluid power equipment.

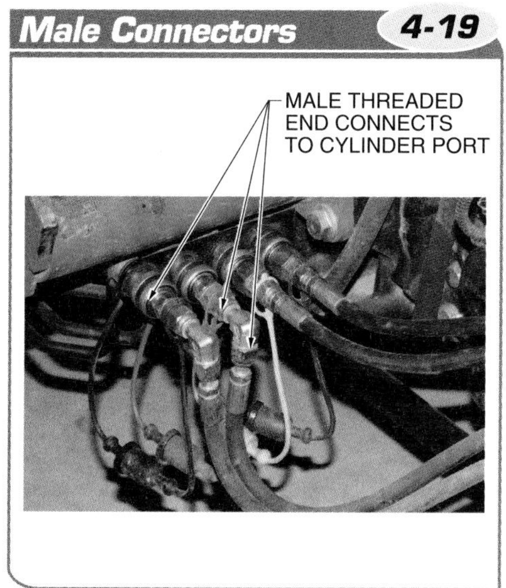

Figure 4-19. A connector that has two male ends typically has one end that attaches to a hydraulic hose while the other end attaches to a threaded port on another.

The first step in establishing the type of connector being used is to establish the thread types and characteristics. There are many different types of connector threads available. The most common types of connector threads include the following:

- straight thread
- American pipe thread
- British parallel pipe thread
- British taper thread
- metric parallel
- metric taper

Connector threads have various characteristics. The different characteristics that connector threads can have include parallel or tapered threads, thread pitch, and thread size.

Parallel or Tapered Threads. Threads can either be parallel or tapered and can typically be determined by visually inspecting the thread or placing the threads vertically against a straight edge. Parallel threads require additional sealing methods such as an O-ring. Tapered threads seal as they are being screwed into the female port. However, this seal is not tight enough alone and requires that a sealant be applied to them before being inserted into the port.

Thread Pitch. Thread pitch for most connectors is found by using a pitch gauge. If a pitch gauge is not available, the thread pitch can be found on most English threads by counting the number of threads in a 1″ length of a threaded device. If the thread is metric, it can be found by measuring the distance from one thread crest to the next crest. **See Figure 4-20.**

Thread Size. Thread size refers to the OD of the threads. For SAE and metric threads, the OD can be found using calipers. For all British threads and NPT, the most common method for determining size is to use a chart or a female connector with a known ID. Another method is to measure the OD and subtract ¼″ from it. For these types of threads, their size is typically referred to as nominal size.

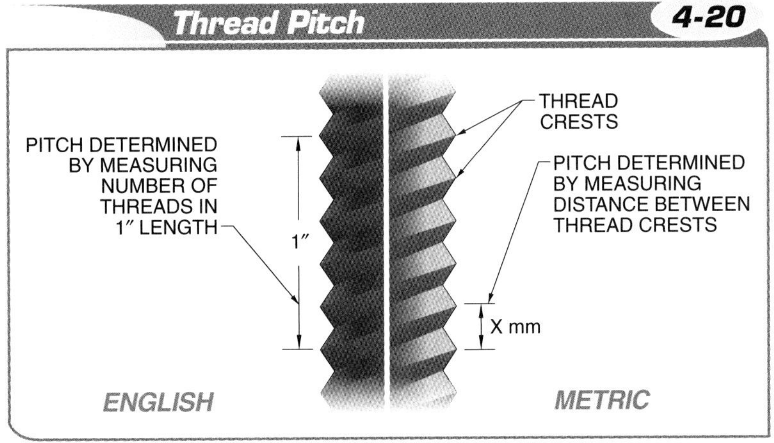

Figure 4-20. Thread pitch can be found on English threads by counting the number of threads in a 1" length of a threaded device and on metric threads by measuring.

TERMS

Pickling is a method of removing scale and rust from metal through chemical treatment.

Thread Types

Each type of thread has different characteristics that separate it from other thread types. Being able to identify which threads are being used will help to ensure a long life of the connector. The most common types of threads are Society of Automotive Engineers (SAE) straight threads and national pipe thread/national pipe thread fuel (NPT/NPTF). Throughout the world, British standard pipe tapered (BSPT) and British standard pipe parallel (BSPP) are the two most popular types of connector thread types and can be found on many types of equipment.

BSPP and NPT/NPTF are similar and can be accidentally interchanged. They have different flank angles. BSPP and BSPT have 55° flank angles, while NPT/NPTF have 60° flank angles. They must never be interchanged. **See Figure 4-21.** The other type of thread is metric thread. This is the only thread type that is metric because BSPP and BSPT use nominal sizes when being identified.

Cleaning Conductors before Installation

One important part of installing any conductor is to verify that the conductor is free of any contaminants before installation. Contaminants are created in conductors when they are cut to size, shipped, and stored. After any conductor is cut, it must be cleaned. There are many methods for cleaning, and the best method depends on the type of conductor.

For piping and tubing, the ends must be deburred by reaming. Tubes can be cleaned by sandblasting or spraying with pressurized air, which removes most foreign containments. Pipes should be pickled. *Pickling* is a method of removing scale and rust from metal through chemical treatment. Hoses are often cleaned with pressurized air, but to ensure that all contaminants are removed, a hose cleaning device should be used. A hose cleaning device cleans the inside of the hose with detergent and then removes residual cleaning agents with pressurized air. All conductors that will be stored or shipped should be capped to ensure that contaminants do not attach to them.

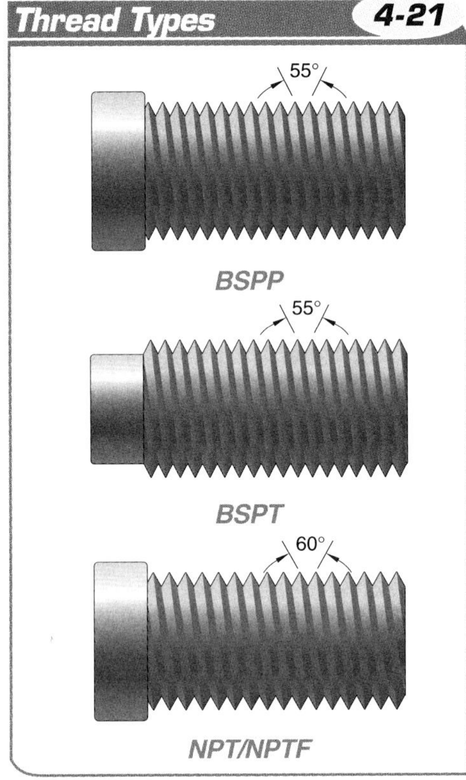

Figure 4-21. The most common types of threads used on hydraulic connectors are British standard pipe parallel thread (BSPP), British standard pipe tapered thread (BSPT), and American national pipe thread/national pipe thread fuel (NPT/NPTF).

Digital Resources

Chapter Review

Fluid Conductors and Connectors — Chapter 4

Name: ______________________________ Date: ______________

MULTIPLE CHOICE

_______________ **1.** Velocity increases ___ times to maintain a constant rate of fluid flow if the diameter of a conductor is halved.

A. two
B. three
C. four
D. does not increase

_______________ **2.** The standard flare angle for hydraulic tubing fittings is ___° from the centerline.

A. 17
B. 24
C. 37
D. 45

_______________ **3.** A ___ mark is a mark used as a guide to join to parts together.

A. ferrule
B. hose
C. pipe
D. witness

_______________ **4.** ___ is the speed of fluid flow through a hydraulic line.

A. Acceleration
B. Pressure
C. Velocity
D. Viscosity

_______________ **5.** Pipes are rated type of scale known as a schedule and range from ___ to ___.

A. 10; 40
B. 40; 80
C. 40; 120
D. 10; 160

_______________ **6.** When performing certain mathematical calculations, the symbol ___ is used to indicate a change in a particular value.

A. Δ
B. ^
C. Σ
D. ~

______________ **7.** A ___ is a tool used to cut threads in steel pipe.
- A. deburring tool
- B. pipe cutter
- C. pipe threader
- D. tap and die

______________ **8.** ___ are/is a passage or series of passages in a fluid power system that is constructed of metal, plastic, or plasticized rubber and conforms to ANSI standards.
- A. Accumulators
- B. Pilot lines
- C. Piping
- D. Valve bodies

______________ **9.** A ___ is a line mark or print on a hose that indicates if a hydraulic hose has been improperly twisted after installation.
- A. layline
- B. pilot line
- C. specification line
- D. witness mark

______________ **10.** ___ is an increase in speed, measured in fps.
- A. Acceleration
- B. Torque
- C. Velocity
- D. Work

COMPLETION

______________ **1.** ___ determines the maximum pressure a pipe can withstand.

______________ **2.** ___ tubing is commonly used in hydraulic systems.

______________ **3.** ___ fittings are typically used in hydraulic equipment when attaching tubing to a hose.

______________ **4.** The velocity of fluid flow increases at any restriction in the piping if the ___ remains the same in the system.

______________ **5.** A(n) ___ is a metal sleeve used for joining one piece of tubing to another.

______________ **6.** A(n) ___ is a piece of pipe that is used to interconnect pipes and allows them to change direction.

______________ **7.** The bending radius of flexible hose must be greater than ___ times its ID.

______________ **8.** The ___ of fluid flow must increase at any restriction to maintain constant flow rate.

______________ **9.** A(n) ___ fitting is a tubing fitting in which the end is spread outward.

______________ **10.** Pressure spikes that occur within a hydraulic system can be as high as ___% of the operating pressure.

______________ **11.** The two types of ___ used in hydraulic systems are black and stainless steel.

______________ **12.** All hydraulic hoses are constructed with a minimum of ___ layers.

______________ **13.** A(n) ___ mark can be used on a nut to avoid under or overtightening certain components.

____________________ 14. ___ is a method of removing scale and rust from metal through chemical treatment.

____________________ 15. Piping is available in standard sizing with ODs up to ___″.

TRUE/FALSE

T F 1. Tubing installed in a hydraulic system cannot be threaded.

T F 2. The lower a pipe schedule number, the smaller the ID of the pipe.

T F 3. Galvanized pipe should not be used with hydraulic systems.

T F 4. Hydraulic and pneumatic pipes each require the same amount of support in a fluid power system.

T F 5. A 45° flare angle should not be used for high-pressure hydraulic systems.

T F 6. Pipes that are bolted to flanges do not require welding at the pipe joint.

T F 7. Flow velocity in a pipe is related to the area of its inside diameter.

T F 8. When applying pipe thread sealant, the last few threads at the end of the pipe should not have any sealant applied to them.

T F 9. Piping sections in a hydraulic system can only be connected together through welding.

T F 10. Once a fluid is flowing, its speed never changes.

T F 11. A nonpositive seal is a type of seal that does not allow any amount of fluid leakage.

T F 12. Tubing is more commonly used in hydraulic systems than pneumatic systems.

T F 13. Fluid travels at a higher velocity through large diameter piping than it does through low diameter piping.

T F 14. Straight pipe threads are preferred over tapered pipe threads because of the reduced possibility of leakage.

T F 15. The difference between fluid flow (gpm) and velocity (fps) is that fluid flow is distance per unit of time, while velocity is volume per unit of time.

SHORT ANSWER

1. List at least five advantages of using tubing over piping in a hydraulic system.

2. List six common types of connector threads.

3. Briefly explain why velocity in a hydraulic system should not exceed recommended values.

4. Why must pipe threads for hydraulic equipment not be cut too deep?

5. List three factors that must be considered when selecting pipe thread sealant for a specific application.

Activity 4-1: Pipe Velocity

A 2″ ID pipe that is 10′ in length supplies pressure from a hydraulic power unit must be replaced because it was damaged by a fork-lift truck. The only pipe available is 1.5″ ID pipe and the company is losing thousands of dollars during this downtime. The pump is supplying 12 gpm. The head engineer has asked for the difference in the velocity of the fluid flowing through the pipes to determine if the 1.5″ diameter pipe would work in the system.

1. Determine the difference in velocity of the two pipes.

Activity 4-2: Nomographic Chart Application

Use the nomographic chart to determine the minimum size hose that must be used if the flow rate is 20 gpm and the velocity needs to be 15 fps.

1.

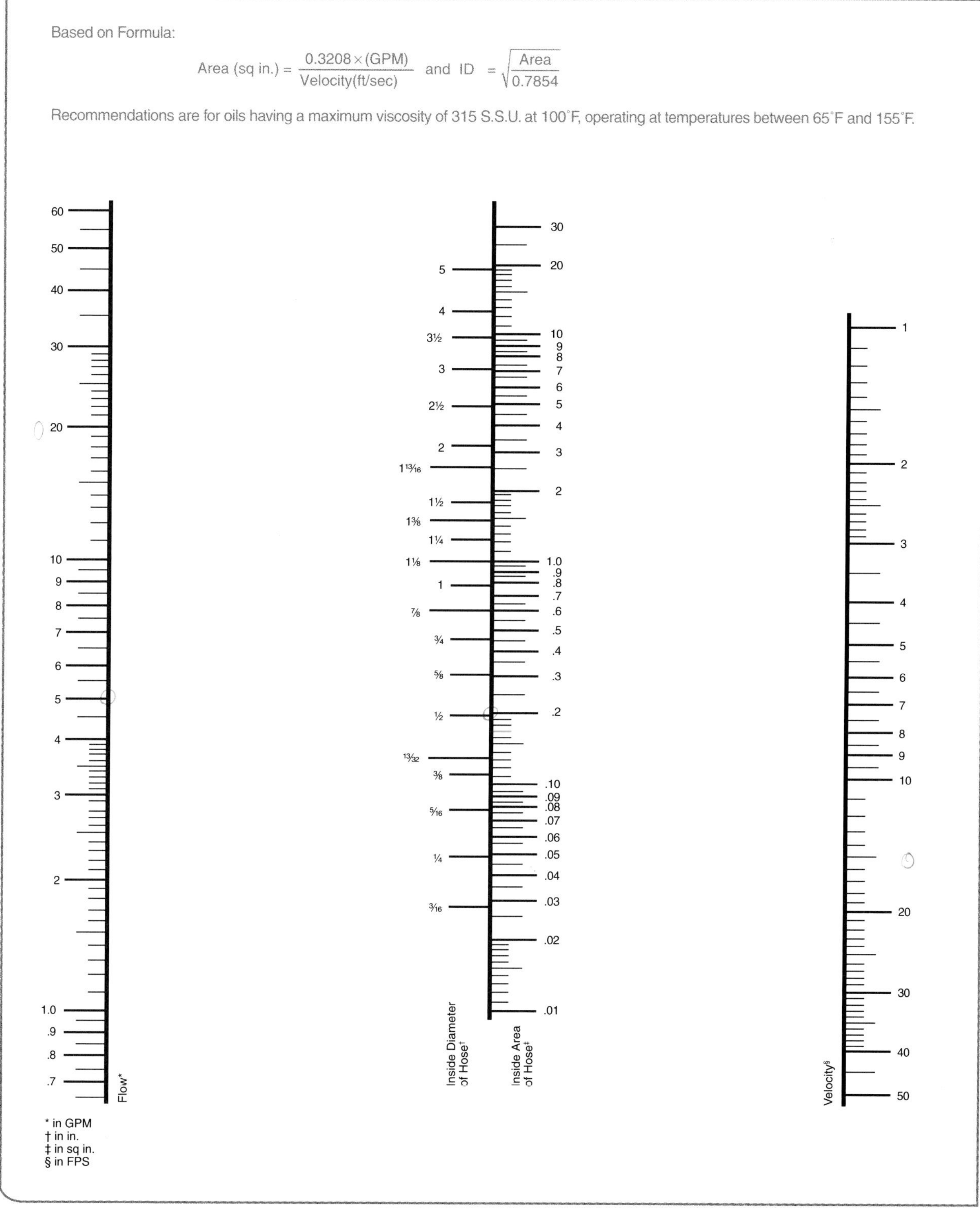
Based on Formula:
Area (sq in.) = 0.3208 × (GPM) / Velocity(ft/sec) and ID = √(Area / 0.7854)
Recommendations are for oils having a maximum viscosity of 315 S.S.U. at 100°F, operating at temperatures between 65°F and 155°F.
Flow*
Inside Diameter of Hose†
Inside Area of Hose‡
Velocity§
* in GPM
† in in.
‡ in sq in.
§ in FPS

Activity 4-3: Building and Testing a FluidSim® Circuit

Log into the FluidSim software and build the fluid power system below by dragging and dropping the correct components and connecting them together. To operate the system, apply the following procedure:

1. Go to File > New.
2. Build the system below using the parts in the Component Library and dragging them onto the white part of the screen.
3. To connect the components, place mouse over connection point, left click, hold, and drag mouse over opposite connection point and release.
4. Once circuit is built, simulate circuit by clicking on the Start button (►) in the toolbar.
5. If an "Open Connection" window appears, click "OK." This is referring to the "T's" connection that is not being used. The program will still operate correctly.
6. If a "There have been warnings. Start simulation anyway?" window appears, select "Yes".
7. Drag mouse over the directional control valve's lever on left. Mouse arrow will turn into a hand with a finger pointing out. When this occurs, left click mouse.

Answer the following questions.

1. Explain what occurs with the simulation.

2. Release mouse, and explain what occurs with the simulation.

Note: Click lever of directional control valve on right. Second cylinder should extend. When releasing mouse, cylinder continues extension because the directional control valve has no spring to return it. To return cylinder, click on far right box of directional control valve. Cylinder should return.

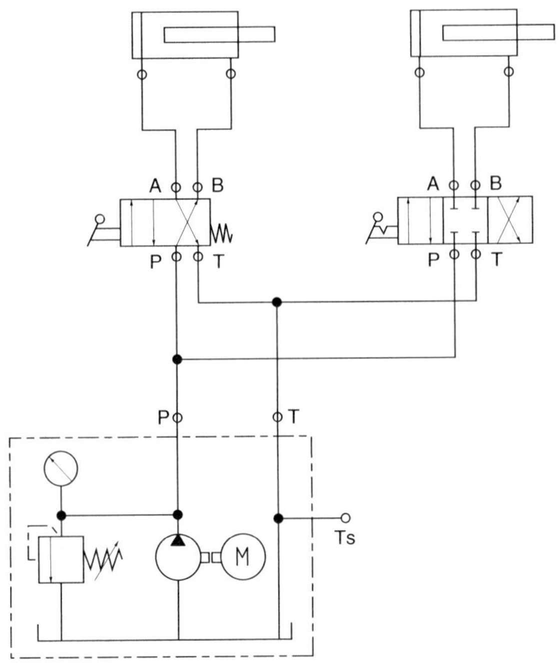

Chapter 5

Hydraulic Pumps

OBJECTIVES

- Describe the basic steps for creating fluid flow that are common to all hydraulic pumps.
- Describe how hydraulic pumps are rated and distinguish the different ratings.
- Distinguish between the different types of hydraulic pumps that are used in hydraulic systems.
- Describe the specific actions that the different types of hydraulic pumps use to create fluid flow and pressure.
- Describe the different types of variable pump controls.

INTRODUCTION

All devices that accomplish work must have a power source. For hydraulic systems, the hydraulic power source is a pump. A pump acts in the same manner as an electric generator. While an electric generator changes rotating mechanical energy into electric energy, a pump changes mechanical energy into hydraulic energy, which is the energy used in a hydraulic system. All hydraulic systems have a pump, and although there are many different types of pumps, they all accomplish the goal of moving hydraulic fluid to where it is needed to operate equipment.

TERMS

A **hydraulic pump** is a mechanical device that changes mechanical energy into hydraulic energy (fluid flow).

HYDRAULIC PUMPS

A *hydraulic pump* is a mechanical device that changes mechanical energy into hydraulic energy (fluid flow). Hydraulic pumps create flow by increasing the volume of fluid at their inlet and decreasing the volume of fluid at their outlet. The resistance to fluid flow in the pathway and the load on the system create pressure in the system. The greater the load or system resistances, the higher the pressure in the system.

Hydraulic pumps are used in all hydraulic systems and are classified by the type of pumping mechanisms used. The main categories of pumps are positive-displacement and non-positive-displacement (dynamic) pumps. **See Figure 5-1.** Although categorized as a type of hydraulic pump, dynamic pumps are rarely used for hydraulic applications because they cannot create flow against the pressures found in hydraulic systems. Several factors must be considered when selecting a hydraulic pump for a specific application.

Hydraulic Pump Ratings

Hydraulic pump ratings are assigned by the pump manufacturer and are the main factors in determining pump use. The most common pump ratings include displacement, gallons per minute, pressure, and volumetric efficiency.

All the information needed to service or replace a pump can be found on the pump nameplate. The information is typically provided in the form of a pump model number and must be referenced in the pump's OEM manual. Nameplates are generally affixed to hydraulic pumps to identify the different pump ratings. In addition to pump ratings, the nameplate also identifies the pump manufacturer's name, the pump serial number, pump type, and/or the pump model number.

Nameplates are typically made from stainless steel and affixed to the pump in an easy-to-read location. The pump's nameplate can sometimes break off, so it is important to keep the information in another location in case the pump fails.

Note: Not all hydraulic pumps have nameplates. Relevant pump information can also be obtained through the documents provided by the manufacturer with the pump's original packaging. Such information must never be discarded but saved for future reference.

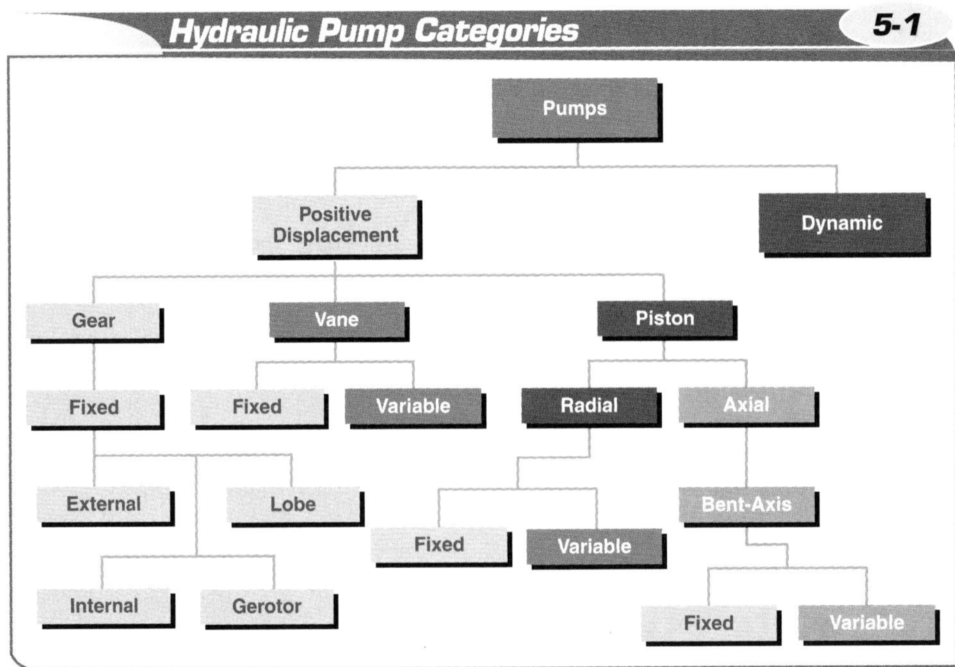

Figure 5-1. The two main types of hydraulic pumps are positive-displacement pumps and dynamic pumps.

Displacement. *Displacement* is the volume of hydraulic fluid moved during each revolution of a pump's shaft. Displacement is rated in cubic inches per revolution of the pump shaft and is supplied by the OEM and found using the pump model number. Displacement per minute represents the amount of fluid a pump displaces in one minute. The amount is based on the displacement of the pump and the rpm of the prime mover and is calculated using the following formula:

$$D_{in^3/min} = rpm \times D$$

where

$D_{in^3/min}$ = displacement per minute

rpm = revolutions per minute of prime mover

D = pump volumetric displacement (in cu in.)

Example: What is the displacement of a pump with revolutions of 1.94 cu in. that has a prime mover with revolutions of 1120 rpm?

$$D_{in^3} = rpm \times D$$

$$D_{in^3} = 1120 \times 1.94$$

$$D_{in^3} = \mathbf{2173\ cu\ in./min}$$

A *prime mover* is a device that supplies rotating mechanical energy to a fluid power system. The two main types of prime movers used in fluid power systems are electric motors and internal combustion engines. **See Figure 5-2.**

Gallons Per Minute. When used in fluid power systems, gallons per minute is the number of gallons of fluid that a pump can force into the system every minute. *Gallons per minute (gpm)* is a measure of fluid flow that is used to measure small volumes of intermittently flowing fluids such as pump discharges. **See Figure 5-3.** Gallons per minute in a pump is typically measured with a flow meter attached to the hydraulic system, but can be calculated if certain variables are known. Gpm in a pump is calculated by applying the following formula:

$$Q = \frac{cu\ in./rev \times rpm}{231}$$

where

Q = gallons per minute

$cu\ in./rev$ = cubic inches of flow per revolution

rpm = revolutions per minute

231 = constant (cu in. in one gal)

Example: What is the gpm of a pump that has a displacement of 4 cu in. that operates at 1725 rpm?

$$Q = \frac{cu\ in./rev \times rpm}{231}$$

$$Q = \frac{4 \times 1725}{231}$$

$$Q = \frac{6900}{231}$$

$$Q = \mathbf{29.87}$$

TERMS

Displacement is the volume of hydraulic fluid moved during each revolution, stroke, of a hydraulic pump shaft.

A **prime mover** is a device that supplies rotating mechanical energy to a fluid power system.

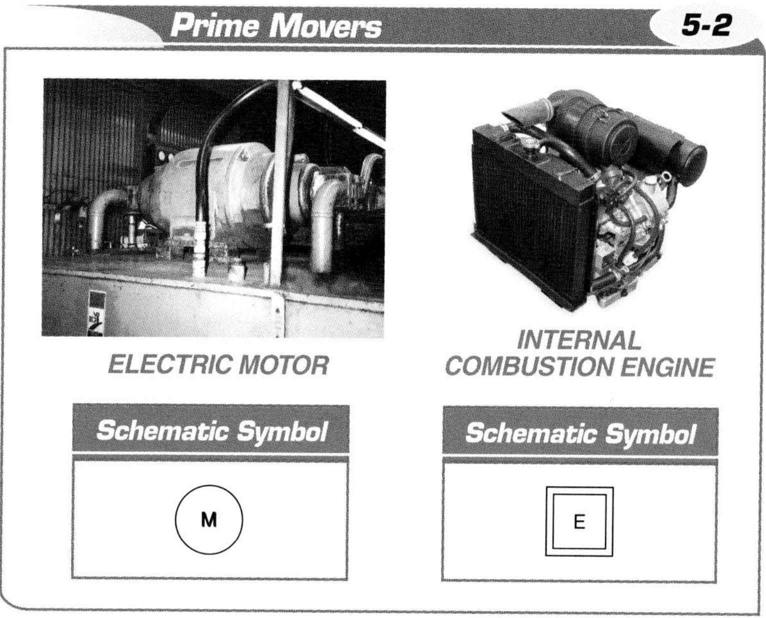

Figure 5-2. The two main types of prime movers used in fluid power systems are electric motors or internal combustion engines.

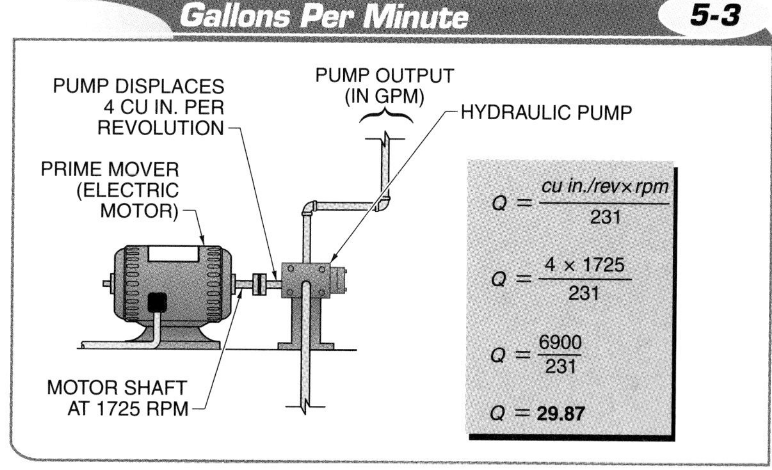

Figure 5-3. Gallons per minute is a measure of fluid flow that is used to measure small volumes of intermittently flowing fluids such as pump discharges.

TERMS

Gallons per minute (gpm) is a measure of fluid flow that is used to measure small volumes of intermittently flowing fluids such as pump discharges.

Pressure rating is the highest amount of pressure at which a pump can create flow against.

Volumetric efficiency is the relationship between actual and theoretical fluid flow, or pump gpm.

A **positive-displacement pump** is a pump that has a positive seal between its inlet and outlet and moves a specific volume of hydraulic oil with each revolution of the shaft.

Pressure Ratings. *Pressure rating* is the highest amount of pressure at which a pump can continually create flow without premature wear. A pump must be specified with a higher pressure rating than the fluid power system's maximum pressure requirement. If a pump's maximum pressure is not properly rated for the application it is used for, the pump will experience premature wear and breakdown.

Volumetric Efficiency. *Volumetric efficiency* is the relationship between actual and theoretical fluid flow, or pump gpm. **See Figure 5-4.** All pumps have internal leakage (slippage). As pressure requirements in a fluid power system increase, the efficiency of the pump decreases. The higher the pump efficiency, the less internal losses there will be inside the pump and less energy will be lost. Volumetric efficiency is calculated by applying the following formula:

$$Eff_V = \left(\frac{P_o}{P_{ro}}\right) \times 100$$

where

Eff_V = volumetric efficiency (in %)

P_o = actual pump output (in gpm)

P_{ro} = rated pump output (in gpm)

Example: What is the pump efficiency of a pump rated for 20 gpm that has actual output of 19 gpm?

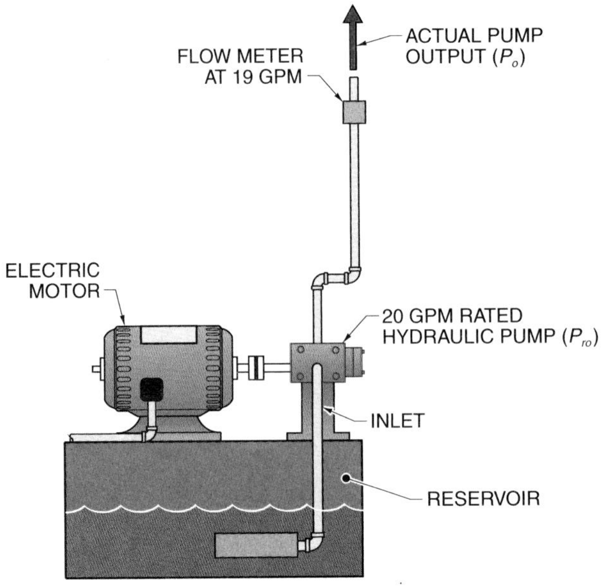

$$Eff_V = \left(\frac{P_o}{P_{ro}}\right) \times 100$$

$$Eff_V = \left(\frac{19}{20}\right) \times 100$$

$$Eff_V = 0.95 \times 100$$

$$Eff_V = \mathbf{95\%}$$

Positive-Displacement Pumps

A *positive-displacement pump* is a pump that has a positive seal between its inlet and outlet and moves a specific volume of hydraulic oil with each revolution of the shaft. Positive-displacement pumps are the only type of pump used in hydraulic systems and equipment. Typical applications for positive-displacement pumps include creating hydraulic flow for hydraulic press equipment, agricultural equipment, construction equipment, flight navigation systems, and robotic systems. **See Figure 5-5.**

Volumetric Efficiency 5-4

Figure 5-4. Volumetric efficiency is the relationship between actual and theoretical fluid flow of the pump.

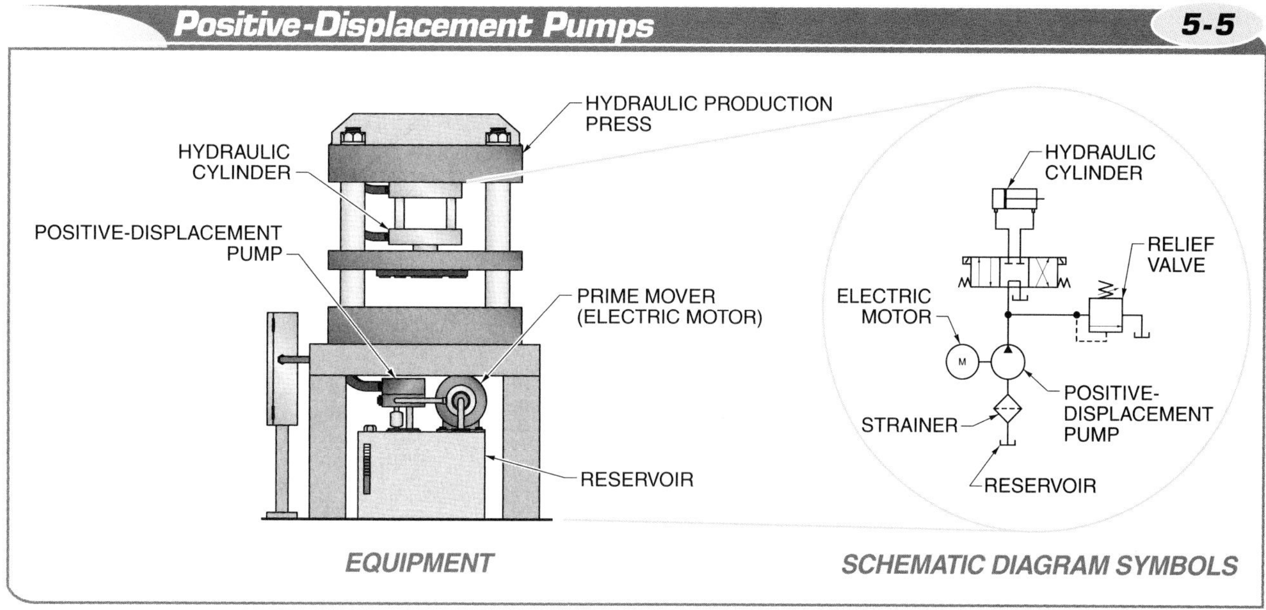

Figure 5-5. A typical application for a positive-displacement pump includes the fluid power systems on a hydraulic production press.

All positive-displacement pumps can operate under a wide range of pressures, have high-pressure capability, can operate in a wide range of environments, and can be small to large in size. Positive-displacement pumps are typically categorized as fixed or variable.

Fixed-Displacement Pumps. A *fixed-displacement pump* is a positive-displacement pump where the fluid flow rate (gpm) cannot be changed. Fixed-displacement pumps are rated by pressure rating, volumetric displacement per rotation, maximum pressure rating, minimum and maximum operating speeds, overall efficiency, noise level, mounting options, maximum input power (horsepower and kilowatts), torque range, and fluid cleanliness requirements. Hydraulic fixed-displacement pumps cannot vary the amount of fluid flow that they produce during operation without changing the speed at which they operate.

A second method used to change the amount of fluid flow produced by the pump is to change the pump's internal components. Typical applications for fixed-displacement pumps include hydraulic systems that do not require variation in system pressure, such as mobile hydraulic equipment. **See Figure 5-6.**

Figure 5-6. Fixed-displacement pumps are available in various sizes with different displacement ratings.

TERMS

A **fixed-displacement pump** is a positive-displacement pump where the fluid flow rate (gpm) cannot be changed.

A **variable-displacement pump** is a positive-displacement pump that can have its flow rate (gpm) changed.

Fixed-displacement pumps are more economical than variable-displacement pumps because they have less initial cost. Although fixed pumps are less electrically efficient, they create a constant flow at a high pressure (with some controls), which means that there is a higher electrical cost for a fixed pump over its lifetime.

Variable-Displacement Pumps. A *variable-displacement pump* is a positive-displacement pump that can have its flow rate (gpm) changed. The variable amount of hydraulic fluid that the pump moves is dictated by the demand of the fluid power system it is installed in. A variable-displacement pump varies the amount of hydraulic fluid with movable internal components while the rpm of the prime mover remains fixed.

Like fixed-displacement pumps, variable-displacement pumps are rated by maximum pressure rating, minimum and maximum operating speeds, overall efficiency, noise level, mounting options, maximum input power (horsepower and kilowatts), torque range, and fluid cleanliness requirements. However, unlike fixed-displacement pumps, variable-displacement pumps are also rated in maximum volumetric displacement per rotation and minimum and maximum pressure compensated ranges. Excessive pressure may cause the pump housing to rupture or the pump seals to burst. For maximum safety, the hydraulic pump selected for a fluid power system must be rated to withstand a pressures that are higher than the highest anticipated pressure in the system. This prevents damage and possible failure if an overpressure condition occurs. Typical applications for variable-displacement pumps include systems that operate with varied system fluid flow, such as industrial robotic systems. **See Figure 5-7.**

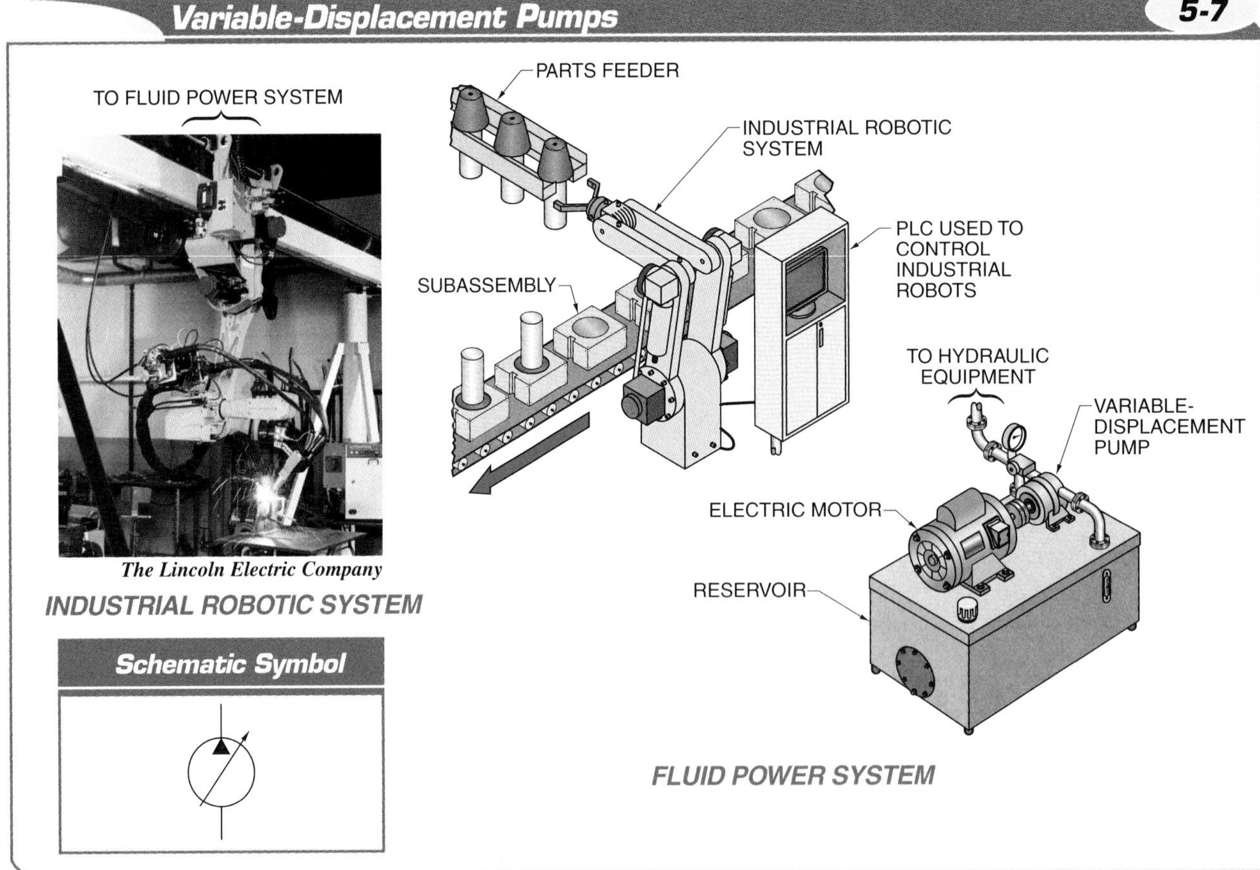

Figure 5-7. Typical applications for variable-displacement pumps include systems that operate with varied system fluid flow, such as industrial robots.

Positive-Displacement Pump Operation

All positive-displacement pumps operate similarly to create fluid flow. Therefore, the principles of positive-displacement pump operation can be applied directly to the different types of positive-displacement pumps encountered in the field. All hydraulic positive-displacement pumps have similar parts such as a shaft, pump housing, inlet port, and outlet port. **See Figure 5-8.** A positive-displacement pump operates in four basic steps:

1. The pump creates a vacuum by increasing the volume at its inlet, which is connected to the reservoir. Vacuum is created in a confined space that has less pressure than atmospheric pressure. The vacuum in a pump is created when the pump rotates and an increased volume is created at its inlet. The pressure in the tank is at atmospheric pressure, which is higher than the vacuum created at the inlet. Atmospheric pressure forces the fluid to flow from the tank into the inlet of the pump.
2. Once the fluid enters the pump through the inlet, the pump traps the fluid through a sealing method. A *seal* is an airproof and/or fluidproof joint between two members. The fluid travels through the pump, in decreasing volume, towards the outlet side of the pump.
3. Once the trapped fluid moves toward the outlet side of the pump, the sealed chamber opens, decreases its volume, and releases the fluid into the pump outlet.
4. The sealed chamber then closes, preventing fluid in the outlet side of the pump from slipping to the inlet side of the pump. At the same time, more fluid is forced out of the outlet and into the hydraulic equipment.

TECH FACT

Positive displacement, as it relates to hydraulic pumps, means that regardless of the load or speed involved with a hydraulic system, the pump will always displace the same amount of fluid per shaft revolution.

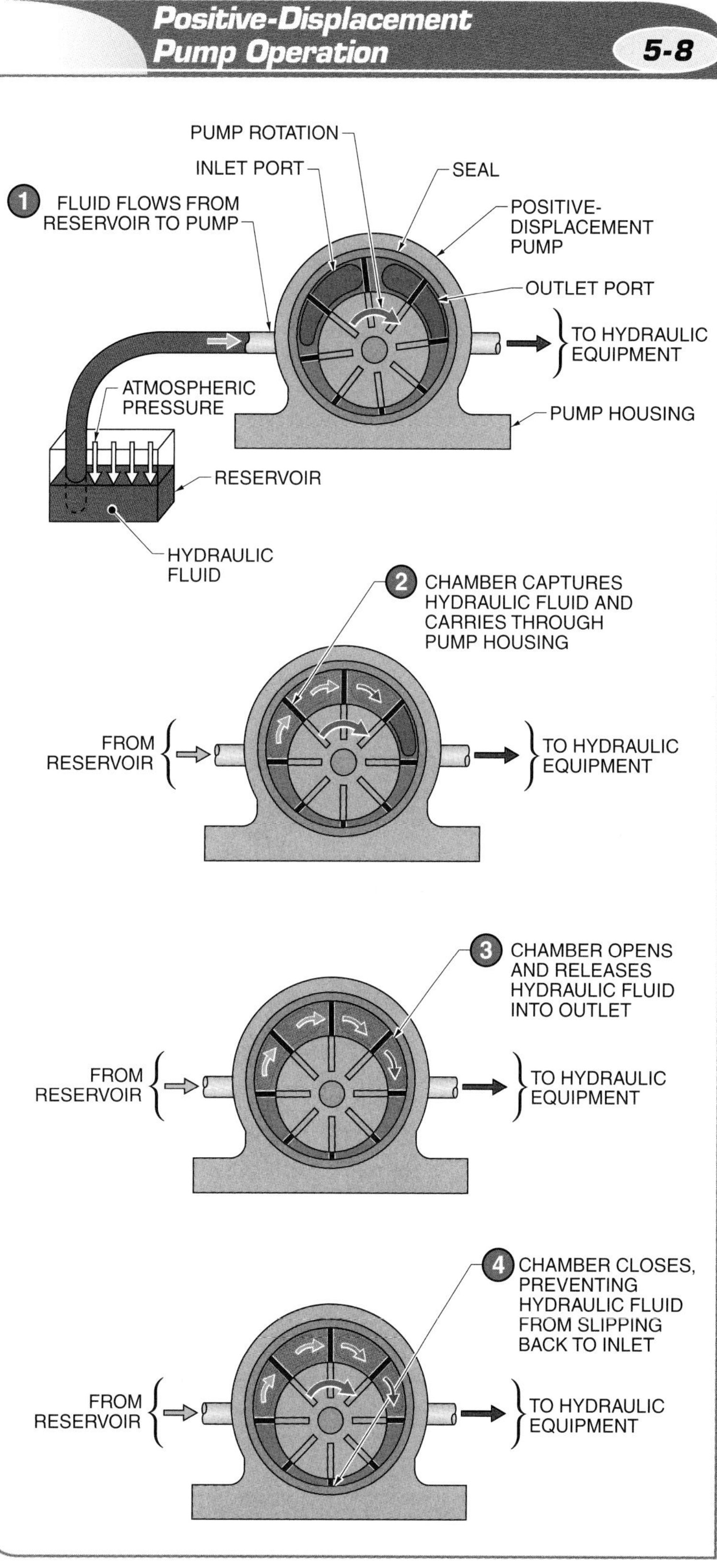

Figure 5-8. All positive-displacement pumps follow the same basic operational steps to create fluid flow.

TERMS

A **seal** is an airproof and/or fluidproof joint between two members.

A **gear pump** is a hydraulic pump that consists of gears that mesh together in various manners to create fluid flow.

An **external gear pump** is a gear pump that consists of two externally toothed gears that form a seal within the pump housing.

HYDRAULIC PUMP TYPES

Hydraulic pump types include gear, vane, and piston pumps. All hydraulic pumps create fluid flow by following four basic steps, with some differences according to pump type. For example, a gear pump uses meshing gears to create fluid flow, while a piston pump uses the extension and retraction of multiple pistons.

Gear Pumps

A *gear pump* is a hydraulic pump that consists of gears that mesh together in various manners to create fluid flow. Gear pumps may be external, lobe, internal, or gerotor pumps. The teeth on the meshing gears create a vacuum at the inlet and then push the fluid to the outlet by using a decreasing volume. **See Figure 5-9.** Gear pumps are ideal for equipment that operates in dirty environments, such as mobile hydraulic construction equipment, because a small amount of dirt in the system will not affect the overall performance of the pump.

Gear pumps consist of two meshing gears enclosed in a close-fitting housing.

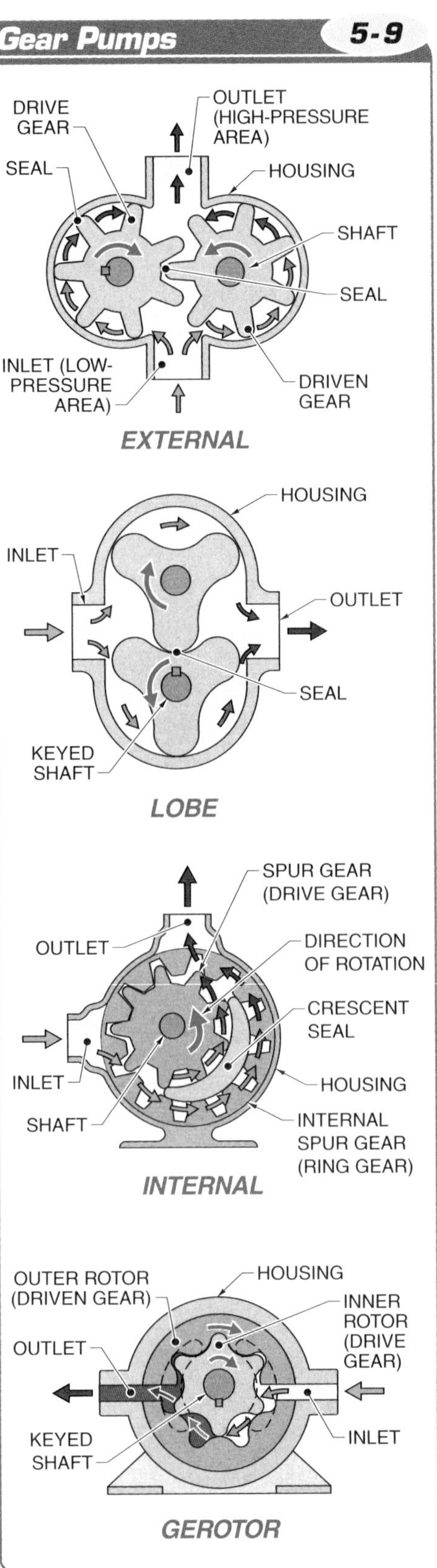

Figure 5-9. A gear pump consists of various types of meshing gears.

All gear pumps are fixed-displacement pumps. The most practical method used to change the flow of a gear pump is to replace its meshing gears with those of a different size. Each gear pump has specific applications, and each has advantages and disadvantages. Advantages of using gear pumps are as follows:

- low cost
- ability to handle dirt better than other types of pumps
- easier to repair than other pumps
- can be used with a wide range of hydraulic fluid viscosities
- reduced sensitivity to pump cavitation
- low noise levels
- output that is more predictable because it is linear with the speed of the prime mover.

Disadvantages include that they are not as efficient as other types of pumps, have fixed clearance ends, and can only accomplish volume control by changing the pump gears or changing the speed of the prime mover. Gear pumps can be found in low volume transfer applications such as hydraulic aerial lifts, log splitters, dump trucks, trailers, earthmovers, trucks, buses, and machine tool equipment.

External Gear Pumps. An *external gear pump* is a gear pump that consists of two externally toothed gears that form a seal within the pump housing. External gear pumps have two equal-sized gears, the drive gear and driven gear, which rotate to cause fluid to flow into the system. **See Figure 5-10.** External gear pumps operate in four basic steps:

1. As the drive gear rotates, it turns the driven gear. This causes both gears to move away from the inlet. This movement creates a vacuum on the inlet side of the pump as the gear teeth pull apart. The atmospheric pressure then pushes the fluid from the reservoir into the inlet.
2. As both gears rotate away from the inlet, the gear teeth move closer to the pump housing internal wall. This traps the fluid and forces it between the gear teeth and the internal wall toward the discharge.
3. As the gear teeth reach the outlet, they pull away from the internal wall. The fluid is then released. As the gear teeth start moving toward each other again, all fluid is forced out of the pump by decreasing volume.
4. As the gear teeth mesh back together, they form a seal that does not allow most of the fluid to flow back into the inlet. More fluid moves into the outlet area each time fluid is forced out of the gear teeth. This forces the fluid through the outlet port and into the hydraulic system.

Lobe Pumps. A *lobe pump* is a positive-displacement pump that has two external-driven, intermeshing, lobe-shaped gears. A lobe pump operates similar to an external gear pump. **See Figure 5-11.** The main advantage of having two driven lobes is that the two lobes do not have to make contact with each other, thereby reducing the amount of operational noise and wear on the pump. A lobe pump requires timing gears to drive the lobe-shaped driven gears.

Although lobe pumps can be used in hydraulic applications, they are typically used in pneumatic applications or to move products such as slurries, pastes, and solids. Lobe pumps are commonly used in industries such as pulp and paper processing chemical refining, food and beverage processing, pharmaceutical production, and biotechnology. Lobe pumps operate in four basic steps:

1. As the lobes unmesh, they create a vacuum by increasing the volume on the inlet side of the pump.
2. The fluid flows into the pump housing and is trapped by the lobes as they rotate.
3. The fluid travels around the interior of the pump housing in the gaps between the lobes and starts to leave the trapped areas.
4. The meshing of the lobes forces the fluid through the outlet into the system.

TECH FACT

Common external gear pump applications include pumping acids, fuels, and lube oils; metering chemical additives and polymers; mixing and blending chemicals; transferring low-volume fluid; and operating industrial and mobile hydraulic equipment, such as log splitters.

TERMS

A **lobe pump** is a positive-displacement pump that has two external-driven, intermeshing, lobe-shaped gears.

An **internal gear pump** is a gear pump that consists of a small external drive gear mounted inside a large internal spur gear, also called a ring gear.

A **crescent seal** is a crescent-moon-shaped seal between the gears and between the inlet and outlet sides of an internal gear pump.

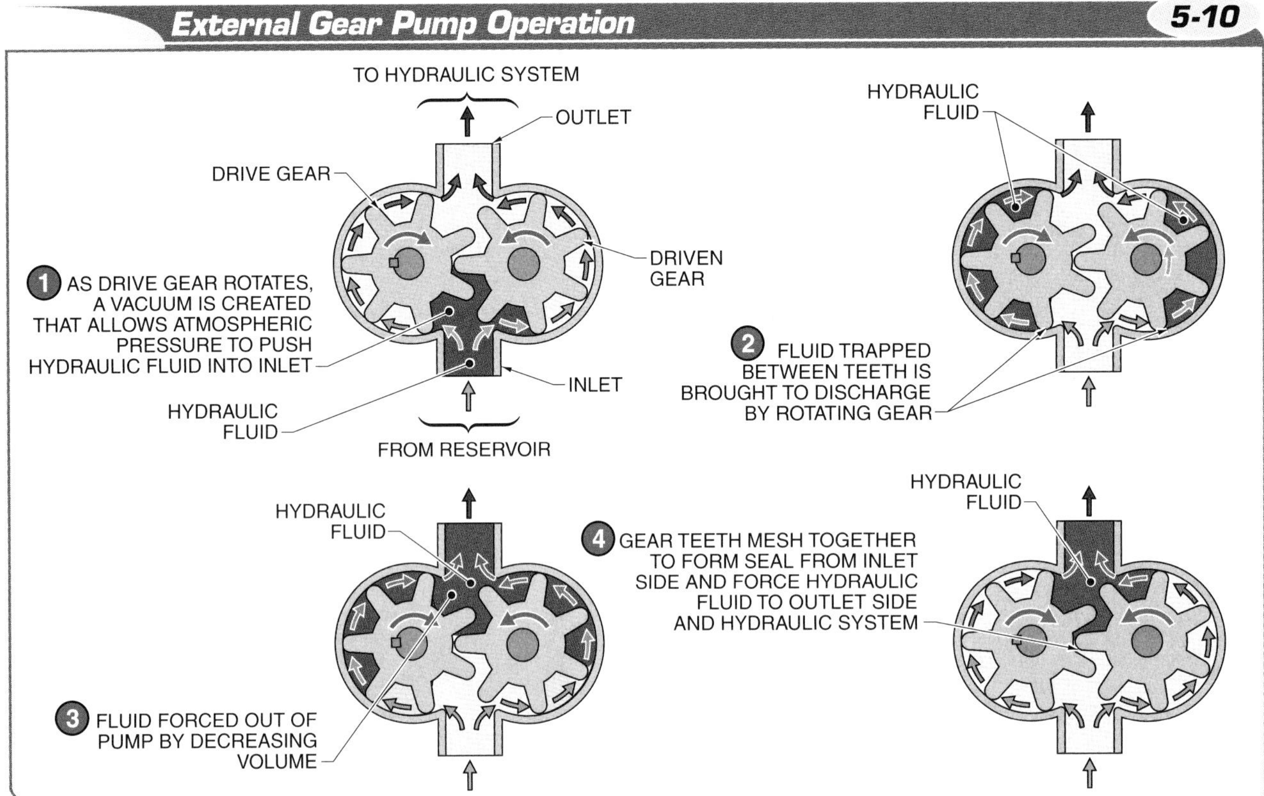

Figure 5-10. An external gear pump consists of meshing gears that form a seal with the pump housing and operates similar to the four basic steps of a positive-displacement pump.

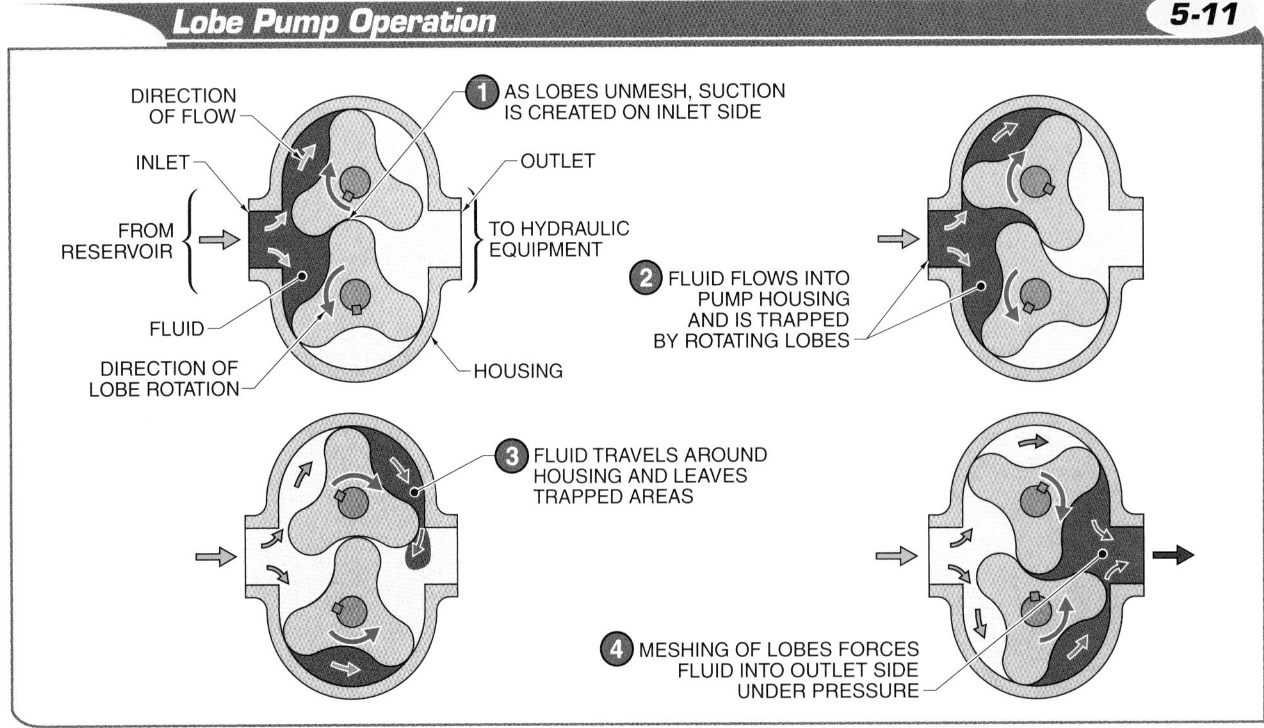

Figure 5-11. A lobe pump has two external-driven gears and operates similar to an external gear pump.

Internal Gear Pumps. An *internal gear pump* is a gear pump that consists of a small external drive gear mounted inside a large internal spur gear (ring gear). The two gears rotate in the same direction. **See Figure 5-12.** A crescent seal separates the low- and high-pressure areas of the pump. A *crescent seal* is a crescent-moon-shaped seal between the gears and between the inlet and outlet sides of an internal gear pump. Hydraulic fluid is trapped as the gears rotate and is discharged through the pump outlet.

Internal gear pumps can pump fluids with a wide range of viscosity and can operate at temperatures up to 750°F. Because there are only two moving parts, internal gear pumps are reliable and easy to maintain. Internal gear pumps are typically used in industrial production facilities and in automotive automatic transmissions. Internal gear pumps operate in three basic steps:

1. As the motor turns the external toothed gear, the gear teeth unmesh, creating an increasing volume. Atmospheric pressure then pushes the fluid from the reservoir into the inlet of the pump.
2. The fluid becomes trapped in the cavities of the unmeshed gears. As the gears rotate, the crescent seal separates the internal gear and the external gear. The fluid continues to move as the two gears continue to rotate.
3. As the two gears reach the end of the crescent seal, the gears begin to mesh again, decreasing the volume. The decrease in volume forces the fluid out of the cavities between the teeth and causes the fluid to flow through the outlet side of the pump and into the hydraulic system.

TECH FACT

In addition to hydraulic oil, internal gear pumps are used to pump fuel oil, lube oil, resins, polymers, alcohols, solvents, asphalt, bitumen, tar, polyurethane foam, products, paint, inks, pigments, soaps, surfactants, and glycol.

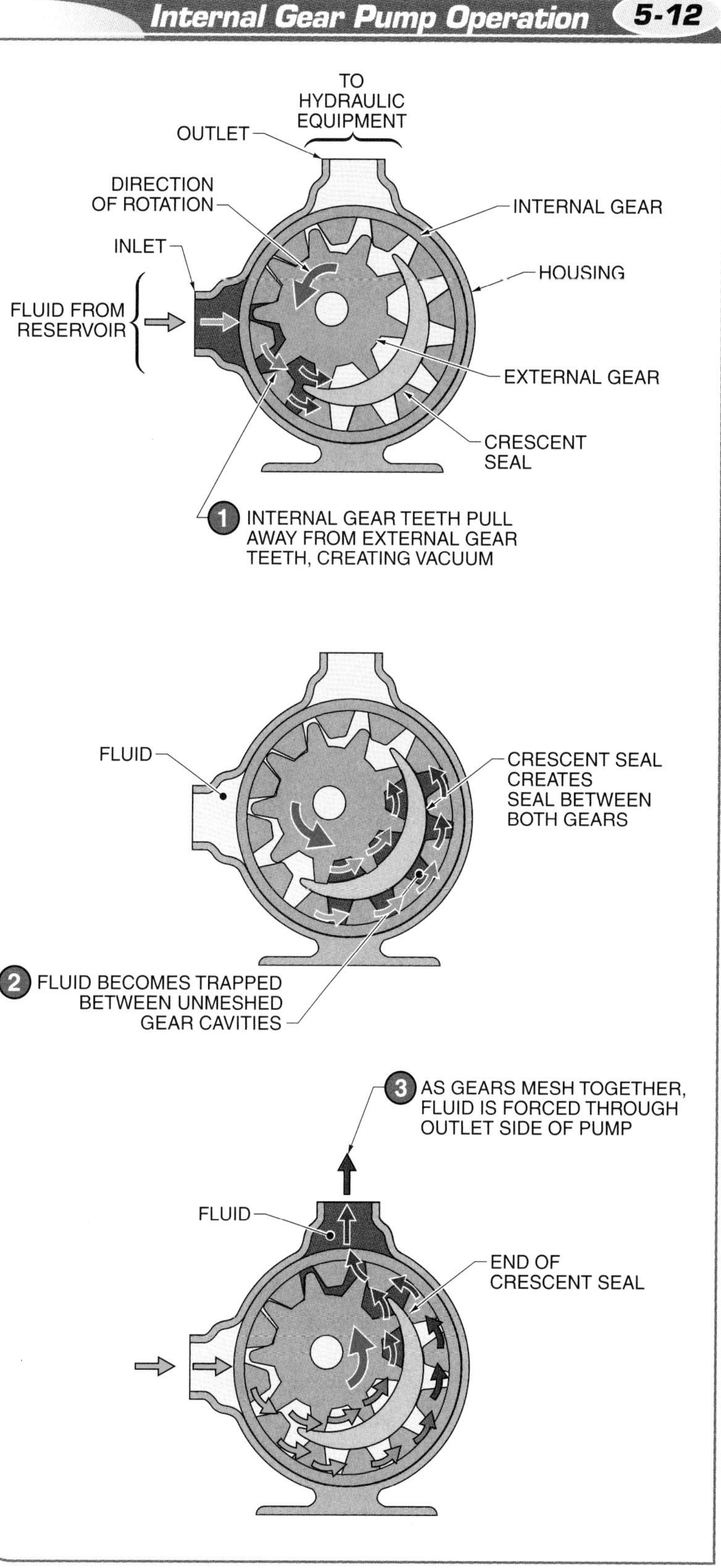

Figure 5-12. An internal gear pump consists of a small external drive gear mounted inside a large internal gear.

TERMS

A **gerotor pump** is a gear pump that has an inner rotor that meshes with the gear teeth of an outer rotor.

Gerotor Pumps. A *gerotor pump* is a gear pump that has an inner rotor that meshes with the gear teeth of an outer rotor. The inner rotor has one less gear tooth than the outer rotor and both rotors rotate in the same direction. Gerotor pump operation is similar to internal gear pump operation without the crescent seal. **See Figure 5-13.**

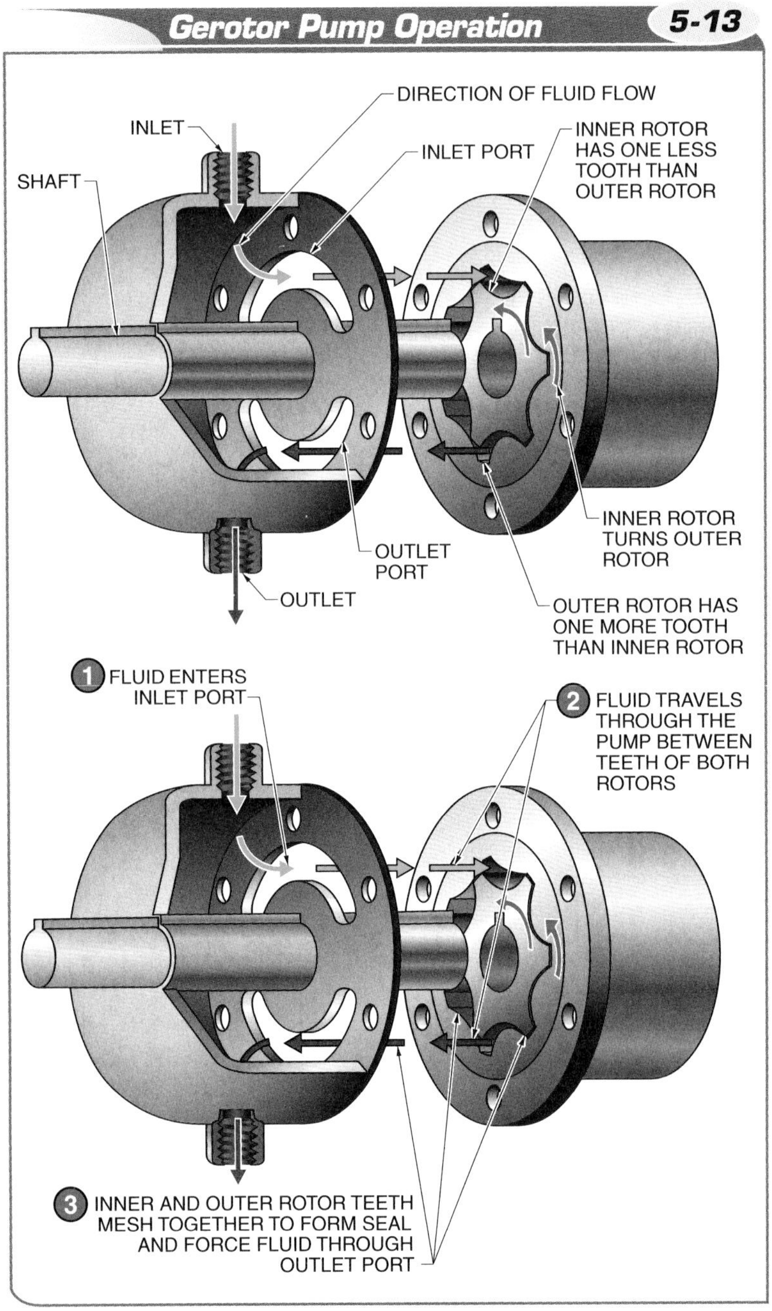

Gerotor Pump Operation 5-13

DIRECTION OF FLUID FLOW
INLET
INLET PORT
INNER ROTOR HAS ONE LESS TOOTH THAN OUTER ROTOR
SHAFT
INNER ROTOR TURNS OUTER ROTOR
OUTLET PORT
OUTLET
OUTER ROTOR HAS ONE MORE TOOTH THAN INNER ROTOR
1 FLUID ENTERS INLET PORT
2 FLUID TRAVELS THROUGH THE PUMP BETWEEN TEETH OF BOTH ROTORS
3 INNER AND OUTER ROTOR TEETH MESH TOGETHER TO FORM SEAL AND FORCE FLUID THROUGH OUTLET PORT

Figure 5-13. Gerotor pump operation is similar to internal gear pump operation, with the inner rotor having one less gear tooth than the outer rotor.

Gerotor pumps are used for low- to moderate-pressure hydraulic applications of less than 1000 psi, such as trash compactors, hydraulic lifts, and hydraulic elevators. Because gerotor pumps are susceptible to problems with dirt, they can only be used in applications that have clean oil. Due to these considerations, gerotor pumps are used in clean, low-pressure industrial and commercial applications. Gerotor pumps operate in three basic steps:

1. Fluid enters the inlet port between the inner rotor and outer rotor.
2. Fluid travels through the pump between the teeth of the two rotors.
3. The inner rotor and outer rotor teeth mesh to form a seal between the inlet and outlet ports, forcing the fluid through the outlet port.

Gear Pump Assembly

Gear pumps can sometimes be repaired in the field. For a technician to accomplish this, they must have an understanding of how the pump is assembled. Gear pump assemblies consist of major parts (a frame, gears, housing, and a shaft) and minor parts (O-rings, backup rings, and seals) that may need to be repaired, replaced, or refurbished. Many gear pump manufacturers supply step-by-step procedures for the assembly and disassembly of their pumps. This allows a technician to repair a gear pump in the field, rather than removing it from the hydraulic equipment and sending it out for repair. Gear pump manufacturers also supply part numbers and/or descriptions for every part that may need to be replaced. **See Figure 5-14.**

For example, the two gears in a gear pump can be replaced by applying the following procedure:

1. Remove the pump from its motor coupling by removing the bolts.
2. Remove all bolts from the front of the pump that hold the pump assembly together.

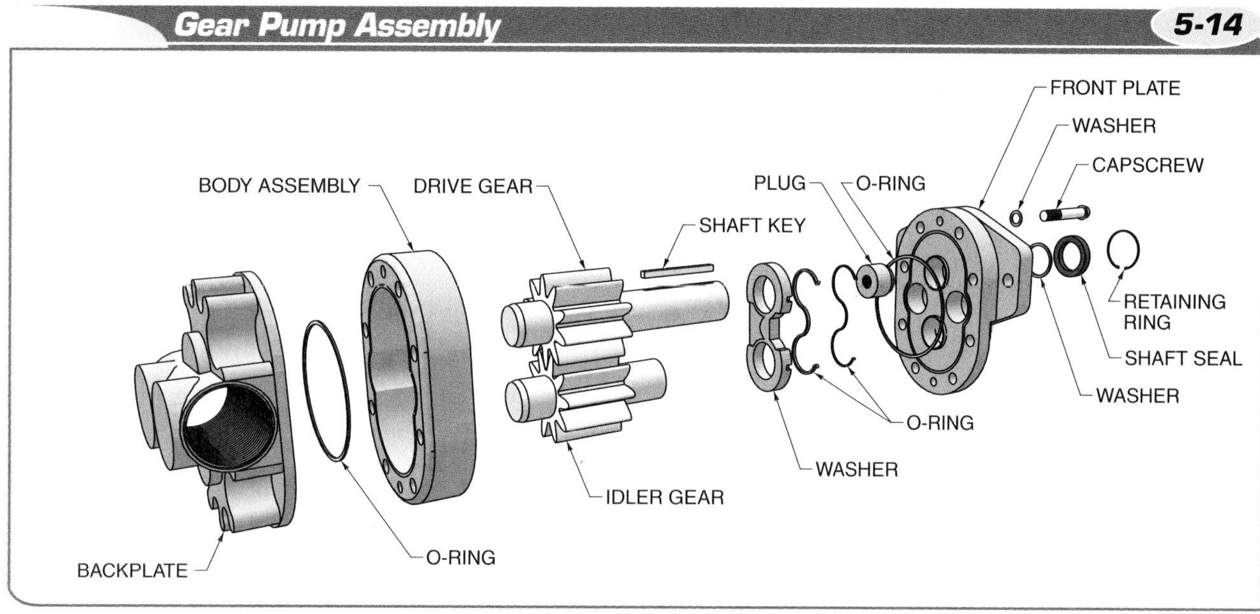

Figure 5-14. Gear pump manufacturers typically provide assembly diagrams for each specific pump.

3. Disassemble pump.
   a. Detach the front plate assembly from the pump.
   b. Remove the two gears from the body housing.
   c. Detach the body housing from the back plate assembly.
4. Place all parts in a straight line in the exact order that they were disassembled.
5. Inspect all seals that connect the three main sections together.
   a. If the seals are good, proceed to step 6.
   b. If the seals are worn or damaged, replace them.
6. Inspect the ball bearings for wear and lubrication. Replace or regrease as required.
7. Replace the two gears with new gears in the order that they have been laid out.
8. Reassemble the gear pump by working backwards from step 7.

Vane Pumps

A *vane pump* is a hydraulic pump that creates a vacuum by rotating a rotor inside a cam ring while trapping fluid between vanes that expand and retract from the rotor while moving the fluid toward the output. **See Figure 5-15.** There are a number of different types of vane pumps including unbalanced vane pumps; variable-displacement, pressure-compensated vane pumps; and balanced vane pumps.

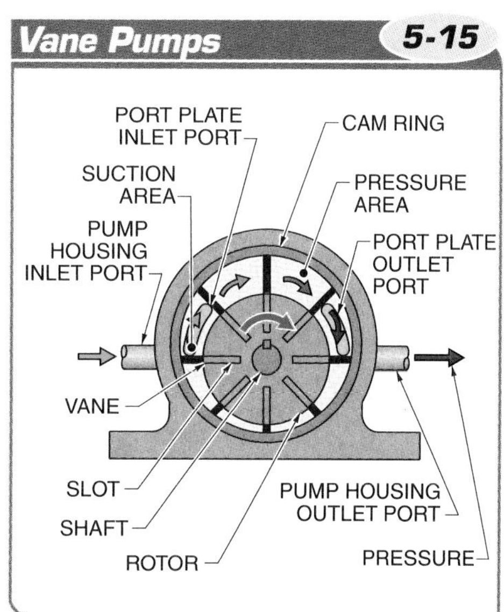

Figure 5-15. A vane pump contains vanes in an offset rotor and rotates the rotor to produce the flow of hydraulic fluid.

TERMS

A **vane pump** is a hydraulic pump that creates a vacuum by rotating a rotor inside a cam ring while trapping fluid between vanes that extend and retract from the rotor while moving the fluid toward the output.

An **unbalanced vane pump** is a fixed- or variable-displacement hydraulic pump in which the pumping action occurs in the chambers on one side of the rotor and shaft.

Each type of vane pump follows four basic operational steps. However, all vane pumps have similar parts such as the shaft, cam ring, vanes, rotor, inlet port, outlet port, and pump body.

Vane pumps are known for their ease of maintenance and good suction characteristics over the life of the pump. Because of a smoother, nonpulsing flow rate, vane pumps run quietly and efficiently. Vane pumps can operate at temperatures ranging from –25°F to 500°F. Vane pumps extend their vanes through centrifugal force or through a mechanical means such as a spring.

Unbalanced Vane Pumps. An *unbalanced vane pump* is a fixed- or variable-displacement hydraulic pump in which the pumping action occurs in the chambers on one side of the rotor and shaft. Unbalanced vane pumps have the simplest design of the various types of vane pumps.

An unbalanced vane pump is typically used in low-pressure applications. **See Figure 5-16.** A high pressure differential between the inlet and outlet of the pump creates a load on the rotor that is attached to the shaft. Larger bearings and shafts must be used because of this extra load, which limits the size of unbalanced vane pumps. Unbalanced vane pumps operate in four basic steps:

1. The offset rotor rotates with vanes contacting the cam ring, creating a vacuum at the inlet of the pump. As the rotor rotates, the vanes extend outward by centrifugal force, creating a confined space for the fluid. The shape of the inlet allows the pump to pull more fluid into the confined space.
2. As the fluid becomes trapped between the vanes and the cam ring, it is forced toward the outlet of the pump.
3. As the trapped fluid gets closer to the outlet side of the pump, it is released from its confined space. The outlet is also shaped to allow more fluid to move through the pump. At this point, the vanes are forced back into the rotor by the cam ring, making a smaller confined space for the fluid.
4. As the fluid is forced out of the confined space, the rotor and the cam ring provide a leakproof seal that does not allow the fluid to slip back to the inlet. At the same time, more fluid is forced out of the pump, forcing fluid into the system.

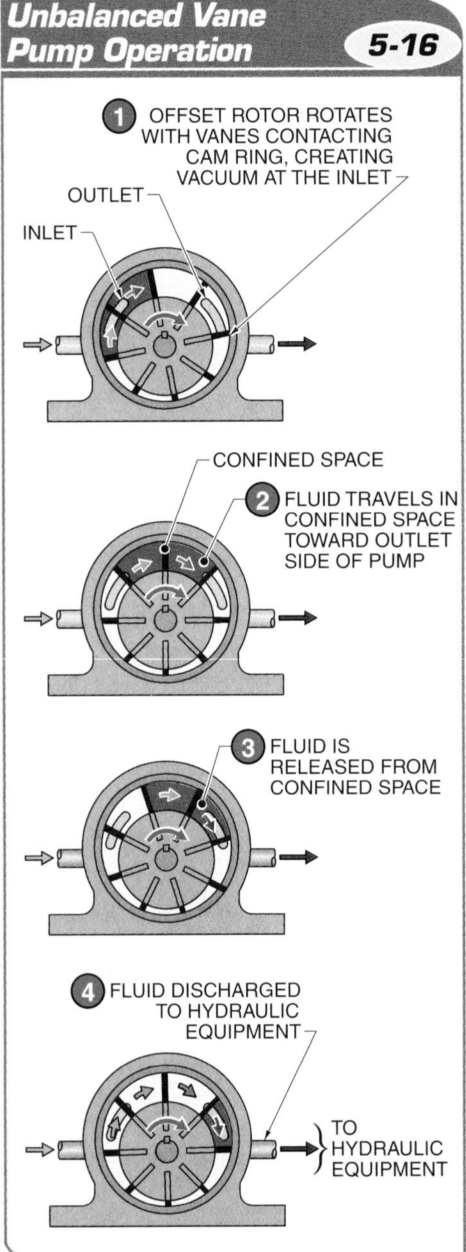

Figure 5-16. Unbalanced vane pumps operate similar to the four basic steps of a positive-displacement pump.

Variable-Displacement, Pressure-Compensated Vane Pumps. A *variable-displacement, pressure-compensated vane pump* is a pump that automatically adjusts the amount of volume it displaces per rotation by centering the rotor when the pressure in the system starts to build. **See Figure 5-17.** A variable-displacement, pressure-compensated vane pump operates by adjusting a cam ring to allow the volume per revolution to change according to system pressure. This type of pump protects itself against excessive pressure by reducing power consumption as the flow rate decreases. When pressure reaches a certain value, the compensator spring force equals the hydraulic piston force. As pressure continues to increase, the compensator spring is compressed until concentricity on the compensator ring is achieved. Maximum pressure is then achieved. At this point, the pump is protected because it produces no more flow, resulting in no power loss and no fluid heating. A thrust block is used to ensure smooth movement and correct placement of the cam ring.

Variable-displacement, pressure-compensated vane pumps have ease of maintenance over the life of the pump. Setting the maximum fluid flow of a variable-displacement, pressure-compensated vane pump is performed using the following procedure:

1. Connect the outlet side of the pump directly to a pressure gauge.
2. Turn the pump ON and record the pressure when the pump rotor becomes centered.
3. Use the pressure adjustment screw to set the desired fluid flow.

TERMS

A **variable-displacement, pressure-compensated vane pump** is a pump that automatically adjusts the amount of volume it displaces per rotation by centering the rotor when the pressure in the system starts to build.

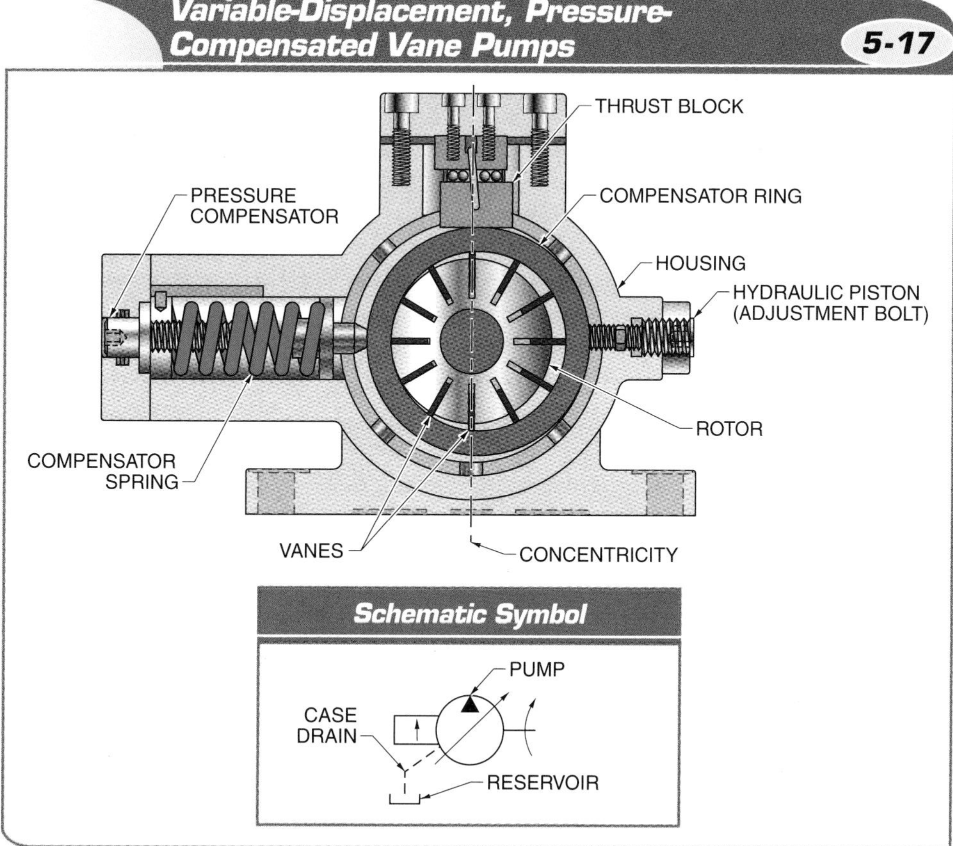

Figure 5-17. A variable-displacement, pressure-compensated vane pump is a pump that automatically adjusts the amount of volume it displaces per rotation by centering the rotor when the pressure in the system starts to build.

TERMS

A **balanced vane pump** is a pump that consists of a cam ring, rotor, vanes, and a port plate with opposing inlet and outlet ports.

A **cartridge assembly** is a cartridge located in a vane pump that houses the vanes, rotor, and cam ring, which are all placed between two end plates.

A **piston pump** is a hydraulic pump in which fluid flow is produced by reciprocating pistons.

Balanced Vane Pump. A *balanced vane pump* is a pump that consists of a cam ring, rotor, vanes, and a port plate with opposing inlet and outlet ports. **See Figure 5-18.** This creates a balanced load on the pump bearings and seals. The two inlets and two outlets are set 180° apart from each other. This helps to prolong the life of the shaft bearings and allows the pump to run at higher speeds and at higher pressure ratings than an unbalanced vane pump. A balanced vane pump also has an elliptical cam ring.

Balanced Vane Pumps 5-18

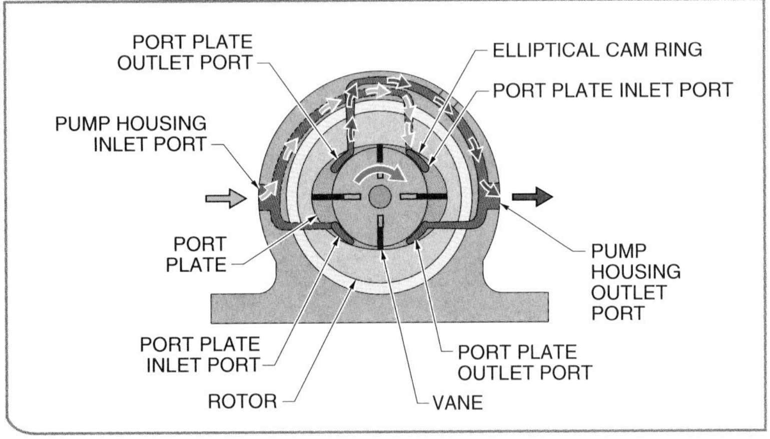

Figure 5-18. A balanced vane pump has two inlets and outlets at opposite sides of the pump and contains an elliptical cam ring.

A *cartridge assembly* is a cartridge located in a vane pump that houses the vanes, rotor, and cam ring, which are all placed between two end plates. **See Figure 5-19.** A cartridge assembly allows a technician to service the pump quickly because it allows for replacement of the entire assembly when the internal parts need to be replaced. Additionally, changing the size of the cartridge assembly can change the displacement of the pump.

Vane Pump Assembly

Like gear pumps, many types of vane pumps can be repaired in the field. Vane pump assemblies are more complicated than gear pump assemblies and require more time to repair. Many manufacturers provide detailed assembly and disassembly instructions for vane pumps that make it possible to repair, replace, or refurbish any part of a vane pump. It is also common for vane pump assemblies to have rebuild kits that can be ordered and kept onsite for quicker repair time. Usually specific assembly and disassembly instructions are used because vane pumps may have a cartridge that holds the vanes. **See Figure 5-20.**

Cartridge Assemblies 5-19

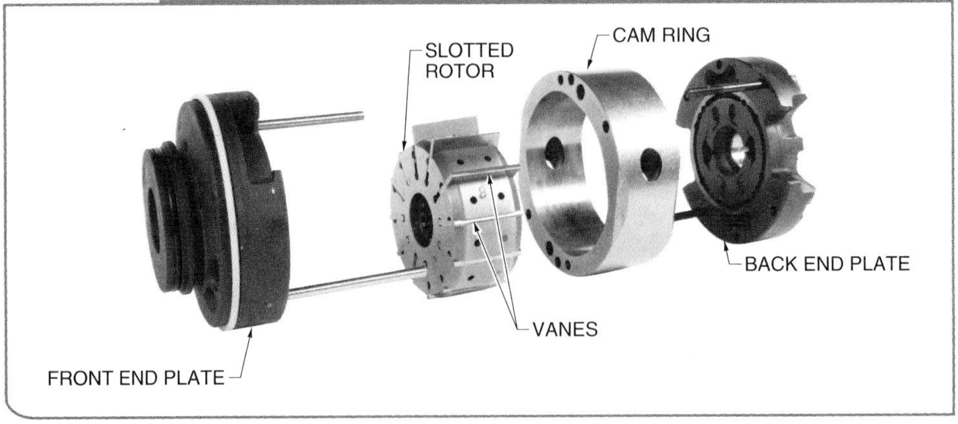

Figure 5-19. A cartridge assembly is located in a vane pump and houses the vanes, rotor, and cam ring, which are all placed between two plates.

Vane Pump Assembly **5-20**

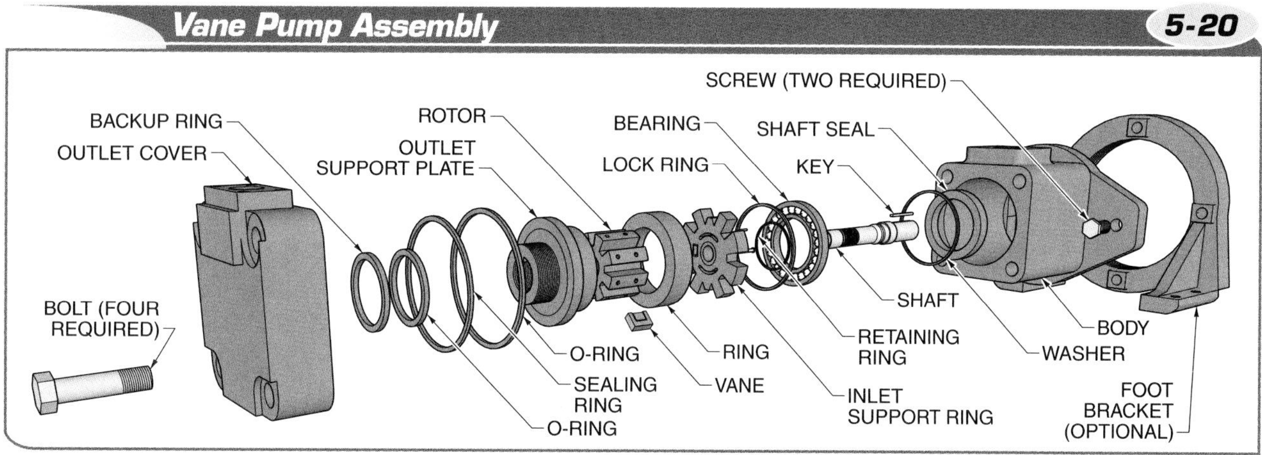

Figure 5-20. Vane pump manufacturers typically provide assembly diagrams for each specific pump.

For example, a cam ring in a vane pump is replaced by applying the following procedure:

1. Remove the pump from its foot bracket by removing the bolts.
2. Remove four bolts from the back of the pump.
3. Disassemble the pump into three sections.
   a. One section is the back housing with O-rings.
   b. Another section is the cartridge assembly.
   c. The third section is the front housing with the shaft.
4. Place all parts in a straight line in the exact order that they were disassembled.
5. Inspect all the seals that connect the three main sections together.
   a. If the seals are good, go to step 6.
   b. If the seals are worn or damaged, replace them.
6. Inspect the ball bearings for wear and lubrication. Replace or regrease as required.
7. Remove two screws holding the vane cartridge together.
8. Place all parts from the vane cartridge in a straight line in the exact order that they were disassembled.
9. Remove the cam ring and the rotor, making sure that the vanes do not fall out of the rotor.
10. Remove all vanes from the rotor and inspect for premature wear.
11. Replace the damaged cam ring with a new cam ring.
12. Reassemble the vane pump by working backwards from step 11.

Piston Pumps

A *piston pump* is a hydraulic pump in which fluid flow is produced by reciprocating pistons. Piston pumps are either fixed or variable displacement. They use a rotating internal piston assembly to create a vacuum as the pistons pull away from the inlet. The piston assembly then forces fluid out when the pistons are pushed toward the outlet.

A bent-axis piston pump is the most durable type of hydraulic pump and can operate at pressures of up to 10,000 psi. Piston pumps are typically used in applications such as small loaders. Types of piston pumps include inline axial, bent axial, variable-axis, variable-bent-axis, radial, and variable radial piston pumps.

TECH FACT

To achieve a uniform volumetric flow rate of hydraulic fluid, hydraulic piston machines, such as piston pumps, are designed with an odd number of pistons. For example, many hydraulic piston pumps are designed to house seven or nine pistons.

TERMS

An **inline axial piston pump** is a piston pump that consists of pistons in a rotating piston block parallel to the drive shaft.

A **swash plate** is an angled plate in contact with the piston heads that moves the pistons in the cylinders of a pump.

A **pressure-compensated variable-displacement inline axial piston pump** is a piston pump in which the angle of the swash plate can be varied.

Inline Axial Piston Pumps. An *inline axial piston pump* is a piston pump that consists of pistons in a rotating piston block parallel to the drive shaft. Inline axial piston pumps create smooth fluid flow for piston pumps. **See Figure 5-21.** Axial piston pumps consist of a number of pistons, a piston block, piston shoes, a swash plate, and a drive shaft. A *swash plate* is an angled plate in contact with the piston heads that causes the pistons in the cylinders of a pump to extend and retract. Typical applications for piston pumps include high-pressure devices, such as large press machines, and heavy-duty construction and industrial equipment. Axial piston pumps operate in four basic steps:

1. As the drive shaft rotates, the piston block rotates in the same direction. This pulls a piston from the number of pistons out and creates suction. The piston is pulled because it is attached to the swash plate and the swash plate is in a fixed, slanted position. The farther back the piston moves, the more volume of fluid it will move. The length of the piston is the main determinant to the amount of gpm a pump can produce.
2. As the piston moves through the first half of the pump, it pulls more fluid every degree it turns, trapping more fluid in the piston barrel.
3. When the piston reaches the halfway point of a cycle (180°), the piston pushes fluid out of the piston barrel. As the shaft continues to rotate, more fluid is forced out of the piston barrel.
4. As the piston completes a 360° cycle, all fluid is pushed out of the piston barrel, creating a leakproof seal that will not allow the oil to reenter the inlet. At the same time, the next piston forces fluid out of its barrel, which forces fluid flow in the hydraulic system.

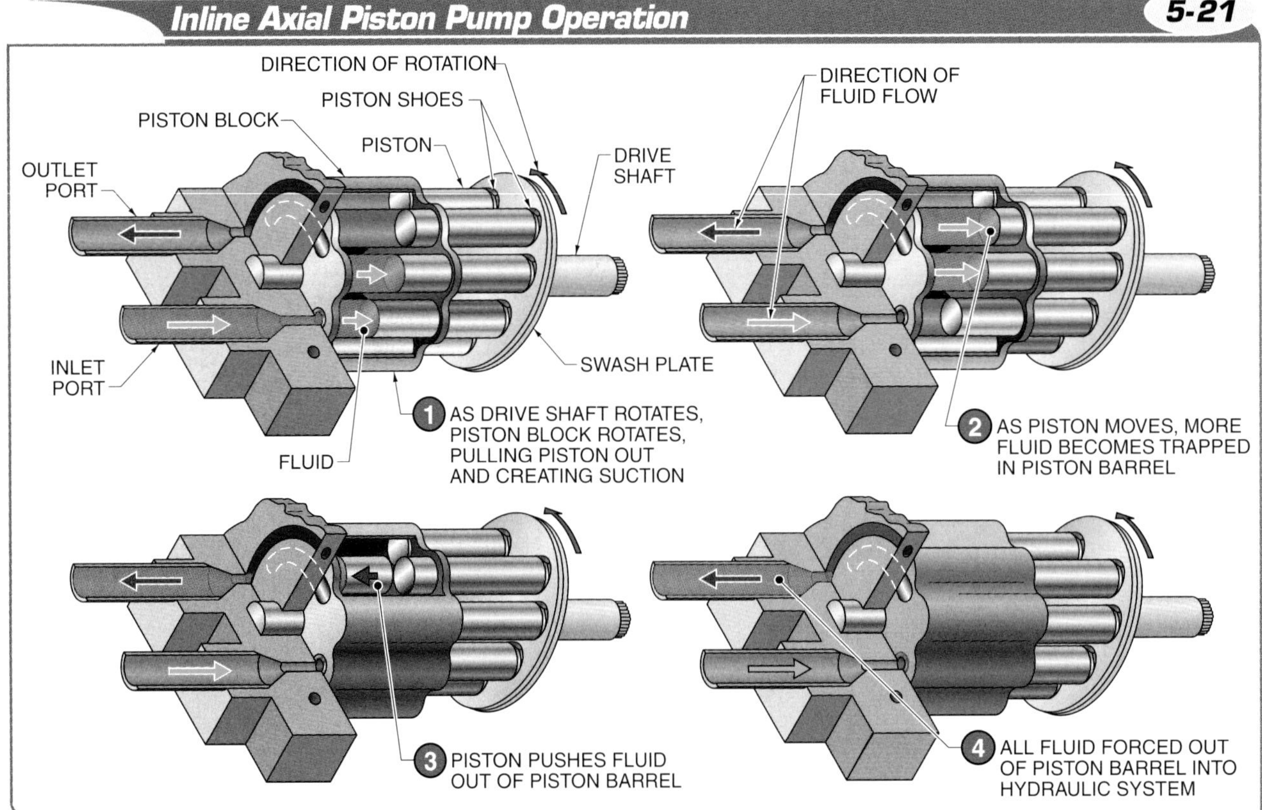

Figure 5-21. Inline axial piston pumps consist of a number of pistons, a piston block, piston shoes, a swash plate, and a shaft and operate with four basic steps.

Hydraulic pumps are used to provide fluid power to concrete compression testers.

Pressure-Compensated Variable-Displacement Inline Axial Piston Pumps. A *pressure-compensated variable-displacement inline axial piston pump* is a piston pump in which the angle of the swash plate can be varied. **See Figure 5-22.** Pressure-compensated variable-displacement inline axial piston pumps have a simple design. They are also reliable and durable. Pressure-compensated variable-displacement inline axial piston pumps include applications where a large amount of varied pressure is required such as backhoes, hydraulic cranes, heavy-duty presses, and balers. They are also used in applications where heat buildup can affect pump performance.

A pressure-compensated variable-displacement inline axial piston pump works under the same principles as an axial piston pump with the exception of the variability of the angle of the swash plate. When the angle of the swash plate is varied, it changes the distance of how far back a piston pulls. This causes the piston to allow more or less fluid into its barrel, varying the amount of gpm that the pump produces.

TECH FACT

The pistons in an axial piston pump reciprocate parallel to the centerline of the drive shaft of the piston block. Rotary shaft motion is converted into axial reciprocating motion. Most axial piston pumps contain multiple pistons and use check valves or port plates to direct fluid flow from the inlet port to the outlet port.

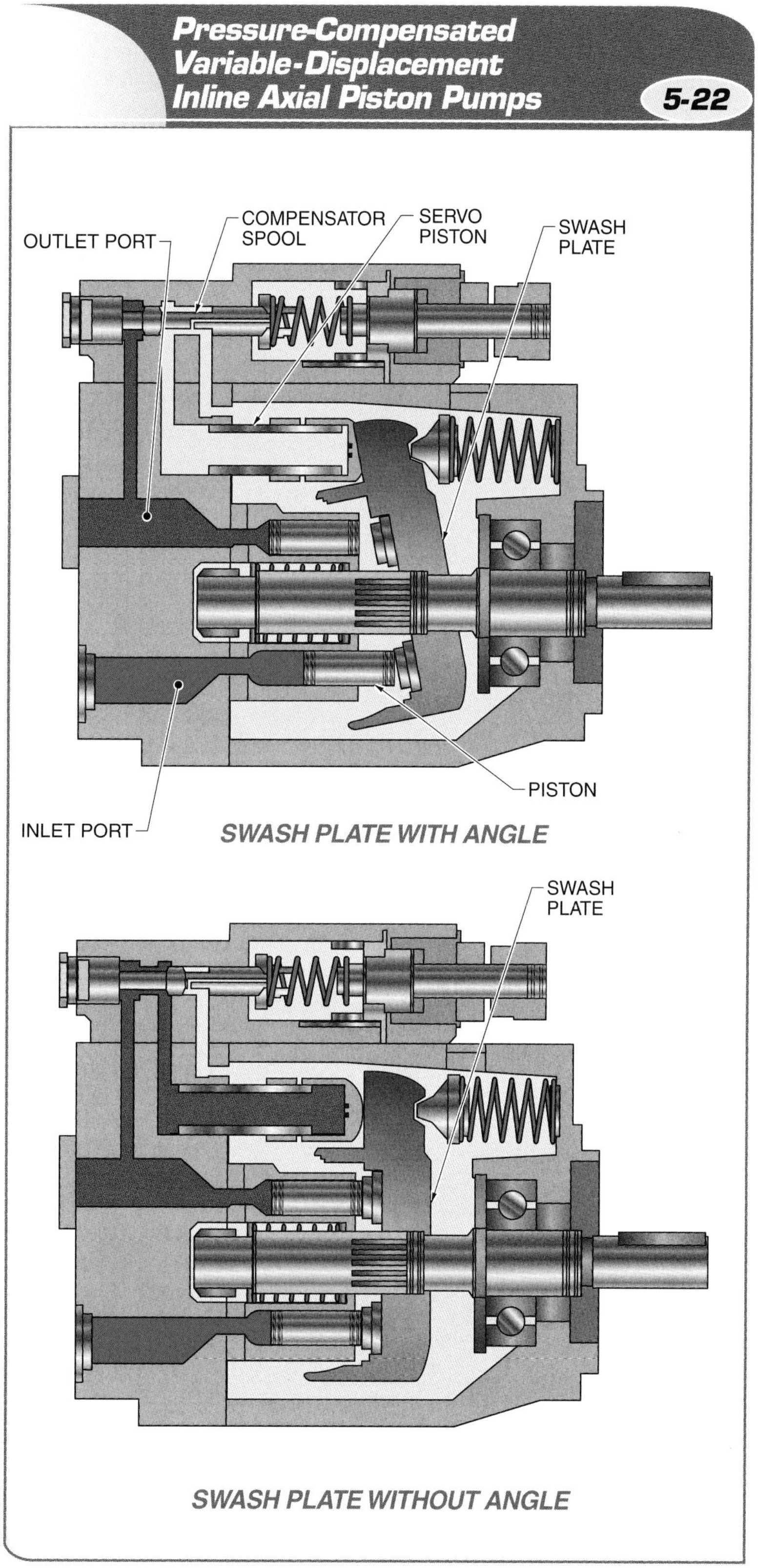

Pressure-Compensated Variable-Displacement Inline Axial Piston Pumps 5-22

OUTLET PORT
COMPENSATOR SPOOL
SERVO PISTON
SWASH PLATE
PISTON
INLET PORT

SWASH PLATE WITH ANGLE

SWASH PLATE

SWASH PLATE WITHOUT ANGLE

Figure 5-22. A pressure-compensated variable-displacement inline axial piston pump has a swash plate at an angle that can be varied, thereby varying the amount of fluid flow (gpm).

TERMS

A **bent-axis piston pump** is a piston pump in which the pistons and cylinders are at an angle to the drive shaft and thrust plate.

A **radial piston pump** is a piston pump that consists of pistons located perpendicular to the pump shaft.

The most common method to vary the angle of a swash plate is through internal pilot pressure. As the pressure in the fluid power system begins to reach the set pressure of the pump, pressure from the internal pilot lines begins to push on the pilot valve attached to the swash plate.

Maximum pilot pressure is set with a setscrew that adjusts a control spring. As the swash plate begins to move, the distance the pistons pull back into the barrel changes. When the swash plate is vertical, there is no fluid flow produced by the pump. However, the prime mover still rotates at the same rpm, which saves energy because there is no load on the motor and there is no energy being wasted as heat from fluid moving through the pressure relief valve.

Bent-Axis Piston Pumps. A *bent-axis piston pump* is a piston pump in which the pistons and cylinders are at an angle to the drive shaft and thrust plate. Bent-axis piston pumps operate similarly to axial piston pumps, but rather than the swash plate being at an angle (offset), the pistons and piston block are at an angle (offset). The angle at which the pistons and piston block are offset determines the amount of fluid that each piston can take in. Thus, the angle at which the pistons and piston block are set determines the amount of fluid flow. **See Figure 5-23.**

Bent-axis piston pumps can be either fixed or variable. Fixed bent-axis piston pumps work by rotation from the prime mover that the angled piston is attached to. As they rotate, the pistons extend and retract, creating fluid flow. Variable bent-axis piston pumps work by adjusting the angle at which the pistons and the piston block sit. Typical applications for bent-axis piston pumps include mobile and industrial equipment where a high-pressure rating is required and space is limited.

Radial Piston Pumps. A *radial piston pump* is a piston pump that consists of a cylinder barrel, pistons with shoes, a ring, and a valve block located perpendicular to the pump shaft. **See Figure 5-24.** Radial piston pumps are high-pressure hydraulic pumps, capable of operating at 10,000 psi. Radial piston pumps are used because of the design of their pistons and barrel, which allow for a short stroke. Typical applications for radial piston pumps include equipment that uses a heavy fluid for low fluid flow and high pressure, such as plastics injection molding machines or die-casting machines for metals.

Radial piston pumps are classified as cam or rotating piston pumps. In a cam pump, a rotating internal cam moves the pistons in cylinders. The cam is shaped to push the pistons out during one half of the cam rotation and allow the pistons to retract during the other half. There are also variable pressure-compensated radial piston pumps.

A variable pressure-compensated radial piston pump works similar to a variable axial model by adjusting the stroke of the pistons to adjust the amount of fluid flow as pressure increases. The amount of fluid flow is adjusted by centering the cam ring and controlling the distance that the pistons extend and retract. In a rotating piston pump, pistons are housed in a rotating piston block that is offset inside the pump housing and rotates around a fixed shaft. Fluid enters the pump inlet as the pistons extend and is discharged from the pump outlet as the pistons retract.

Radial piston pumps operate on the same basic principles as axial piston pumps but are built with the pistons lying flat and facing inward toward the shaft. The inlet and outlet are located close to the shaft, and the piston block is off-center inside the cam ring. As the shaft rotates, the pistons extend and retract to complete the four basic operational steps of positive-displacement pumps.

TECH FACT

Bent-axis hydraulic pumps can have its input shaft and pistons arranged coaxially or its input shaft mounted on an angle to the piston bores. Bent-axis hydraulic pumps tend to be more volumetrically efficient than other pumps, but they also can be slightly larger for a given capacity, and their shape can cause installation difficulties in some applications.

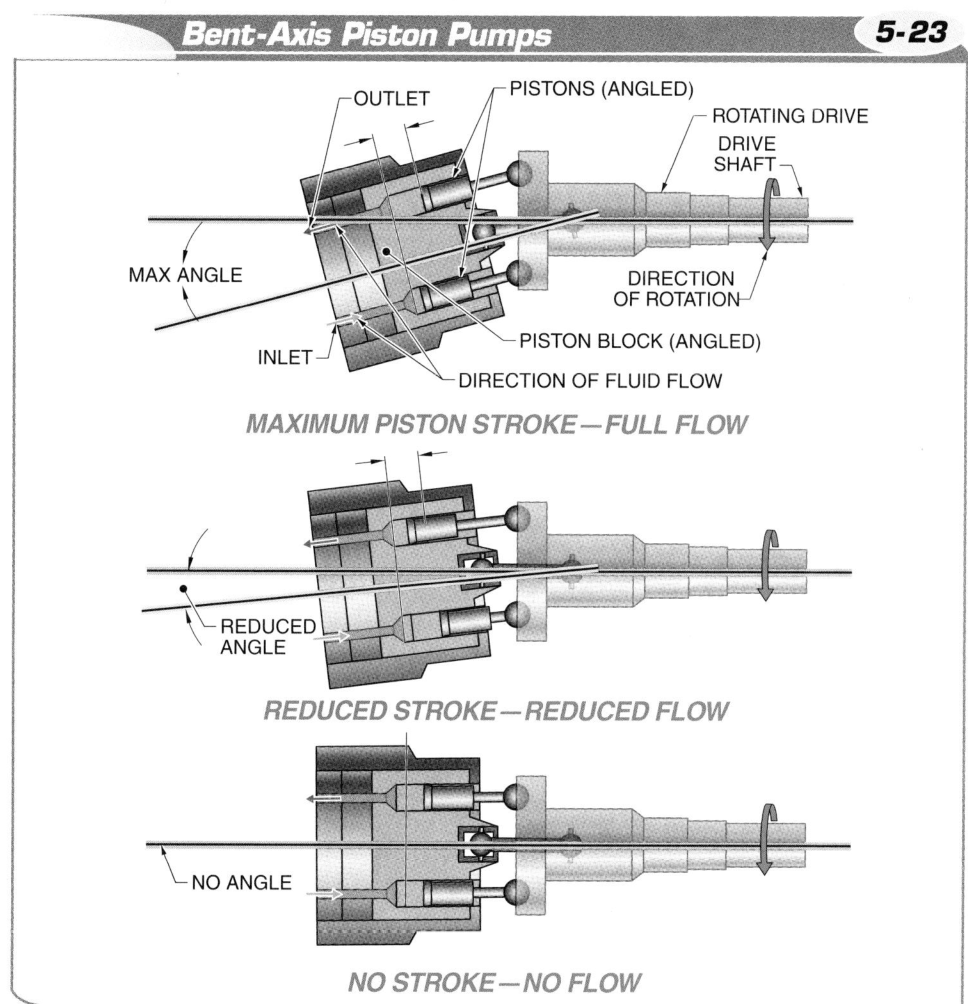

Figure 5-23. Bent-axis piston pumps operate in the same manner as an axial piston pump, but rather than the swash plate being at an angle, the pistons and piston block are at an angle.

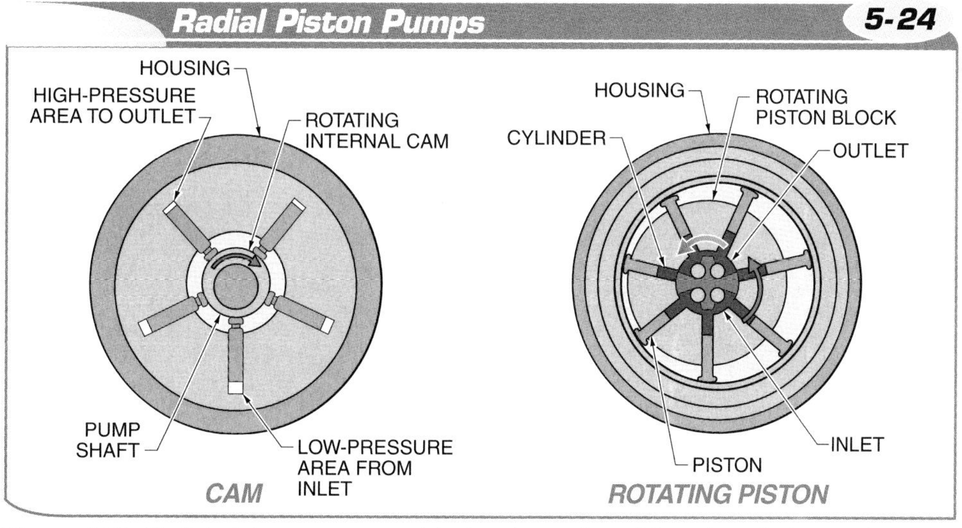

Figure 5-24. Radial piston pumps consist of reciprocating pistons in cylinders and can be classified as cam or rotating piston pumps.

TERMS

Cavitation is a localized gaseous condition within a stream of fluid, which occurs when pressure is reduced to vapor pressure.

Implosion is an inward bursting.

Pseudocavitation is artificial cavitation caused by air being allowed into the pump suction line.

Variable-Displacement Pump Control Methods. Although pressure compensated variable displacement pumps are the most common type of displacement control methods, there are other methods of controlling displacement of a vane or piston pump. The most common types of variable displacement control include load sensing, torque or horsepower limiter, and electronic displacement.

Load Sensing—Load sensing controls a variable displacement pump by sensing an increase of a load in the system by monitoring the pressure drop across an orifice. If the load side pressure of the orifice increases, the pump varies the displacement of the pump to increase the pressure on the supply (pump) side of the orifice. **See Figure 5-25.**

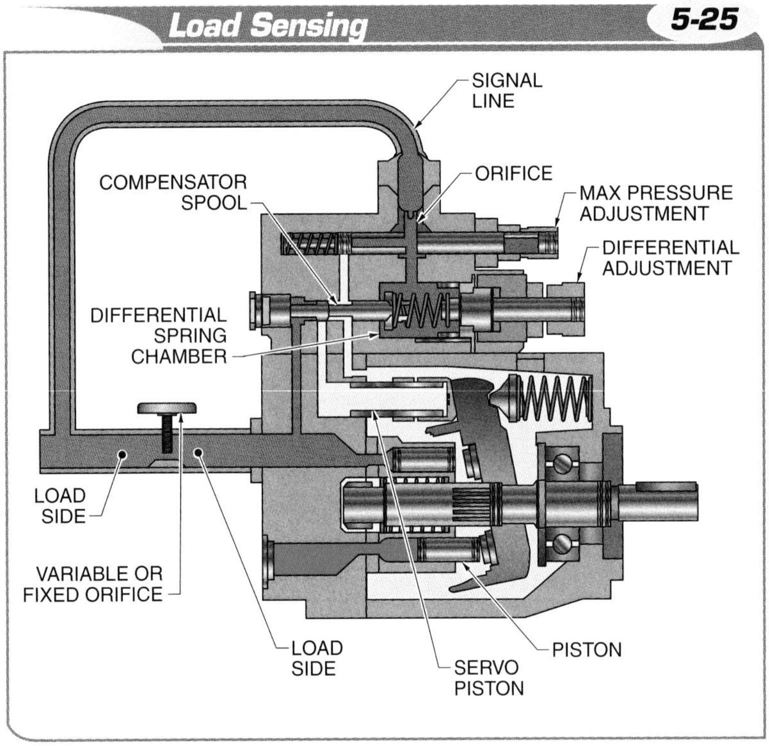

Load Sensing 5-25

SIGNAL LINE
ORIFICE
COMPENSATOR SPOOL
MAX PRESSURE ADJUSTMENT
DIFFERENTIAL ADJUSTMENT
DIFFERENTIAL SPRING CHAMBER
LOAD SIDE
VARIABLE OR FIXED ORIFICE
LOAD SIDE
SERVO PISTON
PISTON

Figure 5-25. Load sensing controls a variable displacement pump by sensing an increase of a load in the system by monitoring the pressure drop across an orifice.

Torque or Horsepower Limit Control—Torque or horsepower limiter control varies the displacement to maintain a constant horsepower or torque input when the speed of the prime mover does not change. The concept of this type of control is when higher flow is required pressure drops to supply the needed flow. When higher pressure is needed, the flow is reduced to supply the needed pressure. With this concept, the same amount of torque and horsepower is required from the prime mover when the needs of pressure and flow in the system change. Because torque and horsepower have a direct relationship and the rpm of the prime mover does not change, this type of control limits both the horsepower and the torque. If this type of control is used in mobile hydraulics, it is referred to as torque control. If this type of control is used in industrial hydraulics, it is referred to as horsepower control.

Electronic Displacement—Electronic control controls the displacement of a pump using an electrical signal and is accomplished with either a proportional or a servo attachment. **See Figure 5-26.** The electrical signal is transmitted from an industrial computer that typically uses other electrical signals from the system to calculate how much flow is required in the system. The industrial computer then sends the signal to the proportional valve, which adjusts the displacement of the pump.

Cavitation. *Cavitation* is a localized gaseous condition within a stream of fluid, which occurs when pressure is reduced to vapor pressure. *Implosion* is an inward bursting. **See Figure 5-27.** *Pseudocavitation* is artificial cavitation caused by air being allowed into the pump suction line. Pseudocavitation is caused by low reservoir fluid, contaminated fluid, or leaking pump suction lines. Pseudocavitation is indicated by a loud noise for an extended period of time.

Pseudocavitation occurs when the inlet port of a pump is restricted. An indication of pump cavitation is a high shrieking sound or a sound similar to loose marbles or ball bearings in the pump. Pseudocavitation is normally created when the suction line is damaged, plugged, or collapsed. Pseudocavitation may also be caused by an increase in pump rpm that requires more fluid than the system piping allows, fluids with an increased viscosity due to lower ambient temperatures, or an increase in the viscosity of a fluid in a system when the system has a long suction line.

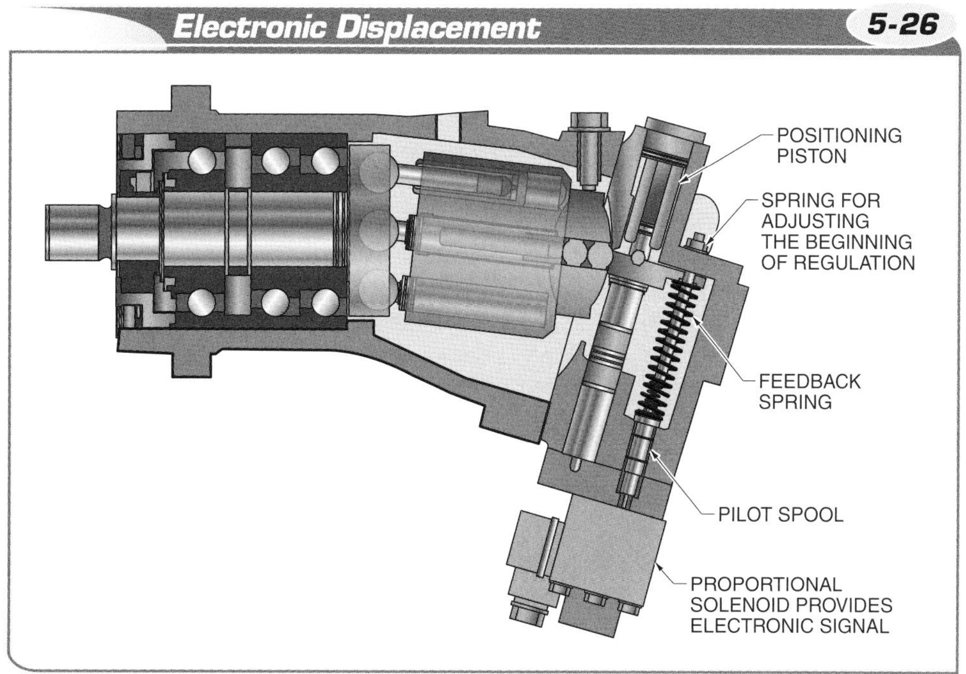

Figure 5-26. Electronic control controls the displacement of a pump using an electrical signal and is accomplished with either a proportional or a servo attachment.

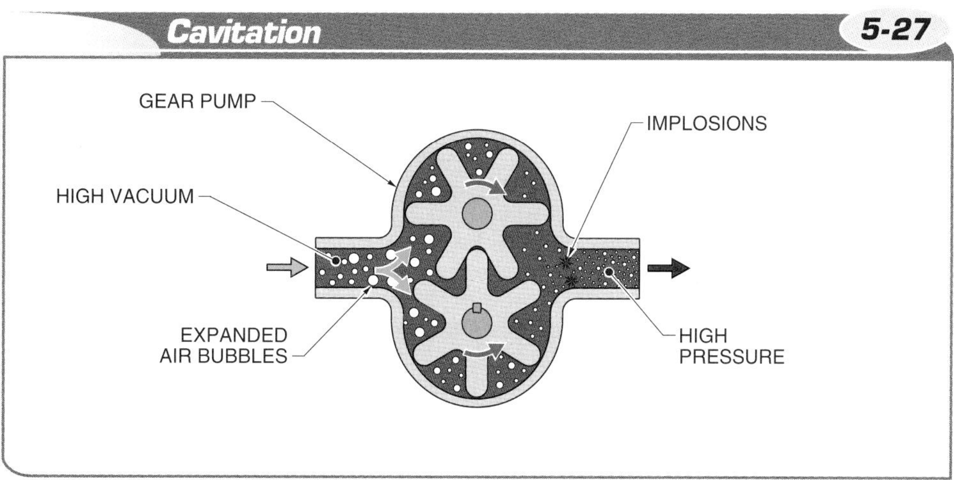

Figure 5-27. Cavitation occurs as gas bubbles expand in a vacuum and implode when entering a pressurized area.

With cavitation, as the pump pulls against a fluid that does not flow, a greater vacuum is created. Any microscopic air or gas within the fluid expands. Expanded bubbles on the inlet side collapse rapidly on the outlet side of the pump. The small but tremendous implosions can cause extensive damage to pump parts. Theoretically, an air bubble exposed to a 5000 psi cavitation may create an implosion pressure of 75,000 psi and travel at a speed of 600 fps to 4000 fps.

Double and Triple Pumps

Hydraulic pumps are available with double and triple pump styles. These pump styles have two or three pumping units inside a single pump housing on the same shaft. The pumping units may or may not be the same size. The advantage of these systems is that one prime mover is used to rotate two or three pumps at the same time. Double and triple pumps can provide greater flow by connecting the pumps in series or they can supply two or three different hydraulic systems using only one prime mover. Each pump requires its own relief valve. Only one pump unit needs to be installed when using double or triple pumps, which simplifies the installation process. Pumps that are connected together can deliver greater volumes of hydraulic fluid and produce higher pressures than a single pump.

In addition to high cost, the main disadvantage of using double or triple pumps is that if one of the pump units breaks down, the entire hydraulic system must be shutdown while the pump is repaired. Double and triple pumps are used in mobile and industrial applications such as hydraulic log splitters, shear presses, underground well drilling machines, and trash compactors. For example, a small pump can provide hydraulic fluid to a piece of equipment while the larger pump can unload fluid to the reservoir. The most common types of double and triple pumps are gear and vane pumps. **See Figure 5-28.**

Hydraulic Pump Schematic Symbols

Hydraulic pump schematic symbols are used to determine general information about the pump used in a system. **See Figure 5-29.** While hydraulic pump schematic symbols do not provide direct information on pump type, such as piston, gear, or vane, they do provide information on whether the pump is unidirectional or bidirectional by using arrows. These symbols also provide information on whether the pump is fixed or variable.

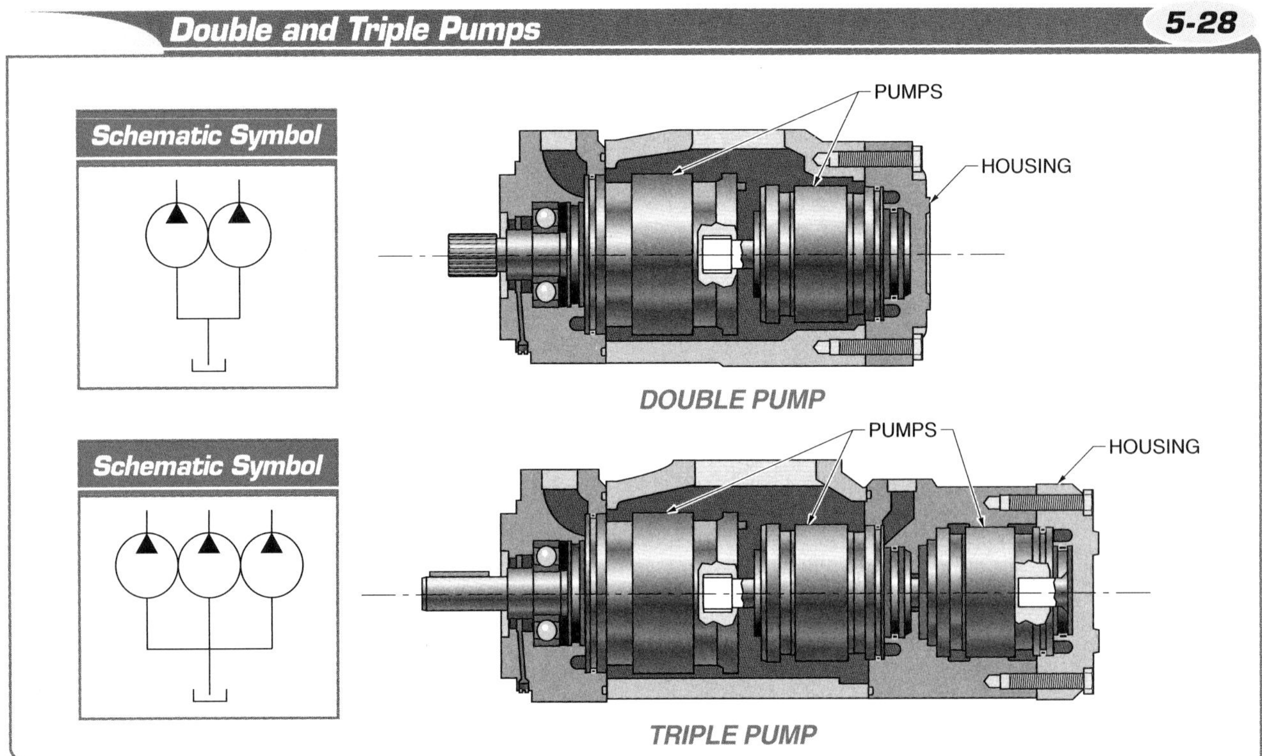

Figure 5-28. Hydraulic pumps are available in both double and triple pump designs.

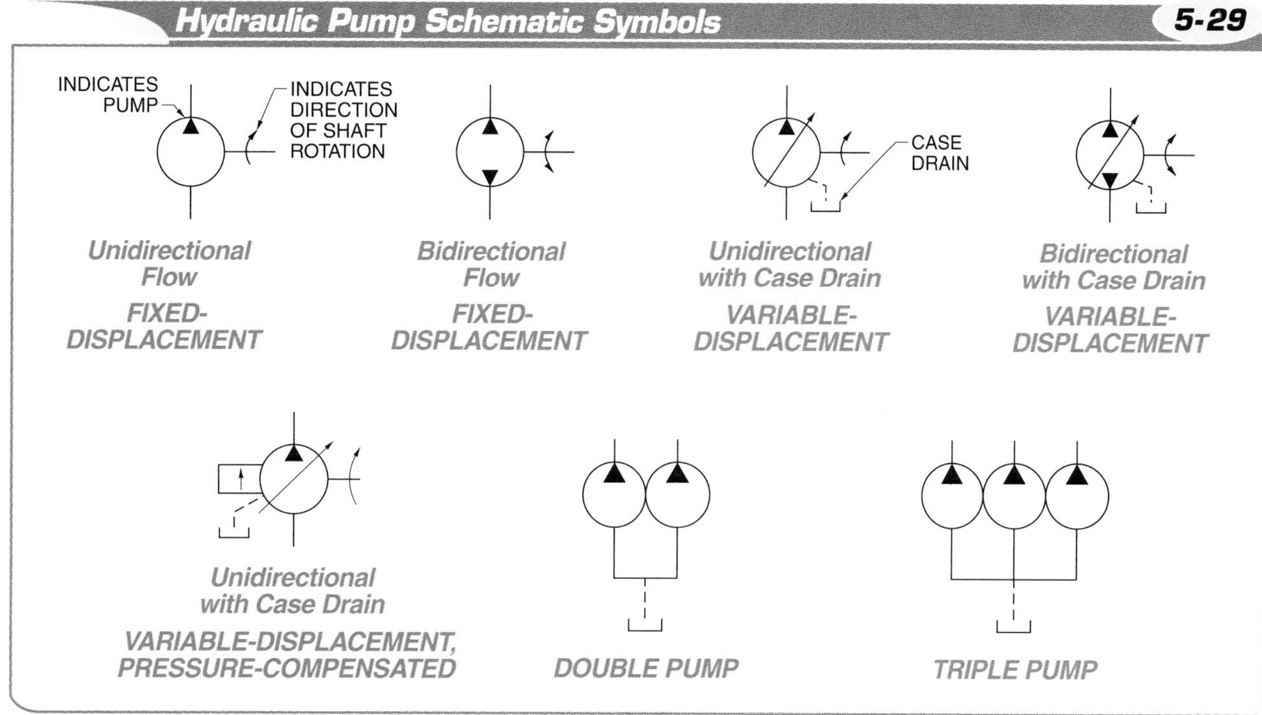

Figure 5-29. Hydraulic pump schematic symbols are used to determine general information about the type of pump used in a system.

HYDRAULIC PUMP SELECTION CONSIDERATIONS

All hydraulic pumps create fluid flow by using variations of the four basic operational steps of a positive-displacement pump. Each type of pump has applications where it functions better. In addition to a facility's replacement parts inventory, there are a number of variables to consider:

- **Equipment cost** - Typically, the least expensive pump that accomplishes the facility requirements is installed.
- **Operating environment** - The type of operating environment that the fluid power system is installed in can determine which type of pump to use. Facilities that operate in clean, dirty, or excessively loud environments can require certain types of pumps.
- **Pressure operating range** - A pump with a maximum pressure rating that is higher than that of the application requirements is typically installed.
- **Efficiency** - Pump efficiency must be taken into consideration. Certain applications may require high efficiency pumps.
- **Pump design** - Pumps can be fixed or variable. If fluid flow needs to be adjusted, a variable pump must be used.
- **Design complexity** - Design complexity is associated with equipment cost. A less complex pump, such as a gear pump, has lower cost than that of a more complex pump, such as a piston pump.
- **Loudness** - Although not as expensive as piston or vane pumps, gear pumps have the highest level of loudness when in operation.

A pump comparison chart can help show the differences between different types of pumps.

Pump Comparison

	Gear	Vane	Piston
Equipment cost	Least expensive	Intermediate	Most expensive
Operating environment	Least clean	Clean	Most clean
Pressure operating range	1500 to 3000*	200 to 3000*	3000 to 12,000*
Efficiency	Least	Intermediate	Most
Pump design	Fixed only	Fixed or variable	Fixed or variable
Design complexity	Least	Intermediate	Most
Loudness	100†	70†	80†

* in psi
† in dB

Digital Resources

Chapter Review

Chapter 5 Hydraulic Pumps

Name: ______________________ Date: ______________

MULTIPLE CHOICE

_______________ 1. Fixed-displacement pumps are rated by fluid flow and ___.
- A. electrical requirements
- B. housing size
- C. pressure rating
- D. speed

_______________ 2. Hydraulic pump types include gear, vane, and ___ pumps.
- A. centrifugal
- B. gerotor
- C. lobe
- D. piston

_______________ 3. ___ is a localized gaseous condition within a stream of fluid that occurs when pressure is reduced to vapor pressure.
- A. Cavitation
- B. Implosion
- C. Pseudocavitation
- D. all of the above

_______________ 4. A lobe pump requires ___ to drive the lobe-shaped driven gears.
- A. lobe gears
- B. pistons
- C. timing belts
- D. timing gears

_______________ 5. ___ is the volume of hydraulic fluid moved during each revolution of the shaft of a pump.
- A. Displacement
- B. Gallons per minute
- C. Revolutions per minute
- D. Volumetric efficiency

_______________ 6. A ___ pump has an inner rotor that meshes with the gear teeth of an outer rotor.
- A. gerotor
- B. lobe
- C. piston
- D. vane

_______________ 7. A variable-displacement pump varies the amount of hydraulic fluid with movable internal components while the ___ of the prime mover remains fixed.
- A. displacement
- B. gpm
- C. rpm
- D. volumetric efficiency

_______________ **8.** Bent-axis piston pumps are different from axial piston pumps because their ___ are at an angle (offset).

A. pistons and piston block
B. swash plates
C. pistons, piston block, and swash plate
D. none of the above

_______________ **9.** A(n) ___ plate is an angled plate used to extend and retract the pistons in an axial piston pump.

A. axial
B. cylinder barrel
C. piston
D. swash

_______________ **10.** The main categories of hydraulic pumps are dynamic and ___ pumps.

A. fixed
B. nonpositive-displacement
C. positive-displacement
D. variable

_______________ **11.** The most common method to vary the angle of a swash plate is through ___.

A. internal pilot pressure
B. the speed of the prime mover
C. the rotating vanes
D. the cartridge assembly

_______________ **12.** A ___ is part of a vane pump and houses the vanes, rotor, and cam ring, which are all placed between two end plates.

A. cam ring
B. cartridge assembly
C. cartridge fitting
D. slotted rotor

_______________ **13.** Gear pumps may be external, internal, lobe, or ___.

A. centrifugal
B. gerotor
C. piston
D. vane

_______________ **14.** ___ is the highest amount of pressure at which a pump can create fluid flow.

A. Displacement
B. Gallons per minute
C. Pressure rating
D. Volumetric efficiency

_______________ **15.** A(n) ___ vane pump allows adjustment of the flow rate of the pump.

A. axial
B. balanced
C. variable-displacement
D. variable-displacement, pressure-compensated

COMPLETION

_______________ 1. A(n) ___ is an airproof and/or fluidproof joint between two members.

_______________ 2. Vane pump types include unbalanced, balanced, and ___.

_______________ 3. A(n) ___ is a crescent-moon-shaped seal between the gears and between the inlet and outlet sides of an internal gear pump.

_______________ 4. A(n) ___ pump has pistons in a rotating cylinder block parallel to the drive shaft.

_______________ 5. A(n) ___ piston pump has pistons and cylinders that are at an angle to the drive shaft and thrust plate.

_______________ 6. A(n) ___ allows a technician to replace an entire assembly in a vane pump when the internal parts need to be replaced.

_______________ 7. ___ is the relationship between actual and theoretical fluid flow.

_______________ 8. Pseudocavitation is caused by ___ being allowed into the suction line.

_______________ 9. ___ are generally affixed to a pump to identify the different pump ratings.

_______________ 10. A(n) ___ vane pump is a fixed- or variable-displacement hydraulic pump in which the pumping action occurs in the chambers on one side of the rotor and shaft.

_______________ 11. A(n) ___-displacement piston pump can have the angle of its swash plate varied.

_______________ 12. A(n) ___ pump uses gears that mesh together in various manners in a close-fitting housing to create fluid flow.

_______________ 13. A(n) ___-displacement pump cannot have its fluid flow (gpm) changed.

_______________ 14. ___ pumps are known for their dry priming capability, ease of maintenance, and good suction characteristics over the life of the pump.

_______________ 15. A(n) ___ consists of a cam ring, rotor, vanes, and a port plate with opposing inlet and outlet ports set 180° apart from each other.

_______________ 16. ___ is the volume of hydraulic fluid moved during each revolution of the shaft of a pump.

_______________ 17. A(n) ___ is a mechanical device that changes mechanical energy into hydraulic energy (fluid flow).

_______________ 18. A(n) ___ piston pump has pistons located perpendicular to the pump shaft.

_______________ 19. Pseudocavitation occurs when the ___ port of a pump is restricted.

_______________ 20. A(n) ___ pump has a positive seal between its inlet and outlet and moves a specific volume of hydraulic fluid with each revolution of the shaft.

TRUE/FALSE

T F 1. Although lobe pumps can be used in pneumatic applications, they are typically used in hydraulic applications to move products such as slurries, pastes, and solids.

T F 2. Hydraulic fixed-displacement pumps cannot vary the amount of fluid flow that they produce during operation without changing the speed at which they operate.

T F 3. External gear pumps have two different size gears.

T F 4. A balanced vane pump has an elliptical cam ring.

T F 5. Displacement is the fluid flow that a pump can produce.

T F 6. Axial piston pumps create smooth fluid flow for piston pumps.

T F 7. Changing the angle of the swash plate in a variable-displacement piston pump does not affect the distance of how far back a piston pulls.

T F 8. Gear pumps operate through the extension and retraction of multiple pistons.

T F 9. All bent-axis piston pumps are variable.

T F 10. Dynamic pumps include gear, vane, and piston pumps.

T F 11. Pseudocavitation can be caused by low reservoir fluid.

T F 12. Gear pumps have the highest level of loudness while in operation.

T F 13. Hydraulic pump schematic symbols provide direct information on pump type.

T F 14. Vane pumps are typically used in applications that require the movement of high-viscosity fluids.

T F 15. A radial piston pump consists of pistons located parallel to the pump shaft.

T F 16. Radial piston pumps can be classified as cam or rotating piston pumps.

T F 17. All hydraulic pumps have nameplates affixed to them.

T F 18. The length of the piston is the main determinant to the amount of gpm that an axial piston pump can produce.

T F 19. Gear pumps are ideal for equipment that operates in dirty environments.

T F 20. A bent-axis piston pump can operate at pressures above 10,000 psi.

T F 21. The rotor of a variable-displacement, pressure-compensated, vane pump is centered with a crescent seal.

T F 22. All gear pumps are fixed-displacement pumps.

T F 23. Unbalanced vane pumps have the simplest design of the various types of vane pumps.

T F 24. Gerotor pumps are used for high-pressure applications of greater than 1000 psi.

T F 25. Positive-displacement pumps are typically classified as fixed or variable.

MATCHING

Cartridge Assemblies

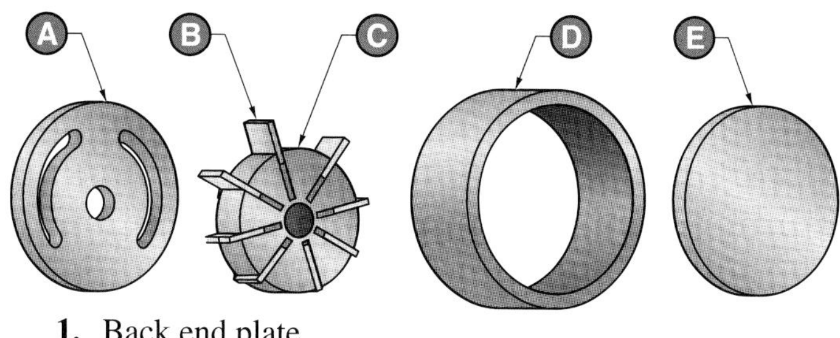

_______________ 1. Back end plate

_______________ 2. Cam ring

_______________ 3. Front end plate

_______________ 4. Slotted rotor

_______________ 5. Vane

Schematic Symbols

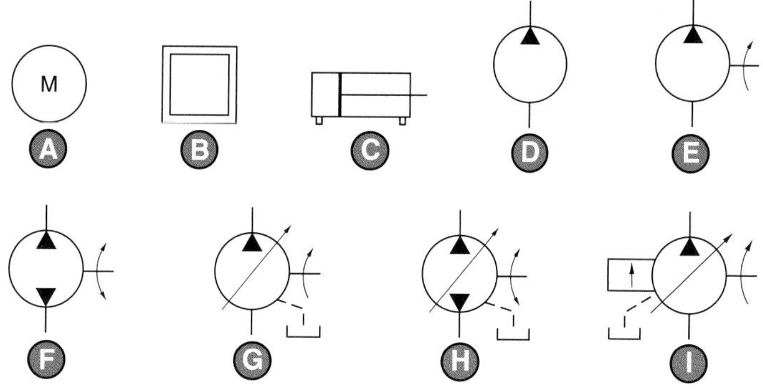

_______________ 1. Hydraulic pump—bidirectional flow—fixed-displacement

_______________ 2. Hydraulic pump—bidirectional flow with case drain—variable-displacement

_______________ 3. Electric motor

_______________ 4. Double-acting cylinder

_______________ 5. Hydraulic pump, unidirectional flow

_______________ 6. Internal combustion engine

_______________ 7. Hydraulic pump—unidirectional flow—fixed-displacement

_______________ 8. Hydraulic pump—unidirectional flow with case drain—variable-displacement

_______________ 9. Hydraulic pump—unidirectional flow with case drain—variable-displacement, pressure compensated

Gear Pumps

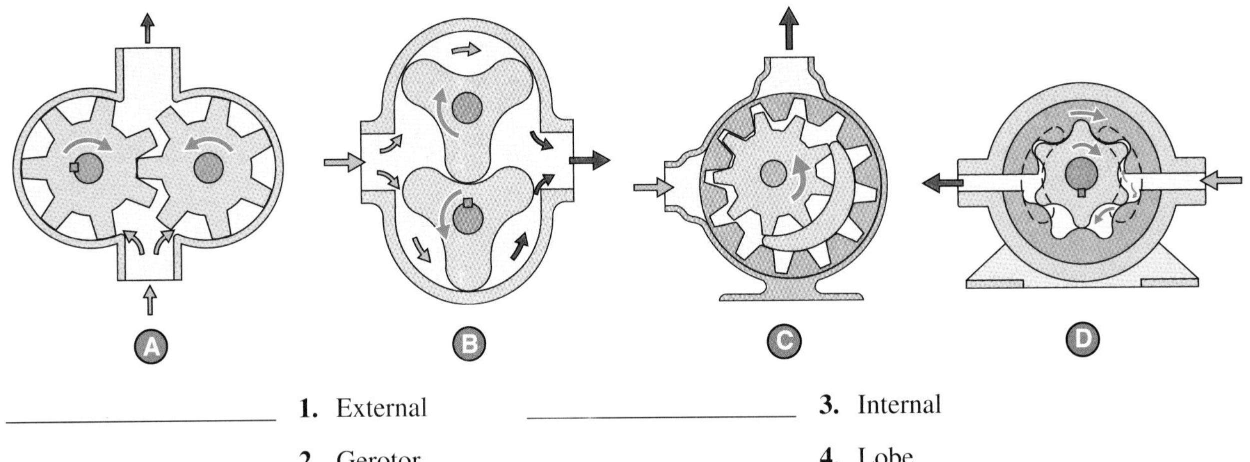

______________ 1. External

______________ 2. Gerotor

______________ 3. Internal

______________ 4. Lobe

Balanced Vane Pumps

______________ 1. Elliptical cam ring

______________ 2. Port plate

______________ 3. Port plate inlet port

______________ 4. Port plate outlet port

______________ 5. Pump housing inlet port

______________ 6. Pump housing outlet port

______________ 7. Rotor

______________ 8. Vane

SHORT ANSWER

1. List and describe the four main pump ratings.

Activities

Activity 5-1: Pump Identification Information

A technician must be able to use a pump nameplate to identify the specific information of different pump parts. The hydraulic pump on a mobile hydraulic system has failed. The company that manufactured the pump no longer exists, but the original operator's manual is available.

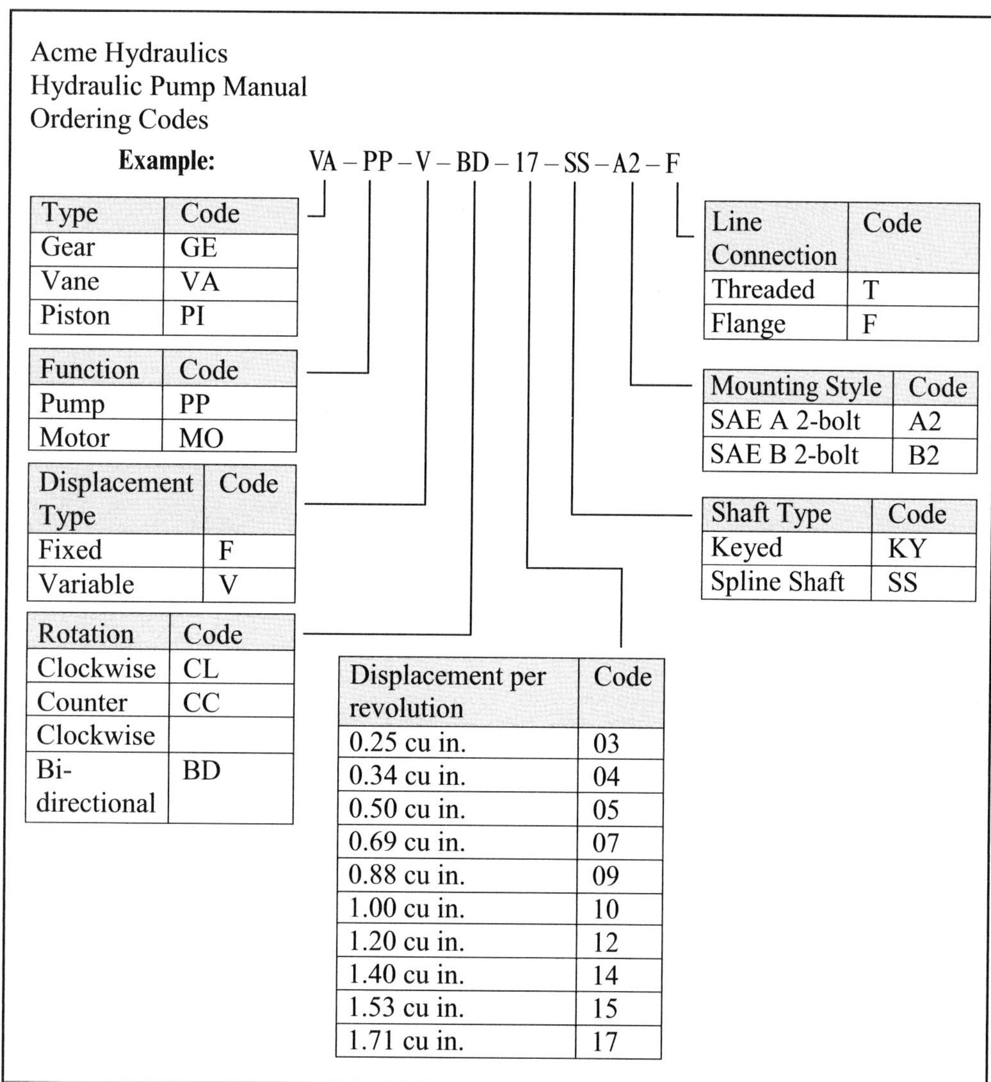

Acme Hydraulics
Hydraulic Pump Manual
Ordering Codes

Example: VA – PP – V – BD – 17 – SS – A2 – F

Type	Code
Gear	GE
Vane	VA
Piston	PI

Function	Code
Pump	PP
Motor	MO

Displacement Type	Code
Fixed	F
Variable	V

Rotation	Code
Clockwise	CL
Counter Clockwise	CC
Bi-directional	BD

Displacement per revolution	Code
0.25 cu in.	03
0.34 cu in.	04
0.50 cu in.	05
0.69 cu in.	07
0.88 cu in.	09
1.00 cu in.	10
1.20 cu in.	12
1.40 cu in.	14
1.53 cu in.	15
1.71 cu in.	17

Shaft Type	Code
Keyed	KY
Spline Shaft	SS

Mounting Style	Code
SAE A 2-bolt	A2
SAE B 2-bolt	B2

Line Connection	Code
Threaded	T
Flange	F

1. List the pump description required for ordering a replacement pump for the existing pump GE-PP-F-CC-07-KY-A2-T.

Activity 5-2: Vane Pump Repair

A vane pump used in a hydraulic press has failed. The rotor has been damaged from contaminated hydraulic fluid and must be replaced.

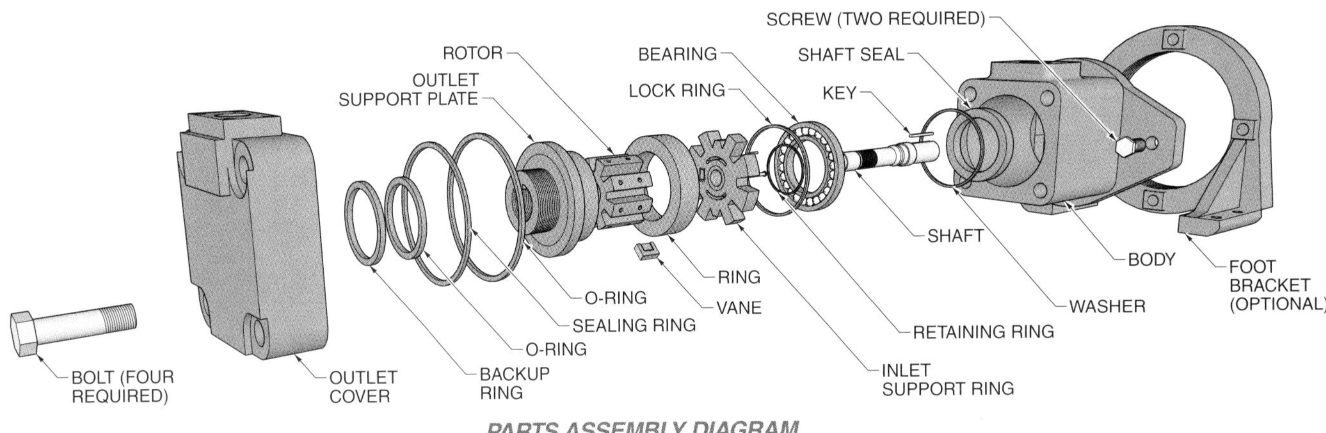

PARTS ASSEMBLY DIAGRAM

1. Using the parts assembly diagram, list the steps required to disassemble the pump and replace the damaged rotor.

Activity 5-3: Vane Pump Cartridge Analysis

Below is a chart for two vane pump cartridges commonly used in industrial applications. Both vane pumps were tested to see how much fluid flow each produced at specific pressures under standard conditions. The standard conditions for both vane pump cartridges are 150 Saybolt Universal Seconds (SUS) hydraulic fluid at 120°F.

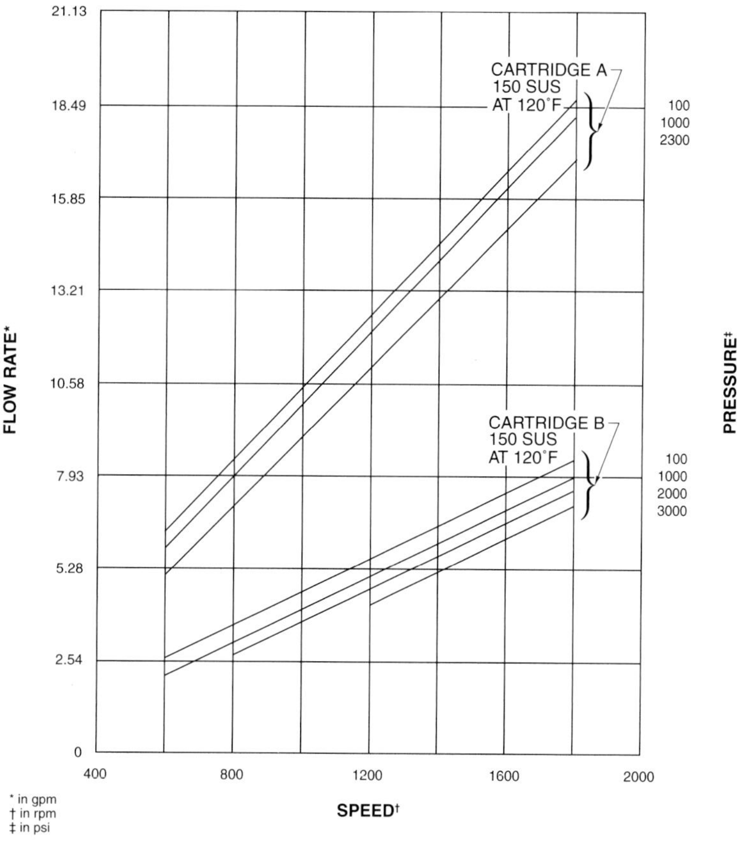

VANE PUMP CARTRIDGE ANALYSIS CHART

Use the Vane Pump Cartridge Analysis chart and information from the chapter to answer the following questions:

_______________ **1.** If the pump with cartridge A installed is operating at 1000 psi and the prime mover is rotating at 1200 rpm, what should a flow meter read if installed within 6′ of the pump discharge?

_______________ **2.** If the pump with cartridge A installed is operating at 2300 psi and the prime mover is turning at 1700 rpm, what should a flow meter read if installed within 6′ of the pump discharge?

_______________ **3.** If the pump with cartridge B installed is operating at 2000 psi and the prime mover is rotating at 1100 rpm, what should a flow meter read if installed within 6′ of the pump discharge?

_______________ **4.** If the pump with cartridge B installed is operating at 3000 psi and the prime mover is rotating at 1400 rpm, what should a flow meter read if installed within 6′ of the pump discharge?

_______________ **5.** If the pump with cartridge A installed is operating at 1000 psi, the prime mover is rotating at 1700 rpm, and the flow meter reads 13.5 gpm, what is the volumetric efficiency of the pump?

_______________ **6.** The pump with cartridge A installed is operating at 2300 psi, the prime mover is rotating at 1600 rpm, and the cartridge is 2.01 cu in. Place a mark on the chart where that should be represented.

Activity 5-4: Pump Analysis

Refer to Hydraulic System on page 153 to answer the following questions.

_______________ **1.** The pump displacement in Hydraulic System is rated at 2.56 cu in. and is rotating at 1200 rpm. If flow meter FM1 shows a reading of 12.25 gpm, what is the pump efficiency of the pump to FM1?

_______________ **2.** The pump displacement in Hydraulic System is rated at 1.34 cu in. and is rotating at 1000 rpm. If flow meter FM2 shows a reading 5.2 gpm what is the pump efficiency of the pump to FM2?

_______________ **3.** The pump displacement in Hydraulic System is rated at 1.34 cu in. and is rotating at 1000 rpm. If flow meter FM3 shows a reading 4.2 gpm what is the pump efficiency of the pump to FM3?

_______________ **4.** What is the difference in pump efficiency between FM2 and FM3?

5. List at least two possible reasons for the difference in flow from FM2 and FM3.

6. The technician responsible for inspecting Hydraulic System notices that the pump efficiency is down 12% at each flow meter. List at least five possible reasons for lower pump efficiency. *Note:* Assume that the system could possibly be damaged.

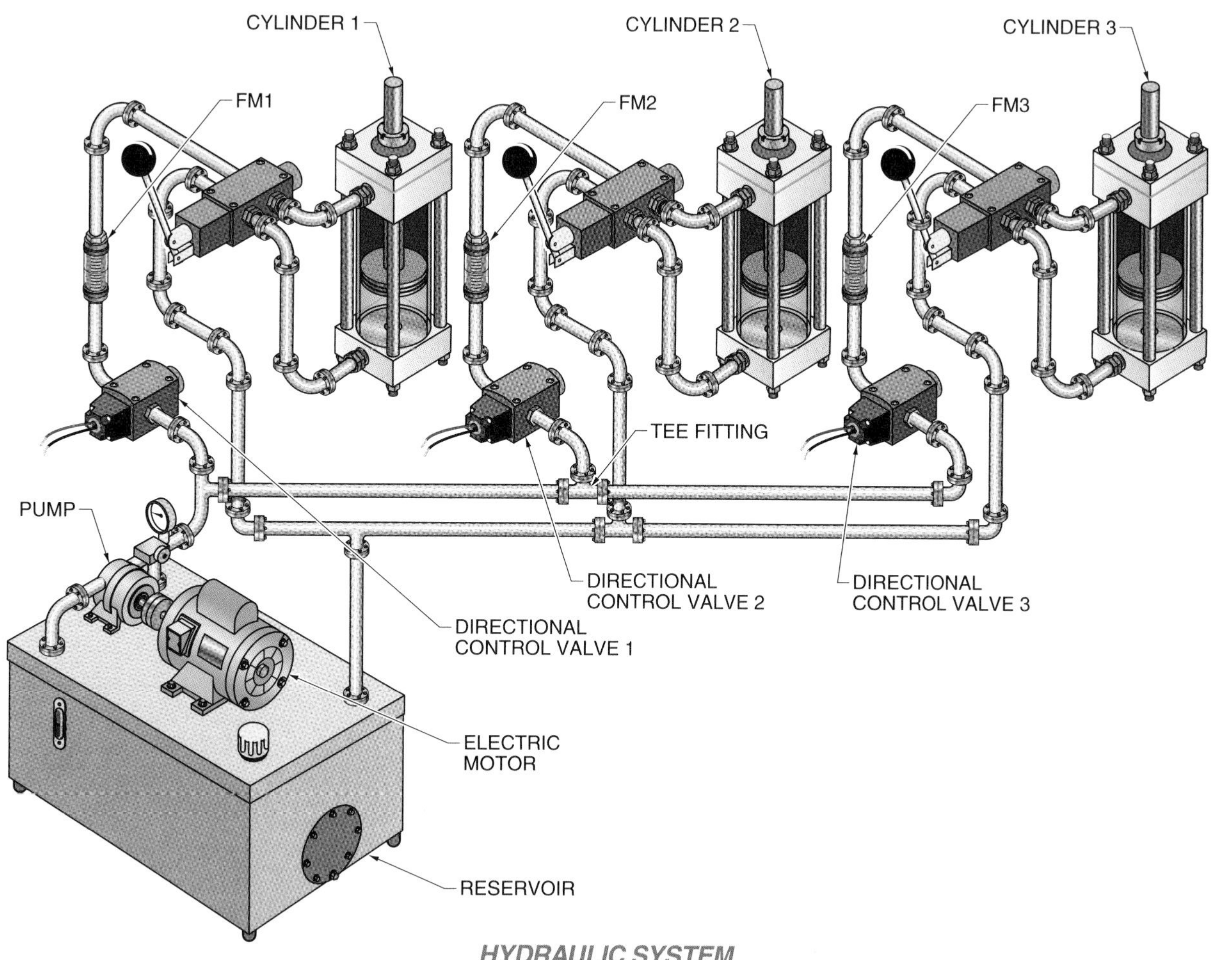

HYDRAULIC SYSTEM

Activity 5-5: Pump Flow Rate

1. Open FluidSIM software program and build system below.
2. After system is built, double left click on power unit to ensure that the flow is set to US gal/min. If not, select correct setting.
3. Simulate circuit, activate directional control valve by left clicking on lever, and monitor the speed of the cylinders extension.
4. Left click once on power unit and change flow rate to 0.45 gallons.
5. Activate directional control valve and monitor the speed of the cylinder.
6. Left click on the power unit and change the flow rate to 4.98 gallons.
7. Activate directional control valve and monitor the speed of the cylinder.

Answer the following questions:

1. What was the outcome of the change of pump flow rate?

2. List different methods of changing the pump flow rate.

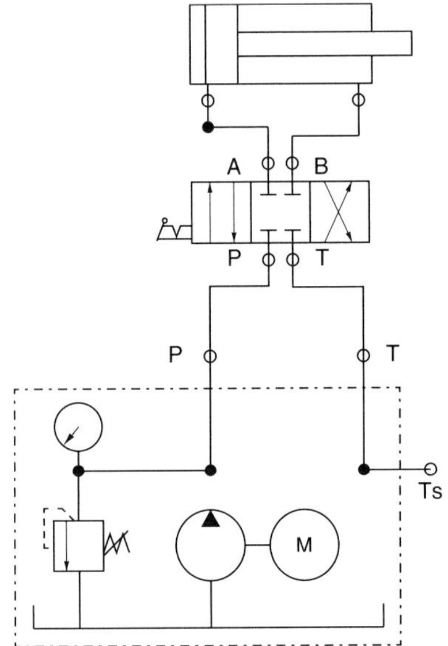

Chapter 6

Directional Control

OBJECTIVES

- Identify the different types of check valves.
- Identify the different directional control valve designs.
- Identify the different parts of directional control valves.
- Describe two- or three-position, two-, three-, and four-way directional control valves.
- Identify the four center positions used in three-position, four-way directional control valves.
- Describe check valve operation.
- Identify the different types of directional control valve spool actuators.
- Identify the main types of logic valves.

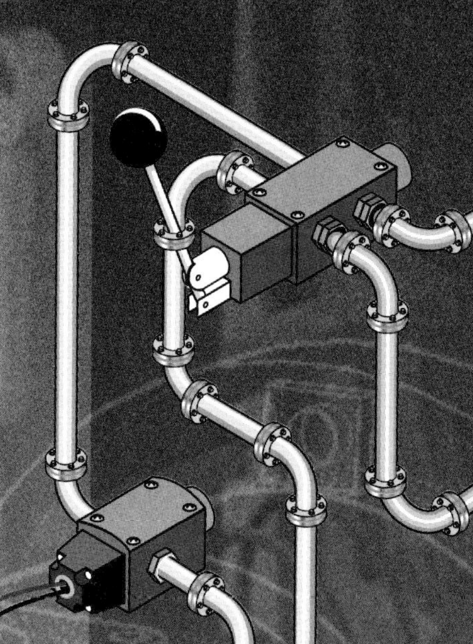

INTRODUCTION

The direction of fluid flow is one of the most important aspects of a hydraulic system. If a hydraulic system is to function in the required manner, proper directional control devices must be installed. Fluid flow is typically controlled by valves. Valve selection is based on type, size, actuating method, and automatic control capability. The main types of control valves are directional control valves and check valves.

Directional control valves determine the path of fluid flow within the hydraulic system. They also establish the direction and the motion of actuators, such as cylinders and motors. Check valves restrict the direction of fluid flow and are used to bypass certain components within a hydraulic system. The control of fluid flow can be accomplished with two-way, three-way, four-way, AND logic, or shuttle (OR logic) valves.

TERMS

A **direct-acting check valve** is a check valve that is directly activated or moved by fluid flow from the primary port.

A **ball check valve** is a check valve that uses a ball located between the biasing spring and the seat to block fluid flow.

CHECK VALVES

A check valve is a valve that allows fluid flow in one direction and stops fluid flow in the opposite direction. Check valves also prevent the backflow of fluid within the system. A check valve acts like a one-way directional control valve or the fluid power equivalent of an electric diode. Check valves are common to most hydraulic systems and are typically used to allow fluid flow to bypass components. When a bypass check valve is used, it is usually built directly into the valve that is doing the work. Check valves can also be used to isolate sections of a system or system components. The two main types of check valves are direct-acting check valves and pilot-operated check valves.

Direct-Acting Check Valves

A *direct-acting check valve* is a check valve that is directly activated or moved by fluid flow from the primary port. A direct-acting check valve consists of an inlet port, an outlet port, a valve body, a moveable part, and usually a biasing spring. The movable part is typically a ball or a poppet. The biasing spring is not adjustable. The pressure required to overcome the spring energy depends on the rating of the spring, in pounds.

When fluid flows from one direction, the force created by the pressure unseats the ball or poppet from the port and fluid is allowed to pass. When fluid flows from the opposite direction, the ball or poppet pushes against the port and stops fluid flow. The three different types of direct-acting check valves are ball, poppet, and right angle. **See Figure 6-1.**

Ball Check Valves. A *ball check valve* is a check valve that uses a ball located between the biasing spring and the seat to block fluid flow. Ball check valves are used in hydraulic applications to bypass components that control the flow rate to a cylinder, such as in an industrial band saw.

Ball check valves are often used to allow fluid flow to bypass other hydraulic components. This is done by connecting a ball check valve in parallel with the component to be bypassed. In one direction, fluid flow is not allowed through the check valve and has to flow into the component. In the opposite direction, fluid flow is allowed through the check valve and can bypass the component. Advantages of a ball check valve include low cost, a wide range of cracking pressures, in-line or cartridge construction, and low amounts of leakage when blocking fluid flow.

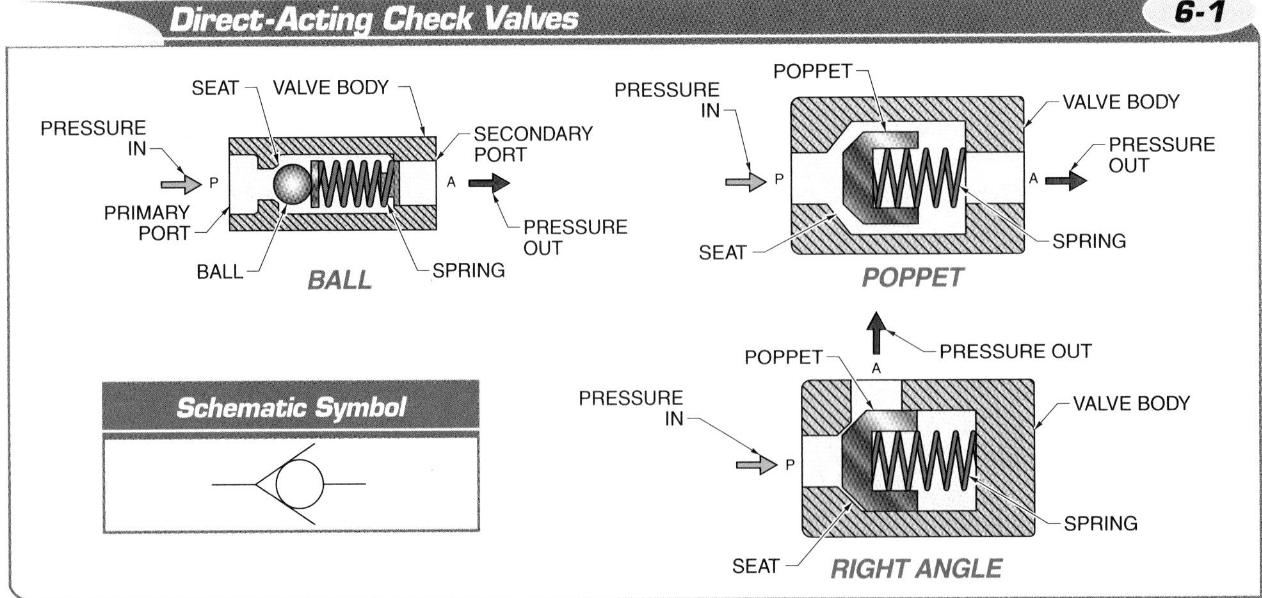

Figure 6-1. The three different types of direct-acting check valves are ball, poppet, and right angle.

Poppet Check Valves. A *poppet check valve* is a check valve that uses a poppet located between the biasing spring and the seat to block fluid flow. *Cracking pressure* is the amount of pressure required to slightly unseat a ball or a poppet and start to allow fluid flow through the check valve. Cracking pressure in a poppet check valve is typically 5 psi, although poppet check valves with cracking pressures as high as 65 psi are available. Poppet check valves are available with a range of flow rates up to several hundred gpm.

An advantage of poppet check valves over ball check valves is that they can be designed with an orifice through them for use as a safety bypass. A disadvantage of poppet check valves is that the poppet can become worn in the same area through repeated use, which allows leaks to develop at the point where the poppet fits against the seat. Leaks can create unwanted heat in the system, unnecessary movement of the different actuators in the system, bypass of filters, and possible safety issues when a cylinder needs to hold a load.

Right Angle Check Valves. A *right angle check valve* is a check valve that has its inlet and outlet ports set at a right angle to each other. It is installed mainly in hydraulic systems, rather than pneumatic systems.

A right angle check valve usually has a poppet to restrict fluid flow in one direction. Fluid flow in the opposite direction forces the poppet out of the path of fluid flow, which allows for higher flow rates with lower pressure drop. Cracking pressure in a right angle check valve is typically 5 psi, although valves with cracking pressures as high as 20 psi are available. Right angle check valves are available with a range of flow rates up to 400 gpm.

A *restriction (orifice) check valve* is a type of poppet check valve with an orifice placed in the center of the poppet to permit restricted fluid flow through the valve in the NC position. An *orifice* is a restricted passage in a fluid power line or component and is used to control fluid flow or to create a pressure differential. Orifices are sized to control the flow rate at a specific inlet pressure. Fluid flow increases through an orifice if the pressure differential across the orifice increases.

Restriction check valves are used in applications that require controlling the rate of compression and decompression, such as plastics compression molding machines. They also extend component life by reducing system shock caused by high flow rates and high volumes of pressurized fluid. **See Figure 6-2.**

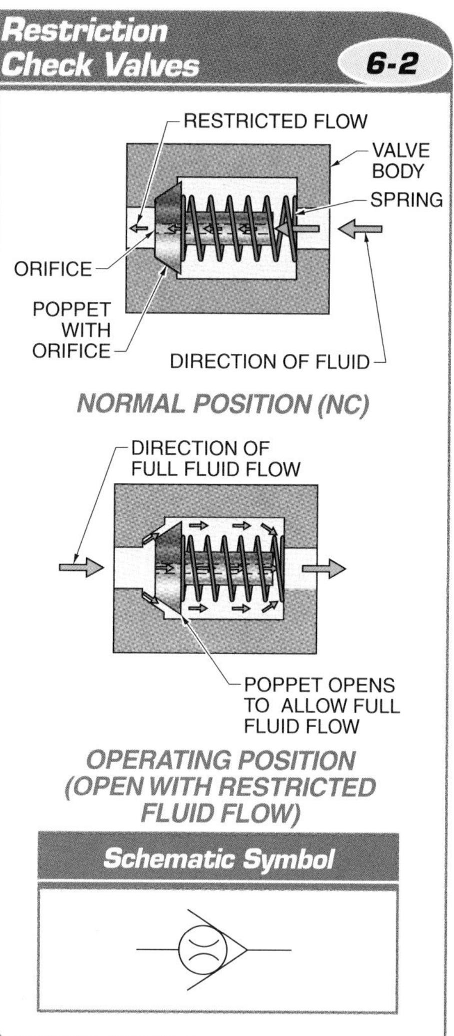

Figure 6-2. A restriction check valve has an orifice placed in the center of the poppet to permit restricted fluid flow through the valve in the NC position.

TECH FACT

Right angle check valves are common in older hydraulic systems because they were originally less expensive to manufacture.

TERMS

A **poppet check valve** is a check valve that uses a poppet located between the biasing spring and the seat to block fluid flow.

Cracking pressure is the amount of pressure required to slightly unseat a ball or a poppet and start to allow fluid flow through the valve.

A **right angle check valve** is a check valve that has its inlet and outlet ports set at a right angle to each other.

A **restriction (orifice) check valve** is a type of poppet check valve with an orifice placed in the center of the poppet to permit restricted fluid flow through the valve in the NC position.

An **orifice** is a restricted passage in a fluid power line or component and is used to control fluid flow or to create a pressure differential.

TERMS

A **pilot-operated check valve** is a check valve that operates with a pilot line that allows or stops fluid flow in both directions when activated.

A **pilot line** is a line used to transmit pressure in a hydraulic system for control purposes.

Direct-Acting Check Valve Installation. Direct-acting check valves are installed in most hydraulic systems and are used for two main applications. The first application is maintaining pressure control. For example, if gravity forces fluid back to a reservoir that is below a system not in operation, the seals in the system can be damaged. A direct-acting check valve keeps the majority of the system primed and prevents fluid from draining out of the system, which saves start-up time and cost.

The second application of a direct-acting check valve is as a bypass to change fluid flow and pressure in one direction but not in the other. For example, when using pressure and flow control valves, some pipes have fluid flow going in both directions at different times. When the flow rate needs to be slow in one of those directions but not the other, a bypass check valve is placed parallel with the flow control valve. This allows the flow rate to remain at the unrestricted rate in the opposite direction. **See Figure 6-3.**

Pilot-Operated Check Valves

A *pilot-operated check valve* is a check valve that operates with a pilot line that allows or stops fluid flow in both directions when activated. Pilot-operated check valves consist of all the same components as direct-acting check valves, but they also have a pilot line, sometimes a drain line, and a plunger or an internal spring.

A *pilot line* is a line used to transmit pressure in a hydraulic system for control purposes. On schematic diagrams, pilot lines are indicated by dashed lines while pressurized lines are solid. Pilot-operated check valves typically use a poppet to seal the port from fluid flow. Pilot-operated check valves are mainly used in hydraulic systems and can be pilot-to-open or pilot-to-close.

Pilot-to-Open Check Valves. A pilot-to-open check valve allows fluid flow in one direction and restricts fluid flow in the opposite direction until it receives a pressure signal through a pilot line applied to the port against the piston. **See Figure 6-4.** When the check valve gets a pilot pressure signal, it opens and allows fluid flow in both directions.

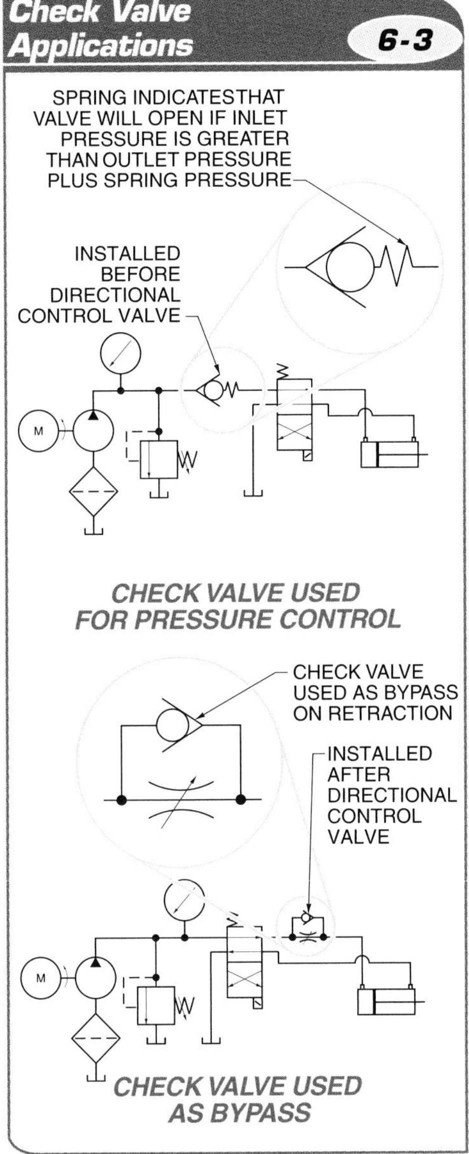

Figure 6-3. Check valves are mainly used for pressure control and to bypass certain system components.

TECH FACT

When a check valve gets dirty, the ball or poppet cannot seal completely against the seat, causing fluid leakage through the valve. The only way a check valve can be cleaned is by removing it from the system and cleaning it manually. To save cost and time, a dirty check valve can be replaced with a new one.

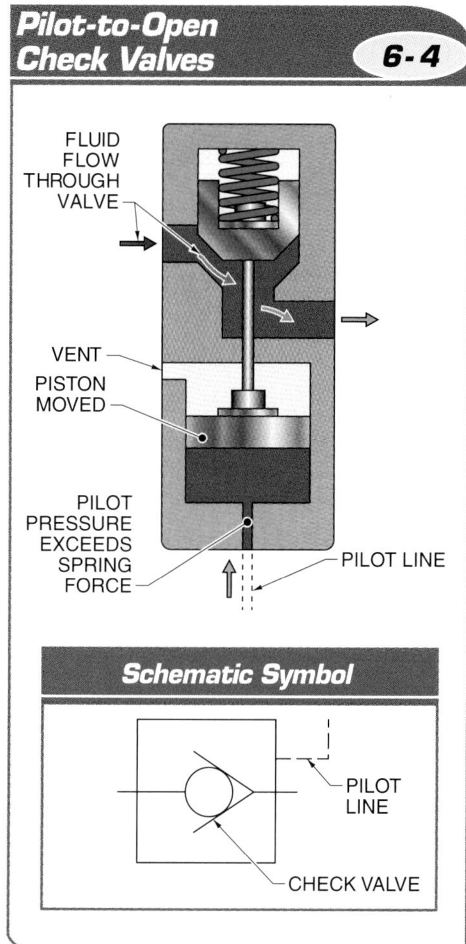

Figure 6-4. A pilot-to-open check valve allows fluid flow in one direction and stops fluid flow in the opposite direction until it receives a pressure signal through a pilot line applied to the piston port.

When the pilot pressure signal is received, a piston in the valve unseats the poppet, which allows fluid flow in both directions. When the pilot pressure signal is not present, the spring forces the piston to reseat the poppet. This valve has a drain line so the fluid that leaks by the piston does not stop the pilot-operated check valve from actuating. Common cracking pressures for pilot-to-open check valves are 4 psi, 30 psi, 75 psi, and 150 psi. The most common cracking pressure is 30 psi.

A pilot-to-open check valve is used in different applications. It can be used to hold a cylinder in place if there is a load on the cylinder. In this case, the system usually includes a double-acting cylinder and a three-position, four-way directional control valve with a tandem-center position.

When the directional control valve is in the center position, the piston and the load are locked and held in place by the check valve. When it is in the right-side position, the check valve allows the piston to retract. When it is in the left-side position, as soon as there is enough pressure in the line that is attached to the pilot line, the valve opens and the piston extends. The load will stay suspended as long as the seals in the check valve do not fail. **See Figure 6-5.**

This type of system can be used in applications that have a heavy load attached to the cylinder head, such as with metal shear press machines or large steam vents.

Pilot-to-Close Check Valves. A pilot-to-close check valve allows fluid flow in one direction. The valve is closed when the pilot pressure signal is sent and fluid flow is not allowed in either direction. Pilot-to-close check valves are typically used in applications that use a regenerative system to increase the speed of a cylinder, such as with braking systems on garbage trucks or grapples on a hydraulic crane. The best method to differentiate between a pilot-to-open and pilot-to-close check valve is to refer to the schematic symbol. **See Figure 6-6.**

Pilot-Operated Check Valve Installation. A pilot-operated check valve is usually designed as a cartridge. The cartridge connects to a standard valve body, which connects to the means of transmission through a subplate, stack valve system, or manifold. The three connections that must be made to the pilot-operated check valve include connections to the inlet port, outlet port, and pilot line. Some pilot-operated check valves require a fourth, smaller connection to a drain line.

A cartridge valve has an advantage over an in-line valve. If a cartridge valve fails, it can be replaced without disconnecting the entire valve body, stack valve, or manifold from the system.

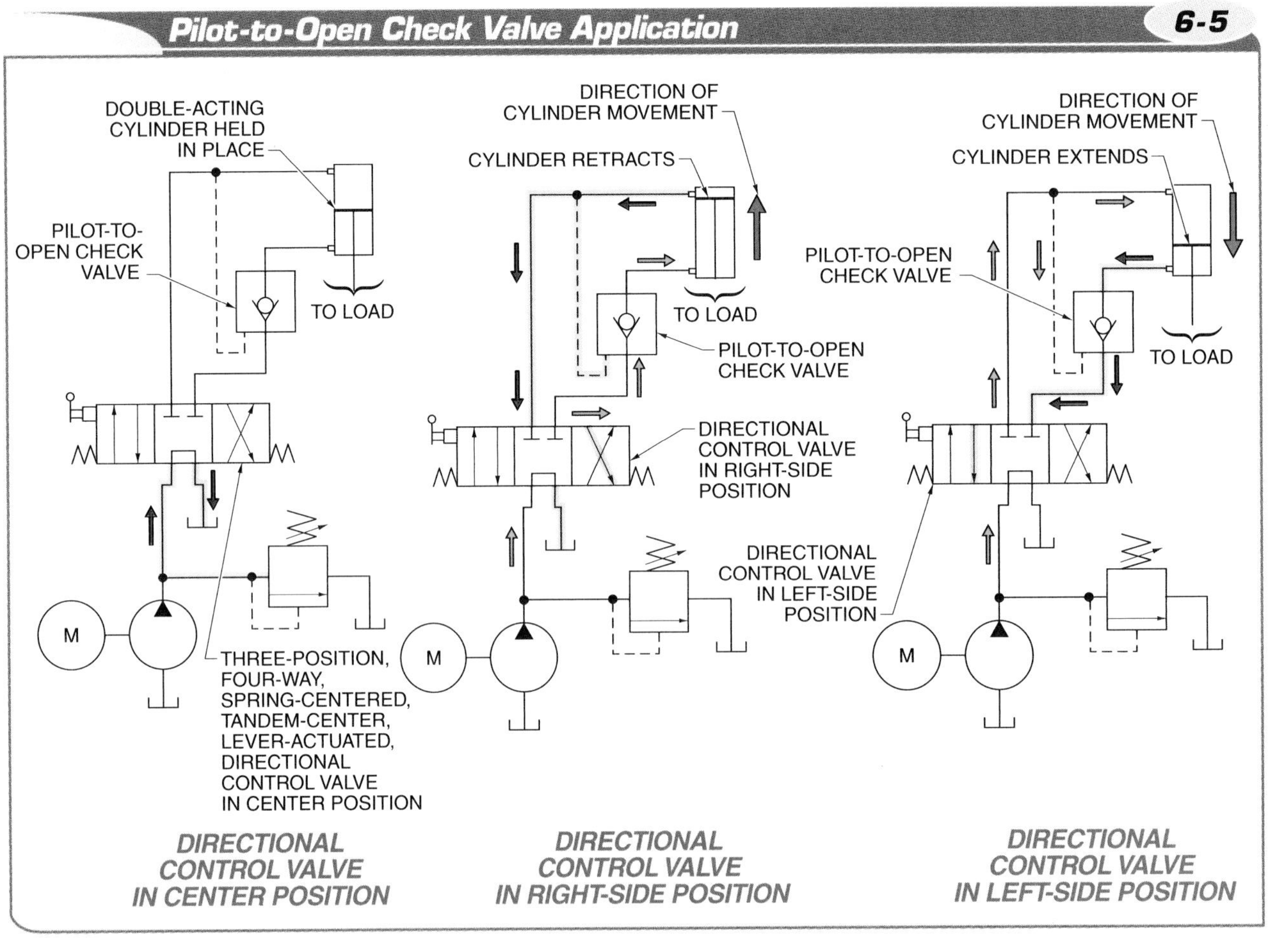

Pilot-to-Open Check Valve Application
6-5
DOUBLE-ACTING
CYLINDER HELD
IN PLACE
PILOT-TO-
OPEN CHECK
VALVE
TO LOAD
M
THREE-POSITION,
FOUR-WAY,
SPRING-CENTERED,
TANDEM-CENTER,
LEVER-ACTUATED,
DIRECTIONAL
CONTROL VALVE
IN CENTER POSITION
DIRECTIONAL
CONTROL VALVE
IN CENTER POSITION
DIRECTION OF
CYLINDER MOVEMENT
CYLINDER RETRACTS
TO LOAD
PILOT-TO-OPEN
CHECK VALVE
DIRECTIONAL
CONTROL VALVE
IN RIGHT-SIDE
POSITION
M
DIRECTIONAL
CONTROL VALVE
IN RIGHT-SIDE POSITION
DIRECTION OF
CYLINDER MOVEMENT
CYLINDER EXTENDS
PILOT-TO-OPEN
CHECK VALVE
TO LOAD
DIRECTIONAL
CONTROL VALVE
IN LEFT-SIDE
POSITION
M
DIRECTIONAL
CONTROL VALVE
IN LEFT-SIDE POSITION

Figure 6-5. A pilot-to-open check valve is often used to hold a double-acting cylinder in place when the cylinder has a load suspended on it.

Pilot-to-Close Check Valves
6-6
SPRING
POPPET
NO FLUID
FLOW
THROUGH
VALVE
PILOT LINE
PISTON
MOVED
VENT
Schematic Symbol
PILOT
LINE
CHECK VALVE

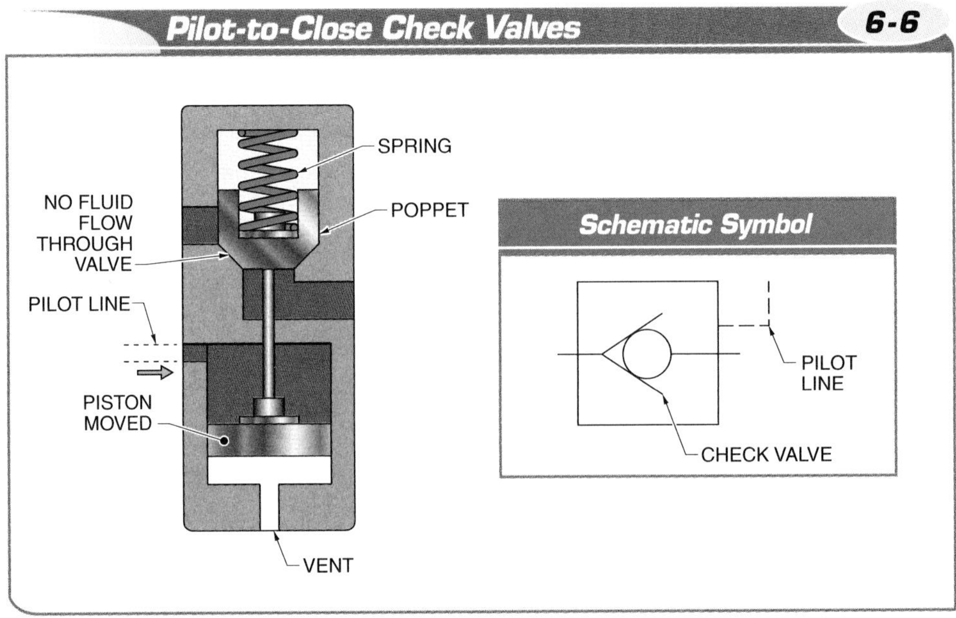

Figure 6-6. A pilot-to-close check valve allows fluid flow in one direction until the pilot signal is sent to close the valve.

DIRECTIONAL CONTROL VALVES

A *directional control valve* is a valve that allows or prevents fluid flow to specific piping or system actuators. A *spool* is an internal component of a directional control valve that is used to control fluid flow and to connect internal passages and ports. All directional control valves consist of a valve body, a spool or poppet, spool actuator and ports to connect the pump, the reservoir, and the actuator to the directional control valve. **See Figure 6-7.**

Directional control valves are available in many different designs. Common designs include two-position, two-way directional control valves; two-position, three-way directional control valves; two-position, four-way directional control valves; and three-position, four-way directional control valves. For proper operation, directional control valves must be correctly mounted within the hydraulic system. Schematic diagrams are used to depict the operation of each design.

Directional Control Valve Schematic Diagrams

Schematic diagrams of directional control valves are used for installation and troubleshooting purposes. Newly installed directional control valves typically have a schematic symbol printed or embossed on the valve body. It is important to have access to the original schematic diagrams so that operating characteristics of the design of the installed directional control valves can be determined if necessary.

A directional control valve schematic diagram consists of position boxes, valve ports, internal arrows, spool actuator symbols, and the name of the directional control valve design. **See Figure 6-8.**

TERMS

A **directional control valve** is a valve that directs or prevents fluid flow to specific piping or system actuators.

A **spool** is an internal component of a directional control valve that is used to control fluid flow and to connect internal passages and ports.

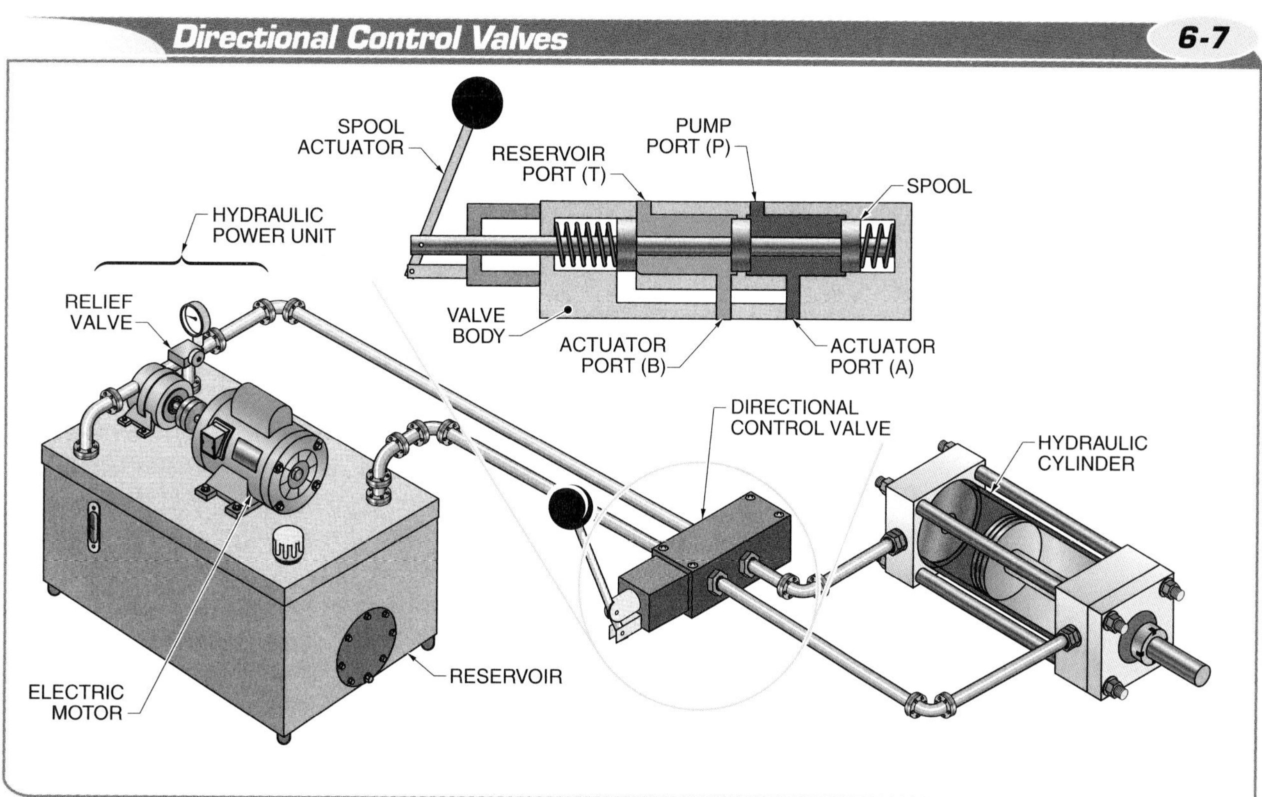

Figure 6-7. The primary function of a directional control valve is to direct or prevent fluid flow to specific piping or system actuators.

TERMS

A **position box** is a symbol that represents of the number of different positions that the spool of a directional control valve is capable of moving into.

A **position** is a schematic representation of the direction in which a spool forces fluid to flow.

A **way** is the number of piping ports in a directional control valve.

A **port** is the section of a directional control valve that connects a pipe, hose, or tube to the internal passages of the directional control valve.

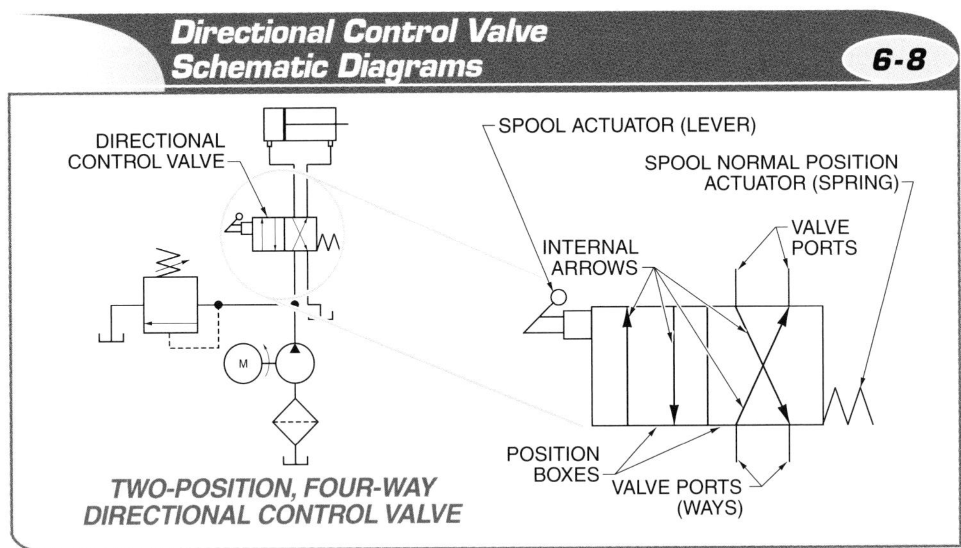

Figure 6-8. A schematic diagram can be used to determine the operating characteristics of a directional control valve.

Position Boxes. A *position box* is a symbol that represents of the number of different positions that the spool of a directional control valve is capable of moving into. Every directional control valve has at least two positions. A *position* is a schematic representation of the direction in which a spool forces fluid to flow. Positions on directional control valves are represented in schematic symbols by the number of position boxes.

Most directional control valves used in commercial and industrial applications usually have two, three, or four positions. They are also available with five and six positions, but these designs are usually used in special applications, such as large material handling equipment in steel mills or quarries.

Prince Manufacturing Corporation

Depending on the application, some directional control valves used in hydraulic systems can have more than one handle.

Ways. A *way* is the number of piping ports in a directional control valve. For example, a valve with two ports is a two-way valve. A *port* is the section of a directional control valve that connects a pipe, hose, or tube to the internal passages of the directional control valve. Every directional control valve has ports to provide the controlled entry and exit of fluid.

On a schematic symbol, valve port depictions serve two purposes. They depict the location of the port and they show the normal (or starting) position of the valve when the system is idle. The three different types of ports are pressure (P), tank (or reservoir) (T), and actuator (A and B). Port P receives fluid flow from the pump, port T returns fluid flow to the reservoir, and ports A and B allow fluid flow in and out of the actuator. Most directional control valves used in commercial and industrial applications are two-way, three-way, or four-way valves. **See Figure 6-9.**

TECH FACT

Five-way valves are sometimes used in mobile power equipment for steering control valves. An additional port is used for load sensing.

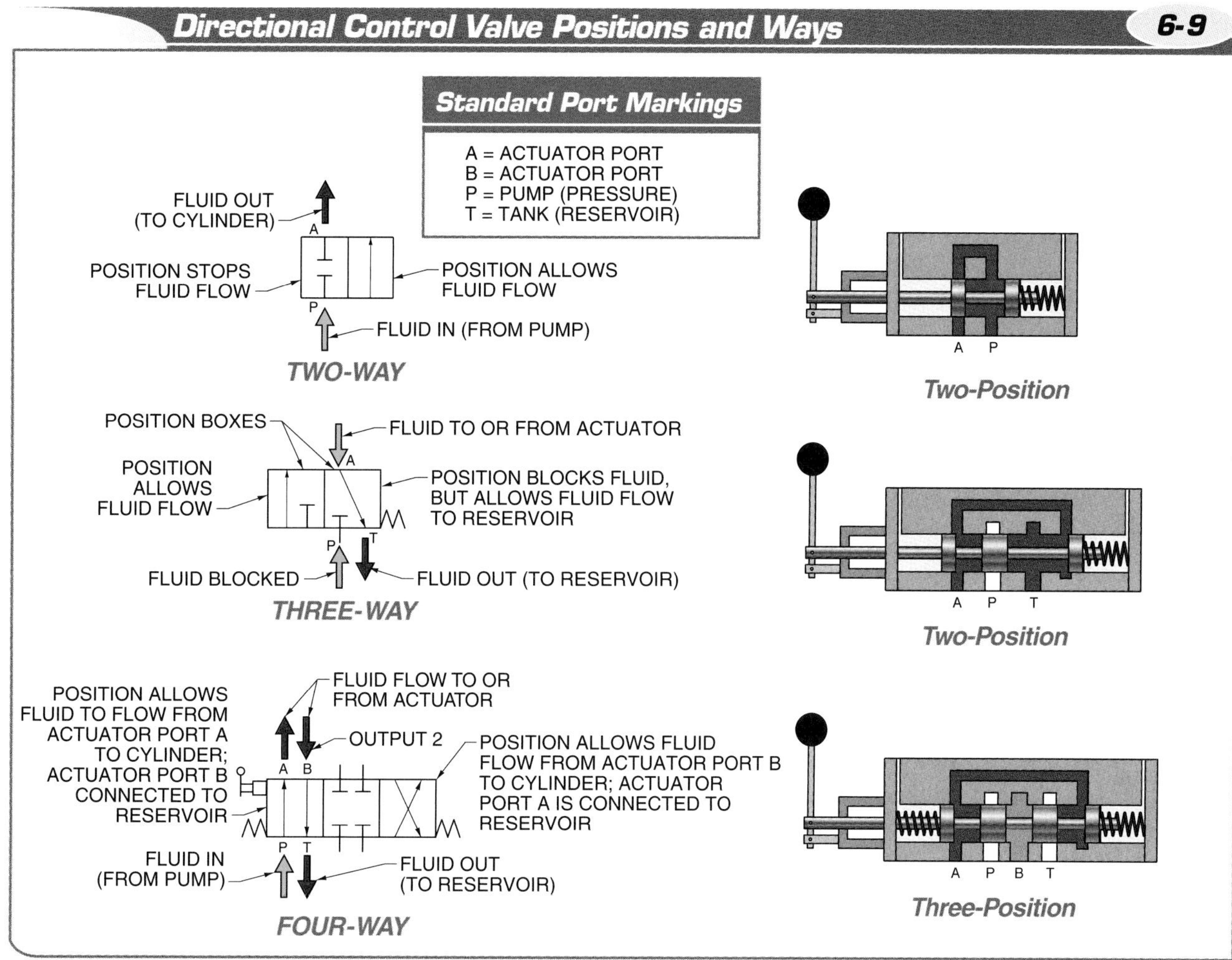

Figure 6-9. A way is the number of valve ports, and position is the number of different positions that a spool of a directional control valve is capable of moving into in a directional control valve.

Internal Arrows. In a directional control valve schematic symbol, internal arrows inside the position boxes depict the direction of fluid flow for each spool position. Internal arrows on hydraulic schematic diagrams are always colored solid.

Position boxes may have an internal T-shaped symbol that depicts where fluid is blocked by the spool in the valve. **See Figure 6-10.** In some directional control valves, fluid flow, or valve ports, are blocked when the spool is in a certain position.

Note: On pneumatic schematic symbols, arrows are sometimes outlined.

The direction of internal arrows on a schematic symbol can help a technician troubleshoot a hydraulic system. For example, if the system does not function correctly after the valve has been replaced, it may be because the new directional control valve does not have the same type of spool. The best troubleshooting method is to compare the schematic symbols of each directional control valve and verify whether or not the internal arrows match.

TECH FACT

Schematic symbols used with directional control valves indicate the function and methods of operation. These symbols make the function of directional control valves simple to understand and are not limited by language barriers when describing hydraulic systems.

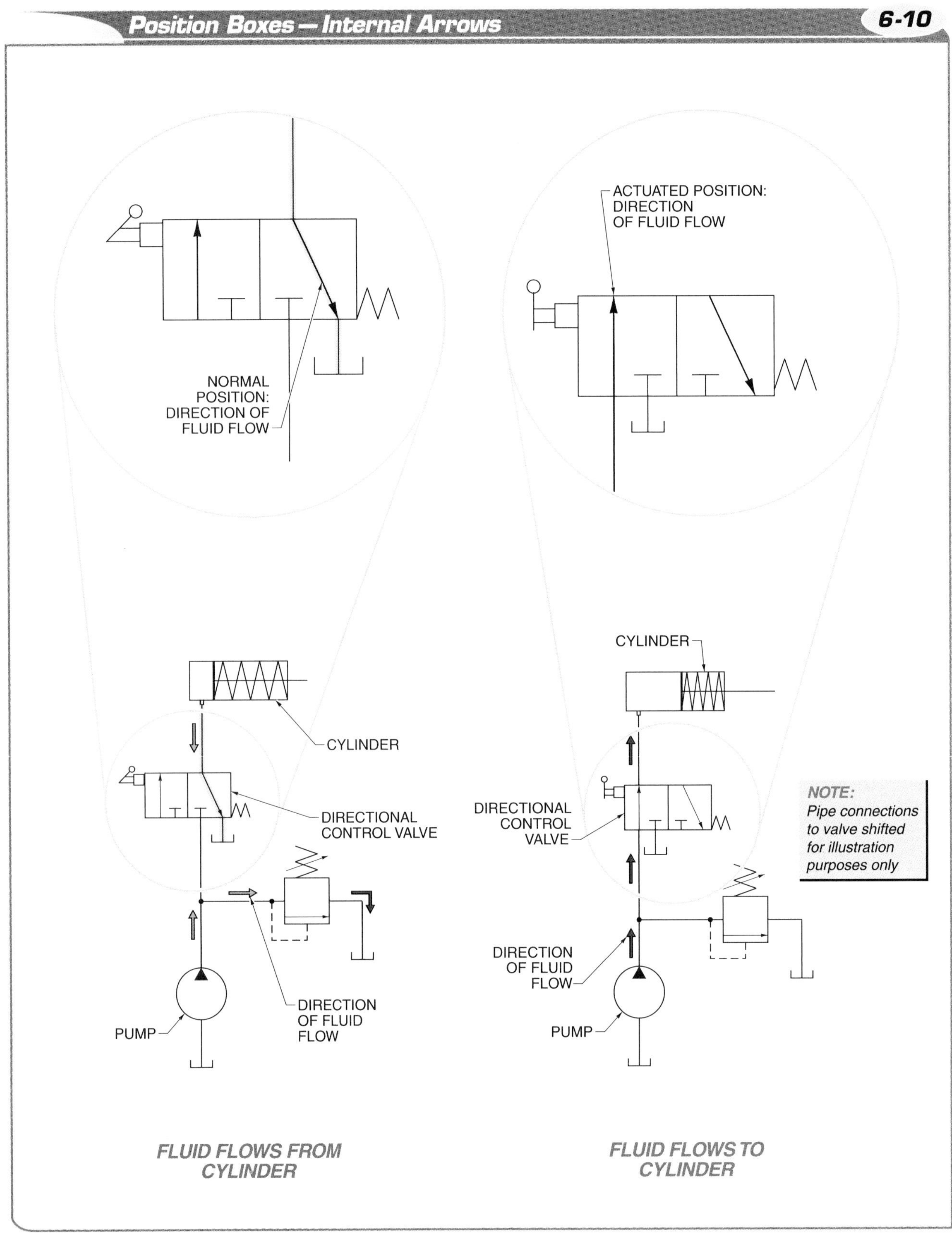

Figure 6-10. In a schematic diagram, internal arrows inside the position boxes depict the direction of fluid flow for each spool position.

Naming Standards. On a schematic symbol, naming standards are used to describe the directional control valve design. Naming standards include the number of positions, the number of ways, the types of directional control valve actuators, and the normal spool position. Schematic symbols are used to indicate the type of directional control valve actuator and the type of spool positioning method, where applicable. **See Appendix. See Figure 6-11.**

Normal Positions. Many two-position directional control valves are available as either normally open (NO) or normally closed (NC). A *normally open (NO) valve* is a valve that allows fluid flow from the pump port to an actuator port in the spring-actuated (normal) position. A *normally closed (NC) valve* is a valve that does not allow fluid flow from the pump port to an actuator port in the spring-actuated (normal) position. On a schematic symbol, a NO valve is indicated by arrows going from port P to an actuator port. An NC valve is indicated by arrows going from port P to a blocked port.

Two-Position, Two-Way Directional Control Valves

A *two-position, two-way directional control valve* is a directional control valve with two positions and two ports. One position allows fluid flow through the valve and the other position stops fluid flow completely. Two-position, two-way directional control valves are used in hydraulic systems to operate as ON/OFF switches. They can be used to control or isolate different parts of a hydraulic system. For example, they can turn on and off separate components of systems that are powered by the same hydraulic power unit. **See Figure 6-12.**

They can be either NO or NC. The normal position can be determined by locating the valve ports on a schematic diagram.

TECH FACT

When replacing a directional control valve, it is important to note the flow paths for each position of the valve so that it can be replaced with the same operating valve.

TERMS

A **normally open (NO) valve** is a valve that allows fluid flow from the pump port to an actuator port in the spring-actuated (normal) position.

A **normally closed (NC) valve** is a valve that does not allow fluid flow from the pump port to an actuator port in the spring-actuated (normal) position.

A **two-position, two-way directional control valve** is a directional control valve with two positions and two ports.

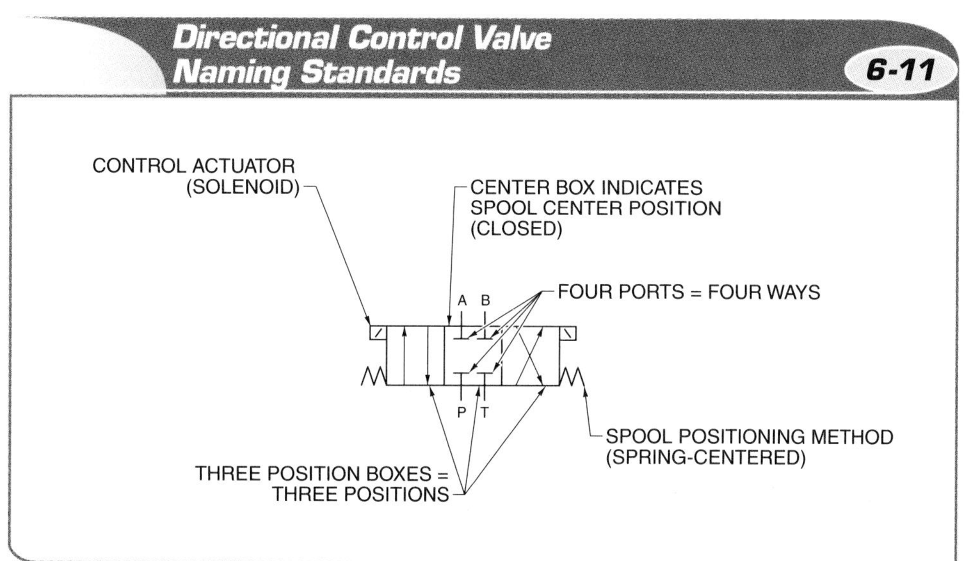

Figure 6-11. Naming standards include the number of positions, the number of ways, the types of directional control valve actuators, and the normal spool center position.

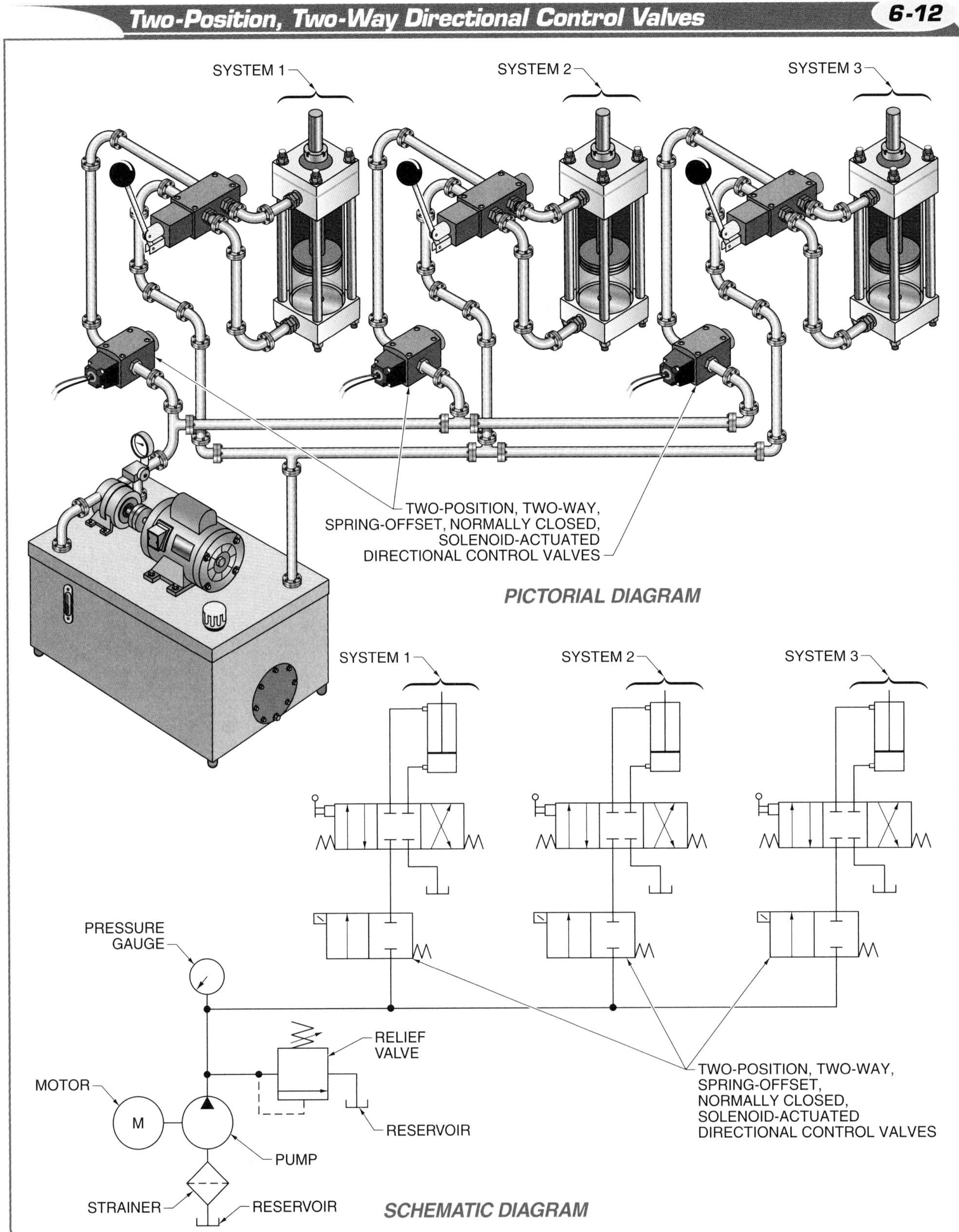

Figure 6-12. Two-position, two-way directional control valves can turn on and off separate hydraulic systems that are powered by the same hydraulic power unit.

Two-Position, Three-Way Directional Control Valves

A *two-position, three-way directional control valve* is a directional control valve that has two positions and three ports. A two-position, three-way directional control valve has a pressure port (P), a tank port (T), and an actuator port (A). One position allows fluid flow from port P to port A, and the other position allows fluid flow from port A to port T. **See Figure 6-13.**

Two-position, three-way directional control valves are usually used in applications with a spring-offset spool because spring pressure allows return fluid flow. Another common application for a two-position, three-way directional control valve is in a forklift or an automobile lift, where gravity can retract the cylinder after it has been extended.

TECH FACT

If there is a need for a two-position, three-way directional control valve in a hydraulic system but none are available, a two-position, four-way directional control valve can be used with port B plugged.

Two-Position, Four-Way Directional Control Valves

A *two-position, four-way directional control valve* is a directional control valve with two positions and four ports. **See Figure 6-14.** A two-position, four-way directional control valve has a pressure port (P), a tank port (T), and two actuator ports (A and B). In one position, the spool allows fluid flow from port P to port A, while also allowing fluid flow from port B to port T. In the other position, the spool allows fluid flow from port P to port B, while also allowing fluid to flow from port A to port T.

A two-position, four-way directional control valve can be used in different control systems. For example, it can be used to control the extension and retraction of a double-acting cylinder, to control the direction of rotation of a hydraulic motor, or to control cylinders that move heavy loads, such as cylinders in hydraulic automotive lifts, hydraulic cranes, scrap metal balers, and bridge lifts.

TERMS

A **two-position, three-way directional control valve** is a directional control valve that has two positions and three ports.

A **two-position, four-way directional control valve** is a directional control valve with two positions and four ports.

A **three-position, four-way directional control valve** is a directional control valve with three positions and four ports.

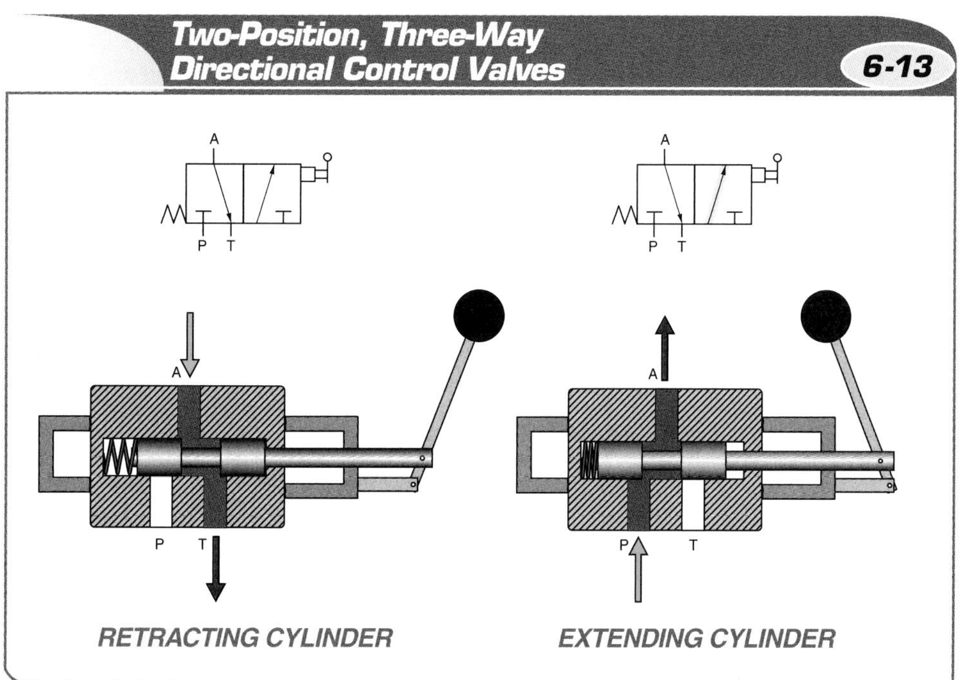

Figure 6-13. A two-position, three-way directional control valve has two positions, one to allow fluid flow from the cylinder port to the tank port and the other to allow fluid flow from the pump to the cylinder port.

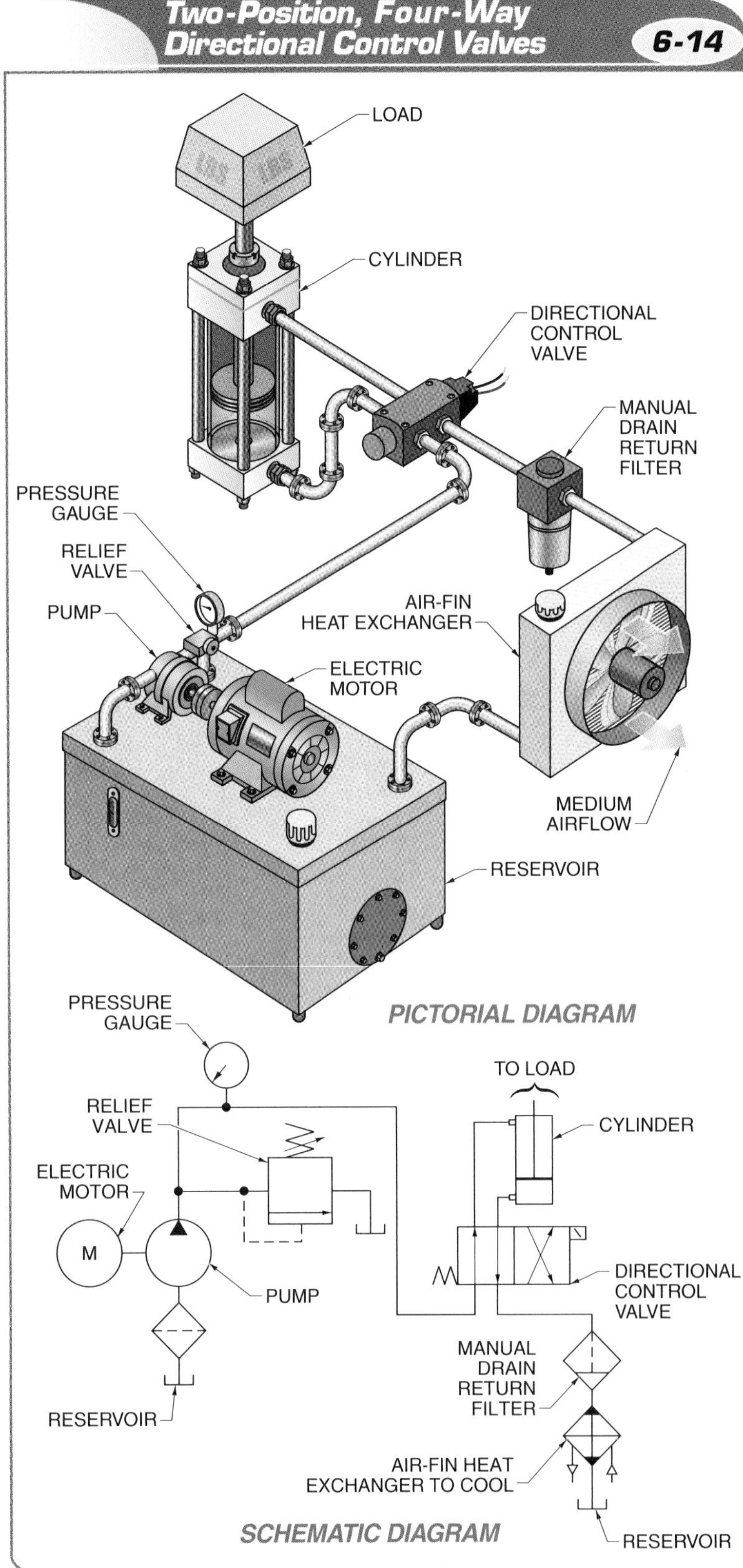

Figure 6-14. Two-position, four-way directional control valves are typically used to control cylinders that lift heavy loads.

Three-Position, Four-Way Directional Control Valves

A *three-position, four-way directional control valve* is a directional control valve with three positions and four ports. A three-position, four-way directional control valve has a pressure port (P), a tank port (T), and two actuator ports (A and B).

Two and three-way directional control valves only control one direction of fluid flow at a time, but three-position, four-way directional control valves can control two directions of fluid flow at a time. A three-position, four-way directional control valve is used to control the extension and retraction of a double-acting cylinder and the direction of rotation of a hydraulic motor.

The advantage of a three-position, four-way directional control valve is that it has four different center positions that can perform four different operations. The four center positions are open, closed, tandem, and float. **See Figure 6-15.** Each center position allows a hydraulic system to operate differently, depending on the application requirements. The center position is always the normal position, or the starting position, of the directional control valve when the system is idle.

Manufacturers design directional control valves so that spools with different designs spools with different center designs can be installed in a directional control valve body. The ability to change the directional control valve spool is an advantage because of reduced cost and less equipment downtime. For example, rather than replace the entire open-center position of a three-position, four-way directional control valve in a hydraulic lift, the spool can be changed to a closed-center position.

TECH FACT

Regenerative-center positions are used in more specialized hydraulic applications, such as production-press equipment cylinders that must retract as fast as possible.

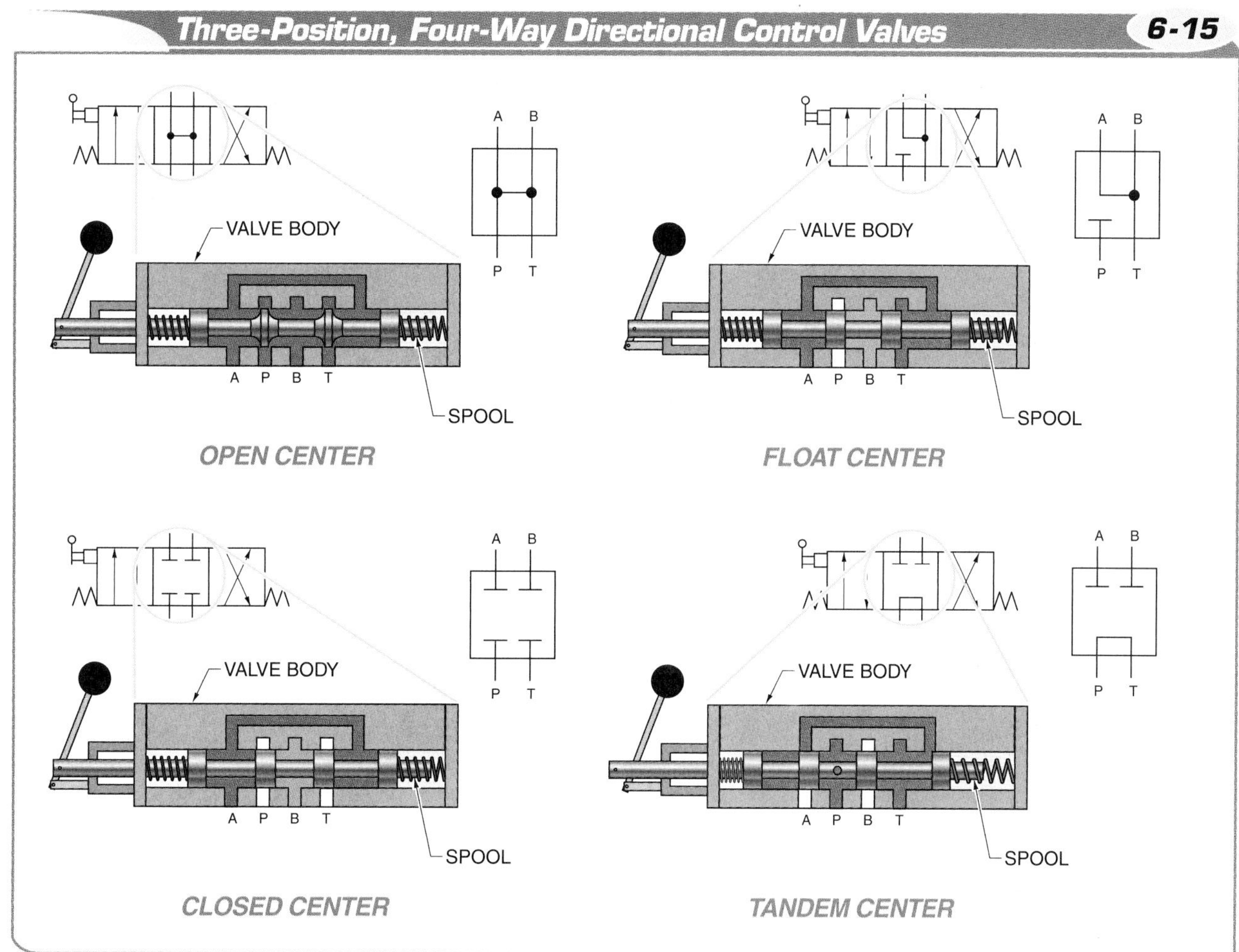

Figure 6-15. An advantage of three-position, four-way directional control valves is that they have four different center positions that can perform four different operations.

Open-Center Positions. The open-center position on a three-position, four-way directional control valve is the position where the pressure port (P), the tank port (T), and both actuator ports (A and B) are connected together. **See Figure 6-16.** The open-center position allows fluid to return to the reservoir when the system is idle, which prevents heat buildup and inefficiency in the system. Open-center positions are typically used in applications where there is only one cylinder in the hydraulic system.

Closed-Center Positions. The closed-center position on a three-position, four-way directional control valve is the position where the pressure port (P), tank port (T), and both actuator ports (A and B) are all closed off to each other. **See Figure 6-17.** In a closed-center position, the cylinder cannot move because there is no fluid flow to or from the cylinder. Fluid flow created by a variable-displacement, pressure-compensated pump either goes through a relief valve or is automatically reduced to no flow.

A closed-center position is used in industrial facilities that may have more than one hydraulic system but only one hydraulic power unit. A closed-center position also allows a hydraulic system to operate with more than one actuator and with each actuator operating independently of the others.

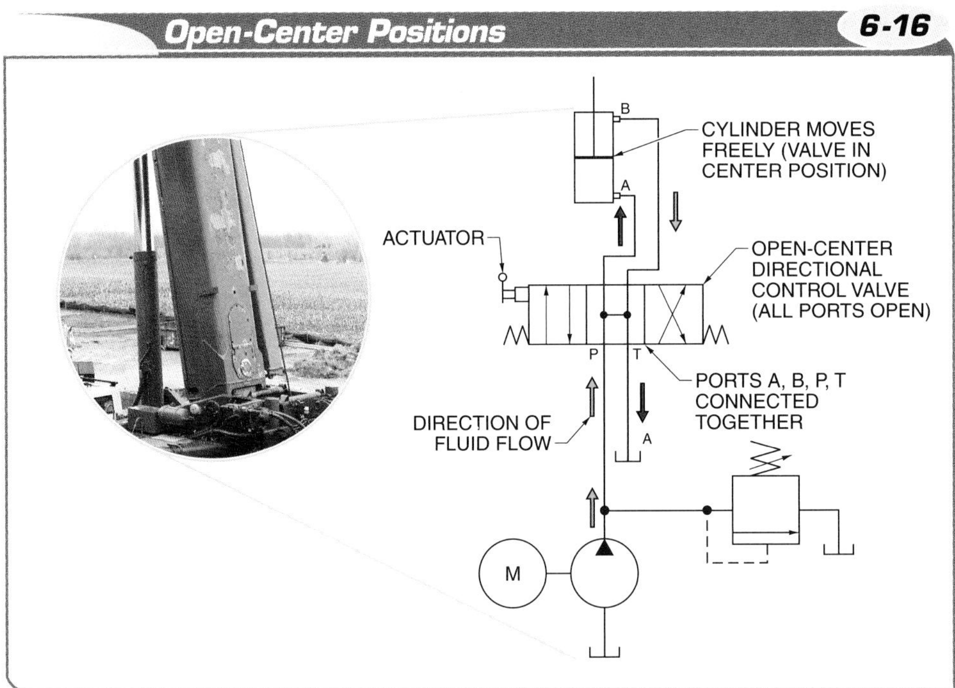

Figure 6-16. Open-center positions are typically used in applications where there is only one cylinder in the hydraulic system.

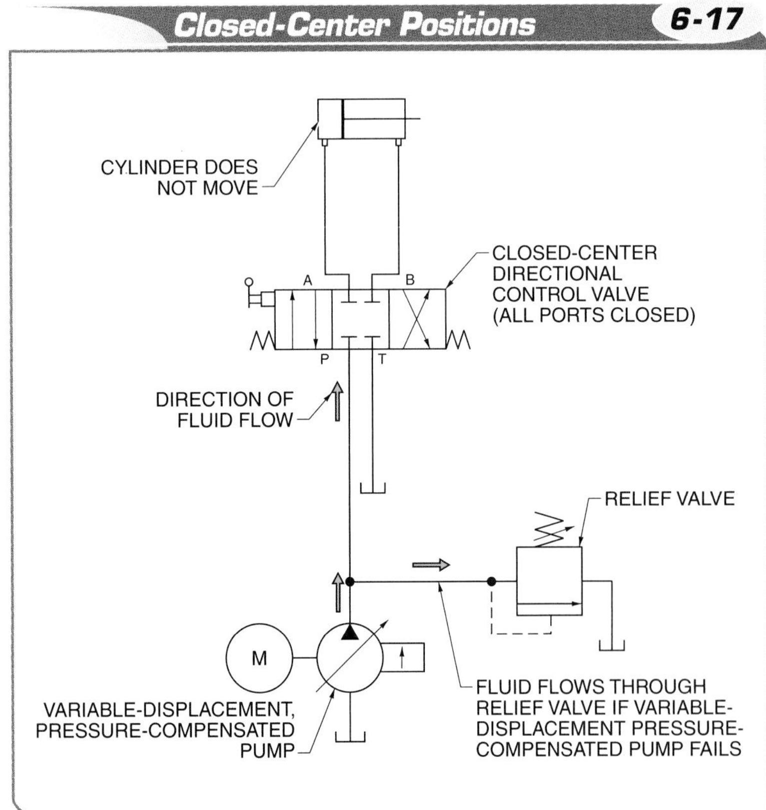

Figure 6-17. When using a closed-center position, fluid flow created by a variable-displacement, pressure-compensated pump either goes through a relief valve or is automatically reduced to no flow.

When using the closed-center position with a fixed-displacement pump, the relief valve must be in good working condition. Relief valves should be tested whenever there is downtime. Also, any routine or preventive maintenance that is recommended by the manufacturer should be performed.

If the relief valve fails, unlimited resistance to fluid flow is created by the closed-center position, causing a hazardous working condition. As the pump works to overcome the unlimited resistance, pressure will increase until the hydraulic system fails. A failed relief valve will cause damage to hydraulic system devices and create a hazard to personnel in the form of overpressurized hydraulic equipment that may explode. A relative valve is used as additional protection for variable-displacement pumps with a closed center.

TECH FACT

Although a closed-center position stops all fluid flow through a directional control valve, there can be residual fluid leakage that can cause the cylinder to drift if a worn directional control valve is held in the center position for too long.

Tandem-Center Positions. The tandem-center position on a three-position, four-way directional control valve has the pressure port (P) and the tank port (T) connected to each other, with both of the actuator ports (A and B) blocked. Tandem-center positions are used to hold the piston in a cylinder in the desired position while allowing fluid flow to be directed back to the reservoir without the need to activate a relief valve. **See Figure 6-18.**

The tandem-center position is used to hold the cylinder in a stable position while fluid flows back to the reservoir from the pump under 150 psi, depending on the distance between the pump and the valve and the resistance to fluid flow in the piping. Because the pump is operating under low pressure, fluid flowing directly to the reservoir prevents heat buildup in the system and can increase the life of the pump and piping. This is sometimes referred to as unloading with a tandem center. A tandem-center position can be used if there are cylinders connected in a series, as with many mobile hydraulic systems such as forklift trucks or hydraulic cranes.

Float-Center Positions. The float-center position on a three-position, four-way directional control valve has the two actuator ports (A and B) connected to the tank port (T). The float-center position of a directional control valve allows a cylinder to move freely while the valve is in the center position and still directing fluid back to the reservoir. Fluid flow is stopped at the pump port (P). **See Figure 6-19.**

Float-center position directional control valves are common for controlling a hydraulic motor because they can slow the motor down gradually rather than bringing it to an abrupt stop when changing directions. For example, when a hydraulic motor is being used as a fan, a float-center position is commonly used to help the motor slow down before it stops. Sometimes the float-center position is in mobile hydraulic valves as the fourth position.

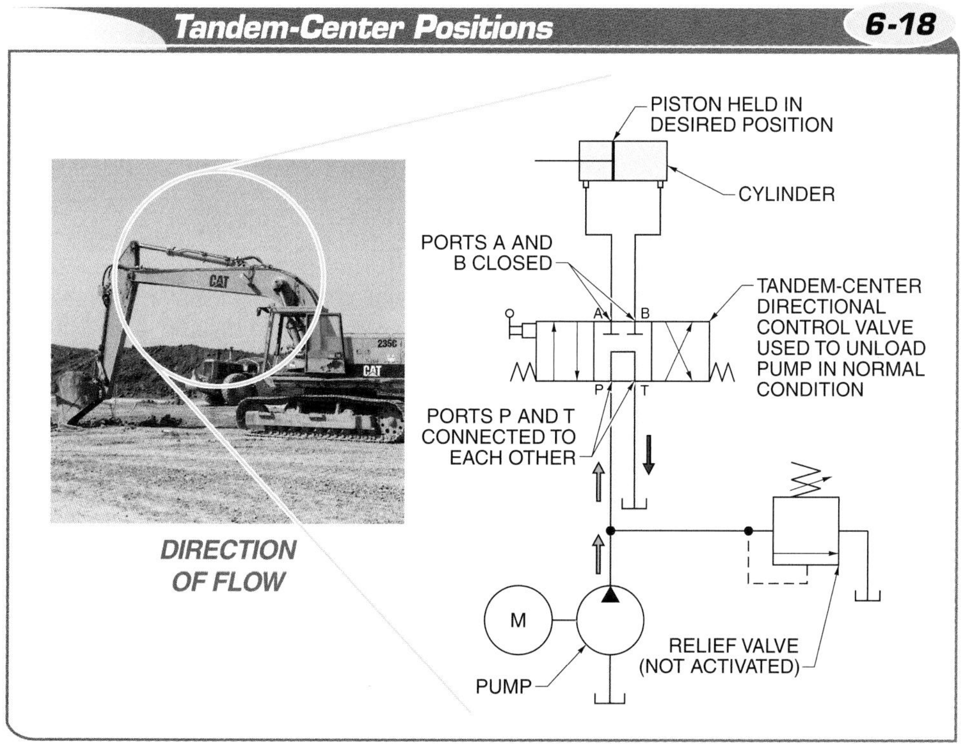

Figure 6-18. Tandem-center positions are used to hold a cylinder in the desired position while allowing fluid flow to be directed back to the tank without the need to activate a relief valve.

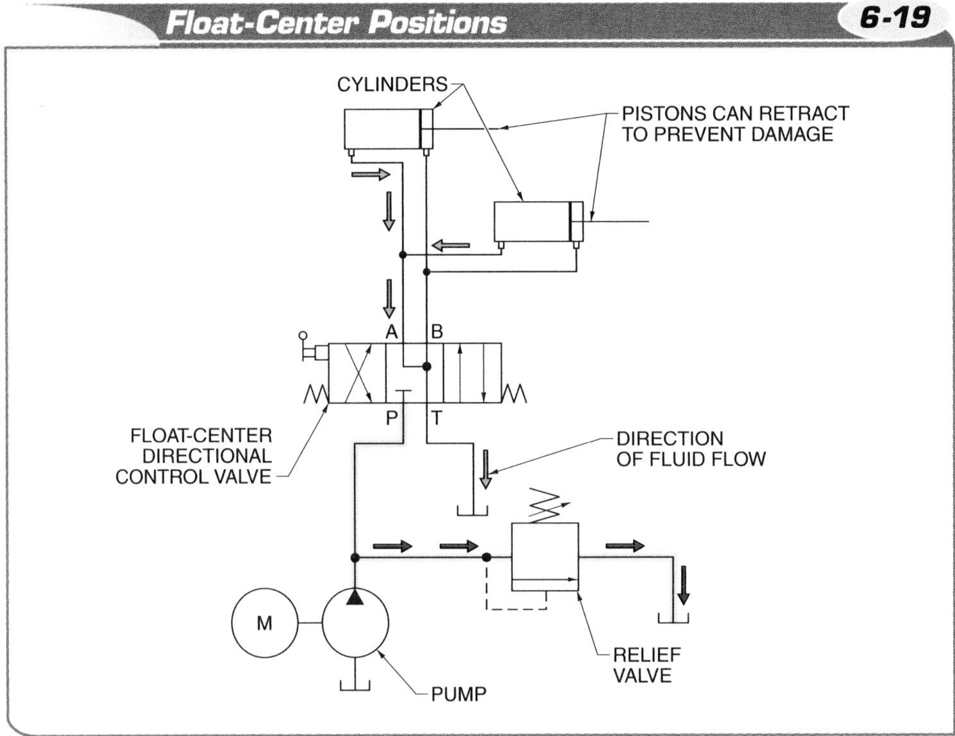

Figure 6-19. Float-center positions allow the piston in a cylinder to move while it is in the center position and can still direct fluid flow back to the tank.

Center Position Selection. There are two variables that can help determine the type of center position a directional control valve requires. The first variable is the type of pump that is installed in the system. If the system has a variable-displacement, pressure-compensated pump, a closed-center position or a float-center position is used because the pump automatically lowers the system flow while maintaining set pressure when in the center position. If the system does not have a variable-displacement, pressure-compensated pump, an open-center position or a tandem-center position can be used so that the hydraulic fluid can be returned to the reservoir.

The second variable is the position that the cylinder needs to be in when the directional control valve is in the center position. When in the float-center position or open-center position, the piston can move. When in the tandem-center position or closed-center position, the piston cannot move.

When selecting which center position to use in a hydraulic system, there are several other variables that should also be considered, including the type of pump, type of cylinder, type of directional control valve actuator, size of the system, system pressure requirements, the size of the means of transmission, and the mounting conditions.

When ordering a specific directional control valve from a specific manufacturer, it is important to understand their ordering codes. Every manufacturer has a different coding system used for ordering their directional control valves.

TECH FACT

Directional control valves can shift instantly from fully open to fully closed, causing the velocity of hydraulic fluid to rapidly accelerate and decelerate. Under certain conditions, a change in velocity can cause fluid hammer, which sounds like a hammer striking the inside of a pipe or reservoir.

Directional Control Valve Mounting Methods

The two methods of mounting a directional control valve that are commonly used in hydraulic systems are subplate mounting and in-line mounting. *Subplate mounting* is a mounting method where a directional control valve is mounted to a plate that attaches to system piping. **See Figure 6-20.**

Subplate Mounting 6-20

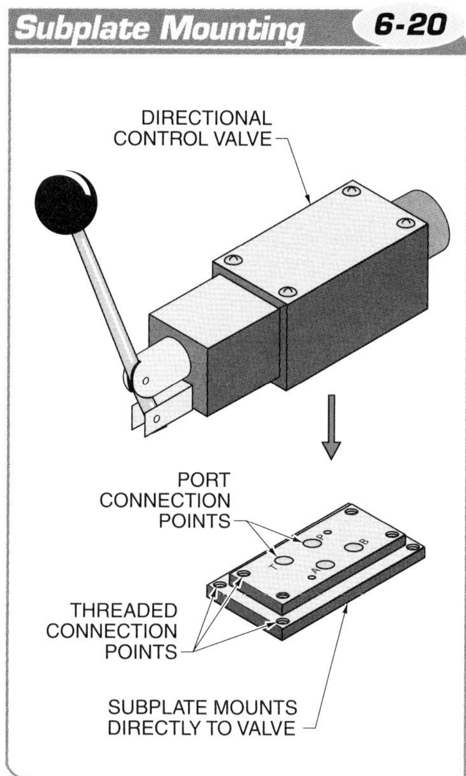

Figure 6-20. Subplate mounting allows for a directional control valve to be mounted to a plate that attaches directly to system piping.

Most subplate designs are available with two different types of mounting holes. One type of mounting hole allows the subplate to be mounted securely within the hydraulic system, while the other type has screw threads that are used to attach the directional control valve body to the subplate.

The subplate has valve ports through which fluid flows to and from the directional control valve. These valve ports have an O-ring that prevents leakage. The bottom or the side of the subplate has threaded connections for the means of transmission and is usually labeled to indicate what the port connects to on the valve. Typically, a threaded connection is attached to a metal fitting that allows for easier connection to a hose, tube, or pipe. Although subplate mounting is more costly than screwing the pipe into threads in the valve body, it does have advantages such as easy replacement. When a valve needs to be replaced, it can be done so in a timely manner, which quickly allows the machine back into service. In addition, the valve body will not be distorted by screwing a pipe directly into it and leaks around the threaded connections are eliminated.

Subplates are available in common sizes and styles that are standardized by ANSI, in collaboration with the International Organization for Standardization (ISO), and the National Fluid Power Association (NFPA). The seven different sizes of subplates are D02, D03, D05, D05H, D07, D08, D10. Each size of subplate has different designs for the location of ports to the directional control valve. **See Appendix.**

A *manifold subplate* is a subplate that can be used for mounting multiple directional control valves of the same configuration. The advantage of a manifold subplate is that several directional control valves can be mounted and connected in a smaller space. Also, only one pressure port and one tank port need to be connected into the subplate to allow all of the directional control valves to receive and return fluid flow.

Manifold subplates reduce the amount of space and piping that a hydraulic system requires. They are commonly used where several cylinders are controlled in close proximity, and can connect up to 16 cylinders for use in a hydraulic system. Although manifold subplates are available in standard sizes, they are usually custom-made.

The other type of mounting method is in-line mounting. *In-line mounting* is a mounting method that uses tubing or threaded pipes that connect directly to the directional control valve. In some applications, the directional control valves are attached to a secure location, but in other applications, the piping holds them in place.

TERMS

Subplate mounting is a mounting method where a directional control valve is mounted to a plate that attaches to system piping.

A **manifold subplate** is a subplate that can be used for mounting multiple directional control valves of the same configuration.

In-line mounting is a mounting method that uses tubing or threaded pipes that connect directly to the directional control valve.

TERMS

A **directional control valve actuator** is a mechanism that is used to move the position of the spool in a directional control valve.

A **manual actuator** is a type of directional control valve actuator that requires a person for actuation.

An **automated actuator** is a type of directional control valve actuator that can be actuated without a person being involved.

A **spring actuator** is a mechanism that uses one or more springs to move a directional control valve spool to normal position.

In-line mounting is used with valves that are located in areas away from other hydraulic components. When a valve is mounted in-line, unions should be used on all pipes going to or from the valves. This allows the valve to be replaced if it fails or malfunctions. In-line mounting is not as common as subplate mounting, but it is often used with other types of hydraulic valves.

DIRECTIONAL CONTROL VALVE ACTUATORS

A *directional control valve actuator* is a mechanism that is used to move the position of the spool position in a directional control valve. There are several common actuators used with directional control valves. All directional control valve actuators are used to move the spool into the proper position per the application requirements, but they all do it differently.

Prince Manufacturing Corporation

Some designs of directional control valves have a joystick as the actuating mechanism.

Directional control valve actuators are categorized as manual and automated. A *manual actuator* is a type of directional control valve actuator that requires a person for actuation. Manual actuators include pushbuttons, hand levers, foot pedals, and manual overrides. Manual actuators are typically used in applications such as tow-truck pulley systems, trash compactors, and hydraulic automobile lifts.

An *automated actuator* is a type of directional control valve actuator that can be actuated without a person being involved. Automated actuators are typically used in applications such as in hydraulic punch presses on production manufacturing lines.

While some directional control valves are either automated or manual, certain valve designs use both automatic and manual actuators to position the spool. Directional control valves that have a combination of automated and manual actuators are typically used in specific applications, such as in a wheelchair lift on a bus. For example, if a solenoid-controlled actuator malfunctions, a manual override mechanism can be actuated and the lift can complete its cycle. Automated actuator control methods include spring, pilot-pressure, solenoid, detent (mechanical) actuators, and proportional actuators. **See Appendix.**

Spring Actuators

A *spring actuator* is a mechanism that uses one or more springs to move a directional control valve spool to normal position. **See Figure 6-21.**

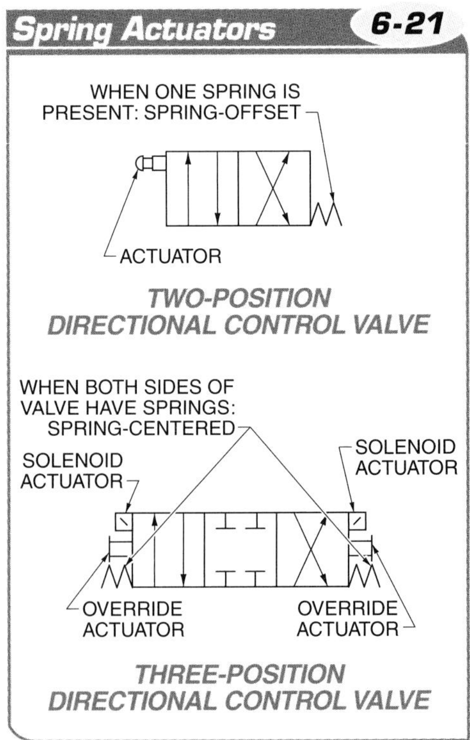

Figure 6-21. Spring actuators are used to shift the spool of a directional control valve to normal position.

In a two-position directional control valve, during system idle or startup, the spring is placed on one side to make sure that the spool returns to its normal position when the valve actuator is completely deactivated. On a three-position directional control valve, both sides of the spool have springs to center it for normal position. These are used to center the spool because the center position of a spring-centered directional control valve is always the normal position.

Pilot Pressure Actuators

A *pilot pressure actuator* is a mechanism that uses fluid pressure to shift a directional control valve spool. **See Figure 6-22.** *Pilot pressure* is a valve control method that uses fluid pressure from somewhere in the hydraulic system to control or shift a valve. Pilot pressure is commonly used to control hydraulic systems because it does not require electrical, mechanical, or manual actuation.

TERMS

A **pilot pressure actuator** is a mechanism that uses fluid pressure to shift a directional control valve spool.

Pilot pressure is a valve control method that uses fluid pressure from somewhere in the hydraulic system to control or shift a valve.

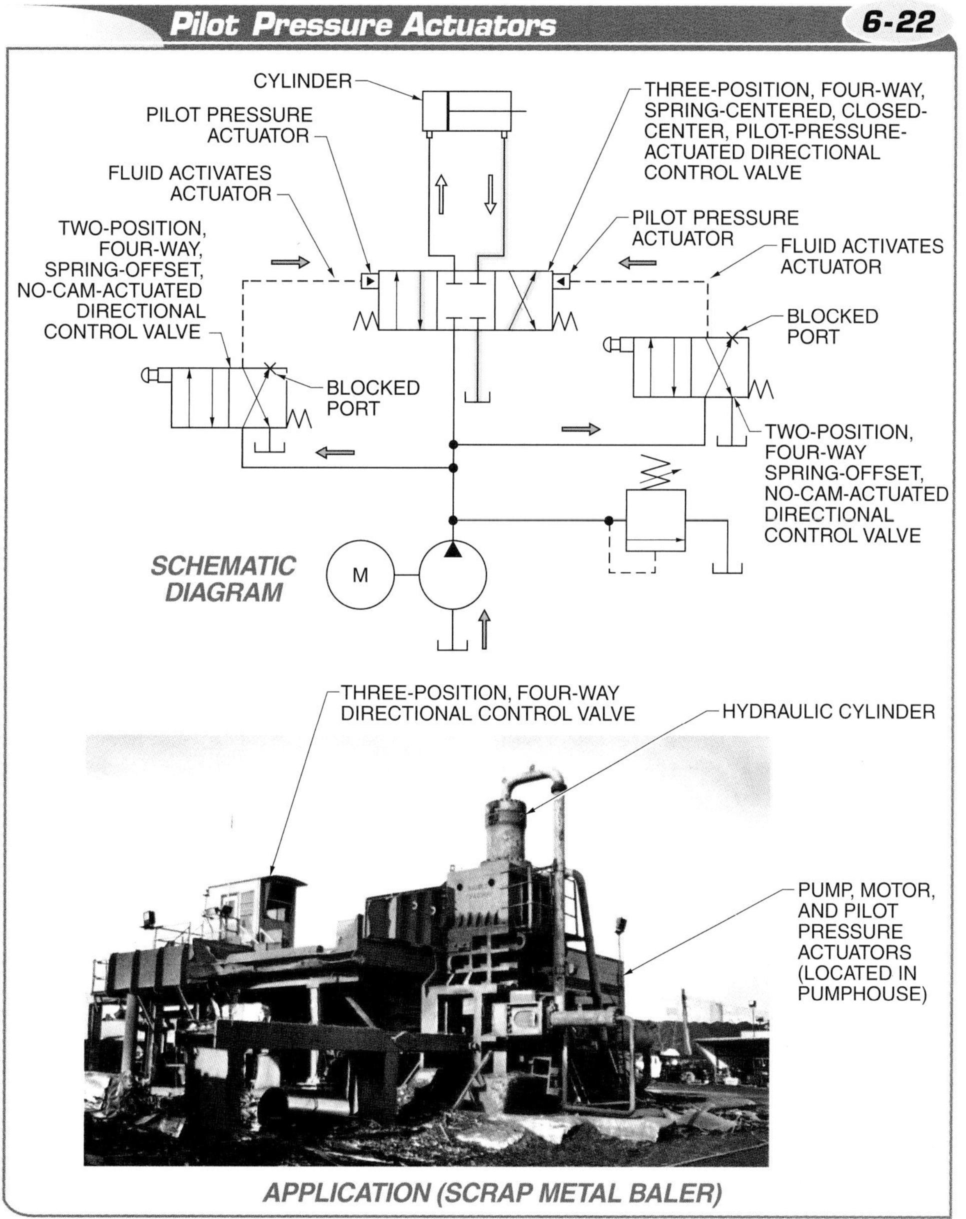

Figure 6-22. A pilot pressure actuator uses fluid pressure to actuate a directional control valve spool.

TERMS

A **solenoid** is an electrical control device that converts electrical energy into linear mechanical energy when a current passes through a magnetic coil in the solenoid.

A **solenoid actuator** is an actuator that uses electricity to actuate a directional control valve spool.

A **piggyback valve** is a combination valve that consists of a small solenoid-actuated directional control valve that is used to control the main pilot-operated valve.

A **detent lock** is an actuator that is used to hold a directional control valve spool in selected positions.

An **automatic detent** is a detent that releases a spool to its original position when force at the detent reaches a certain point.

The advantage to using pilot pressure is that it provides a means of actuation that can shift a spool in the directional control valve when there is a large amount of fluid flow in the system. Common applications for pilot pressure actuators are on the large industrial cylinders that are used to compress steel in scrap metal balers, cut steel bars in shear press machines, and shred metal in industrial shredders.

Solenoid Actuators

A *solenoid* is an electrical control device that converts electrical energy into linear mechanical energy when a current passes through a magnetic coil in the solenoid. Solenoids are used to control one circuit with another and consist of a hollow coil that contains a metal rod to open or close electrical contacts. A *solenoid actuator* is an actuator that uses electricity to actuate a directional control valve spool. Solenoid actuators are used when electronic control of a hydraulic system is desired.

When electricity is supplied to a solenoid, it produces a magnetic field and turns its coil into an electromagnet. The coil in a solenoid-actuated directional control valve is attached to a plunger. The plunger is pulled into the electromagnetic field and moves the spool into the desired position.

In hydraulic systems, solenoid actuators are used to control fluid flow in two-position valves or to control the position of the spool in three-position valves. Directional control valves can be single- or double-solenoid actuated. **See Figure 6-23.** Solenoid actuators are typically used in newer hydraulic equipment, such as hydraulic presses, stamping machines, large paper cutters, and steel coil rollers.

TECH FACT

Solenoids operate similar to mechanical actuators because they work against a pushpin, which is sealed to prevent external oil leakage. Solenoids are also designed in a manner that dissipates heat created by the current in the electrical conductors that form the internal coil.

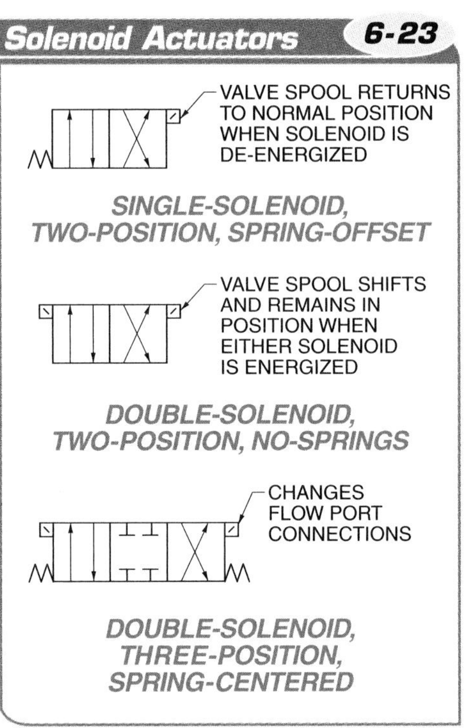

Figure 6-23. Directional control valves can be single- or double-solenoid actuated.

Note: Solenoid-actuated directional control valves should always be mounted horizontally to help prevent the valve spool from drifting out of the normal position.

Solenoid-Actuated, Pilot-Operated Directional Control Valves (Piggyback Valves). Although solenoids are efficient and quick acting, they have limited strength and are typically only used in directional control valves that have small spools. To overcome the limited strength, combination valves such as piggyback valves are used.

A *piggyback valve* is a combination valve that consists of a small solenoid-actuated directional control valve that is used to control the main pilot-operated valve. Typical pilot operation consists of a piggyback design where a smaller solenoid-actuated directional control valve is positioned on top of the main valve. When shifting is required, fluid flow from a piggyback valve is directed as pilot pressure to either side of the main directional control valve spool. **See Figure 6-24.**

Solenoid-Actuated, Pilot-Operated Directional Control Valves 6-24

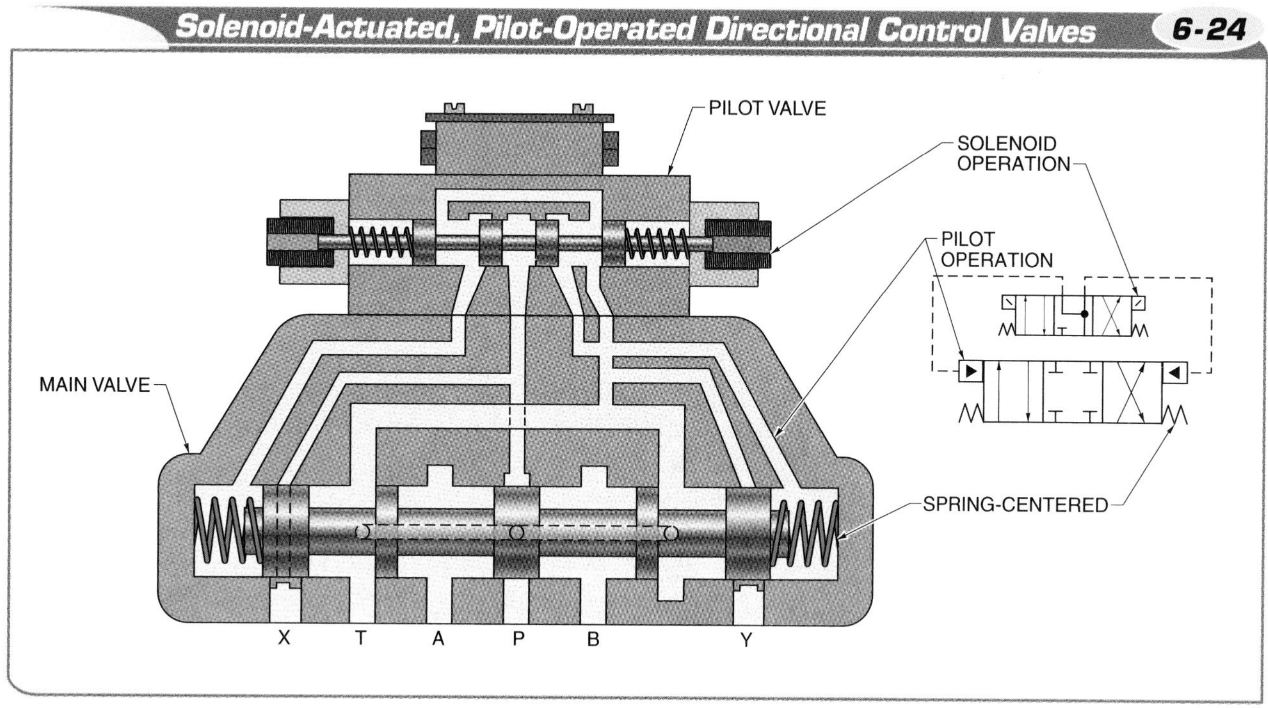

Figure 6-24. When shifting is required, fluid flow from a piggyback valve is directed as pilot pressure to either side of the main directional control valve spool.

System pressure or pilot pressure is often used in combination with solenoids because many directional control valves require more pressure to operate the spool than a solenoid can produce. For example, when fluid flow in a hydraulic system is greater than 25 gpm, solenoids alone are not strong enough to move the spool.

Piggyback valves are used to account for the weight of the main spool in large valve designs. Solenoid-actuated, pilot-operated directional control valves are used when fluid flow is more than 25 gpm and solenoid control is required.

A solenoid-actuated, pilot-operated directional control valve operates by allowing the solenoids to control pilot signals that move the main valve's spool. For example, when the solenoid receives an electrical signal, it shifts the spool away from the center position. Pilot pressure at the pressure port of the pilot valve is allowed to flow through its actuator port to one side of the spool in the main valve. This shifts the main spool and allows system pressure to flow from the main valve's pressure port to its actuator port and actuate the cylinder or motor.

Solenoid-actuated, pilot-operated directional control valves are commonly used in industry to control hydraulic systems with electrical controls. The two actuator ports from the pilot-operated directional control valve are attached to a larger directional control valve and used to control the movement of the spool. A typical application for a solenoid-actuated, pilot-operated directional control valve is in automatic hydraulic compression presses of 500 t and larger.

Pilot Chokes. A *pilot choke* is an assembly that is mounted between a main valve and a pilot valve that is used to slow or briefly delay reversals in spool movement. Solenoid-actuated, pilot-operated directional control valves are available with a pilot choke that can control the speed at which the main spool shifts in a two-stage valve. This prevents quick reversal of the actuator, preventing pressure spikes and premature component wear.

TERMS

A **pilot choke** is an assembly that is mounted between a main valve and a pilot valve that is used to slow or briefly delay reversals in spool movement.

TERMS

A **pilot piston** is a device that allows a directional control valve spool to shift quickly by placing the pilot piston in a pilot-pressure chamber near the spool in the main valve.

A **detent** is an actuator that is used to hold a directional control valve spool in selected positions.

An **automatic detent** is a detent that releases a spool to its original position when force at the detent reaches a certain point.

A **programmable logic controller (PLC)** is a solid-state control device that can be programmed and reprogrammed to automatically control electrical systems in commercial and industrial facilities.

A **proportional actuator** is a type of actuator that uses analog signaling to actuate a proportional directional control valve with a spool that can be placed in an infinite number of positions.

Pilot chokes can be internal or external. Internal pilot chokes are sometimes incorporated into the main valve or inserted into the pressure port on the main valve. This type of control is typically fixed so it cannot be used to adjust fluid flow. External pilot chokes are stacked between the main valve and the pilot valve and meter the fluid in the pilot lines. External pilot chokes typically allow fluid flow to be adjusted. **See Figure 6-25.** *Note:* Meter-out control is the most common type of control.

Pilot Pistons. Another option available in solenoid-actuated, pilot-operated directional control valves is a pilot piston. A *pilot piston* is a device that allows a directional control valve spool to shift quickly by placing the pilot piston in a pilot-pressure chamber near the spool in the main valve. Pilot pistons are typically small pistons placed at the end of the pilot line from the pilot-operated directional control valve and are attached to the spool. Since the pilot piston has a smaller surface area than the spool, it takes less fluid volume to move the spool, which increases the speed of spool shifting. **See Figure 6-26.**

Detent Actuators

A *detent* is an actuator that is used to hold a directional control valve spool in selected positions. A detent is typically used in hydraulic applications where the spool must be held in a certain position, such as in elevators or log splitters. A detent operates by locking the valve spool in a desired position and not allowing it to move back into its normal position until it is manually moved. **See Figure 6-27.**

An *automatic detent* is a detent that releases a spool to its original position when force at the detent reaches a certain point. Automatic detents are known as kick-out or pressure-release detents. The pressure point at which an automatic detent releases can be fixed or variable, depending on the type of valve installed in the system. Automatic detents are commonly used in forestry equipment, waste recycling facilities, and agricultural equipment.

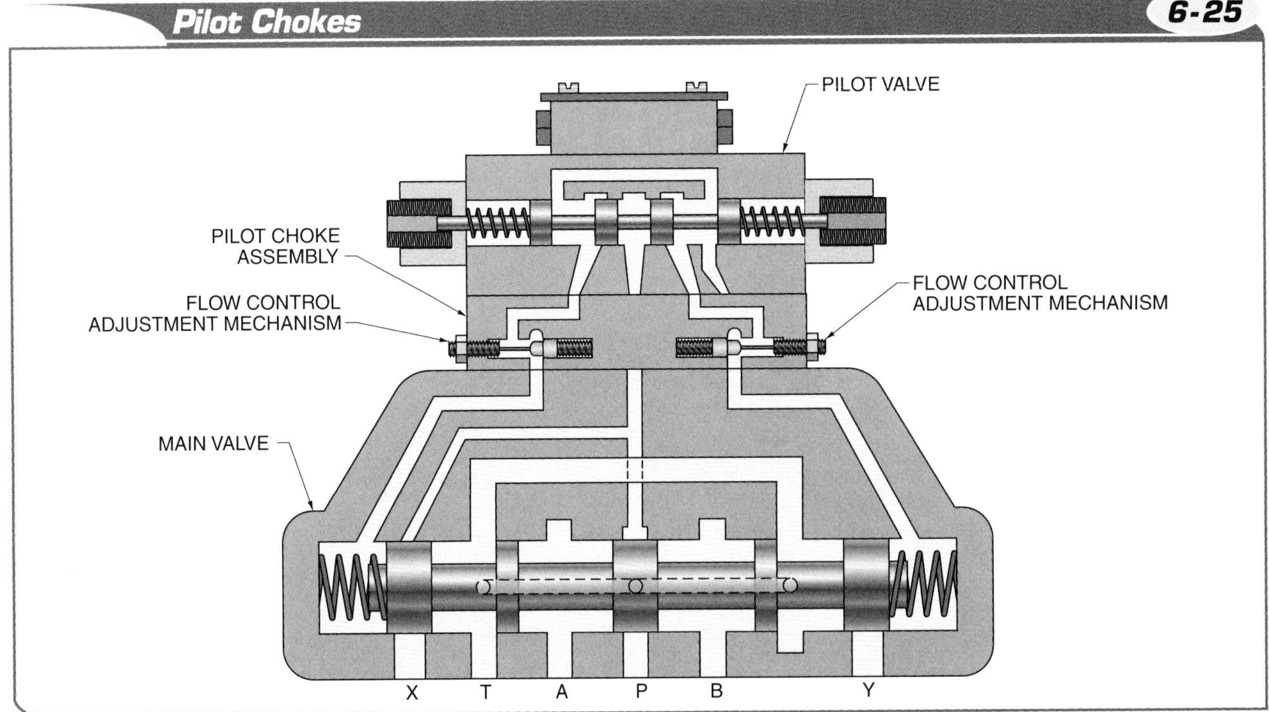

Figure 6-25. A pilot choke is mounted between a main valve and a pilot valve that is used to slow or briefly delay reversals in spool movement.

Pilot Pistons 6-26

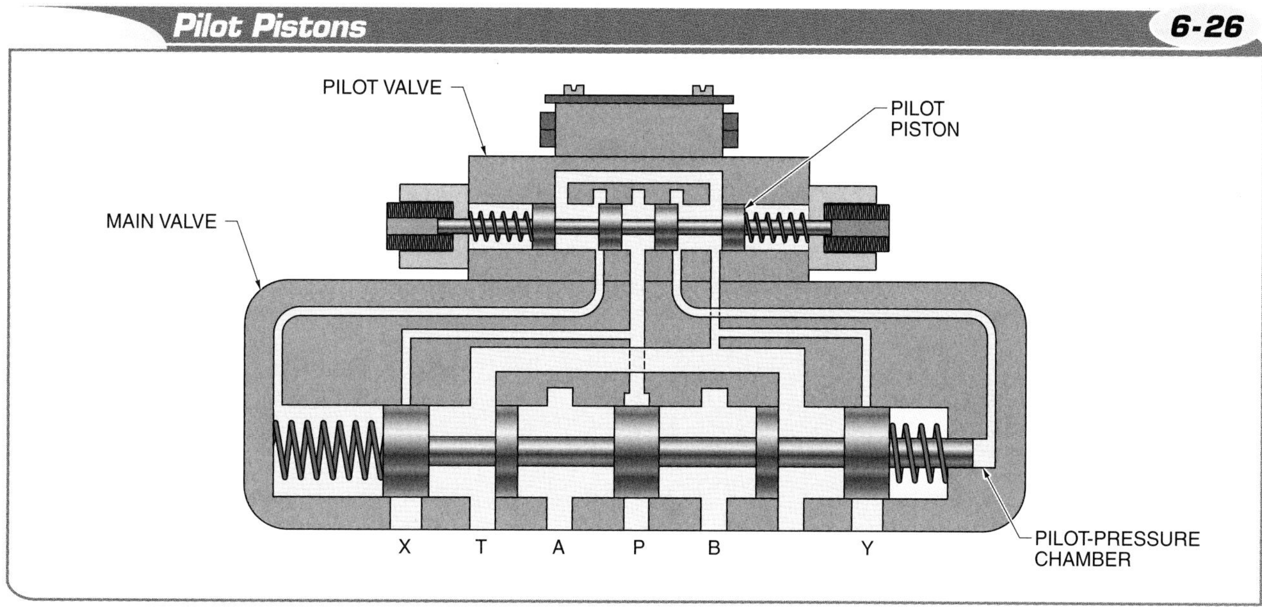

Figure 6-26. A pilot piston allows a directional control valve spool to shift quickly by placing the pilot piston in a pilot-pressure chamber near the spool in the main valve.

Directional Control Valve Detents 6-27

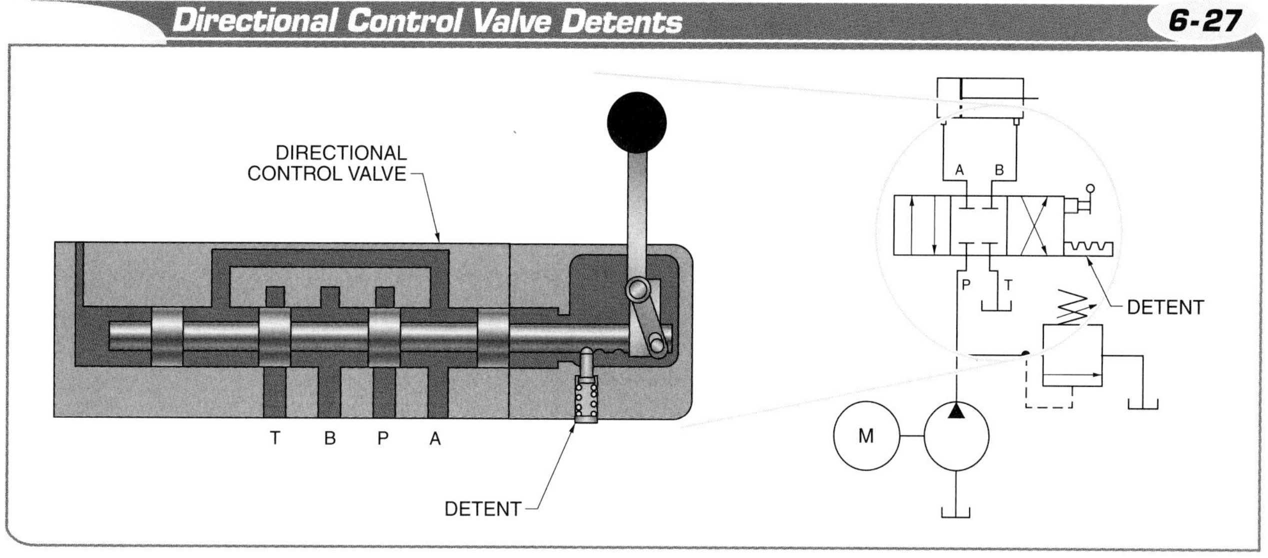

Figure 6-27. A detent operates by locking the spool in a desired position and not allowing it to move back into its normal position until it is manually repositioned.

Proportional Actuators

A *programmable logic controller (PLC)* is a solid-state control device that can be programmed and reprogrammed to automatically control electrical systems in commercial and industrial facilities. A *proportional actuator* is a type of actuator that uses analog signaling to actuate a proportional directional control valve with a spool that can be placed in an infinite number of positions.

For example, one method to signal a proportional directional control valve is with a joystick that controls the precise proportional distance a rod extends. This controls the position of the spool, by controlling the amount of fluid flow that is allowed into the cylinder. Using a joystick to signal a proportional directional control valve provides a great amount of flexibility when controlling a cylinder.

TERMS

A **logic valve** is a valve that uses pressure signals at its input ports to determine when fluid flow will occur from its output ports.

An **AND logic valve** is a type of logic valve that allows fluid flow out of the valve when pressure is present on input 1 and input 2.

Two types of proportional control are variable pressure control and variable electrical control. Variable pressure control uses fluid pressure to move the spool into the desired position. Variable electrical control, which is standard in industry, uses a varying electrical signal to control the position of the spool.

LOGIC VALVES

A *logic valve* is a valve that uses pressure signals at its input ports to determine when fluid flow will occur from its output ports. Logic valves are controlled by pilot pressure. The two types of logic valves commonly used in hydraulic systems are shuttle valves (OR logic valves) and AND logic valves. AND logic valves commonly consist of two or more valves.

AND Logic Valves

An *AND logic valve* is a type of logic valve that allows fluid flow out of the valve when pressure is present on input 1 and input 2. It consists of two input ports and one output port. When pressure is present at the two input ports, an AND logic valve allows fluid flow out of the valve to the hydraulic system.

An AND logic valve has a spool between the input ports that requires the same amount of pressure from both sides of the valve to allow fluid flow to a predetermined part of the hydraulic system. If only one side of the valve is pressurized or has higher pressure than the opposite side, the spool shifts and does not allow fluid to pass through the valve. The operation of an AND logic valve can be understood through the use of a logic chart or truth table. **See Figure 6-28.**

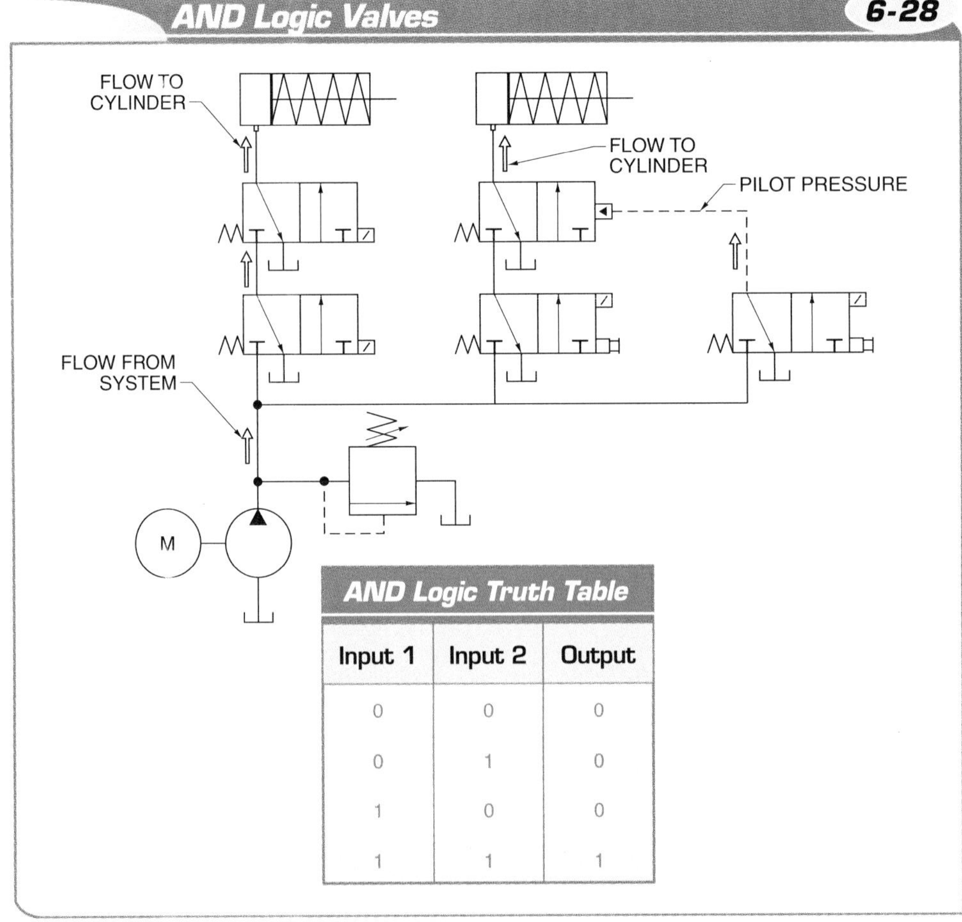

AND Logic Truth Table

Input 1	Input 2	Output
0	0	0
0	1	0
1	0	0
1	1	1

Figure 6-28. In an AND hydraulic circuit, pressure must be present from the hydraulic system (input 1) and through the pilot line (input 2) before the cylinder (output) can operate.

AND logic valves are used in applications that have two inputs and a single output. These valves are connected one after another from an actuator port to a pressure port. This allows the circuit to only operate when all valves have been actuated. The operation of an AND logic circuit can be understood through the use of a logic chart or truth table. This ensures that the operator's hands are not in the path of the shear blade when the press cycles. The logic that is performed by an AND logic valve can also be performed by two separate two-position, three-way directional control valves.

Shuttle Valves

A *shuttle valve* (OR logic valve) is a logic valve that permits fluid flow from the highest pressure of two different input signals. A shuttle valve can have a similar configuration as an AND logic valve, has two input ports and one output port, but operates differently. Fluid will flow to a desired part of the hydraulic system if either input port is pressurized or if both input ports are pressurized. Fluid flow is at no-flow condition if neither input port is pressurized.

Shuttle valves are used for applications such as switching flow from a pump that has failed or malfunctioned to a backup pump. For example, if production equipment is controlled from one of two control stations and the pump at the start-up station fails, the shuttle valve switches operation to the pump at the other control station. Shuttle valves can be used to ensure that the highest system working pressure is used. The valve shifts so that the highest pressure is sent to the next part of the circuit. The operation of a shuttle valve can be understood through the use of a logic chart, or truth table. **See Figure 6-29.**

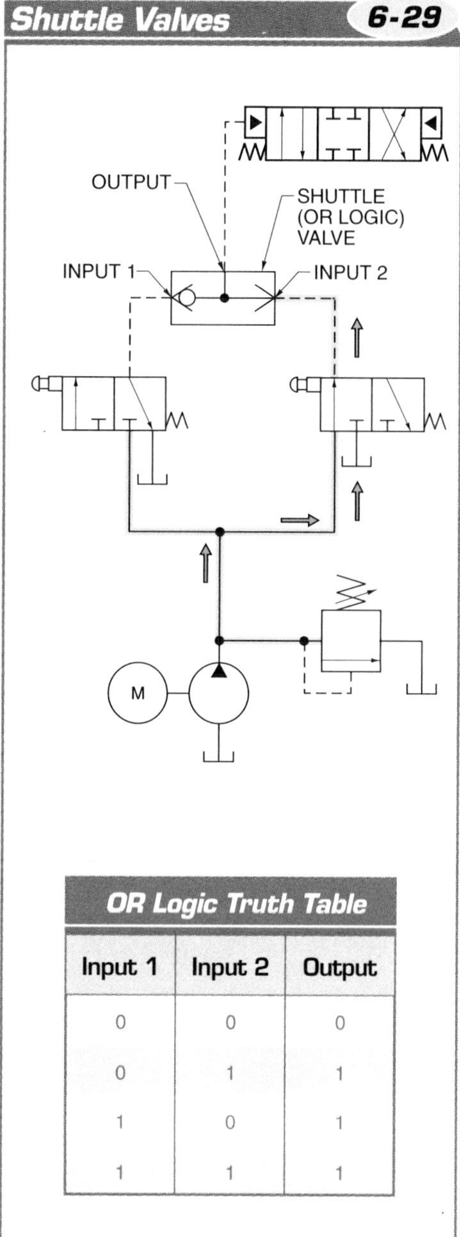

OR Logic Truth Table

Input 1	Input 2	Output
0	0	0
0	1	1
1	0	1
1	1	1

Figure 6-29. A shuttle (OR logic) valve permits a hydraulic system to operate from either of two different hydraulic signals.

TERMS

A **shuttle valve** (OR logic valve) is a logic valve that permits fluid flow from the highest pressure of two different input signals.

Chapter Review

Name: ______________________ Date: ____________

MULTIPLE CHOICE

________ **1.** In a directional control valve, port ___ receives fluid flow from the pump, while port ___ returns fluid flow to the reservoir.

A. A; B
B. B; A
C. P; T
D. T; P

________ **2.** A(n) ___-center position directional control valve allows a hydraulic system to operate with more than one actuator and with each actuator operating independently of the others.

A. closed
B. float
C. open
D. tandem

________ **3.** A(n) ___ is an internal component of a directional control valve that is used to control fluid flow and to connect internal passages and ports.

A. actuator
B. piston
C. plunger
D. spool

________ **4.** There are ___ different subplate designs.

A. two
B. seven
C. nine
D. ten

________ **5.** A(n) ___-center position directional control valve is used to hold the piston in a cylinder in the desired position while allowing fluid flow to be directed back to the reservoir without the need to activate a relief valve.

A. closed
B. float
C. open
D. tandem

________ **6.** ___ mounting is a method that uses tubing or threaded pipes that connect directly to a directional control valve.

A. In-line
B. Manifold subplate
C. Pipe
D. Valve

_______________ **7.** A directional control valve ___ is a mechanism that is used to move the position of the spool in a directional control valve.

A. actuator
B. handle
C. positioner
D. slide

_______________ **8.** Solenoids alone are not strong enough to move a directional control valve spool when fluid flow is greater than ___ gpm.

A. 20
B. 25
C. 35
D. 50

_______________ **9.** In a directional control valve schematic symbol, ___ the position boxes depict the direction of fluid flow for each spool position.

A. external arrows outside
B. internal arrows inside
C. diagonal lines outside
D. T-shaped symbols

_______________ **10.** A hydraulic system that does not have a variable-displacement, pressure-compensated pump typically uses a(n) ___-center or ___-center position directional control valve so that the hydraulic fluid can be returned to the reservoir.

A. closed; float
B. float; tandem
C. open; closed
D. open; tandem

_______________ **11.** ___ pressure is the amount of pressure required to slightly unseat a ball or poppet and start to allow fluid flow through a check valve.

A. Cracking
B. Operating
C. Pilot
D. Poppet

_______________ **12.** A(n) ___ is the schematic representation of the direction in which a spool forces fluid to flow.

A. actuator
B. position
C. valve port
D. way

_______________ **13.** ___ pressure is initiated when the fluid pressure in a fluid power system shifts the spool into its desired position.

A. Pilot
B. PLC
C. Solenoid
D. Spring

______________ 14. A ___ check valve allows fluid flow in one direction and restricts fluid flow in the opposite direction until it receives a pressure signal through a pilot line applied to the port against the piston.
A. direct-acting
B. pilot-to-close
C. pilot-to-open
D. solenoid-actuated

______________ 15. The advantage of a three-position, four-way directional control valve is that it has ___ different center positions that can perform ___ different operations.
A. three; three
B. three; four
C. four; three
D. four; four

COMPLETION

______________ 1. A(n) ___ valve allows or prevents fluid flow to specific piping or system actuators.

______________ 2. A(n) ___ subplate can be used to mount multiple directional control valves of the same configuration.

______________ 3. A(n) ___ is an actuator that is used to hold a directional control valve spool in selected positions.

______________ 4. A(n) ___ is the section of a directional control valve that connects a pipe, hose, or tube to the internal passages of the directional control valve.

______________ 5. ___ and ___ logic valves have two inputs and a single output.

______________ 6. A(n) ___ valve allows fluid flow from the pump port to an actuator port in the spring-actuated (normal) position.

______________ 7. A T-shape symbol within the schematic symbol of a directional control valve is used to represent a(n) ___.

______________ 8. In a(n) ___-center position directional control valve, the cylinder cannot move because there is no fluid flow to or from the cylinder.

______________ 9. Directional control valve actuators are categorized as ___ and ___.

______________ 10. Right angle check valves are available with a range of flow rates up to ___ gpm.

______________ 11. ___ mounting is a mounting method where a directional control valve is mounted to a subplate that attaches to system piping.

______________ 12. The ___-center position is used to hold a cylinder in a stable position while fluid flows back to the reservoir from the pump under low pressure.

______________ 13. A(n) ___ is a solid-state control device that can be programmed and reprogrammed to automatically control electrical systems in commercial and industrial facilities.

______________ **14.** A(n) ___ valve is a combination valve that consists of a small solenoid-actuated directional control valve that is used to control the main pilot-operated valve.

______________ **15.** A(n) ___ check valve is directly activated or moved by fluid flow from a primary port.

______________ **16.** A(n) ___ valve allows fluid flow in one direction and stops fluid flow in the opposite direction.

______________ **17.** A(n) ___ is a symbol that represents the number of different positions that the spool in a directional control valve is capable of moving into.

______________ **18.** ___ symbols are used to indicate the type of directional control valve actuator and the type of spool positioning method, where applicable.

______________ **19.** A(n) ___ is an electrical control device that converts electrical energy into linear mechanical energy when current passes through a magnetic coil located in the device.

______________ **20.** A(n) ___ valve has its inlet and outlet ports set at a right angle to each other.

TRUE/FALSE

T F **1.** Two-position, two-way directional control valves can turn on and off separate components of systems that are powered by the same hydraulic power unit.

T F **2.** Most directional control valves used in industrial applications have two, three, or four positions, but some can have five or six positions.

T F **3.** Every manufacturer has a different coding system used for ordering their directional control valves.

T F **4.** Three-position, four-way directional control valves can only control one direction of fluid flow at a time.

T F **5.** Internal arrows on a hydraulic schematic diagram can be either colored solid or outlined.

T F **6.** On a three-position directional control valve, both sides of the spool have springs to center the spool for normal condition.

T F **7.** An AND logic valve allows fluid flow out of the valve when pressure is present on input 1 and input 2.

T F **8.** A shuttle valve is also known as an AND logic valve.

T F **9.** Right angle check valves are installed mainly in hydraulic systems, rather than pneumatic systems.

T F **10.** A shuttle valve operates differently than an OR logic valve.

T F **11.** Pilot-pressure actuators are used when electronic control of a hydraulic system is desired.

T F **12.** On schematic diagrams, pilot lines are indicated by dotted lines.

T F 13. Open-center position directional control valves are typically used in applications where there is only one cylinder in the hydraulic system.

T F 14. Pilot pressure does not require electrical, mechanical, or manual actuation.

T F 15. A position is the number of piping ports in a directional control valve.

T F 16. Solenoid-actuated directional control valves are always double-solenoid actuated.

T F 17. A shuttle valve is a type of logic valve.

T F 18. Newly installed directional control valves typically have a schematic symbol printed or embossed on the valve body.

T F 19. The pressure point at which an automatic detent releases can be fixed or variable.

T F 20. Directional control valve actuators can be automated or manual but never a combination of both.

T F 21. The spring pressure of the biasing spring in a direct-acting check valve is adjustable.

T F 22. When a directional control valve is in the closed-center or tandem-center position, the piston in a hydraulic cylinder can move.

T F 23. Two-position, two-way directional control valves are always normally closed (NC).

T F 24. Two- and three-way directional control valves only control one direction of fluid flow at a time.

T F 25. Manifold subplates reduce the amount of space and piping that a hydraulic system requires.

SHORT ANSWER

1. List the four common directional control valve center positions.

2. List four directional control valve automated actuator control methods.

3. Briefly describe the two main applications that direct-acting check valves are used for in hydraulic systems.

4. Describe the main advantage of using a cartridge valve.

MATCHING

Center Positions

_______ 1. Closed
_______ 2. Float
_______ 3. Open
_______ 4. Tandem

A B P T — A
A B P T — B
A B P T — C
A B P T — D

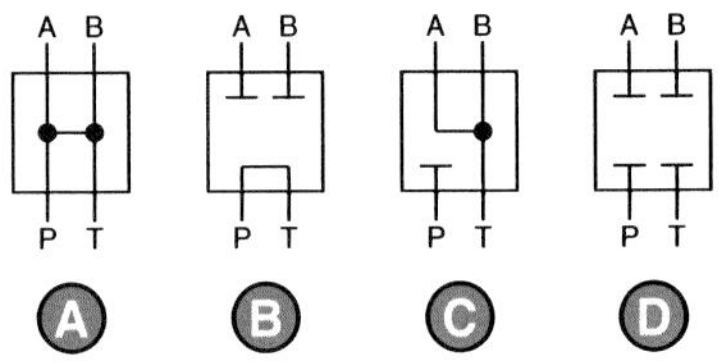

CENTER POSITIONS

Schematic Symbols

_______ 1. Check valve
_______ 2. Check valve used as bypass
_______ 3. Check valve used for pressure control
_______ 4. Three-position, four-way, spring-centered, NC solenoid-actuated directional control valve
_______ 5. Two-position, four-way, double solenoid-actuated directional control valve
_______ 6. Pilot-to-close check valve
_______ 7. Pilot-to-open check valve
_______ 8. Restriction check valve
_______ 9. Two-position, four-way, spring-offset, solenoid-actuated directional control valve
_______ 10. Two-position, three-way, spring-offset, lever-actuated directional control valve

A B C D E F
G H I J

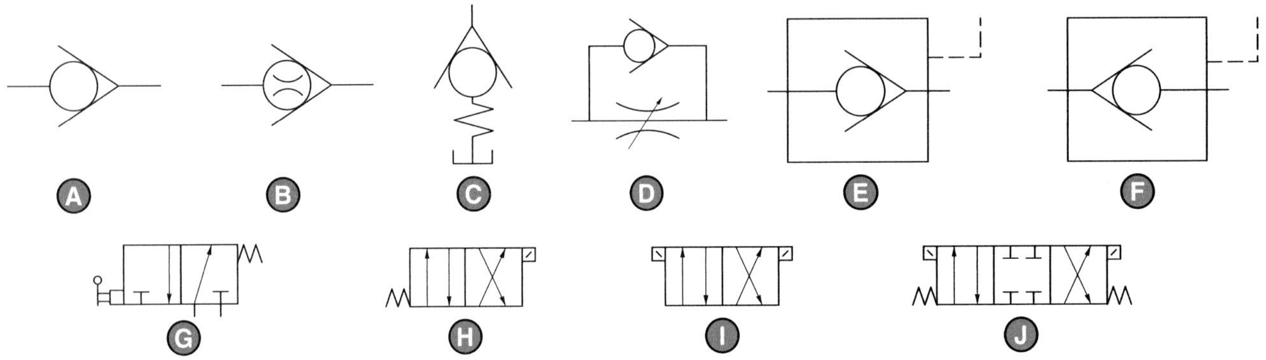

SCHEMATIC SYMBOLS

Activity 6-1: Directional Control Valve Ordering Specifications

A door controlled by a hydraulic system has malfunctioned and it has been determined that the directional control valve is the cause. However, there are no directional control valves remaining in the spare parts inventory. In order to reduce cost, the maintenance manager asks the technician to compare the prices of directional control valves from different manufacturers before ordering a replacement valve. The technician must use the existing directional control valve's model number to determine the type and parameters of the replacement valve to be ordered. The model number of the existing valve is the following:

DC-IT-1-0-4-P3A-SP3-CL-FF-ST-FFM-1

1. Use Directional Control Valve Ordering Code Chart and list the information that is required to determine the type and parameters of the directional control valve that is to be replaced.

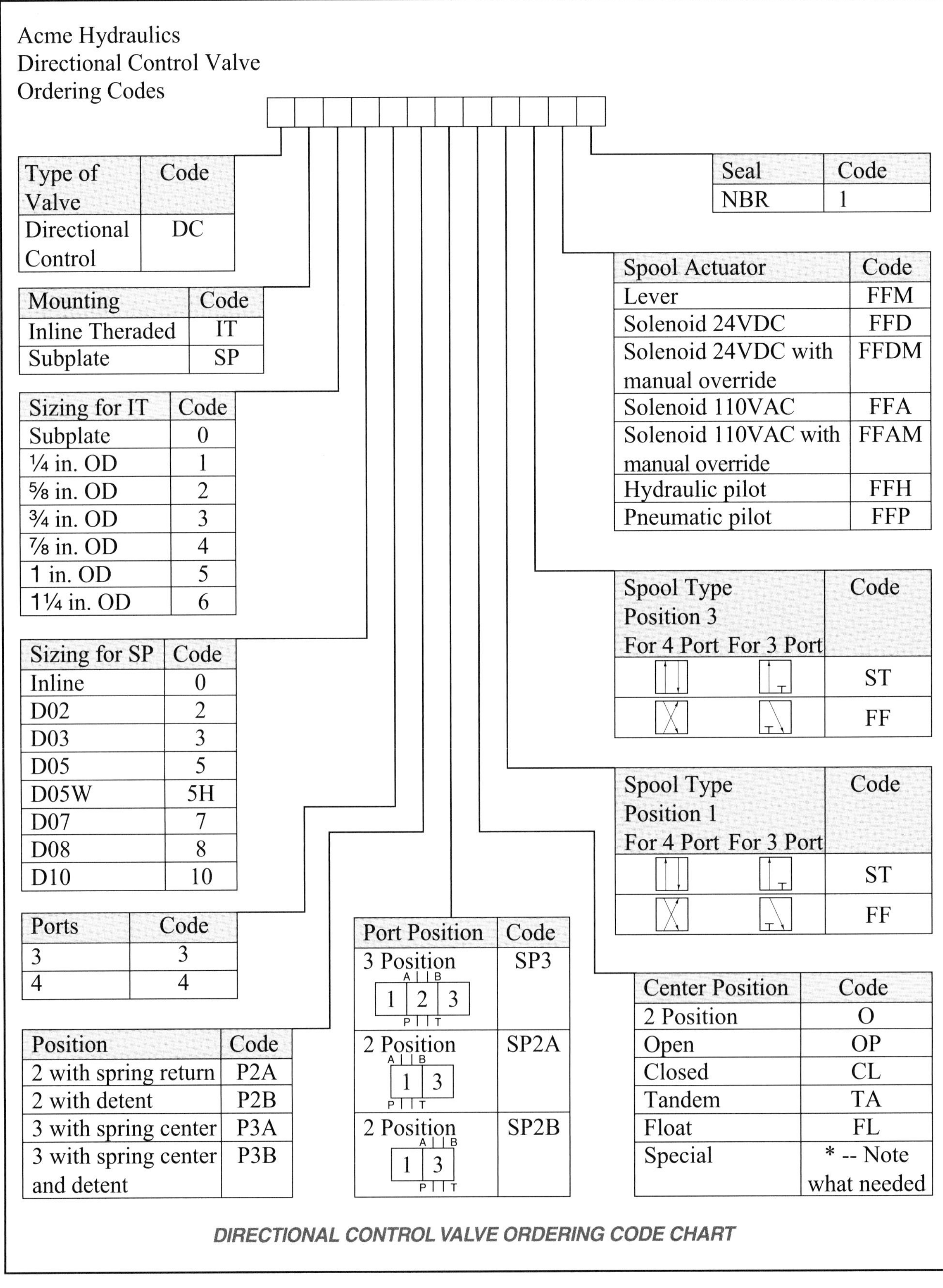

Acme Hydraulics
Directional Control Valve
Ordering Codes

Type of Valve	Code
Directional Control	DC

Mounting	Code
Inline Theraded	IT
Subplate	SP

Sizing for IT	Code
Subplate	0
¼ in. OD	1
⅝ in. OD	2
¾ in. OD	3
⅞ in. OD	4
1 in. OD	5
1¼ in. OD	6

Sizing for SP	Code
Inline	0
D02	2
D03	3
D05	5
D05W	5H
D07	7
D08	8
D10	10

Ports	Code
3	3
4	4

Position	Code
2 with spring return	P2A
2 with detent	P2B
3 with spring center	P3A
3 with spring center and detent	P3B

Port Position	Code
3 Position	SP3
2 Position	SP2A
2 Position	SP2B

Seal	Code
NBR	1

Spool Actuator	Code
Lever	FFM
Solenoid 24VDC	FFD
Solenoid 24VDC with manual override	FFDM
Solenoid 110VAC	FFA
Solenoid 110VAC with manual override	FFAM
Hydraulic pilot	FFH
Pneumatic pilot	FFP

Spool Type Position 3 For 4 Port	For 3 Port	Code
		ST
		FF

Spool Type Position 1 For 4 Port	For 3 Port	Code
		ST
		FF

Center Position	Code
2 Position	O
Open	OP
Closed	CL
Tandem	TA
Float	FL
Special	* -- Note what needed

DIRECTIONAL CONTROL VALVE ORDERING CODE CHART

Use Directional Control Valve Ordering Code Chart and information from the appendix to draw the corresponding schematic symbols in the space provided for each directional control valve model number listed.

2. *DC-SP-0-5-4-P2A-SP2A-0-FF-ST-FFM-1*

3. *DC-SP-0-10-4-P3A-SP3-TA-FF-ST-FFH-1*

4. *DC-IT-2-0-3-P2B-SP2B-0-FF-ST-FFD-1*

5. *DC-IT-3-0-4-P2A-SP2A-0-FF-ST-FFH-1*

6. *DC-SP-3-5-4-P3B-SP3-OP-ST-FF-FFA-1*

7. *DC-IT-1-0-4-P3B-SP3-FL-ST-FF-FFP-1*

8. *DC-SP-0-3-4-P3A-SP3-CL-ST-FF-FFDM-1*

Activity 6-2: Ram Cylinder Control

In a steel factory, a single-acting ram cylinder is used to open a portion of the roof for ventilation during certain production processes. A pilot-operated check valve is used to hold the ram cylinder in place when it is extended. The maintenance manager decides that the roof must be closed with an electrical control, rather than using the lever-actuated directional control valve.

1. Design a system using the symbols below to meet the following parameters:

- A lever-actuated directional control valve will raise the roof.
- A pilot-operated check valve will hold the cylinder in the extended position.
- A solenoid-actuated directional control valve will allow fluid flow to exit the cylinder, allowing gravity to retract the roof.

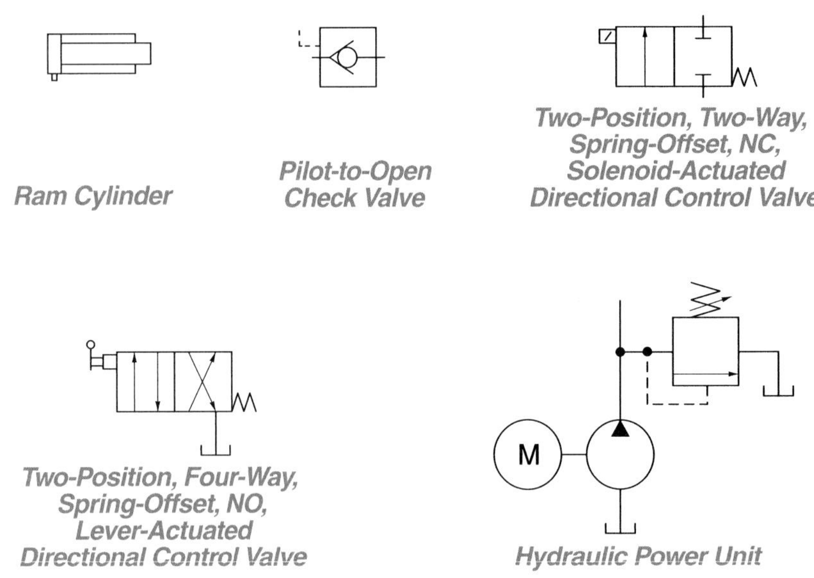

SCHEMATIC SYMBOLS

Activity 6-3: Hydraulic Power Unit with Lever-Actuated Directional Control Valve

A warehouse has a hydraulic system with a double-acting cylinder that is used to push boxes filled with product off a conveyor and onto a platform. A small hydraulic power unit dedicated for this task with a two-position, four-way, lever-actuated directional control valve controls the hydraulic system.

The warehouse has decided to make the system more energy efficient and reduce costs. It has been decided to replace the two-position, four-way, lever-actuated directional control valve with a three-position, four-way, tandem-center, lever-actuated directional control valve. This will allow the pump in the system to run at a lower pressure during the machine's idle time and reduce costs by saving electricity.

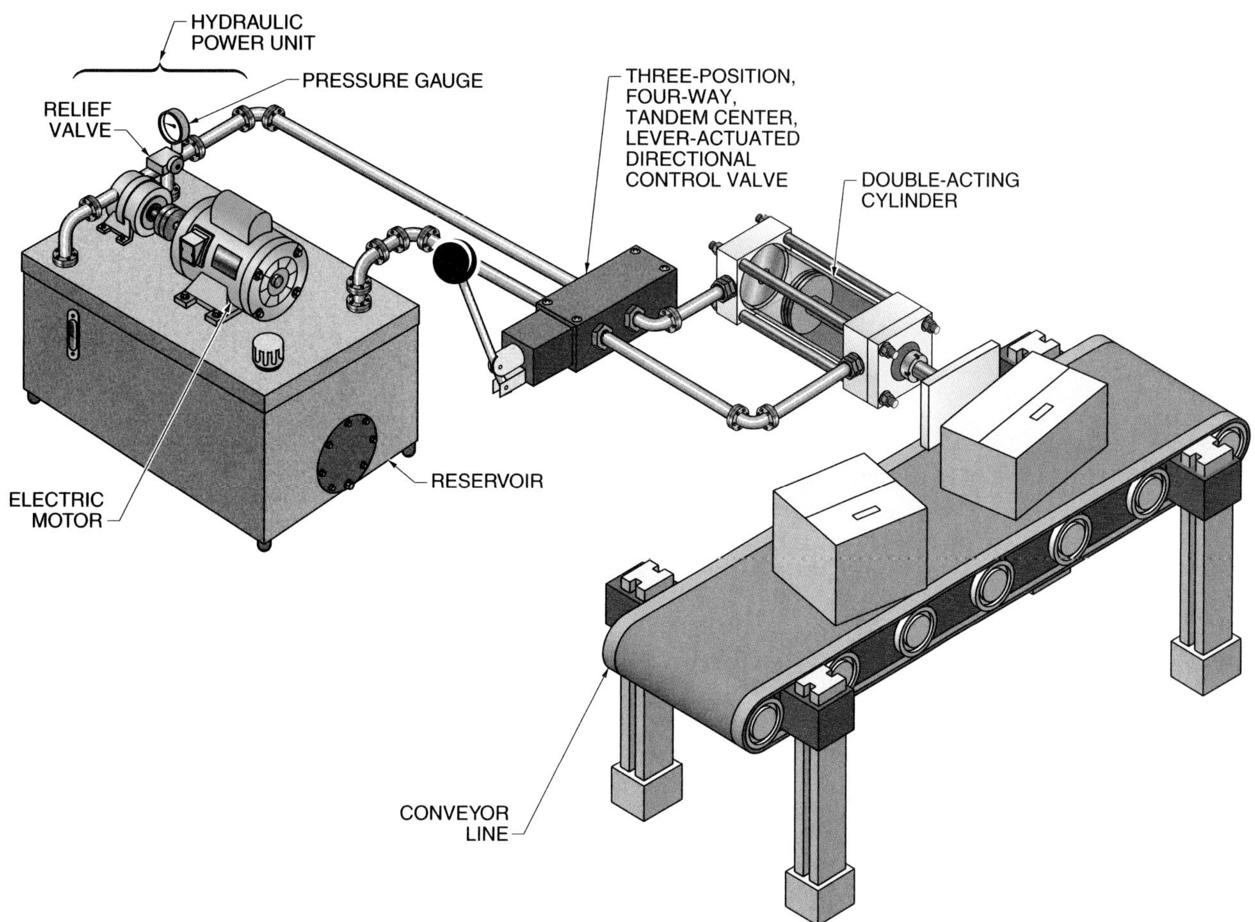

HYDRAULIC POWER UNIT WITH LEVER-ACTUATED DIRECTIONAL CONTROL VALVE

1. Sketch a schematic diagram of the hydraulic system with a three-position, four-way, closed-center, lever-actuated directional control valve with the crossover on the left position. Then, use the FluidSIM® software included on the CD-ROM to build the hydraulic system and verify that it will operate properly.

FluidSIM® Activity

Activity 6-4: Directional Control Valve Selection

Using FluidSIM®, replace the current directional control valve from the hydraulic system in Activity 6-3 with the one listed in each question and run the simulation.

1. System has a three-position, four-way, closed-center, lever-actuated directional control valve with the crossover on the right position. What happens with the cylinder?

2. System has a three-position, four-way, float-center, lever-actuated directional control valve with the crossover on the left position. What happens with the cylinder?

3. System has a three-position, four-way, float-center, lever-actuated directional control valve with the crossover on the right position. What happens with the cylinder?

Chapter 7

Flow Control

OBJECTIVES

- Describe the difference between a fixed orifice and a variable orifice.
- Describe the different types of flow control valves.
- Describe the variables that affect fluid flow.
- Distinguish between meter-in, meter-out, and bleed-off systems.
- Describe the operation and applications of flow dividers.

INTRODUCTION

Flow control is one of the most important aspects of a hydraulic system. If proper flow control devices are not selected and installed, a hydraulic system will not function as intended. In some hydraulic systems, the speed of a piston in a cylinder needs to be slow in order to protect the load from damage. In other hydraulic systems, piston speed needs to be as fast as possible to maintain a high level of production. There are also hydraulic systems where piston speed must change during operation. Flow control valves are used to control the flow rate of fluid. Metering systems are used to control flow rate and piston speed.

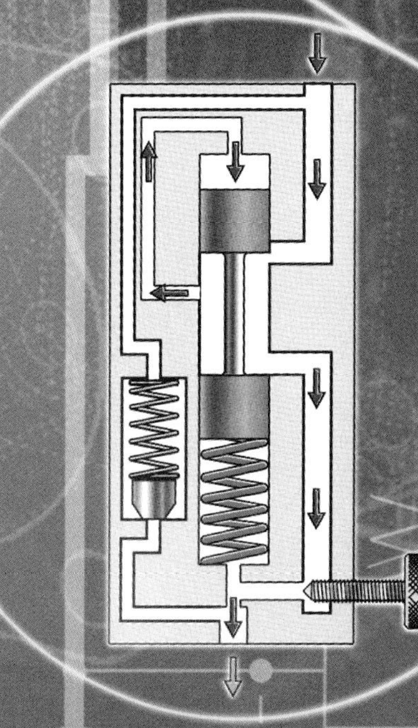

HYDRAULIC FLOW CONTROL

Fluid flow must be controlled to properly operate hydraulic equipment and machinery. Without flow control valves, fluid flow rate would be determined solely by the pump. In a hydraulic system, fluid flow rate controls the speed of a piston in a cylinder. The ability to control fluid flow rate, and therefore piston speed, allows hydraulic systems to be integrated into other industrial systems. The factors to consider when controlling fluid flow are the diameters of any orifices, the style of flow control valve to use for an application, the speed required for a piston in a specific application, and the pressure differential across the flow control valve (constant or variable).

Pressure Differential

In a hydraulic system, the pressure differential across an orifice or a flow control valve affects fluid flow, thus affecting actuator speed. The greater the pressure differential across an orifice, the greater the flow rate through the orifice or flow control valve. Any change in pressure before or after a flow control valve affects fluid flow through it, which also results in a change in actuator speed.

The main variable that affects pressure differential is the weight of a load. If the load on a cylinder increases, the pressure differential across the flow control valve decreases and the flow out of the valve also decreases. This reduces piston speed. To increase piston speed, the orifice is partially opened.

If the load on the cylinder is reduced, piston speed increases. When piston speed is too fast, it can be corrected by using a pressure-compensated flow control valve.

A pressure-compensated flow control valve compensates for the changing pressures in a hydraulic system. **See Figure 7-1.** The amount of fluid flow through an orifice is directly proportional to the pressure differential across the orifice. When there is a pressure differential in a flow control valve, piston speed is determined by applying the following procedure:

1. Calculate the pressure needed to move a piston when there is a load on the cylinder.

$$P_F = \frac{F}{A_{pf}}$$

where

P_F = pressure created by the load on the cylinder (in psi)

F = load on the cylinder (in lb)

A_{pf} = piston face area (in sq in.)

2. Calculate the pressure differential in the flow control valve.

$$P_D = P_{BV} - P_F$$

where

P_D = pressure differential (in psi)

P_{BV} = pressure before the flow control valve (in psi)

P_F = pressure created by the load on the cylinder (in psi)

3. Calculate the flow rate through the flow control valve.

$$V_Q = \frac{P_D}{P_Q}$$

where

V_Q = flow rate through the flow control valve (in gpm)

P_D = pressure differential (in psi)

P_Q = pressure per gallon of fluid flow through the flow control valve (in psi)

4. Calculate piston speed.

$$P_s = \frac{(V_Q \times 19.25)}{A_{pf}}$$

where

P_s = piston speed (in fpm)

V_Q = flow rate through the flow control valve (in gpm)

A_{pf} = piston face area (in sq in.)

TECH FACT

Large mobile hydraulic systems, such as those in off-road loaders in excavation and mining operations, use double-acting cylinders to control the lift and tilt system of the loader bucket.

Example: How fast does a piston with an area of 4.91″ and a system pressure of 750 psi move when there is a load of 250 lb on the cylinder? (and one gallon of fluid flow through the flow control valve is 75 psi).

1. Calculate the pressure needed to move a piston with a load on the cylinder.

$$P_F = \frac{F}{A_{pf}}$$

$$P_F = \frac{250}{4.91}$$

$$P_F = 50.92 \text{ psi}$$

2. Calculate the pressure differential through the flow control valve.

$$P_D = P_{BV} - P_F$$

$$P_D = 750 - 50.92$$

$$P_D = 699.08 \text{ psi}$$

3. Calculate the flow rate through the flow control valve.

$$V_Q = \frac{P_D}{P_Q}$$

$$V_Q = \frac{699.08}{75}$$

$$V_Q = 9.32 \text{ gpm}$$

4. Calculate piston speed.

$$P_s = \frac{(V_Q \times 19.25)}{A_{pf}}$$

$$P_s = \frac{(9.32 \times 19.25)}{4.91}$$

$$P_s = \frac{179.41}{4.91}$$

$$P_s = \mathbf{36.53\ fpm}$$

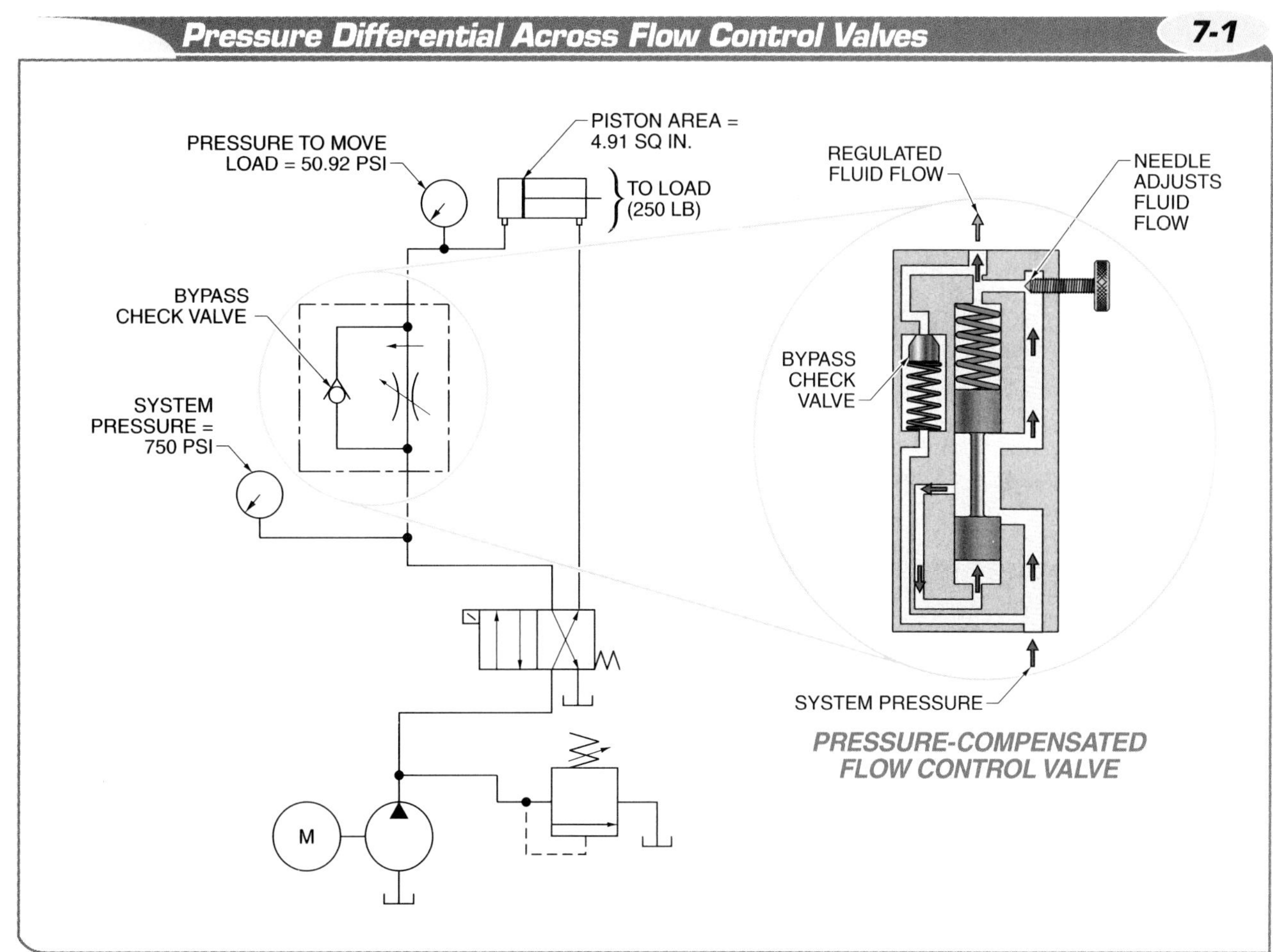

Figure 7-1. A pressure-compensated flow control valve compensates for the changing pressures in a hydraulic system.

TERMS

A **fixed orifice** is an orifice that cannot be adjusted.

A **variable orifice** is an orifice that allows an adjustable amount of fluid flow.

A **flow control valve** is a valve whose primary function is to control the rate of fluid flow.

Metering is the controlling of the rate of fluid flow and how the fluid flow is being accomplished.

The lower the pressure differential, the lower the flow rate. The heavier the load on the cylinder, the lower the pressure differential, and the slower the piston moves. The lighter the load on the cylinder, the faster the piston moves. The other variable that affects the pressure differential is when the relief valve is adjusted. A relief valve is typically manually adjusted.

Example: What is piston speed when the relief valve is set to 500 psi and the load on the cylinder is 250 lb? (and one gallon of fluid flow through the flow control valve is 75 psi).

1. Calculate the pressure needed to move a piston with a load on the cylinder.

$$P_F = \frac{F}{A_{pf}}$$

$$P_F = \frac{250}{4.91}$$

$$P_F = 50.92 \text{ psi}$$

2. Calculate the pressure differential through the flow control valve.

$$P_D = P_{BV} - P_F$$
$$P_D = 500 - 50.92$$
$$P_D = 449.08 \text{ psi}$$

3. Calculate the flow rate through the flow control valve.

$$V_Q = \frac{P_D}{P_Q}$$

$$V_Q = \frac{449.08}{50.92}$$

$$V_Q = 8.82 \text{ gpm}$$

4. Calculate piston speed.

$$P_s = \frac{(V_Q \times 19.25)}{A_{pf}}$$

$$P_s = \frac{8.82 \times 19.25}{4.91}$$

$$P_s = \frac{169.79}{4.91}$$

$$P_s = \mathbf{34.58 \text{ fpm}}$$

When system pressure is decreased, piston speed is decreased. If system pressure is increased, piston speed is increased.

Orifices

An orifice is an opening through which fluid flows. An orifice is always smaller than the inside diameter of a pipe, tubing, hose, or internal passage. For example, the orifice through the center of a poppet in a restriction check valve has a considerably smaller diameter than the port that the poppet will seal. Fluid flow through an orifice is affected by the size of the orifice, the pressure differential across the orifice, and the temperature of the hydraulic fluid. The two types of orifices that are available in hydraulic systems are fixed orifices and variable orifices.

A *fixed orifice* is an orifice that cannot be adjusted. The diameter of a fixed orifice cannot be adjusted. For example, a fixed orifice is the orifice in the center of a poppet in a restriction check valve. This design allows maximum fluid flow in one direction and restricts it in the opposite direction.

A *variable orifice* is an orifice that allows an adjustable amount of fluid flow. **See Figure 7-2.** The diameter of a variable orifice can be adjusted by turning a handwheel, handle, or knob attached to the valve stem. The schematic symbol for a variable orifice is the same as the symbol for a fixed orifice but with a diagonal arrow through it. A variable orifice is a valve.

Flow Control Valves

A *flow control valve* is a valve whose primary function is to control the rate (gpm) of fluid flow. *Note*: Flow control valves can refer to both a valve having a variable orifice or a variable orifice with a check valve. These are typically needle valves, so it is common to refer to valves with a variable orifice as needle valves and valves with a variable orifice and check valve as flow control valves. The terms have become interchangeable in industry. A technician should always ensure that the description of the device is understood. *Metering* is the controlling of the rate of fluid flow and how the fluid flow is being accomplished.

When a flow control valve controls fluid flow, it causes internal resistance that increases pressure from the valve back to the pump. The increased pressure opens the relief valve, allowing excess fluid flow to return to the reservoir while metered (reduced) fluid flow continues to a branch system or hydraulic actuator. In some cases, the excess fluid flow is used in another part of the hydraulic system rather than returning to the reservoir. Noncompensated flow control in a hydraulic system can be accomplished with globe valves, gate valves, ball valves, or needle valves. **See Figure 7-3.** Most hydraulic systems use needle valves. Most hydraulic systems use needle valves to meter fluid flow instead of globe, gate, and ball valves, which are typically used as shut-off valves and only intended to start and stop the flow of hydraulic fluid. These valves have two connections, an inlet and an outlet, and are also referred to as two-way valves.

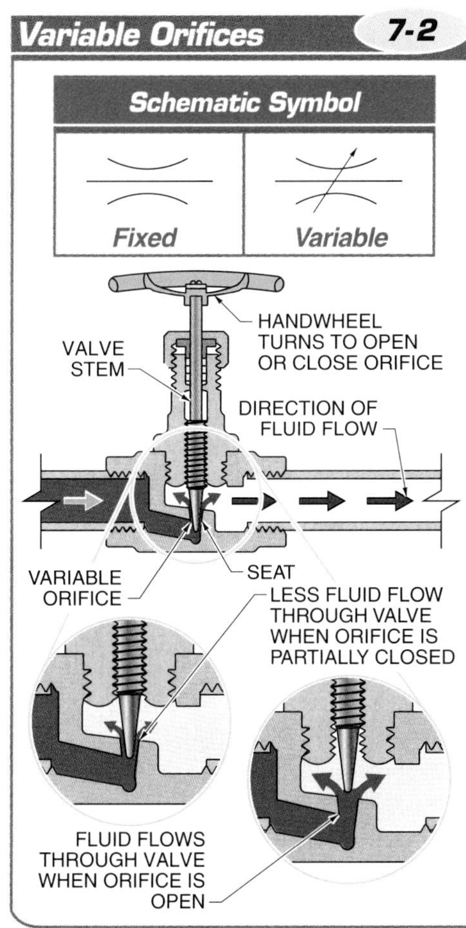

Figure 7-2. A variable orifice allows a variable amount of fluid flow.

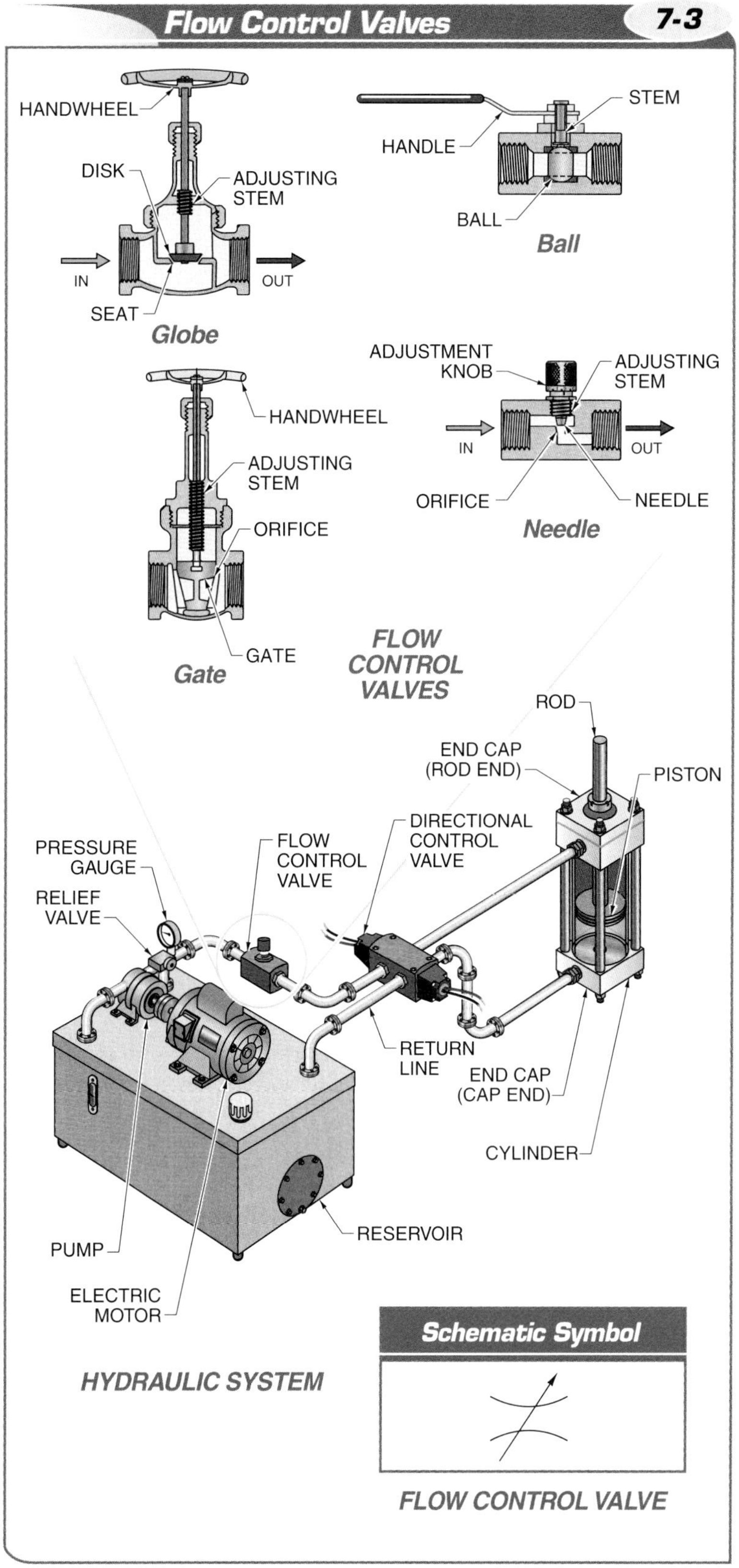

Figure 7-3. Variable flow control in a hydraulic system can be accomplished with globe valves, gate valves, ball valves, or needle valves.

TERMS

A **globe valve** is a valve with a disk that is raised or lowered over an orifice.

A **gate valve** is a two-position valve with an internal gate between the two seats of the orifice.

Globe Valves. A *globe valve* is a valve with a disk that is raised or lowered over an orifice. **See Figure 7-4.** Globe valves are designed as shut-off valves for low-pressure service, typically no higher than 150 psi. They have seats that are slightly smaller than the means of transmission. They do not have a straight path for fluid flow, which must make two 90° turns to pass in one direction through the valve. The opening between the seat and the disk is controlled to meter fluid flow from no-to full-flow operations. An arrow placed on the side of the valve body indicates the direction of fluid flow through a globe valve.

Gate Valves. A *gate valve* is a two-position valve with an internal gate between the two seats of the orifice. **See Figure 7-5.** Gate valves are generally used for full-flow or no-flow operations and are not designed for restricting fluid flow. Gate valves have ports that are the same size as the means of transmission. This causes the hydraulic fluid to flow in a straight path through the valve, allowing only a slight pressure drop in the system when the valve is fully open. Some gate valves tend to provide smooth fluid flow, however, vibration and erratic fluid flow can occur when a gate valve is used in a partially or completely open position.

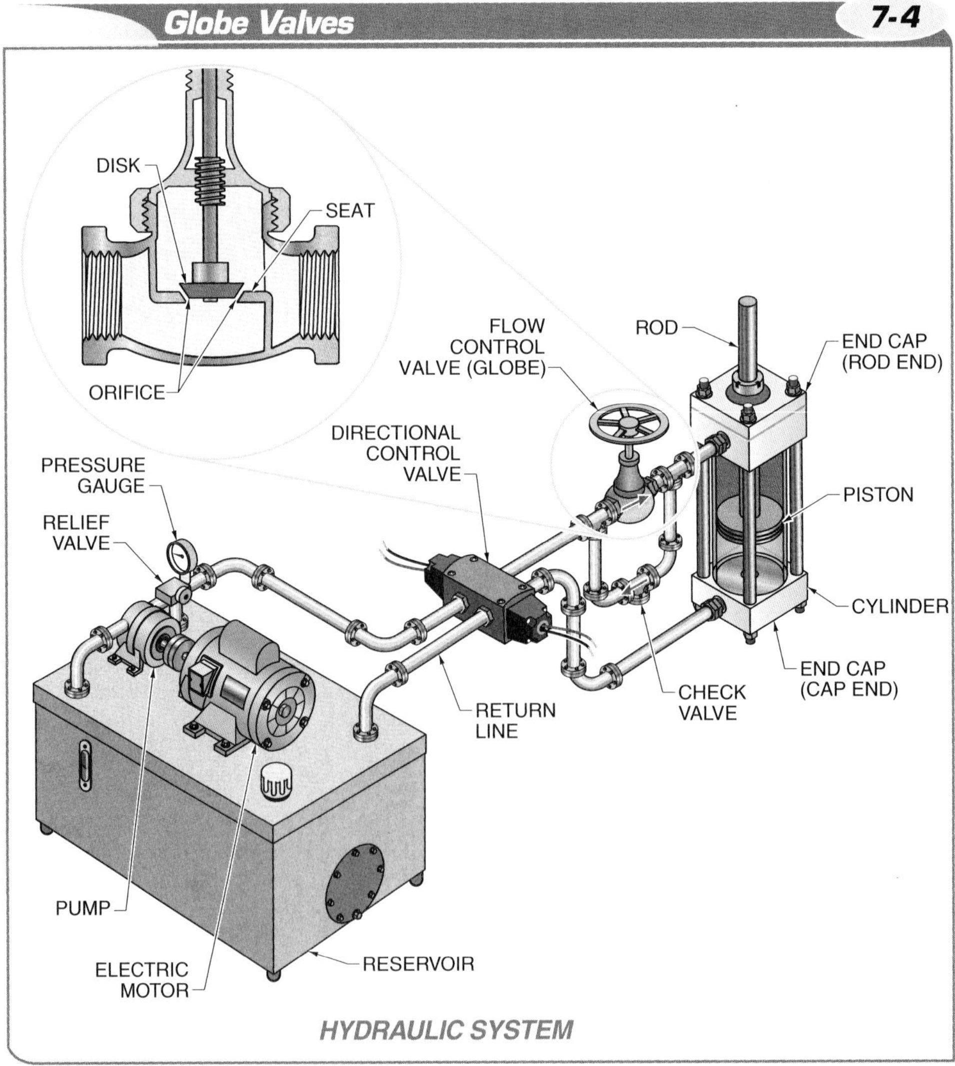

Figure 7-4. A globe valve has a disk that is raised or lowered over an orifice.

Gate valves are used for applications with pressures of 2000 psi to 5000 psi. They can also allow fluid flow in both directions, unless otherwise marked. Gate valves can be either rising-stem or nonrising-stem.

TECH FACT

In the United States, "American style" ball valves are used, which are closed when the handle is perpendicular to the piping or ports. In Europe, "European-style" ball valves are used, which are closed when the handle is parallel to the piping or ports. The style of the ball valve that is fitted into the equipment should be indicated to avoid misuse.

Parker Motion & Control

Flow control valves are adjusted manually by turning an adjustment knob.

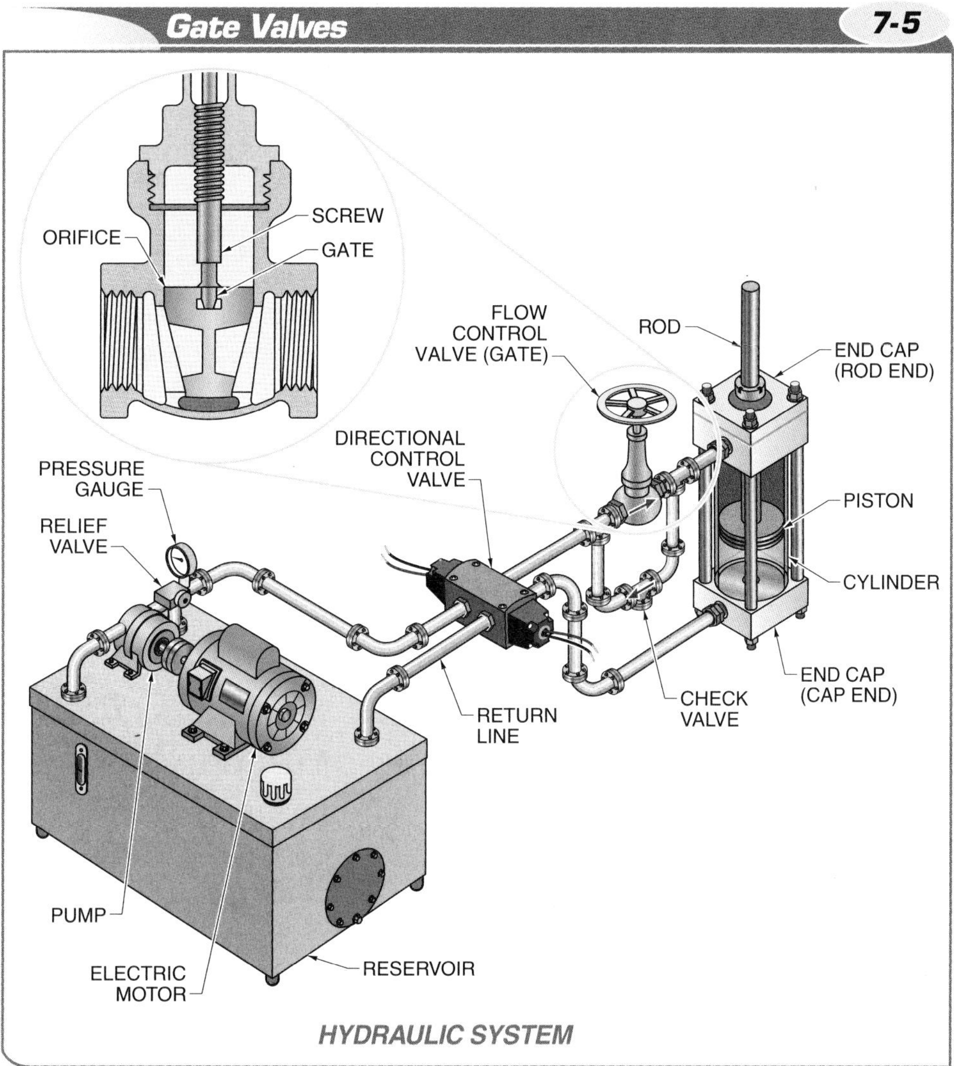

Figure 7-5. A gate valve has an internal gate that slides over an orifice.

TERMS

A **ball valve** is an infinite-position flow control valve with a ball that has an orifice through the center to allow fluid flow.

A **needle valve** is an infinite-position flow control valve that has a narrowly tapered stem or needle positioned in-line with an orifice that is the same size as the stem or needle.

Ball Valves. A *ball valve* is an infinite-position flow control valve with a ball that has an orifice through the center to allow fluid flow. **See Figure 7-6.** Ball valves allow hydraulic fluid to flow in a straight path, which causes less internal resistance and wastes less energy.

Ball valves used in hydraulic systems are often referred to as high-pressure ball valves or hydraulic ball valves. When a ball valve is used in a hydraulic system, it should always be ensured that it is rated for use with high pressures and is compatible with hydraulic fluid. Ball valves can be rated as high as 12,000 psi. Ball valves are also commonly used in pneumatic systems.

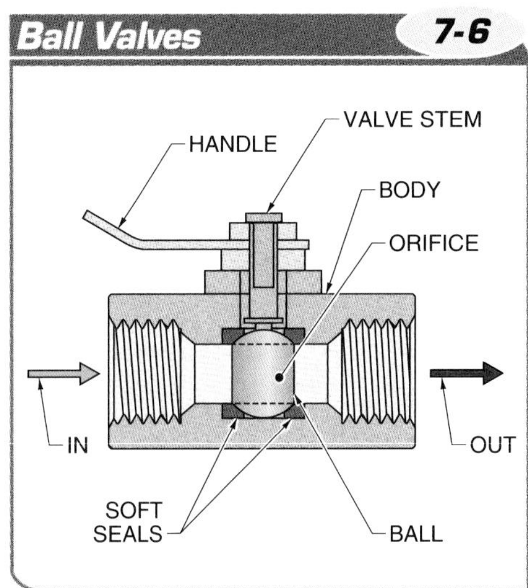

Figure 7-6. A ball valve has a ball with a hole drilled through the center to allow fluid flow.

A ball valve is manually controlled with a handle that is connected to the valve with a valve stem. The handle is moved a quarter of a turn to open or close the valve. In the completely open position, the orifice in the ball is aligned with the ports to allow fluid flow. In the completely closed position, the orifice in the ball is perpendicular to the ports to stop fluid flow. If required, the orifice in the ball can also be moved to allow partial fluid flow. The ball rides on soft seals for leakproof sealing when closed.

Needle Valves. A *needle valve* is an infinite-position flow control valve that has a narrowly tapered stem or needle positioned in-line with an orifice that is the same size as the stem or needle. **See Figure 7-7.** Needle valves are flow control valves used in speed-control applications. Fluid flow takes two 90° bends through a needle valve, allowing the needle to create a pressure differential. Some needle valves are available with color-coded valve stems, which allow the technician to view any adjustments that are made. Needle valves are the most frequently used valves in industrial hydraulic applications.

Needle valves are available with built-in check valves for applications that require fluid flow in the opposite direction of their metered flow. The check valve directs fluid flow around the needle valve. Hydraulic fluid passes unrestricted through the check valve when the flow is reversed. An arrow on the side of the needle and check valve combination indicates the direction of metered fluid flow.

The greater the pressure differential across a needle valve, the greater the flow rate through it. Fluid flow through a needle valve is affected by any change in pressure before or after the needle valve, which results in a change in piston speed. For example, an increase in the load on a cylinder decreases the pressure differential and reduces fluid flow through the needle valve. This then reduces piston speed. To return the cylinder to its original speed, the needle valve needs to be adjusted.

To increase piston speed, the flow control valve can be partially opened. Piston speed can become too fast if the load on the cylinder is reduced. For example, if a cylinder is used to move a specific load at a specific speed that is set by a needle valve, it will extend at the same speed every time. However, if the weight of that load is reduced, the cylinder will extend faster. To return the cylinder to its original speed, the needle valve needs to be adjusted. To avoid adjusting a needle valve when the pressure differential across it changes, a pressure compensated flow control valve can be used.

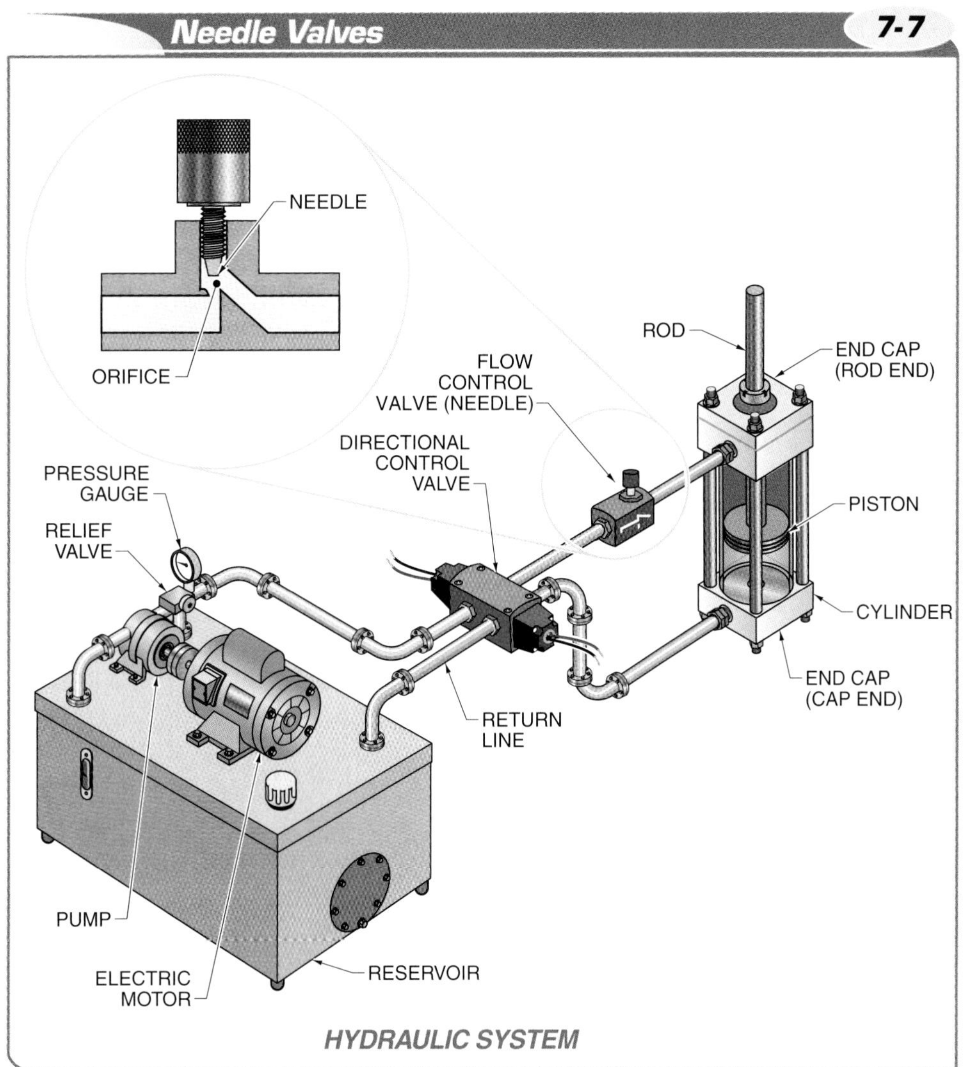

Figure 7-7. A needle valve has a narrowly tapered stem or needle positioned in-line with an orifice that is the same size as the stem or needle.

TERMS

A **pressure-compensated flow control valve** is a flow control valve that changes flow due to changes in pressure before or after the value to keep flow constant from the orifice.

Pressure-Compensated Flow Control Valves. A *pressure-compensated flow control valve* is a flow control valve that changes flow due to changes in pressure before or after the valve to keep flow constant from the orifice. **See Figure 7-8.** The size of the orifice in the valve is determined by a needle valve, while a spring and spool compensate for any change in pressure. Stable fluid flow occurs when the spool, assisted by the spring and controlled orifice, maintains a constant pressure differential across the valve.

When consistent piston speed is necessary and it is important to avoid changing the speed during the course of operation, a pressure-compensated flow control valve is used. Piston speed is affected by changes in system pressure before the flow control valve or in the load on the cylinder. When pressure increases before the valve, it pushes on the end of the spool and against spring tension. As the spool compresses the spring, the land of the spool closes the passageway to the orifice of the valve. Because the spool is restricting fluid flow, pressure before the valve increases more, causing the relief valve to crack open. The unwanted fluid flow returns to the reservoir through the relief valve.

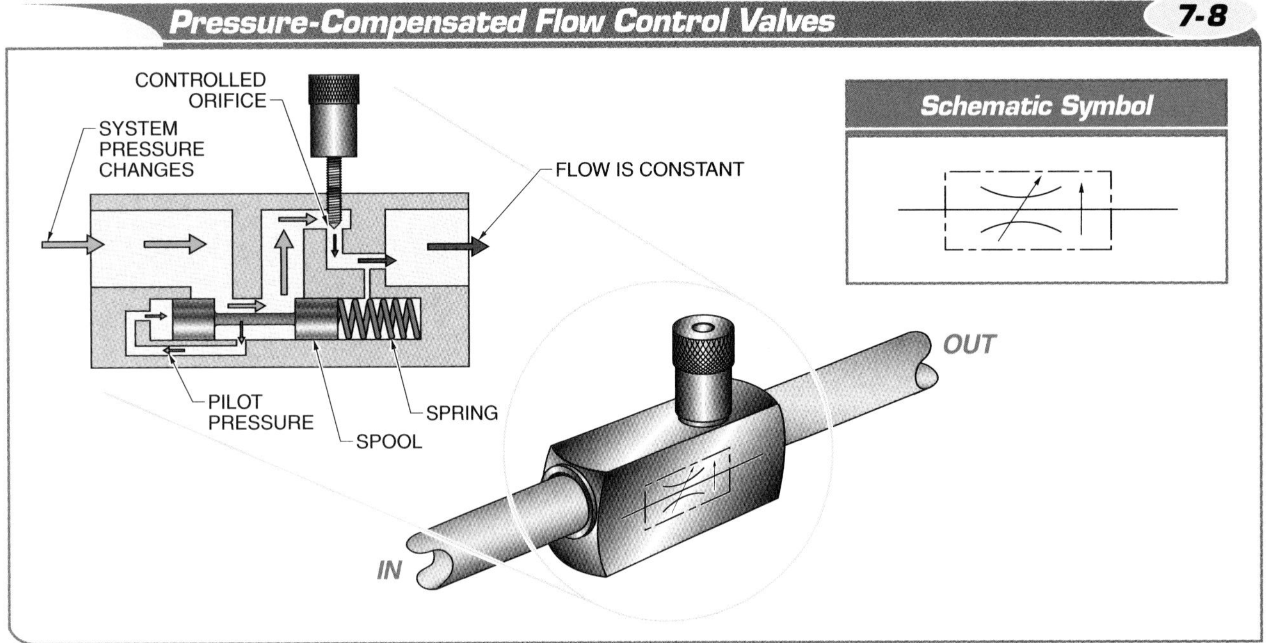

Figure 7-8. A pressure-compensated flow control valve changes flow due to changes in pressure before or after the valve to keep flow constant from the valve.

TERMS

A **temperature- and pressure-compensated flow control valve** is a flow control valve that compensates for changes in hydraulic fluid temperature and pressure.

If the weight of the load on the cylinder increases, the pressure after the valve increases, increasing the pressure on the spring side of the spool and lowering the pressure differential and flow out of the valve.

Pressure-compensated flow control valves are used in applications such as elevating platforms, lift tables, pulp and paper mills, and machine tool equipment. The symbol for a pressure-compensated flow control valve is similar to the flow control valve symbol but has an upward pointing arrow.

Temperature- and Pressure-Compensated Flow Control Valves. A *temperature- and pressure-compensated flow control valve* is a flow control valve that compensates for changes in hydraulic fluid temperature and pressure. **See Figure 7-9.** Hydraulic fluid viscosity changes as temperature changes, which affects flow rate and piston speed. The higher the temperature, the easier it is for hydraulic fluid to flow. The lower the temperature, the more difficult it is for hydraulic fluid to flow.

In some hydraulic systems, such as outdoor hydraulic systems, the temperature can vary greatly throughout the day. To compensate for differences in temperature, temperature- and pressure-compensated flow control valves are used. They are similar to pressure-compensated flow control valves, with a needle valve and a spring and spool, but have an added component that compensates for hydraulic fluid temperature variation.

The most common method is to give the needle a bimetallic stem. The bimetallic stem expands when the temperature increases or contracts when the temperature decreases. As hydraulic fluid temperature increases, it flows more easily. However, the expansion of the bimetallic stem narrows the orifice, maintaining a set amount of fluid flow. As hydraulic fluid temperature decreases, it flows slower and with more difficulty. The contraction of the bimetallic stem enlarges the orifice, maintaining a set amount of fluid flow.

TECH FACT

Dirt can easily clog up a needle valve, which needs to be very precise. Typically, the valve is replaced when it becomes clogged. In some instances, the valve can be opened as much as possible to allow increased fluid flow to unclog it. If this is done, the needle valve must be readjusted afterward to control piston speed.

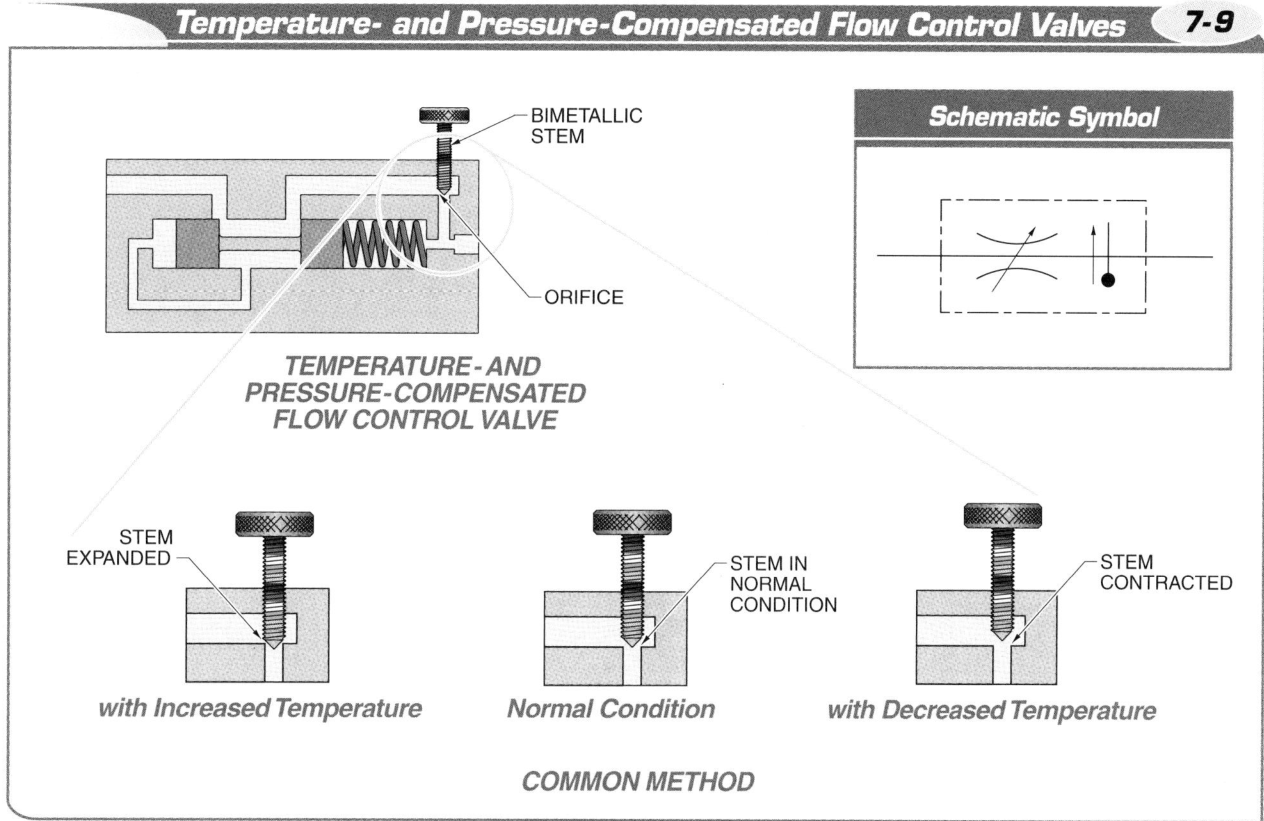

Figure 7-9. A temperature- and pressure-compensated flow control valve compensates for changes in hydraulic fluid temperature and pressure.

Temperature- and pressure-compensated flow control valves are typically used in applications where hydraulic equipment is located outdoors, in areas with equipment in both outdoor and indoor locations, or in facilities with wide fluctuations in temperature. The symbol for a temperature- and pressure-compensated flow control valve is similar to the symbol for a pressure-compensated flow control valve, but has a thermometer symbol adjacent to the upright arrow.

TECH FACT

The type of hydraulic fluid that is used in hydraulic systems with temperature- and pressure-compensated flow control valves must be specified because the valve stem portion is bimetallic and sensitive to different fluid properties. A bimetallic valve stem will degrade prematurely if used with an improper hydraulic fluid.

The control of piston speed in a hydraulic cylinder is critical in applications such as aerial man lifts.

TERMS

Meter-in is a type of flow control where a needle valve controls the fluid flow into an actuator to control its speed.

Meter-out is a type of flow control where a needle valve controls the fluid flow out of an actuator to control its speed.

Piston Speed

In a hydraulic system, fluid flow must move the actuator at the proper speed. Therefore, it is important to calculate piston speed. Piston speed is affected by the amount that an orifice in a flow control valve is open to allow fluid flow. If the pump cannot create enough fluid flow to move the piston at the proper speed, the entire operation may be too slow or may be inoperable.

When calculating the piston speed of a cylinder with no load, the two variables that are taken into consideration are the flow rate and the piston face area. **See Figure 7-10.** Depending on the length of the rod, piston speed can be measured either in feet per minute (fpm) or inches per minute (ipm). Piston speed in fpm is calculated by applying the following formula:

$$P_s = \frac{(Q \times 19.25)}{A_{pf}}$$

where

P_s = piston speed (in fpm)
Q = flow rate (in gpm)
19.25 = constant
A_{pf} = piston face area (in sq in.)

Example: If a pump supplies 15 gpm and a needle valve is used to restrict fluid flow down to 10 gpm before it reaches a piston with a diameter of 4″ at what speed does the piston extend?

$$P_s = \frac{(Q \times 19.25)}{A_{pf}}$$

$$P_s = \frac{(10 \times 19.25)}{12.57}$$

$$P_s = \frac{192.5}{12.57}$$

$$P_s = \mathbf{15.31\ fpm}$$

When working with short rods, piston speed can also be calculated in ipm. Piston speed in ipm is calculated by applying the following formula:

$$P_s = \frac{(Q \times 231)}{A_{pf}}$$

where

P_s = piston speed (in ipm)
Q = flow rate (in gpm)
231 = constant
A_{pf} = piston face area (in sq in.)

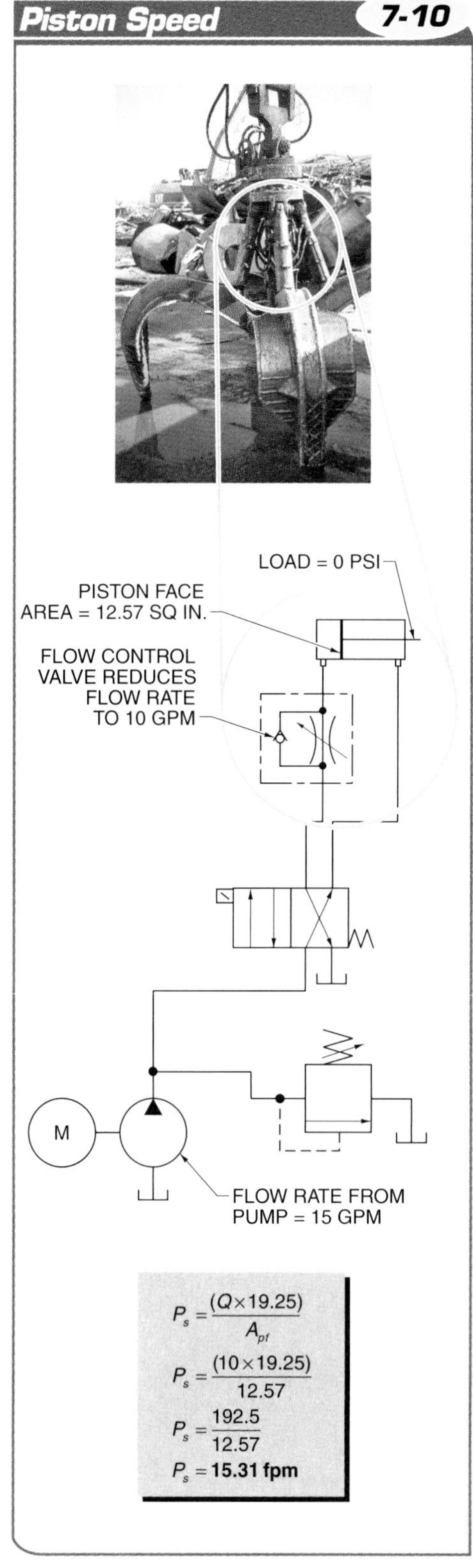

Figure 7-10. When calculating the piston speed of a cylinder with no load, the two variables that are taken into consideration are the flow rate and the piston face area.

Example: If a pump supplies 10 gpm and a needle valve is used to restrict fluid flow down to 7 gpm before it reaches a piston with a diameter of 3.75″ at what speed does the piston extend?

$$P_s = \frac{(Q \times 231)}{A_{pf}}$$

$$P_s = \frac{(7 \times 231)}{11.05}$$

$$P_s = \frac{1617}{11.05}$$

$$P_s = \mathbf{146.33\ ipm}$$

To increase or decrease piston speed, the amount that the needle valve is opened should be adjusted accordingly.

METERING FLUID FLOW

Separate flow control valves can control piston speed by metering fluid flow during extension and retraction. Fluid flow can be metered to single-acting and double-acting cylinders, motors, oscillators, and the pilot lines of directional control valves and other devices.

In many hydraulic systems, fluid flow is metered in only one direction. This is accomplished with a combination of a flow control valve and a check valve. **See Figure 7-11.** By placing a flow control valve and a check valve in reverse parallel with each other, it is possible to meter fluid flow in one direction and allow unrestricted or full flow in the opposite direction. This combination is usually referred to as a flow control valve with a check or internal check. Flow control valves with checks are used in most applications where a flow control valve is installed. Anytime full flow is required in the opposite direction of metered flow, flow control valves with check are used.

Flow control valves with checks can be used in meter-in or meter-out applications. *Meter-in* is a type of flow control where a needle valve controls the fluid flow into an actuator to control its speed. An internal check valve is used to allow unrestricted flow in the opposite direction, out of the actuator. *Meter-out* is a type of flow control where a needle valve controls the fluid flow out of an actuator to control its speed. Typically, an internal check valve is used to allow unrestricted flow in the opposite direction, into the actuator.

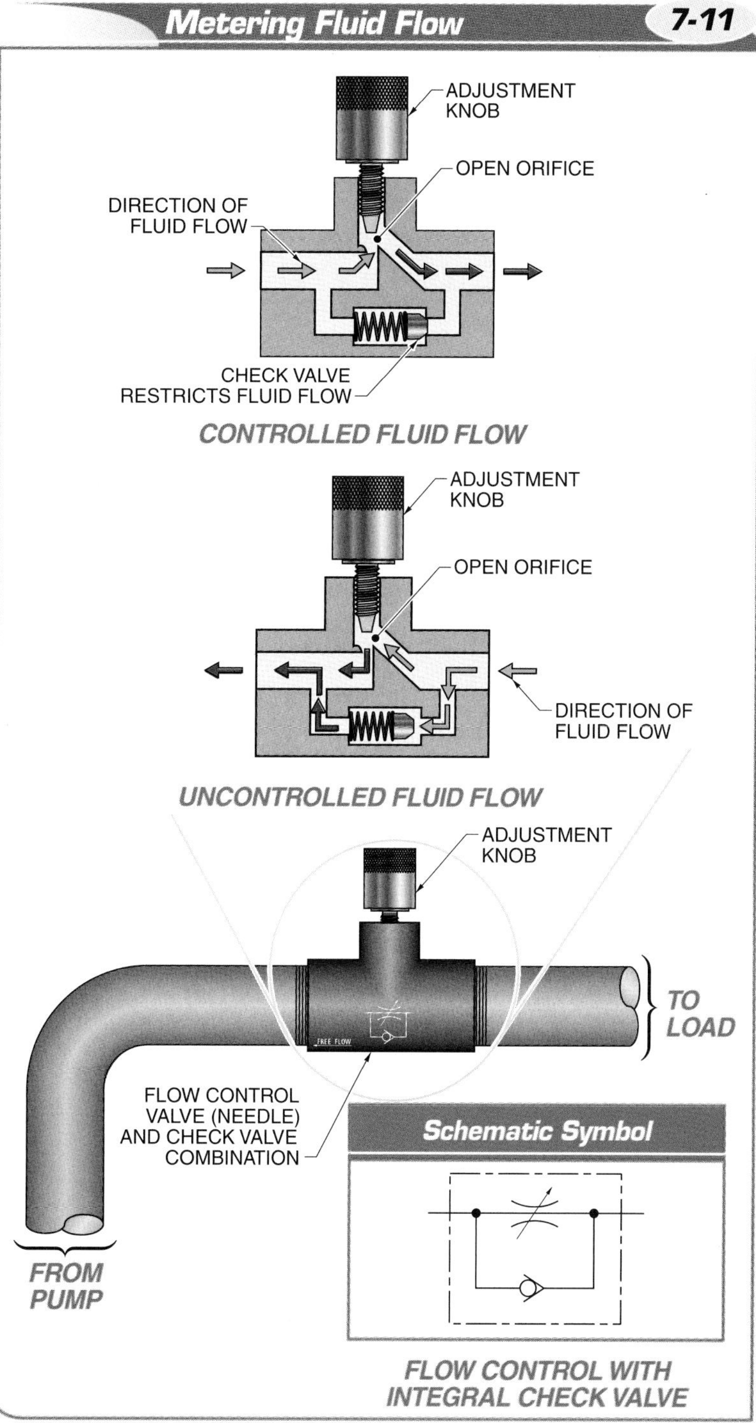

Figure 7-11. Metering fluid flow in one direction is accomplished with a combination of a flow control valve and a check valve.

TERMS

Bleed-off is a type of flow control where a needle valve (no check) controls the fluid flow to the reservoir in a parallel leg of the system to control the speed of an actuator.

A **single-acting cylinder** is a cylinder that uses fluid flow for extension of movement and spring, gravity, or other mechanical means for retraction.

Bleed-off is a type of flow control where a needle valve (no check) controls the fluid flow to the reservoir in a parallel leg of the system to control the speed of an actuator. Bleed-off metering uses a flow control valve without a check valve. Bleed-off systems are used when less flow control accuracy and more system efficiency is required.

Metering Single-Acting Cylinders

A *single-acting cylinder* is a cylinder that uses fluid flow for extension of movement and spring, gravity, or other mechanical means for retraction. **See Figure 7-12.** When hydraulic fluid flows into the cylinder, the piston and rod extend. When hydraulic fluid flows out of the cylinder, the piston and rod retract. A flow control valve with a bypass is used to control one direction of piston movement and speed in a single-acting cylinder.

In this application, meter-in flow control is used to meter fluid flow into the single-acting cylinder to control the speed of the piston during extension. When the piston and rod retract, hydraulic fluid flows unrestricted out of the cylinder through the check valve of the flow control valve. This allows the spring or load to retract the piston and rod as fast as possible. Meter-in systems are typically used for applications that need to extend a heavy load but retract no load.

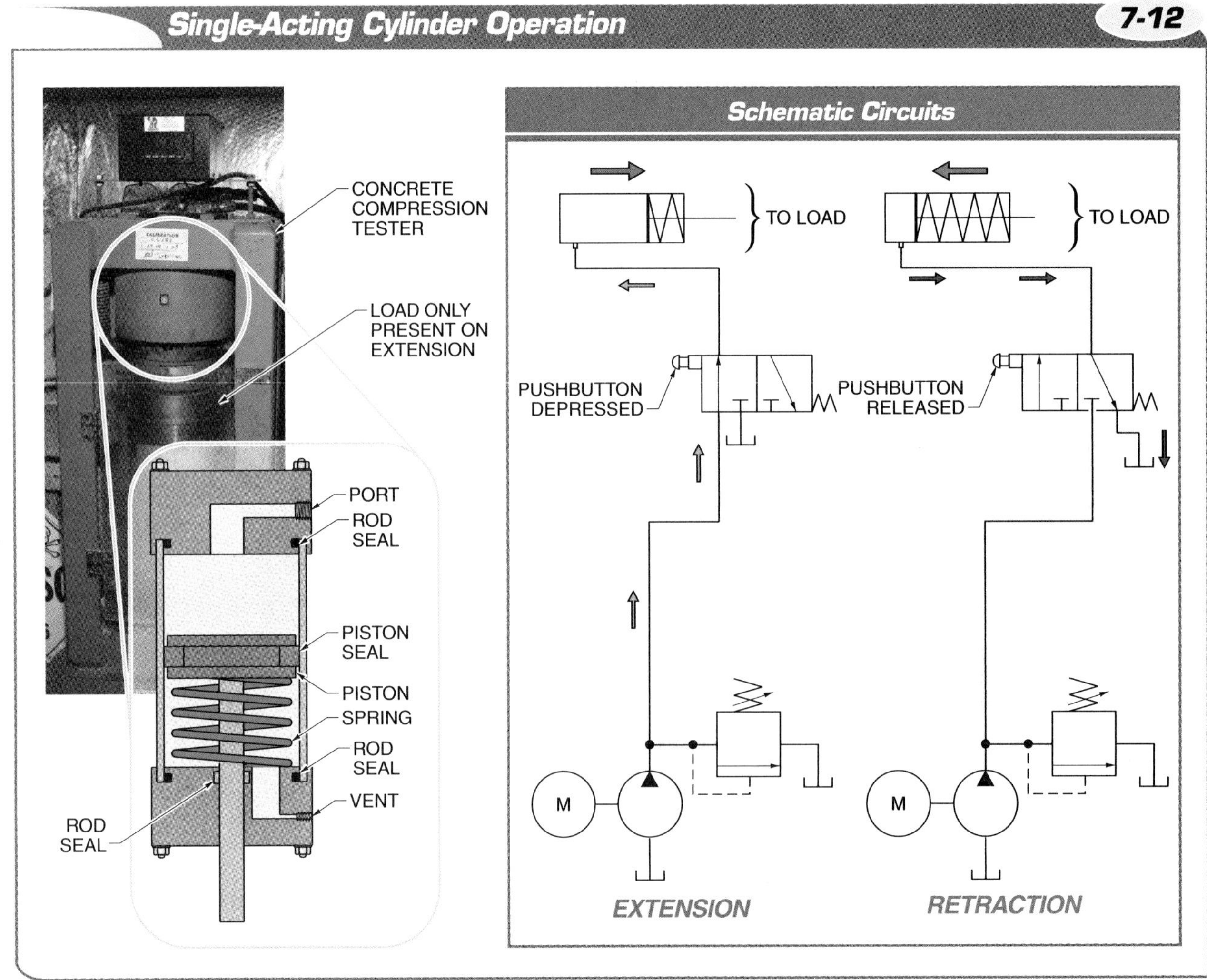

Figure 7-12. A single-acting cylinder uses metered fluid flow for extension, and spring, gravity, or other mechanical means for retraction.

In this application, meter-out flow control is used to meter fluid flow out of the single-acting cylinder to control the speed of the piston during retraction. When the piston and rod extend, hydraulic fluid flows unrestricted into the cylinder through the bypass of the flow control valve. This allows the cylinder to extend the piston and rod as fast as possible. Meter-out is typically used in applications that require high accuracy unless pressure conditions do not allow such use.

To control the piston speed of a single-acting cylinder in both directions, a meter-in and meter-out system is used. This system uses two flow control valves with check valves that are placed directly after each other between the directional control valve and the cylinder. During extension, hydraulic fluid flows unrestricted through the check valves of the first flow control valve and is metered in by the second flow control valve. During retraction, hydraulic fluid flows unrestricted through the check valves of the second flow control valve and is metered out by the first flow control valve.

Another common method of controlling piston speed during extension and retraction is to place the meter-in flow control valve between the cylinder and the directional control valve. The meter-out flow control valve is placed between the directional control valve and the reservoir. **See Figure 7-13.**

Metering Double-Acting Cylinders

A *double-acting cylinder* is a cylinder that has fluid pressure flow alternately to both sides of the cylinder to extend and retract the piston. A double-acting cylinder actuates when pressure develops on one side of the piston. If the pressure is greater on that side of the piston than on the other side of the piston, a pressure differential develops. This causes fluid flow to extend or retract the piston and the rod.

The speed that the piston and rod move is based on the maximum fluid flow allowed into the cylinder, which is determined by the pressure differential across the piston. To control the piston speed of a double-acting cylinder, a flow control valve with a check valve is used to control the flow rate at the cap-end port or the rod-end port of the cylinder. Metering a double-acting cylinder is different than metering a single-acting cylinder because fluid flow can be controlled at two different ports.

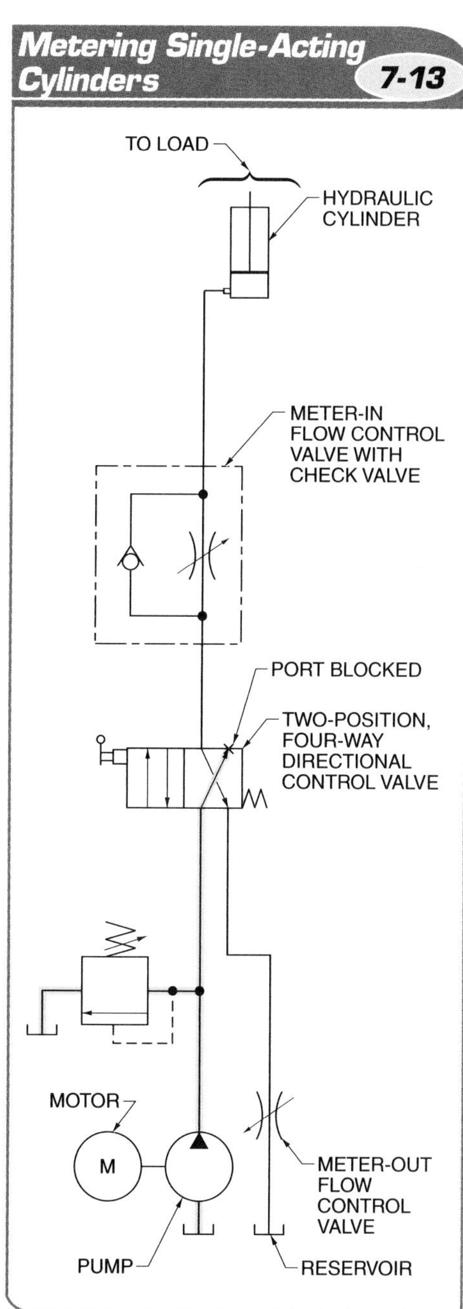

Figure 7-13. A single-acting cylinder uses a meter-in systems and meter-out flow control valve to control piston speed in both directions.

TERMS

A **double-acting cylinder** is a cylinder that has fluid pressure flow alternately to both sides of the cylinder to extend and retract the piston.

Either port of a double-acting cylinder can function as an input or output, depending on whether the cylinder is extending or retracting. **See Figure 7-14.** During extension, fluid flow enters the cap end of the cylinder with enough pressure to extend the piston and rod. Fluid flow then exits the rod end of the cylinder without restricting the movement of the piston and rod. During retraction, fluid flow enters the rod end of the cylinder with enough pressure to retract the piston and rod. Fluid flow then exits the cap end of the cylinder without restricting the movement of the piston and rod.

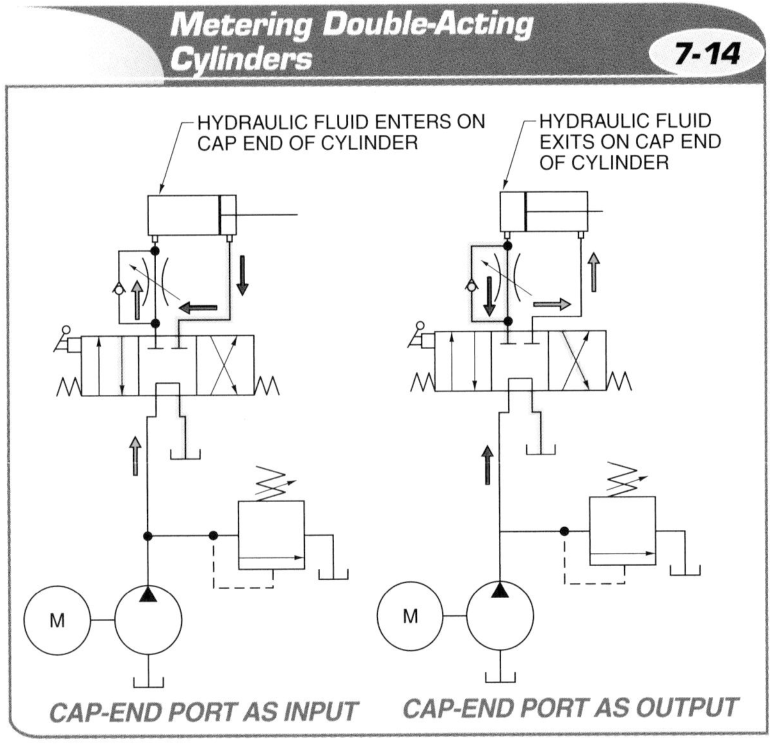

Figure 7-14. Either port of a double-acting cylinder can function as an input or output, depending on whether the cylinder is extending or retracting.

There are three methods of metering a double-acting cylinder. A flow control valve with a bypass is needed for each method. The three methods of metering a double-acting cylinder are controlling the piston speed only on extension, controlling the piston speed only on retraction, and controlling the piston speed on both extension and retraction.

Metering In Double-Acting Cylinders. When using meter-in flow control to control the speed of a piston during extension only, a flow control valve with a check valve is placed between the directional control valve and the cap-end port of the cylinder. Fluid flow is metered as it enters the cap-end port during extension. During retraction, the hydraulic fluid flows unrestricted through the check valve when it exits the cap end.

To control the piston speed only during retraction, the flow control valve is attached to the rod end of the cylinder. Fluid flow is metered as it enters the rod-end port during retraction. During extension, the hydraulic fluid flows unrestricted through the check valve when it exits the rod end.

To control piston speed in both directions using meter-in, two flow control valves with bypasses are placed between the directional control valve and the cylinder. **See Figure 7-15.** One is attached to the cap-end port and the other is attached to the rod-end port.

During extension, fluid flow is metered by the cap-end flow control valve into the cap end of the cylinder. Hydraulic fluid flows out of the rod end, bypassing the rod-end flow control valve unrestricted. During retraction, fluid flow is metered by the rod-end flow control valve into the rod end of the cylinder. Hydraulic fluid flows out of the cap end, bypassing the cap-end flow control valve unrestricted.

Metering Out Double-Acting Cylinders. When using meter-out flow control to control the speed of a piston during extension, a flow control valve with a check valve is placed between the directional control valve and the rod-end port of the cylinder. A meter-out system operates differently from a meter-in system to control piston speed in a double-acting cylinder. The same flow control valves are used in both systems, but are reversed.

If the flow control valve were completely closed, the hydraulic fluid could not flow out of the cylinder and the piston would be unable to move. Regulating the size of the orifice in the flow control valve will meter the flow rate out of the cylinder and control piston speed.

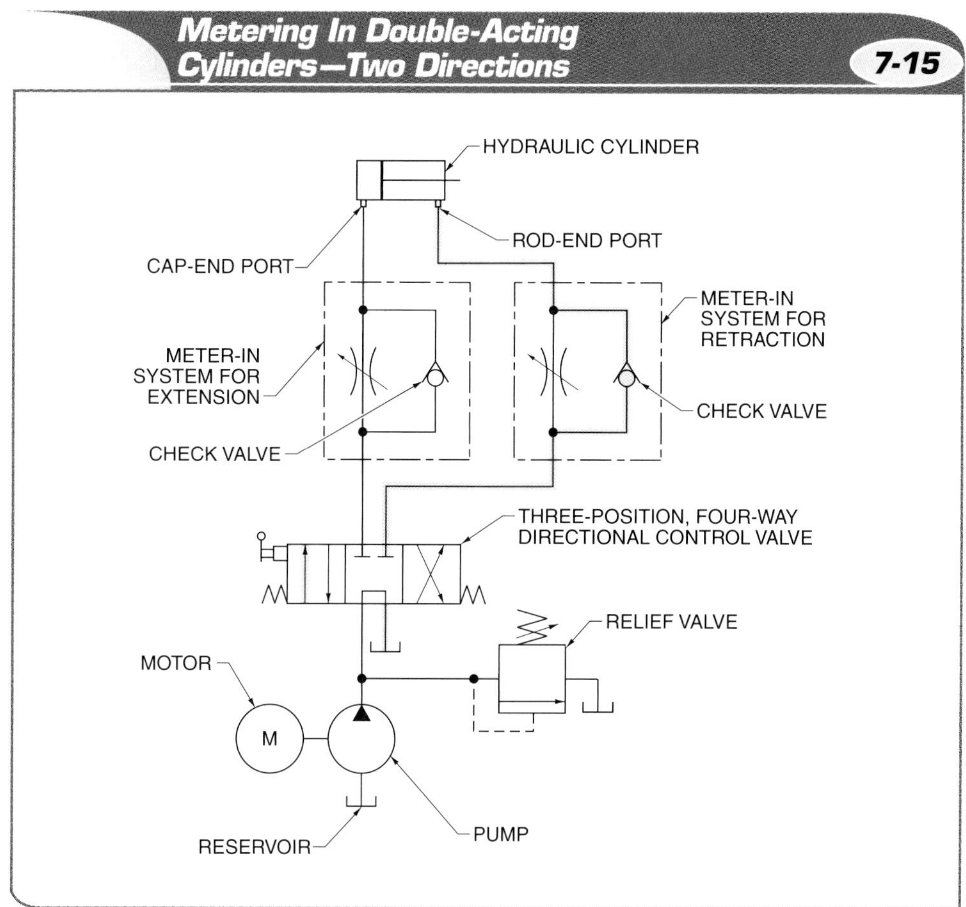

Figure 7-15. To control piston speed in both directions using meter-in, two flow control valves with check valves are placed between the directional control valve and the cylinder.

To control the piston speed only during extension, a flow control valve is attached to the rod end of the cylinder. Fluid flow is metered as it exits the rod-end port during extension. During retraction, the hydraulic fluid flows unrestricted through the check valve when it enters the rod end.

To control the piston speed only during retraction, a flow control valve is attached to the cap end of the cylinder. Fluid flow is metered as it exits the cap-end port during retraction. During extension, the hydraulic fluid flows unrestricted through the check valve when it enters the cap end.

To control piston speed in both directions using a meter-out system, two flow control valves with check valves are placed between the directional control valve and the cylinder. One is attached to the cap-end port and the other is attached to the rod-end port.

A double-acting hydraulic cylinder can be identified by the presence of hose installations on each end of the cylinder barrel.

During extension, hydraulic fluid flows unrestricted through the cap-end bypass into the cap end of the cylinder. Hydraulic fluid in the rod end is then metered out through the rod-end flow control valve. During retraction, hydraulic fluid flows unrestricted through the rod-end check valve into the rod end of the cylinder. Hydraulic fluid on the cap end is then metered out through the cap-end flow control valve.

Meter-out systems are typically used when a change from high pressure to low pressure may occur, such as a drill press breaking through a workpiece. Meter-out systems help prevent uncontrolled movement of a cylinder by metering hydraulic fluid that is exiting the cylinder. Meter-out systems are more commonly used in industrial hydraulic applications than meter-in systems. On a schematic diagram, the bypass check valve arrow points towards the cylinder for meter-in systems and away from the cylinder for meter-out systems. **See Figure 7-16.**

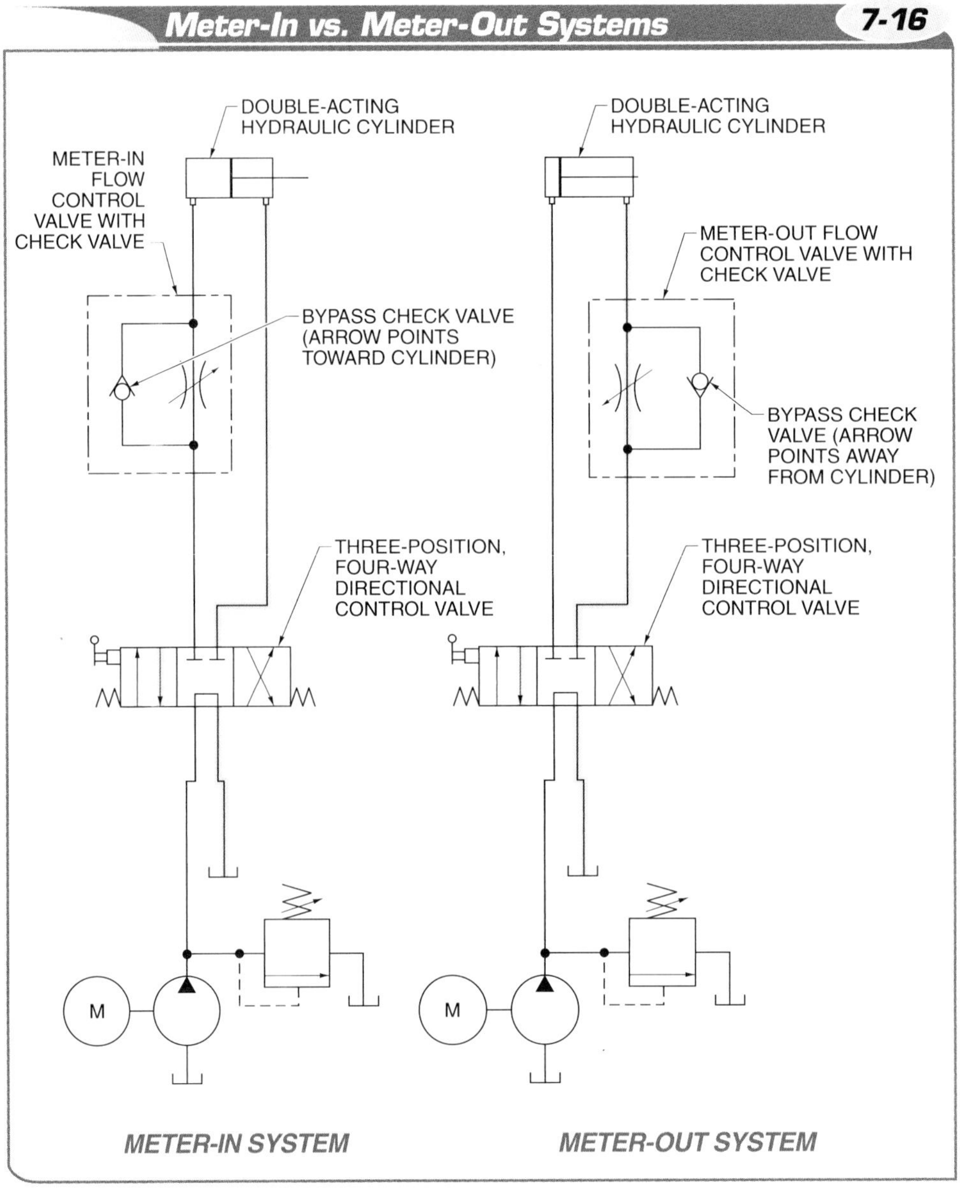

Meter-In vs. Meter-Out Systems **7-16**

Figure 7-16. On a schematic diagram, the bypass check valve arrow points toward the cylinder for meter-in systems and away from the cylinder for meter-out systems.

Bleed-Off Systems

Bleed-off meters flow fluid to and from a cylinder by returning (bleeding off) some of the hydraulic fluid to the reservoir. Flow control valves used for bleed-off do not require a check valve. **See Figure 7-17.** The flow control valve is connected to the line entering the cylinder and to the reservoir. Bleed-off systems take excess fluid flow from the pump and direct it through the flow control valve to the reservoir. For example, if there is a flow rate of 12 gpm at the pump and 4 gpm is sent to the reservoir, then there is a flow rate of 8 gpm to the cylinder. Bleed-off systems can be used in applications such as reciprocating grinding operations and the vertical lifting of a load.

Bleed-off systems used for flow control indirectly control piston speed but have better efficiency than meter-in or meter-out systems. However, they are less accurate for piston speed control than meter-in or meter-out systems because they cannot compensate for pressure fluctuations within the hydraulic system. They are also not as accurate because the metered fluid flow goes to the reservoir rather than to the cylinder. Bleed-off systems should be used only when there is a constant load, precision piston speed accuracy is not required, and there is no chance of uncontrolled cylinder movement.

TECH FACT

Meter-in and meter-out functions are not typically performed at the same time on the same piece of hydraulic equipment because piston speed is controlled in only one direction. However, meter-in and meter-out functions can be performed at the same time on specially designed high-performance hydraulic equipment that requires precise piston-speed control.

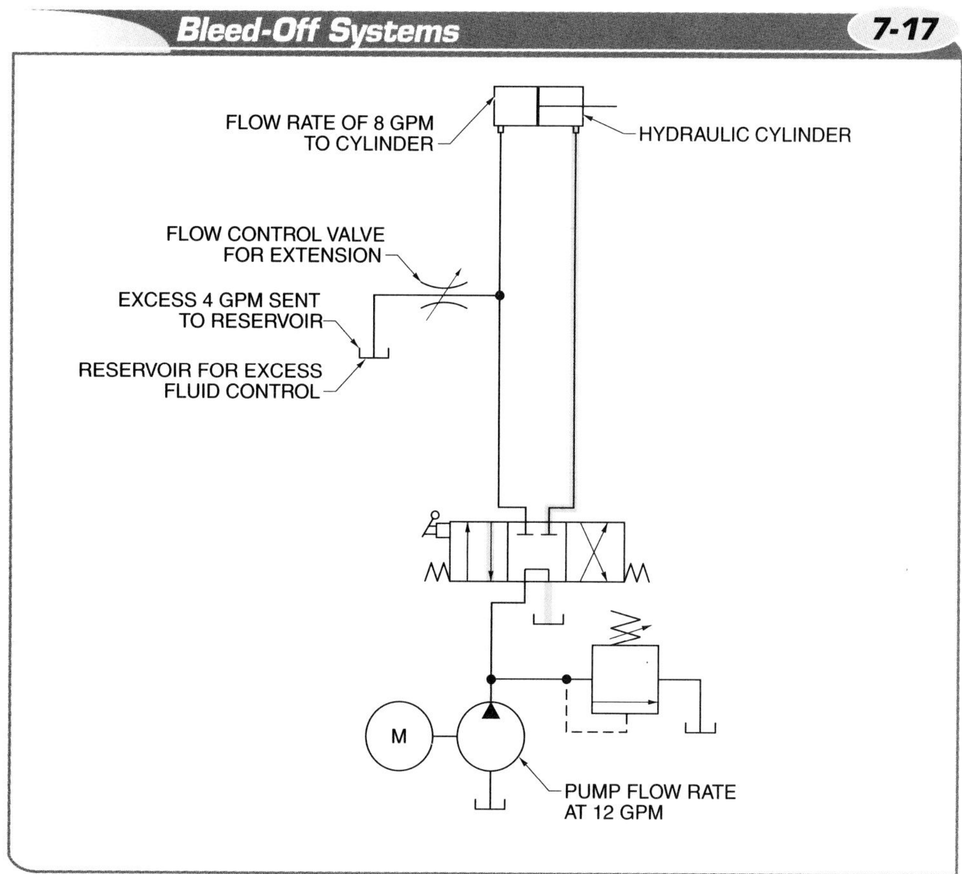

Figure 7-17. A bleed-off system uses a flow control valve without a bypass.

FLOW DIVIDERS

A *flow divider* is a valve that divides flow from a single source equally to two or more sections or components of a hydraulic system. The purpose of a flow divider is to maintain constant flow to two or more sections or components of a system even if each system is operating at a different pressure. The two types of flow dividers are spool-type and rotary-type.

Spool-Type Flow Dividers

A spool-type flow divider uses a spool to divide the flow from a pump between two outlets using a spool for the proper balance. The two types of spool flow dividers are proportional and priority. Spool flow dividers are used in applications such as door drives, lifting platforms, and work access platforms.

Proportional Flow Dividers. A proportional flow divider divides flow through two different orifices and does not change flow rate even if there is pressure fluctuation from one of the outlet systems. The size of the orifice determines how much flow can go through each outlet. If both orifices are the same size, the flow is at a ratio of 1:1. If one orifice is larger than the other, the proportions can vary. There are standard ratios available such as 3:1 and 3:2, but they can also be customized for almost any desired ratio. A common application for proportional flow dividers is to ensure that all two cylinders extend at the same speed. **See Figure 7-18.**

Priority Flow Dividers. A priority flow divider is a flow divider that sends fluid flow to the controlled flow (CF) port and any flow left over to the excess flow (EF) port. The CF port is connected to the part of the system that is vital to the production of the product or is needed to ensure that the machine is running safely. The EF port is connected to either a secondary operation or back to tank and only supplies flow after the needs of the CF have been met. **See Figure 7-19.** This valve is commonly found in mobile power equipment because the speed of the pump can be determined by the speed of the engine. The priority flow divider allows the designer to prioritize what needs flow first in the system. Priority flow dividers can be fixed or variable depending on the needs of the system.

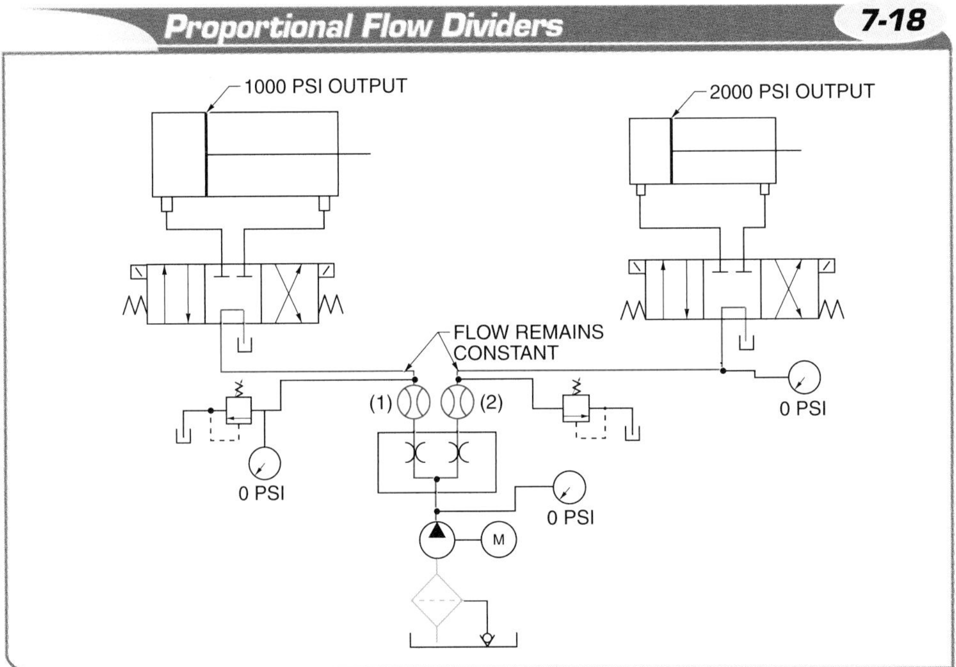

Figure 7-18. A proportional flow divider divides flow through two different orifices and does not change flow rate even if there is pressure fluctuation from one of the outlet systems.

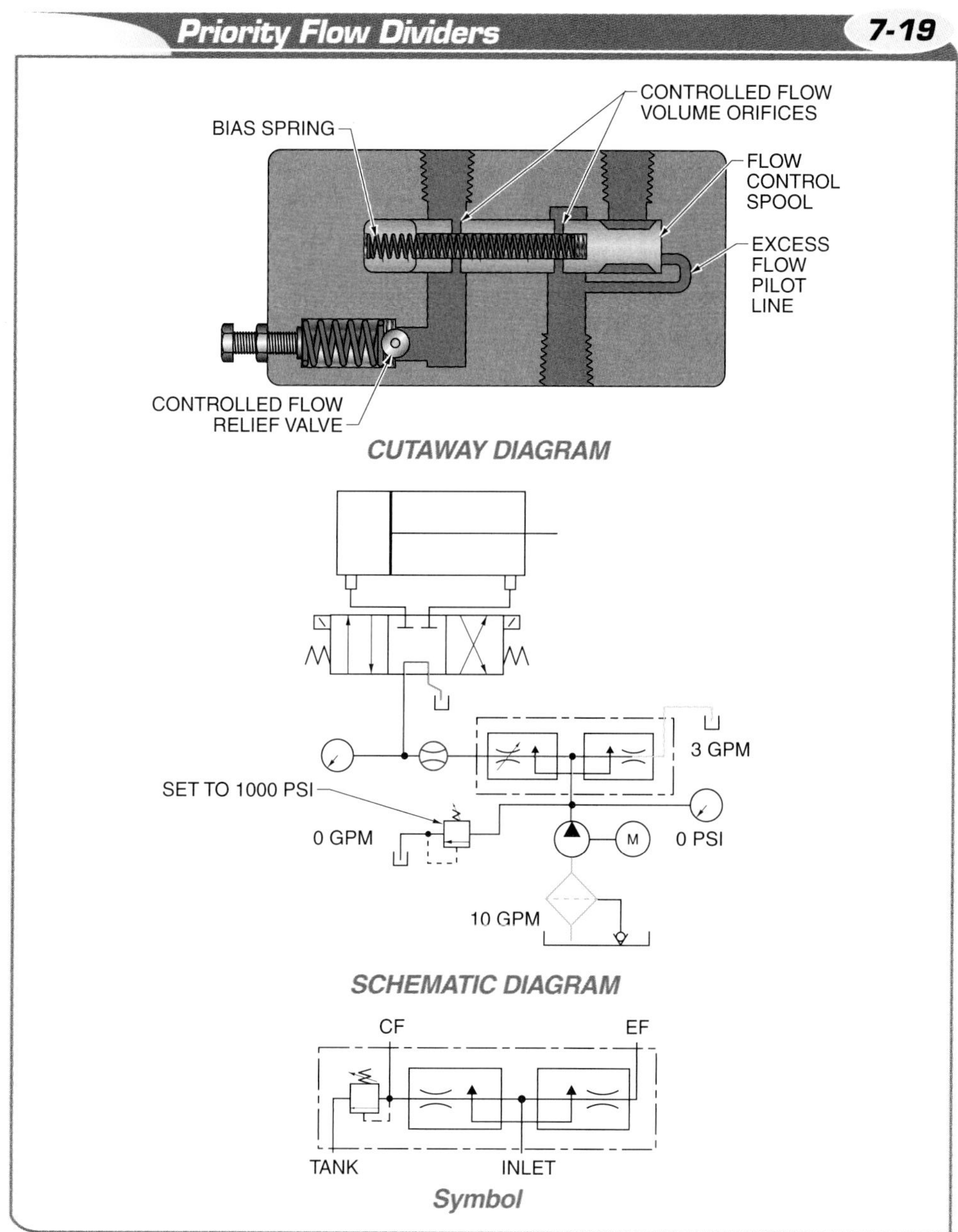

Figure 7-19. A priority flow divider is a flow divider that sends fluid flow to the controlled flow (CF) port and any flow left over to the excess flow (EF) port.

Rotary-Type Flow Dividers

A rotary (motor) flow divider has one or more motors attached to a shaft that rotates them all at the same speed. If all motors have the same displacement, then they supply the same amount of fluid flow to each part of the system. **See Figure 7-20.** If each motor has a different displacement, then the divider supplies a different amount of fluid flow to each part of the system. The advantage of this type of divider is that there is less heat loss because motors are mechanically linked. This makes rotary flow dividers more energy efficient. Also, there can be up to eight motors attached to the shaft, so it will divide into eight separate flows that are not affected by each other. Rotary flow dividers are used in applications such as fuel distribution systems, lube systems, forklift trucks, container handlers, cranes, lifts, and many types of multiple-function machines.

Rotary Flow Dividers 7-20

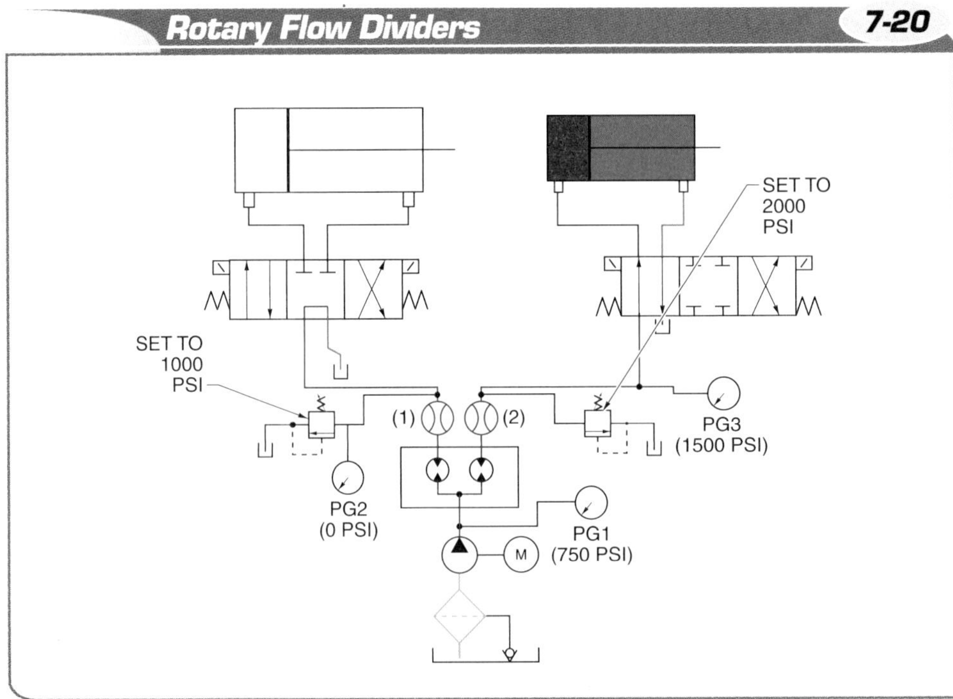

Figure 7-20. A rotary flow divider, or motor flow divider, has one or more motors attached to a shaft that rotates them all at the same speed and is used to regulate fluid flow within the system.

Digital Resources

Chapter Review

Chapter 7 Flow Control

Name: ______________________ Date: ______________

MULTIPLE CHOICE

______________ **1.** A ___ valve is typically used in speed control applications.
- A. ball
- B. gate
- C. globe
- D. needle

______________ **2.** ___ valves are typically used for full-flow or no-flow operations and are not designed for restricting fluid flow.
- A. Ball
- B. Gate
- C. Globe
- D. Needle

______________ **3.** Depending on the length of a rod, piston speed can be measured in ___ or ___.
- A. fpm; ipm
- B. fpm; rpm
- C. gpm; rpm
- D. ipm; rpm

______________ **4.** Gate valves are used for applications with pressures of ___ psi to ___ psi.
- A. 500; 2000
- B. 1000; 3000
- C. 2000; 5000
- D. 5000; 10,000

______________ **5.** A ___ system is typically used for applications that need to extend a heavy load but retract no load.
- A. bleed-off
- B. meter-in
- C. meter-out
- D. all of the above

______________ **6.** Fluid flow must make two 90° turns to pass in one direction through a ___ valve.
- A. ball
- B. gate
- C. globe
- D. none of the above

______________ **7.** A double-acting cylinder can be metered by controlling piston speed on ___.
- A. extension only
- B. extension and retraction
- C. retraction only
- D. all of the above

_______________ 8. A ___ valve has a disk that is raised or lowered over an orifice.
A. ball
B. gate
C. globe
D. needle

_______________ 9. On schematic diagrams, the bypass check valve arrow points ___ the cylinder for meter-in systems.
A. away from
B. towards
C. in either direction of
D. none of the above

_______________ 10. A flow control valve and a check valve in reverse parallel with each other is also referred to as a ___.
A. flow check valve
B. flow control valve with a check or internal check
C. meter-in system
D. meter-out system

_______________ 11. Ball valves can be rated as high as ___ psi.
A. 5000
B. 6000
C. 12,000
D. 21,500

_______________ 12. When piston speed is too high, it can be corrected by using a ___ valve.
A. directional control
B. flow control
C. relief
D. none of the above

_______________ 13. ___ is the controlling of the rate of fluid flow and how the fluid flow is being accomplished.
A. Bleeding
B. Choking
C. Flowing
D. Metering

_______________ 14. To control the piston speed of a double-acting cylinder, a flow control valve with a check valve is used to control the flow rate at ___ of the cylinder.
A. only the cap-end port
B. the cap-end port or the rod-end port
C. only the rod-end port
D. none of the above

_______________ 15. The main variable that affects pressure differential in a hydraulic system is the ___.
A. flow rate (gpm) of the pump
B. speed of piston extension
C. speed of piston retraction
D. weight of a load

COMPLETION

_______________ 1. A meter-in system is used to control piston speed of a single-acting cylinder during ___.

_______________ 2. ___ systems are typically used in double-acting cylinders when an abrupt change from high pressure to low pressure may occur.

_______________ 3. Fluid flow is usually metered in only one direction by using a flow control valve with a(n) ___.

_______________ 4. Fluid flow takes a(n) ___ turn through a flow control valve to create a pressure differential.

_______________ 5. A(n) ___ cylinder uses fluid flow for extension and a spring, gravity, or other mechanical means for retraction.

_______________ 6. A(n) ___ valve compensates for changes in hydraulic fluid temperature and pressure.

_______________ 7. A(n) ___ cylinder has fluid pressure flow alternately to both sides of the cylinder to extend and retract the piston.

_______________ 8. Most flow control valves used in hydraulic systems are ___ valves.

_______________ 9. When calculating the piston speed of a cylinder with no load, the two variables that must be considered are ___ and ___.

_______________ 10. ___ is controlling of the rate of fluid flow.

_______________ 11. A variable orifice in a flow control valve is also referred to as ___.

_______________ 12. When a load on a double-acting cylinder remains constant, a meter-___ system has the most accurate speed control.

_______________ 13. Some types of ___ valves have color-coded valve stems.

_______________ 14. ___ flow control valves operate by maintaining a constant pressure differential across the valve's orifice.

_______________ 15. A(n) ___ valve is a two-position valve with an internal gate between the two seats of the orifice.

_______________ 16. When using meter-in flow control to control the speed of the piston during extension only, a flow control valve with a check valve is placed between the directional control valve and the ___-end port of the cylinder.

_______________ 17. ___ is a type of flow control where a needle valve (no check) controls the flow of fluid to the reservoir in a parallel leg of the system to control the speed of an actuator.

_______________ 18. The most common method of compensating for changes in hydraulic fluid temperature is to use a flow control valve that has a(n) ___ through the path of fluid flow.

_______________ 19. The primary function of a(n) ___ valve is to control the rate of fluid flow.

_______________ 20. If the load on a hydraulic cylinder is reduced, piston speed ___.

TRUE/FALSE

T F 1. Meter-in systems are used only for controlling piston speed.

T F 2. Bleed-off systems are used when more flow control accuracy and less system efficiency is required.

T F 3. To control the piston speed of a double-acting cylinder during retraction by using meter-in, a flow control valve is attached to the rod end of the cylinder.

T F 4. Gate valve have ports that are the same size as the means of transmission.

T F 5. Piston speed is not affected by the amount that an orifice in a flow control valve is open to allow fluid flow.

T F 6. The metering of double-acting and single-acting cylinders is performed differently.

T F 7. Meter-in systems are more commonly used than meter-out systems in double-acting cylinders.

T F 8. A variable orifice is a valve.

T F 9. The difference between a meter-in system and a meter-out system on a double-acting cylinder is that the flow control valves must be reversed.

T F 10. Gate valves can allow fluid flow in both directions, unless otherwise marked.

T F 11. The greater the pressure differential across a needle valve, the less the flow rate through the valve.

T F 12. A ball valve has a ball with an orifice through its center to allow fluid flow.

T F 13. A bleed-off system uses a flow-control valve without a check valve.

T F 14. Gate valves are designed for low-pressure service, typically no higher than 150 psi.

T F 15. Either port of a double-acting cylinder can function as an input or an output, depending on whether the cylinder is extending or retracting.

T F 16. Meter-out is used in applications that require high accuracy unless pressure conditions do not allow such use.

T F 17. Needle valves can have built-in check valves for applications that require fluid flow in the opposite direction of the needle valve's metered flow.

T F 18. Bleed-off systems used for flow control have less efficiency than meter-in or meter-out systems.

T F 19. Fluid flow can only be metered while using double-acting cylinders.

T F 20. Bleed-off meters fluid flow to and from a cylinder by returning (bleeding-off) some of the hydraulic fluid to the reservoir.

T F 21. An orifice is smaller than the inside diameter of a pipe, hose, tube, or internal passage.

T F 22. All gate valves have a nonrising-stem design.

T F **23.** If the load on a cylinder increases, the pressure differential across the flow control valve decreases and fluid flow out of the flow control valve also decreases.

T F **24.** Needle valves have an infinite number of positions for flow control.

T F **25.** A bleed-off system uses a flow control valve with a check valve as a bypass.

SHORT ANSWER

1. Describe how a bimetallic stem is used in a flow control valve to compensate for changes in the temperature of hydraulic fluid.

2. List the three factors to be considered when controlling fluid flow in a hydraulic system.

3. Describe the differences between meter-in and meter-out.

MATCHING

Flow Control Valves

____________________ **1.** Ball

____________________ **2.** Gate

____________________ **3.** Globe

____________________ **4.** Needle

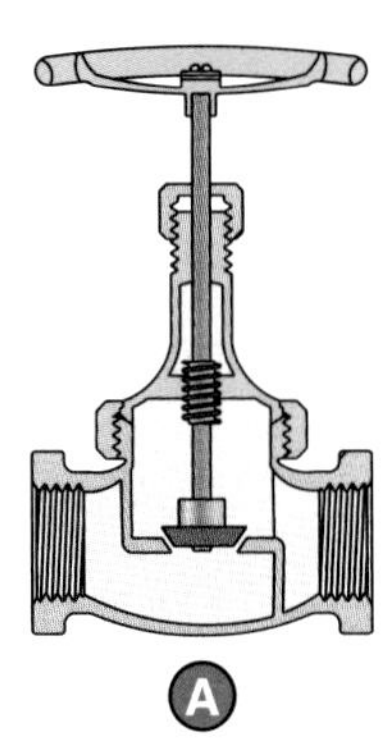

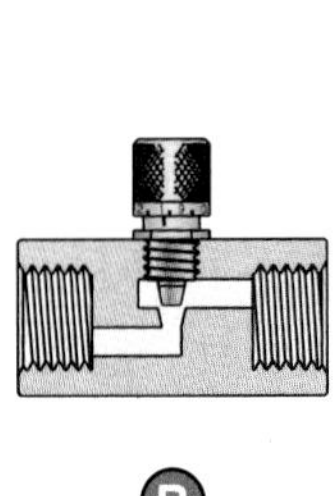

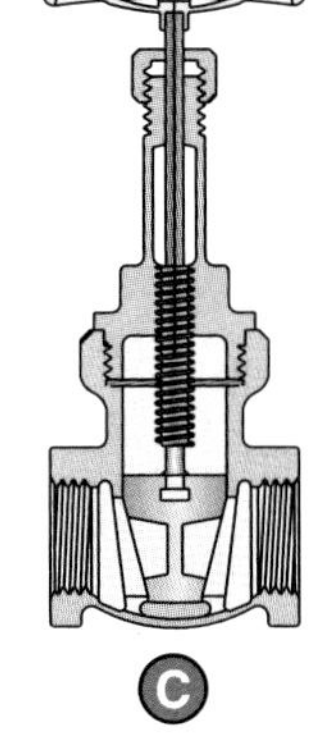

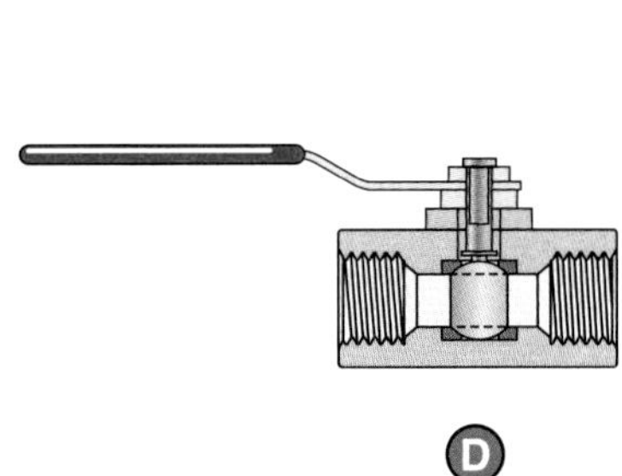

FLOW CONTROL VALVES

Schematic Symbols

Note: Some symbols are used more than once.

____________________ **1.** Ball valve

____________________ **2.** Flow control valve with fixed orifice

____________________ **3.** Flow control valve with variable orifice

____________________ **4.** Gate valve

____________________ **5.** Globe valve

____________________ **6.** Needle valve

____________________ **7.** Pressure-compensated flow control valve

____________________ **8.** Temperature- and pressure-compensated flow control valve

A

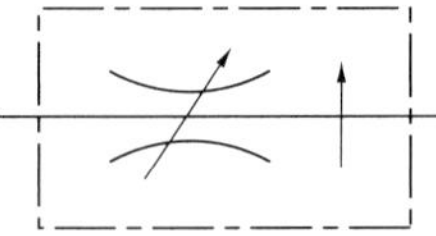

C

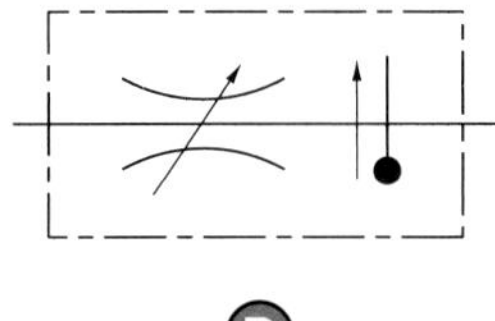

SCHEMATIC SYMBOLS

Activity 7-1: Flow Control Valve Specification

An industrial manufacturing facility performed an energy audit on all of their hydraulic equipment and determined that the piping on the conveyor lift is too small for the amount of fluid flow required. To be able to manage the proper amount of fluid flow, the 3/8″ diameter pipe must be replaced with 1/2″ diameter pipe. To complete this project, the existing in-line flow control valve (Order Number PCBIT55S2) must be replaced with an in-line flow control valve that can be used with 1/2″ diameter pipe. Refer to Flow Control Valve Selection Chart to determine the Order Number for the replacement in-line flow control valve.

1. What is the order number for the in-line flow control valve?

Acme Flow Control

Compensated	Code
Pressure Compensated	PC
Temperature Pressure Compensated	TPC
Non-compensated	NC

Bypass	Code
Bypass check valve	B
No bypass	0

Mounting	Code
Subplate	S
Inline Threaded	IT

Size and Max Pressure	Code
3/8 in – 550 psi	55
1/2 in – 700 psi	70
3/4 in – 1200 psi	120

Material	Code
Steel	S

Flow Rate	Code
* 0.1 – 3 gpm	1
* 0.1 – 5 gpm	2
** 0.1 – 15 gpm	01
** 0.1 – 20 gpm	02
*** 0.5 – 25 gpm	001
*** 0.5 – 30 gpm	002
* 3/8 only ** 1/2 only *** 3/4 only	

FLOW CONTROL VALVE SELECTION CHART

Activity 7-2: Hydraulic Punch Press System

An industrial manufacturing facility has a hydraulic punch press that is controlled by a meter-out. The speed of the cylinder in the hydraulic punch press must be controlled on cylinder extension. Refer to Hydraulic Punch Press System and Flow Control Valve Selection Chart to answer the following questions:

1. What is the order number for the flow control valve?

2. Complete the pictorial diagram below by placing a flow control valve and additional piping in its proper location.

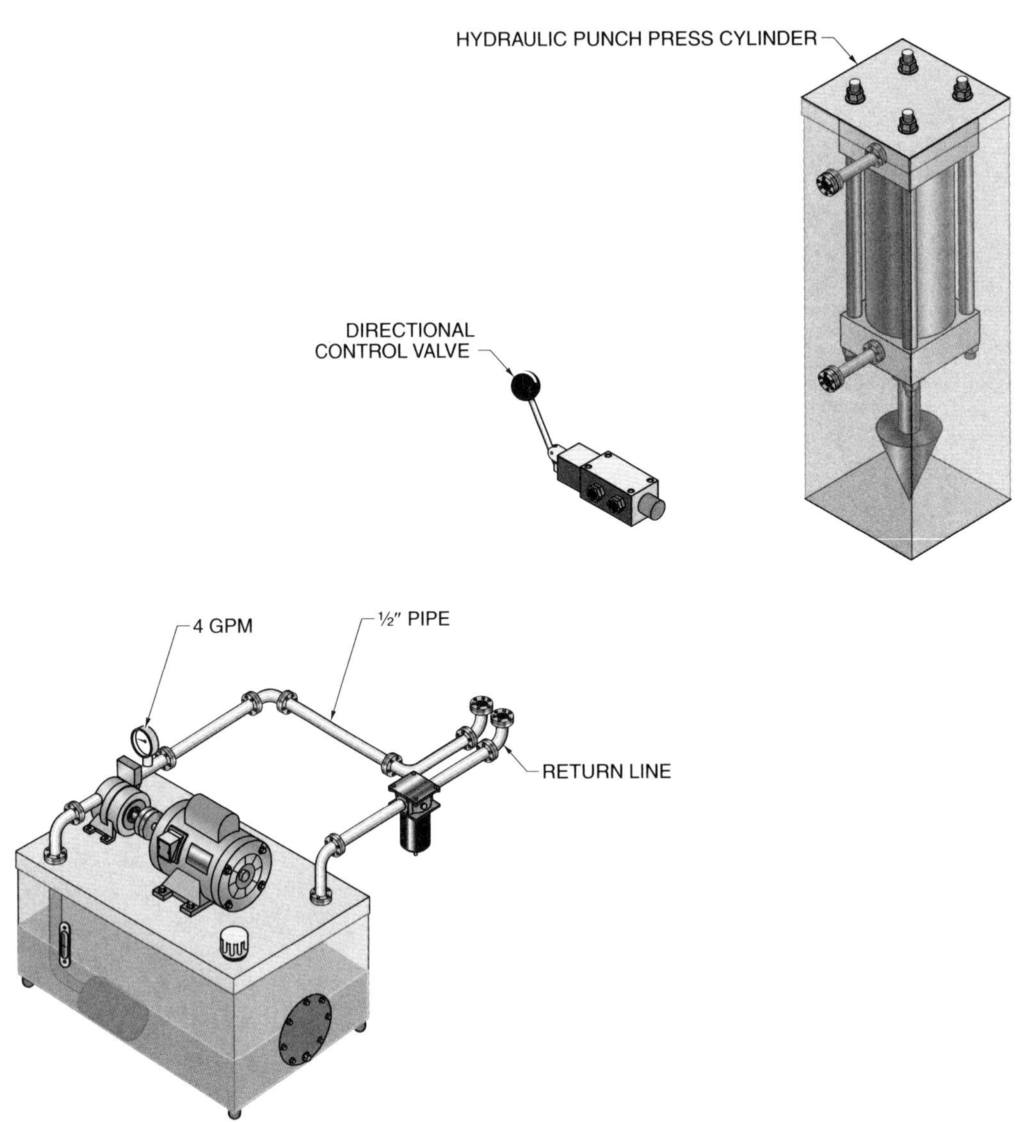

HYDRAULIC PUNCH PRESS SYSTEM

3. List the procedure to add a flow meter into the hydraulic punch press system. *Note:* All pipes are threaded.

Activity 7-3: Hydraulic Clamping Press System

A production manufacturing facility purchases a new hydraulic clamping press. To be compatible with the existing production manufacturing equipment, the clamping cylinder on the new hydraulic clamping press must extend at 150 ipm.

1. Refer to Hydraulic Clamping Press System to determine if the size of the flow control valve orifice needs to be increased or decreased.

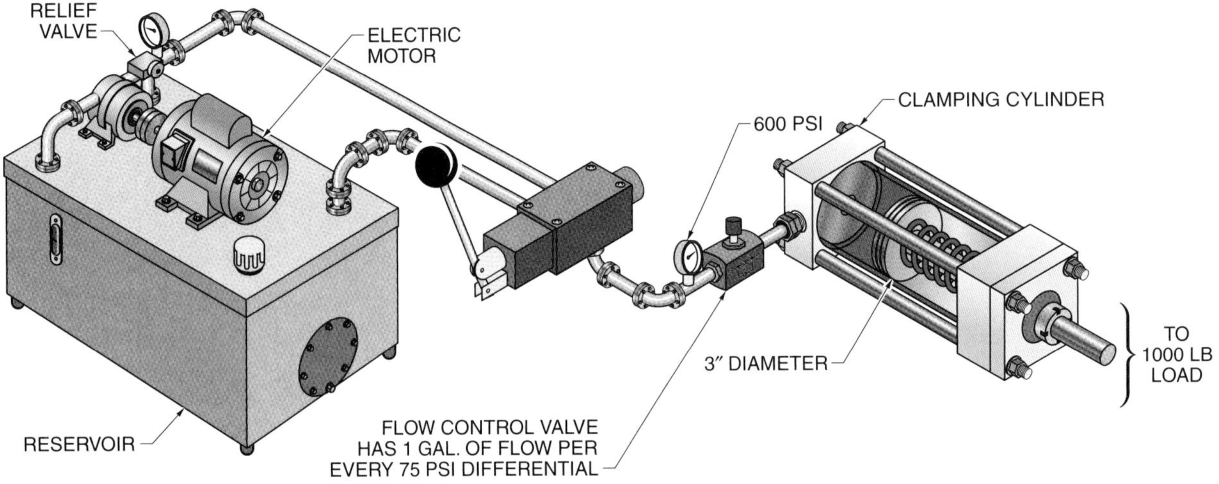

HYDRAULIC CLAMPING PRESS SYSTEM

FluidSIM® Activity — Flow Control, Chapter 7

Activity 7-4: Hydraulic Shear Press Cylinder Control

The speed of a cylinder used in a hydraulic shear press has always been controlled by the flow from the pump. However, the operation of the press needs to be modified to control the speed of cylinder while it extends to shear the metal.

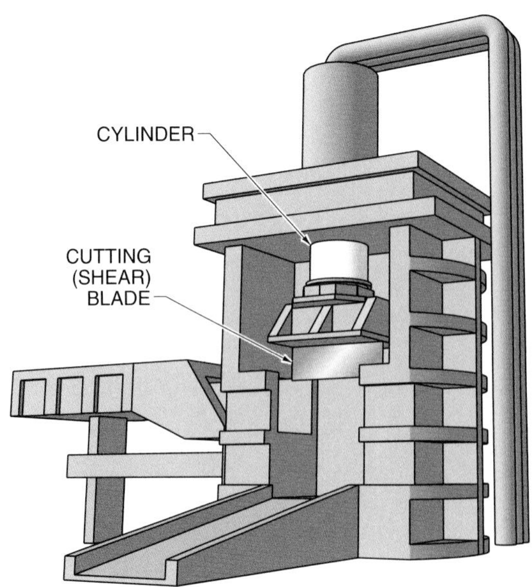

HYDRAULIC SHEAR PRESS

1. Sketch a schematic diagram that shows a meter-out system for controlling the speed of the double-acting cylinder on the extension. Then, use FluidSIM® to build the system and ensure that it will operate properly. A two-position, four-way, spring-return, lever-actuated directional control valve with crossover in the right position controls the cylinder. To adjust the flow control setting, place the computer mouse over the flow control while in simulation mode and click once.

2. Adjust the flow control opening to 1.5 and press the enter key on the computer keyboard. Briefly describe what occurs with the cylinder when the system is operated.

3. Adjust the flow control opening to 8.5 and press the enter key on the computer keyboard. Briefly describe what occurs with the cylinder when the system is operated.

Chapter 8

Hydraulic Actuators

OBJECTIVES

- Describe the different types of hydraulic cylinders and their applications.
- Describe the different types of seals.
- Identify the different types of cylinder mounting methods.
- Describe the different types of hydraulic motors and their applications.

INTRODUCTION

Hydraulic systems have actuators that perform work. The two most common types of hydraulic actuators used in commercial and industrial applications are hydraulic cylinders and hydraulic motors. Seals can be made from different materials and are used to help contain pressure and prevent leaks within a hydraulic cylinder. There are also several different types of hydraulic motors used in hydraulic systems.

TERMS

A **hydraulic cylinder** is an actuator that moves its shaft in a straight line called a stroke.

A **ram cylinder** is a single-acting cylinder that has a large piston and a rod with the same diameter as the piston, and is capable of producing large amounts of linear force during extension.

HYDRAULIC CYLINDERS

A *hydraulic cylinder* is an actuator that moves its shaft in a straight line called a stroke. The main function of a hydraulic cylinder is to convert hydraulic fluid pressure into a mechanical linear force. There are different types and sizes of hydraulic cylinders that perform push and pull operations using linear force.

Hydraulic cylinders are easily identified in a hydraulic system because they have unique features such as a cylinder barrel, a piston and rod, cap-end and rod-end ports, cylinder seals, and end caps. The rod protrudes from the cylinder barrel and is kept clean from dirt and other debris by rod wipers.

Hydraulic cylinders can be mounted in hydraulic systems using different mounting methods. The mounting method used depends on the needs of the hydraulic system and the type of hydraulic cylinder. The different types of hydraulic cylinders are categorized as ram, single-acting, and double-acting. **See Figure 8-1.**

Figure 8-1. Hydraulic cylinders are categorized as ram, single-acting, and double-acting.

Ram Cylinders

A *ram cylinder* is a single-acting cylinder that has a large piston and a rod with the same diameter as the piston, and is capable of producing large amounts of linear force during extension. A ram cylinder consists of a cylinder barrel, a cap-end end cap and port, a piston and rod, a ring stop, and rod seals. **See Figure 8-2.** It is the simplest design of all hydraulic cylinders because it has only one chamber in the cylinder barrel and provides force in only one direction. Ram cylinders are often used in applications such as plastics extrusion presses, hydraulic jacks, and metals processing equipment.

As hydraulic fluid enters the cap-end port, the piston and rod extend. The rod wipers keep dirt out of the cylinder barrel and off the rod as it retracts. When the rod reaches the end of its extension stroke, the ring stop prevents it from overextending. Rod seals form a seal that prevents hydraulic fluid leaks. A ram cylinder retracts using gravity, the weight of the load, or both.

Single-Acting Cylinders

A single-acting cylinder is a hydraulic cylinder that uses metered fluid flow for one direction of movement and spring, gravity, or other mechanical means for the other direction. Single-acting cylinders are used in many hydraulic systems. They are typically less expensive than double-acting cylinders due to installation and product costs. Although a two-position, three-way directional control valve can be used to control a single-acting cylinder, a two-position, four-way directional control valve with one cylinder port plugged is typically used. This is often the case because suppliers usually only stock two-position, four-way directional control valves due to their flexibility for controlling single-acting cylinders and other types of actuators.

Most single-acting cylinders have a rod-end vent that allows air to enter and escape from the rod end of the cylinder. **See Figure 8-3.** This prevents pressure buildup during extension or vacuum during retraction. When there is pressure buildup or vacuum in a hydraulic cylinder, the piston and rod require more energy to move to their desired position. Also, it is common to have the vent piped back to tank, so any fluid that leaks past the piston will go back to tank. Single-acting cylinders are available in several variations including single-acting, spring-return cylinders and telescoping ram cylinders.

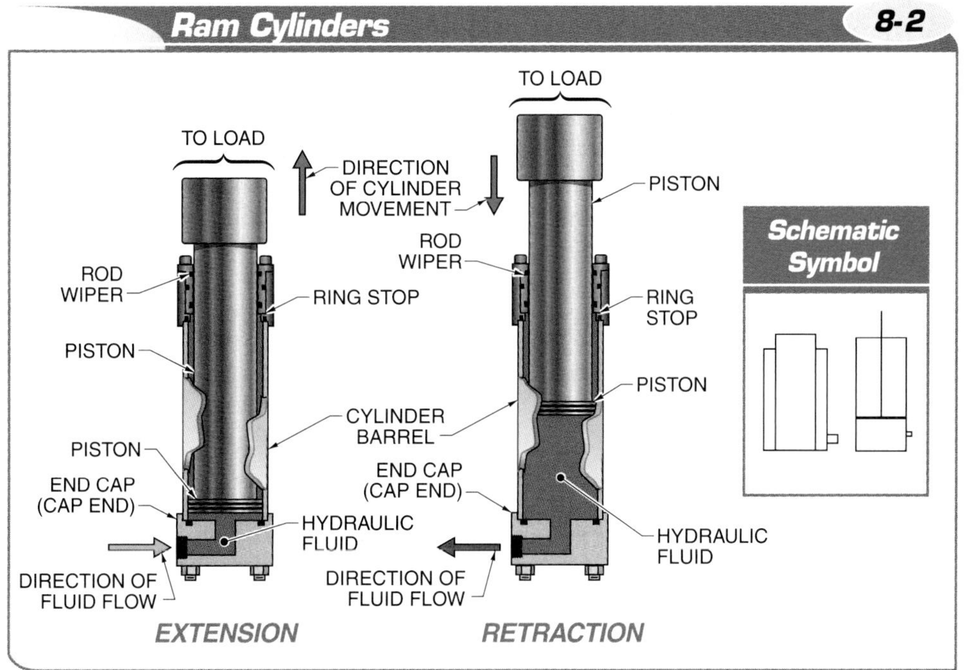

Figure 8-2. A ram cylinder has the simplest design of all hydraulic cylinders because it has only one chamber and provides force in only one direction.

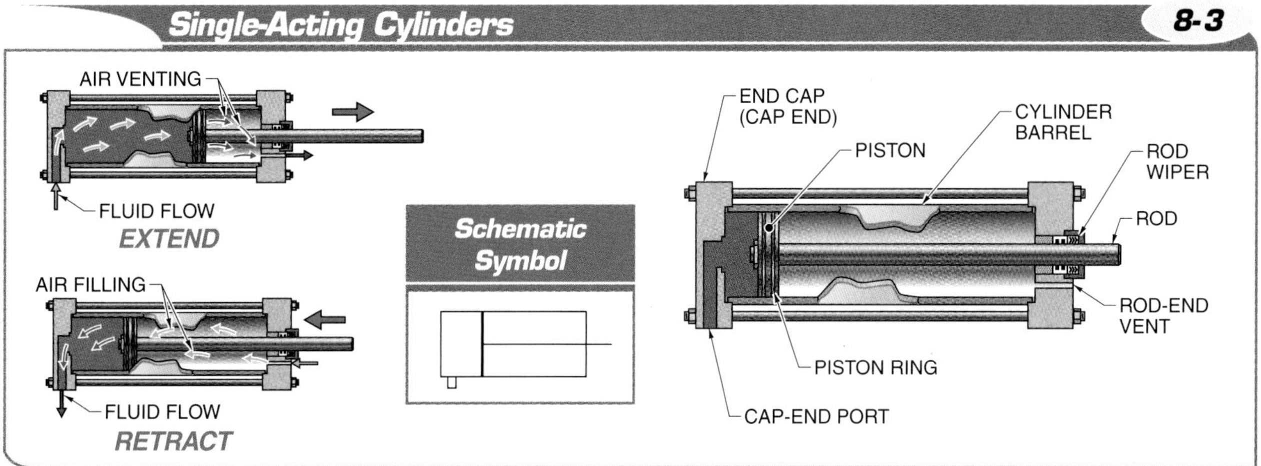

Figure 8-3. Most single-acting cylinders have a rod-end vent that allows air to enter and vent from the rod end of the cylinder.

TERMS

A **single-acting, spring-return cylinder** is a hydraulic cylinder that uses fluid flow for extension and a spring for reaction.

A **telescoping ram cylinder** is a ram cylinder that extends its rod to a length longer than the housing of the cylinder by extending it in stages.

Single-Acting, Spring-Return Cylinders. A *single-acting, spring-return cylinder* is a hydraulic cylinder that uses fluid flow for extension and a spring for retraction. **See Figure 8-4.** The piston and rod are retracted when the cylinder is in its normal position. A single-acting, spring-return cylinder operates through the following procedure:

1. Hydraulic fluid enters the cap-end port, forcing the piston and rod to extend from the cylinder barrel.
2. As the piston and rod extend, air escapes out of the rod-end vent to prevent pressure from building up on the rod side of the piston.
3. Fluid flow to the cylinder is stopped. The cylinder is connected to the reservoir, allowing the spring to retract the piston and rod.
4. As the spring returns the piston and rod to their retracted position, air fills the rod end of the cylinder and assists with retraction.

Common applications for single-acting, spring-return cylinders are in production manufacturing facilities where hydraulic systems are the main method used for handling product. For example, hydraulic systems with single-acting, spring-return cylinders are used to position products along a production line for labeling. Another common application for a single-acting, spring-return cylinder is clamping a workpiece into place for it to be drilled or stamped.

Telescoping Ram Cylinders. A *telescoping ram cylinder* is a ram cylinder that extends its rod to a length longer than the housing of the cylinder by extending it in stages. **See Figure 8-5.** Each stage of the rod is a set length. As hydraulic fluid enters the cylinder, the first stage extends. Once the first stage has extended to its set length, the next stage extends. After the rod has extended to the required length, it retracts as hydraulic fluid flows out of the cylinder. The stages collapse in reverse order due to the weight of the load, such as the bed of a dump truck.

Telescoping ram cylinders are available with rods that have two to five stages and can extend up to 30′. They rarely extend facing downward because the telescoping ram would be unable to retract. There are also double-acting telescoping ram cylinders that use rod-end ports to assist gravity and the weight of the load during retraction. Telescoping ram cylinders are typically used in hydraulic equipment such as elevators, dump truck trailers, and forklift trucks.

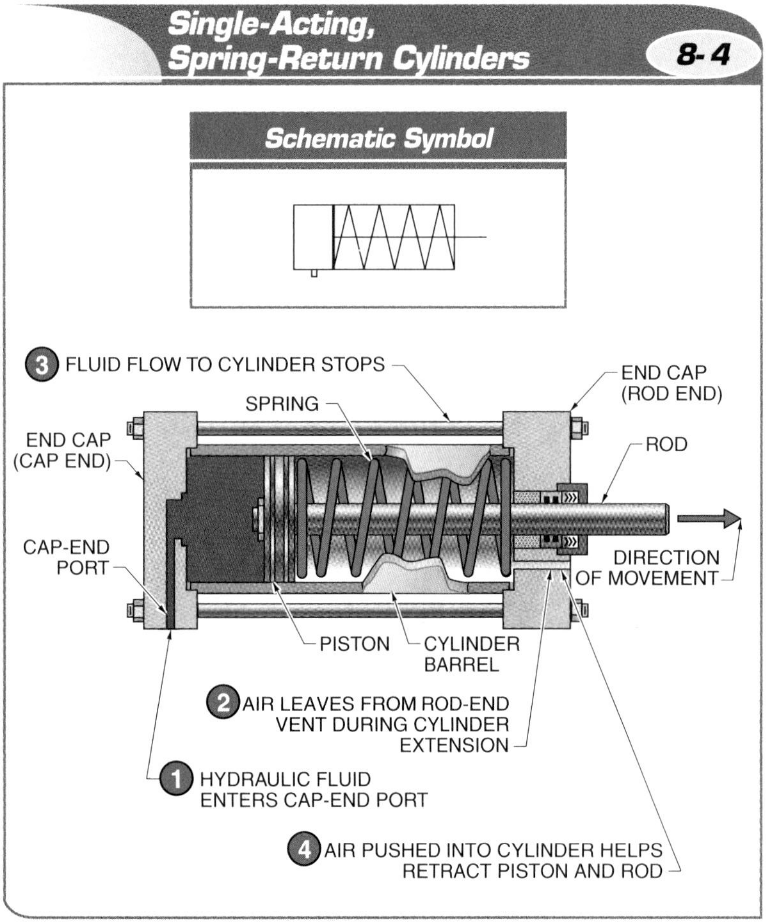

Figure 8-4. A single-acting, spring-return cylinder uses fluid flow for extension and a spring for retraction.

TECH FACT

A hydraulic cylinder can be identified as single-acting or double-acting by the number of hoses attached to the cylinder barrel. A single-acting cylinder has one hose, while a double-acting cylinder has two hoses.

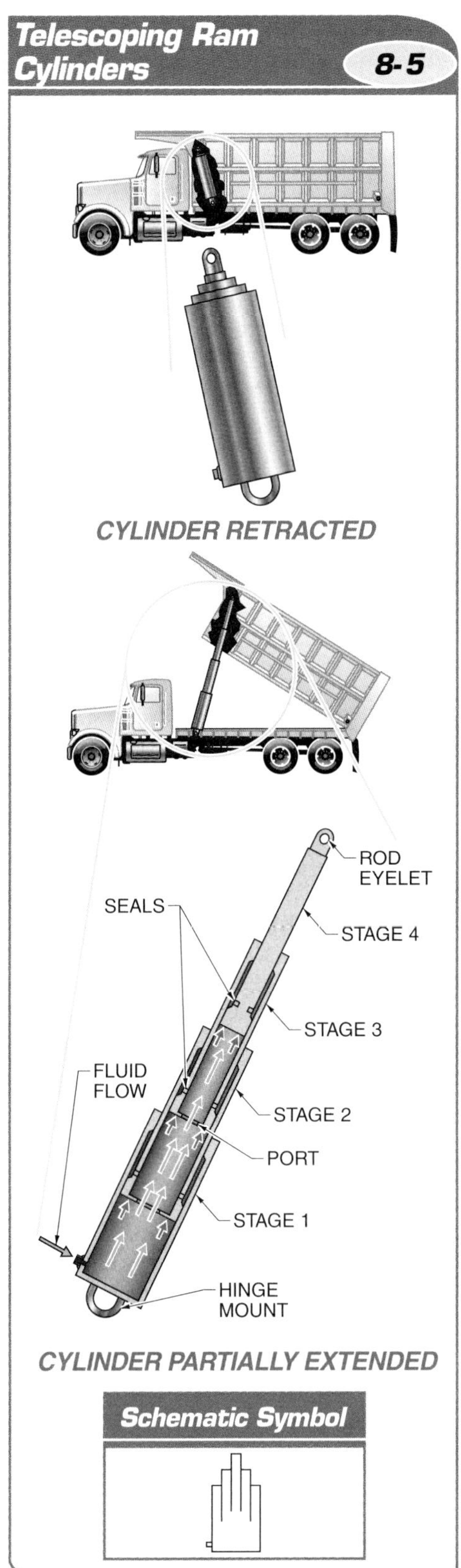

Figure 8-5. A telescoping ram cylinder is a hydraulic cylinder that extends its rod in stages.

Double-Acting Cylinders

A *double-acting cylinder* is a type of cylinder that uses fluid flow through two ports to extend and retract its rod and create force in both directions. Double-acting cylinders are the most common of all hydraulic cylinders. They are able to produce force in both directions by applying hydraulic fluid pressure to either side of the piston. **See Figure 8-6.** Double-acting cylinders are available in a variety of lengths and diameters and can be mounted using different methods.

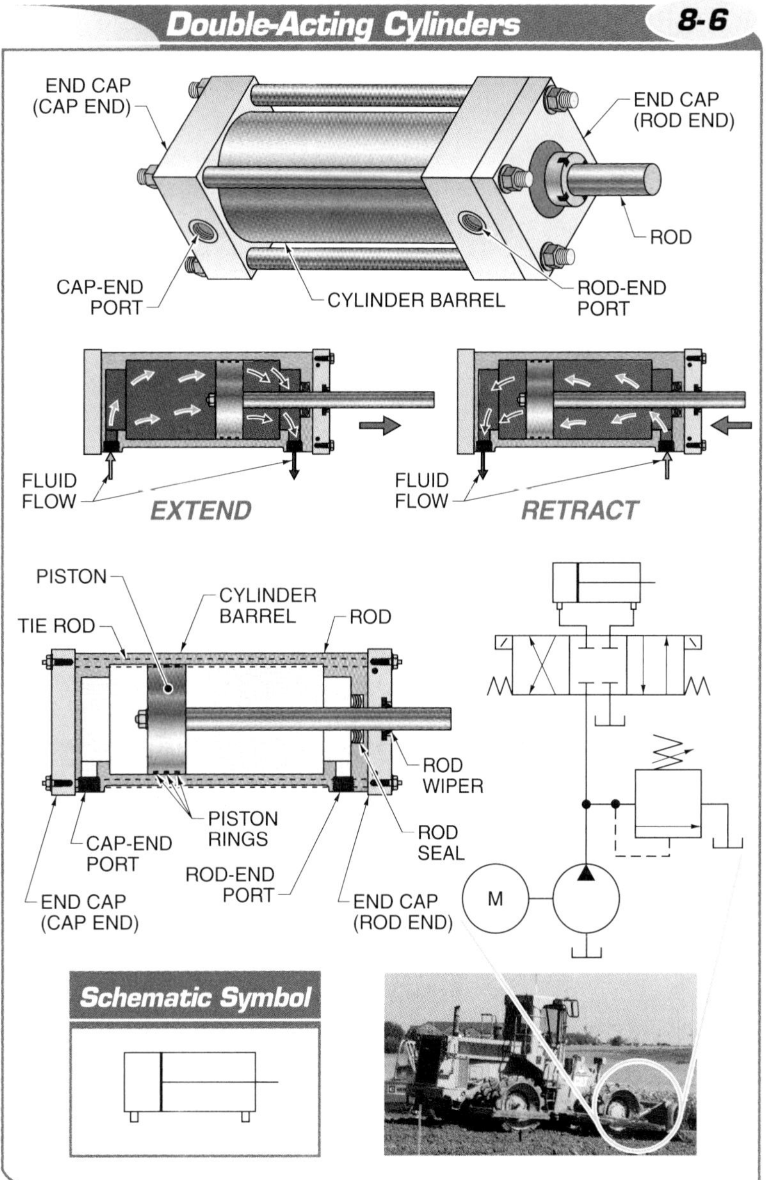

Figure 8-6. Double-acting cylinders are able to produce force in both directions by applying hydraulic fluid pressure to either side of the piston.

TERMS

A **double-rod cylinder** is a hydraulic cylinder that has a single piston and rod that protrudes from both end caps of the cylinder.

A **tandem cylinder** is a hydraulic cylinder that consists of two or more in-line cylinders with their rods connected to form a common rod.

A **duplex cylinder** is a hydraulic cylinder that consists of two or more in-line cylinders that do no have their rods connected to form two or more cylinders in one housing.

The directional control of a double-acting cylinder is performed by a directional control valve that has two cylinder ports. There are several different types of double-acting cylinders including double-rod cylinders, tandem cylinders, and duplex cylinders. Accessories for double-acting cylinders are used for various applications. For example, double-acting cylinders can be fitted with cylinder cushions and cylinder stop tubes and used in applications such as presses, log-splitters, and robotics.

Double-Rod Cylinders. A *double-rod cylinder* is a hydraulic cylinder that has a single piston and rod that protrudes from both end caps of the cylinder. **See Figure 8-7.** A double-rod cylinder has the same amount of surface area on both sides of the piston. This will produce the same amount of force during extension and retraction. A double-rod cylinder can also provide equal piston speed in both directions and can be more productive than a single-rod cylinder. The control of a double-rod cylinder is the same as with a double-acting cylinder.

Telescoping cylinders are typically used with mobile hydraulic equipment such as dump truck trailers.

Double-rod cylinders are used in applications such as opening and closing filters and filter tanks for cleaning and maintenance. They are also used in push and pull operations or where equal piston speed and force is required in both directions. They can be placed adjacent to identical workstations so one side of the rod is retracting while the other side is extending. For example, double-rod cylinders can be installed in applications that have product moving along more than one conveyor system.

Tandem Cylinders. A *tandem cylinder* is a hydraulic cylinder that consists of two or more in-line cylinders with their rods connected to form a common rod. **See Figure 8-8.** Tandem cylinders have two pistons on the same rod and have four ports. Two of the ports always work together as the inlets, while the other two ports work together as the outlets.

A tandem cylinder is used when a single large-diameter cylinder cannot be used because of space considerations. Since there are two pistons inside the cylinder barrel, the barrel must be longer than a regular double-acting cylinder to extend to the same length. Tandem cylinders have fast extension and retraction speeds and are used in applications that require short strokes and high force. Typical applications of tandem cylinders include hydraulic presses and forming machines that require a high amount of force but have limited space for a large diameter cylinder to operate.

Duplex Cylinders. A *duplex cylinder* is a hydraulic cylinder that consists of two or more in-line cylinders that do not have their rods connected to form two or more cylinders in one housing. Duplex cylinders have the same assembly as tandem cylinders, but the rods are not connected together. **See Figure 8-9.** Duplex cylinders can be used to accomplish and maintain different fixed stroke lengths. They are used in applications such as two-post hydraulic lifts, which are used to lift vehicles in automotive repair facilities.

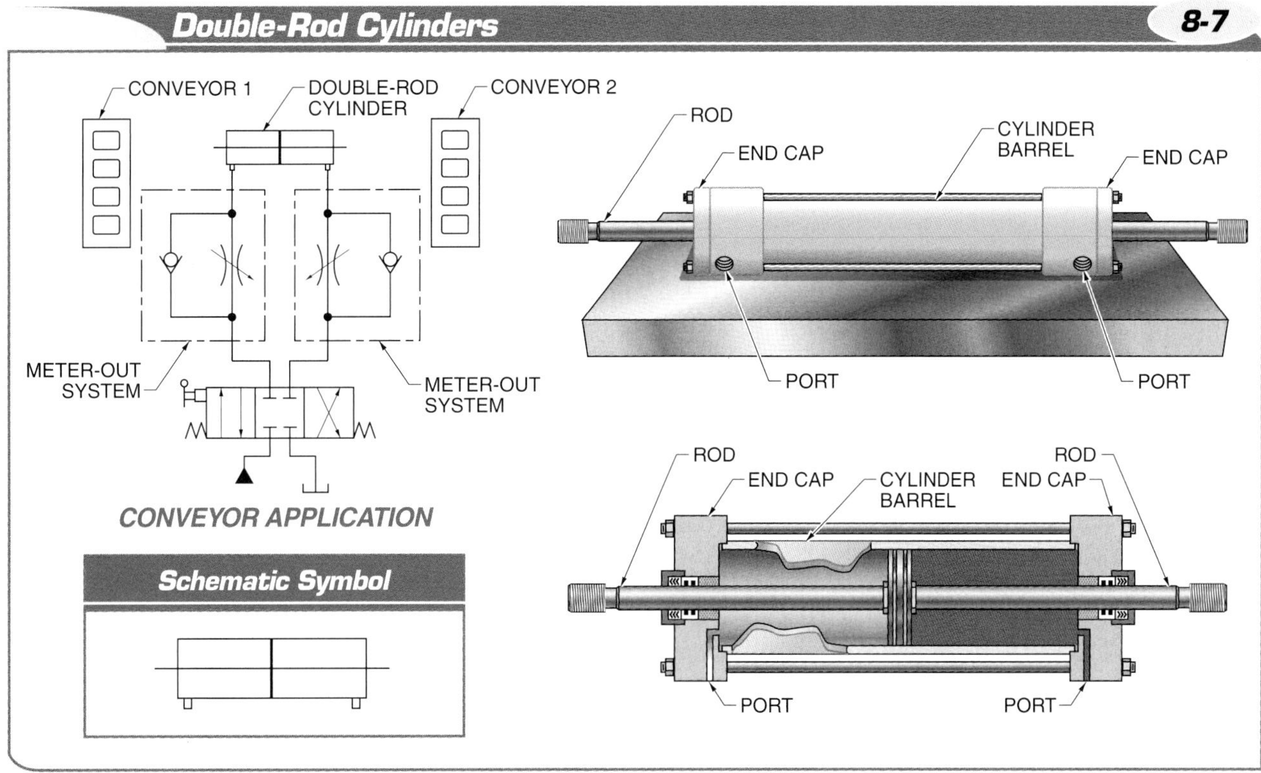

Figure 8-7. A double-rod cylinder can provide equal force and piston speed in both directions.

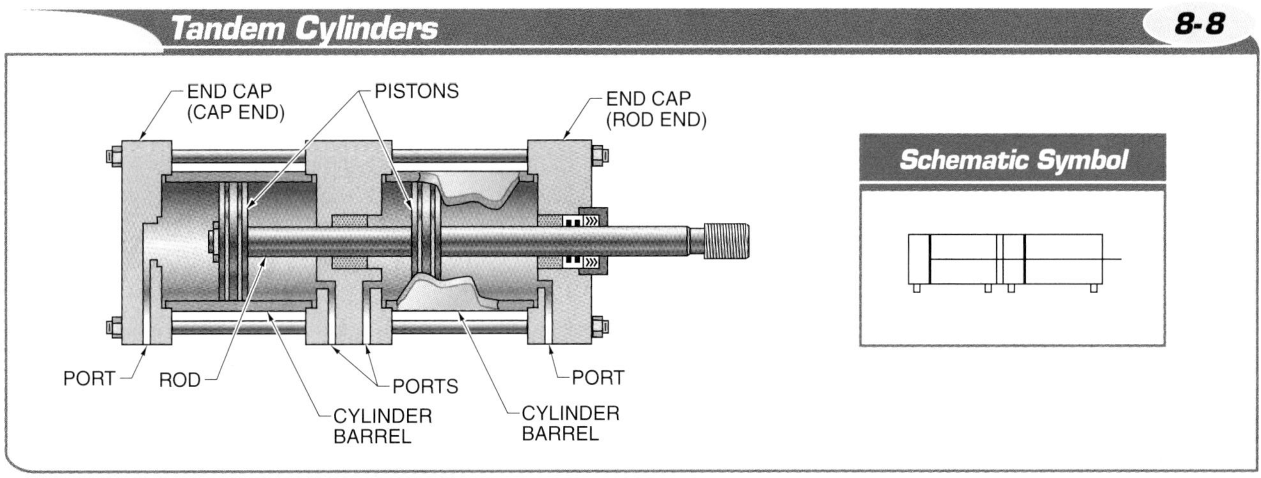

Figure 8-8. Tandem cylinders consist of two or more in-line cylinder barrels with their rods connected to form a common rod.

Cushions. A *cylinder cushion* is a tapered plug attached to the piston or rod that fills an exit hole on either end of the cylinder and is used to prevent the piston from colliding with the inside surfaces of the end caps. When a rod moves, it is stopped at the end of its stroke by one of the caps at the end of the cylinder. The cylinder is stopped by the rod-end end cap on the extend stroke and the cap-end end cap on the return stroke. When a cylinder is being used in a high-speed application, abruptly stopping the cylinder as it moves at a fast speed can damage its cap- or rod-end end cap. To prevent damage to the end caps, a cushion can be installed. The two types of cushions are variable cushions and fixed cushions.

TERMS

A **cylinder cushion** is a tapered plug attached to the piston or rod that fills an exit hole on either end of the cylinder and is used to prevent the piston from colliding with the inside surfaces of the end caps.

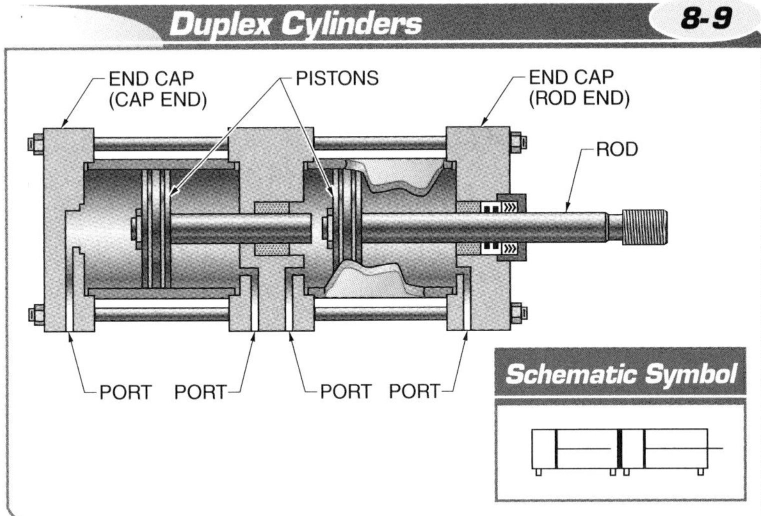

Figure 8-9. Duplex cylinders have the same assembly as tandem cylinders, but the rods are not connected.

Cylinder stop tubes are often used with hydraulic equipment that has long, horizontal rods.

TERMS

A **variable cushion** is a needle flow control valve with a check valve in the rod-end end cap, cap-end end cap, or both and a tapered plug attached to the piston or rod to slow the piston at the end of its stroke.

A **fixed cushion** is a tapered plug attached to the piston or rod that fills an exit hole on either end of the cylinder.

A *variable cushion* is a needle flow control valve with a check valve in the rod-end end cap, cap-end end cap, or both and a tapered plug attached to the piston or rod to slow the piston at the end of its stroke. Variable cushions prevent the piston from colliding with the inside surfaces of the cylinder end caps. Variable cushions are also adjustable. **See Figure 8-10.** A variable cushion can be used on either end cap of the cylinder and operates through the following procedure:

1. As the piston gets close to the end of its stroke, the cushion attached to the piston or rod fills the space of the exit hole.
2. As the exit hole is filled by the cushion, all remaining fluid is forced to flow through the needle valve on the same end of the cylinder.
3. Fluid flow through the needle valve causes the piston speed to slow and allows the cushion to absorb any residual force, preventing it from slamming into the inside surface of the end cap.
4. When fluid enters the cylinder to push the piston in the opposite direction (return stroke), fluid flows through a check valve, bypassing the needle valve. This prevents the cylinder from being too slow at the beginning of its stroke.

Needle valves are adjusted with an adjustment screw that can rotate more than 360°. The only method to adjust the cylinder cushion is to operate the cylinder and adjust the needle valve adjustment screws with a flathead screwdriver until the cylinder cushion is set to the desired setting.

A *fixed cushion* is a tapered plug attached to the piston or rod that fills an exit hole on either end of the cylinder. A fixed cushion operates in the same manner as a variable cushion, but does not have an adjustable needle valve. The amount and speed of the cushion is fixed and cannot be adjusted.

TECH FACT

One method to determine if a cylinder cushion has malfunctioned is by the sound of the cylinder during operation. When the cylinder cushion has malfunctioned, the piston slams against the cylinder cap end hard enough that a banging sound can easily be heard. Also, if the needle valve gets clogged with large particle debris such metal chips, the cylinder will not be able to complete its stroke and can cause a problem within the hydraulic system.

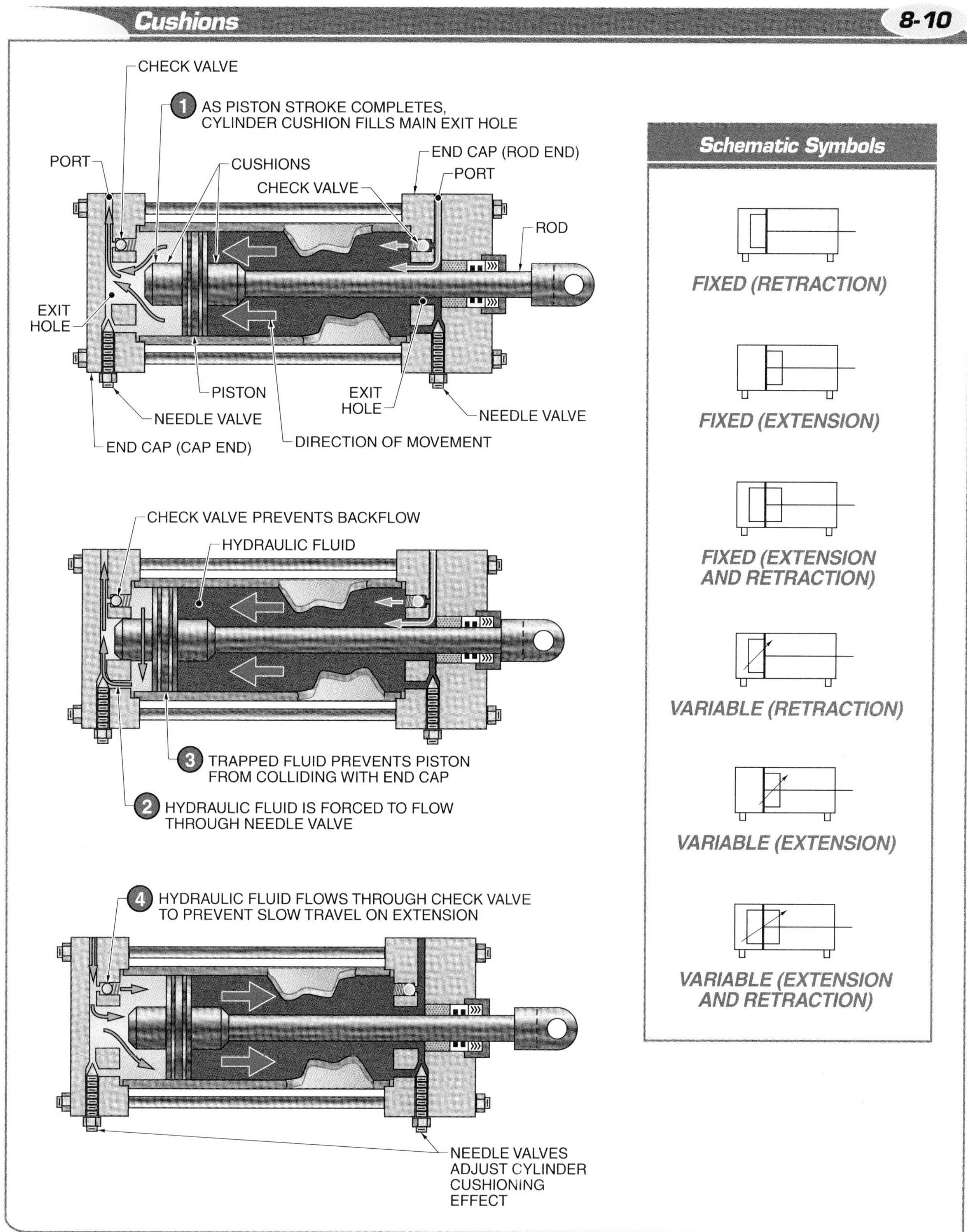

Figure 8-10. A cushion is a tapered plug attached to the piston and rod and includes a needle valve with a check valve in the end cap at the rod end, cap end, or both.

TERMS

A **stop tube** is a short hollow metal tube attached to the rod end of a hydraulic cylinder that changes the piston stroke.

A **piston stroke** is the distance the piston and rod of the cylinder travel.

A **positive seal** is a seal that does not allow any hydraulic fluid to pass.

A **nonpositive seal** is a seal that allows a small amount of hydraulic fluid to pass to provide lubrication between mating surfaces.

A **static seal** is a device that creates a positive seal between parts that do not move relative to one another.

A **dynamic seal** is a device that creates a positive seal between parts that move relative to one another.

Variable and fixed cushions can be located on the cap-end side of the piston, the rod-end side of the piston, or both. The selection of the type cushions is often determined when ordering the cylinder.

Stop Tubes. Cylinders with long rods sometimes have stop tubes installed. A *stop tube* is a short hollow metal tube attached to the rod end of a hydraulic cylinder that changes the piston stroke. A *piston stroke* is the distance the piston and rod of the cylinder travel. For example, a cylinder with a stroke of 4″ will have a piston and rod that travel 4″ inside the cylinder barrel. The rod will also travel 4″ from the beginning of the stroke to the end of the stroke. **See Figure 8-11.** A long rod can sag under its own weight. A stop tube prevents the rod from sagging by moving the piston away from the end cap, which provides a length of straight rod between the piston and rod steady bearing. It also prevents the piston from pivoting and destroying the piston rings.

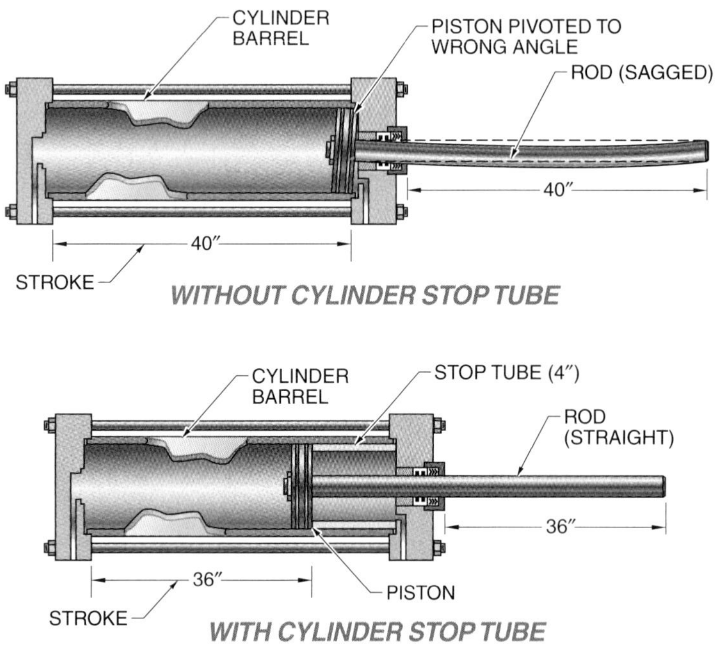

Figure 8-11. Stop tubes prevent rods from sagging and pistons from pivoting.

Stop tubes are used only on cylinders with long horizontal rods over 40″ in length. The length of the stop tube increases by 1″ for every 10″ over 40″. A stop tube reduces the piston stroke. To maintain the same piston stroke, the cylinder must be sized to accommodate the stop tube length and still allow the required stroke needed for the application.

Seals

A seal is a device that creates contact between the components of a hydraulic cylinder to help contain pressure and prevent leaks. **See Figure 8-12.** If system pressure is too high, the cylinder can leak hydraulic fluid, decreasing the efficiency of the system or completely stopping the cylinder from operating. Seals are installed between the metal parts of cylinders or other hydraulic components to prevent leakage between the parts. Seals can be categorized as positive or nonpositive and static or dynamic.

A *positive seal* is a seal that does not allow any hydraulic fluid to pass. A *nonpositive seal* is a seal that allows a small amount of hydraulic fluid to pass to provide lubrication between mating surfaces. A *static seal* is a device that creates a positive seal between parts that do not move relative to one another. A *dynamic seal* is a device that creates a positive seal between parts that move relative to one another.

A hydraulic cylinder uses both static and dynamic seals. Static seals are located at the rod end and cap end between the cylinder barrel and the end caps. Dynamic seals are located around the circumference of the piston. Piston seals, in the form of piston rings, use hydraulic fluid pressure to create a dynamic seal. They also prevent the hydraulic fluid from leaking between the two sides of the piston. A rod seal is located where the rod enters and exits the end cap. It prevents hydraulic fluid leakage at the point where the rod exits the cylinder.

TECH FACT

Most hydraulic applications do not require the use of a stop tube. Typically, when a stop tube is required, it will be specified by the equipment manufacturer.

Seals are made from different types of materials that can withstand typical system operating pressures, temperatures, and hydraulic fluid additives. These materials include plastics, elastoplastics, and elastomers. Some seals are also made from soft metals. Common seals include O-rings, Quad-rings®, and compression seals.

O-Rings. An *O-ring* is a molded synthetic rubber seal with a circular cross section. O-rings are the most commonly used type of static seal. While typically used for static applications, they can also be used for dynamic applications in low-pressure operations, such as short-stroke force multiplication systems. However, O-rings are not typically used with rotating parts or in applications where they could be exposed to vibration.

The use of O-rings as seals depends on the smoothness of the moving parts and the closeness of their fit for the best service life. Normally, well-fitted O-rings are compressed about 10% when installed in the groove between the inside of the cylinder barrel and the piston. As pressure builds, the O-ring distorts to completely fill the empty space at one end of its groove. This distortion provides the dynamic seal.

Excessive fluid pressure can force an O-ring out of its groove and into any space. Using back-up rings in the O-ring groove can help prevent the extrusion of the O-ring. A *back-up ring* is a ring that supports the O-ring and is installed on the side receiving the least amount of pressure. If the O-ring receives pressure from both directions, back-up rings must be installed on both sides of the O-ring. **See Figure 8-13.**

When an O-ring is used as a piston ring, only one can be installed in the groove. Multiple O-rings do not always provide an efficient seal. In some applications, an O-ring is used to hold a plastic band around the circumference of the piston. It keeps the plastic band in contact with the cylinder barrel, which supplies a tight seal between the two surfaces. The advantage of the plastic band is that it does not cause any significant friction.

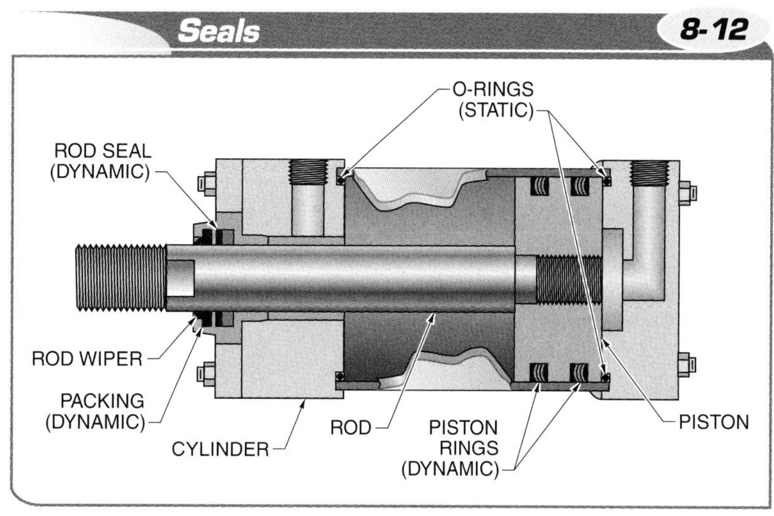

Figure 8-12. A seal creates positive contact between the components of a hydraulic cylinder to help contain pressure and prevent leaks.

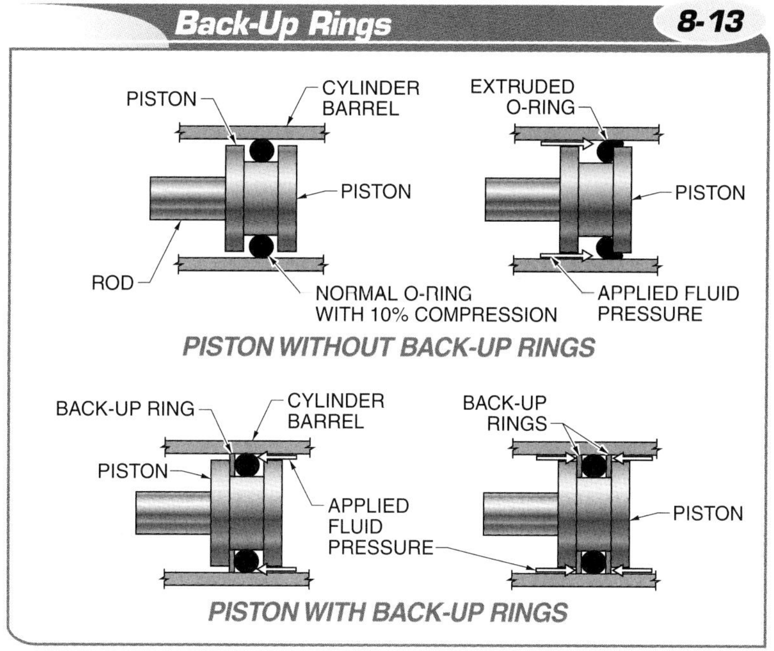

Figure 8-13. A back-up ring is installed on the side of the O-ring receiving the least amount of pressure or on both sides if the O-ring receives pressure from both directions.

Quad-Rings®. A *Quad-ring®* is a molded synthetic rubber seal with a rounded-off, X-shaped cross section. Quad-rings® are similar to O-rings and are used to prevent hydraulic fluid leaks. They are made of flexible, elastomeric materials that can withstand repeated compression without excessive wear and tear.

TERMS

An **O-ring** is a molded synthetic rubber seal with a circular cross section.

A **back-up ring** is a ring that supports the O-ring and is installed on the side receiving the least amount of pressure.

TERMS

A **Quad-ring**® is a molded synthetic rubber seal with a rounded-off, X-shaped cross section.

Resilience is the capability of a material to regain its original shape after being bent, stretched, or compressed.

A **cup seal** is a piston seal that has a sealing edge formed into a lip.

A **V-ring seal** is a cup seal with a V-shaped cross section.

Most Quad-rings® are installed to fit tightly and securely into the groove between the inside of the cylinder barrel and the piston. When pressure is applied to a Quad-ring®, it creates a dynamic seal by pressing against one side of the groove. **See Figure 8-14.** This pressure forces the cylinder seal outward and against the sealing surface of the groove. The more pressure on a Quad-ring®, the greater the sealing force.

Quad-rings® require a back-up ring for pressures of about 1000 psi because of the distortion of the seal under those pressures. The back-up ring prevents the Quad-ring® from being forced out of the cylinder groove. Quad-rings® have twice the sealing power and greater resilience to rolling, twisting, or extruding than O-rings. *Resilience* is the capability of a material to regain its original shape after being bent, stretched, or compressed.

Cup Seals. A *cup seal* is a piston seal that has a sealing edge formed into a lip. Cup seals are typically made of resilient leather, rubber, or synthetic material and are more commonly used in pneumatic applications. When used in hydraulic applications, a cup seal uses the hydraulic fluid pressure on its lip to form a dynamic cylinder seal. As pressure increases, the seal expands as the lips of the seal spread apart. This creates a tighter seal.

Cup seals are available in various shapes to withstand different ranges of pressure. Cup seals are commonly used for sealing rotating or reciprocating shafts, pistons, rods, and pump shafts. Common cup seals include V-ring seals and U-ring seals. **See Figure 8-15.**

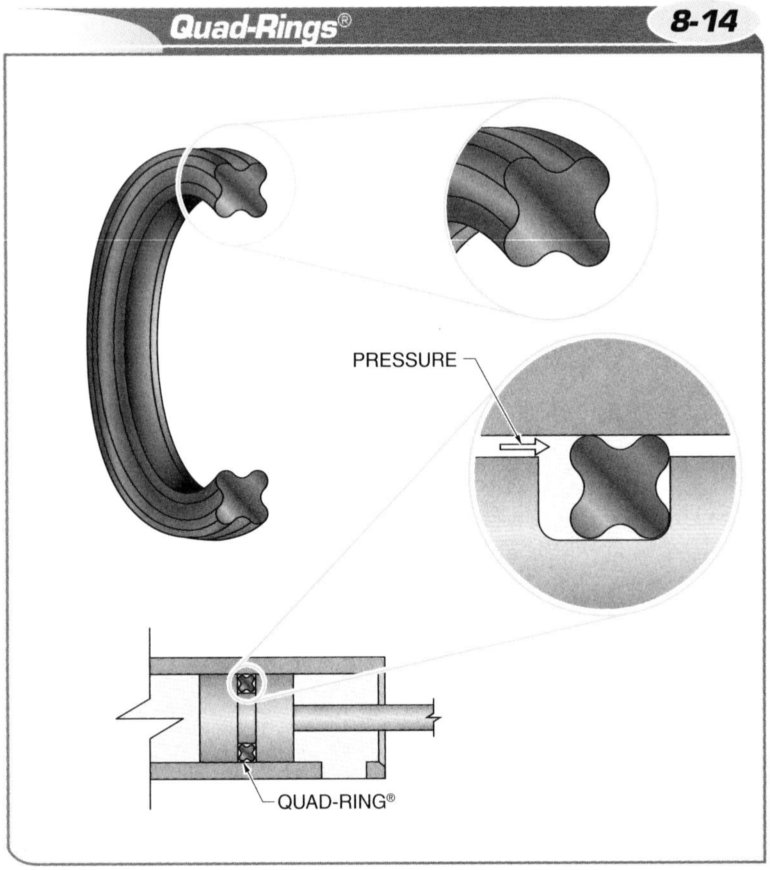

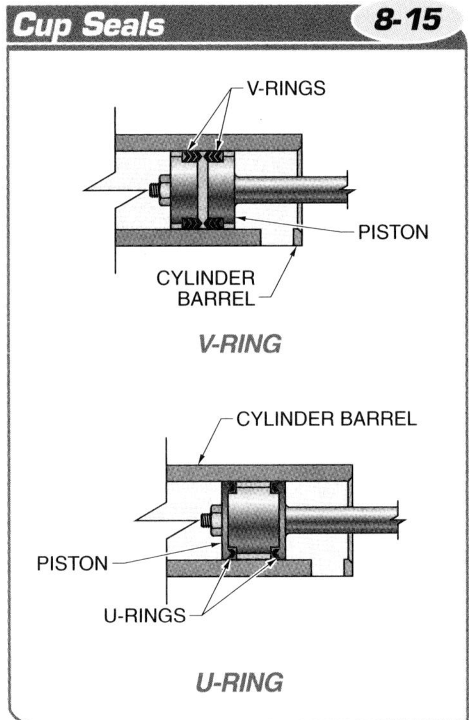

Figure 8-15. Common cup seals include V-ring seals and U-ring seals.

A *V-ring seal* is a cup seal with a V-shaped cross section. V-ring seals are dynamic piston seals that are commonly used in hydraulic systems with severe operating conditions. They are typically made of synthetic rubber or elastoplastics.

Quad-Rings® 8-14

PRESSURE

QUAD-RING®

Figure 8-14. When pressure is applied to a Quad-ring®, it creates a dynamic seal by pressing against one side of the groove.

V-ring seals can be used in combination with dissimilar seals of different materials to provide the best pressure, wear, and friction protection for an application. Because V-ring seals cannot work as stand-alone seals in a hydraulic system, they are usually laid together in stacks ranging from two to eight rings. The more rings in the V-ring stack, the more pressure it can withstand.

A *U-ring seal* is a cup seal with a U-shaped cross section. U-ring seals are dynamic cylinder seals that are commonly used in reciprocating and rotating hydraulic applications. They are made of synthetic rubber, plastics, and elastoplastics.

A U-ring can also be preloaded. A preloaded U-ring has a small O-ring embedded in the U-shaped groove of the ring. The embedded O-ring increases the overall efficiency of the seal by compressing it for a tighter fit. U-ring seals are used in low-pressure or high-pressure hydraulic applications.

Compression Seals. Compression seals are static cylinder seals commonly referred to as gaskets or packing. **See Figure 8-16.** A *gasket* is a seal used between machined parts or around the circumference of flange pipe joints to prevent hydraulic fluid leaks. A gasketed joint is sealed by compressing the gasket material into the imperfections of the mating surfaces of the joint. Gasket material can be plasticized rubber, synthetic leather, or soft metal such as brass or copper. Gasket material must be soft enough to form a proper seal when pressed between two harder materials.

Packing is a bulk deformable material reshaped by compression. It is installed where motion occurs between rigid members of a hydraulic system. Packing must only be compressed to the point where minimal amounts of hydraulic fluid are allowed to leak between the moving and stationary parts. These leaks act as lubricant and coolant for the packing. On large applications, the leakage rate can be as high as 10 drops per minute. On small or light applications, one drop per minute is sufficient.

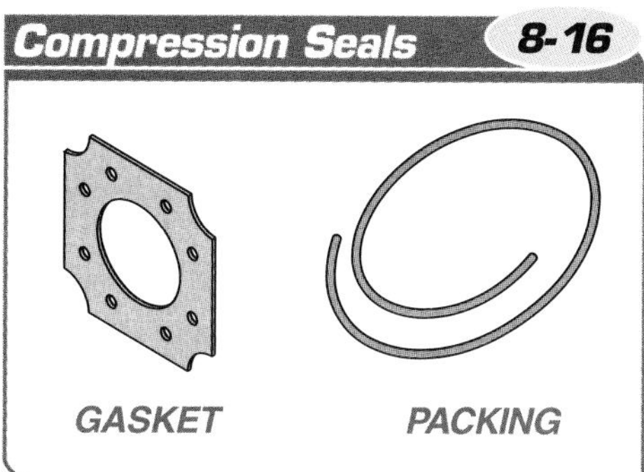

Figure 8-16. Compression seals are static cylinder seals commonly referred to as gaskets or packing.

Packing material must be pliable enough to provide a radial seal when axially compressed and must be frequently adjusted to compensate for wear. Packing material is made of braided, woven, or twisted cotton or flax. Solid lubricants, such as graphite or mica, are sometimes added to the packing to protect the moving parts of a hydraulic system.

Seal Materials. Besides soft metals, seal materials include plastics, elastoplastics, and elastomers. These types of seal materials have different pressure ratings, operating temperature ranges, and hydraulic fluid compatibilities. **See Appendix.**

Plastic seals can be made from polytetrafluoroethylene. *Polytetrafluoroethylene (PTFE),* commonly referred to as Teflon®, is a low-friction plastic material used as sealing in low-pressure hydraulic systems. It can operate in high temperatures and has good compatibility with hydraulic fluids.

Elastoplastic seals can be made from polyurethane. *Polyurethane* is a hard, chemical-resistant plastic used as a sealing material in hydraulic systems. Polyurethane is harder and stronger than other types of plastics, but has less elasticity and flexibility than elastomers. It is suitable for high-pressure applications and has good resistance to abrasion. However, it can be destroyed by exposure to water at temperatures above 140°F and therefore cannot be used in any systems with water-based fluids.

TERMS

A **U-ring seal** is a cup seal with a U-shaped cross section.

A **gasket** is a seal used between machined parts or around the circumference of pipe joints to prevent hydraulic fluid leaks.

Packing is a bulk deformable material reshaped by compression.

Polytetrafluoroethylene (PTFE), commonly referred to as Teflon®, is a low-friction plastic material used as sealing in hydraulic systems.

Polyurethane is a hard, chemical-resistant plastic used as a sealing material in hydraulic systems.

TERMS

Nitrile, (NBR), is an elastomeric compound commonly used as a sealing material in hydraulic systems because it has elasticity, tensile strength, and hardness.

Fluorocarbon (Viton®) is a dense elastomeric compound commonly used as a sealing material in hydraulic systems.

A **rod wiper** is the part of a cylinder that keeps foreign materials that have attached themselves to the rod from entering the cylinder and contaminating the hydraulic fluid.

A **cylinder bellows** is an expandable and retractable protective cover that is attached to the rod.

Elastomeric cylinder seals can be made from nitrile or fluorocarbon. *Nitrile (NBR)*, is an elastomeric compound commonly used as a sealing material in hydraulic systems because it has elasticity, tensile strength, and hardness. Nitrile seals are often standard on hydraulic pumps, motors, cylinders, and valves.

Fluorocarbon (Viton®), is a dense elastomeric compound commonly used as a sealing material in hydraulic systems. O-rings are typically made from a fluorocarbon material. It has similar mechanical properties as nitrile, but it can operate in higher temperatures and has better compatibility with hydraulic fluids. Fluorocarbons can be reinforced with fabric to increase their strength.

Preventing Cylinder Contamination. A *rod wiper* is the part of a cylinder that keeps foreign materials that have attached themselves to the rod from entering the cylinder and contaminating the hydraulic fluid. **See Figure 8-17.** A rod wiper has a lip that wipes foreign materials from a rod with each stroke. Normally, rod wipers are installed under a threshold collar on the outermost portion of the rod end of a hydraulic cylinder. Rod wipers also protect the seal material in the rod-end end cap. Rod wipers are composed of synthetic material and are not designed to seal against pressure.

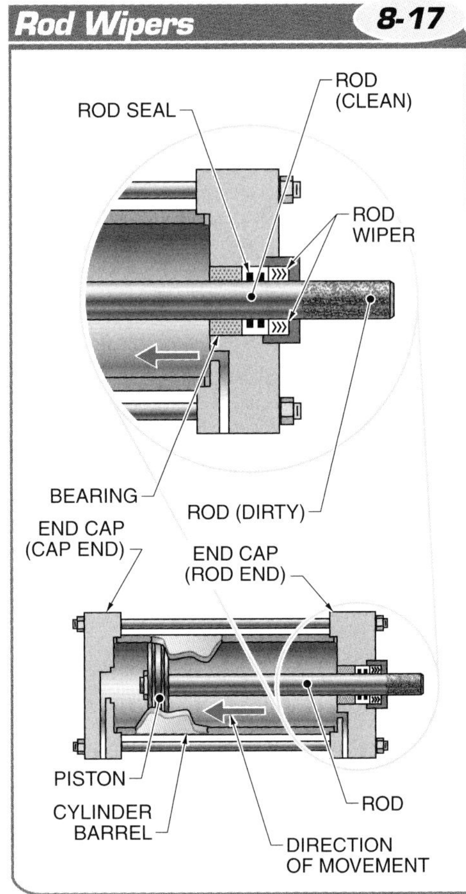

Figure 8-17. A rod wiper prevents foreign materials on the rod from contaminating the hydraulic fluid inside a cylinder.

In extremely dirty operating environments, a rod wiper may not be able to keep all contaminants out of the hydraulic fluid. These operating environments can also cause a rod wiper to wear out quickly. To prevent premature wear on a rod wiper, a cylinder bellows can be used. A *cylinder bellows* is an expandable and retractable protective cover that is attached to the rod. A cylinder bellows can be made of different materials, depending on its application or environment.

In addition to using seal materials, hydraulic rods must be regularly maintained for proper operation.

TECH FACT

When a rod wiper fails, the rod will have streak marks and dark blotches visible on it. In a dirty operating environment, it is good practice to add a rod wiper check into the preventive maintenance plan.

Cylinder Mounting Methods

The National Fluid Power Association (NFPA), the American National Standards Institute (ANSI), and the International Organization for Standardization (ISO) have set mounting method standards for differently sized cylinders used in various applications. For example, the ANSI/NFPA T3.6.7 and ISO 6099:2009 are commonly followed mounting method standards. Manufacturers follow mounting method standards when designing the mounting method of a hydraulic cylinder. This promotes interchangeability between different hydraulic cylinders.

Cylinder mounting methods are categorized as fixed or pivot. **See Figure 8-18.** A *fixed cylinder mount* is a mounting method that holds a hydraulic cylinder rigidly in place, only allowing rod movement.

TERMS

A **fixed cylinder mount** is a mounting method that holds a hydraulic cylinder rigidly in place, only allowing rod movement.

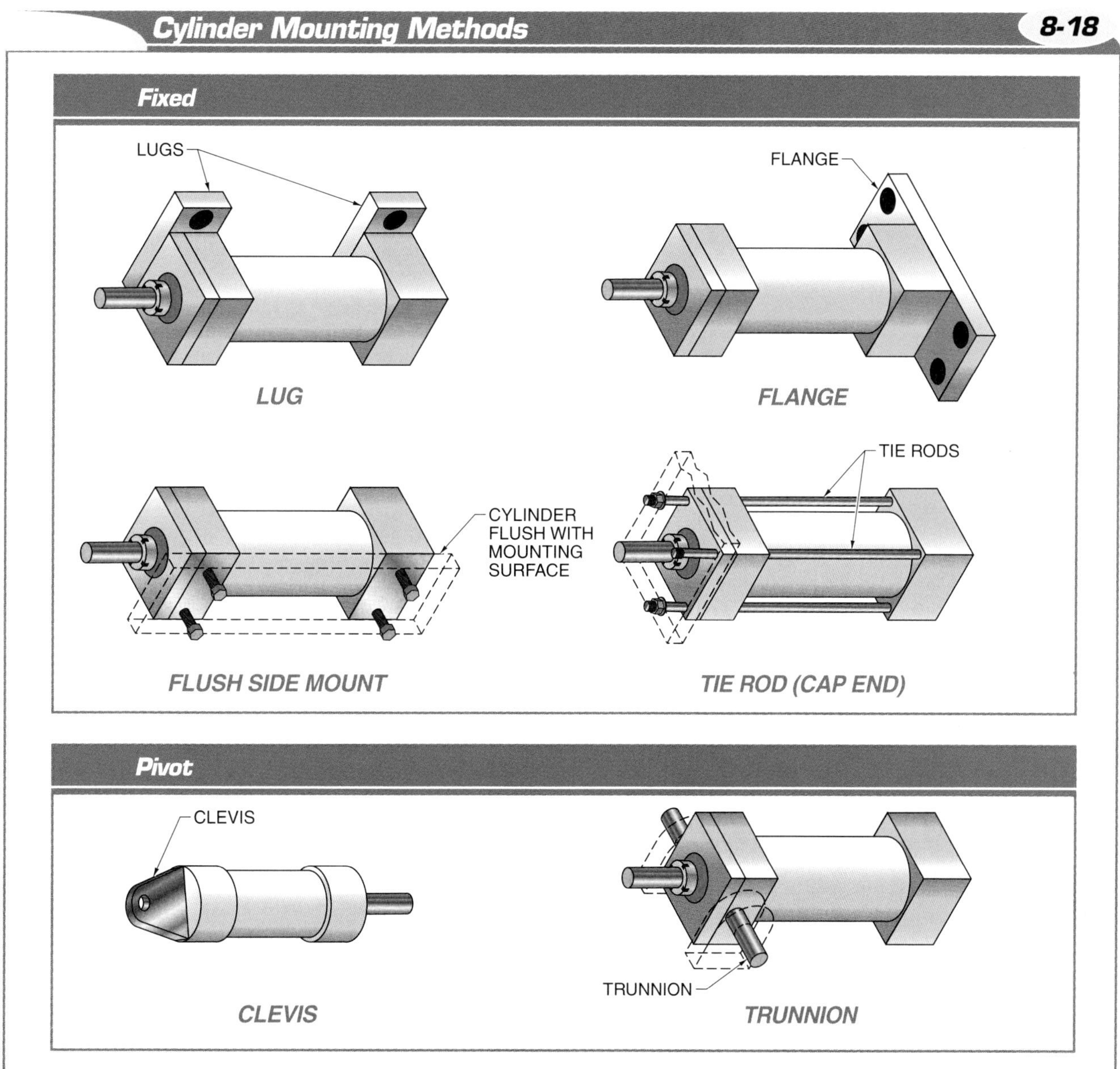

Figure 8-18. Cylinder mounting methods are categorized as fixed or pivot.

Fixed cylinder mounts include lug, flange, flush side mount, and tie rod. A *lug mount* is a fixed cylinder mounting method that has lugs attached to the side of each of the end caps. A *flange mount* is a fixed rectangular-shaped plate that is attached to either end cap in the cylinder and is used to bolt the cylinder into position. A *flush-side mount* is a fixed cylinder mounting method where both end caps in the cylinder have predrilled holes through them for bolting the cylinder into position. A *tie-rod mount* is a fixed cylinder mounting method that has tie rods extending out of the cap end, the rod end, or both.

A *pivot cylinder mount* is a cylinder mounting method that allows the entire hydraulic cylinder to pivot while the rod moves. The two types of pivot cylinder mounts are the clevis and the trunnion. A *clevis* is a steel or iron U-shaped device with holes in the ends that receive a pin. A clevis is attached to the cap end of the cylinder and allows the cylinder to pivot. A clevis mount is the most commonly used mount when a cylinder moves a load in a direction other than horizontal. For example, clevis mounts are used in tractors, snowplows, and bridge cranes.

A *trunnion* is a pivot mechanism consisting of two cylindrical objects that protrude from an end cap. In a hydraulic system, trunnions that allow the cylinder to pivot are attached to both sides of either end cap or to a bracket that fits over the barrel of the cylinder. Trunnion mounted cylinders are typically used in pivot applications such as dump truck beds, mobile conveyors, and large doors.

HYDRAULIC MOTORS

A *hydraulic motor* is a device that converts hydraulic energy into rotating mechanical energy. **See Figure 8-19.** Hydraulic motors can change direction and vary speed quickly and are used to produce large amounts of torque capable of rotating heavy loads. For example, they are used in equipment such as concrete mixing trucks, wood chippers, and industrial shredders.

The construction of hydraulic motors is similar to that of hydraulic pumps. However, the transformation of hydraulic energy to rotating mechanical energy by a hydraulic motor is a reversal of the operation of a hydraulic pump. While hydraulic fluid flows into a hydraulic pump by rotation, the rotation of a hydraulic motor is produced by hydraulic fluid that is forced through it. The type of hydraulic motor used in a hydraulic system depends on its application and the amount of torque needed.

Motor Torque

Torque is a turning or twisting force that causes a shaft, or other object, to rotate. The presence of torque indicates that there is a force present, even without rotation. Torque is measured from the center of the hydraulic motor shaft to the point of force, which is usually the radius of a pulley. This means that the greater the torque, the farther the force is from the center. **See Figure 8-20.**

Torque is equal to the product of force times the distance from the center of the hydraulic motor shaft to the point of force. It is normally measured in pound-inches (lb-in.). Torque is calculated by applying the following formula:

$$T = F \times r$$

where

T = torque (in lb-in.)

F = force (in lb)

r = shaft radius (in in.)

Example: What is the torque required to overcome a 75 lb force connected 3″ from the center of a hydraulic motor shaft?

$$T = F \times r$$

$$T = 75 \times 3$$

$$T = \mathbf{225\ lb\text{-}in.}$$

A motor with a large shaft or pulley would need to apply greater torque than a motor with a small shaft or pulley. For example, if the pulley radius is 5″, the torque required to overcome a 75 lb force is 375 lb-in. (75 × 5 = 375 lb-in.).

TECH FACT

Hydraulic motors work opposite of pumps in that they convert pressurized fluid flow to an equivalent amount of rotating mechanical energy delivered to the shaft.

TERMS

A **lug mount** is a fixed cylinder mounting method that has lugs attached to the side of each of the end caps.

A **flange mount** is a fixed rectangular-shaped plate that is attached to either end cap in the cylinder and is used to bolt the cylinder into position.

A **flush-side mount** is a fixed cylinder mounting method where both end caps in the cylinder have predrilled holes through them for bolting the cylinder into position.

A **tie-rod mount** is a fixed cylinder mounting method that has tie rods extending out of the cap end, the rod end, or both.

A **pivot cylinder mount** is a cylinder mount that allows the entire hydraulic cylinder to pivot while the rod moves.

A **clevis** is a steel or iron U-shaped device with holes in the ends that receive a pin.

A **trunnion** is a pivot mechanism consisting of two cylindrical objects that protrude from the main section of the pivoted object.

A **hydraulic motor** is a device that converts hydraulic energy into rotating mechanical energy.

Torque is a turning or twisting force that causes a shaft, or other object, to rotate.

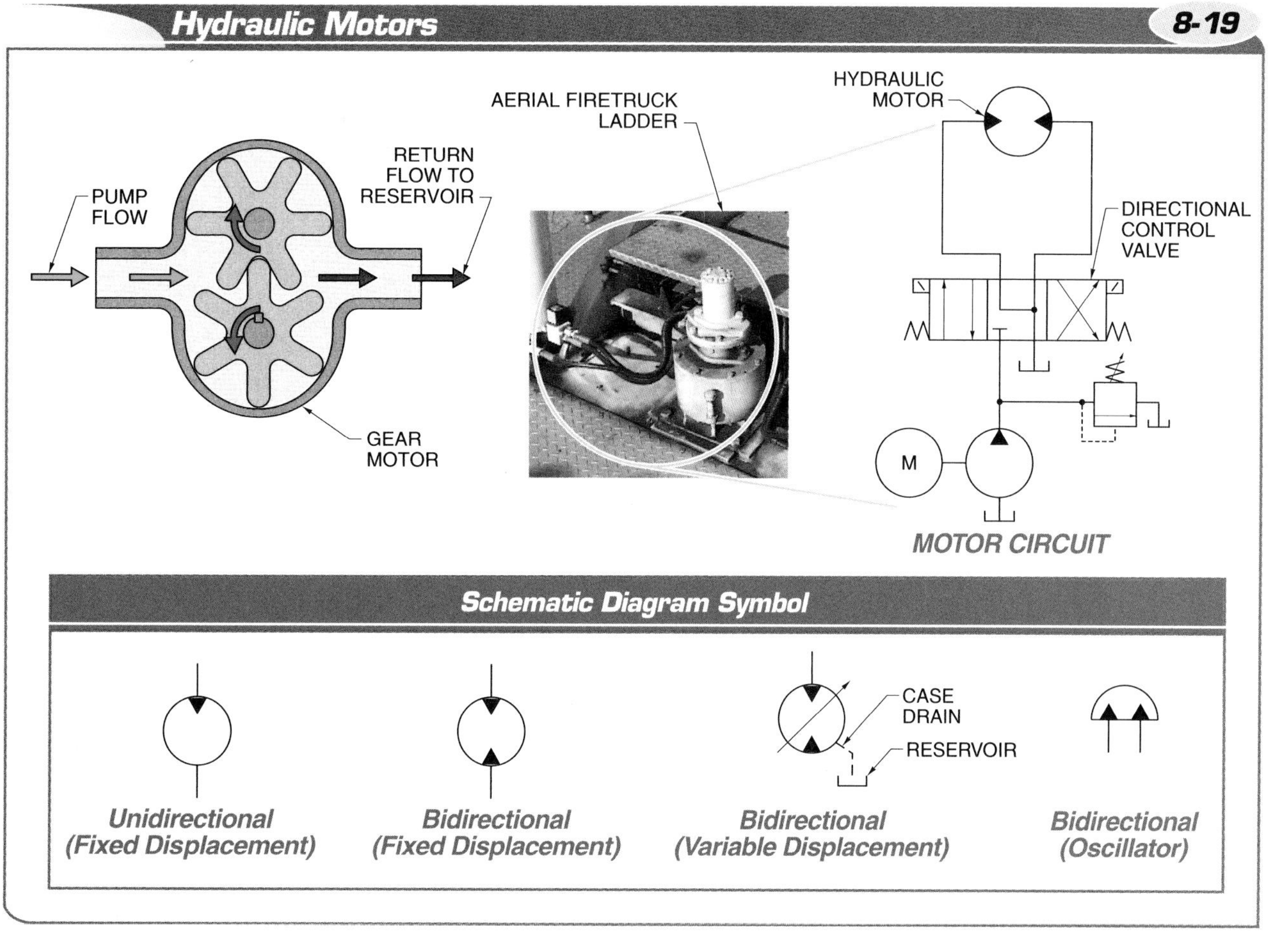

Hydraulic Motors
8-19
PUMP FLOW
RETURN FLOW TO RESERVOIR
GEAR MOTOR
AERIAL FIRETRUCK LADDER
HYDRAULIC MOTOR
DIRECTIONAL CONTROL VALVE
M
MOTOR CIRCUIT
Schematic Diagram Symbol
CASE DRAIN
RESERVOIR
Unidirectional (Fixed Displacement)
Bidirectional (Fixed Displacement)
Bidirectional (Variable Displacement)
Bidirectional (Oscillator)

Figure 8-19. A hydraulic motor converts hydraulic energy into rotating mechanical energy.

Motor Torque
8-20
TORQUE = 225 LB-IN.
3″
PULLEY RADIUS
LOAD
75 LB
T = F × r
T = 75 × 3
T = 225 LB-IN.
T
F
r
TORQUE = 375 LB-IN.
5″
PULLEY RADIUS
LOAD
75 LB
T = F × r
T = 75 × 5
T = 375 LB-IN.

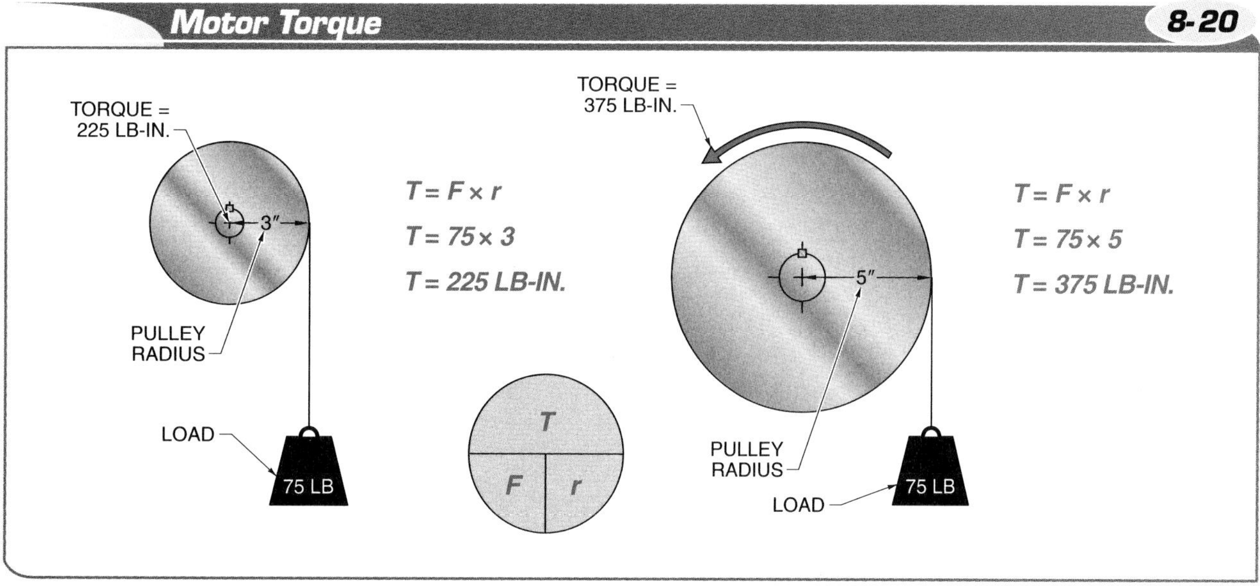

Figure 8-20. The farther a load is from the center of a hydraulic motor shaft, the greater the torque required to move the load.

TERMS

Breakaway torque is the torque required to get a nonmoving load to begin moving.

Starting torque is the torque produced when a motor is started under a load.

Running torque is the torque produced by a rotating motor.

A **gear motor** is a hydraulic motor that has meshing gears, which are rotated by hydraulic fluid flow.

Torque applied by a hydraulic motor can also be calculated with pressure and hydraulic motor displacement in cu in. per revolution. Using these variables, torque is calculated by applying the following formula:

$$T = \frac{p \times D}{2\pi}$$

where

T = torque (in lb-in.)

p = pressure (in psi)

D = hydraulic motor displacement (in cu in. per revolution)

π = 3.1416

Example: What is the available torque delivered by a hydraulic motor with a pressure of 3200 psi and a displacement of 2.146 cu in.?

$$T = \frac{p \times D}{2\pi}$$

$$T = \frac{3200 \times 2.146}{2 \times 3.1416}$$

$$T = \frac{6867}{6.28}$$

$$T = \mathbf{1092.95\ lb\text{-}in.}$$

Torque produced by a hydraulic motor depends on the pressure and displacement in cu in. per revolution. In most cases, if the available torque is not enough, the pressure is increased. In other cases, depending on the motor type, the displacement in cu in. per revolution can be increased. Either of these methods can be necessary to start a hydraulic motor and overcome breakaway, starting, and running torque.

Breakaway, Starting, and Running Torque. *Breakaway torque* is the torque required to get a nonmoving load to begin moving. *Starting torque* is the torque produced when a motor is started under a load. *Running torque* is the torque produced by a rotating motor.

Changing displacement can allow for greater breakaway or starting torque, but it has an adverse effect on motor speed and operating pressure. **See Figure 8-21.** Increasing displacement decreases motor speed and increases available torque. Decreasing displacement increases motor speed and decreases available torque. The available torque of a hydraulic motor is determined by the amount of pressure acting on the area of its rotating parts. The speed of a hydraulic motor is determined by the rate at which hydraulic fluid flows through the motor.

Hydraulic Motor Types

Each hydraulic motor type accomplishes different tasks and has different characteristics. All hydraulic motors produce rotation because of the pressure differential across the internal gears, vanes, or pistons of the motor. The hydraulic motor type used in an application depends on the operating environment, system pressure ratings, cost, motor speed variance, and the ability to change direction of rotation. Hydraulic motor types include gear, vane, and piston. Hydraulic motors are known as rotary actuators.

Gear Motors. A *gear motor* is a hydraulic motor that has meshing gears, which are rotated by hydraulic fluid flow. **See Figure 8-22.** The three different types of gear motors used in hydraulic systems are external gear, internal gear, and gerotor motors. A gear motor operates through the following procedure:

1. Hydraulic fluid flows into the inlet port of the motor at high pressure.
2. Pressurized fluid flow is converted into torque. As the gears in a gear motor rotate, a gear that is keyed with the drive shaft produces torque on it.
3. Hydraulic fluid is discharged through the outlet port at low pressures and back to the reservoir.

An external gear motor is usually bidirectional. It can easily change direction by changing the direction of fluid flow through it. It is the least expensive of all gear motors and has the highest tolerance for solid contaminants. However, it is the least efficient gear motor.

An internal gear motor has an inner gear and an outer gear. Both gears rotate when fluid flows through the motor and back to the reservoir. An internal gear motor is not as common or as efficient as a gerotor motor.

A gerotor motor rotates when hydraulic fluid pressure makes contact with the internal gear at the inlet. Since pressure is less at the outlet, the pressure differential forces the shaft to move, producing rotation. A gerotor motor is more commonly used in a hydraulic system than an internal gear motor because it is capable of producing more torque and is more efficient.

Vane Motors. A *vane motor* is a hydraulic motor that contains a rotor with vanes, which are rotated by fluid flow. When hydraulic fluid enters the inlet of a vane motor, the resulting force causes rotation of the rotor and torque on the drive shaft. **See Figure 8-23.** A vane motor operates through the following procedure:

1. Pressurized hydraulic fluid flows through the inlet port and pushes against the vanes fitted into the rotor, causing rotation of the motor.
2. Rotation of the rotor produces torque on the drive shaft.

Vane motors are often used in applications such as plastics injection molding machines, winch drives, and tunneling machines.

Piston Motors. A *piston motor* is a hydraulic motor that has internal pistons, which are extended by fluid flow to produce rotational movement. **See Figure 8-24.** Piston motors are the most efficient of all hydraulic motors. Piston motors are available as fixed or variable displacement and in bent axial form. Piston motors produce torque through fluid pressure acting on the pistons that are reciprocating inside a piston block.

A variable-displacement piston motor has an adjustable swash plate. This allows the swash plate angle to be changed, which changes the amount that the pistons can extend because of fluid flow. Decreasing the swash plate angle increases piston speed but decreases the amount of torque. Increasing the swash plate angle decreases piston speed but increases the amount of the torque.

Effect on Motor Output from System Changes * — 8-21

Change	Motor Speed	Available Torque
Increase Displacement	Decreases	Increases
Decrease Displacement	Increases	Decreases
Increase Pressure Setting	No effect	Increases
Decrease Pressure Setting	No effect	Decreases
Increase gpm	Increases	No effect
Decrease gpm	Decreases	No effect

* effects from changes assume working loads remain the same

Figure 8-21. Changing displacement can allow for greater torque, but it has an adverse effect on hydraulic system speed and operating pressure.

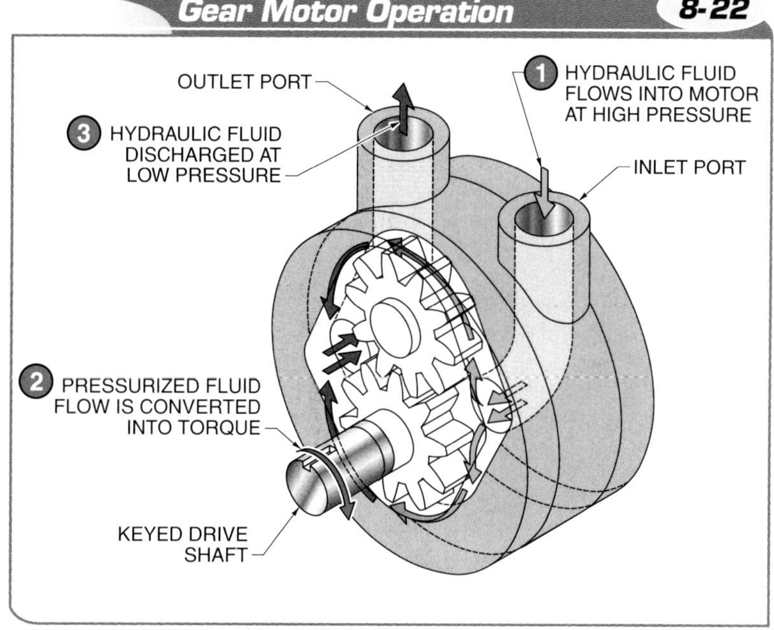

Figure 8-22. A gear motor has meshing gears, which are rotated by hydraulic fluid flow.

The angle of a swash plate is typically adjusted by pressures from somewhere in the system, but can also be adjusted by hand or with an industrial computer such as programmable logic controller (PLC). A variable-displacement piston motor operates through the following procedure:

1. Pressurized hydraulic fluid at the inlet port forces the pistons to extend.
2. The force of the pistons is transmitted to the angled swash plate, causing it to rotate.

TERMS

A **vane motor** is a hydraulic motor that contains a rotor with vanes, which are rotated by fluid flow.

A **piston motor** is a hydraulic motor that has internal pistons, which are extended by fluid flow to produce rotational movement.

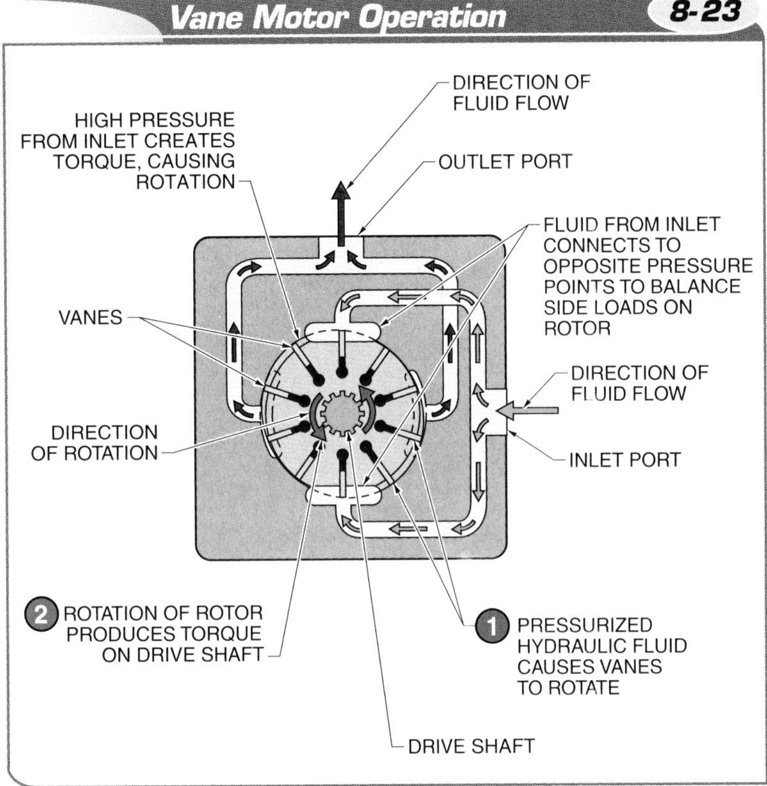

Figure 8-23. When hydraulic fluid enters the inlet of a vane motor, the resulting force causes rotation of the rotor and torque on the drive shaft.

3. The pistons, shoe retainer plate, and piston block rotate together. This rotation produces torque on the drive shaft of the motor.
4. As the swash plate angle changes, it causes the pistons to move back into their cylinders.
5. The hydraulic fluid is pushed out through the outlet port.

All hydraulic motors have internal slippage. Internal slippage causes hydraulic fluid to leak into undesired locations within the motor case. Variable displacement motors have a case drain to assist in removing hydraulic fluid from the cam ring area in the motor case. A *case drain* is a line or passage from the cam ring area in a hydraulic device that carries hydraulic fluid leakage from the device to a low-pressure reservoir. The case drain is a small port in the housing of the motor that must be returned to the reservoir and end above the fluid level as not to allow any back pressure. Case drains must be connected to hydraulic equipment for the system to be operational.

TERMS

A **case drain** is a line or passage from the internal cavity of a hydraulic device that carries hydraulic fluid leakage from the device to a low-pressure reservoir.

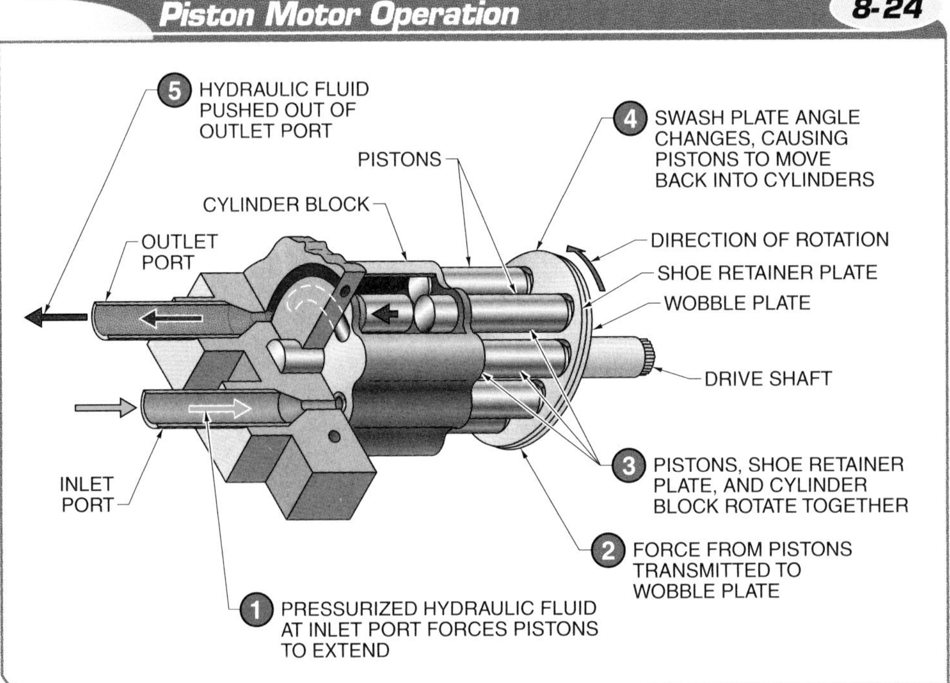

Figure 8-24. A piston motor has internal pistons, which are extended by hydraulic fluid pressure to produce rotational movement.

A fixed-displacement piston motor operates similar to a variable-displacement piston motor. However, a fixed-displacement piston motor has a swash plate angle that cannot be changed, so the speed and the torque cannot be adjusted. Fixed-displacement or variable-displacement piston motors are typically used in equipment such as road paving machines, hydraulic windlasses (mooring), and most hydraulic cranes.

Oscillators. Some hydraulic motors are not used to produce a 360° rotation. An *oscillator* is a motor that moves back and forth over a fixed arc that is less than 360°. The two types of oscillators are fixed and adjustable. A fixed oscillator usually rotates 90°, 180°, 270°, or 360° in either direction. An adjustable oscillator has a rotation range of 0° to 270°.

A *stop* is a stationary barrier used to prevent the continuous rotation of an oscillator. Depending on the setting of the stop, oscillators can provide instant torque at a high rate for moving an object a short distance. Oscillators are used in applications that require clamping, opening and closing, loading, transferring, unloading, or turning. **See Figure 8-25.** There are several types of oscillators used in industrial applications, one common type is the vane oscillator.

A *vane oscillator* is a type of oscillator that has a fixed rotation, typically either 100° or 280°. Vane oscillators can be single- or double-vane. A *single-vane oscillator* is an oscillator that has a cylindrical chamber in which a vane connected to a drive shaft rotates in a 280° arc. The inlet and outlet ports are separated by a stop. Differential pressure applied across the vane rotates the drive shaft until the vane meets the stop. Rotation is reversed by reversing fluid pressure at the inlet and outlet ports.

TERMS

An **oscillator** is a motor that moves back and forth over a fixed arc that is less than 360°.

A **stop** is a stationary barrier used to prevent the continuous rotation of an oscillator.

A **vane oscillator** is a type of oscillator that has a fixed rotation, typically either 100° or 280°.

A **single-vane oscillator** is an oscillator that has a cylindrical chamber in which a vane connected to a drive shaft rotates in a 280° arc.

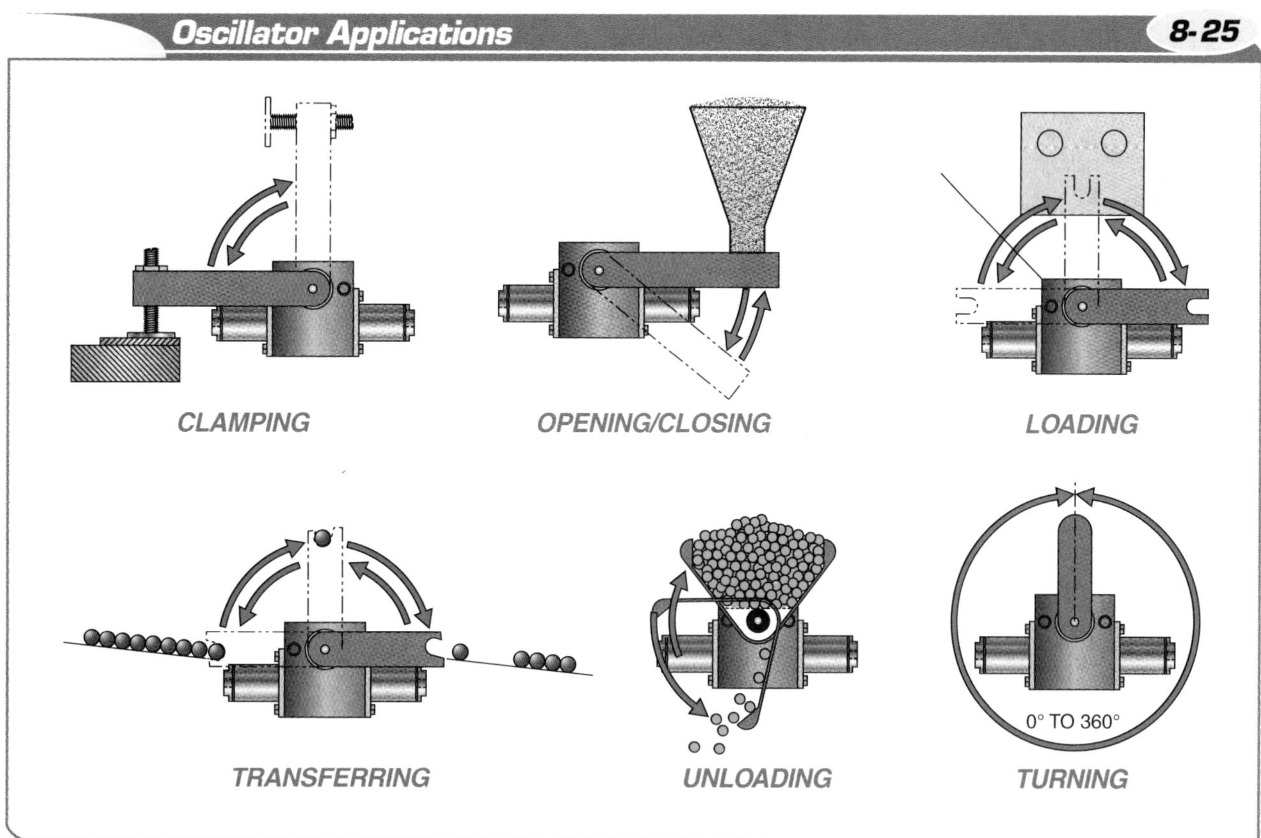

Figure 8-25. Oscillators are used to provide instant torque at a high rate for moving an object a short distance, depending on the setting of the stop.

TERMS

A **double-vane oscillator** is an oscillator that has two diametrically opposed vanes and stops.

A *double-vane oscillator* is an oscillator that has two diametrically opposed vanes and stops. This type of design provides twice the torque in the same space as a single-vane oscillator. However, rotation is generally limited to 100°. Vane oscillators are easy to service because they have fewer parts and less-critical fits than many other types of oscillators.

Valve Actuators. Valve actuators are available as low-torque rack and pinion and high-torque rack and pinion. A rack and pinion valve actuator can have one or two rack gears that mesh with a pinion gear mechanism.As the rack gears move, the pinion gear rotates a specific degree. The pinion gear is connected to a drive shaft, which produces torque as the pinion gear rotates. Multiple-position rack and pinion valve actuators can rotate the output shaft to several positions by varying the pressure of the hydraulic fluid coming through the ports.

Rack and pinion valve actuators can be low torque (for light loads) or high-torque (for heavy loads). **See Figure 8-26.** Rack and pinion valve actuators are typically used in heavy-duty applications because they can handle heavy side and end loads and can accommodate large bearings. They are also used for precision control operations because they provide constant torque output and resistance to drift.

Rack and pinion valve actuators are used as open, closed, or position valves but can also be used for operations such as mixing, dumping, intermittent feeding, screw clamping, continuous rotation, turning, automated transfer, providing constant tension, and material handling. They are also suitable for toggle clamping, indexing, positioning, oscillating, lifting, opening, closing, pushing, pulling, and lowering. For example, they are used in a steel processing facility to tilt electric furnaces and move large coils, turnstiles, and rollover devices.

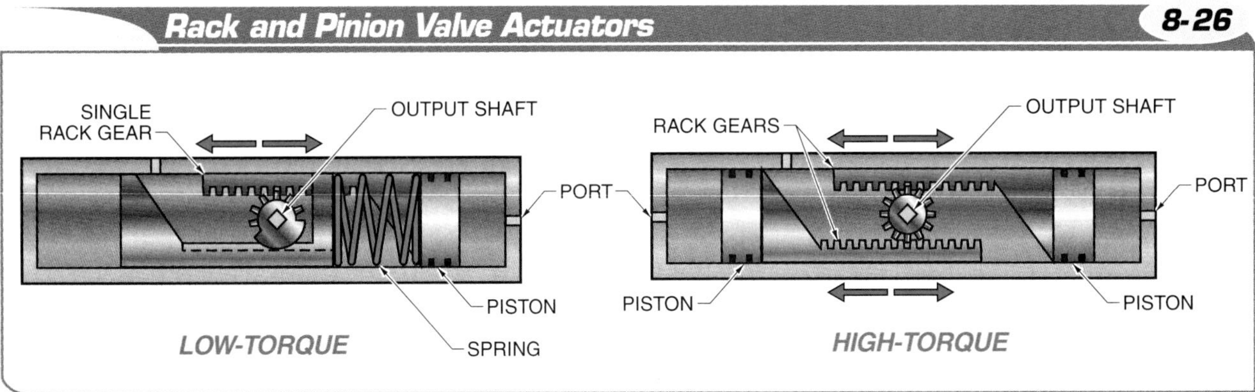

Figure 8-26. Rack and pinion valve actuators can be low-torque or high-torque designs.

Digital Resources

Chapter Review

Hydraulic Actuators

Chapter 8

Name: ______________________________ Date: ______________

MULTIPLE CHOICE

______ **1.** Gear motor types include external gear, internal gear, and ___.
A. capacitor start
B. gerotor
C. piston
D. vane

______ **2.** Well-fitted O-rings are compressed about ___% when installed in the groove between the inside of the cylinder barrel and the piston.
A. 10
B. 15
C. 20
D. 25

______ **3.** A ___ cylinder consists of two or more cylinders where the rods are not connected.
A. double-rod
B. duplex
C. telescoping ram
D. tandem

______ **4.** A(n) ___ is a line or passage from the cam ring area of a hydraulic device that carries hydraulic fluid leakage from the device to a low-pressure reservoir.
A. case drain
B. external drain line
C. outlet port
D. pilot line

______ **5.** ___ torque is the torque required to get a nonmoving load to begin moving.
A. Breakaway
B. Mechanical
C. Running
D. Starting

______ **6.** Polytetrafluoroethylene is also known as ___®.
A. NBR
B. Teflon
C. Torlon
D. Viton

______ **7.** A cylinder used to raise the bed of a dump truck trailer is an example of a ___ cylinder.
A. double-rod
B. duplex
C. telescoping ram
D. tandem

________________ **8.** A(n) ___ ring is a molded seal with a rounded-off, X-shaped cross section.
A. back-up
B. gasket
C. O-
D. Quad-

________________ **9.** A ram cylinder retracts using ___.
A. gravity only
B. the weight of the load only
C. gravity or weight of the load
D. gravity and the weight of the load

________________ **10.** A ___ cylinder consists of two or more in-line cylinders with their rods connected to form a common rod.
A. double-rod
B. duplex
C. tandem
D. telescoping

________________ **11.** A(n) ___ is a molded synthetic rubber seal with a circular cross section.
A. compression seal
B. cup seal
C. O-ring
D. Quad-ring®

________________ **12.** A ___ cylinder is a single-acting cylinder that has a large piston and rod that is the same diameter as the piston and is capable of producing large amounts of linear force during extension.
A. double-acting
B. ram
C. telescoping
D. tandem

________________ **13.** A ___ is a tapered plug attached to a piston or rod that fills an exit hole on either end of the cylinder.
A. cap
B. cushion
C. seal
D. stop tube

________________ **14.** A ___ cylinder has a single piston and rod that protrudes from both end caps of the cylinder barrel.
A. double-rod
B. duplex
C. tandem
D. telescoping

COMPLETION

______________________ **1.** A(n) ___ oscillator has two diametrically opposed vanes and stops.

______________________ **2.** A(n) ___ seal does not allow hydraulic fluid to pass.

______________________ **3.** ___ is a bulk deformable material reshaped by compression.

______________________ **4.** ___ is a turning or twisting force that causes a shaft or other object to rotate.

______________________ **5.** A(n) ___ is a device that creates contact between the components of a hydraulic cylinder to help contain pressure and prevent leaks.

______________________ **6.** A(n) ___ piston motor has an adjustable swash plate.

______________________ **7.** A(n) ___ converts hydraulic energy into rotating mechanical energy.

______________________ **8.** A(n) ___ is the part of the cylinder that prevents foreign materials on a rod from entering the cylinder and contaminating the hydraulic fluid.

______________________ **9.** The two types of pivot cylinder mounts are the ___ and the ___.

______________________ **10.** A(n) ___ cylinder extends its rod in stages.

______________________ **11.** A(n) ___ motor rotates when hydraulic fluid pressure makes contact with the internal gear at the inlet.

______________________ **12.** A(n) ___ cylinder uses fluid flow through two ports to extend and retract the cylinder and create force in both directions.

______________________ **13.** A(n) ___ seal is used to seal nonmoving parts that can be taken apart and reassembled.

______________________ **14.** A(n) ___ ring supports the O-ring in a hydraulic cylinder and is installed on the side receiving the least amount of pressure.

______________________ **15.** ___ torque is produced by a rotating motor.

______________________ **16.** A(n) ___ is an actuator that moves in a straight line called a stroke.

______________________ **17.** ___ is a hard, chemical-resistant plastic used as a sealing material in hydraulic systems.

______________________ **18.** When used on horizontal rods, the length of a cylinder stop tube increases by 1″ for every 10″ over ___″.

______________________ **19.** A(n) ___ cylinder uses fluid flow for extension and a spring for retraction.

______________________ **20.** A(n) ___ cylinder is used in applications where loads are attached to each end of the rod.

TRUE/FALSE

T F **1.** Piston motors are available as fixed or variable displacement and in bent axial form.

T F **2.** Starting torque is produced when a motor is started under load.

T F **3.** A clevis is a type of fixed cylinder mount.

T F **4.** A static seal is used between moving parts to prevent leaks or contamination.

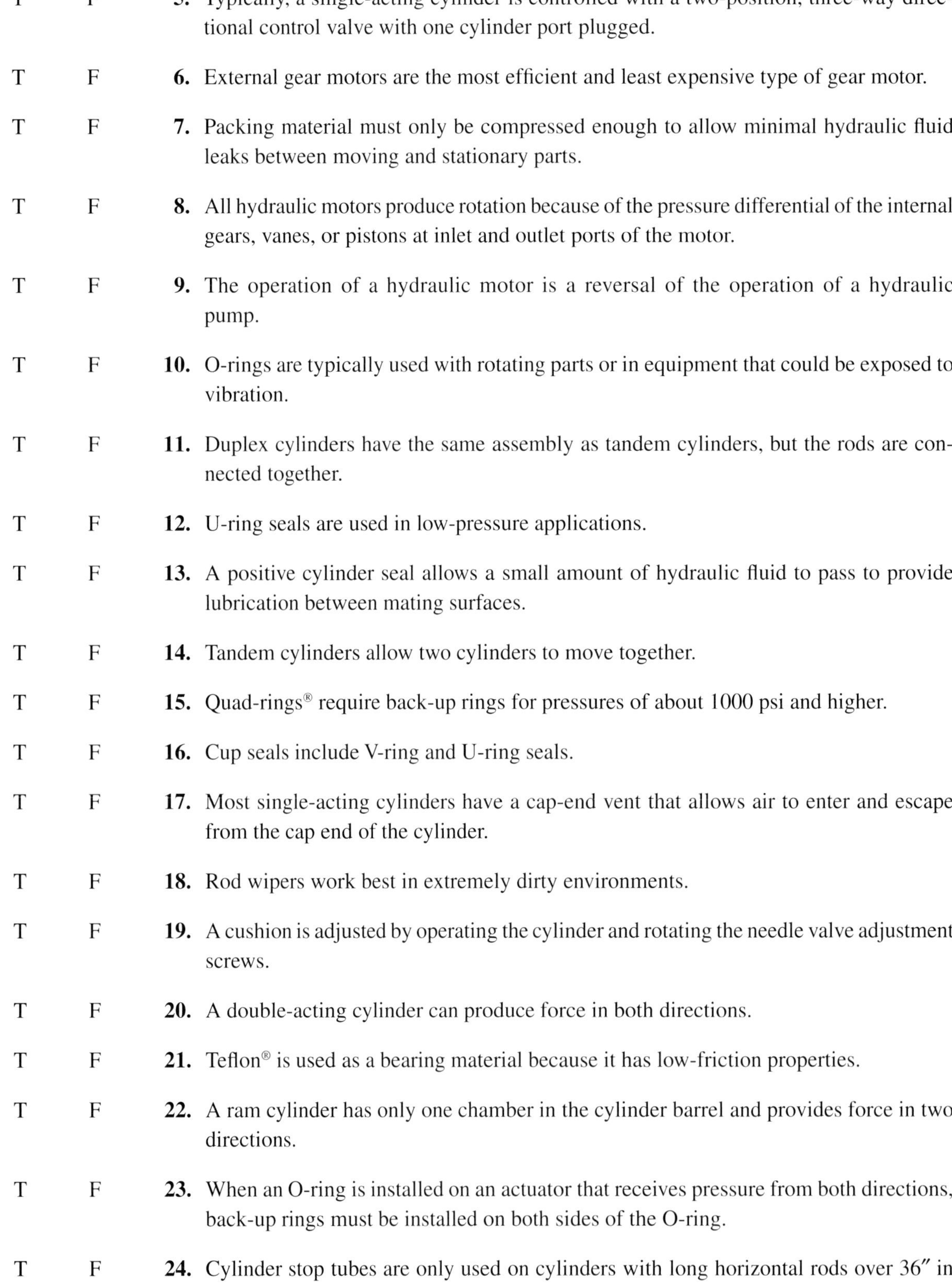

T F **5.** Typically, a single-acting cylinder is controlled with a two-position, three-way directional control valve with one cylinder port plugged.

T F **6.** External gear motors are the most efficient and least expensive type of gear motor.

T F **7.** Packing material must only be compressed enough to allow minimal hydraulic fluid leaks between moving and stationary parts.

T F **8.** All hydraulic motors produce rotation because of the pressure differential of the internal gears, vanes, or pistons at inlet and outlet ports of the motor.

T F **9.** The operation of a hydraulic motor is a reversal of the operation of a hydraulic pump.

T F **10.** O-rings are typically used with rotating parts or in equipment that could be exposed to vibration.

T F **11.** Duplex cylinders have the same assembly as tandem cylinders, but the rods are connected together.

T F **12.** U-ring seals are used in low-pressure applications.

T F **13.** A positive cylinder seal allows a small amount of hydraulic fluid to pass to provide lubrication between mating surfaces.

T F **14.** Tandem cylinders allow two cylinders to move together.

T F **15.** Quad-rings® require back-up rings for pressures of about 1000 psi and higher.

T F **16.** Cup seals include V-ring and U-ring seals.

T F **17.** Most single-acting cylinders have a cap-end vent that allows air to enter and escape from the cap end of the cylinder.

T F **18.** Rod wipers work best in extremely dirty environments.

T F **19.** A cushion is adjusted by operating the cylinder and rotating the needle valve adjustment screws.

T F **20.** A double-acting cylinder can produce force in both directions.

T F **21.** Teflon® is used as a bearing material because it has low-friction properties.

T F **22.** A ram cylinder has only one chamber in the cylinder barrel and provides force in two directions.

T F **23.** When an O-ring is installed on an actuator that receives pressure from both directions, back-up rings must be installed on both sides of the O-ring.

T F **24.** Cylinder stop tubes are only used on cylinders with long horizontal rods over 36″ in length.

T F **25.** Telescoping ram cylinders can have up to five stages and extend up to 30′.

SHORT ANSWER

1. Identify and describe the four types of fixed cylinder mounts.

2. List three types of seal materials.

3. Identify five operations that rack and pinion valve actuators are used for.

4. Describe the operational procedure for a single-acting, spring-return cylinder.

5. Describe the unique features of a hydraulic cylinder.

MATCHING

Schematic Symbols

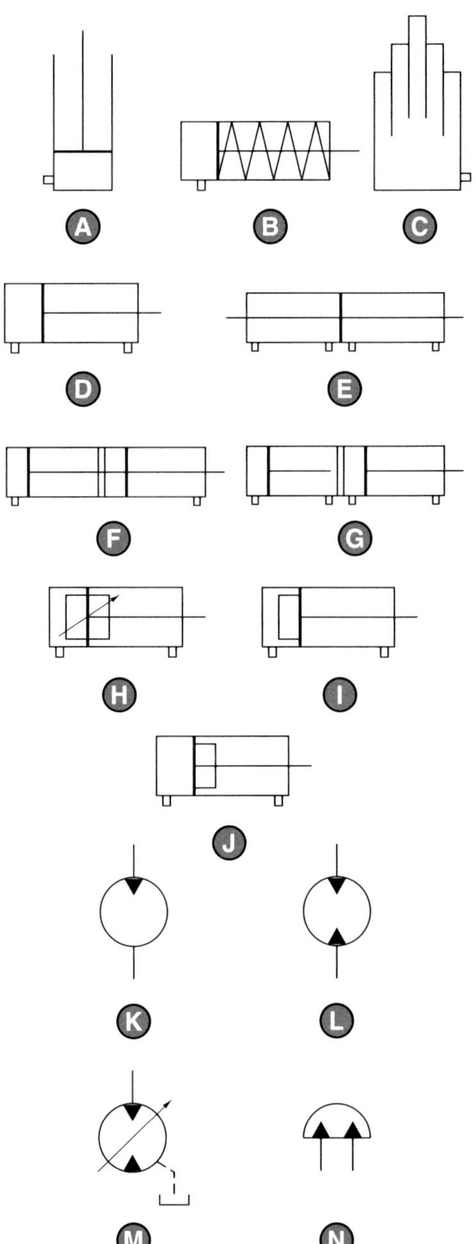

_______________ 1. Bidirectional hydraulic motor

_______________ 2. Bidirectional oscillator

_______________ 3. Bidirectional variable-displacement hydraulic motor

_______________ 4. Double-acting cylinder with cushions on extension and retraction

_______________ 5. Cylinder cushion on extension only

_______________ 6. Cylinder cushion on retraction only

_______________ 7. Double-acting cylinder

_______________ 8. Double-rod cylinder

_______________ 9. Duplex cylinder

_______________ 10. Ram cylinder

_______________ 11. Single-acting spring-return cylinder

_______________ 12. Tandem cylinder

_______________ 13. Telescoping ram cylinder

_______________ 14. Unidirectional hydraulic motor

Activity 8-1: Cylinder Assembly

A tie-rod mounted cylinder on a hydraulic forming press is leaking hydraulic fluid between its cylinder barrel and cap end. The O-ring that is between the cylinder barrel and the cap end must be replaced.

1. Using Parts Assembly Diagram, list the steps required to disassemble the cylinder and replace the damaged O-ring.

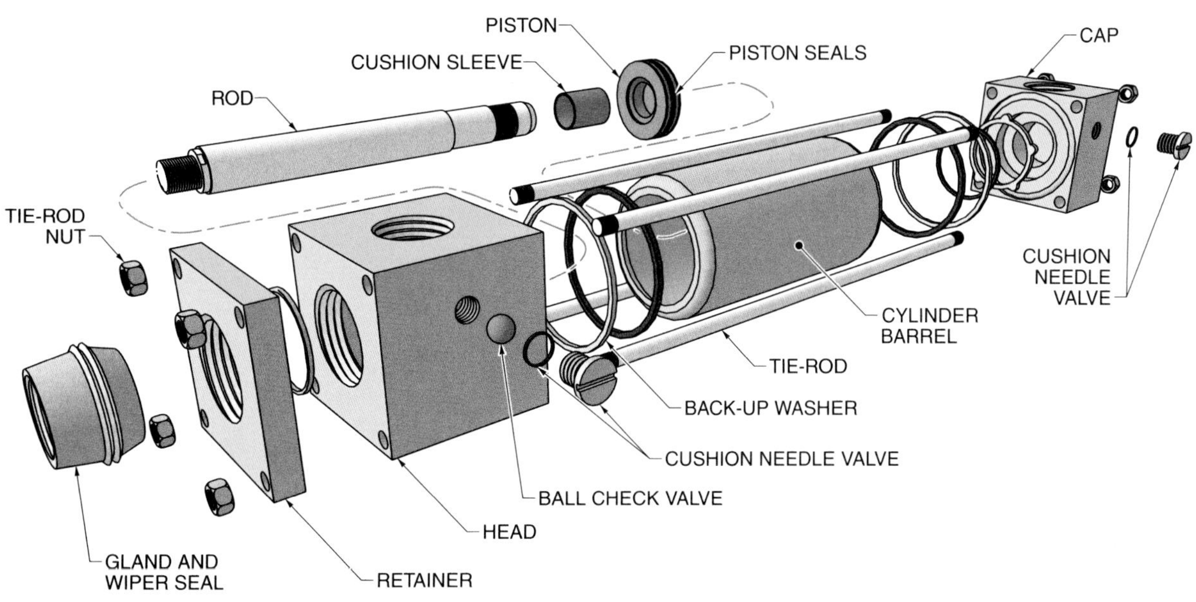

PARTS ASSEMBLY DIAGRAM

Activity 8-2: Cylinder Selection

The cylinder on a hydraulic log splitter (Model Number: 4-NC-SR-RR-H-S-0-1-MT-24) is damaged beyond repair.

1. Use Cylinder Selection Chart and the model number of the damaged cylinder to list all of the required information from the damaged cylinder.

Component Features	Description	Code
Cylinder Bore*	1″–7″	1-7
Cushion	Cushion only on head end	HC
	Cushion only on cap end	CC
	Cushion on both ends	BC
	No cushions	NC
Rod Type	Single Rod	SR
	Double Rod	DR
Mounting Style	Head Tie Rods Extended	HH
	Cap Tie Rods Extended	CC
	Both End Tie Rods Extended	BB
	Head Rectangular Flange	RR
	Cap Rectangular Flange	FF
	Head Square Flange	HF
	Cap Square Flange	FH
	Side Lugs	RF
	Centerline Lugs	FR
	Side Tapped	EE
	Side End Lugs	PP
	Head Trunnion	IK
	Cap Trunnion	KK
	Intermediate Fixed Trunnion	II
	Cap Fixed Clevis	GG
	Cap Detachable Clevis	TH
Piston Seals	Standard Seals	H
Port Type	SAE Thread	S
	Metric Thread	M
Modification†	Bellow	
	Stop tube	SP
	High water content fluid	if non
	Different port placement	0
Rod Diameter*	1″–4″	1-4M
Rod End	Male Threads	MT
	Female Thread	FT
Stroke Length‡	6″–36″ only available in inches	xx

* Half inches are available in 1″– 4″. Place a 5 after the code. For example, a 3.5″ would be 35.
† For any special modifications please note on order form.
‡ For a stroke longer than 36″ contact local sales rep.

CYLINDER SELECTION CHART

Activity 8-3: Cylinder Component Ordering Specifications

To reduce operating cost, a manufacturing facility has decided to rebuild their own hydraulic cylinders in-house rather than send them to be serviced by an outside company. The maintenance department must order the following cylinder components and gather the following information for cylinders that are most commonly used within the facility. Using the Cylinder Specification Tables and four given model numbers, list the required ordering specifications for each cylinder model number.

Bore Size	Torque Requirements for Tie Rods
1	35–40*
1½	60–70*
2	11–12†
2½	11–12†
3½	25–26†
4	25–26†
5	60–64†
6	60–64‡
7	90–94†

* in lb
† in lb-ft
‡ in lb-t

Bore Size	Piston Seal Kit Part Numbers
1	LAK0010
1½	LAK0015
2	LAK0020
2½	LAK0025
3	LAK0030
3½	LAK0035
4	LAK0040
5	LAK0050
6	LAK0060
7	LAK0070

Rod Size	Gland and Wiper Seal Kit Part Numbers
½	MSE0005
1	MSE0010
1½	MSE0015
2	MSE0020
2½	MSE0025
3	MSE0030
3½	MSE0035
4	MSE0040

Bore Size	Body Seal Kit Part Numbers
1	WJK1010
1½	WJK1015
2	WJK1020
2½	WJK1025
3	WJK1030
3½	WJK1035
4	WJK1040
5	WJK1050
6	WJK1060
7	WJK1070

1. **Model Number: 4-NC-SR-RR-H-S-0-1-MT-24**

Tie Rod Torque Requirements ______________________________

Gland and Wiper Seal Kits Part Number ______________________

Piston Seal Kit Part Number _______________________________

Body Seal Kit Part Number ________________________________

2. **Model Number: 6-HC-SR-RR-H-S-0-25-MT-30**

Tie Rod Torque Requirements ______________________________

Gland and Wiper Seal Kits Part Number ______________________

Piston Seal Kit Part Number _______________________________

Body Seal Kit Part Number ________________________________

3. **Model Number: 35-BC-SR-HH-H-S-0-15-MT-12**

Tie Rod Torque Requirements ______________________________

Gland and Wiper Seal Kits Part Number ______________________

Piston Seal Kit Part Number _______________________________

Body Seal Kit Part Number ________________________________

4. **Model Number: 5-CC-SR-PP-H-S-0-15-MT-18**

Tie Rod Torque Requirements ______________________________

Gland and Wiper Seal Kits Part Number ______________________

Piston Seal Kit Part Number _______________________________

Body Seal Kit Part Number ________________________________

Activity 8-4: Hydraulic Motor Seal Replacement

The efficiency of a hydraulic motor for a conveyor line has dropped because of fluid leaks within the motor. It has been determined that the motor must be removed from service and its seals replaced.

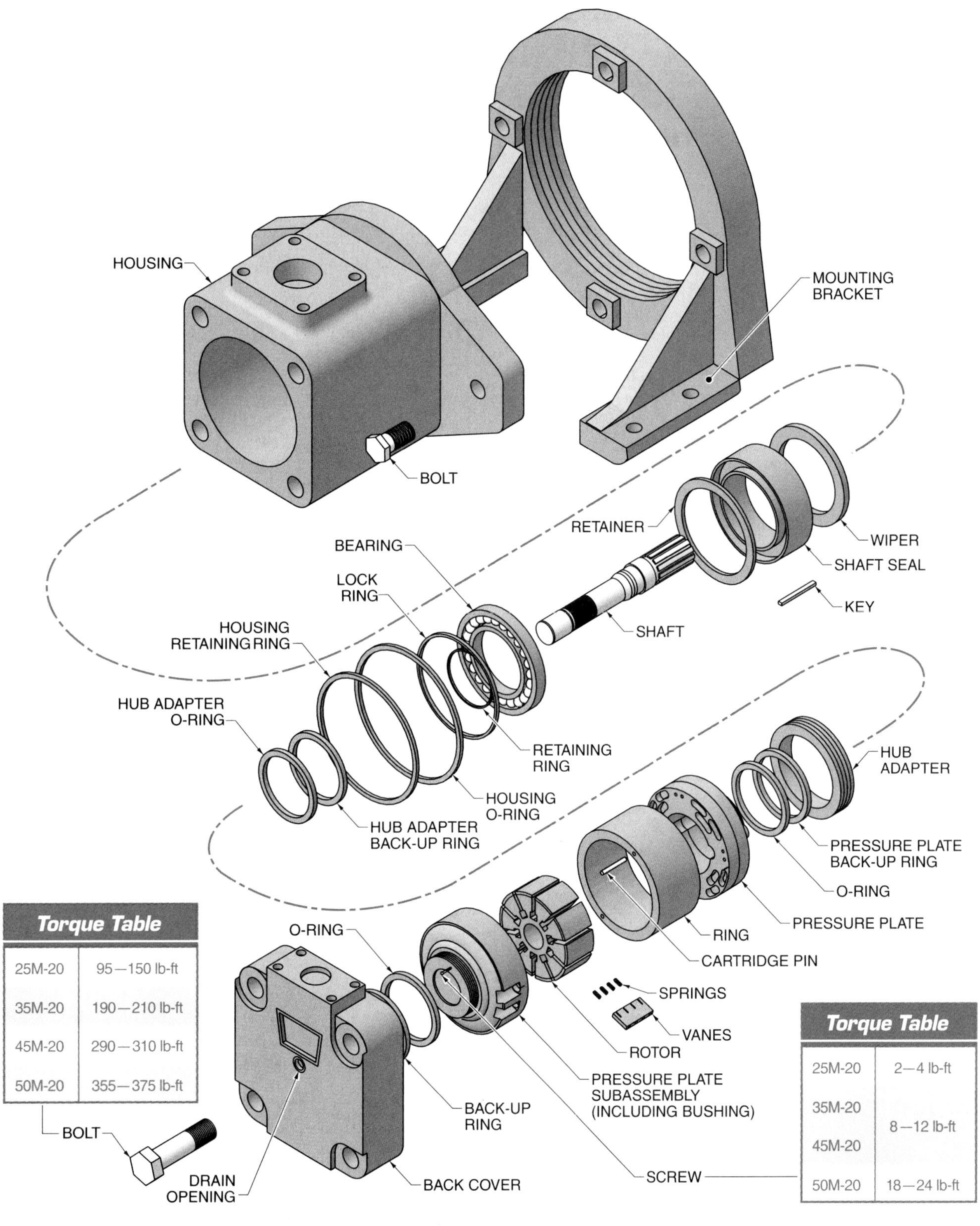

Torque Table (BOLT)

25M-20	95—150 lb-ft
35M-20	190—210 lb-ft
45M-20	290—310 lb-ft
50M-20	355—375 lb-ft

Torque Table (SCREW)

<table>
<tr><td>25M-20</td><td>2—4 lb-ft</td></tr>
<tr><td>35M-20</td><td rowspan="2">8—12 lb-ft</td></tr>
<tr><td>45M-20</td></tr>
<tr><td>50M-20</td><td>18—24 lb-ft</td></tr>
</table>

HYDRAULIC MOTOR ASSEMBLY DIAGRAM

1. Using Hydraulic Motor Assembly Diagram, list all of the seals and other components that must be replaced in the motor once it is out of service.

Activity 8-5: Hydraulic Clamping Machine Application

A single-acting, spring-return cylinder used in a hydraulic clamping system has been having difficulty retracting ever since a larger capacity clamp was installed on the rod. To correct this problem, the original single-acting cylinder is to be replaced with a double-acting cylinder so that cylinder speed can be controlled in both directions. The current hydraulic system uses a two-position, three-way, spring-offset, NC, foot-pedal-actuated directional control valve.

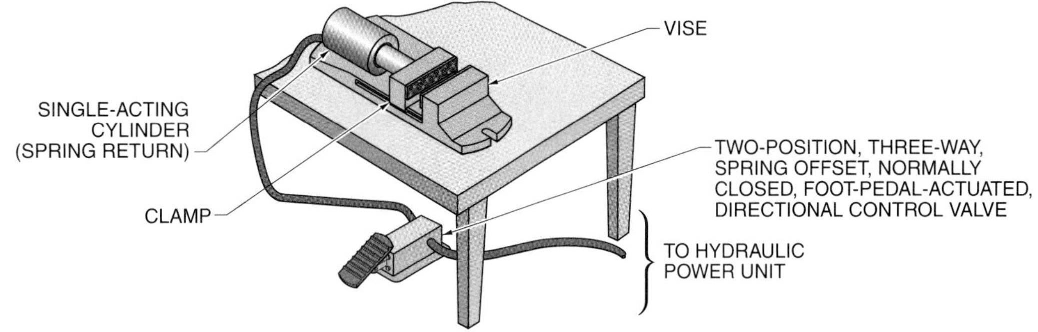

HYDRAULIC CLAMPING MACHINE

1. List all of the new components required to update the current hydraulic system.

2. Sketch the schematic diagram of the hydraulic system with a double-acting cylinder. Then, use the FluidSIM® software included on the CD-ROM to build the hydraulic system and verify that it will operate properly.

Chapter 9

Pressure Control

OBJECTIVES

- Describe the functions and applications of sequence, counterbalance, pressure-reducing, and unloading valves.
- Describe the functions of remote-controlled sequence and remote-controlled counterbalance valves.
- Describe the function and application of brake valves.
- Describe the function of a pilot-operated pressure control valve.

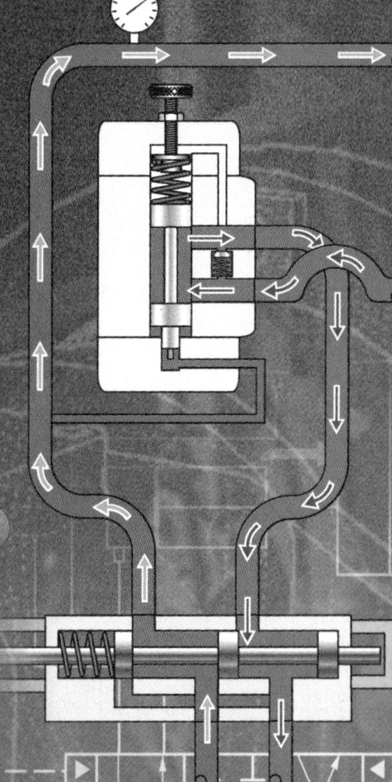

INTRODUCTION

When a hydraulic pump produces fluid flow, pressure increases proportionally with the resistance in the hydraulic system. Pressure control valves allow control of the pressures that hydraulic cylinders use and protect the hydraulic system from excessive pressures. They can also be used to control pressure in specific sections of a system.

In a hydraulic system, a pressure control valve is used to control pressure. It sets the level that pressure can reach. However, a pressure control valve can also be controlled by pressure or do work. A valve that is controlled by pressure is actuated at a set pressure to change the operation of a hydraulic system. Once system pressure is set and controlled, pressure can be used to control the necessary functions of a hydraulic device or system.

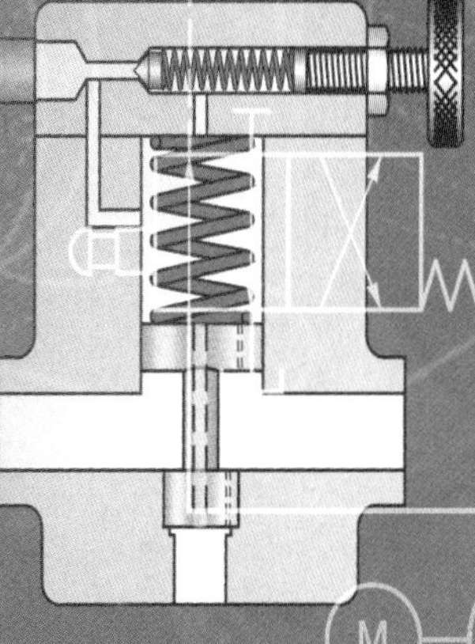

TERMS

A **direct-acting pressure control valve** is a valve that senses pressure through an internal pilot line connected to the inlet or the outlet of the valve in order to direct fluid flow to the bottom of its spool where it works against a biasing spring.

An **internal drain line** is a line that is internally machined into the body of a pressure control valve and used to drain fluid from the top of the spool to the secondary port where the valve is not pressurized.

An **external drain line** is a line that is externally connected to a pressure control valve and reservoir and used to drain fluid from the top of the spool where the secondary port of the valve is pressurized.

A **direct-acting relief valve** is a normally closed hydraulic system safety valve that sets the maximum system pressure and must be capable of allowing full pump flow through to the reservoir.

DIRECT-ACTING PRESSURE CONTROL VALVES

A *direct-acting pressure control valve* is a valve that senses pressure through an internal pilot line connected to the inlet or the outlet of the valve in order to direct fluid flow to the bottom of its spool where it works against a biasing spring. When fluid pressure overcomes the force of the biasing spring, or biasing force, the spool moves to open or close the valve. Internal hydraulic fluid leakage is drained by internal or external drain lines. **See Figure 9-1.** Drain lines are considerably narrower than inlet and outlet lines.

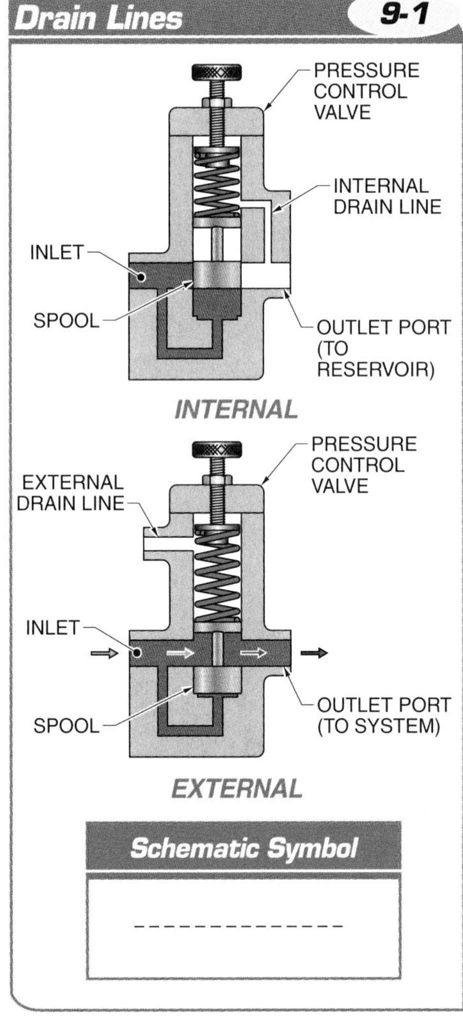

Figure 9-1. Hydraulic fluid can be drained out of a valve body by internal or external drain lines.

An *internal drain line* is a line that is internally machined into the body of a pressure control valve and used to drain fluid from the top of the spool to the secondary port where the valve is not pressurized. Internal drains are used only in valves that do not have pressure on the secondary port of the valve.

An *external drain line* is a line that is externally connected to a pressure control valve and reservoir and used to drain fluid from the top of the spool where the secondary port of the valve is pressurized. External drain lines must end above the fluid level in a reservoir. If an external drain line ends below the fluid level in a reservoir, a backpressure condition will occur. On a schematic diagram, an external drain line appears as a pilot line coming out of the biasing spring and connecting to the reservoir.

When a valve uses fluid pressure to overcome biasing spring pressure, hydraulic fluid can leak into the valve body. Hydraulic fluid can slip past the spool into the area that holds the biasing spring. If it remains in this area, it will change the biasing pressure, which will change the pressure at which the spool moves. To avoid improper operation, hydraulic fluid must be drained from the top of the spool. Direct-acting pressure control valves include relief, sequence, counterbalance, unloading, brake, and pressure-reducing valves.

Direct-Acting Relief Valves

A *direct-acting relief valve* is a normally closed hydraulic system safety valve that sets the maximum system pressure and must be capable of allowing full pump flow through to the reservoir. A direct-acting relief valve is often located within 6′ or less of the hydraulic pump. **See Figure 9-2.** Per OSHA regulations, there must never be any other components connected to a hydraulic system between the pump and its relief valve. Pressure at the inlet port of a direct-acting relief valve is used to overcome the force of the biasing spring. Direct-acting relief valves are always attached to a pump in parallel because when the valve is actuated, hydraulic fluid flows through it to the reservoir.

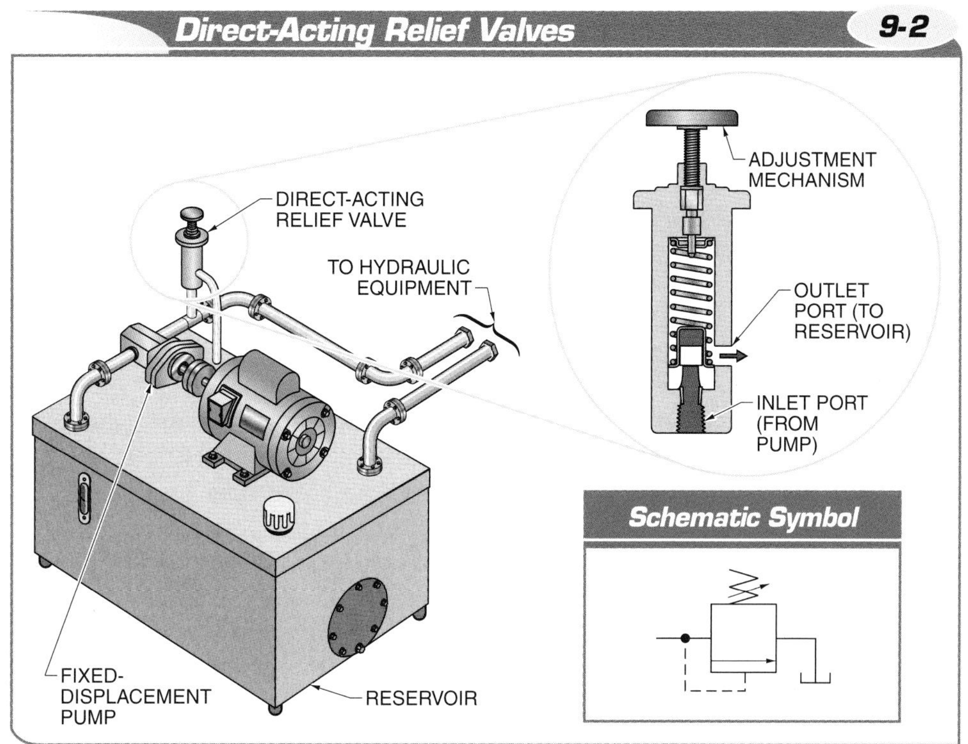

Figure 9-2. A direct-acting relief valve senses pressure at its inlet port and returns hydraulic fluid to the reservoir when there is too much pressure.

TERMS

A **sequence valve** is a normally closed pressure control valve that makes various operations occur in an orderly manner, or one after another.

Sequence Valves

A *sequence valve* is a normally closed pressure control valve that makes various operations occur in an orderly manner, or one after another. A sequence valve uses increasing pressure levels to actuate operations one after another to control when each part of the system will function. A direct-acting sequence valve uses pressure to actuate another operation in the hydraulic system when the pressure reaches a set point.

A sequence valve is installed in a hydraulic system when one operation needs to begin after another has finished, such as when hydraulic cylinders need to be activated in a successive order. Sequence valves are also used to control the order of cylinder operation when the cylinders are actuated from one directional control valve. **See Figure 9-3.** A direct-acting sequence valve requires an external drain line since the outlet port is pressurized when the valve continues with a secondary operation. A sequence valve is also available with a plugged external pilot line.

Parker Motion & Control

Direct-acting sequence valves are often used when hydraulic cylinders are controlled from a main directional control valve.

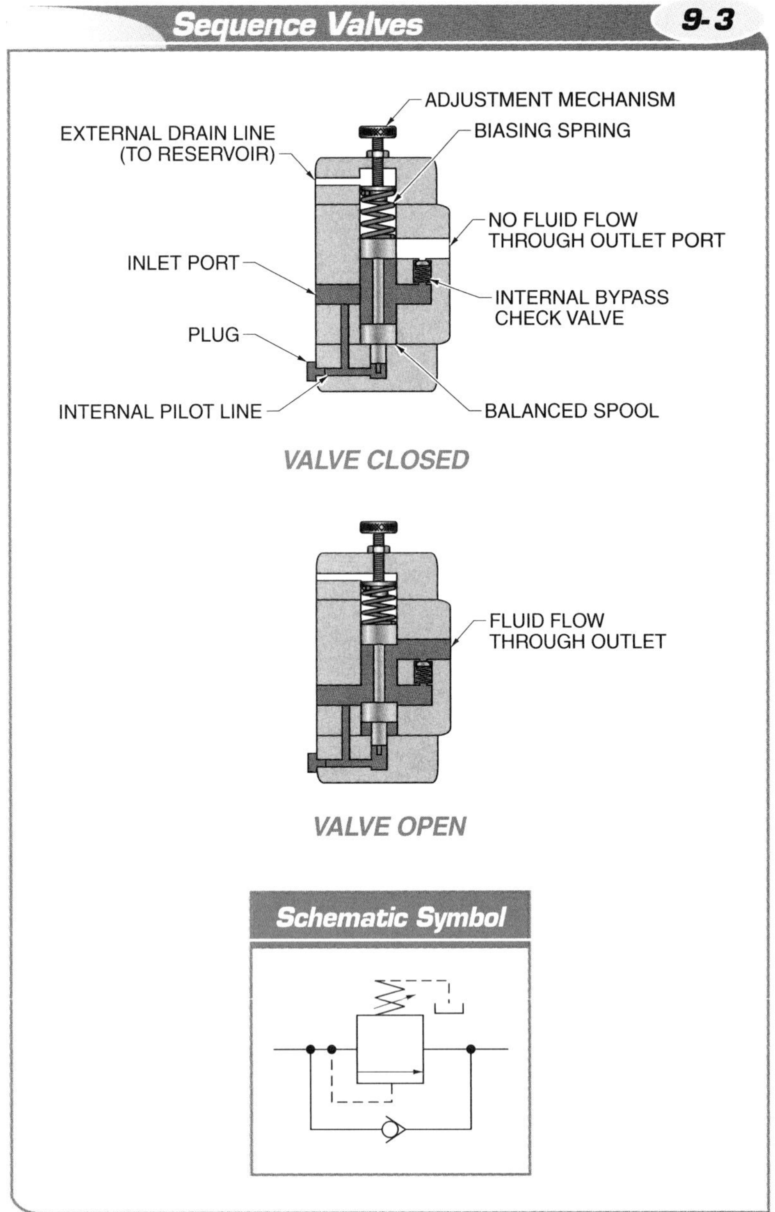

Figure 9-3. A sequence valve has an internal bypass check valve and three ports: an inlet port, an outlet port, and an external drain line.

Sequence valves are commonly installed in hydraulic clamping and bending systems. **See Figure 9-4.** A basic hydraulic clamping and bending system allows the clamping cylinder to extend its rod first. Once that is complete, the rod in the bending cylinder extends. However, unlike during extension, both rods are retracted at the same time. When they are completely retracted, the directional control valve is de-actuated, allowing the spool in the valves to center, completing the clamping and bending operation.

A sequence valve operates a hydraulic clamping and bending system using the following procedure:

1. When the lever of the directional control valve is pulled in one direction, hydraulic fluid is allowed to flow in two directions. In one direction, hydraulic fluid is stopped by the sequence valve that is connected in-line with the cap end of the bending cylinder. In the other direction, hydraulic fluid flows into the cap end of the clamping cylinder and extends its rod.
2. When the rod of the clamping cylinder reaches the workpiece (at end of cylinder stroke), system pressure begins to increase. The sequence valve connected in-line with the bending cylinder is set to actuate at 450 psi.
3. When system pressure reaches 450 psi, hydraulic fluid at the inlet port of the sequence valve flows into the internal pilot line. Fluid pressure through the internal pilot line forces the spool to move and allows fluid flow into the cap end of the bending cylinder. This extends the rod of the bending cylinder and bends the workpiece.
4. When the bending operation is complete, the lever for the directional control valve is actuated and hydraulic fluid is sent in two directions again. Hydraulic fluid flows to the rod end of each cylinder, causing both cylinders to retract at the same time. The hydraulic fluid leaving the cap end of the bending cylinder bypasses the closed sequence valve through its internal check valve.

Direct-acting sequence valves can also be used to control the sequence of rod retraction. In many clamping and bending systems, such as those used for electrical box and panel production, the bending cylinder must be retracted before the clamping cylinder. This ensures that the workpiece remains fixed and that it is bent to correct specifications.

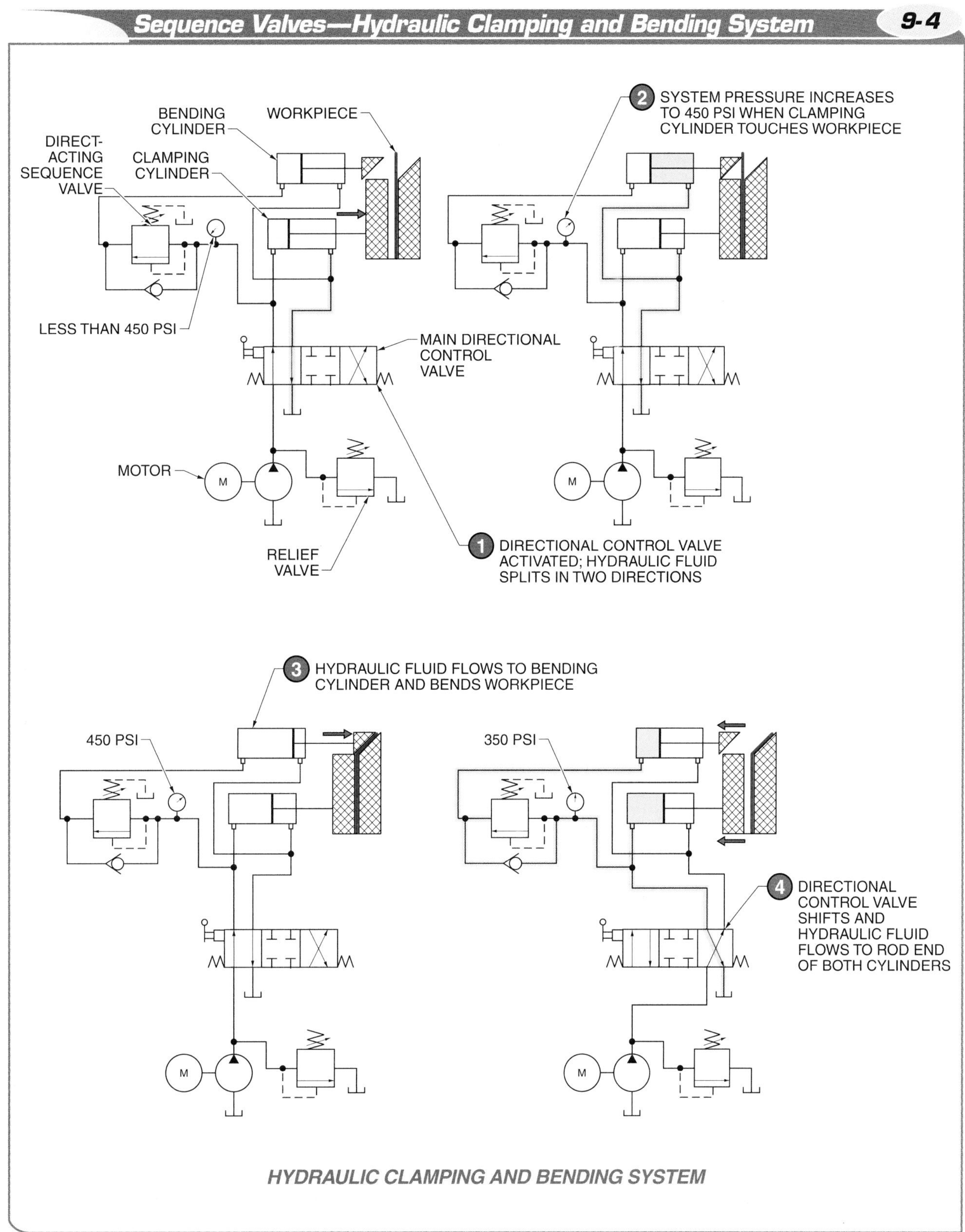

Figure 9-4. A sequence valve is used in a basic hydraulic clamping and bending system and allows the clamping cylinder to extend before the bending cylinder.

TERMS

A **counterbalance valve** is a normally closed pressure control valve that stops a vertically mounted cylinder from running away because of the load weight.

Counterbalance Valves

A *counterbalance valve* is a normally closed pressure control valve that stops a vertically mounted cylinder from running away because of the load weight. A counterbalance valve is sometimes referred to as an internal pilot counterbalance valve. It is used to counteract the pressures created by the weight of a load attached to a cylinder. **See Figure 9-5.** Most direct-acting counterbalance valves have an internal bypass check valve in the main valve body that allows reverse flow through the valve. Hydraulic fluid must bypass the counterbalance valve to retract or extend the cylinder in the opposite direction that the counterbalance valve is being used to control.

TECH FACT

Counterbalance valves are sometimes used in the same applications as pilot-to-open check valves, but are more accurate at controlling pressure.

Pressure reaches the spool of a counterbalance valve by traveling through the internal pilot line from the valve's inlet port. The pressure created in the cylinder by the weight of the load cannot travel through the valve and back to the reservoir until it overcomes the biasing spring force. Once there is enough pressure created, the counterbalance valve opens and allows the fluid in the cylinder to travel through it back to the reservoir.

Counterbalance valves are used when a load will make a piston move faster than the flow rate going to a cylinder dictates, such as with mobile equipment. For example, a counterbalance valve is used to control a boom on a piece of firefighting equipment. The counterbalance valve stops the boom from free-falling by continually opening and closing, thereby creating a condition where the lift can only descend at a slow rate.

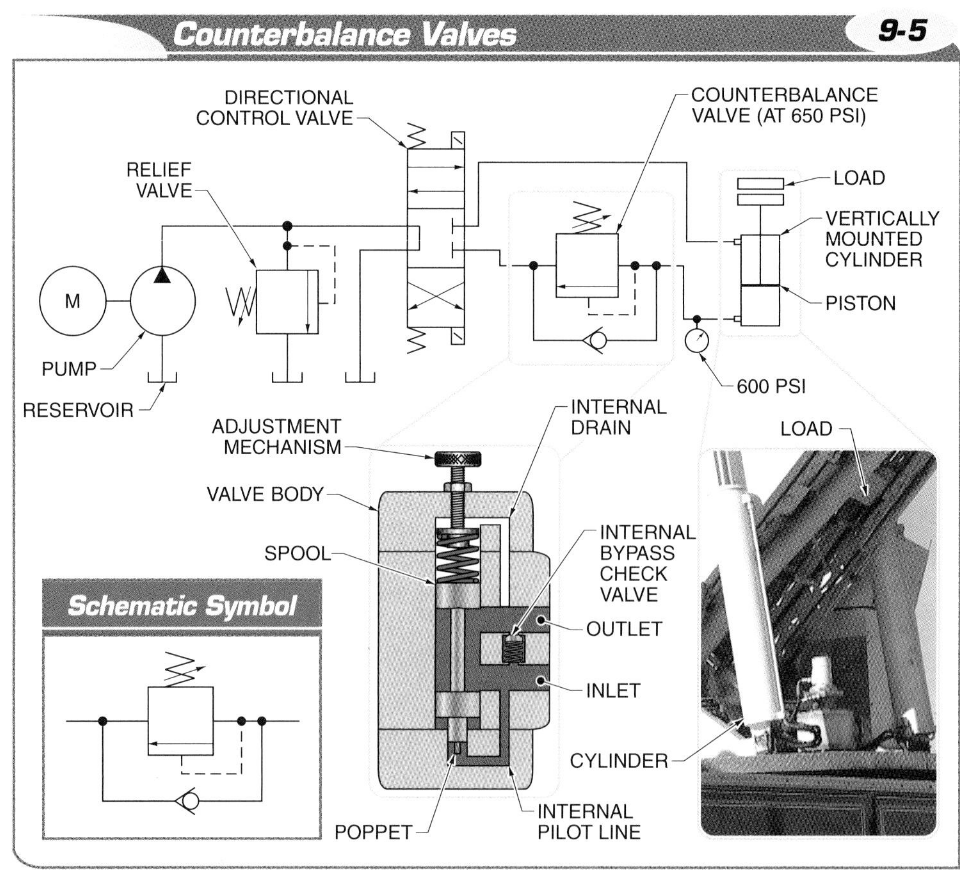

Figure 9-5. A direct-acting counterbalance valve is used to counteract the pressure of a vertical load on a cylinder.

A directional control valve directs fluid flow to bypass the counterbalance valve through the check valve, allowing the rod to extend and raise the boom. If the load on the boom equals a force of 3600 lb and a 2.75″ diameter cylinder is used (piston face area = 6 sq in.), the hydraulic pump must overcome 600 psi (3600 ÷ 6 = 600) to extend the rod.

When the boom is at the desired height, the spool in the main directional control valve is centered to stop fluid flow to the cylinder. The pressure of 600 psi within the cylinder maintains the desired height of the 3600 lb load.

To lower the boom, the directional control valve spool is shifted to direct fluid flow from the cap end of the cylinder to the reservoir. Only 50 psi more is required to overcome the biasing spring in the counterbalance valve. The pressure to overcome the biasing spring (50 psi) is combined with the cylinder pressure (600 psi) for a total of 650 psi on the cap end of the cylinder.

At 650 psi, the counter-balance valve has enough pressure present through the internal pilot line to overcome the biasing spring and open the valve. The spool inside the counterbalance valve moves, allowing controlled retraction of the rod. As hydraulic fluid drains from the counterbalance valve, it is directed back to the reservoir through the directional control valve.

Pressure-Reducing Valves

A *pressure-reducing valve* is a normally open pressure control valve that controls pressure in one leg (one small part) of a system. Because a pressure-reducing valve maintains pressure at its outlet, there does not have to be a hydraulic power unit for every section of the system that requires different pressure settings. Pressure-reducing valves are typically used to regulate pressure in separate sections of a hydraulic system.

A pressure-reducing valve is a normally open pressure control valve that senses pressure at its outlet and controls pressure in one section of a hydraulic system. An internal pilot line is attached to the outlet port of the valve to move the spool when set pressure is reached. **See Figure 9-6.** Pressure is set by a biasing spring that is adjusted with an adjustment mechanism. When system pressure reaches the set level, the spool closes.

A pressure-reducing valve can be used in industrial applications, such as production manufacturing facilities, with applications that have pressure fluctuations between 300 psi and 3000 psi, and to control pressure in pilot lines for control valves. For example, a pressure-reducing valve can be used when a metal workpiece must be stamped three times and each stamping operation must be set at a different pressure. The first stamping operation (Stamp A) is set to Stamp at 1000 psi, the second stamping operation (Stamp B) is set to stamp at 800 psi, and the third stamping operation (Stamp C) is set to stamp at 600 psi.

TERMS

A **pressure-reducing valve** is a normally open pressure control valve that controls pressure in one leg (one small part) of a system.

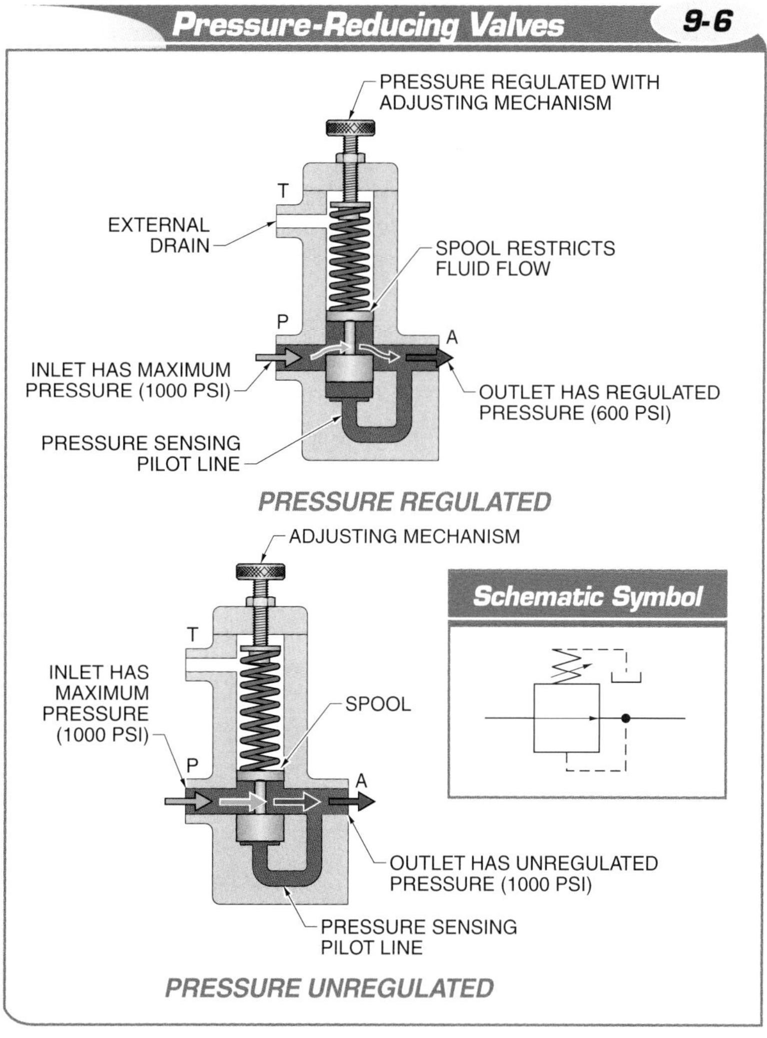

Figure 9-6. A direct-acting pressure-reducing valve senses pressure at its outlet, regardless of its inlet pressure.

TERMS

An **unloading valve** is a normally closed pressure control valve that can set maximum system pressure and unload the pump, which operates under 100 psi, at the same time.

When the metal workpiece moves on the production line conveyor to Stamp A, the directional control valve for the stamping die is manually actuated, and 1000 psi is sent to the cylinder. This is the first stamping operation.

When the metal workpiece reaches Stamp B, the directional control valve for the stamping die is manually actuated again. As the stamping die makes contact with the metal workpiece, resistance and pressure increase. When pressure reaches 800 psi, the pressure on the internal pilot line overcomes the biasing pressure of the spring, causing the spool to close. Closing the spool maintains 800 psi after the pressure-reducing valve. Maintaining this pressure also allows the cylinder to perform the second stamping operation.

When the metal workpiece reaches Stamp C, the directional control valve for the stamping die is manually actuated again. As the stamping die makes contact with the metal workpiece, resistance and pressure increase. When pressure reaches 600 psi, the pressure on the internal pilot line overcomes biasing pressure, causing the spool to close. Closing the spool maintains 600 psi after the pressure-reducing valve. Maintaining this pressure also allows the cylinder to perform the third stamping operation.

Pressure-reducing valves can have an internal bypass check valve to allow reverse fluid flow. Pressure-reducing valves are always installed after the directional control valve. Therefore, an internal bypass check valve must also be installed so that hydraulic fluid can flow out of the actuator and through the valve without being restricted. The accuracy and consistency of system pressure depends on how close the direct-acting pressure-reducing valve is to the cylinder. Pressure-reducing valves must be as close to the point of work as possible.

Unloading Valves

An *unloading valve* is a normally closed pressure control valve that can set maximum system pressure and unload the pump, which operates under 100 psi, at the same time. **See Figure 9-7.** In some applications, an unloading valve is similar to a direct-acting relief valve, except it senses pressure from a remote section of the hydraulic system through a pilot line.

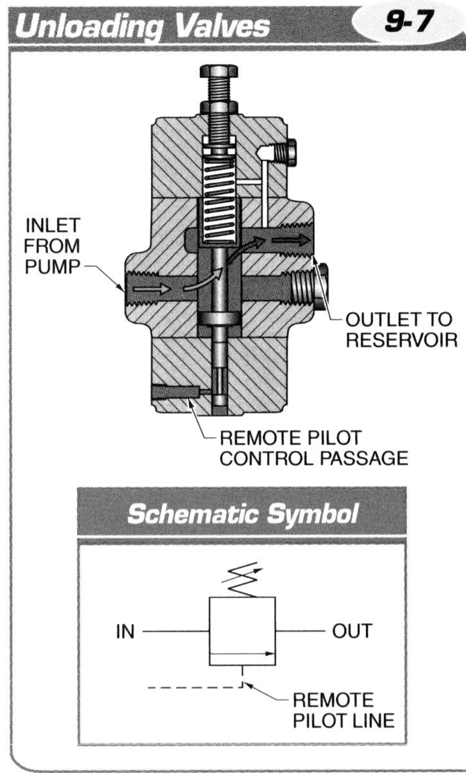

Figure 9-7. An unloading valve returns hydraulic fluid to the reservoir when pressure from a remote section of the hydraulic system is sensed through a pilot line.

One advantage of an unloading valve is that it accurately senses pressure and pressure drops in other sections of the system. This advantage allows for more precise pressure control. When the unloading valve opens to allow the fluid flow of the high-flow, low-pressure pump back to the reservoir, the pump's fluid flow will ideally travel at a pressure lower than 100 psi. Because the fluid flow is returning to the reservoir at low pressure, the pump will encounter less wear, and the electric motor will use a lower percentage of electrical power than normal.

Unloading valves can be used in hydraulic systems that require a rod to quickly extend to the point of work and then slowly extend once it is at the point of work, such as a two-stage log splitter.

For example, an unloading valve is used in a hydraulic system that forms sheet metal, such as a hydraulic press. Quick extension of the rod to the point of work saves time. Slowing extension at the point of work allows the stamping die to stamp the sheet metal with better control, which prevents the workpiece from being punctured. To form sheet metal with both fast and slow rod extension, there must be a large amount of fluid flow and pressure.

Two hydraulic pumps can be used in the system that operates a hydraulic press. One pump produces a high flow rate and operates at low pressure, while the other pump produces a low flow rate and operates at high pressure. These types of hydraulic systems are commonly referred to as high-low systems.

In a high-low system, the unloading valve redirects fluid flow from the high-flow, low-pressure pump when the rod of the cylinder reaches the point of work and pressure climbs. When the unloading valve opens to allow the fluid flow of the high-flow, low-pressure pump back to the reservoir, the pump's fluid flow will ideally travel at a pressure lower than 100 psi. Because the fluid flow is returning to the reservoir at low pressure, the pump will encounter less wear, and the electric motor will use a lower percentage of electrical power than normal. **See Figure 9-8.**

The unloading valve used in a hydraulic press with a high-low system operates using the following procedure:

1. A two-position, four-way directional control valve is actuated and sends a pilot signal that actuates the main directional control valve.
2. When the main directional control valve is actuated to extend the rod, the high-flow and low-flow pumps create fluid flow to move the piston and rod of the cylinder rapidly to the point of work.
3. When the rod reaches the workpiece, pressure increases quickly. When the pressure in the cap end of the cylinder increases to the unloading valve setting (600 psi), the unloading valve will open and allow fluid flow from the high-flow pump to travel to the reservoir under low pressure. The low-flow pump will still create fluid flow going into the system. Because the unloading valve is open, the pressure on that side of the check valve will be less than 100 psi.
4. The isolation check valve closes because of the low pressure on the pump's high-flow side of the check valve. Pressures over 600 psi are created by the resistance of the system for the low-flow pump to work against.

   The speed of the cylinder decreases because the cylinder is only receiving fluid flow from the low-flow pump. The low-flow pump continues to create fluid flow so that the workpiece can be stamped.
5. The maximum pressure that the low-flow pump will have to create fluid flow against is set by the relief valve in the hydraulic system. The relief valve is set so that there is enough pressure to stamp the workpiece. When the directional control valves are shifted to retract the cylinder, pressure drops in the system and closes the unloading valve. The high-flow and low-flow pumps can then both send fluid flow into the system to rapidly retract the rod of the cylinder. Two check valves are placed in the system to isolate the system into different pressure zones.

Another common application for an unloading valve is energy conservation. For example, an unloading valve can be used in a hydraulic press where the rod must extend and hold a load for a certain amount of time. Rather than having a high-pressure pump hold a load on a cylinder, an accumulator is used to maintain pressure in the system while pump pressure is released through the unloading valve. **See Figure 9-9.**

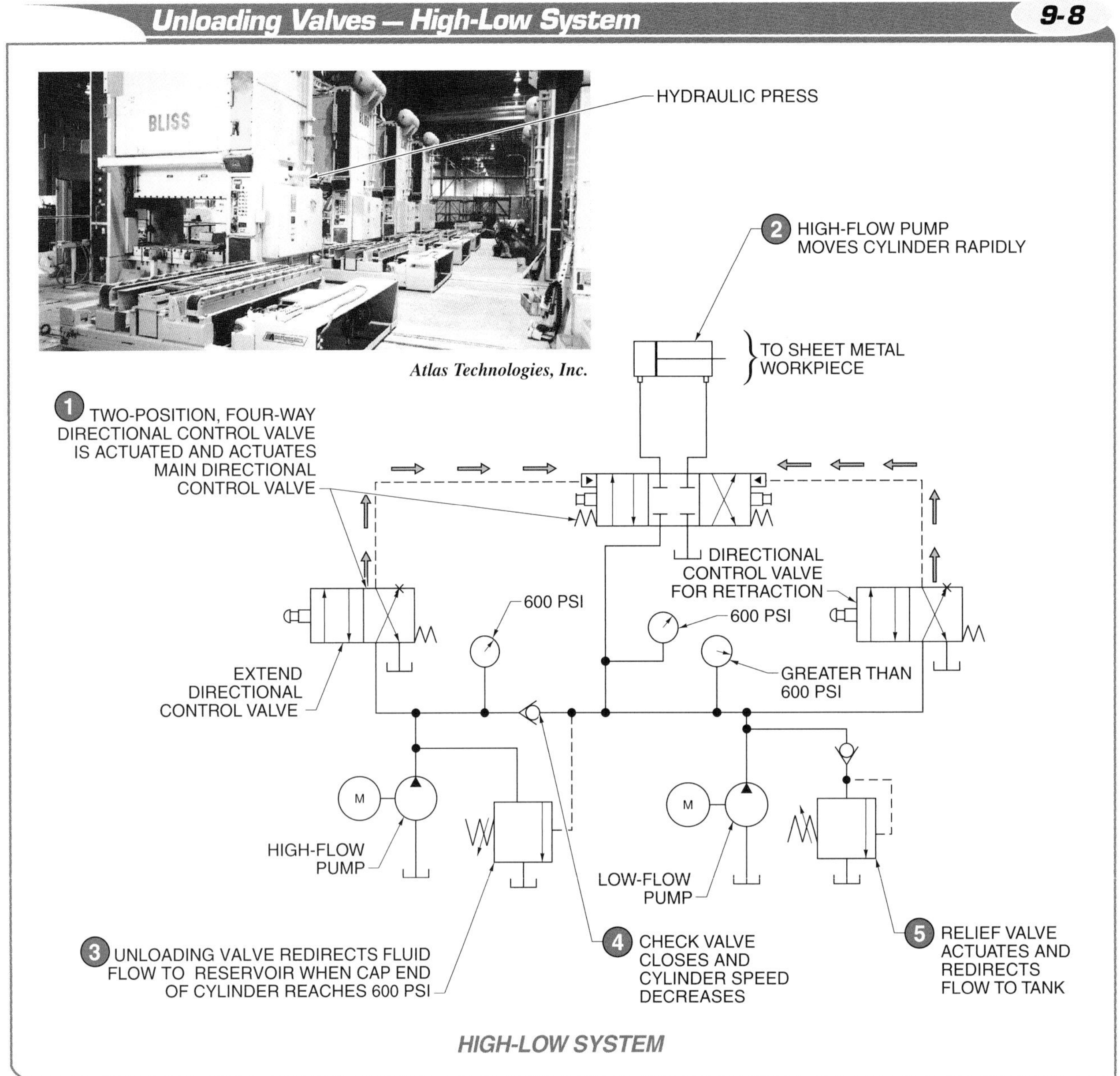

Figure 9-8. An unloading valve is used in a high-low system that operates a hydraulic press.

For example, an unloading valve with a biasing pressure of 1200 psi is installed in a hydraulic press. As the hydraulic press operates, hydraulic fluid flows through the check valve and a two-position, three-way directional control valve. Fluid flow fills and pressurizes the accumulator and cylinder, causing the rod of the cylinder to extend. When the rod is fully extended, system pressure spikes upward. When pressure reaches 1200 psi, the unloading valve opens.

When the unloading valve opens, the check valve closes because the accumulator maintains system pressure at 1200 psi. It also closes because there is more pressure on that side than the pump side of the check valve. The rod is held in place under pressure, pressing the workpiece. The unloading valve is also held open by pressure from the accumulator, which causes hydraulic fluid from the pump to flow into the reservoir under low pressure.

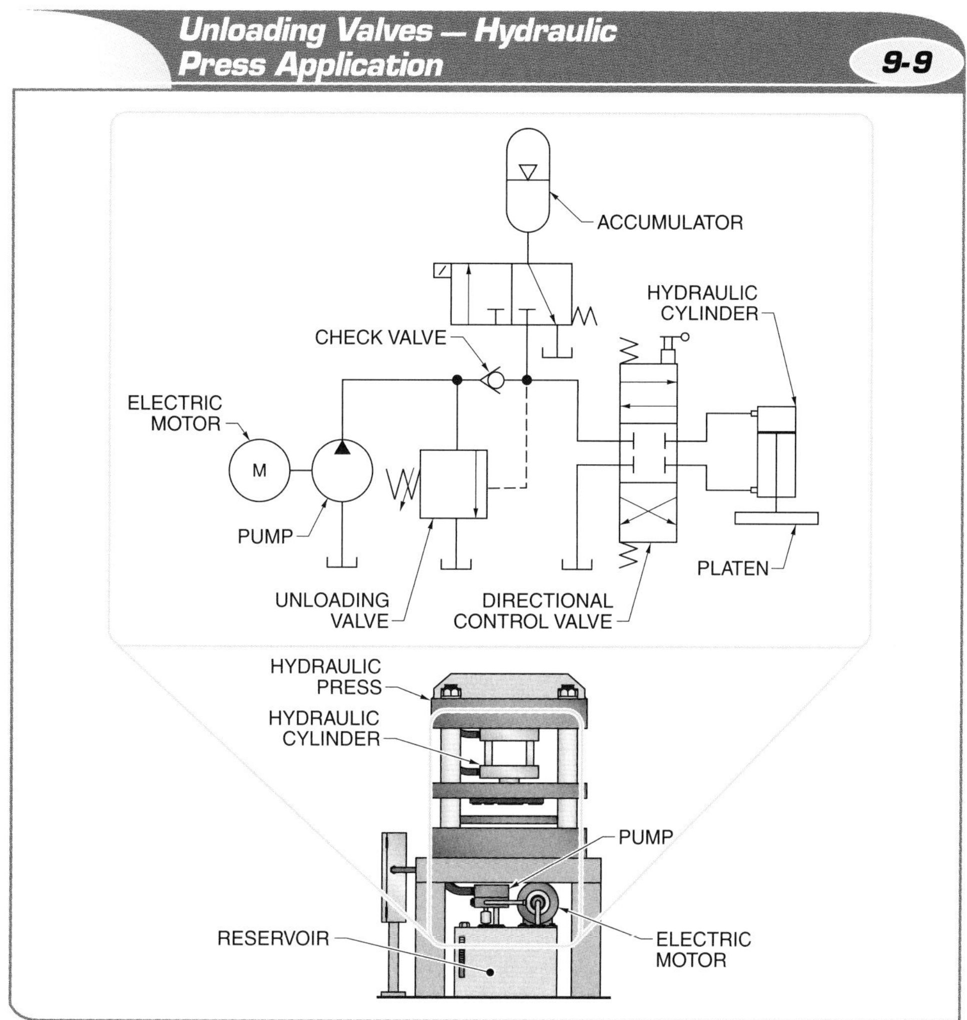

Figure 9-9. An accumulator in a hydraulic press can be used to maintain pressure in the system while pump pressure is released through an unloading valve.

When the directional control valve is shifted to retract the cylinder, the pressure in the accumulator decreases. This opens the check valve and allows fluid flow from the pump to retract the rod. When the rod is completely retracted and the accumulator is charged, the system is ready for the next hydraulic press operation.

TECH FACT

Some hydraulic valve manufacturers design their valves in such a manner that allows the user to convert a valve from one type to another by simply rearranging the top and bottom covers of the valve.

Unloading valves can be used in industrial hydraulic systems with multiple accumulators.

TERMS

A **remote-controlled sequence valve** is a pressure control valve that receives a pilot signal from a remote section of a hydraulic system.

A **remote-controlled counterbalance valve** is a normally closed pressure control valve that uses an external pilot line to sense pressure from a remote section of the hydraulic system and counteract the pressure from a hanging cylinder load.

Remote-Controlled Sequence Valves

A *remote-controlled sequence valve* is a pressure control valve that receives a pilot signal from a remote section of a hydraulic system. Remote-controlled sequence valves operate similarly to direct-acting sequence valves except that they have an external pilot line that can sense pressure in another section of the system rather than at the inlet port. An advantage of a remote-controlled sequence valve is that it can be used when the two actuators are being controlled by different directional control valves.

In hydraulic systems that have more than one cylinder, a remote-controlled sequence valve is sometimes used to prevent multiple cylinders from operating at the same time. For example, a remote-controlled sequence valve is commonly used in a hydraulic clamping and drilling system when the workpiece must be clamped before it is drilled. **See Figure 9-10.** It can be used to sense when the workpiece is properly clamped by cylinder 1 before allowing fluid flow to the drilling actuator's directional control valve. Individual directional control valves are used to control the extension and retraction of each cylinder.

Remote-Controlled Counterbalance Valves

A *remote-controlled counterbalance valve* is a normally closed pressure control valve that uses an external pilot line to sense pressure from a remote section of the hydraulic system and counteract the pressure from a hanging cylinder load. A remote-controlled counterbalance valve is sometimes referred to as an external counterbalance valve.

A remote-controlled counterbalance valve operates like a direct-acting counterbalance valve, but receives its signal from an external pilot line. The external pilot line is usually attached to the pressure line entering the cap end of the cylinder. Pilot pressure from a remote section prevents the back pressure that would otherwise cause the pressure on the rod end of a cylinder to resist the pressure on its cap end.

A remote-controlled counterbalance valve is used in applications where a heavy load is suspended from a vertically mounted cylinder. This load may be an operating mechanism in a hydraulic system. For example, in a concrete compression test machine, a vertically mounted cylinder with an attached compression plate requires a remote-controlled counterbalance valve. **See Figure 9-11.**

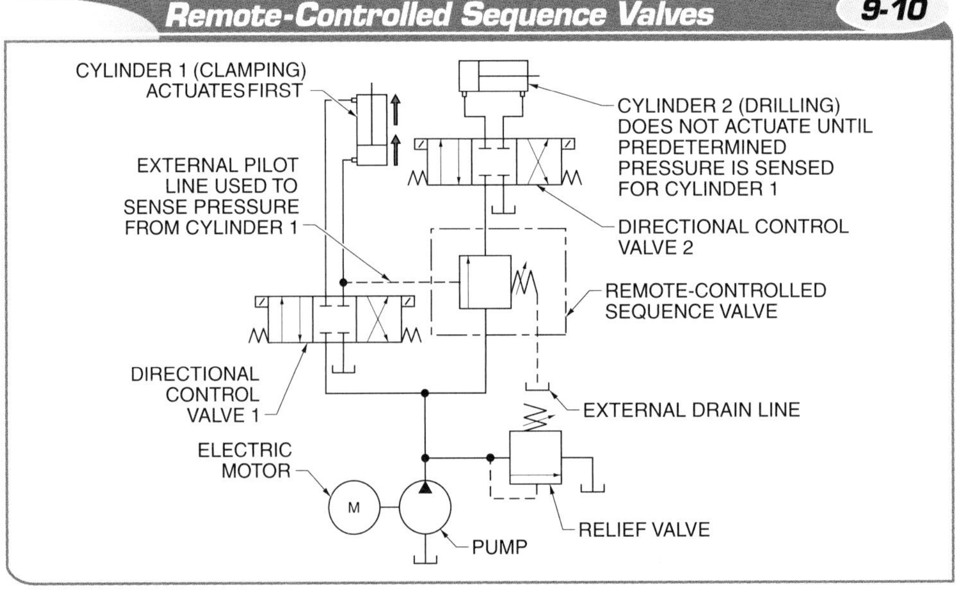

Figure 9-10. Remote-controlled sequence valves can be used to control multiple cylinders in the same system, such as a hydraulic clamping and drilling system.

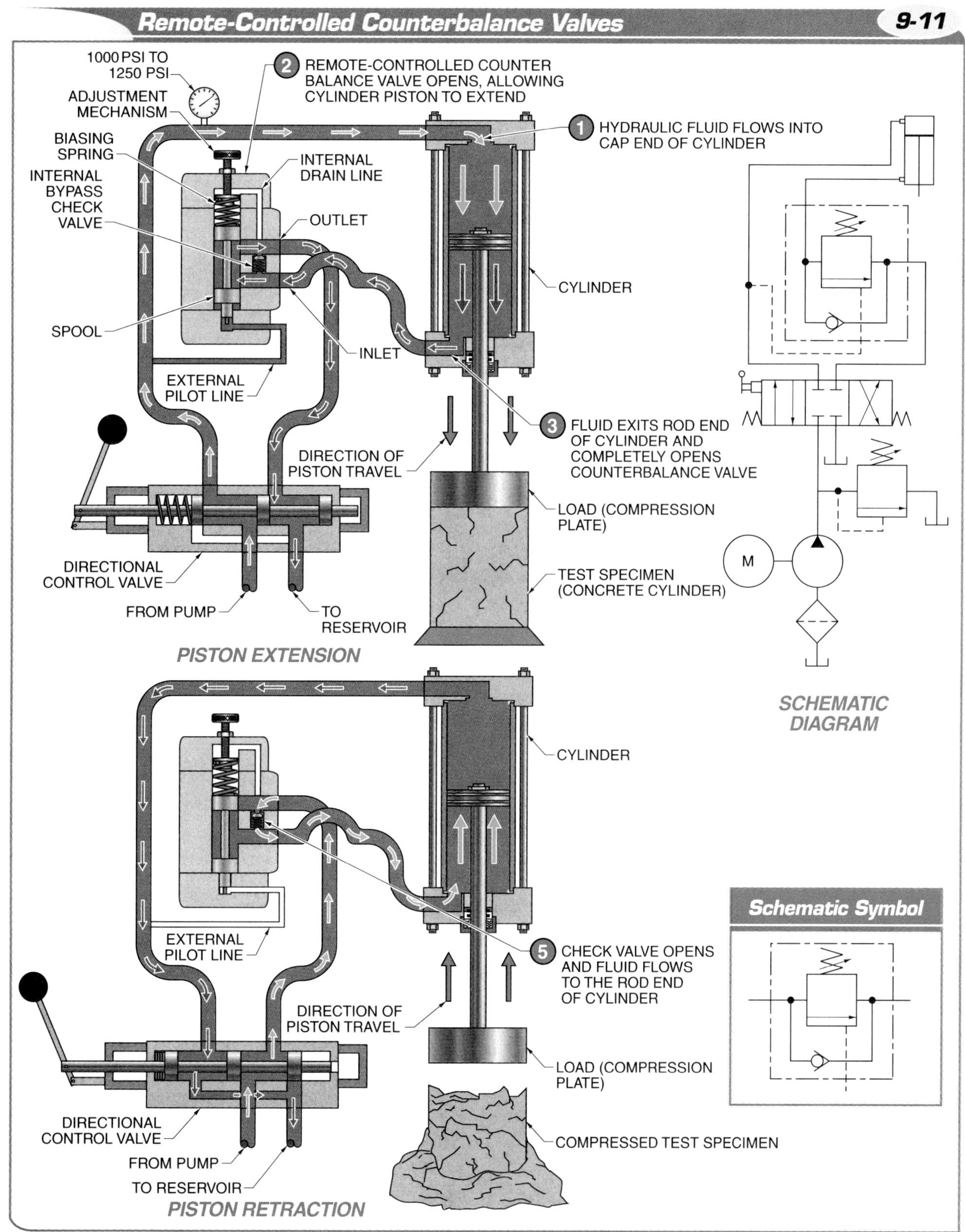

Figure 9-11. Remote-controlled counterbalance valves are typically used in hydraulic systems such as concrete compression test machines.

TERMS

A **brake valve** is a normally closed pressure control valve that prevents a hydraulic motor from running away, or speeding out of control.

A remote-controlled counterbalance valve operates a concrete compression test machine in the following procedure:

1. The directional control valve is actuated when the test specimen, a concrete cylinder, is in place. Hydraulic fluid flows into the cap end of the cylinder. Since hydraulic fluid in the rod end of the cylinder cannot exit, pressure on the cap end starts to build until the biasing pressure of the remote-controlled counterbalance valve is reached.
2. When biasing pressure of 1000 psi is reached, the remote-controlled counterbalance valve opens and allows the rod to slowly extend.
3. When the compression plate reaches the concrete cylinder, pressure begins to increase on the cap end of the cylinder. As pressure increases on the cap end, it also increases in the remote-controlled counterbalance valve of 1250 psi. The pilot line fully opens the internal spool of the remote-controlled counterbalance valve, allowing hydraulic fluid to flow unrestricted from the rod end of the cylinder.
4. Since hydraulic fluid flows unrestricted from the rod end of the cylinder, there is no back pressure working against the extension of the piston. This allows the full force of the cylinder and the weight of the compression plate to compress the concrete cylinder.
5. Once the test specimen is processed, the directional control valve is manually actuated and hydraulic fluid flows through the check valve in the remote-controlled counterbalance valve. This forces the rod to retract.

Counterbalance valves are used in large forklifts to help lower a load slowly or to help hold a load in the up position.

Brake Valves

A *brake valve* is a normally closed pressure control valve that prevents a hydraulic motor from running away, or speeding out of control. A brake valve has both internal pilot lines and remote lines for controlling the speed of the motors in a hydraulic system.

A low setting allows the motor to start at a lower pressure and is controlled by the external pilot line connected to the inlet side of the motor. Because the brake valve is closed, the motor cannot rotate and pressure builds up in the pressure line to the motor. When pressure reaches the set value for the remote pilot line, the brake valve will open. Once the valve opens, the motor begins to rotate. If the motor begins running away, the pressure at the inlet side of the brake valve drops and the brake valve closes again. When the brake valve closes, this creates resistance to the fluid flow coming out of the motor. As the motor begins to stop, the fluid flow coming out of the motor that is working against the closed brake valve generates pressure. This pressure travels through the brake valve's internal pilot line to open the brake valve again. This constant opening and closing of the brake valve prevents the motor from running away. It is also more energy efficient because the brake valve opens at a lower pressure and system pressure does not work against the back pressure created by the brake valve. **See Figure 9-12.**

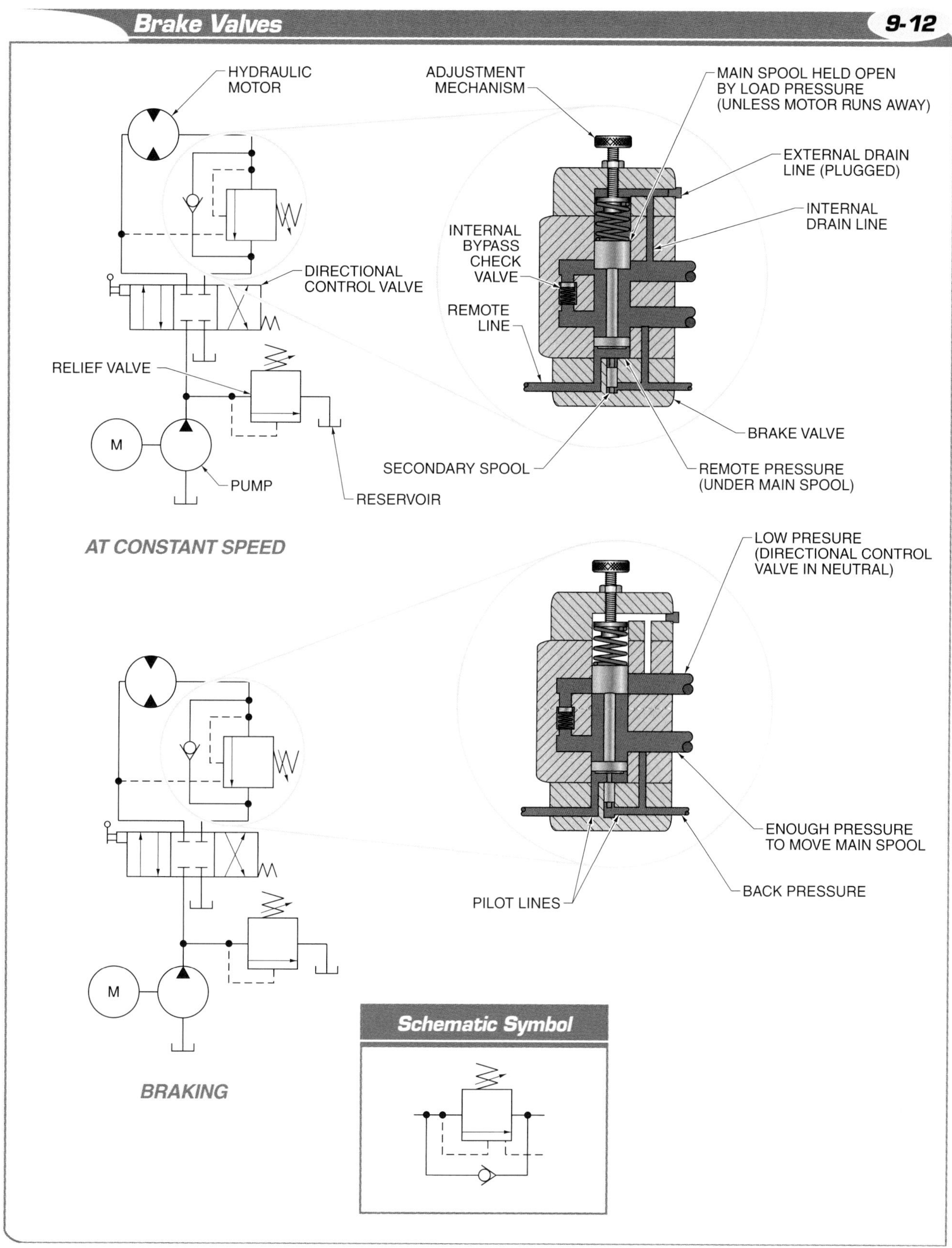

Figure 9-12. A brake valve can be both direct-acting and remote-controlled.

TERMS

A **pilot-operated pressure control valve** is a type of pressure control valve that is remote controlled and, instead of a spring, uses pilot pressure set at a level specified by a pilot valve to bias the main spool.

A **pilot-operated relief valve** is a type of pilot-operated pressure control valve that is used to control the maximum pressure in a hydraulic system.

A **remote-controlled, pilot-operated relief valve** is a type of pilot-operated relief valve that uses one or more external pilot valves to set the pressure(s) at which the relief valve will open.

PILOT-OPERATED PRESSURE CONTROL VALVES

A *pilot-operated pressure control valve* is a type of pressure control valve that is remote controlled and, instead of a spring, uses pilot pressure set at a level specified by a pilot valve to bias the main spool. A main advantage of pilot-operated pressure control valves is that they have low-pressure (under 100 psi) override. Pilot-operated relief valves have several advantages over direct-acting types of pressure control valves, including larger ranges of operating pressure and quicker reaction times.

Pilot-operated pressure control valves are available with remote control capabilities. Because of this, operators can adjust the pressure settings of pilot-operated pressure control valves from other locations.

A pilot-operated pressure control valve has a biasing spring on the main spool that is used to help hold the spool down and the valve closed. Biasing pressure for a light-biasing spring in a pilot-operated pressure control valve is usually 20 psi to 30 psi. A pilot valve is a very small relief valve inside a pilot-operated pressure control valve that is used to set the pressure at which the pressure control valve actuates.

Pilot-Operated Relief Valves

A *pilot-operated relief valve* is a type of pilot-operated pressure control valve that is used to control the maximum pressure in a hydraulic system. Pilot-operated relief valves are used instead of direct-acting relief valves when more energy efficiency is required. **See Figure 9-13.** They also have a quicker reaction time and can withstand a larger range of pressure. Pilot-operated relief valves are used in applications such as cranes, presses, excavation equipment, welding machines, and pipe benders. Pilot-operated relief valves are sometimes called compound two-stage relief valves.

A pilot-operated relief valve consists of a main spool and an adjustable pilot valve that are both internal to the main valve body. The main spool of the valve has an orifice to allow pilot pressure to the top of the spool and is biased by a low-pressure spring. The orifice allows any pressure that is under the spool to travel to the top of the spool to be used for main spool biasing and to travel to the pilot valve. The pilot valve has a dart and biasing spring. In addition, an internal drain line connects the pilot chamber to the reservoir port so when the pilot valve is actuated, hydraulic fluid in the pilot chamber drains to the reservoir. **See Figure 9-14.**

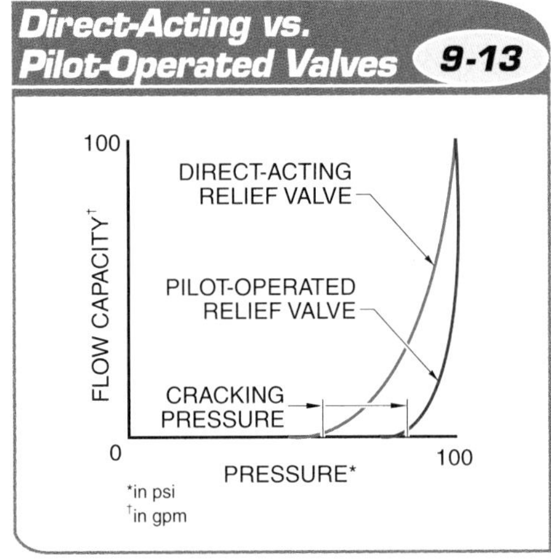

Figure 9-13. Pilot-operated relief valves are more energy efficient than direct-acting relief valves.

A pilot-operated relief valve operates through the following procedure:

1. Hydraulic fluid enters the inlet port of the pilot-operated relief valve.
2. Fluid pressure travels through the orifice in the main spool and fills the chamber above the spool and the pilot chamber.
3. The pressure above the main spool and the added force of the biasing spring hold the main spool in the closed position.
4. The dart of the pilot valve unseats (opens) when the pressure in the pilot chamber reaches the set value of the pilot valve.
5. Once the dart unseats, any fluid above the main spool is allowed to drain to the reservoir through the pilot valve.
6. The main spool shifts upward, allowing hydraulic fluid from the pump to flow directly to the reservoir.
7. Any excess fluid from the main spool or pilot chamber is directed through the pilot valve back to the reservoir.

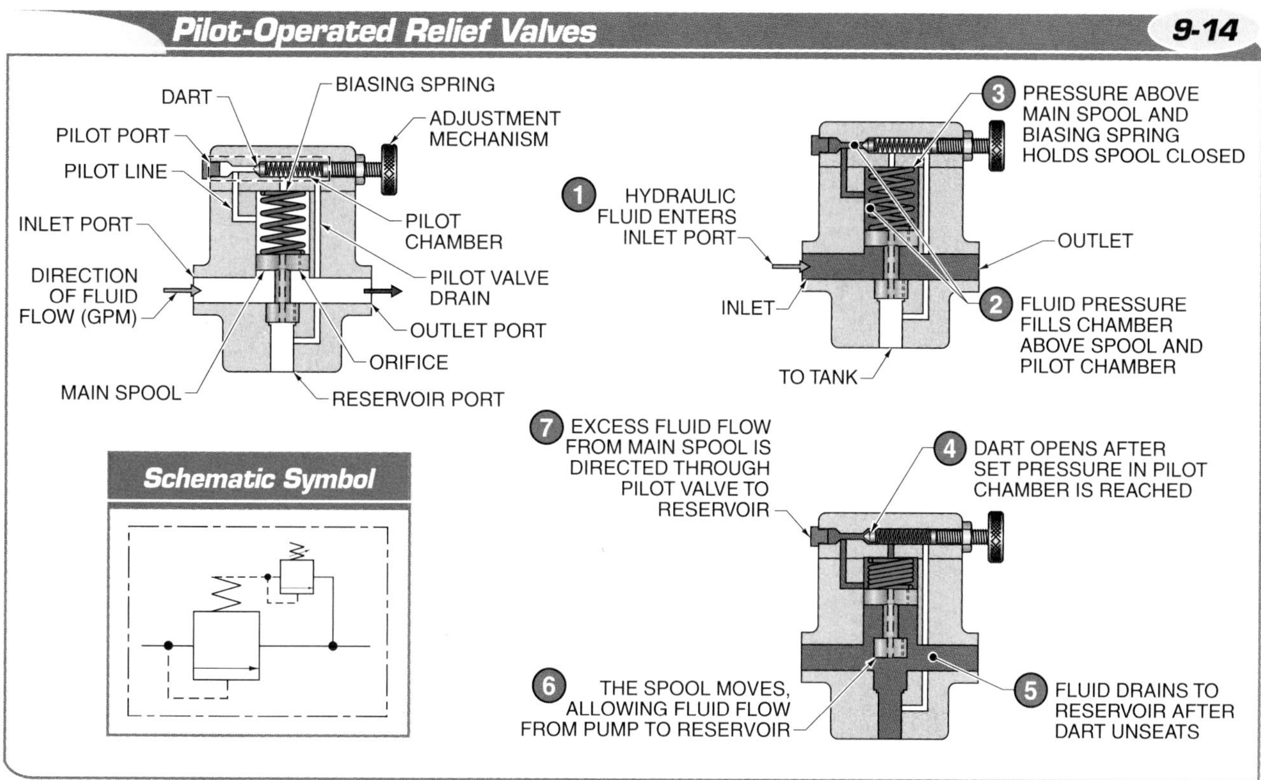

Figure 9-14. Pilot-operated relief valves are used instead of direct-acting relief valves when more precise pressure control is required.

Remote-Controlled, Pilot-Operated Relief Valves. A *remote-controlled, pilot-operated relief valve* is a type of pilot-operated relief valve that uses one or more external pilot valves to set the pressure(s) at which the relief valve will open. The remote-controlled relief valves are connected in parallel and relieve pressure from the pilot side of the main spool. They operate based on the principle that fluid flow will take the path of least resistance.

System pressure is determined by whichever remote pilot relief valve is set at the lowest pressure. For example, if there are two remote pilot relief valves in the system, one set at 1800 psi and the other set at 1600 psi, the pilot-operated relief valve will open at 1600 psi. In addition, if the pilot-operated relief valve is set at 2100 psi, pressure in any of the remote pilot relief valves cannot be higher than 2100 psi.

The advantage of using remote pilot valves is that they can be installed in remote locations, which allows easier adjustment of the pilot-operated relief valve pressure setting. Typically, a pilot-operated relief valve is set at the highest pressure that will ever be required in a hydraulic system. The remote pilot valves are then used to adjust system pressure to any value less than the pressure set on the pilot-operated relief valve. Remote pilot valves can add an extra level of safety to a system by providing more than one path to open the pilot-operated relief valve. **See Figure 9-15.**

Remote-controlled, pilot-operated relief valves are typically used in applications such as process equipment that may be located over several different floors or locations in a facility. Common applications for pilot-operated relief valves include manufacturing systems that require adjustable maximum system pressure. For example, pilot-operated relief valves are used in hydraulic theroforming presses that form different types of plastic sheet material.

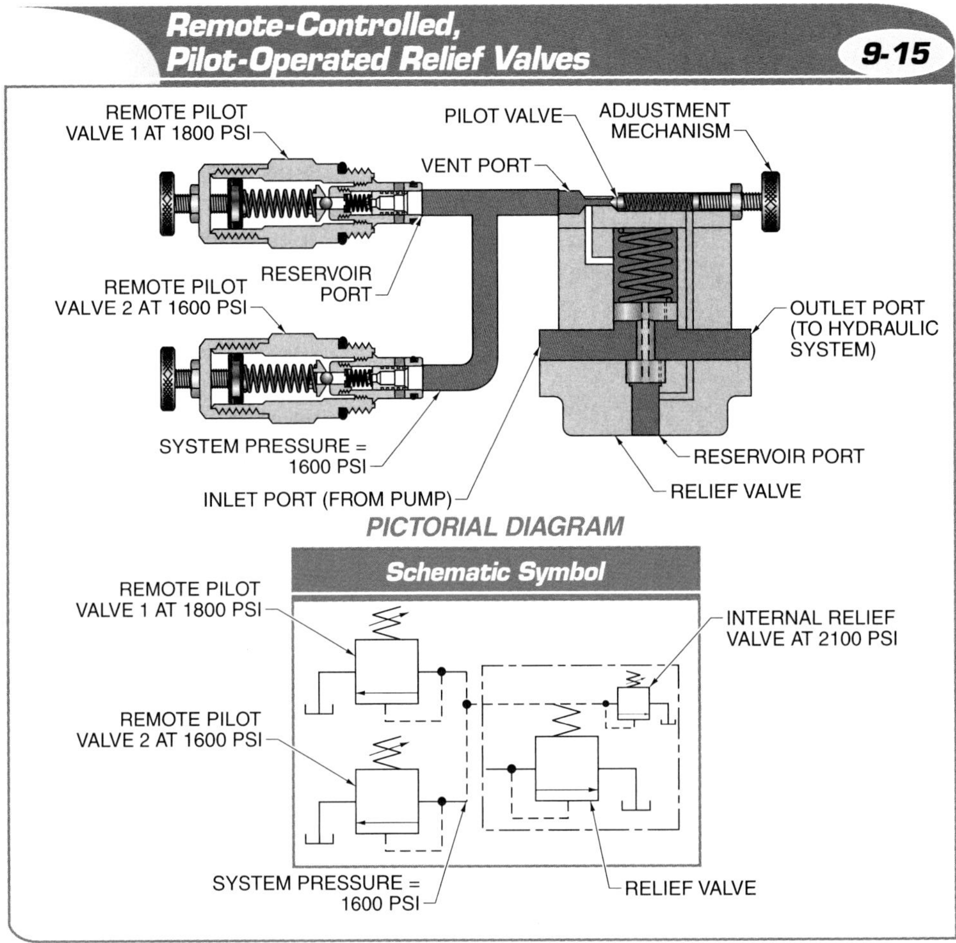

Figure 9-15. A remote-controlled, pilot-operated relief valve can be installed in a remote location and allow adjustment of maximum system pressure for different locations.

Venting Remote-Controlled, Pilot-Operated Relief Valves. The one or more remote-pilot valves connected to a pilot-operated relief valve can be replaced with a two-position, two-way directional control valve. The two-position, two-way directional control valve can be used as an ON/OFF switch that vents hydraulic fluid from the pilot chamber back to the reservoir. **See Figure 9-16.** This is known as a type of pump unloading method.

The two-position, two-way directional control valve remains closed while the hydraulic system is in operation and the pilot-operated relief valve operates normally. When the hydraulic system needs to be in the idle condition, the two-position, two-way directional control valve is opened, allowing hydraulic fluid from the pilot chamber above the main spool to vent back to the reservoir. This decreases the pressure in the pilot chamber, allowing the pressure under the main spool to push it upwards. When the main spool shifts, the pilot-operated relief valve opens, allowing hydraulic fluid to travel to the reservoir under low pressure.

A pilot-operated relief valve can be used to save energy because the motor attached to the hydraulic pump has little load on it since there is no pressure in the system. For example, a pilot-operated relief valve is used in a stand-alone pipe-bending machine that makes bends at set intervals during operation. When the pipe-bending machine is in the idle condition, hydraulic fluid is vented to the reservoir and the low pressures involved help to prevent premature wear on the hydraulic system.

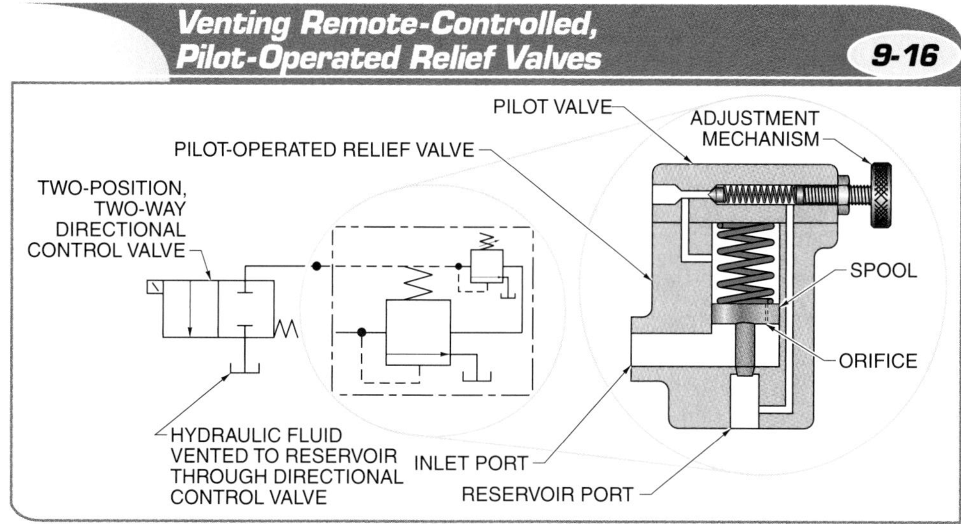

Figure 9-16. A two-position, two-way directional control valve can be used as an ON/OFF switch that vents hydraulic fluid to the reservoir.

Pilot-Operated Pressure-Reducing Valves

A *pilot-operated pressure-reducing valve* is a type of pilot-operated relief valve that is used to control pressures in one leg of a system. Pilot-operated pressure-reducing valves are commonly used in applications such as hydraulic presses, pipe-bending machines, welding machines, and aerial lifts.

Pilot-operated pressure-reducing valves are normally open. Hydraulic fluid from the outlet port travels through the pilot passage and works against the bottom of the spool. Hydraulic fluid from the ports is transmitted through the orifice in the spool to the top of the spool, into the pilot chamber, and to the pilot valve. The main spool of pilot-operated pressure-reducing valves has a low-strength biasing spring to help the spool remain in the open position. Once pressure unseats the dart in the pilot valve, the spool starts to close and stops pressure from increasing in the leg of the system being controlled.

Similar to pilot-operated relief valves, pilot-operated pressure-reducing valves can also have one or more remote pilot valves for more flexible operation. However, unlike a pilot-operated relief valve, when the dart is unseated in a pilot-operated pressure-reducing valve, hydraulic fluid does not drain internally to the reservoir. Instead, hydraulic fluid drains to the reservoir through an external drain line.

Other Pilot-Operated Pressure Control Valves

Counterbalance valves, sequence valves, unloading valves, and brake valves are also available with pilot-operated control. These pilot-operated valves have advantages over direct-acting pressure control valves, including smaller valve bodies, larger ranges of operating pressure, and quicker reaction times. They can also have remote pilot valves that allow them to be adjusted from another location within the system. In addition, the pilot valve has a high-biasing spring. The pressure for the main valve is set by an adjustable mechanism in the pilot valve.

TERMS

A **pilot-operated pressure-reducing valve** is a type of pilot-operated relief valve that is used to control pressures in one leg of a system.

TECH FACT

Pilot-operated pressure-reducing valves must be installed in horizontal piping so that the indicating arrows on the valve housing match with the direction of fluid flow in the hydraulic system.

Digital Resources

Chapter Review

Pressure Control

Chapter 9

Name: ______________________ Date: ______________

MULTIPLE CHOICE

______________ **1.** A remote-controlled, pilot-operated ___ valve is a type of pilot-operated relief valve that uses one or more external pilot valves to set the pressure(s) at which the relief valve will open.

A. counterbalance
B. relief
C. sequence
D. unloading

______________ **2.** A(n) ___ valve is a normally closed pressure control valve that prevents a hydraulic motor from running away, or speeding out of control.

A. brake
B. counterbalance
C. sequence
D. unloading

______________ **3.** A ___ pressure control valve is a type of pressure control valve that is remote controlled and, instead of a spring, uses pilot pressure set at a level specified by a pilot valve to bias the main spool.

A. direct-acting
B. pilot-operated
C. remote-control
D. none of the above

______________ **4.** A(n) ___ valve is a normally closed pressure control valve that stops a vertically mounted cylinder from running away because of the load weight.

A. counterbalance
B. relief
C. sequence
D. unloading

______________ **5.** A remote-controlled ___ valve receives a pilot signal from a remote section of a hydraulic system.

A. pilot
B. relief
C. sequence
D. unloading

______________ **6.** A(n) ___ valve is a normally closed pressure control valve that can set maximum system pressure and unload the pump, which operates under 100 psi, at the same time.

A. counterbalance
B. relief
C. sequence
D. unloading

__________ **7.** A pilot-operated ___ valve is a type of pilot-operated pressure control valve that is used to control the maximum pressure in a hydraulic system.
A. counterbalance
B. relief
C. sequence
D. unloading

__________ **8.** A(n) ___ valve is a normally closed pressure control valve that makes various operations occur in an orderly manner, or one after another.
A. counterbalance
B. relief
C. sequence
D. unloading

__________ **9.** A remote-controlled ___ valve is a normally closed pressure control valve that uses an external pilot line to sense pressure from a remote section of the hydraulic system and counteract the pressure from a hanging cylinder load.
A. counterbalance
B. relief
C. sequence
D. unloading

__________ **10.** A(n) ___ valve is a normally open pressure control valve that controls pressure in one leg (one small part) of a system.
A. counterbalance
B. pressure-reducing
C. sequence
D. unloading

__________ **11.** ___ drains are used only in valves that do not have pressure on the secondary port of the valve.
A. External
B. Internal
C. both A and B
D. none of the above

__________ **12.** A remote-controlled ___ valve operates like a direct-acting counterbalance valve, but receives its signal from an external pilot line.
A. counterbalance
B. pressure-reducing
C. sequence
D. unloading

__________ **13.** Biasing pressure for a light-biasing spring in a pilot-operated pressure control valve is usually ___ psi to ___ psi.
A. 0; 10
B. 10; 20
C. 20; 30
D. 45; 50

______________ **14.** A pilot-operated ___ valve is a type of pilot-operated relief valve that is used to control pressures in one leg of a system.
A. counterbalance
B. pressure-reducing
C. sequence
D. unloading

______________ **15.** A pilot-operated ___ valve can be used to save energy because the motor attached to the hydraulic pump has little load on it since there is no pressure in the system.
A. counterbalance
B. relief
C. sequence
D. unloading

COMPLETION

______________ **1.** Pilot-operated relief valves are used instead of ___ relief valves when more energy efficiency is required.

______________ **2.** A(n) ___ valve is used when a load will make a piston move faster than the flow rate going to a cylinder dictates, such as with mobile equipment.

______________ **3.** A(n) ___ valve has both internal pilot lines and remote lines for controlling the speed of the motors in a hydraulic system.

______________ **4.** In hydraulic systems that have more than one cylinder, a(n) ___ valve is sometimes used to prevent multiple cylinders from operating at the same time.

______________ **5.** ___ valves can be used in hydraulic systems that require a rod to quickly extend to the point of work and then slowly extend once it is at the point of work.

______________ **6.** A(n) ___ valve is installed in a hydraulic system when one operation needs to begin after another has finished.

______________ **7.** A common application for an unloading valve is ___ conservation.

______________ **8.** On a schematic diagram, a(n) ___ drain line appears as a pilot line coming out of the biasing spring and connecting to the reservoir.

______________ **9.** A(n) ___ pressure control valve senses pressure through an internal pilot line connected to the inlet or the outlet of the valve in order to direct fluid flow to the bottom of its spool where it works against a biasing spring.

______________ **10.** A(n) ___ valve operates similarly to a direct-acting sequence valve except that it has an external pilot line that can sense pressure in another section of the system rather than at the inlet port.

______________ **11.** ___ valves are commonly installed in hydraulic clamping and bending systems because their cylinders extend and retract their rods, one after another.

______________ **12.** A(n) ___ valve is a normally open pressure control valve that senses pressure at its outlet and controls pressure in one section of a hydraulic system.

TRUE/FALSE

T F 1. Fluid flow will always take the path of least resistance.

T F 2. A remote-controlled unloading valve is used in applications where a heavy load is suspended from a vertically mounted cylinder.

T F 3. Because a pressure-reducing valve maintains pressure at its outlet, there must be a hydraulic power unit for every section of the system that requires different pressure settings.

T F 4. A main advantage of pilot-operated pressure control valves is that they have low-pressure (under 100 psi) override.

T F 5. Pressure-reducing valves are always installed after the directional control valve.

T F 6. Typically, a pilot-operated relief valve is set at the lowest pressure that will ever be required in a hydraulic system.

T F 7. Pressure-reducing valves must be installed as far from the point of work as possible.

T F 8. Pilot-operated pressure-reducing valves can also have one or more remote pilot valves for more flexible operation.

T F 9. External drain lines must end above the fluid level in a reservoir.

T F 10. Direct-acting relief valves are sometimes called compound two-stage relief valves.

T F 11. Direct-acting relief valves are always attached to a pump in parallel.

T F 12. An advantage of a remote-controlled sequence valve is that it can be used when the two actuators are being controlled by different directional control valves.

T F 13. Pilot-operated pressure control valves are available with remote control capabilities.

T F 14. A direct-acting sequence valve requires an internal drain line since the outlet port is pressurized when the valve continues with a secondary operation.

T F 15. Pilot-operated pressure-reducing valves are normally open.

T F 16. With counterbalance valves, the pressure created in the cylinder by the weight of the load cannot travel through the valve and back to the reservoir until it overcomes the biasing spring force.

T F 17. Per OSHA regulations, there must never be any other components connected to a hydraulic system between the pump and its relief valve.

T F 18. Unloading valves have an internal bypass check valve to allow reverse fluid flow.

T F 19. Most direct-acting counterbalance valves have an internal bypass check valve in the main valve body that allows reverse flow through the valve.

T F 20. Drain lines are considerably larger than inlet and outlet lines.

SHORT ANSWER

1. List two advantages of having fluid return to the reservoir at low pressure (100 psi or less) in a high-low system.

2. List two advantages of a pilot-operated relief valve over a direct-acting pressure control valve.

3. Explain the operation of a high-low system.

MATCHING

Schematic Symbols

_______________ **1.** Brake valve

_______________ **2.** Counterbalance valve

_______________ **3.** Drain line

_______________ **4.** Pilot-operated relief valve

_______________ **5.** Pressure-reducing valve

_______________ **6.** Relief valve

_______________ **7.** Remote-controlled counterbalance valve

_______________ **8.** Remote-controlled, pilot-operated relief valve

_______________ **9.** Sequence valve

_______________ **10.** Unloading valve

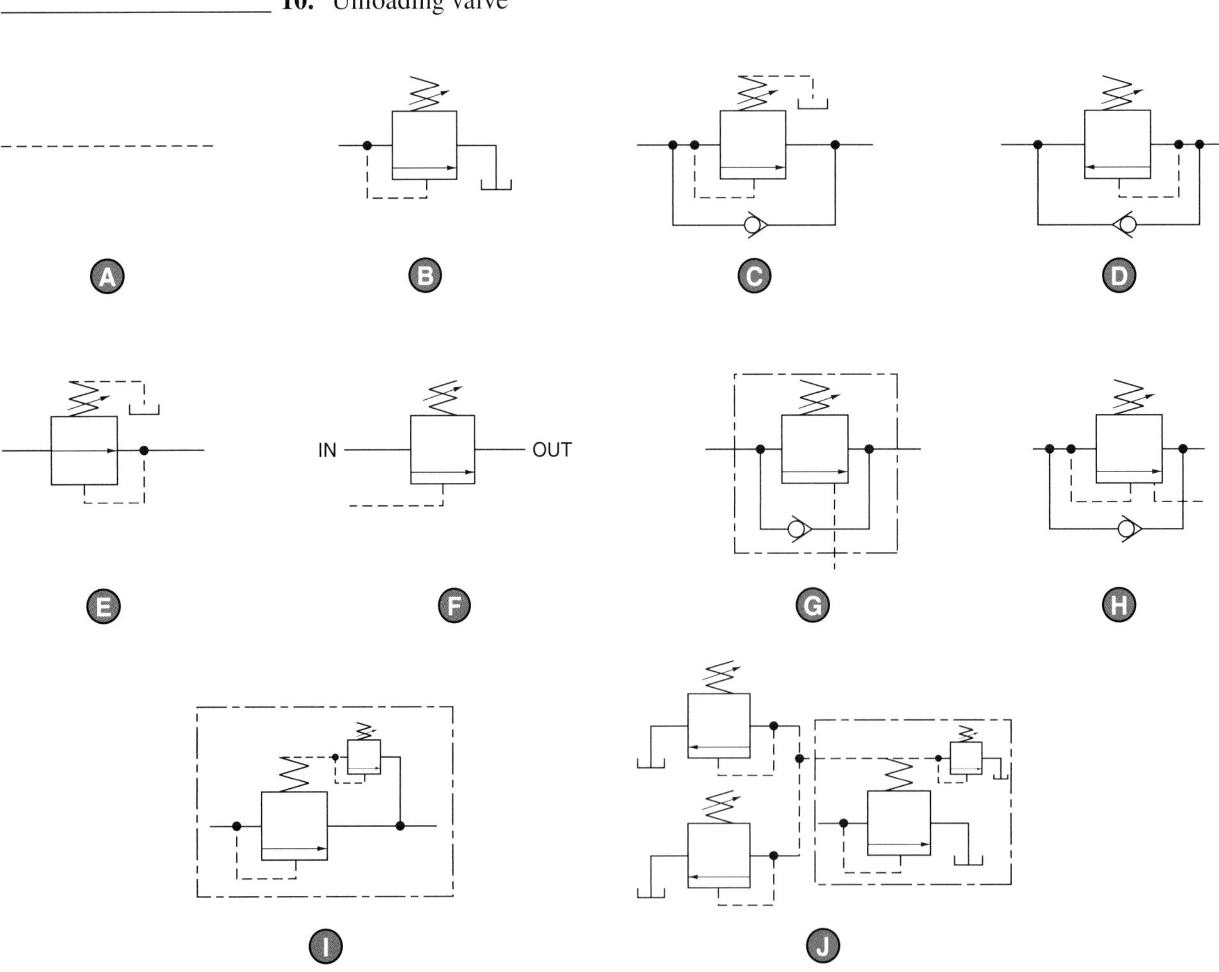

Activity 9-1: Clamping and Bending System

The clamping and bending system allows the clamping cylinder to extend its rod before the bending cylinder extends its rod. Both cylinders retract their rods at the same time. When both rods retract simultaneously, the workpiece has a tendency to fall from the clamping and bending dies.

1. Connect Schematic Symbols to indicate a schematic diagram that will sequence the bending cylinder to retract its rod before the clamping cylinder retracts its rod.

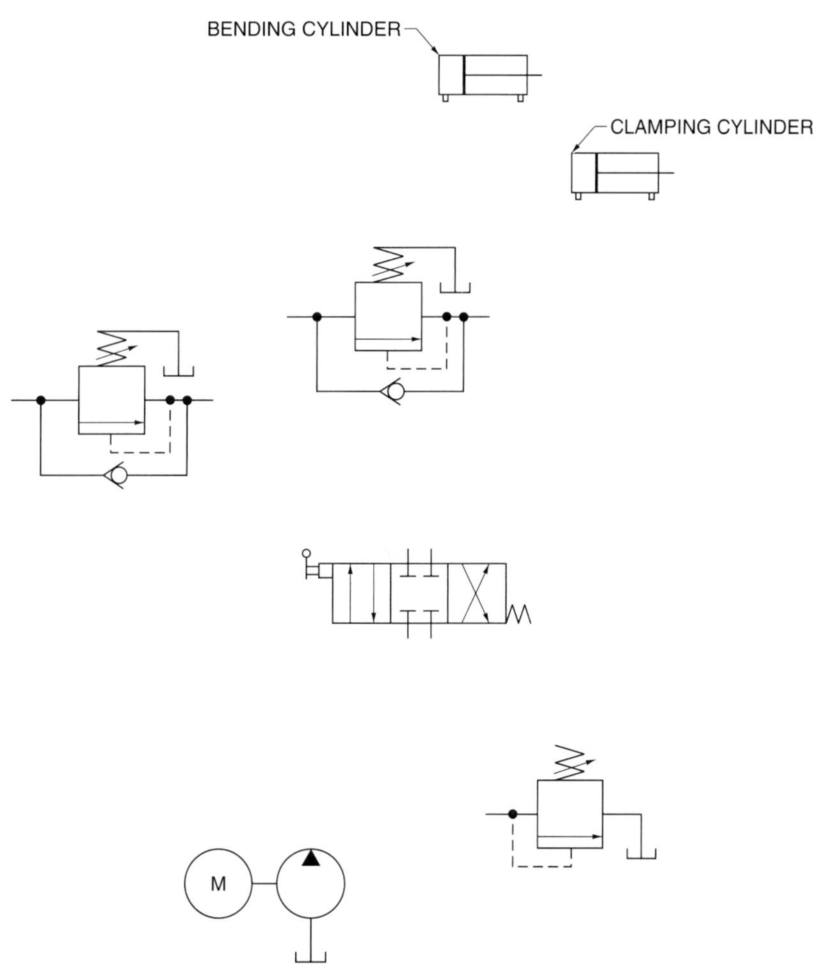

SCHEMATIC SYMBOLS

Activity 9-2: Hydraulic Stage Lifting System

A theater company has hired an engineering consulting firm to design a hydraulic system to lift and hold a section of a stage for a given amount of time. The section of the stage that must be lifted and held weighs 2900 lb. The cylinder has a piston diameter of 2.25 in.

1. Connect Schematic Symbols to indicate a schematic diagram that shows the lifting and holding operation.

_______________ 2. How much pressure is required to lift the load?

_______________ 3. How much pressure is required to hold the load in position?

_______________ 4. What is the amount of spring biasing pressure in the counterbalance valve?

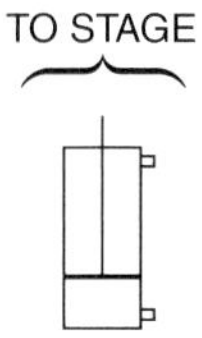

TO STAGE

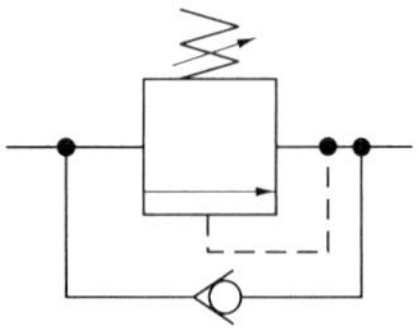

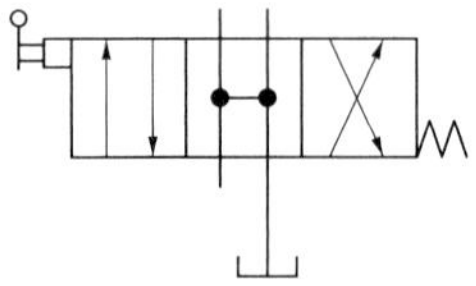

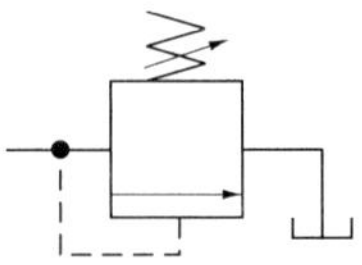

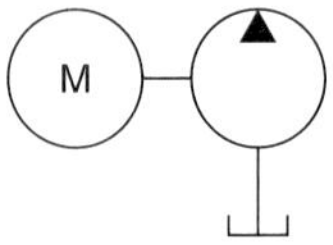

M

Activity 9-3: Pressure Relief Valve Selection

The engineering consulting firm hired by the theater company (in Activity 9-2) needs to determine the pressure relief valve that will best operate in the stage lifting and holding application. *Note:* The relief valve must be able to be locked out and is to be mounted onto the side of the power unit with a size 10 subplate.

1. Select the least expensive pressure relief valve that will meet the needs of the system.

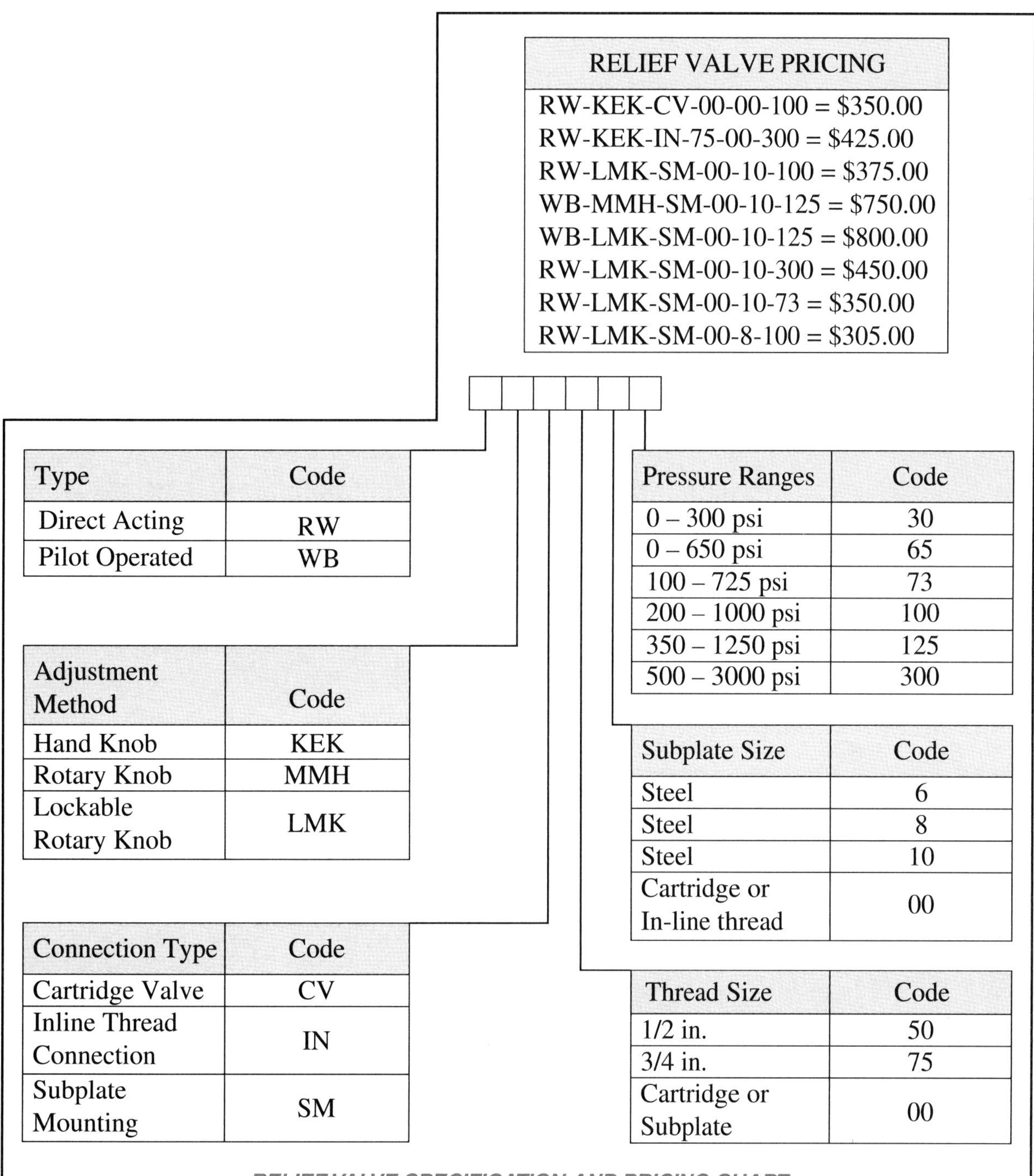

RELIEF VALVE PRICING
RW-KEK-CV-00-00-100 = $350.00
RW-KEK-IN-75-00-300 = $425.00
RW-LMK-SM-00-10-100 = $375.00
WB-MMH-SM-00-10-125 = $750.00
WB-LMK-SM-00-10-125 = $800.00
RW-LMK-SM-00-10-300 = $450.00
RW-LMK-SM-00-10-73 = $350.00
RW-LMK-SM-00-8-100 = $305.00

Type	Code
Direct Acting	RW
Pilot Operated	WB

Adjustment Method	Code
Hand Knob	KEK
Rotary Knob	MMH
Lockable Rotary Knob	LMK

Connection Type	Code
Cartridge Valve	CV
Inline Thread Connection	IN
Subplate Mounting	SM

Pressure Ranges	Code
0 – 300 psi	30
0 – 650 psi	65
100 – 725 psi	73
200 – 1000 psi	100
350 – 1250 psi	125
500 – 3000 psi	300

Subplate Size	Code
Steel	6
Steel	8
Steel	10
Cartridge or In-line thread	00

Thread Size	Code
1/2 in.	50
3/4 in.	75
Cartridge or Subplate	00

RELIEF VALVE SPECIFICATION AND PRICING CHART

Activity 9-4: Hydraulic Motor Mixing Operation

A batch manufacturing facility is reconfiguring one of their mixing operations. They need to ensure that their mixer, which is run by a hydraulic motor, cannot be operated until the hydraulic cylinder has moved the mixer into place.

1. Connect Schematic Symbols for the mixing operation that will accomplish the following:

- The cylinder and hydraulic motor are controlled by separate directional control valves.
- The hydraulic motor cannot start until the cylinder has placed the mixer into position when it is at the end of its stroke.
- The hydraulic motor must have a braking mechanism.

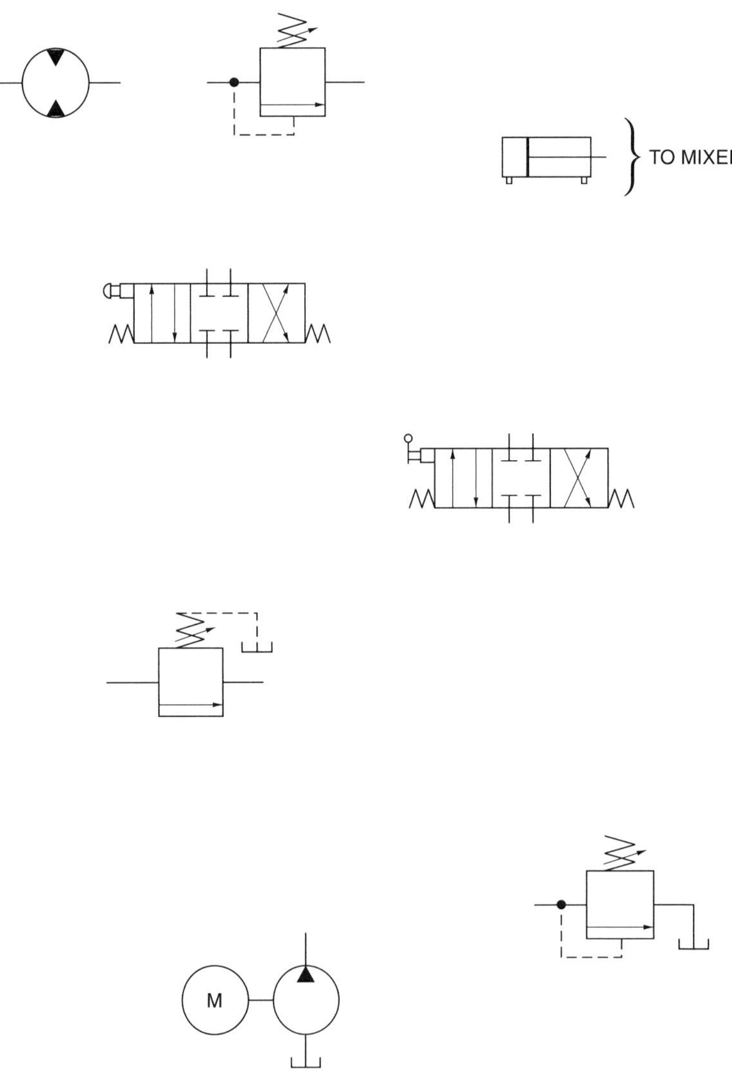

SCHEMATIC SYMBOLS

Activity 9-5: Plastics Thermoforming Press Application

A plastics thermoforming press is having pressure fluctuations when it closes its mold. This problem is caused because the relief valve is installed too far from the cylinder. This creates inconsistent pressure at the cylinder and causes the thermoforming mold to close at different pressures, which damages the workpieces to be formed.

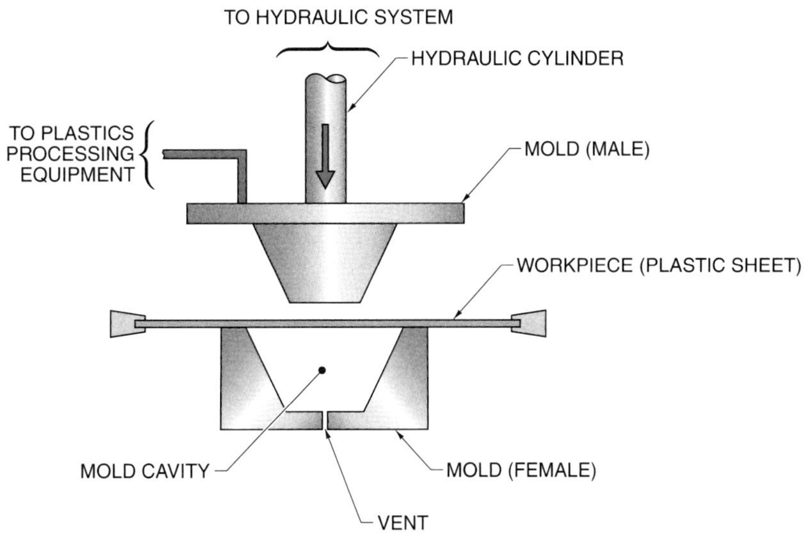

PLASTICS THERMOFORMING PRESS

1. Sketch a schematic diagram of the revised hydraulic system.

Use FluidSIM® to build the revised hydraulic system and verify that it will operate properly. *Note:* If a relief valve and unloading valve are used, the unloading valve needs to be set lower than the relief valve.

2. What happens with the cylinder when an unloading valve (shutoff counteracting valve) is added to the system?

3. What happens with the cylinder when a relief valve is added to the system?

4. What happens with the cylinder when a three-way pressure-reducing valve is added to the system?

Hydraulic Fluid Maintenance

Chapter 10

OBJECTIVES

- Identify contaminant types and sizes.
- Describe common sources of contaminants in a hydraulic system.
- Describe the effects of contaminants in a hydraulic system.
- Describe the methods of filtering contaminants from a hydraulic system.
- Explain how heat is generated in a hydraulic system.
- Describe the methods and equipment used to dissipate heat from hydraulic fluid.
- Identify the components of a reservoir.

INTRODUCTION

Hydraulic fluids are used for several different functions in a hydraulic system. They are used to transmit power to hydraulic cylinders and to cool the system by dissipating heat through reservoirs and cooling devices. Hydraulic fluids also seal system components to prevent leaks and act as a lubricant for any moving surfaces. Hydraulic fluids travel through filters to have contaminants removed and spend time in the reservoir to allow contaminants to settle out. Hydraulic fluids must be treated and conditioned to meet specifications that allow the system to operate at peak efficiency.

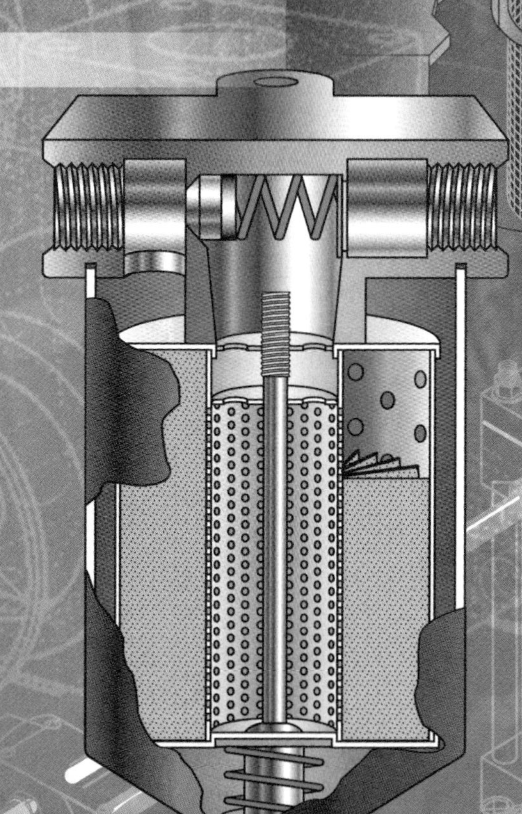

TERMS

A **contaminant** is a foreign substance in the hydraulic fluid that is not meant to be a part of fluid.

A **micron (µm)** is one-millionth of a meter and 310-millionths of an inch.

CONTAMINANTS

A *contaminant* is a foreign substance in the hydraulic fluid that is not meant to be a part of fluid. Common contaminants include dust, dirt, metal, and/or rubberized plastic chips. All hydraulic systems generate various amounts of different contaminants during normal operation.

Early hydraulic systems did not filter contaminants because the clearance between moving parts was large enough to allow most contaminants to travel through the system. As fluid power technology advanced however, the requirement to keep even the smallest contaminants out of a hydraulic system became a necessity.

All hydraulic systems are affected by contaminants. Contaminants come from different sources and are found in various sizes. Hydraulic fluid contaminants are introduced into a system by handling hydraulic fluid, new equipment added to the system, troubleshooting, or normal operation. **See Figure 10-1.** Different types of protection from contaminants are available to help prevent damage to system pumps, valves, and actuators.

Because of the presence of dirt or dust in their operating environments, many types of hydraulic equipment are susceptible to contamination.

Sizing Contaminants

Contaminants in a hydraulic system are rated in micrometers, also known as microns. A *micron (µm)* is one-millionth of a meter and 39-millionths of an inch. Micron-sized particles are invisible to the human eye. Items under 40 µm in diameter that are invisible to the human eye include bacteria (2 µm), red blood cells (7 µm), and white blood cells (25 µm). Even contaminants smaller than 40 µm can have degrading effects on high-tolerance hydraulic valves and pumps. Items above 40 µm that are visible to the human eye include human hair (74 µm to 100 µm) and table salt (105 µm). **See Figure 10-2.**

Contaminant Sources

Contaminants in hydraulic systems can come from various sources. Hydraulic systems are usually found in environments such as mills and factories where dirt and dust are constantly present. Contaminants are also entrained into the hydraulic fluid being used. Other contaminant sources include dirt and dust present in new hydraulic systems, particles from component wear, weld scale, insect and rodent parts, and dirt and dust accidentally introduced during mandatory maintenance or repairs.

New Hydraulic Systems. New hydraulic systems are not clean enough to be used for normal, first time operation. Many contaminants are often introduced in the production, assembly, and transport of a new hydraulic system. During production and assembly, system components such as reservoirs, hoses, and valves easily collect contaminants.

Although these components are typically cleaned after assembly and before transport, they can collect contaminants because they are typically coated with a thin layer of machine oil to prevent corrosion. Atmospheric dust, dirt, wood particles (sawdust), cardboard particles, metal chips, paper, and other solid debris can become embedded in the machine oil and adhere to the coated parts causing contamination.

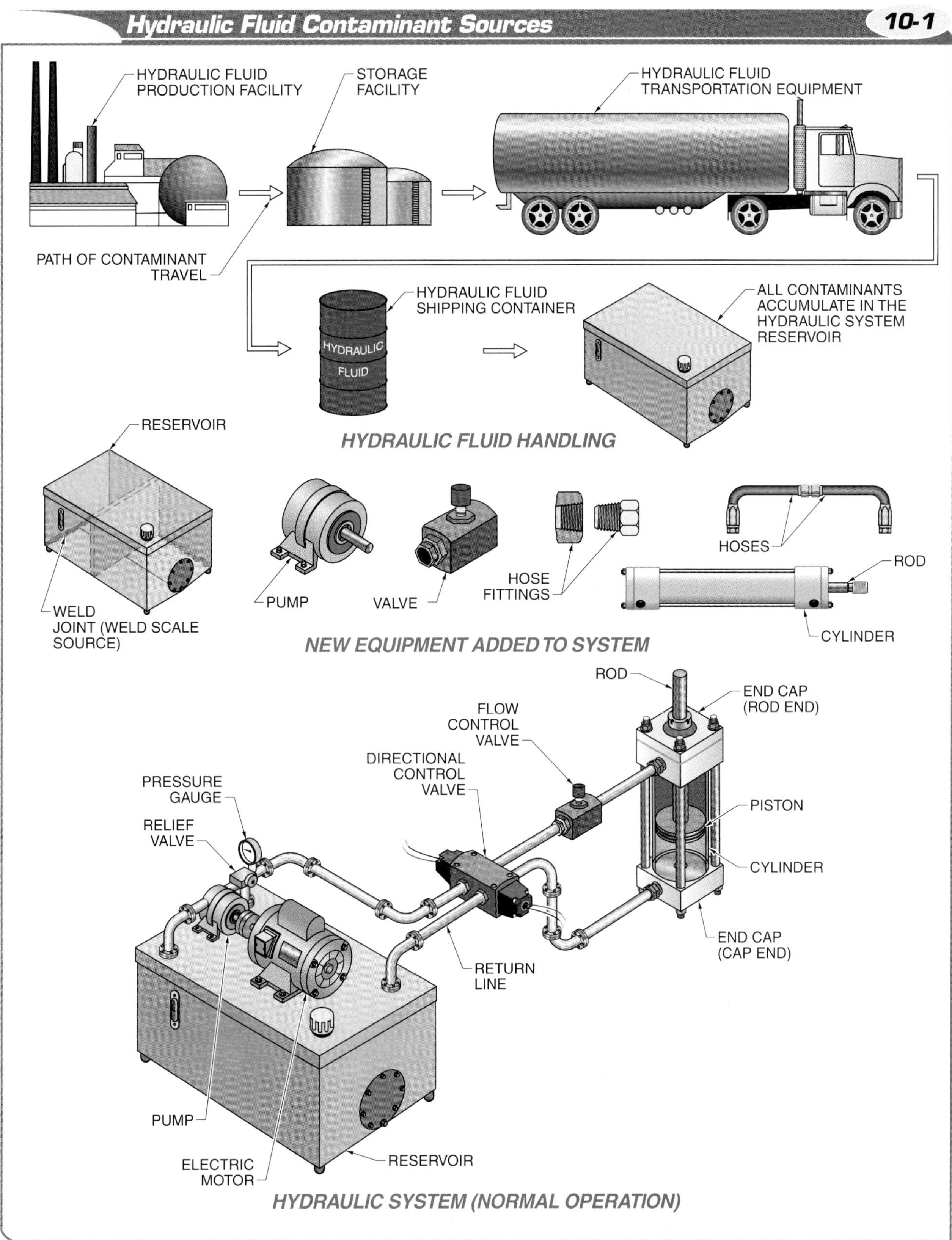

Figure 10-1. Hydraulic fluid contaminants are introduced into a system by handling hydraulic fluid, new equipment added to the system, troubleshooting, or normal operation.

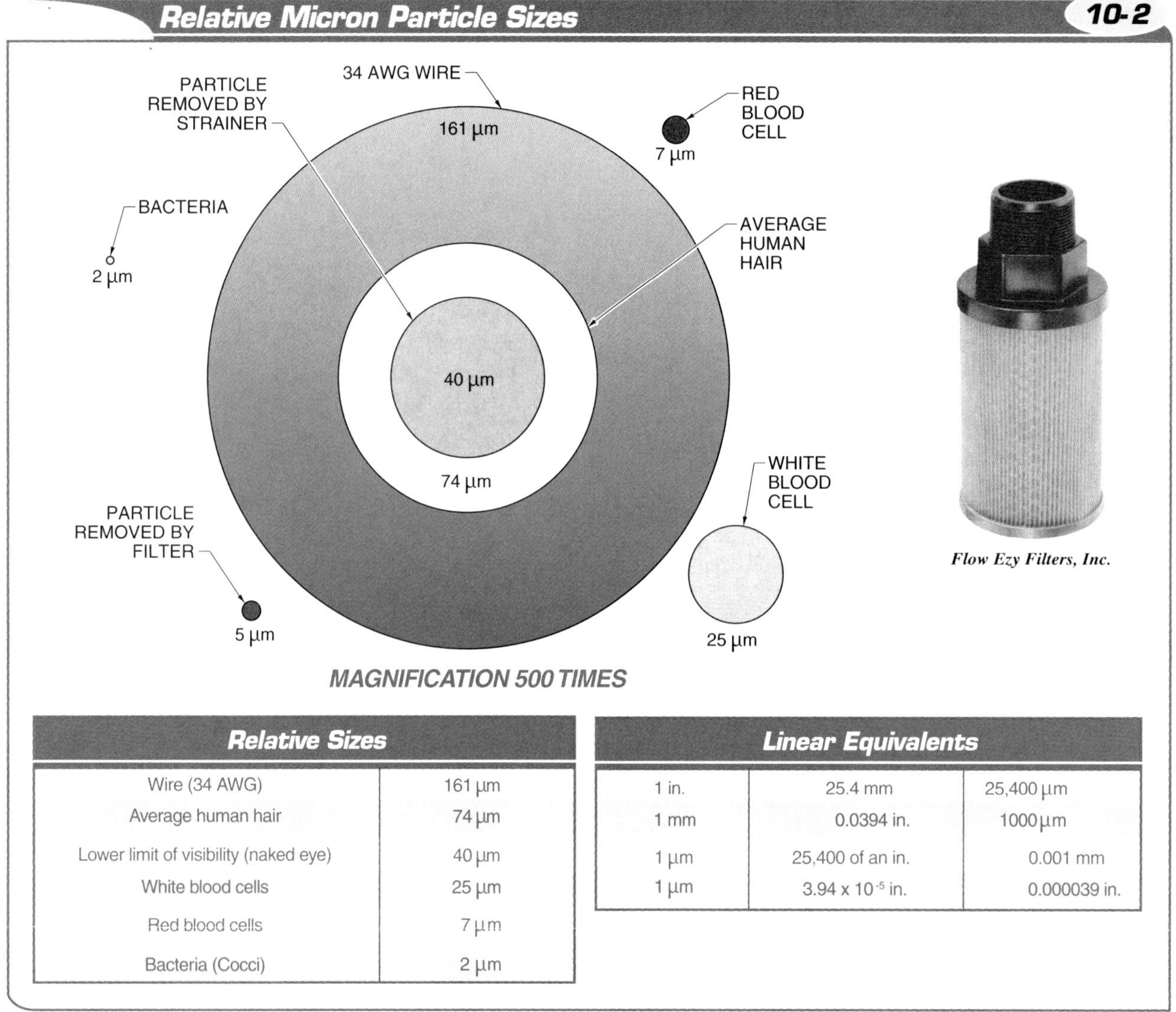

Relative Sizes

Wire (34 AWG)	161 µm
Average human hair	74 µm
Lower limit of visibility (naked eye)	40 µm
White blood cells	25 µm
Red blood cells	7 µm
Bacteria (Cocci)	2 µm

Linear Equivalents

1 in.	25.4 mm	25,400 µm
1 mm	0.0394 in.	1000 µm
1 µm	25,400 of an in.	0.001 mm
1 µm	3.94×10^{-5} in.	0.000039 in.

Figure 10-2. Micron-sized particles can be both visible and invisible to the human eye.

Contaminants are also collected during the transport of assembled hydraulic systems. Hydraulic systems are often shipped in containers such as cardboard boxes, wooden crates, and pallets. Particles and dust from the shipping container often collect in or on the hydraulic components. During the installation of a new hydraulic system, keeping the system clean must not be overlooked.

Contaminants can enter a hydraulic system through new hydraulic fluid. Hydraulic fluid is shipped to factories in various sizes including 1 qt, 1 gal., 5 gal., or 55 gal. containers. The hydraulic fluid collects contaminants that are in the shipping containers because the containers are often not cleaned before being filled.

Contaminants can enter a new hydraulic system when it is filled with hydraulic fluid for the first time. Hydraulic fluid must be filtered before it enters the reservoir. Whether the hydraulic fluid placed into the hydraulic system is new or recycled, it must be filtered when it is transferred from its original shipping container to the reservoir. This is usually done with a portable filtration unit. **See Figure 10-3.** However, recycled hydraulic fluid that meets the specifications for new hydraulic fluid can also be used, which lowers operating costs and is more environmentally friendly.

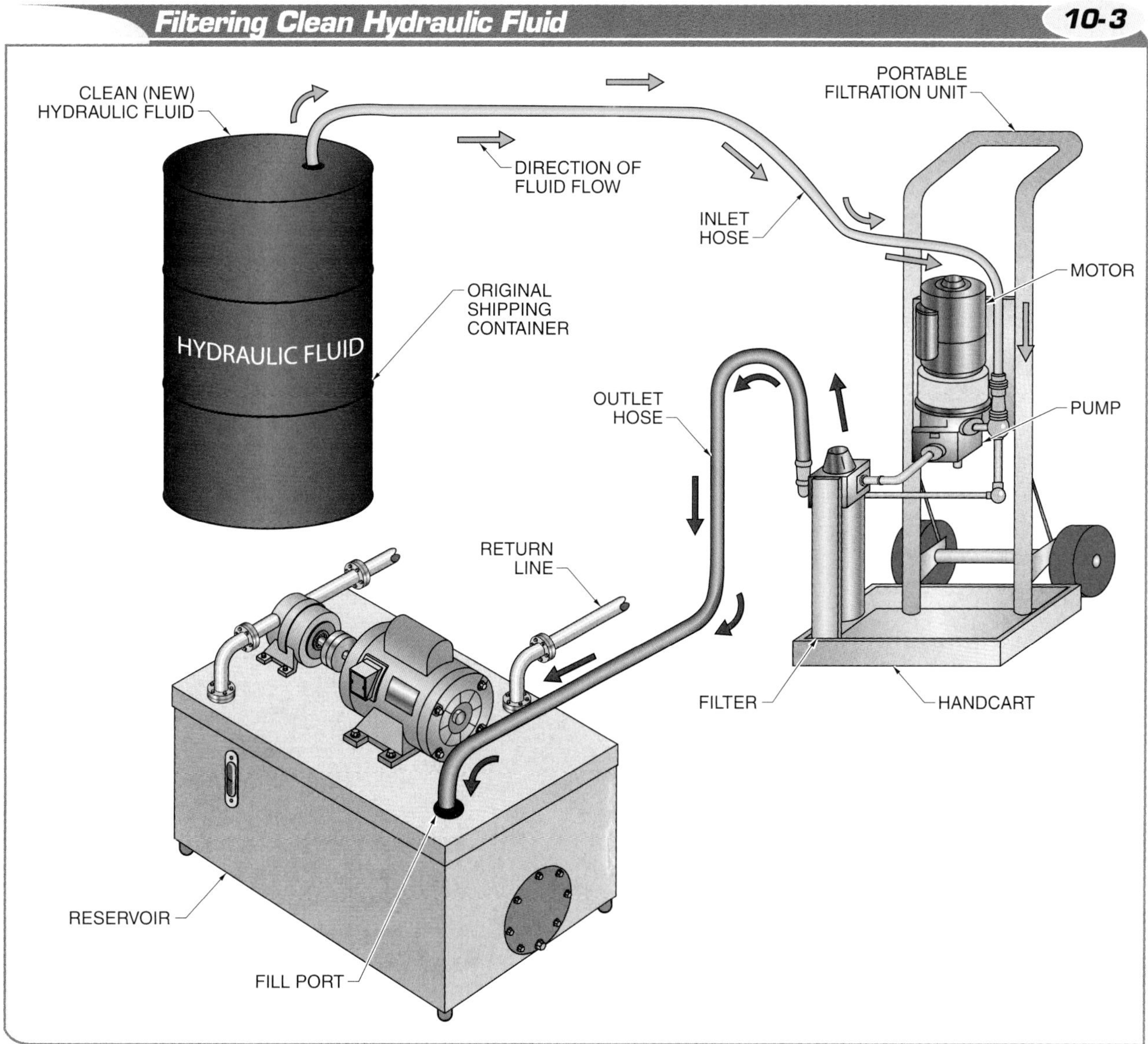

Figure 10-3. Hydraulic fluid must be filtered when it is transferred from its original shipping container to the reservoir.

Portable Filtration Units. A *portable filtration unit* is a nonfixed filtration machine that filters hydraulic fluid as it is being transferred from a storage vessel to the reservoir. Portable filtration units are sometimes referred to as off-line filters and are attached to portable devices, such as handcarts. A portable filtration unit is connected to a hydraulic system that is not in use and used to filter most of the contaminants of a certain size and larger from the hydraulic fluid.

A portable filtration unit is also used to transfer hydraulic fluid from the original shipping container to the reservoir of the hydraulic system. A portable filtration unit can easily be transported to the location of any mobile hydraulic system being used, such as those found in agriculture, excavation, and road construction. Because portable filtration units are mobile, they can be used with different pieces of equipment.

TERMS

A **portable filtration unit** is a nonfixed filtration machine that filters hydraulic fluid as it is being transferred from a storage vessel to the reservoir.

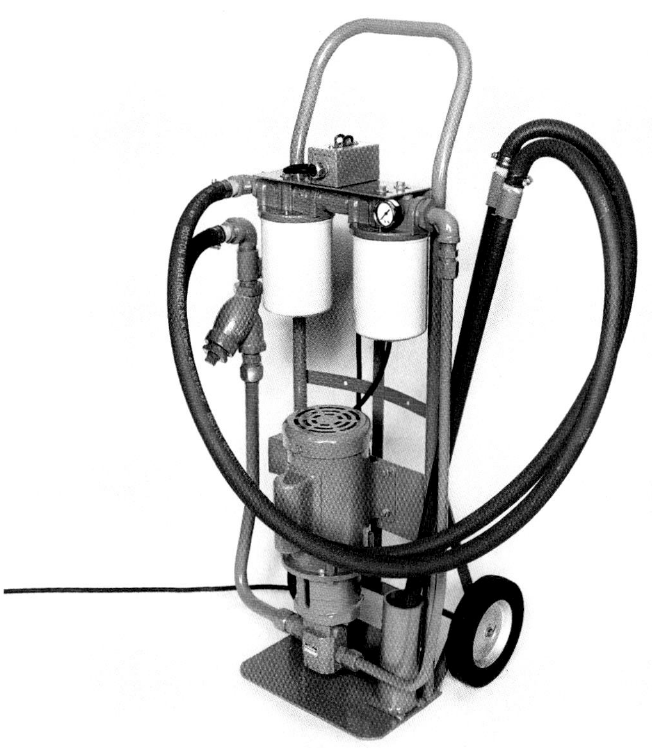

Flow Ezy Filters, Inc.

Portable filtration units are typically used to filter new hydraulic fluid before it is added to a hydraulic system.

TERMS

A **breather/filler cap assembly** is a device used to allow air in and out of a reservoir as the pump delivers hydraulic fluid into the system.

Implementing a cleaning procedure for new hydraulic equipment before it is put into service can lessen contamination. All components should be cleaned with soap and water, acid cleaner, alkaline cleaner, solvent, ultrasonic cleaners, or mechanical cleaners. Most facilities that use a large amount of hydraulic equipment have standard procedures to ensure a minimum cleanliness level for components and new hydraulic fluid.

Added System Contaminants. Contaminants that accumulate in the hydraulic system come from several different sources in the operating environment. Common sources of contaminants include the following:

- Airborne contaminants—A *breather/filler cap assembly* is a device used to allow air in and out of a reservoir as the pump delivers hydraulic fluid into the system. The breather/filler cap has a filter to prevent air contaminants from entering the system. Sometimes the breather/filler cap is temporarily removed when new fluid is added, leaving the hydraulic fluid exposed to any airborne contaminants.
- Maintenance and repairs—All hydraulic systems that are continuously operated require regular repairs. When repairs are performed, contaminants can enter the hydraulic system. For example, when a valve fails, the hoses that are attached to it must be disconnected and placed aside. There must be a clean location for the hoses to be placed. If not, one must be created. When the hoses are reconnected to the new valve, any dirt or other contaminants that they have collected often enter the hydraulic system. Also, when new hoses are assembled or built, they are usually prepared in a maintenance area. When a hose is made, it must be flushed. Typically, a set flushing procedure is applied. A flushing procedure eliminates any contaminants trapped when preparing the hose. **See Appendix.**
- Transportation—Hydraulic equipment often collects contaminants such as dust, dirt, and small wood particles from shipping crates and pallets during transportation from the equipment manufacturer to the end user's location. Equipment should be thoroughly cleaned before startup. Most facilities that use a large amount of hydraulic equipment have a set cleaning procedure for all incoming equipment.

System-Created Contaminants. Contaminants are also created within a hydraulic system. Hydraulic equipment that operates continuously eventually begins to wear. The friction created between the internal surfaces of pumps, cylinders, and valves are all sources of contaminants for a hydraulic system. For example, as a pump begins to wear from normal use, micron-sized metal fragments are created and added to the hydraulic fluid.

Another source of contaminants from inside a hydraulic system are from seals that begin to wear down. The constant linear actuation of cylinders causes cylinder seal material to degrade and enter the hydraulic fluid. The flow of hydraulic fluid through a pipe also causes contaminants to move throughout the hydraulic system.

Contaminants will always be created internally. Even good preventive maintenance plans cannot keep all of the contaminants out of a hydraulic system. For facilities that use hydraulic equipment, the best practice is to keep as many of the external contaminants out of the system as possible and control the number of internal contaminants being created.

Contaminant Effects

Once contaminants are in a hydraulic system, the effects can include overheated hydraulic fluid, premature component failure, and complete system failure. As solid contaminants accumulate in a reservoir, the amount of clean hydraulic fluid decreases and begins to overheat.

Contaminants in the clearance between moving parts in a hydraulic system can cause abrasion. *Abrasion* is the wearing of a material due to grinding or friction. Abrasion damages moving surfaces and can create additional abrasive contaminants. Hydraulic fluid prevents abrasion because it lubricates components that have moving parts, such as spools inside of valves.

However, abrasion can cause the lubrication that is between the moving parts, such as a piston and cylinder barrel, to break down. When lubrication breaks down, it causes friction and heat between the moving parts. This leads to premature component failure. The three sizes of contaminant particles that affect the lubrication of moving hydraulic parts are contaminants that are smaller than, equal to, or greater than the size of the clearance between moving parts. **See Figure 10-4.**

Varnish is a coating of a sludge-like residue that results from the breakdown of hydraulic fluid from the heat and pressure within a hydraulic system. Varnish can cause premature breakdown and failure of hydraulic components by clogging orifices and increasing the thickness of mating surfaces. Contaminants smaller than the size of the clearance still allow the lubrication of moving parts, but can cause abrasion. Contaminants equal to the size of the clearance have the same effect on lubrication as contaminants smaller than the clearance. However, contaminants with sizes equal to the clearance have the potential to seize (lock) a moving part. Contaminants that have a size greater than the clearance can block lubrication from entering the moving parts altogether or can break up into smaller contaminants. All three sizes of contaminants can cause a hydraulic system to fail.

TERMS

Abrasion is the wearing of a material due to grinding or friction.

Varnish is a coating of a sludge-like residue that results from the breakdown of hydraulic fluid from the heat and pressure within a hydraulic system.

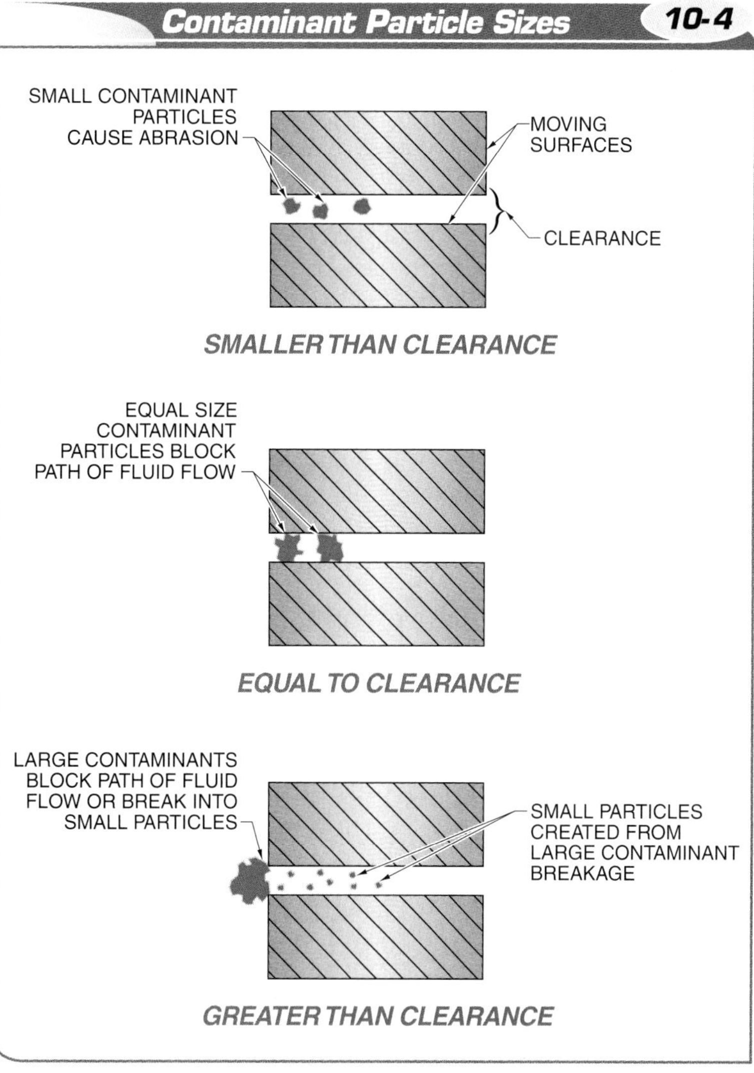

Figure 10-4. Contaminants of different sizes can cause equipment lubrication problems in the form of abrasion or blocked fluid flow.

TERMS

A **filter** is a porous device used to separate solid material from a liquid or gas.

A **strainer** is a screen, mesh, or perforated obstruction used to separate a solid from a liquid.

A **nominal rating** is a value that indicates the average size or larger contaminant that a filter can trap.

An **absolute rating** is a value that indicates the diameter of the largest hard, sphere-shaped particle that can pass through a filter under specific test conditions.

A **surface-type element** is a filter element that is composed of a single closely woven fabric or cellulose layer.

A **depth-type element** is a filter element that is composed of multiple layers of fabric or threaded material.

CONTAMINANT PROTECTION

Protecting a hydraulic system from contaminants is necessary to maintain the maximum operational life of the system. Filters and strainers are devices that help to clean or maintain the cleanliness of hydraulic fluid. **See Figure 10-5.**

A *filter* is a porous device used to separate solid material from a liquid or gas. A *strainer* is a screen, mesh, or perforated obstruction used to separate a solid from a liquid. Strainers typically have openings larger than 50 μm. All filters and strainers have performance ratings to help determine the size of solid contaminants that can be separated from the hydraulic fluid. Various hydraulic components also require specific hydraulic fluid cleanliness ratings.

Filter and Strainer Ratings

Hydraulic filters and strainers are rated as nominal or absolute. A *nominal rating* is a value that indicates the average size or larger contaminant that a filter can trap. For example, a filter with a nominal rating of 30 μm will trap at least one particle that is 30 μm or larger. However, a nominal-rated filter of 30 μm may allow a 40 μm contaminant to pass but can trap some 20 μm contaminants.

All filters are nominal rated, and other filters may be absolute rated. An *absolute rating* is a value that indicates the diameter of the largest hard, sphere-shaped particle that can pass through a filter under specific test conditions.

Note: Both nominal and absolute ratings must only be used to compare filters from the same OEM because there is no industry standard to compare these types of filter ratings. Terminology can be interpreted by filter OEMs differently and makes it impossible to compare filters that are created by different OEMs using these ratings.

Materials used for filters include cellulose, cloth, metallic thread elements, fiberglass, and perforated stainless steel. The threads of a strainer are set vertically and horizontally in equal amounts and are counted per square inch. For example, a 200-mesh filter has 200 vertical threads and 200 horizontal threads per square inch.

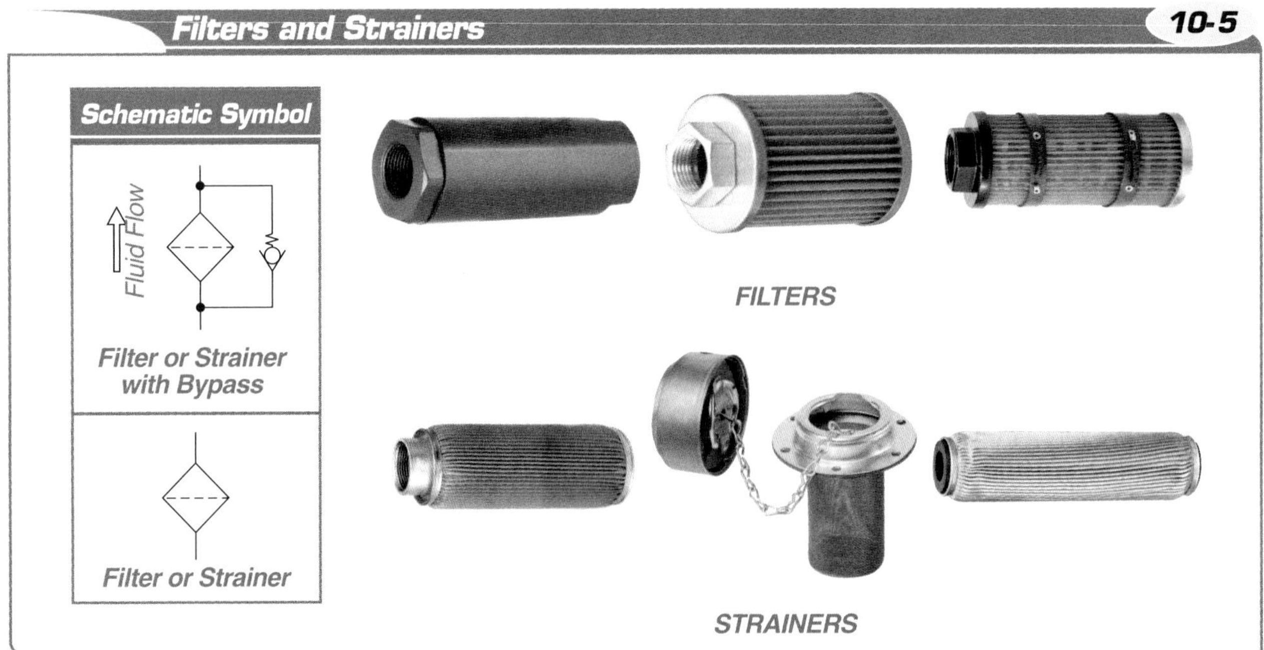

Flo Ezy Filters, Inc.

Figure 10-5. Filters and strainers are devices that help to clean or maintain the cleanliness of hydraulic fluid.

Filter elements are either surface-type or depth-type. A *surface-type element* is a filter element that is composed of a single closely woven fabric or cellulose layer. This allows nearly uniform sizes for the passage of hydraulic fluid. Surface-type elements are the most common type of filter elements and are used in filters with nominal ratings.

A *depth-type element* is a filter element that is composed of multiple layers of fabric or threaded material. This allows the hydraulic fluid passages to vary in size. The degree of filtration depends on the flow rate of the system. Depth-type elements are typically used in filters with absolute ratings and are more efficient.

Pieces of contaminants larger than the filter rating can still fit through the filter element. For example, particles that are long but have smaller diameters than the uniform hydraulic fluid passages may fit through. To prevent this, absolute-rated filters with depth-type elements can be installed. Absolute-rated filters with depth-type elements have 99% efficiency, while nominal-rated filters with surface-type elements have 50% to 75% efficiency. Most absolute-rated filters use fiberglass as a filtering media. **See Figure 10-6.**

A strainer uses a surface-type element because the element traps contaminants as they attempt to pass through the strainer mesh element. A strainer mesh element is removable and typically housed in a rigid plastic or metal housing that attaches directly to the hydraulic system. Opening sizes depend on the mesh size that is set when the strainer is manufactured. For example, a 200-mesh strainer has a opening size of 74 µm.

Strainer mesh is given an absolute rating because the opening size can be accurately controlled. A higher strainer mesh number indicates a smaller opening size. Most strainers remove particles through one layer of material. However, strainers with more than one layer of material are available. Strainers are used because their strainer mesh elements are not as fine as filter elements and they offer less resistance to fluid flow. Strainers can be cleaned periodically and reused, unlike filter elements, which must be discarded.

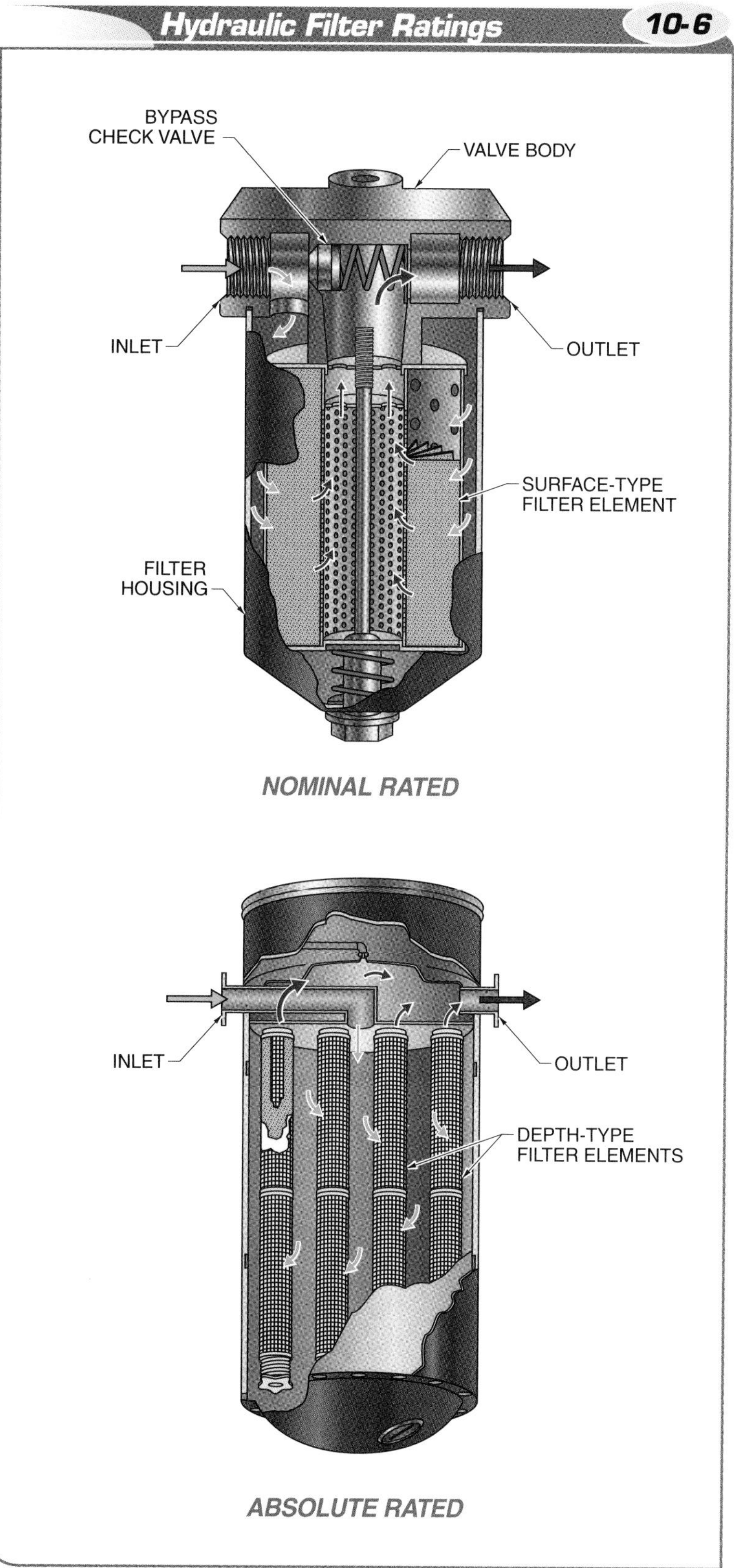

Figure 10-6. Hydraulic filters are rated as nominal and absolute, with absolute being the most efficient at removing contaminants.

TERMS

A **beta ratio** is a number that represents the number of particles at a given size or larger that can pass through a filter.

Filter Efficiency. While an absolute rating of a filter indicates the diameter of the largest circular particle that will pass through a filter under laboratory test conditions, it does not provide the actual efficiency of a filter or strainer. The efficiency of a filter or strainer is determined by the filter's beta ratio. A *beta ratio* is a number that represents the number of particles at a given size or larger that can pass through a filter. The beta ratio is determined by using ISO Standard 16889, *Multipass Filter Test*, which includes testing the number of contaminants of a given size upstream from the filter compared to the number of contaminants of the same size downstream from the filter. The beta ratio can be calculated by applying the following formula:

$$beta\ ratio_x = \frac{N_{UP}}{N_{DP}}$$

where

beta ratio = filter or strainer efficiency
x = contaminant size (in μm)
N_{UP} = number of upstream particles
N_{DP} = number of downstream particles

Example: What is the beta ratio of a filter that is in a system with 100 particles (> 5 μm) upstream of its location and 5 particles (> 5 μm) downstream of its location?

$$beta\ ratio_x = \frac{N_{UP}}{N_{DP}}$$

$$beta\ ratio\ 5\mu m = \frac{100}{5}$$

$$beta\ ratio\ 5\mu m = \mathbf{20}$$

The beta ratio then correlates to the filter's removable efficiency at that specific micron rating. For example, if a filter has a beta ratio of 20, that means the filter will remove 95% of the contaminants that are 5 μm or larger in size. **See Figure 10-7.** Most filter manufacturers typically use a beta ratio of 200 (99.5% efficiency) as their rated micron level.

Hydraulic System Filtration

Hydraulic filters and strainers are typically installed in several different locations within a system. Ideally, a filter should be placed before every component. Because this is not practical, filters of various types and ratings are often installed strategically within a system to offer the most economically productive results. Some filters are equipped with a bypass check valve that is typically set between 15 psi and 20 psi of pressure drop. The bypass check valve opens when the filter creates more than a 15 psi to 20 psi pressure drop. Filters and strainers are placed in standard locations such as at suction lines, pressure lines, and return lines. **See Figure 10-8.**

Filter Efficiency 10-7

Beta Ratio	Filter Removable Efficiency*
2	50
5	80
10	90
20	95
75	98.6
100	99
200	99.5
1000	99.9

* in %

Figure 10-7. The efficiency of a filter or strainer is determined by the filter's beta ratio.

Suction Filters. To protect the pump from contaminants that are in the reservoir, a strainer is attached at the pump suction line. Strainers are sometimes referred to as inlet strainers, suction line filters, or sump strainers. Strainers are used to protect pumps from particle contamination. They are submerged in hydraulic fluid inside the reservoir and placed at the end of the suction line before the pump.

In some applications, more than one strainer may be necessary. Multiple strainers can be added to help filter out additional contaminants. Poor maintenance of the strainer deprives the pump of hydraulic fluid and prevents its lubrication. This causes the pump to generate high temperatures and allows more contaminants into the hydraulic fluid. Strainer ratings can range from 22 μm to 500 μm.

A *suction filter* is a type of filter that is designed to operate in a vacuum and is connected to the hydraulic system before the pump and outside of the reservoir to protect the pump from contaminants in the reservoir. When a strainer alone is not capable of keeping small contaminants out of the system, suction filters are used to filter hydraulic fluid to prevent premature pump failure. **See Figure 10-9.** They are often installed before piston or vane pumps. Replacement of suction filter elements must be performed in accordance with the manufacturer's recommendations.

Pressure Filters. A *pressure filter* is a type of filter that is designed to operate in high pressures and is placed before one or more hydraulic system components to protect them from contaminants in the hydraulic fluid. **See Figure 10-10.** They are placed in pressure lines between the pump and components or between valves and actuators in a system. Pressure filters placed between components have the advantage of filtering out particles introduced upstream by a deteriorating component. Pressure filters typically have a bypass in order to withstand the full pressure differential of a hydraulic system without collapsing.

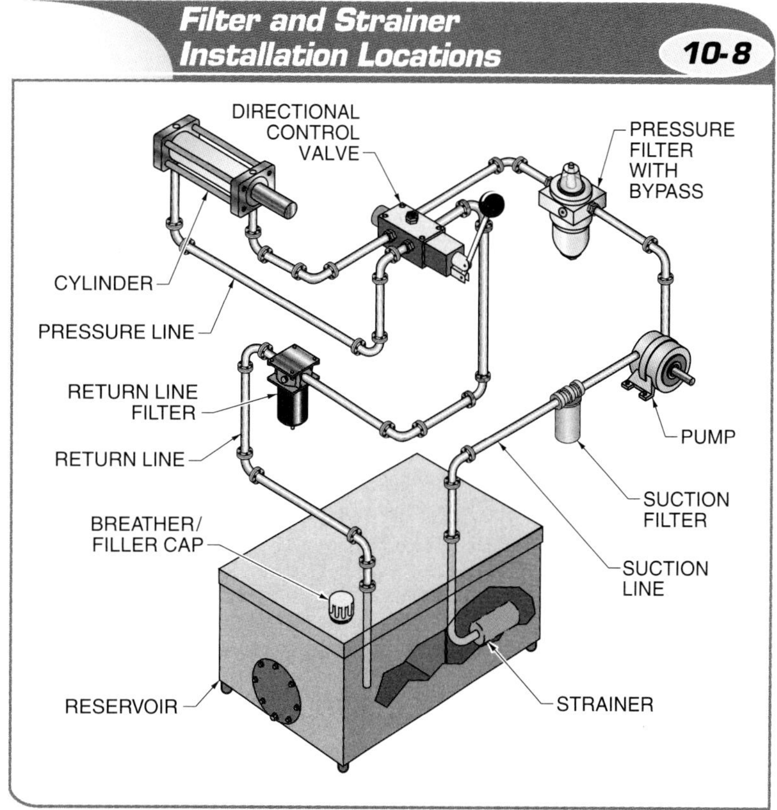

Figure 10-8. Filters and strainers are placed within a hydraulic system at the pump suction lines, pressure lines, and return lines.

TECH FACT

Many hydraulic fluid manufacturers now produce ecologically friendly, vegetable oil-based hydraulic fluids.

TERMS

A **suction filter** is a type of filter that is designed to operate in a vacuum and is connected to the hydraulic system before the pump and outside of the reservoir to protect the pump from contaminants in the reservoir.

A **pressure filter** is a type of filter that is designed to operate in high pressures and is placed before one or more hydraulic system components to protect them from contaminants in the hydraulic fluid.

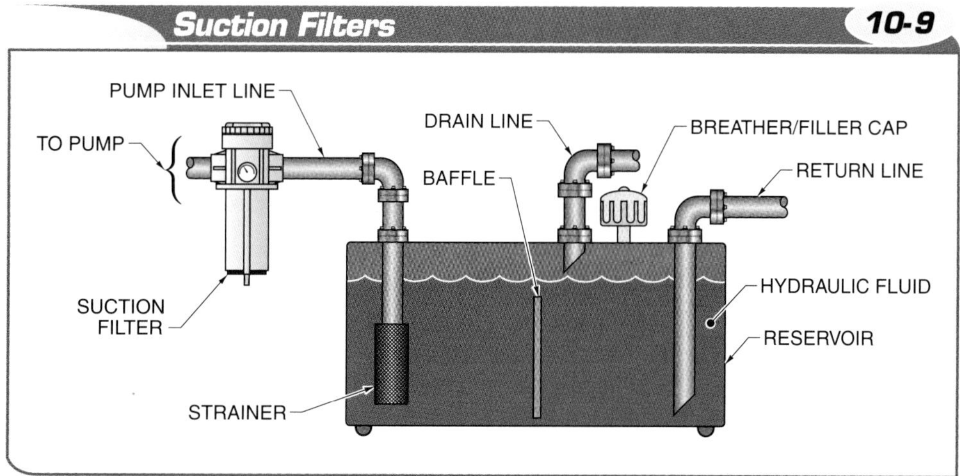

Figure 10-9. Suction filters are used when a strainer alone is not capable of keeping small contaminants out of the system and preventing premature pump failure.

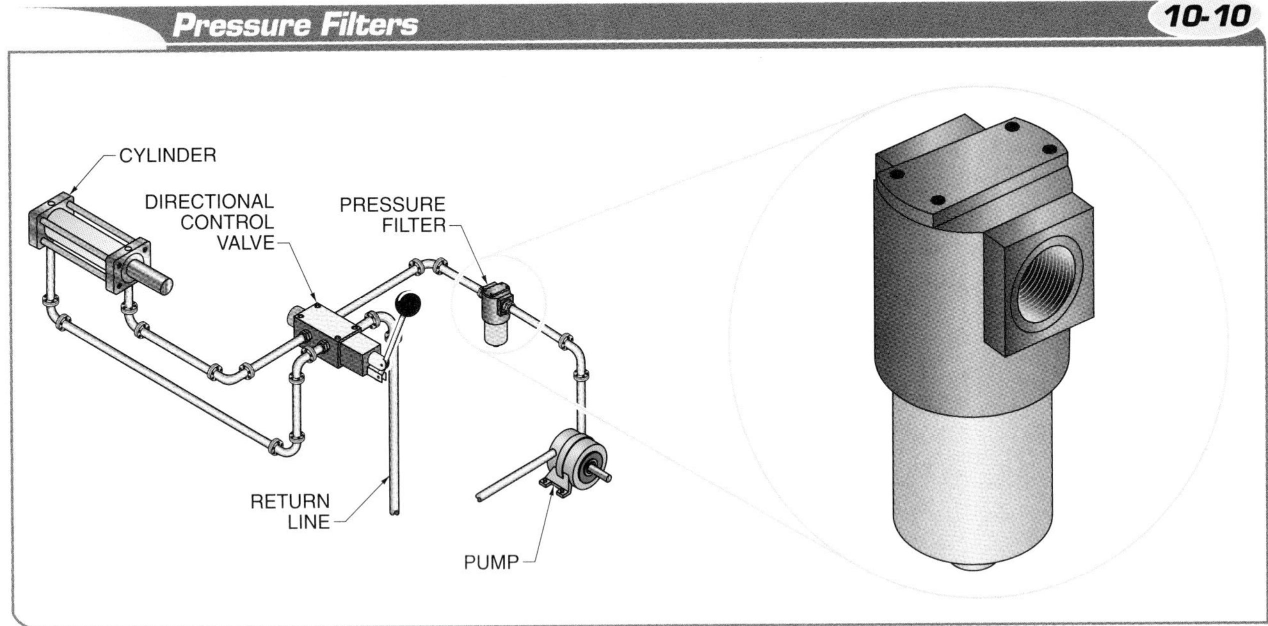

Figure 10-10. Pressure filters are placed before hydraulic components to protect them from contaminants in the system.

TERMS

A **bypass filter** is a hydraulic filter equipped with a bypass check valve to allow hydraulic fluid to bypass the filter element as the element becomes clogged.

A **return line filter** is a type of filter that is designed to operate in low pressures and is connected to a main return line of the hydraulic system before the line enters the reservoir.

Some pressure filters are capable of handling bidirectional fluid flow. Bidirectional fluid flow pressure filters may be used between a directional control valve and a cylinder. Most pressure filter elements range from 3 µm to 40 µm and can filter out fine particles. If a filter element becomes overcontaminated, particles can push through the filter element and it may collapse or tear. Pressure filters are typically used in hydraulic systems such as those used in paper mills, steel mills, plastics production facilities, and sawmills.

Bypass Filters. A *bypass filter* is a hydraulic filter equipped with a bypass check valve to allow hydraulic fluid to bypass the filter element as the element becomes clogged. If a bypass filter element is not replaced, it becomes clogged. This can cause extra resistance in the bypass line and forces the pump to work harder. Pressure will also build up on the inlet side of the filter and eventually cause the filter element to collapse or tear. A bypass check valve helps prevent damage.

Most bypass filters are color-coded to indicate three different filter conditions. **See Figure 10-11.** The first condition is green, which indicates that the filter is "good" or "clean" and in operating condition. The second condition is yellow, which indicates that the filter element must be replaced or cleaned. The third condition is red, which indicates that the filter is in "bypass mode" or "filter failed."

Some bypass filters also have a lamp, buzzer, or automatic shutoff mechanism to shut the system down when the filter fails or reaches the third condition. The activation of a lamp, buzzer, or automatic shutoff mechanism indicates that the hydraulic fluid flowing into the system is not properly filtered, which can quickly cause problems in the system. When the indicator is activated during the second or third condition, the filter element must be replaced as soon as possible.

Return Line Filters. A *return line filter* is a type of filter that is designed to operate in low pressures and is connected to a main return line of the hydraulic system before the line enters the reservoir. **See Figure 10-12.** Return line filters provide filtration for the hydraulic fluid before it is returned to the reservoir and operate under pressure that is typically less than 100 psi. Return line filters have ratings that range from 5 µm to 40 µm.

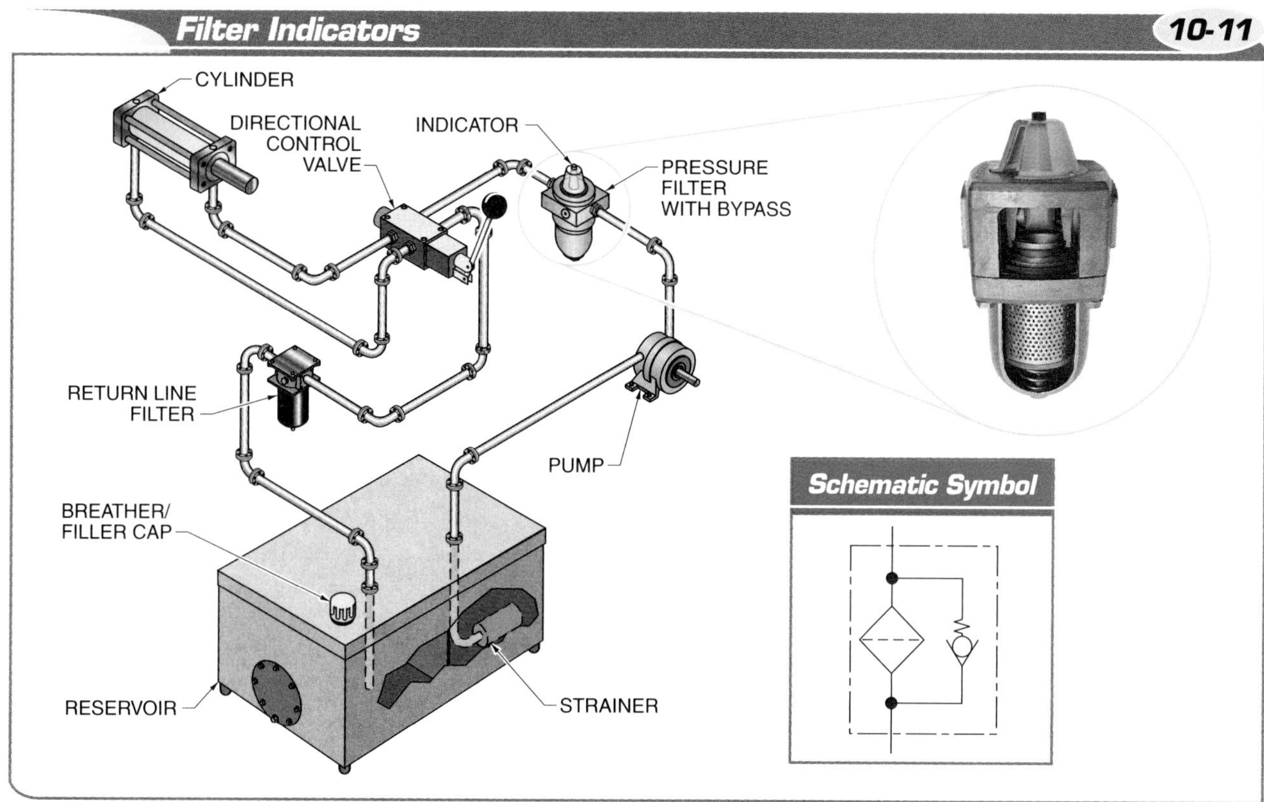

Figure 10-11. Most bypass filters have color-coded filter condition indicators that indicate three different filter conditions.

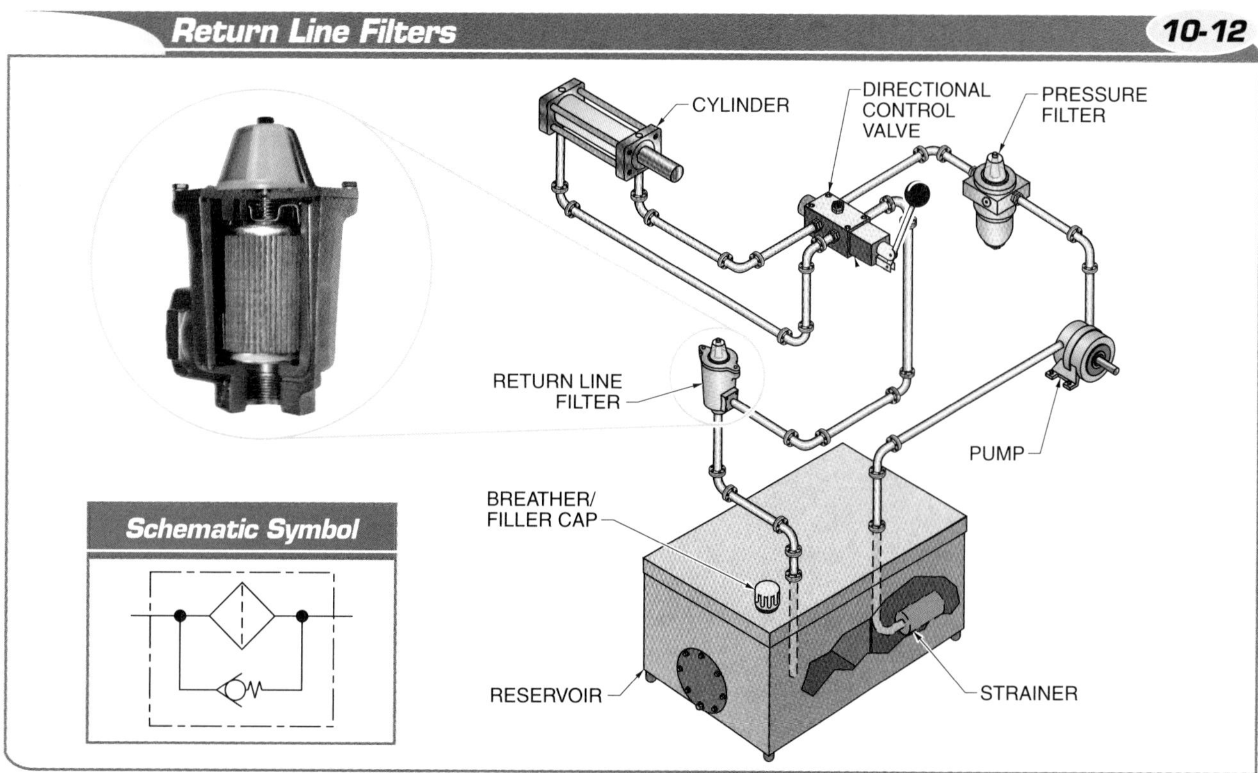

Figure 10-12. Return line filters provide filtration to the hydraulic fluid before it is returned to the reservoir and operate under pressure that is typically less than 100 psi.

TERMS

Back pressure is the level of pressure on the return line of a device or hydraulic system.

The pressure that return line filters operate under, back pressure, is lower than other pressure filters. *Back pressure* is the level of pressure on the return line of a device or hydraulic system. Return line filters have low operating back pressure and are installed in hydraulic systems to filter out contaminants. However, poor maintenance on return line filters increases back pressure and adversely affects system components, because of excess contamination in the hydraulic fluid and decreased system efficiency. Also, return line filters sometimes need to be sized higher than pressure filters because they can have full system flow from the pump and hydraulic fluid returning from the system such as when a large single-acting cylinder is retracting and all fluid from the pump is going back to tank. In this case, there is more fluid flowing through the return filter than would ever flow through the other filters in the system.

Hydraulic Fluid Cleanliness Ratings

Hydraulic fluid cleanliness is rated by International Standards Organization (ISO) Standard 4406:1999. This standard uses a three-number code to determine the cleanliness of hydraulic fluid. The hydraulic fluid is examined under a microscope and the contaminant particles are counted. The first number indicates the number of particles less than or equal to 4 μm/mL of hydraulic fluid. The second number indicates the number of particles less than or equal to 6 μm/mL. The third number indicates the number of particles less than or equal to 14 μm/mL.

Hydraulic filters typically have labels affixed near the filter installation point that provide recommended filter replacement cycles.

Many hydraulic components require a specific hydraulic fluid cleanliness. For example, a gear pump requires hydraulic fluid with a cleanliness rating that conforms to an ISO Code rating of 19/17/14, or 5000 particles 4 μm or less in size, 1300 particles 6 μm or less in size, and 160 particles 14 μm or less in size for each milliliter of hydraulic fluid. **See Figure 10-13.**

In addition to microscopic inspection, a chemical analysis of hydraulic fluid is performed to identify the presence of acids and/or chemical breakdown. Portable test kits are frequently used by fluid power technicians on location to test critical hydraulic systems. Otherwise, testing laboratories that specialize in hydraulic fluid analysis perform the chemical analysis. When a testing laboratory performs the chemical analysis, it may give a report that lists recommendations. The report may recommend completely flushing the system, using more efficient filters, and/or changing maintenance procedures, such as using lint-free rags for cleaning.

Chemical analysis reports of hydraulic fluid are typically electronic. A report lists hundreds of different characteristics of a hydraulic fluid sample, such as amounts of chemicals, metals, and other contaminants within the sample. Most hydraulic fluid analysis reports also include supporting graphs and/or photos. **See Figure 10-14.**

Some hydraulic equipment only specifies a hydraulic fluid cleanliness ISO rating with two-digit numbers. For example, a directional control valve might require hydraulic fluid with a cleanliness rating that conforms to an ISO C rating of 17/14. The number 17 represents the number of particles at 6 μm and the number 14 represents the number of particles 14 μm. When only two-digit numbers are used, the number of particles at 4 μm will not commonly affect the component.

TECH FACT

Most hydraulic fluids are made from petroleum-based products. Other hydraulic fluids are made from either synthetic (such as phosphate ester) or water glycol-based products.

Hydraulic Fluid Cleanliness Standards 10-13

ISO Code	Particles/Milliliter		
	>4 µm	>6 µm	>14 µm
23/21/18	40,000 to 80,000	10,000 to 20,000	6300 to 2500
22/20/18	20,000 to 40,000	5,000 to 10,000	6300 to 2500
22/20/17	20,000 to 40,000	5,000 to 10,000	640 to 1300
22/20/16	20,000 to 40,000	5,000 to 10,000	320 to 640
21/19/16	10,000 to 20,000	2500 to 5000	320 to 640
20/18/15	5000 to 10,000	1300 to 2500	160 to 320
19/17/14	2500 to 5000	640 to 1300	80 to 160
18/16/13	1300 to 2500	320 to 640	40 to 80
17/15/12	640 to 1300	160 to 320	20 to 40
16/14/12	320 to 640	80 to 160	20 to 40
16/14/11	320 to 640	80 to 160	10 to 20
15/13/10	160 to 320	40 to 80	5 to 10
14/12/9	80 to 160	20 to 40	2.5 to 5
13/11/8	40 to 80	10 to 20	1.3 to 2.5
12/10/8	20 to 40	6 to 10	1.3 to 2.5
12/10/7	20 to 40	6 to 10	0.64 to 1.3
12/10/6	20 to 40	6 to 10	0.32 to 0.64

Cleanliness Required for Typical Hydraulic Components

Component	ISO Rating
Servovalves	16/14/11
Proportional valves	17/15/12
Vane and piston pumps and motors	18/16/13
Directional and pressure control valves	18/16/13
Flow control valves and cylinders	20/18/15
New unused fluid	20/18/15

Figure 10-13. Many hydraulic components require a specific hydraulic fluid cleanliness, which is rated by International Standards Organization (ISO) Standard 4406:1999.

Chemical Analysis of Hydraulic Fluid 10-14

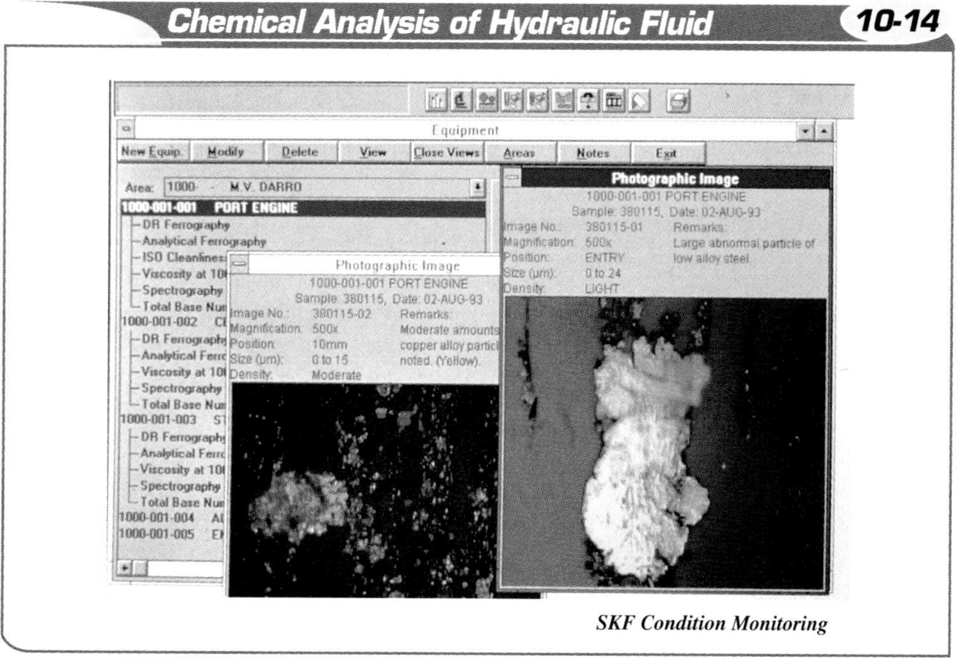

SKF Condition Monitoring

Figure 10-14. In addition to a list of contaminants present in a hydraulic fluid sample, most hydraulic fluid analysis reports also include supporting graphs and/or photos.

TERMS

A **heat exchanger** is a mechanical device used to transfer heat from one fluid to another fluid.

HYDRAULIC SYSTEM HEAT

The operation of a hydraulic system generates heat. In small hydraulic systems, this heat is usually dissipated through the reservoir. In large hydraulic systems, however, the amount of heat generated cannot be completely dissipated through the reservoir. When heat cannot be properly dissipated, heat exchangers must be placed in the system to help cool the hydraulic fluid by removing damaging heat from the system.

Hydraulic System Heat Generation

Any component within a hydraulic system that has fluid flow traveling through it offers resistance, and resistance generates heat. Flow control valves, pressure control valves, orifices and passages in control valves, and bends in hoses and pipes are all areas where resistance to fluid flow creates heat.

Heat is also generated at the hydraulic pump. When the pump begins to operate under pressure, heat is generated because more energy is exchanged between the electric motor and the hydraulic pump. Hydraulic fluid is not efficient at dissipating heat, but it is efficient at transmitting it. As heat is generated in a hydraulic system, it is transmitted through the hydraulic fluid to the reservoir.

Hydraulic filters are sometimes fitted with an analog gauge that indicates when the filter needs to be replaced.

Heat Exchangers

A *heat exchanger* is a mechanical device used to transfer heat from one fluid to another fluid. Heat exchangers are normally constructed of tubes with one fluid on the inside of the tubes and the other fluid on the outside of the tubes. Heat exchangers are low-pressure and are either air-cooled or water-cooled.

Heat exchangers have separate heating and/or cooling systems that are attached to the hydraulic system. These systems typically operate when the hydraulic system is operating. In many hydraulic systems, the heat exchanger is not required to be in constant operation. A temperature control system that is actuated by a thermostat attached to the reservoir controls the heat exchanger and its system operations.

Air-Cooled Heat Exchangers. Air-cooled heat exchangers, also called oil coolers, allow hot hydraulic fluid to be pumped through tubes that are attached to sheet metal cooling fins. **See Figure 10-15.** The hot hydraulic fluid is cooled by ambient air that is blown over and through the heat exchanger tubes and fins by a blower. Air-cooled heat exchangers operate similar to automobile radiators, dissipating heat through convection. They are used in mobile hydraulic systems where water is not readily available, and when its not cost-effective to use water for cooling.

Hydraulic fluid flows through an air-cooled heat exchanger to increase the amount of fluid that is exposed to the air. The more fluid that is exposed to the air, the more the hydraulic fluid is efficiently cooled. Air-cooled heat exchangers work best when the ambient temperature is lower than 80°F. When the ambient temperature is above 80°F, an air-cooled heat exchanger loses some of its efficiency and will show this by having longer running times.

TECH FACT

If further cooling of a hydraulic system is needed and the system already has a heat exchanger in place, a second heat exchanger or a larger-capacity heat exchanger can be installed. A second heat exchanger should be identical to the existing one.

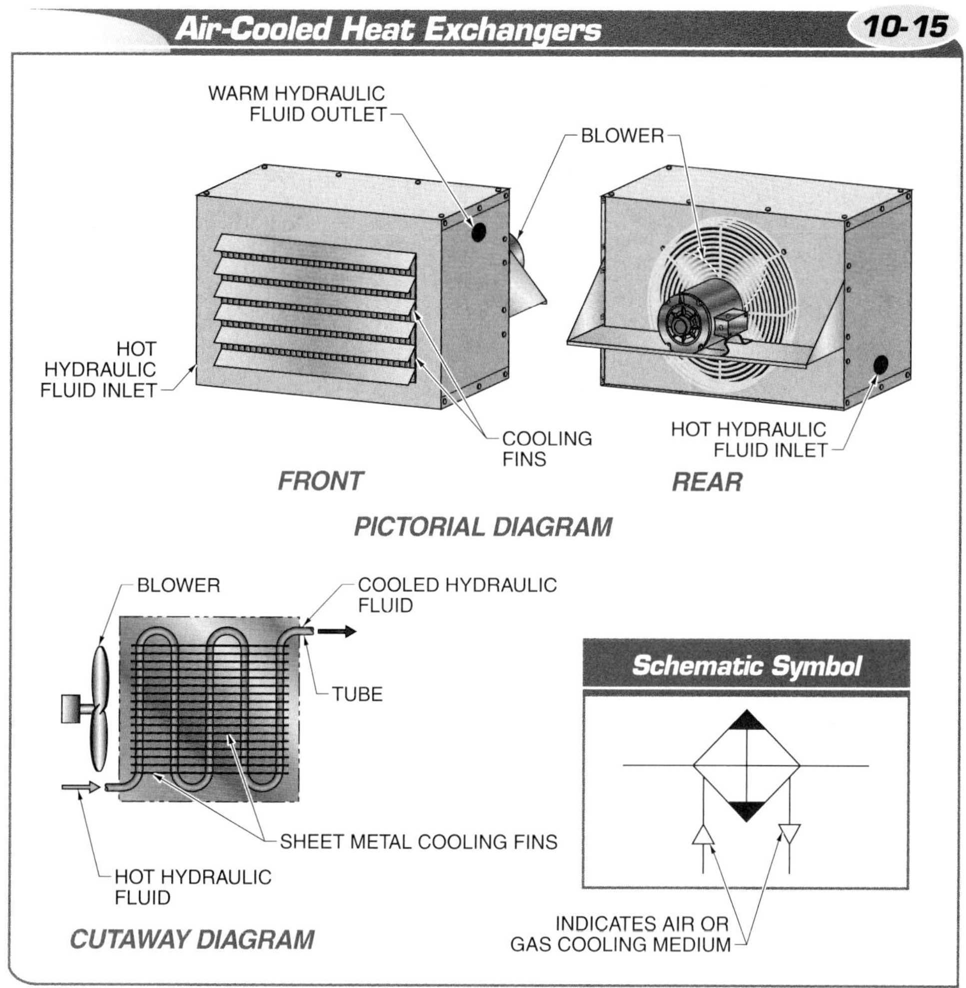

Figure 10-15. Air-cooled heat exchangers allow hot hydraulic fluid to be pumped through tubes that are attached to sheet metal cooling fins.

Water-Cooled Heat Exchangers. Water-cooled heat exchangers cool hot hydraulic fluid as it is pumped through tubes that are surrounded by flowing cool water. The circulating cool water removes unwanted heat from the hydraulic fluid. Water-cooled heat exchangers consist of a shell, tube sheets that create header boxes, cooling tubes, baffles, inlets for hydraulic fluid and water, and outlets for hydraulic fluid and water. **See Figure 10-16.**

When water is available for the hydraulic system, a water-cooled heat exchanger is the preferred choice because it is more efficient than an air-cooled heat exchanger and is less affected by ambient temperatures.

Water-cooled heat exchangers are typically located on the low-pressure return line of a hydraulic system so that their maximum operating pressure is 100 psi, with the possibility that pressures can go up to 200 psi. The return line is less likely to develop a leak because of its continuous high-volume, low-pressure fluid flow. Water-cooled heat exchangers must be protected from hydraulic fluid pressure surges by using a relief valve. A water-cooled heat exchanger operates in the following procedure:

1. Hot hydraulic fluid flows into its inlet and through the individual cooling tubes toward the hydraulic fluid outlet.
2. At the same time, cool water flows into the inlet at the top of the shell in the opposite direction that the hydraulic fluid is flowing, traveling towards the water outlet.

3. Heat from the hydraulic fluid is transmitted through the cooling tubes to the water through conduction.
4. As the temperature of the water rises, the warm water exits the heat exchanger through its outlet. The cooled hydraulic fluid exits out the hydraulic fluid outlet.

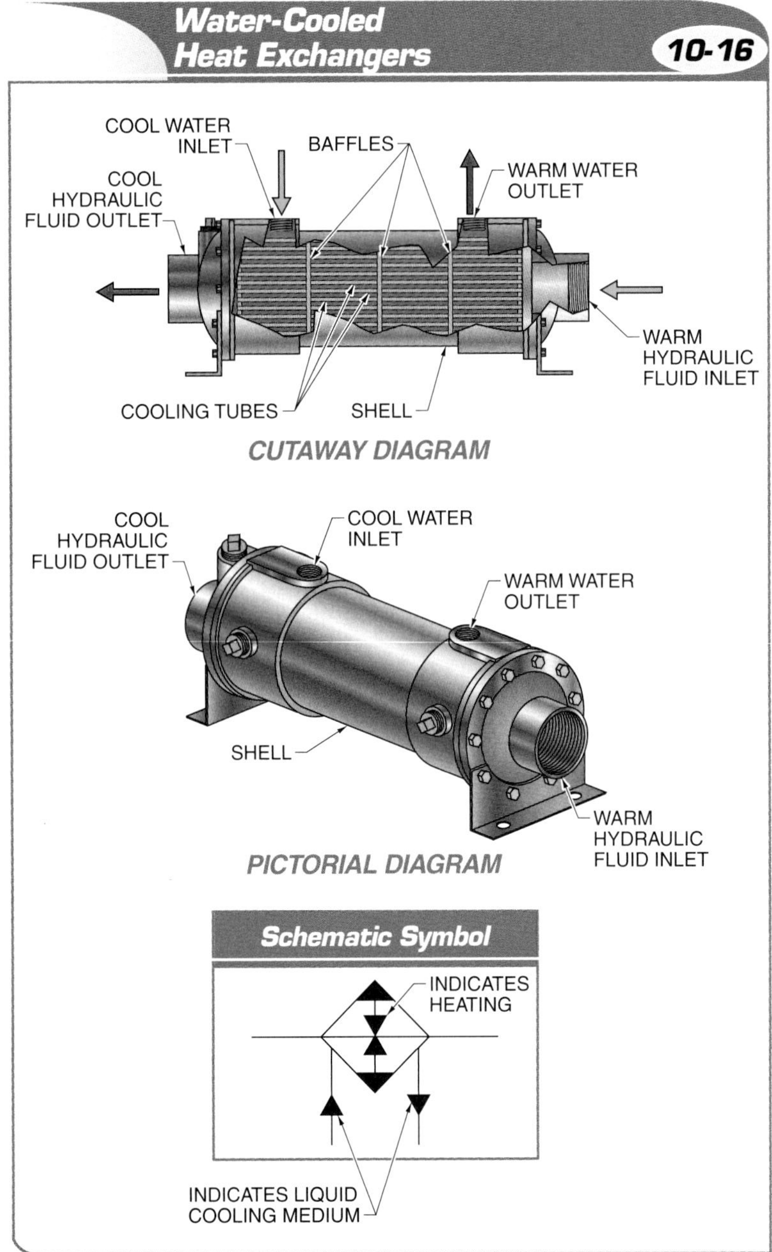

Figure 10-16. Water-cooled heat exchangers allow hot hydraulic fluid to be pumped through the tubes, while cool water circulates over the tubes.

The cooling tubes in the water-cooled heat exchanger must be made from a metallic material that easily conducts thermal energy. Most water-cooled heat exchangers have specialty-metal cooling tubes such as stainless steel or copper. The water and hydraulic fluid travel in opposite directions (counter flow) to help enhance heat transfer from the hydraulic fluid to the water.

Water-cooled heat exchangers typically receive water from a source that is not filtered well. It is necessary to monitor the temperature of the hydraulic fluid for high temperatures, a sign that the heat exchanger is clogged. When the water-cooled heat exchanger is clogged, it must be disconnected from the system and cleaned. In some cases, the cooling tubes are fixed in place and have to be cleaned with a wire bottlebrush. In other cases, the cooling tubes can be removed and cleaned by an easier method, such as power washing.

Heating Hydraulic Fluid. Some hydraulic systems are installed in a cool environment, such as an outdoor location during the winter months. If such a system is not operated for an extended time, it may be necessary to heat the hydraulic fluid. Although hydraulic heaters are not as common as hydraulic coolers, in some systems they are just as important. When hydraulic fluid is below the recommended temperature, its viscosity is much higher. High viscosity causes the system to operate considerably slower and increases the amount of power that must be generated to overcome resistance.

Water-cooled heat exchangers may have their operation reversed to warm hydraulic fluids rather than cool them. By circulating warm water around the tubes, hydraulic fluid within the tubes can be heated. An electric heater, called an immersion heater, is used to heat the hydraulic fluid in a reservoir. An electric heater is submerged in the reservoir below the hydraulic fluid level and near the pump inlet. Electric heaters produce a low-heat output of about 10 W to warm the hydraulic fluid but prevent it from burning. Electric heaters used with hydraulic fluid typically have steel heating elements.

RESERVOIRS

A *reservoir* is a tank that stores hydraulic fluid that is not in use by the hydraulic system. Reservoirs also help clean and cool hydraulic fluid, separate out entrained air and water from hydraulic fluid, and send hydraulic fluid to the pump. Breather/filler cap assemblies are used to add hydraulic fluid to the system. Other reservoir components include sight gauges, cleanout covers, drain plugs, and permanent chip magnets. Most reservoirs are constructed of welded steel plates with end plate extensions that support the unit.

Hydraulic Fluid Storage

The main part of a reservoir allows the storage of hydraulic fluid. The capacity (in gal.) of a reservoir should be at least a minimum of three times the flow rate (gpm) of the pump. All types of reservoirs accomplish the same tasks, although there are four designs typically in use. The four typical reservoir designs available are flattop (conventional), vertically mounted, overhead, and L-shaped. **See Figure 10-17.**

Flattop reservoirs are the most common type of reservoir design. They are sometimes referred to as conventional reservoirs. In this design, the pump and the motor typically sit on top of the reservoir while hydraulic fluid is received from and sent to the hydraulic system.

Due to their configurations, vertically mounted reservoirs offer several advantages over the other designs, including reduced occupied floor space, more efficient hydraulic fluid cooling, vibration dampening, less chance of pump cavitation or aeration, and reduced drive force stresses. In a vertically mounted reservoir, the pump is placed inside the sealed reservoir, which allows for a cleaner system. All reservoirs must be supported a minimum distance 6″ to 7″ from the floor so they can be drained and so that they will not prematurely rust.

Overhead reservoirs are used for large systems where it may be necessary to have more than one hydraulic pump connected to the same reservoir. The motor and the pump are both located below the reservoir. Overhead reservoirs help prevent pseudocavitation because the hydraulic fluid flows freely into the inlet of the pump, reducing the chance that air will enter the pump housing.

TERMS

A **reservoir** is a tank that stores hydraulic fluid that is not in use by the hydraulic system.

One function of a reservoir is to store hydraulic fluid when it is not in use by the hydraulic system.

TECH FACT

Hydraulic fluid that is not properly filtered can cause a buildup of solids in the reservoir and reduce its capacity.

L-shaped reservoirs have a shelf that protrudes from the bottom of the reservoir where the pump and the motor sit, forming an L-shape. L-shaped reservoirs are used for larger units that typically operate at 100 HP and higher. An advantage of an L-shaped reservoir is that the pump does not have to work as hard to receive hydraulic fluid, which reduces the chance of the pump receiving air with the hydraulic fluid and causing pseudocavitation. An L-shaped reservoir also allows easier access to the interior of the reservoir from the top when the strainer needs to be replaced.

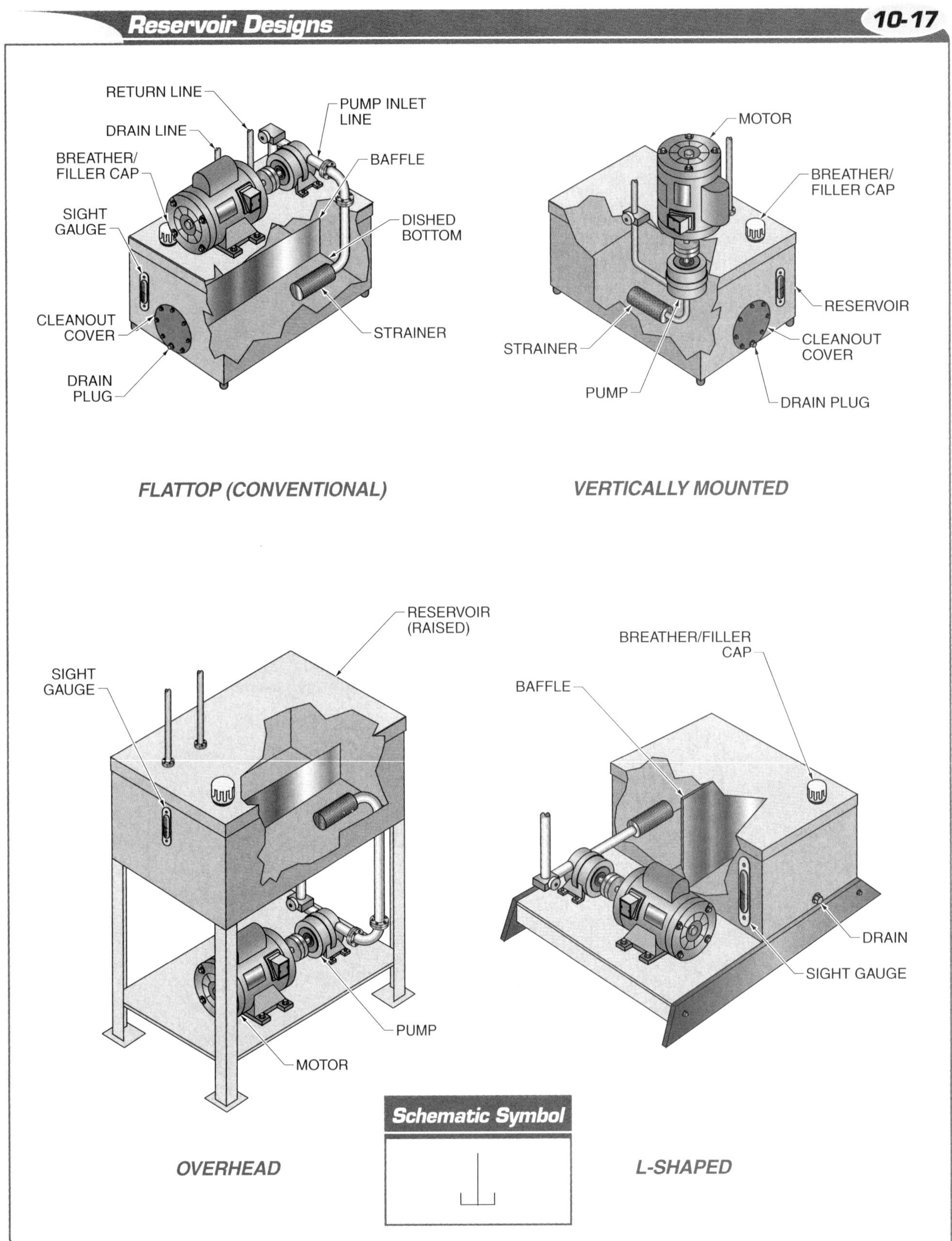

Figure 10-17. Reservoir designs include flattop (conventional), vertically mounted, overhead, and L-shaped.

Breather/Filler Cap Assemblies

A breather/filler cap assembly is a device used to allow air in and out of a reservoir as hydraulic fluid is pumped into the system. **See Figure 10-18.** The breather/filler cap assembly is also the main entrance for atmospheric air to enter the reservoir. When the pump is creating a vacuum, atmospheric pressure is forcing hydraulic fluid into the pump. The breather/filler cap assembly allows air to push into the reservoir so it can force the hydraulic fluid to move into the pump. Air is also allowed to exit the reservoir through the breather/filler cap assembly.

The breather/filler cap assembly must be large enough to withstand the airflow required to maintain atmospheric pressure. The higher the airflow rate, the larger the breather/filler cap assembly required. If air enters the reservoir through return or drain lines, it can exit the reservoir through the breather/filler cap assembly.

The breather/filler cap assembly allows the level of the hydraulic fluid in the reservoir to increase or decrease, depending on the amount delivered to the system. The breather/filler cap assembly is used to refill a reservoir when the amount of hydraulic fluid decreases. All breather/filler cap assemblies are equipped with a filter, which must be regularly maintained, and a removable strainer basket. The breather cap should never be removed from the reservoir without leaving the strainer basket in the filler opening. If left open, undesirable contaminants, such as dirt, debris, insects, and rodents, can enter the reservoir, causing contamination throughout the system.

Hydraulic Fluid Cooling and Cleaning

A reservoir also cools and cleans the hydraulic fluid used in the system. All reservoirs have a baffle that separates the drain and return lines from the pump inlet line. A *baffle* is an artificial surface, usually a plate or wall, for deflecting, retarding, or regulating the flow of hydraulic fluid. **See Figure 10-19.** A baffle is used because it does not allow the hydraulic fluid returning from the system to be pulled directly into the suction line.

When hydraulic fluid in the reservoir takes an indirect path to the pump inlet line, it allows the hydraulic fluid to cool off and separate from any air, which delays reuse. As the hydraulic fluid moves inside the reservoir, most of its heat can be dissipated through the walls of the reservoir. Any remaining contaminants in the hydraulic fluid begin to settle to the bottom of the reservoir. A reservoir must be properly sized and regularly cleaned in order to provide hydraulic fluid cleaning and cooling.

TERMS

A **baffle** is an artificial surface, usually a plate or wall, for deflecting, retarding, or regulating the flow of hydraulic fluid.

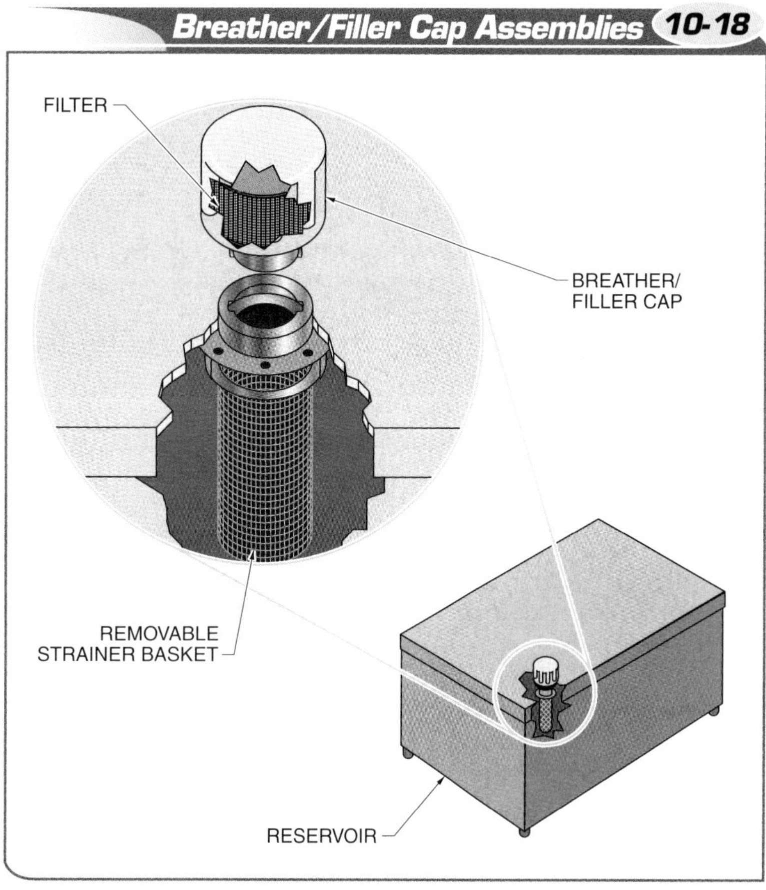

Figure 10-18. A breather/filler cap assembly allows air in and out of a reservoir as hydraulic fluid is pumped into the system.

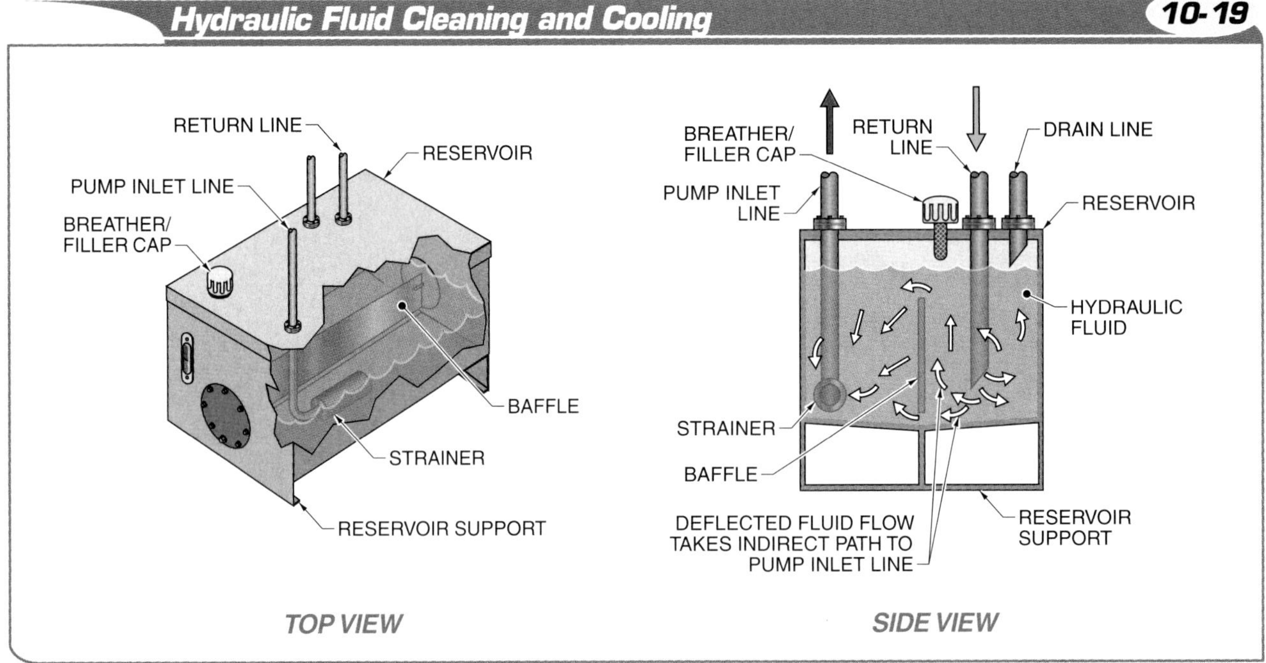

Figure 10-19. A reservoir cools and cleans hydraulic fluid in the system using a baffle that separates the drain and return lines from the pump inlet line.

TERMS

A **sight gauge** is a device used to allow a visual inspection of the hydraulic fluid level in a reservoir at any given time.

Receiving Hydraulic Fluid from the System

A reservoir receives hydraulic fluid used in the system from two separate lines, the return lines and the drain lines. Used hydraulic fluid cannot be released into the operating environment, so it must be returned to the reservoir through the return line. The return line must be below the hydraulic fluid level inside of the reservoir. A return line located above the hydraulic fluid level will cause the hydraulic fluid to bubble and foam as it returns under pressure. It is estimated that it costs 10 times more to extract contaminants from hydraulic fluid than it does to prevent them from entering a hydraulic system.

Drain lines are attached to hydraulic components such as pumps, motors, sequence valves, and pressure-reducing valves. The hydraulic fluid in the drain line returns to the reservoir under little to no pressure. Unlike the return line, the drain line must end above the hydraulic fluid level so it can be allowed to drain. If the drain line is located below the hydraulic fluid level, there will not be enough pressure to force the hydraulic fluid out of the drain line and it will create back pressure.

Sending Hydraulic Fluid to the Pump

After hydraulic fluid has been added to the reservoir or returned from the system through the drain line or the return line, it is ready to be sent to the pump. The pump inlet line is located on the opposite side of the baffle as the drain and return lines. It usually has a strainer attached to its end because it connects directly to the inlet of the pump. The pump inlet line and its strainer must be located below the hydraulic fluid level in the reservoir or air may enter the pump, which can cause pseudocavitation.

Other Reservoir Components

Other reservoir components include sight gauges, cleanout covers, drain plugs, and permanent chip magnets. **See Figure 10-20.** A *sight gauge* is a device used to allow a visual inspection of the hydraulic fluid level in a reservoir at any given time. A sight gauge can also be used to send a signal to a warning device if the hydraulic fluid level drops below recommended operating requirements.

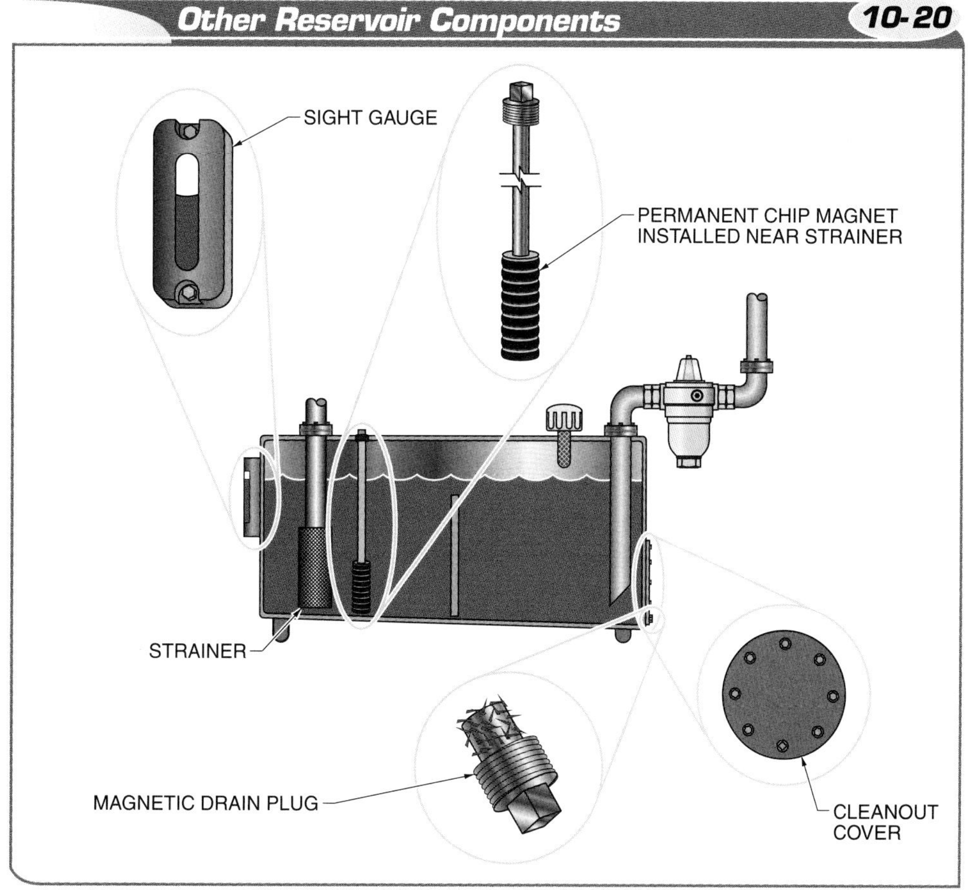

Figure 10-20. Other reservoir components include a sight gauge, cleanout cover, drain plug, and permanent chip magnet.

TERMS

A **cleanout cover** is a device used to access a reservoir when it requires cleaning from the build-up of solid materials.

A **drain plug** is a threaded device installed at the lowest point of a reservoir to allow for the removal of hydraulic fluid.

A **permanent chip magnet** is a magnetic device placed in a reservoir to attract and hold ferrous metal particles that have contaminated the system but not been recovered by system filters or strainers.

A *cleanout cover* is a device used to access a reservoir when it requires cleaning from the build-up of solid materials. Sometimes there are two cleanout covers located on opposite sides of the baffle. This allows both sides of the reservoir to be thoroughly cleaned.

A *drain plug* is a threaded device installed at the lowest point of a reservoir to allow for the removal of hydraulic fluid. A *permanent chip magnet* is a magnetic device placed in a reservoir to attract and hold ferrous metal particles that have contaminated the system but not been recovered by system filters or strainers. Permanent chip magnets are typically placed in the reservoir in the form of magnetic drain plugs or magnetic rings that are placed near the strainer.

Chapter Review

Name: ______________________________ Date: ______________

MULTIPLE CHOICE

______ 1. When heat cannot be properly dissipated, ___ must be placed in the system to help cool the hydraulic fluid by removing damaging heat from the system.

A. cooling fans
B. heat exchangers
C. intercoolers
D. vents

______ 2. A ___ filter is placed before components to protect them from contaminants in hydraulic fluid.

A. basket
B. pressure line
C. reservoir
D. suction line

______ 3. ___ is a coating of a sludge-like residue that results from the breakdown of hydraulic fluid from the heat and pressure within a hydraulic system.

A. Abrasion
B. Agglomeration
C. Cavitation
D. Varnish

______ 4. ___ pressure is the level of pressure on the return line of a device or hydraulic system.

A. Back
B. Bypass
C. Return line
D. Operating

______ 5. A ___ is an artificial surface, usually a plate or wall, for deflecting, retarding, or regulating the flow of hydraulic fluid.

A. baffle
B. cover
C. reservoir
D. support plate

______ 6. Air-cooled heat exchangers work best when the ambient temperature is lower than ___°F.

A. 50
B. 75
C. 80
D. 100

______ 7. A ___ is used to allow air in and out of a reservoir as hydraulic fluid is pumped into the system.

A. breather/filler cap assembly
B. cleanout cover
C. pump inlet line
D. return line

_______________ **8.** A(n) ___ is a holding tank for hydraulic fluid that is not in use by the hydraulic system.
- A. baffle
- B. heat exchanger
- C. intercooler
- D. reservoir

_______________ **9.** Water-cooled heat exchangers have a maximum operating pressure up to ___ psi.
- A. 150
- B. 200
- C. 250
- D. 300

_______________ **10.** Hydraulic fluid cleanliness is rated by standards developed by the ___.
- A. American Society of Testing Materials (ASTM)
- B. International Fluid Power Society (IFPS)
- C. International Standards Organization (ISO)
- D. National Fluid Power Association (NFPA)

_______________ **11.** A ___ filtration unit is a nonfixed filtration machine that filters hydraulic fluid as it is being transferred from a storage vessel to the reservoir.
- A. handheld
- B. manual
- C. mobile
- D. portable

_______________ **12.** Strainer ratings can range from ___ µm to ___ µm.
- A. 10; 100
- B. 22; 500
- C. 74; 200
- D. 100; 200

_______________ **13.** Absolute-rated filters with depth-type elements are close to ___% efficient.
- A. 50
- B. 75
- C. 95
- D. 99

_______________ **14.** Contaminants in a hydraulic system are rated in ___.
- A. inches (in.)
- B. millimeters (mm)
- C. micrometers (µm)
- D. nanometers (nm)

_______________ **15.** Strainers typically have openings larger than ___.
- A. 0.005 in.
- B. 0.05 mm
- C. 50 nm
- D. 50 µm

COMPLETION

_______________ **1.** Electric heaters produce a low-heat output of about ___ W to warm the hydraulic fluid but prevent it from burning.

_______________ **2.** A(n) ___ helps clean and cool hydraulic fluid, separate out entrained air and water from hydraulic fluid, and send hydraulic fluid to the pump.

_______________ **3.** A(n) ___ heat exchanger operates similar to an automobile radiator by dissipating heat through convection.

_______________ **4.** A(n) ___ filter is designed to operate in a vacuum and is connected to the hydraulic system before the pump and outside of the reservoir to protect the pump from contaminants in the reservoir.

_______________ **5.** A(n) ___ is a device used to allow a visual inspection of the hydraulic fluid level in a reservoir at any given time.

_______________ **6.** A(n) ___ is placed in a reservoir to attract and hold ferrous metal particles that have contaminated the system but not been recovered by system filters or strainers.

_______________ **7.** A(n) ___ is a device used to access a reservoir when it requires cleaning from the buildup of solid materials.

_______________ **8.** It is estimated that it costs ___ times more to extract contaminants from hydraulic fluid than it does to prevent them from entering a hydraulic system.

_______________ **9.** A(n) ___ is commonly used to heat the hydraulic fluid in a reservoir.

_______________ **10.** A(n) ___ is a foreign substance in the hydraulic fluid that is not meant to be part of the fluid.

_______________ **11.** A(n) ___ is a filter placed in front of the return line before it enters the reservoir.

_______________ **12.** A(n) ___ rating is a value that indicates the diameter of the largest hard, sphere-shaped particle that can pass through a filter under specific test conditions.

_______________ **13.** A(n) ___ is used to allow air in and out of a reservoir as the pump delivers hydraulic fluid into the system.

_______________ **14.** A(n) ___ is a mechanical device used to transfer heat from one fluid to another fluid.

_______________ **15.** A(n) ___-type filter element is composed of multiple layers of fabric or threaded material.

_______________ **16.** A(n) ___ filter is equipped with a bypass check valve to allow hydraulic fluid to bypass the filter element as the element becomes clogged.

_______________ **17.** A(n) ___ filtration unit is sometimes referred to as an off-line filter.

_______________ **18.** A(n) ___ is a screen, mesh, or perforated obstruction used to separate a solid from a liquid.

_______________ **19.** Some filters are equipped with a bypass check valve that is typically set between ___ psi and ___ psi of pressure drop.

_______________ **20.** ___ is the reduction of the dimensions of a material due to grinding.

TRUE/FALSE

T F 1. Heat exchangers can be either air-cooled or water-cooled.

T F 2. A breather cap should always be removed from a reservoir along with the strainer basket in the filler opening.

T F 3. A nominal rating indicates the average size or larger contaminant that a filter can trap.

T F 4. All reservoirs must be supported a minimum distance 12″ to 18″ from the floor so they can be drained and so that they will not prematurely rust.

T F 5. Most absolute-rated filters use fiberglass as a filtering media.

T F 6. Water-cooled heat exchangers are less efficient than air-cooled heat exchangers.

T F 7. Hydraulic fluid in a drain line returns to the reservoir under little to no pressure.

T F 8. The lower the airflow rate, the larger the breather/filler cap assembly required.

T F 9. Strainers can be cleaned periodically and reused, unlike filter elements, which must be discarded.

T F 10. All hydraulic systems generate various amounts of contaminants during normal operation.

T F 11. A 200-mesh filter has 100 vertical threads and 100 horizontal threads per square inch.

T F 12. Contaminants that are smaller than the clearance of moving surfaces in a hydraulic system cannot cause system failure.

T F 13. Nominal-rated filters with surface-type elements are 50% to 75% efficient.

T F 14. The capacity (in gal.) of a reservoir should be at least a minimum of two times the flow rate (gpm) of the pump.

T F 15. Although recycled hydraulic fluid must be filtered before it enters a reservoir, new hydraulic fluid from the manufacturer does not need to be filtered.

T F 16. Most water-cooled heat exchangers have specialty-metal cooling tubes such as stainless steel or copper.

T F 17. Most bypass filters are color-coded to indicate three different filter conditions.

T F 18. When a new hydraulic hose is assembled or built, it is not necessary to apply a flushing procedure.

T F 19. A higher strainer mesh number indicates a smaller opening size.

T F 20. A strainer is a porous device used to separate solid material from a liquid or a gas.

T F 21. New hydraulic systems are typically clean enough to be used for normal, first time operation.

T F 22. Strainers are typically submerged in hydraulic fluid inside a reservoir and placed at the end of the suction line before a pump.

T F 23. Hydraulic system contaminants are only created externally.

T F 24. A surface-type element is a filter element composed of a single closely woven fabric or cellulose layer.

T F 25. Implementing a cleaning procedure for new hydraulic equipment can lessen contamination.

SHORT ANSWER

1. Describe the color-coding system used with bypass filters.

2. Describe how hydraulic fluid contaminants can be introduced into a system.

3. Describe the effects that contaminants can have on a hydraulic system.

4. Identify four common contaminants found in hydraulic fluid.

5. Describe what each number represents in the three-number code from the industry standard that is used to determine the cleanliness of hydraulic fluid.

MATCHING

Reservoir Designs

__________ 1. Flattop (conventional)

__________ 2. L-shaped

__________ 3. Overhead

__________ 4. Vertically mounted

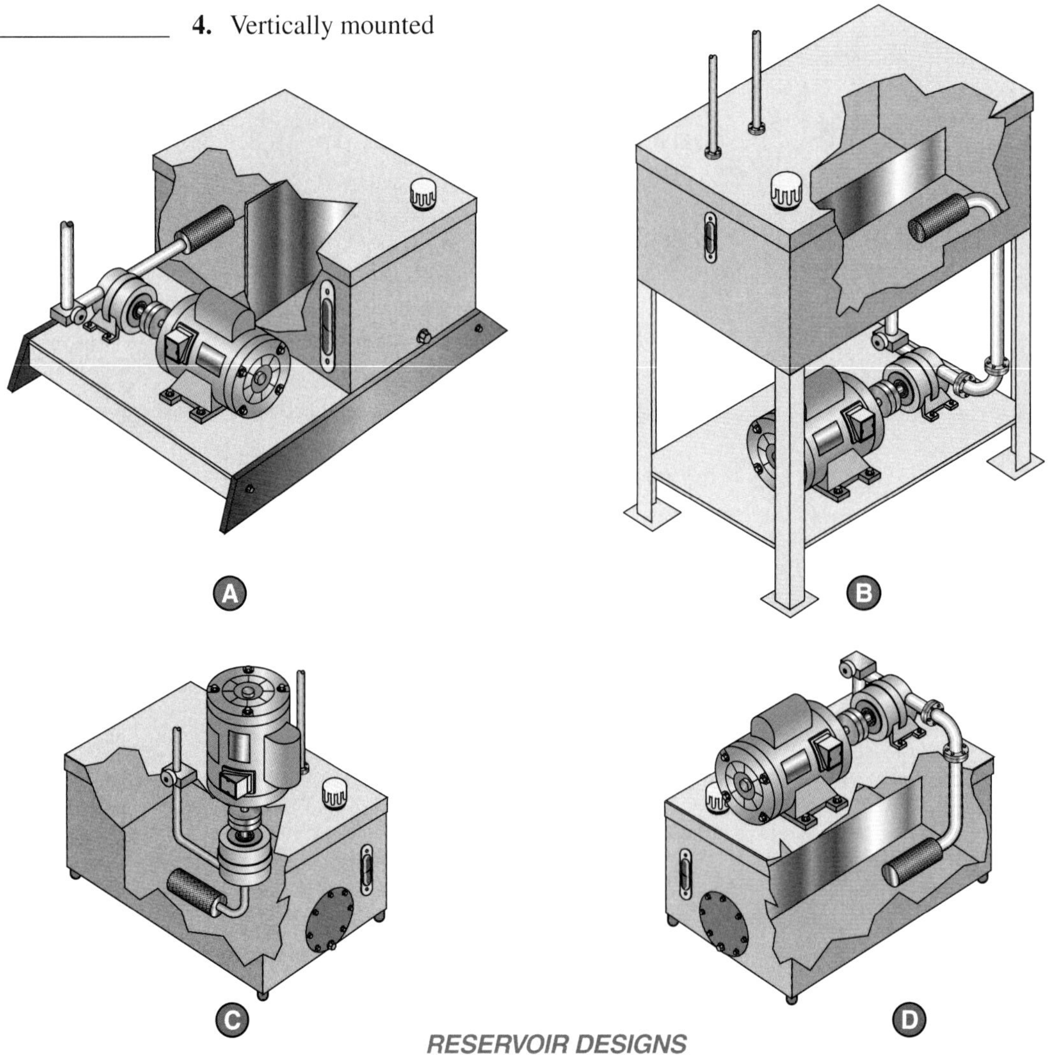

A

B

C

D

RESERVOIR DESIGNS

Schematic Symbols

_______________ **1.** Air-cooled heat exchanger

_______________ **2.** Bypass filter

_______________ **3.** Filter or strainer

_______________ **4.** Pressure filter

_______________ **5.** Reservoir

_______________ **6.** Return line filter

_______________ **7.** Water-cooled heat exchanger

A B C D

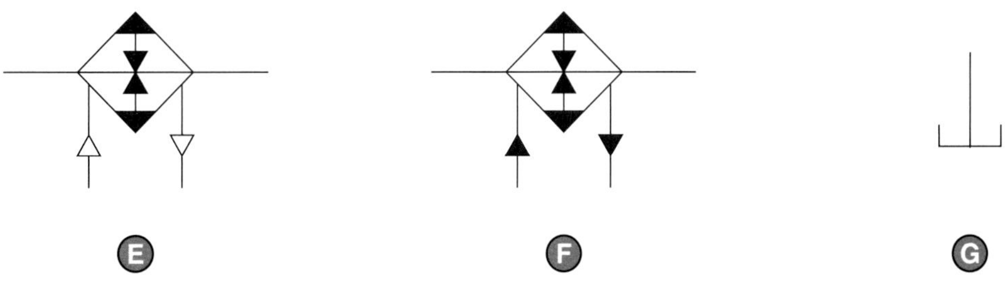

SCHEMATIC SYMBOLS

Activity 10-1: Filter and Strainer Locations

1. Redraw the schematic diagram of Clamping and Bending Machine and add schematic symbols for piping and filters or strainers to the locations where they are required.

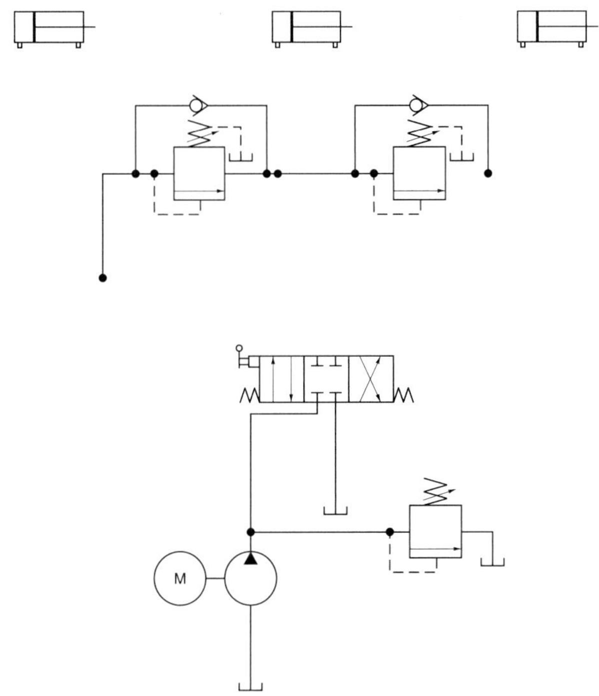

CLAMPING AND BENDING MACHINE

Activity 10-2: Manually Cleaning Water-Cooled Heat Exchanger

A hydraulic system with a water-cooled heat exchanger overheats every 6 months to 7 months. When this occurs, the water-cooled heat exchanger must be cleaned because the water that is used to cool the hydraulic fluid is unfiltered and the water tubes in the heat exchanger become clogged with contaminants such as calcium deposits, rust particles, dust, and sludge. The manufacturer's instructions for the water-cooled heat exchanger includes the following information:

To clean the water tubes, flush out with clean water or a good quality commercial cleaner. Use a rod to remove any contaminants that have become lodged in the tubes.

1. Using the information from the manufacturer's instructions and Water-Cooled Heat Exchanger, write a procedure that can be used to ensure that the water-cooled heat exchanger is properly removed from the hydraulic system, cleaned, and reinstalled.

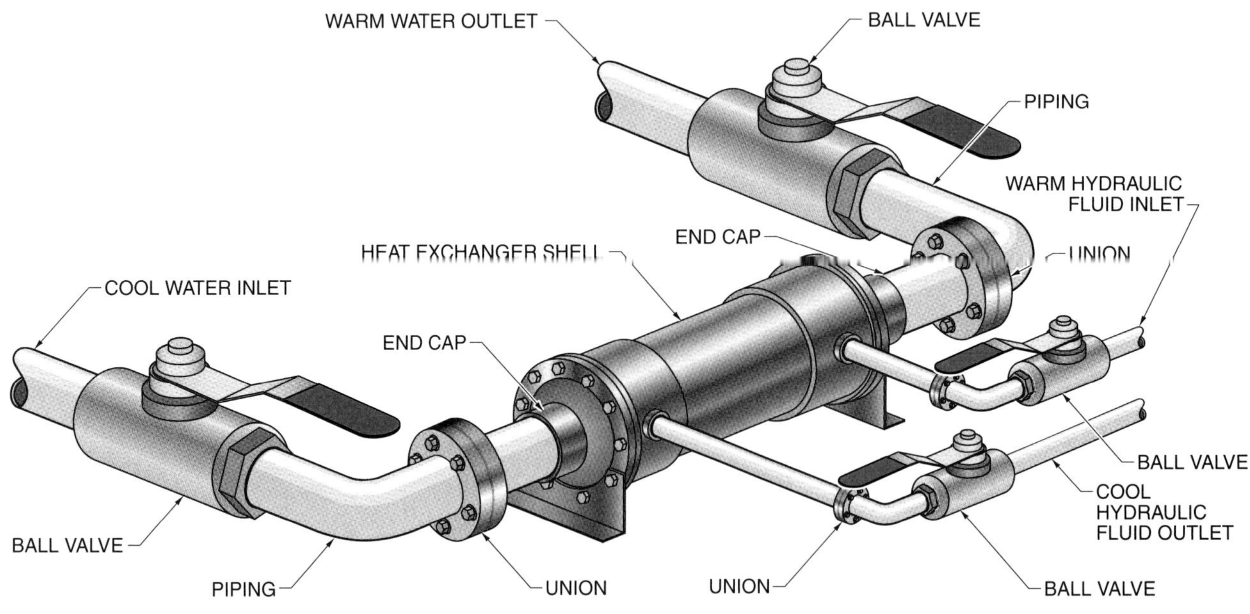

WATER-COOLED HEAT EXCHANGER

Activity 10-3: Strainer Specification

A factory has been experiencing random hydraulic equipment breakdown since the hydraulic system was originally installed. After the pump failed, the company's maintenance department decided to hire a hydraulic system consultant to troubleshoot the system to determine if there was a specific cause for the random equipment failures. After the consultant inspected the system, hydraulic fluid, and repair logbooks, it was determined that metal contaminants in the system were the main cause of the problems. The consultant recommended replacing the 100-mesh strainer in the reservoir with a 60-mesh strainer with magnets attached to it to correct the problems. Using Original Strainer Specifications and Replacement Strainer Specifications, determine the ordering code (model number, wire mesh code, and flow rate code) of the replacement strainer.

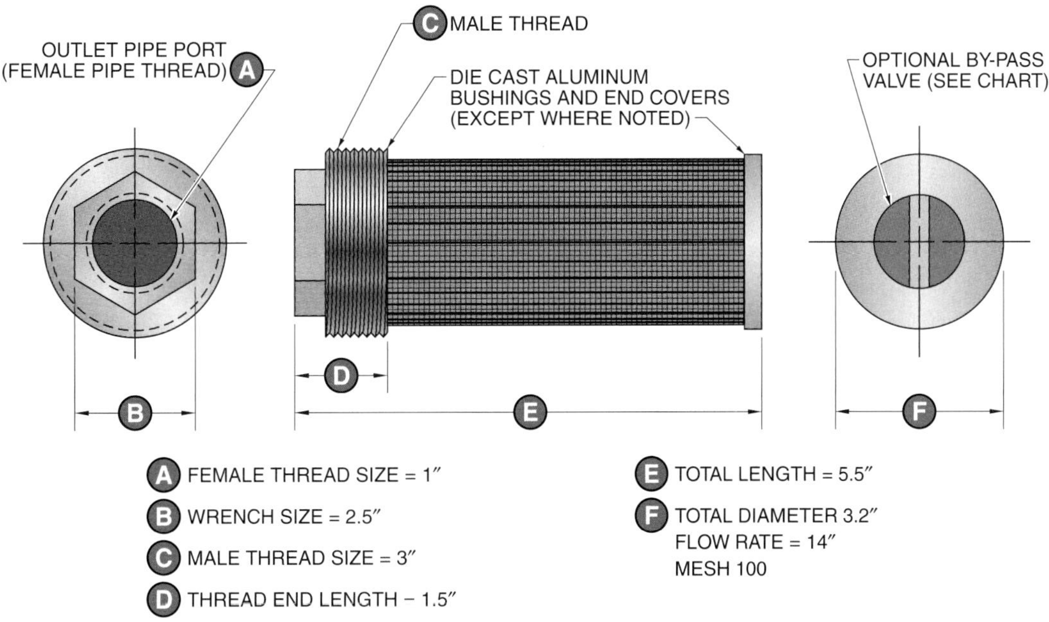

ORIGINAL STRAINER SPECIFICATIONS

Model Number	Female Thread Size*
6075	¾
6100	1
6125	1¼

Code	Wire Mesh
J	200
B	100
E	75
C	60
K	30

Code	Flow Rate Max†
B	10
K	15

* in in.
† in gpm

OUTLET PIPE PORT
MAGNET WITH POLARIZING STEEL END PLATES
4″ D
0.62″
WIRE CLOTH ELEMENTS
A

REPLACEMENT STRAINER SPECIFICATIONS

1. Replacement strainer ordering code: _ _ _ _ - _ - _

Activity 10-4: Hydraulic Fluid Cleanliness and Filtration Requirements

Two hydraulic systems have had recurring problems with component failures because of hydraulic fluid cleanliness. In order to troubleshoot each hydraulic system, the filtration requirements and hydraulic fluid cleanliness for each system must be determined. Use Hydraulic System 1, ISO Code Chart, and Hydraulic System 2 to answer the following questions.

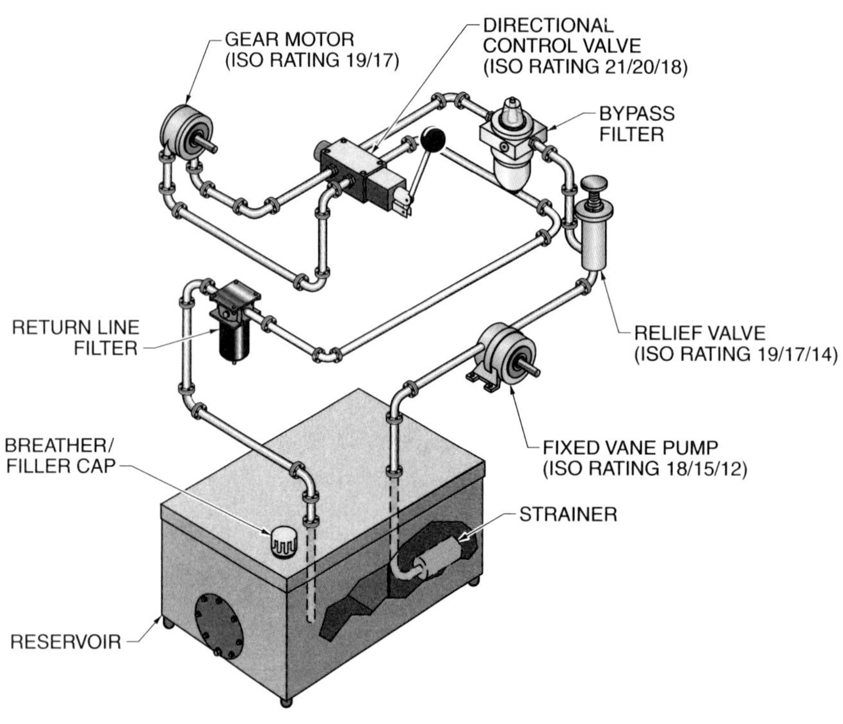

HYDRAULIC SYSTEM 1

ISO Code Chart

ISO Code	Contaminant Size with Beta Ratio of 200*
21/18/15	21/18/15
20/17/14	20/17/14
19/16/13	19/16/13
18/15/12	18/15/12

* in μm

________________ 1. Which component in Hydraulic System 1 is most affected by hydraulic fluid cleanliness?

________________ 2. Use ISO Code Chart to determine the size of contaminants (in μm) that the strainer should trap to protect the pump.

________________ 3. Use ISO Code Chart to determine the size of contaminants (in μm) that the filter should trap to protect the directional control valve and motor.

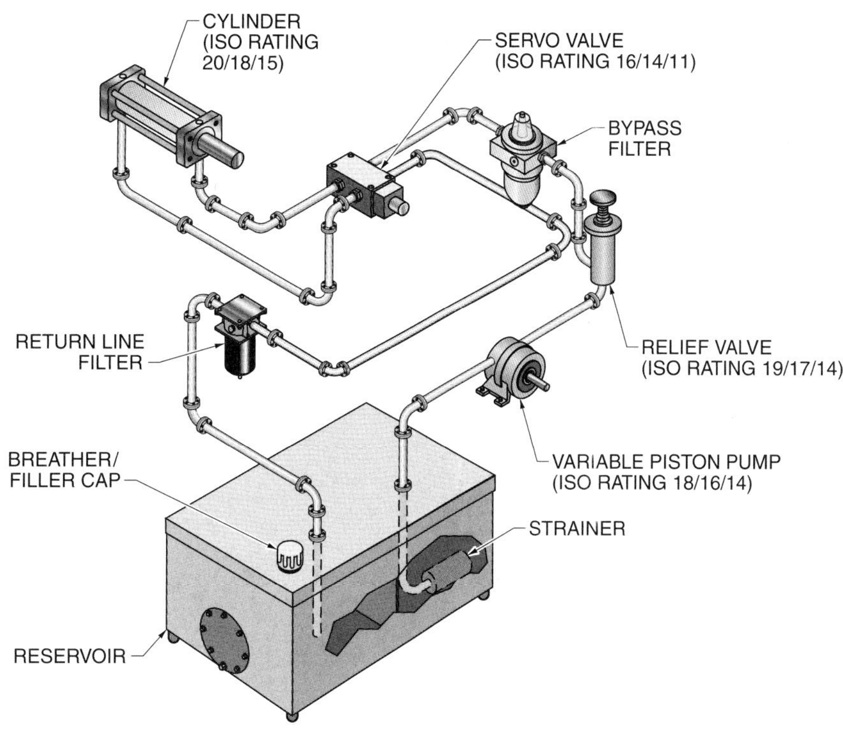

HYDRAULIC SYSTEM 2

__________ **4.** Which component in Hydraulic System 2 is most affected by hydraulic fluid cleanliness?

__________ **5.** Use ISO Code Chart to determine the size of contaminants (in μm) that the strainer should trap to protect the pump.

__________ **6.** Use ISO Code Chart to determine the size of contaminants (in μm) that the filter should trap to protect the servo valve and cylinder.

__________ **7.** What is the amount of 15 μm contaminants (per mL) that the directional control valve can withstand in a filter or strainer with an ISO rating of 22/20/16?

__________ **8.** What is the amount of 5 μm contaminants (per mL) that the pump can withstand in a filter or strainer with an ISO rating of 17/15/12?

__________ **9.** What is the amount of 2 μm contaminants (per mL) that the cylinder can withstand in a filter or strainer with an ISO rating of 20/18/15?

__________ **10.** What is the amount of 15 μm contaminants (per mL) that the relief valve can withstand in a filter or strainer with an ISO rating of 18/16/13?

__________ **11.** What is the amount of 2 μm contaminants (per mL) that the servo valve can withstand in a filter or strainer with an ISO rating of 16/14/11?

Activity 10-5: Hydraulic Filters

The hydraulic system below does not have any filters included in the drawing. The plant engineer wants the plans redrawn to include all required filters. Redraw the diagram and add all required filters. Assume the two directional control valves are 100′ apart. *Note*: Bypass check valves can be added by placing check valves with springs in parallel with filters.

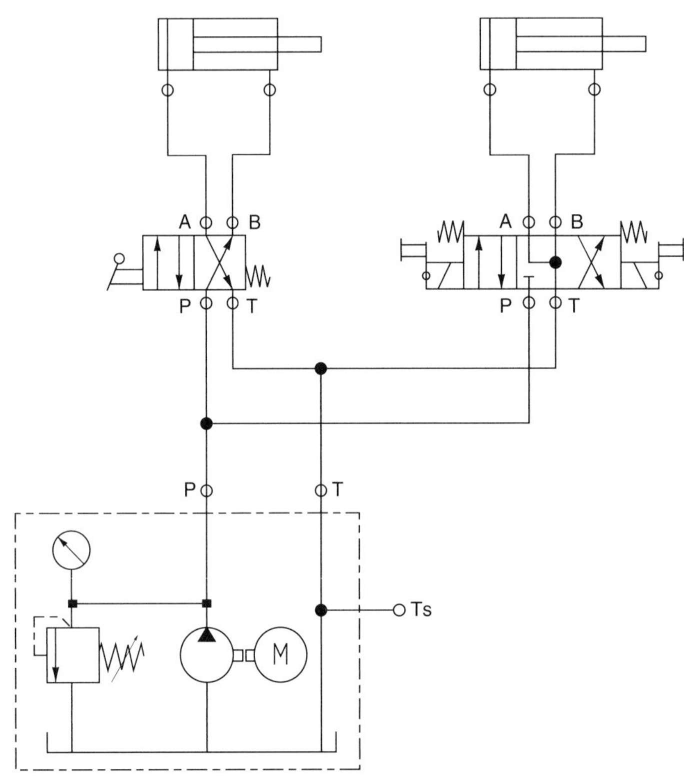

Chapter 11

Pneumatic System Fundamentals

OBJECTIVES

- Describe the basic physical properties of a gas.
- Calculate the volume of a receiver.
- Describe absolute temperature.
- Convert temperature to the Rankine scale from the Fahrenheit scale.
- Describe the gas laws.
- List the advantages and disadvantages of using pneumatic systems.
- Define airflow.

INTRODUCTION

Pneumatic systems are fluid power systems that use gas, typically in the form of air, to transmit energy. Pneumatic systems are widely used in industry to accomplish work such as precision drilling and industrial food processing. Pneumatic systems can operate faster than hydraulic systems, but they cannot create the force needed to move very heavy loads.

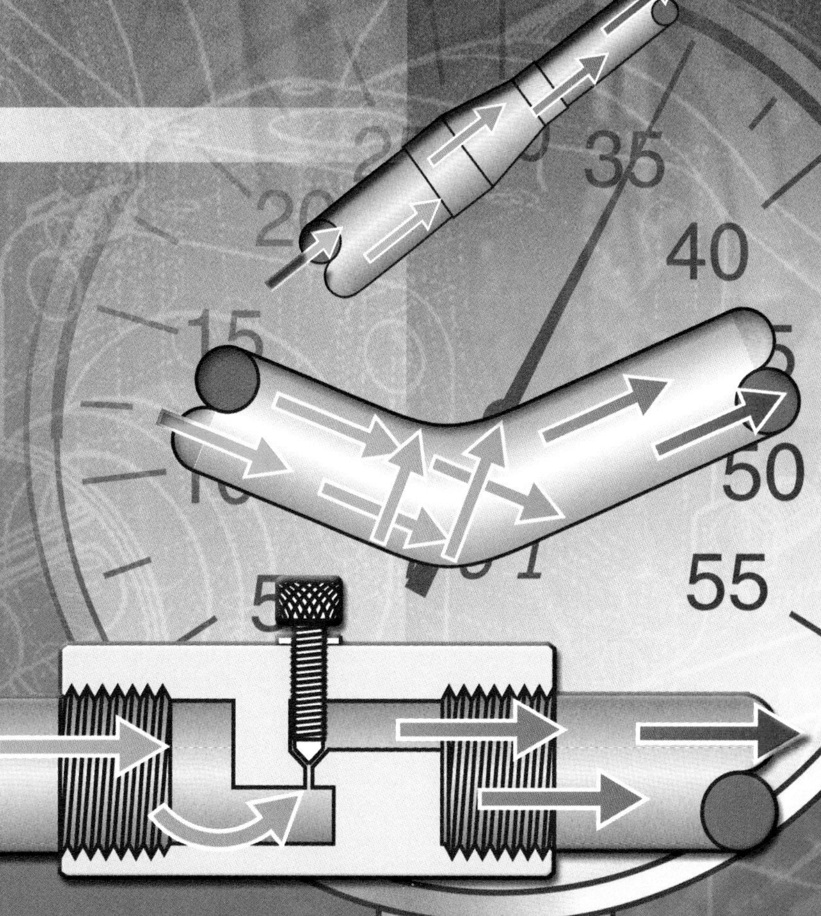

TERMS

Volume is the size of a space or chamber measured in cubic units.

BASIC GAS PHYSICS

Pneumatic systems use gas under pressure to create movement. Pneumatic systems operate on the principles of fluid mechanics, which state that a fluid can flow, has no definite shape, and increases in volume with an increase in temperature. The attraction between the molecules of a substance determines whether the substance is a solid, liquid, or gas. **See Figure 11-1.**

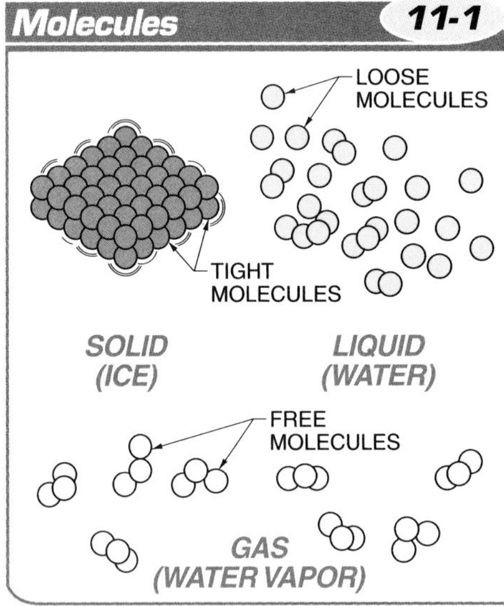

Figure 11-1. The attraction between the molecules of a substance determines whether the substance is solid, liquid, or gas.

The molecules of gases move around quickly and float freely in a state of chaos, unlike the molecules of solids and liquids, which have stronger attractions to each other. The term "gas" is derived from the Greek word for "chaos." Because gas molecules are not strongly attracted to each other and move around in a chaotic state, they are easily compressed. Gases have the following characteristics:

- Gases are shapeless and thus take the shape of their containers.
- Gases have a lower viscosity than liquids because there is no attraction or repulsion between similar gas molecules or atoms. The only interaction between gas molecules is when they collide.
- Gases can be compressed inside a closed container. When gases are compressed, the gas molecules collide with greater frequency, increasing the pressure and temperature in the container.
- Gases can also decompress inside a container if the container expands. When a container expands, the gas molecules collide with less frequency, decreasing the pressure and temperature in the container.
- When gases are heated inside a container, the pressure in the container increases because the gas molecules collide more frequently with each other and with the surface of the container.

There are several scientific laws that determine the properties of a gas. The gas laws include Boyle's law, Charles's law, and Gay-Lussac's law, as well as the combined gas law. To help understand the gas laws, it is necessary to understand volume as it relates to a gas in a container using absolute pressure and absolute temperature.

Volume

Volume is the size of a space or chamber measured in cubic units. Because a gas takes the shape of the container it is in, the volume of the gas, unlike a solid or liquid, is determined by the volume of the container. The smaller the container a gas is in, the less volume of gas is present. The larger the container a gas is in, the more volume of gas is present. **See Figure 11-2.**

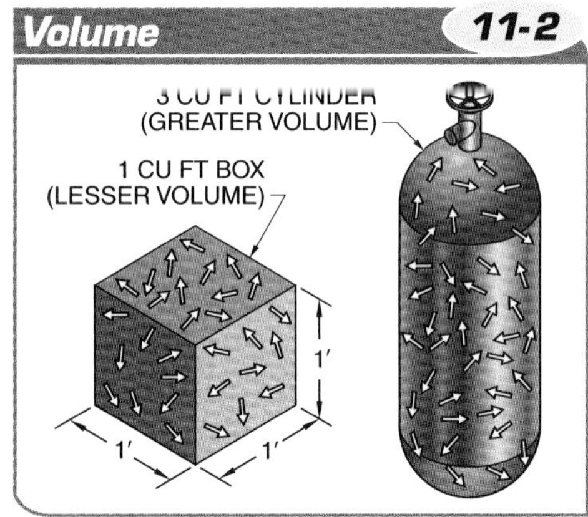

Figure 11-2. Volume is the size of a space or container measured in cubic units.

Absolute Pressure

Absolute pressure is the sum of gauge pressure and atmospheric pressure (14.7 psi) and is expressed in pounds per square inch absolute (psia). Gauge pressure and absolute pressure can be easily converted from one to the other because gauge pressure is always 14.7 psi less than absolute pressure. For example, if absolute pressure is 84 psia, then gauge pressure is 69.3 psig (84 – 14.7 = 69.3). Alternatively, absolute pressure is always 14.7 psi greater than gauge pressure. For example, if gauge pressure is 81 psig, then absolute pressure is 95.7 psia (81 + 14.7 = 95.7).

Absolute Temperature

Absolute temperature is any temperature on a scale that begins with absolute zero. *Absolute zero* is the hypothetical temperature at which molecular motion ceases, precisely –273.15°C (–459.67°F). The absolute temperature scale is based on the Rankine scale (°R). While molecules are still in motion at 0°F, molecules do not move at 0°R. A comparison of the Rankine and Fahrenheit scales shows that the temperature in °R is always approximately 460° greater than the temperature in °F. **See Figure 11-3.**

Temperature in Fahrenheit is converted to Rankine by applying the following formula:

$$T_R = 460 + T_F$$

where

T_R = temperature (in °R)

T_F = temperature (in °F)

Example: What is the equivalent Rankine temperature of 96°F?

$$T_R = 460 + T_F$$
$$T_R = 460 + 96$$
$$T_R = \mathbf{556°R}$$

Boyle's Law

Boyle's law is a gas law that states that when temperature remains constant, there is an inverse relationship between pressure and the volume of a gas. This means that when the pressure applied to a gas increases, its volume decreases. Also, when the pressure applied to a gas decreases, its volume increases. Boyle's law is expressed by the following formula:

$$V_1 = \frac{P_2 \times V_2}{P_1}$$

where

V_1 = volume of the first condition (in cu in.)

P_2 = absolute pressure of the second condition (in psia)

V_2 = volume of the second condition (in cu in.)

P_1 = absolute pressure of the first condition (in psia)

A *receiver* is a specialized vessel that holds gas for future use in a pneumatic system. According to Boyle's law, if the pressure in a receiver increases and the temperature remains constant, the volume of the gas in the receiver decreases. If the pressure decreases and the temperature remains constant, then the volume of the gas increases.

TERMS

Absolute pressure is the sum of gauge pressure and atmospheric pressure (14.7 psi) and is expressed in pounds per square inch absolute (psia).

Absolute temperature is any temperature on a scale that begins with absolute zero.

Absolute zero is the hypothetical temperature at which molecular motion ceases, precisely –273.15°C (–459.67°F).

Boyle's law is a gas law that states that when temperature remains constant, there is an inverse relationship between pressure and the volume of a gas.

A **receiver** is a specialized vessel that holds gas for future use in a pneumatic system.

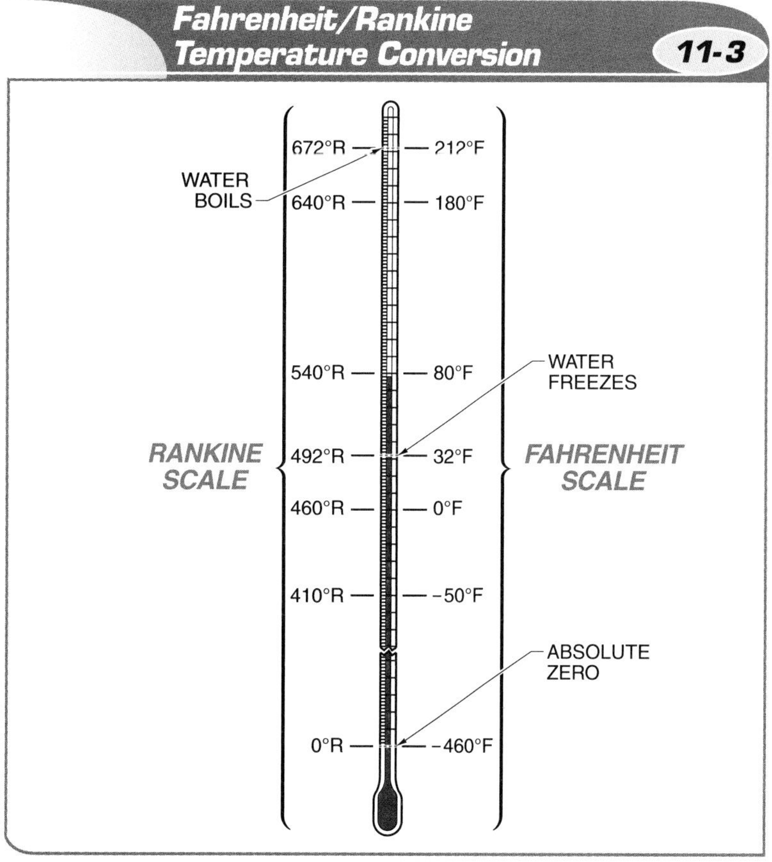

Figure 11-3. Absolute temperature (°R) is always approximately 460° greater than the temperature in degrees Fahrenheit (°F).

TERMS

Charles's law is a gas law that states that when the pressure on a gas is constant, the volume of the gas has a direct relationship with its absolute temperature.

Example—Calculating Volume: The pressure in a receiver is increased from 5 psi to 35 psi. If the new volume is calculated to be 120 cu in., what was the original volume? **See Figure 11-4.** *Note:* Convert psi to psia.

$$V_1 = \frac{P_2 \times V_2}{P_1}$$

$$V_1 = \frac{(35 + 14.7) \times 120}{5 + 14.7}$$

$$V_1 = \frac{49.7 \times 120}{19.7}$$

$$V_1 = \frac{5964}{19.7}$$

$$V_1 = \mathbf{302.74\ cu\ in.}$$

TECH FACT

Irish chemist Robert Boyle performed the first quantitative experiments on gases in 1662. Boyle performed his experiments by setting up a J-shaped tube with a closed end in the multistory entryway of his home. He studied the relationships between the pressure of the trapped gas and its volume in the J-shaped tube. Boyle's law is sometimes referred to as the Boyle-Mariotte law.

Atlas Copco

Air compressors that are installed in outdoor locations must be carefully monitored for pressure and volume due to changes in the ambient temperature.

Charles's Law

Charles's law is a gas law that states that when the pressure of a gas is constant, the volume of the gas has a direct relationship with its absolute temperature. If heat is applied to a gas, the volume of the gas increases. If heat is taken away from a gas, the volume decreases. Charles's law is expressed by the following formula:

$$V_2 = \frac{V_1 \times T_2}{T_1}$$

where

V_2 = volume of the second condition (in cu ft)

V_1 = volume of the first condition (in cu ft)

T_2 = absolute temperature of the second condition (in °R)

T_1 = absolute temperature of the first condition (in °R)

If heat is applied to the gas inside a receiver and the pressure remains constant, then the volume of the gas increases. If the gas cools and the pressure remains constant, then its volume decreases.

Example—Calculating Volume: If 45 cu ft of a gas is cooled from 105°F to 70°F, what is the new volume of the gas? **See Figure 11-5.** *Note:* Convert °F to °R.

$$V_2 = \frac{V_1 \times T_2}{T_1}$$

$$V_2 = \frac{45 \times (460 + 70)}{460 + 105}$$

$$V_2 = \frac{45 \times 530}{565}$$

$$V_2 = \frac{23{,}850}{565}$$

$$V_2 = \mathbf{42.21\ cu\ ft}$$

TECH FACT

Charles's law is named after the French physicist Jacques Charles. Jacques Charles was known for his experiments with gas and was the first person to successfully fly a hot-air balloon in 1783.

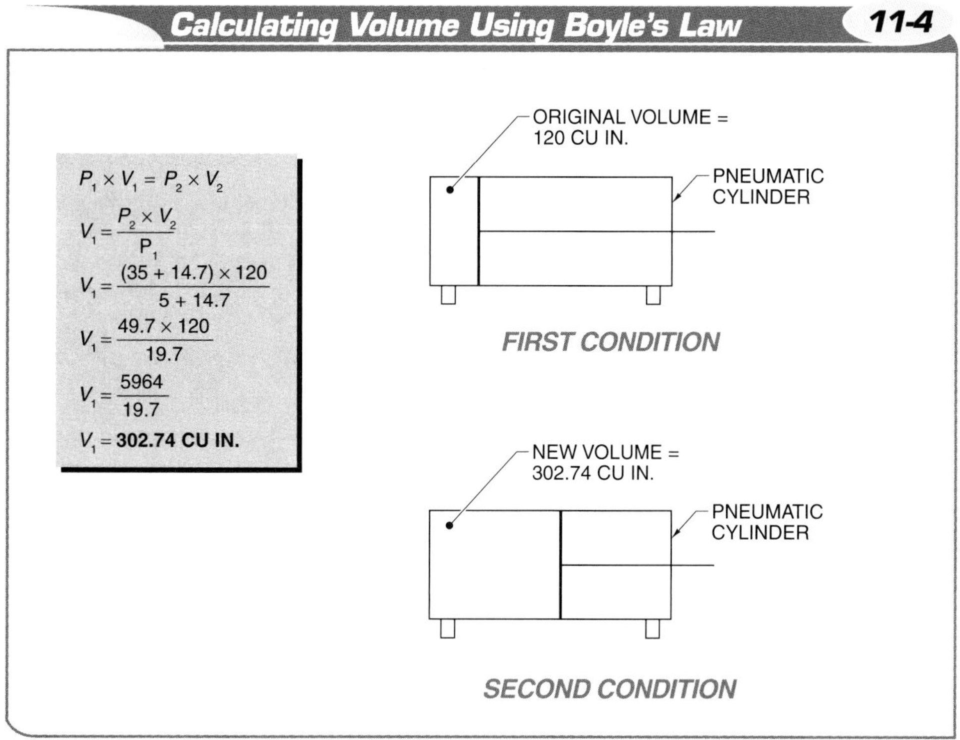

Figure 11-4. Boyle's law states that when temperature remains constant, there is an inverse relationship between pressure and the volume of a gas.

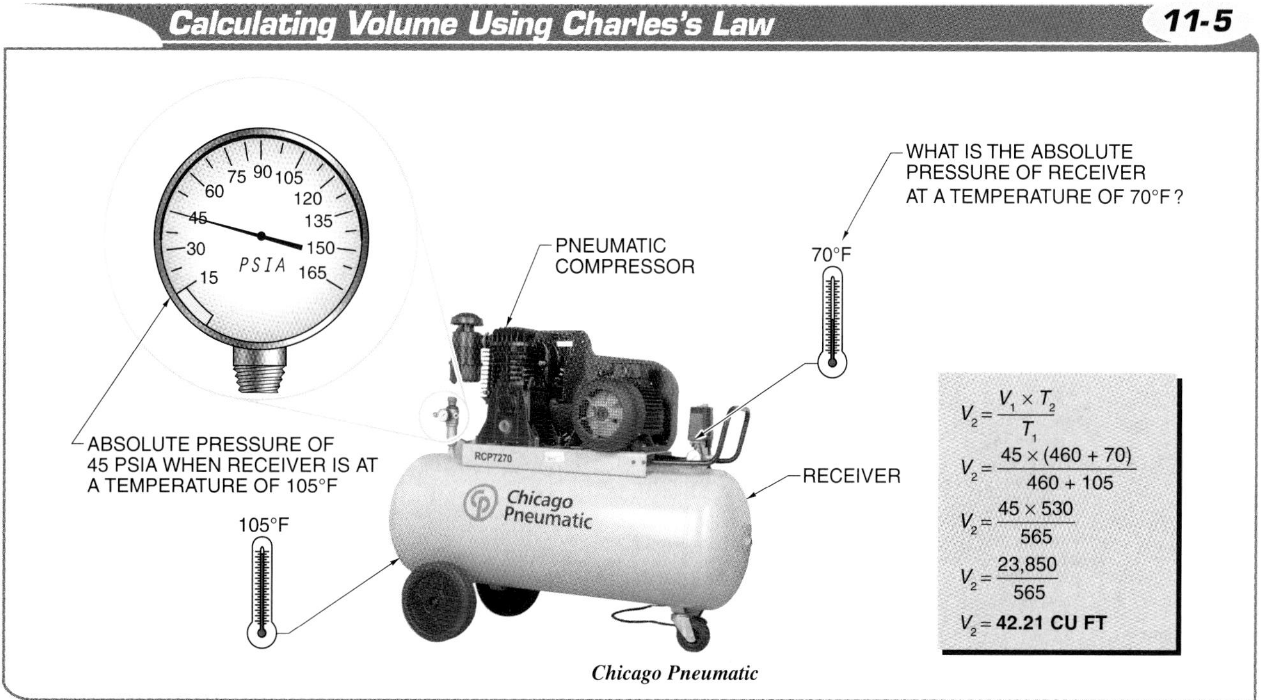

Figure 11-5. Charles's law states that when the pressure of a gas is constant, the volume of the gas has a direct relationship with its absolute temperature.

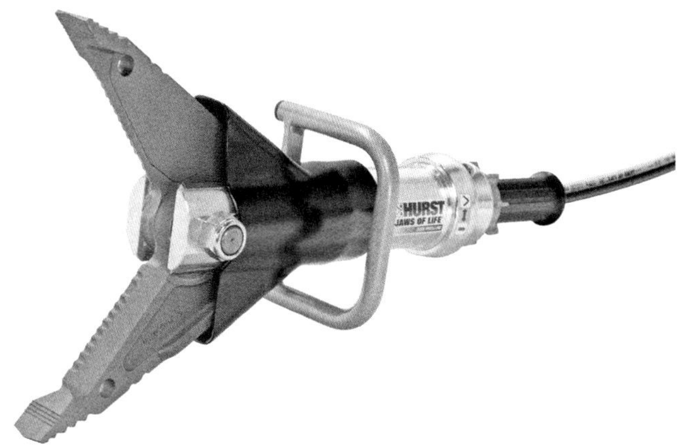

Hurst Jaws of Life

Pneumatic power is used for equipment such as heavy-duty rescue equipment.

TERMS

Gay-Lussac's law is a gas law that states that when the volume of a gas is held constant, the pressure exerted by the gas has a direct relationship to its absolute temperature.

Gay-Lussac's Law

Gay-Lussac's law is a gas law that states that when the volume of a gas is held constant, the pressure exerted by the gas has a direct relationship to its absolute temperature. Gay-Lussac's law can be applied to find pressure when there are temperature changes, or to find temperature when there are pressure changes. Gay-Lussac's law is expressed by the following formula:

$$P_2 = \frac{P_1 \times T_2}{T_1}$$

where

P_2 = pressure of the second condition (in psia)

P_1 = pressure of the first condition (in psia)

T_2 = temperature of the second condition (in °R)

T_1 = temperature of the first condition (in °R)

When heat is applied to a gas in a receiver and the volume of the gas is constant, the pressure of the gas increases. When the gas in the receiver cools and the volume of the gas is constant, the pressure of the gas decreases.

Example—Calculating Pressure: What is the new pressure in a receiver that had a pressure of 120 psi at 75°F if the temperature is increased to 110°F? (Any unknown value can be found by rearranging the given formula.) **See Figure 11-6.** *Note:* Convert psi to psia and °F to °R.

$$P_2 = \frac{P_1 \times T_2}{T_1}$$

$$P_2 = \frac{(120 + 14.7) \times (110 + 460)}{75 + 460}$$

$$P_2 = \frac{134.7 \times 570}{535}$$

$$P_2 = \frac{76{,}779}{535}$$

$$P_2 = \mathbf{143.51\ psia}$$

Note: 143.51 psia – 14.7 = 128.81 psi

TECH FACT

The French scientist Joseph Louis Gay-Lussac based his gas laws on earlier work performed by Jacques Charles and gave him credit when he first published his gas laws in 1802.

Combined Gas Law

The relationship between pressure, volume, and temperature is best determined when the formulas from Boyle's, Charles's, and Gay-Lussac's laws are combined. The combined gas law covers the values regarding the relationships between the pressure, volume, and temperature of a gas. The combined gas law is expressed by the following formula:

$$V_1 = \frac{P_2 \times V_2 \times T_1}{T_2 \times P_1}$$

where

V_1 = volume of the first condition (in cu in.)

P_2 = absolute pressure of the second condition (in psia)

V_2 = volume of the second condition (in cu in.)

T_1 = absolute temperature of the first condition (in °R)

T_2 = absolute temperature of the second condition (in °R)

P_1 = absolute pressure of the first condition (in psia)

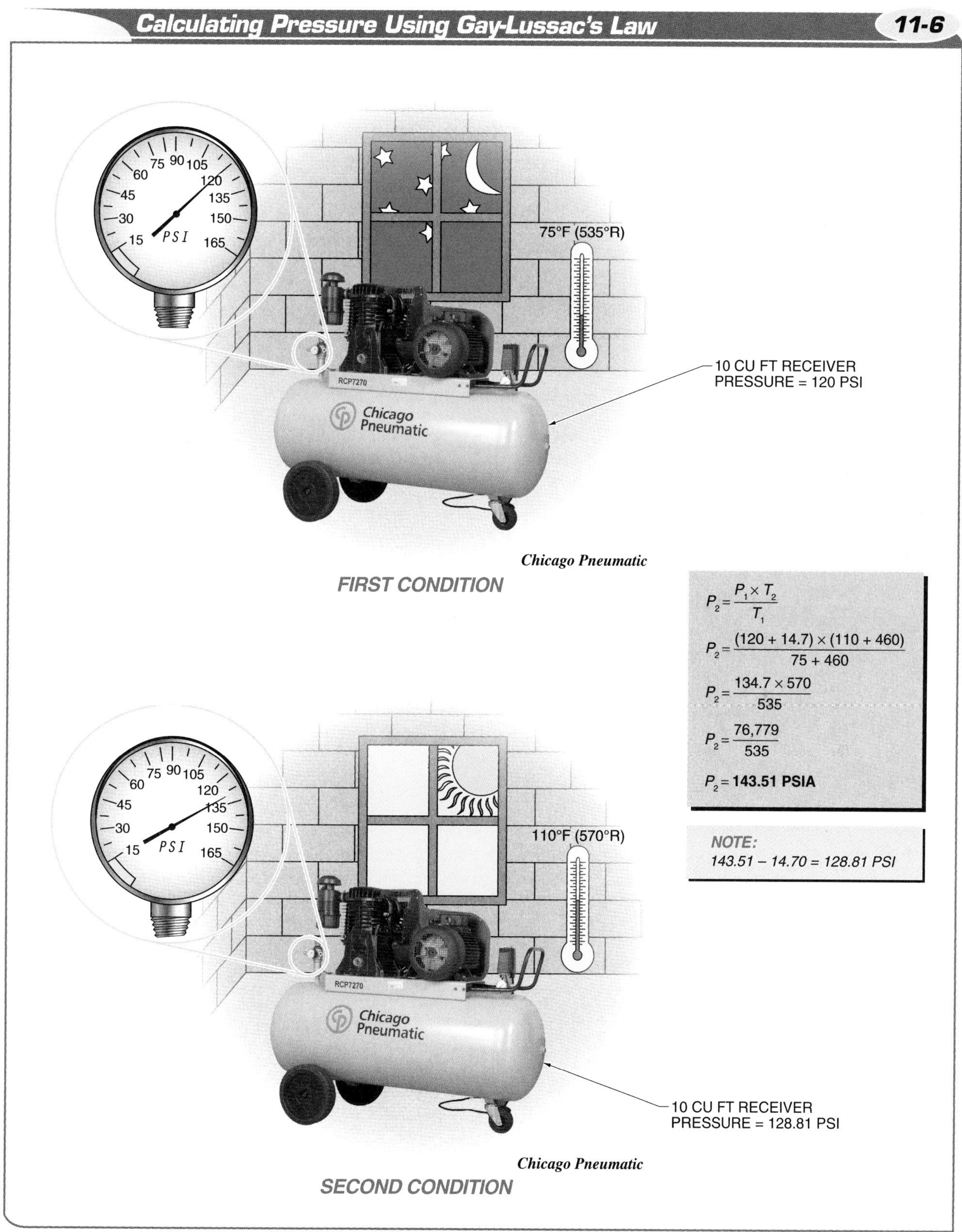

Figure 11-6. Gay-Lussac's law states that when the volume of a gas is held constant, the pressure exerted by the gas has a direct relationship to its absolute temperature.

As with the other gas laws, any unknown value can be found by rearranging the given formula. For example, if the two pressures, the two temperatures, and the new volume are known, the formula can be rearranged to solve for the original volume.

Example—Calculating Volume: A pneumatic system compressor increases its pressure from 110 psi to 135 psi. Temperature also increases from 65°F to 98°F. The new volume is calculated to be 155 cu in. What was the original volume? **See Figure 11-7.** *Note:* Convert psi to psia and °F to °R.

$$V_1 = \frac{P_2 \times V_2 \times T_1}{T_2 \times P_1}$$

$$V_1 = \frac{(135 + 14.7) \times 155 \times (65 + 460)}{(98 + 460) \times (110 + 14.7)}$$

$$V_1 = \frac{149.7 \times 155 \times 525}{558 \times 124.7}$$

$$V_1 = \frac{12{,}181{,}838}{69{,}583}$$

$$V_1 = \mathbf{175.07\ cu\ in.}$$

PNEUMATIC SYSTEMS

A pneumatic system is a system that uses gas under pressure to create movement. The air compressors in pneumatic systems are typically rated in standard cubic feet per minute (SCFM). The cylinders operated by pneumatic systems are generally used for low- to medium-force applications. Pneumatic systems are similar to hydraulic systems, but cannot handle heavy loads or generate as much force as hydraulic systems. However, pneumatic systems can operate at higher speeds than hydraulic systems.

Pneumatic systems are used in production manufacturing facilities, construction equipment, vehicle-service equipment, industrial equipment, maintenance equipment, and marine applications. **See Figure 11-8.** Common pneumatic tools include pneumatic torque wrenches, grinders, nail guns, and impact wrenches.

When proper precautions are taken, contamination levels are low in pneumatic equipment as opposed to hydraulic equipment because pneumatic systems use compressed air instead of hydraulic fluid. Pneumatic systems are used in food, pharmaceutical, and other related processing facilities for their cleanliness. Pneumatic tools are also used for medical and dental applications due to their low levels of contamination.

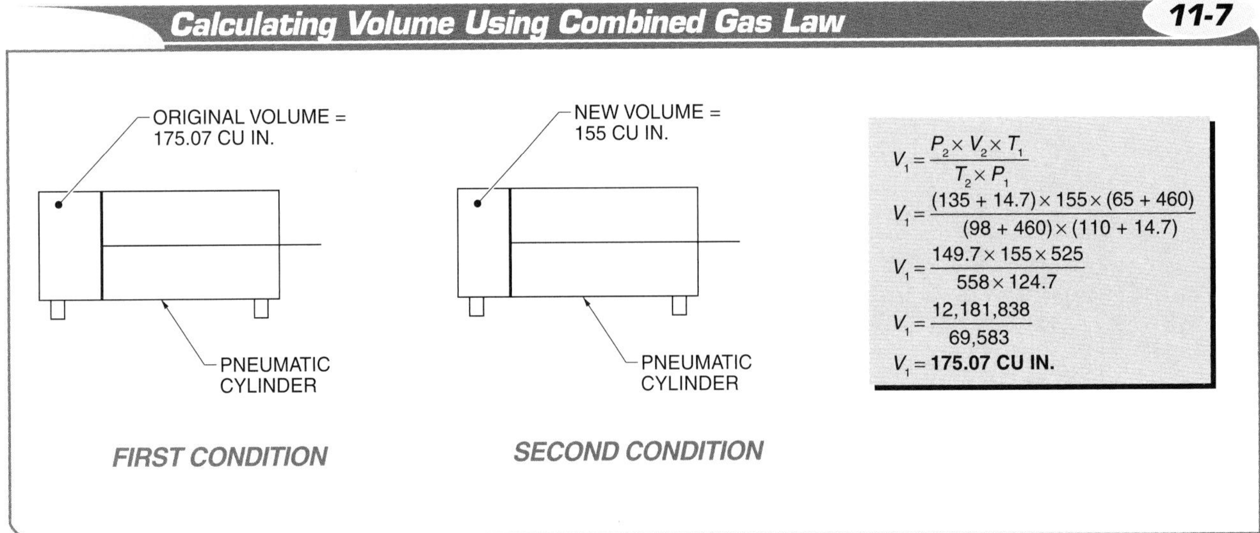

Figure 11-7. The combined gas law can be used to solve for any value, pressure, volume, or temperature, in any condition.

The production manufacturing industry is the largest user of pneumatic systems. Pneumatic systems are used in production manufacturing facilities to move conveyors, to rotate assembly fixtures, to provide vibration for feeding mechanisms, or to rotate mixing and agitating shafts. Pneumatic systems are also used to assemble products, to open or close machine safety guards, to operate automatic cutting blades, or to inject ink into a printer.

Pneumatic System Applications **11-8**

Chicago Pneumatic
PRODUCTION MANUFACTURING FACILITIES

Chicago Pneumatic
CONSTRUCTION EQUIPMENT

Chicago Pneumatic
VEHICLE-SERVICE EQUIPMENT

Saylor-Beall Manufacturing Company
INDUSTRIAL EQUIPMENT

Chicago Pneumatic
MAINTENANCE EQUIPMENT

Chicago Pneumatic
MARINE APPLICATION

Figure 11-8. Pneumatic systems are used in production manufacturing facilities, construction equipment, vehicle service equipment, industrial equipment, maintenance equipment, and marine applications.

TERMS

Air is a gas primarily composed of nitrogen and oxygen, with traces of other gases.

A **bellows** is a device that expands to draw air in through a flapper valve and contracts to expel the air through a nozzle.

An **air compressor** is a device that increases the pressure of a gas by mechanically decreasing its volume.

The gas that pneumatic systems typically use is air. *Air* is a gas primarily composed of nitrogen and oxygen, with traces of other gases. Air does not have to be separately manufactured when it is used as a fluid in a pneumatic system. Air is brought from and exhausted to the atmosphere and does not need to be piped back to a receiver. Also, air can be more environmentally friendly than hydraulic fluids. This makes using air an advantage of a pneumatic system.

There are also some disadvantages to pneumatic systems. Air is not capable of creating and/or sustaining the amount of power that hydraulic systems can provide. Air cannot be used to lift or push extremely heavy weights. For example, air cannot be used to operate medium- to large-crushing mechanisms.

Air compressors create flow for the pneumatic system. Air compressors also maintain pressure in a system by continuously creating a flow of air into the system.

Compressing Air

There are many types of gases, such as hydrogen, oxygen, nitrogen, carbon dioxide, propane, and butane. However, only certain gases are used for pneumatic systems. The most common gas used in pneumatic systems is air. Air composition is 78% nitrogen (N), 20.95% oxygen (O_2), 0.93% argon (Ar), and 0.12% carbon dioxide (CO_2) and trace amounts of other gases (including water vapor). **See Figure 11-9.**

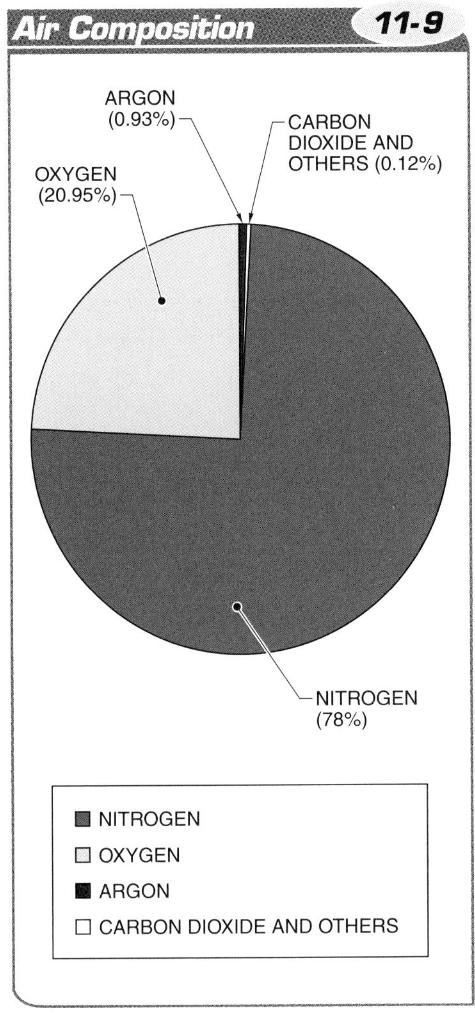

Figure 11-9. Air composition is 78% nitrogen (N), 20.95% oxygen (O_2), 0.93% argon (Ar), and 0.12% carbon dioxide (CO_2) and trace amounts of other gases.

BELLOWS

One of the earliest forms of a pneumatic system that used compressed air was the bellows. A *bellows* is a device that expands to draw air in through a flapper valve and contracts to expel the air through a nozzle. A bellows consists of two wooden handles and a flexible leather cover. It uses air from a sealed container to start or increase the heat of a fire.

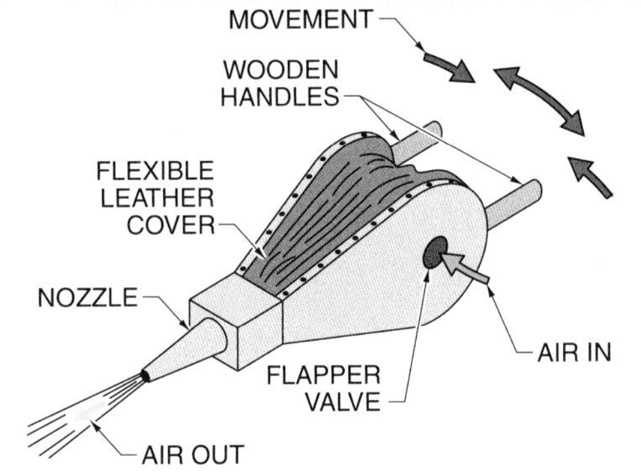

When air is compressed, it is trapped in a cylinder chamber and the volume of the chamber is decreased through mechanical means. For example, a manual air pump compresses air in a cylinder chamber. Manual air pumps are used for inflating devices with inflatable bladders that require a high internal air pressure, such as a basketball. **See Figure 11-10.**

When the handle of the manual air pump is pulled upwards, air fills the cylinder. When the handle is pushed downwards, the piston compresses the air in the cylinder chamber. The compressed air flows through a hose to the basketball.

Air compressors operate similarly, except in larger volumes. An *air compressor* is a device that increases the pressure of a gas by mechanically decreasing its volume. Air compressors are used to compress air for use in pneumatic systems. **See Figure 11-11.** The atmosphere pushes air into the cylinder as the piston retracts and then compresses the air as the piston extends.

The cylinder inlet allows air to be pushed into the cylinder as a vacuum is created inside the cylinder by the piston. The vacuum is created by increasing the volume in the cylinder chamber. Atmospheric pressure then forces air to enter the cylinder through the inlet. As the piston extends, the intake valve blocks the air from escaping through the inlet, and all of the air is forced past the open discharge valve through the outlet. Once the air has been compressed, the resulting airflow is ready to accomplish work.

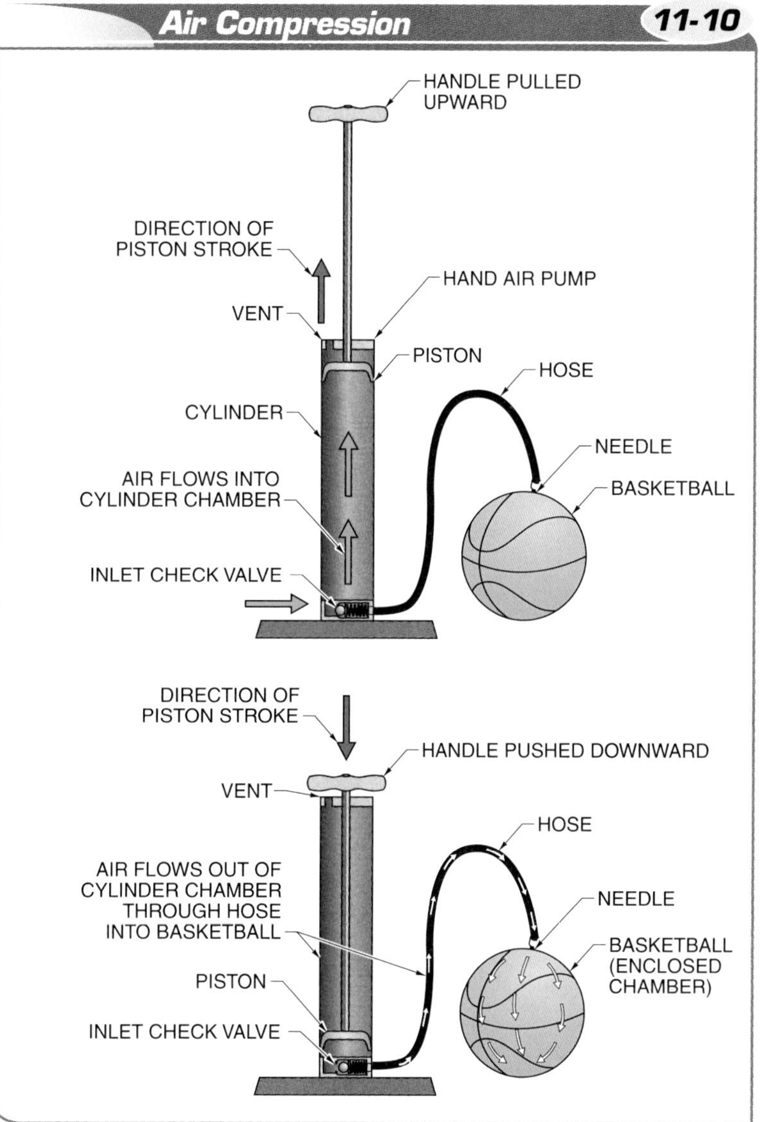

Figure 11-10. A manual air pump compresses air into an enclosed chamber.

Airflow

Airflow is the movement of air in a pneumatic system through pipes, tubing, and hoses that transmit energy produced by the air compressor to the location where it is needed to accomplish work. Airflow occurs when there is a pressure differential between two points in a closed system. Airflow and pressure differential are directly proportional. A higher pressure differential results in more airflow. A lower pressure differential results in less airflow.

For example, it is easy to inflate a completely deflated basketball using a manual air pump because there is a large pressure differential. As the basketball begins to fill with air and the pressure starts to increase, the pressure differential decreases. It becomes more difficult to create airflow into the basketball as the pressure in the basketball increases. Eventually, the pressure in the basketball will be the same as the pressure that the pump can produce. When there is no pressure differential, airflow stops because the piston in the pump cannot be forced down without damaging the ball.

TERMS

Airflow is the movement of air in a pneumatic system through pipes, tubes, and hoses that transmit energy produced by the air compressor to the location where it is needed to accomplish work.

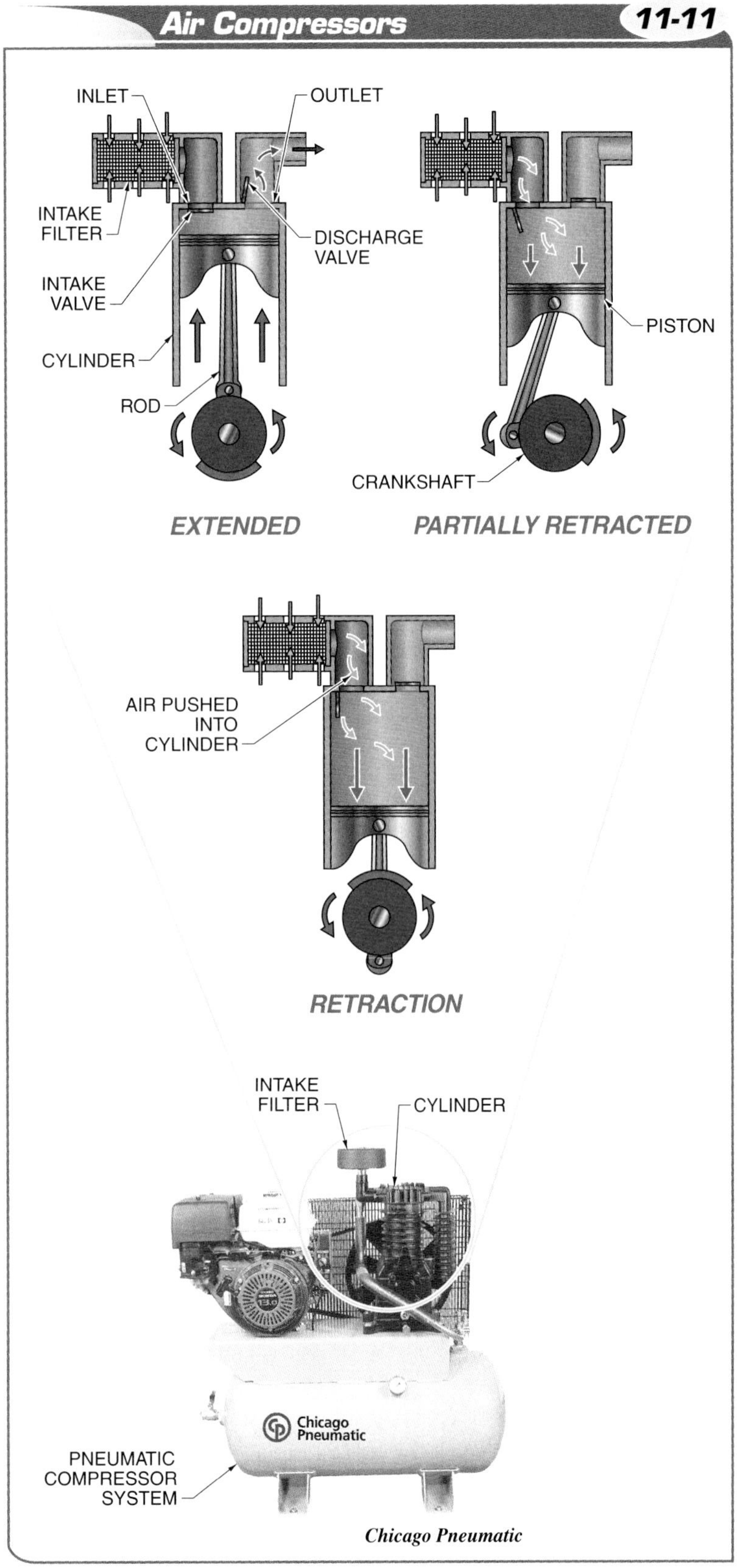

Air Compressors 11-11

INLET, OUTLET, INTAKE FILTER, DISCHARGE VALVE, INTAKE VALVE, CYLINDER, ROD, PISTON, CRANKSHAFT

EXTENDED — PARTIALLY RETRACTED

AIR PUSHED INTO CYLINDER

RETRACTION

INTAKE FILTER, CYLINDER, PNEUMATIC COMPRESSOR SYSTEM

Chicago Pneumatic

Figure 11-11. An air compressor allows air into the cylinder as the piston retracts, and then compresses the air as the piston extends.

The same principle can be applied to pneumatic systems. Rather than pumping a basketball, an air compressor fills the receiver and system piping with air. Initially, the air compressor easily fills the receiver. But as the pressure differential begins to decrease, the air compressor must work harder. Once the receiver is filled, the air compressor stops. When pneumatic systems must perform work, the receiver and/or system piping releases their stored energy, and air flows through the supply line because of a pressure differential in the piping.

A large pipe called the supply line distributes air from the receiver throughout the pneumatic system to one or more pieces of pneumatic equipment. Sometimes in a pneumatic system, one piece of pneumatic equipment is operating and another piece of pneumatic equipment is idle. The operating equipment has a greater pressure differential because it has flow. Because of the greater pressure differential, the supply line to the operating equipment has more airflow than the supply line to the idle equipment. **See Figure 11-12.**

Airflow Rate. Air in piping flows to its destination because of the pressure differential between the compressor(s) and the destination of the air. When there is airflow, air molecules do not float in a chaotic manner, but in a straight path through a pipe. The air closest to the pipe wall flows the slowest because of the resistance created between it and the pipe wall. The air in the center of the pipe flows the fastest because it has the least amount of resistance.

Rating flow for pneumatic systems is more difficult than with hydraulic systems because air is compressible. Airflow is rated as the amount of cubic feet that can pass a given point in one minute, or cubic feet per minute (cfm). One disadvantage to using this type of unit for measuring flow is that the volume of the air changes when it is compressed, becoming smaller under pressure. For example, 5 cfm could have a various range of volumes, depending on the amount of pressure it is operating under.

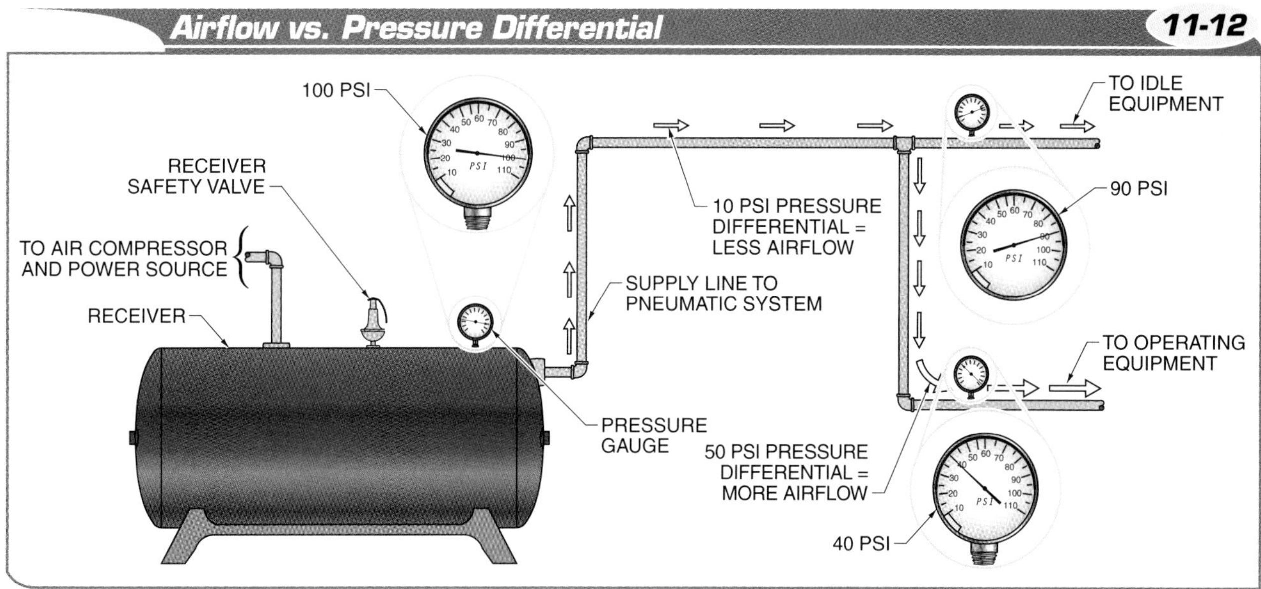

Figure 11-12. When the pressure differential increases in a pneumatic system, there is also an increase in airflow.

Cfm Ratings

Pneumatic system air compressors are rated in standard cubic feet per minute (scfm). The number of cubic feet per minute is determined by the amount of air pushed into the inlet of the air compressor. Since air can have a various range of volumes once it has been compressed, standard cubic feet per minute is used as a unit of measure because it is the only method that gives a consistent rating.

The number of standard cubic feet per minute does not change once system pressure begins to increase. For this reason, many air compressors are rated at two different pressures. For example, an air compressor can have a rating of 5.2 cfm at 40 psi and 4.0 cfm at 90 psi. Other air compressors have ratings of cubic feet per minute at maximum pressure.

Standard Cubic Feet per Minute. The air that is pushed into the inlet of the air compressor can vary at each facility. The condition of the air can also change from day to day. Because the condition of air constantly varies, standard cubic feet per minute measurement is used. *Standard cubic feet per minute (scfm)* is a unit of measure that assumes a constant atmospheric environment for air before it is pushed into the system.

There are differences between standard cubic feet per minute measurements and cubic feet per minute measurements. In a standard cubic feet per minute measurement, it is assumed that the air in the environment is at 68°F, 36% humidity, and has a pressure of 14.7 psia (sea level). When using a cubic feet per minute measurement, the airflow rate is measured at different pressures and is lower when the pneumatic system is operating at higher pressures.

System Air Speed. The speed of air in a system is incredibly fast and is measured in feet per second (fps). The top speed of air is equal to the speed of sound (1125 fps). The fastest air speed in any pneumatic system, however, should be no more than 600 fps. Airflow rate (cfm) is determined by the pressure differential between two points in a pneumatic system. The higher the flow rate, the faster air flows to its destination. The lower the flow rate, the slower air flows to its destination.

Another variable for the speed of air in a system is the size of the pneumatic pipe. The larger the diameter of the piping, the slower the air speed (velocity). The smaller diameter of the piping, the faster the air velocity.

TERMS

Standard cubic feet per minute (scfm) is a unit of measure that assumes a constant atmospheric environment for air before it is pushed into the system.

Restrictions to Airflow. Restrictions to airflow in a pneumatic system make it more difficult to send a volume of air at the required flow rate (cfm) to its destination. **See Figure 11-13.** Pneumatic systems have several different restrictions to airflow, including the following:

- resistance between airflow and piping walls
- bends in the piping
- reduction in piping diameter
- reduction in orifice sizes in pneumatic components

Pneumatic systems are not considered energy efficient, and it is important to minimize waste whenever possible. When airflow is restricted, the wasted energy is converted into heat. Heat in a pneumatic system is one of the biggest wastes of energy.

It is not possible to eliminate all sources of restriction in a pneumatic system. However, the following steps can be taken to help lower the amount of energy wasted:

- Install properly sized piping to keep velocities under 600 fps.
- Avoid bends in piping whenever possible. In some systems, 90° bends can be replaced with two 45° bends to decrease the restrictions to airflow and increase laminar flow.
- Locate and eliminate leaks in system piping, components, and valving.
- Control the speed of the air compressor when it is first turned on by using a speed control device, such as a variable frequency drive (VFD), to force the speed of the electric motor to increase over a predetermined time.
- Replace compressors and aftercoolers with new, energy-efficient models.

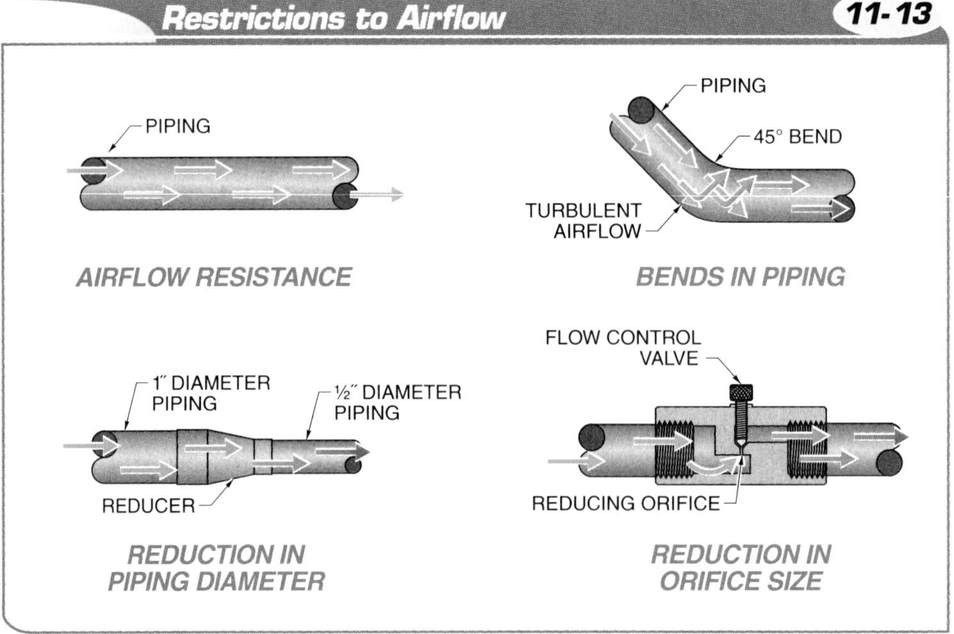

Figure 11-13. Restrictions to airflow in a pneumatic system reduce the volume of air and its speed.

Digital Resources

Chapter Review

Pneumatic System Fundamentals

Chapter 11

Name: ______________________________ Date: ______________

MULTIPLE CHOICE

______________ **1.** A(n) ___ is a device that increases the pressure of a gas by mechanically decreasing its volume.

A. air compressor
B. cylinder
C. piston
D. receiver

______________ **2.** The fastest air speed in any pneumatic system should be no more than ___ fps.

A. 200
B. 400
C. 600
D. 1000

______________ **3.** Absolute ___ is the hypothetical temperature at which molecular motion ceases, precisely –273.15°C (–459.67°F).

A. airflow
B. pressure
C. temperature
D. zero

______________ **4.** A(n) ___ is a specialized vessel that provides storage of gas for use in a pneumatic system.

A. air tank
B. cylinder
C. receiver
D. reservoir

______________ **5.** ___ law states that when the pressure of a gas is constant, the volume of the gas has a direct relationship with its absolute temperature.

A. Boyle's
B. Charles's
C. Gay-Lussac's
D. The combined gas

______________ **6.** ___ law that states that when the volume of a gas is held constant, the pressure exerted by the gas has a direct relationship to its absolute temperature.

A. Boyle's
B. Charles's
C. Gay-Lussac's
D. The combined gas

_______________ **7.** In a standard, cubic feet per minute is measurement; it is assumed that air in the environment is at ___°F, and ___% humidity at 14.7 psia (sea level).

A. 68; 36
B. 70; 40
C. 72; 40
D. 75; 36

_______________ **8.** ___ law states that when temperature remains constant, there is an inverse relationship between pressure and the volume of a gas.

A. Boyle's
B. Charles's
C. Gay-Lussac's
D. The combined gas

_______________ **9.** Temperature in °R is always approximately ___° greater than the temperature in °F.

A. 32
B. 160
C. 360
D. 460

_______________ **10.** Air compressors in pneumatic systems are rated in ___.

A. cubic inches (cu in.)
B. cubic inches per minute (cim)
C. pounds per square inch (psi)
D. standard cubic feet per minute (scfm).

COMPLETION

_______________ **1.** ___ is the movement of air in a pneumatic system through pipes, tubing, and hoses that transmit energy produced by the air compressor to the location where it is needed to accomplish work.

_______________ **2.** ___ is a unit of measure that assumes a constant atmospheric environment for air before it is pushed into a system.

_______________ **3.** A(n) ___ system uses a gas under pressure to create movement.

_______________ **4.** When ___ is compressed, it is trapped in a cylinder chamber and the volume of the chamber is decreased through mechanical means.

_______________ **5.** Absolute pressure is always ___ psi greater than gauge pressure.

_______________ **6.** Pneumatic systems operate on the principles of ___, which state that a fluid can flow, has no definite shape, and increases in volume with an increase in temperature.

_______________ **7.** ___ is a gas primarily composed of nitrogen and oxygen, with traces of other gases.

_______________ **8.** In a pneumatic system, ___ is rated as the amount of cubic feet that can pass a given point in one minute, or cubic feet per minute (cfm).

______________ 9. Absolute pressure is the sum of gauge pressure and ___ pressure.

______________ 10. The number of cubic feet per minute in an air compressor is determined by the amount of air pushed into the ___ of the air compressor.

______________ 11. The ___ gas law covers the values regarding the relationships between the pressure, volume, and temperature of a gas.

______________ 12. The term "gas" is derived from the Greek word for "___."

______________ 13. ___ temperature is any temperature on a scale that begins with absolute zero.

______________ 14. ___ gas law can be applied to find pressure when there are temperature changes, or to find temperature when there are pressure changes.

______________ 15. Molecules have no movement at ___°R.

TRUE/FALSE

T F 1. The larger the diameter of the piping in a pneumatic system, the slower the air speed.

T F 2. Cylinders operated by pneumatic systems are generally used for low- to medium-force applications.

T F 3. If the gas inside of a receiver cools and the pressure remains constant, then its volume decreases.

T F 4. The gas that pneumatic systems typically use is a mixture of only nitrogen and oxygen.

T F 5. The top speed of air is equal to the speed of sound.

T F 6. In pneumatic system piping, the air closest to the pipe wall flows the fastest because of the resistance created between it and the pipe wall.

T F 7. Airflow and pressure differential are directly proportional.

T F 8. If the pressure increases and the temperature remains constant, then the volume of a gas increases.

T F 9. The volume of a gas is determined by the volume of the container that it is in.

T F 10. The relationship between pressure, volume, and temperature is best determined when the formulas from Boyle's, Charles's, and Gay-Lussac's laws are combined.

T F 11. Molecules continue to be in motion at 0°F and at 0°R.

T F 12. The smaller the diameter of the piping in a pneumatic system, the faster the air speed.

T F 13. A high pressure differential within a pneumatic system results in more airflow.

T F 14. Bends in pneumatic piping systems should be avoided whenever possible.

T F 15. When airflow is restricted, wasted energy is converted into cool air.

T F **16.** As the pressure differential in a pneumatic system begins to decrease, the air compressor that is part of the system must work harder.

T F **17.** Pneumatic tools are often used for medical and dental applications due to their low levels of contamination.

T F **18.** Gas molecules are not easily compressed.

T F **19.** Airflow occurs when there is no pressure differential between two points in a closed system.

T F **20.** With all gas laws, any unknown value can be found by rearranging the given formula.

T F **21.** Pneumatic systems are considered energy efficient.

T F **22.** If heat is applied to the gas inside a receiver and the pressure remains constant, then the volume of the gas inside the receiver increases.

T F **23.** Many air compressors are rated at one pressure because the number of standard cubic feet per minute can change once system pressure begins to increase.

T F **24.** When proper precautions are taken, contamination levels are low in pneumatic equipment as opposed to hydraulic equipment because pneumatic systems use compressed air instead of hydraulic fluid.

T F **25.** Gauge pressure is always 14.7 psi greater than absolute pressure.

SHORT ANSWER

1. Describe the main disadvantage of using a pneumatic system over a hydraulic system.

2. List four restrictions to airflow in a pneumatic system.

3. List six characteristics of a gas.

4. List four steps that can be taken to help reduce the amount of energy wasted in a pneumatic system.

5. Describe the main difference between standard cubic feet per minute measurements and cubic feet per minute measurements.

Activities

Pneumatic System Fundamentals

Chapter 11

Activity 11-1: Pneumatic System Pressure

A production manufacturing facility with two existing paint lines has installed a third paint line. Each paint line requires pneumatic power to operate. The third paint line will be located in an area of the facility that does not have a pneumatic power source. A 1″ inside diameter pressurized pneumatic pipeline (composed of a 2′ long section, 50′ long section, and 200′ long section) must be installed and routed to the third paint line from the existing receiver. Use Pneumatic Paint Line System to answer the following questions:

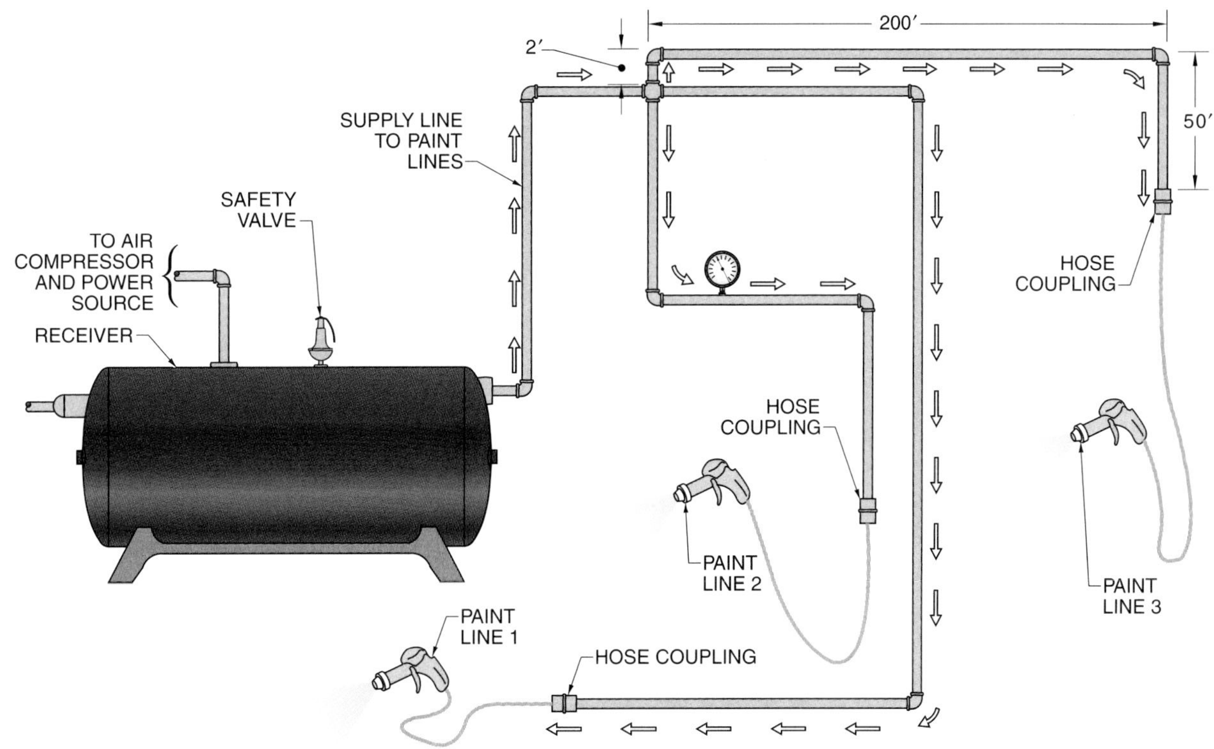

PNEUMATIC PAINT LINE SYSTEM

____________________ **1.** If the total volume of the receiver and original pipeline sections (V_1) is 8000 cu in., what is the total volume (V_T) of the receiver, original pipelines, and three additional pipeline sections?

____________________ **2.** Assuming that the ambient temperature remains constant and that the pressure in the receiver is 110 psi, what occurs with system pressure when the valve to the new section is open?

____________________ **3.** How much pressure will be in the system once the valve is opened?

4. What overall affect will adding pressurized lines have on the existing pneumatic compression system?

Activity 11-2: Increasing Pneumatic System Efficiency

A plastics injection molding facility must increase the energy efficiency of their pneumatic system. The current system is a grid system.

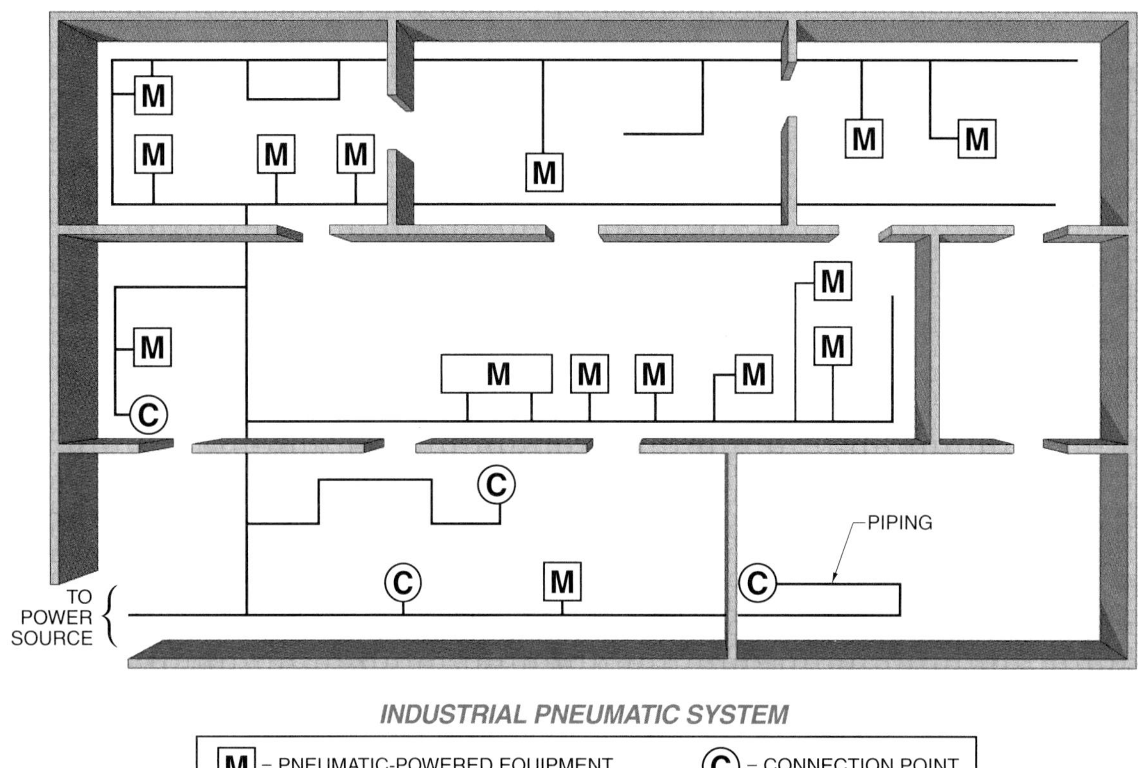

INDUSTRIAL PNEUMATIC SYSTEM

1. Redraw Industrial Pneumatic System so that it is more energy efficient.

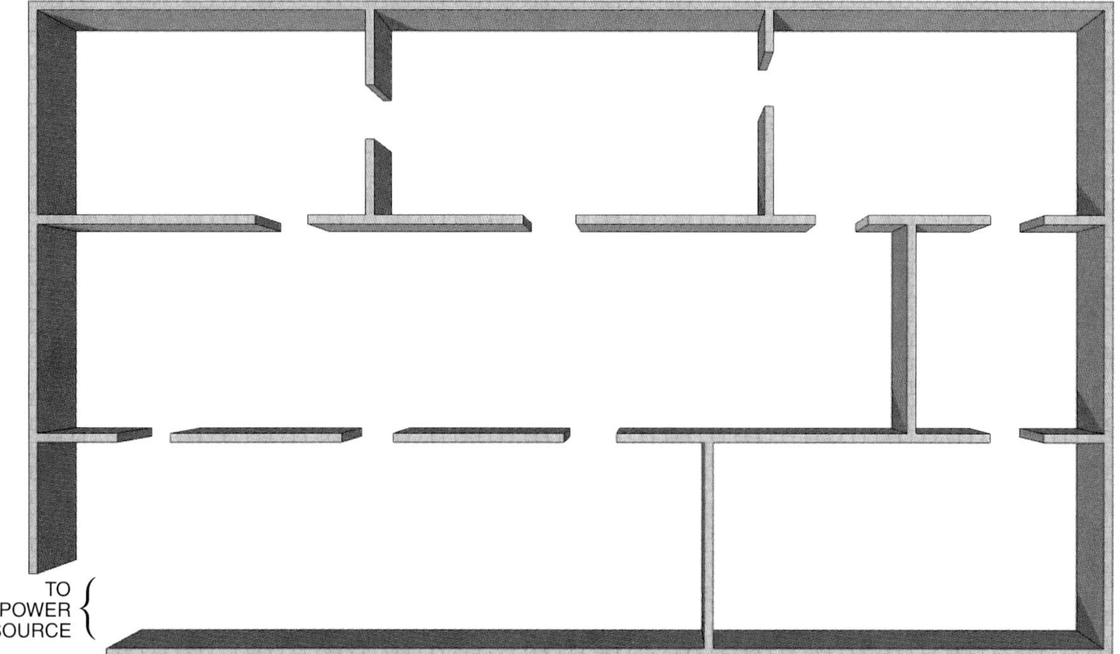

2. What could be done to the new system to keep water condensation out of the system?

Activity 11-3: Calculating Pressure

1. What is the new pressure in a receiver that currently has pressure of 160 psi at 60°F if the temperature is increased to 80°F? *Note:* Convert psi to psia and °F to °R.

Activity 11-4: Calculating Volume

1. An accumulator increases its pressure from 135 psi to 175 psi. The temperature also increases from 60°F to 82°F. The new volume is calculated to be 152 cu in. What was the original volume of the accumulator? *Note:* Convert psi to psia and °F to °R.

Pneumatic System Compression and Control

Chapter 12

OBJECTIVES

- Describe the different types of air compressors.
- Define the different types of control valves used to control compressed air.
- Describe the different types of pneumatic cylinders.
- Describe the different types of air motors.
- Explain how the different types of vacuum cups are used.

INTRODUCTION

A compressor, aftercooler, and the pneumatic components they supply with compressed air are vital to the operation of equipment for production manufacturing, material handling, assembly, and maintenance. Various flow, directions, and pressure control valves are used to operate pneumatic systems. Properly operated pneumatic systems run industrial and commercial equipment efficiently and dependably.

PNEUMATIC SYSTEMS

Pneumatic systems use gas under pressure to accomplish work. The main components of a pneumatic system are a power source (prime mover), an air compressor, and a receiver. **See Figure 12-1.**

Facilities that use pneumatic systems have at least one compressor. A facility with a single compressor system uses a large centralized loop system. A facility with multiple compressors uses smaller decentralized loop systems that provide the entire facility with compressed air, depending on its size and needs. **See Figure 12-2.**

An advantage of installing a single large compressor is that compressed air can be easily piped to where it is needed in a facility. A disadvantage of using a single large compressor is that it supplies the entire facility with compressed air, requiring the entire facility to be shutdown if there is a problem.

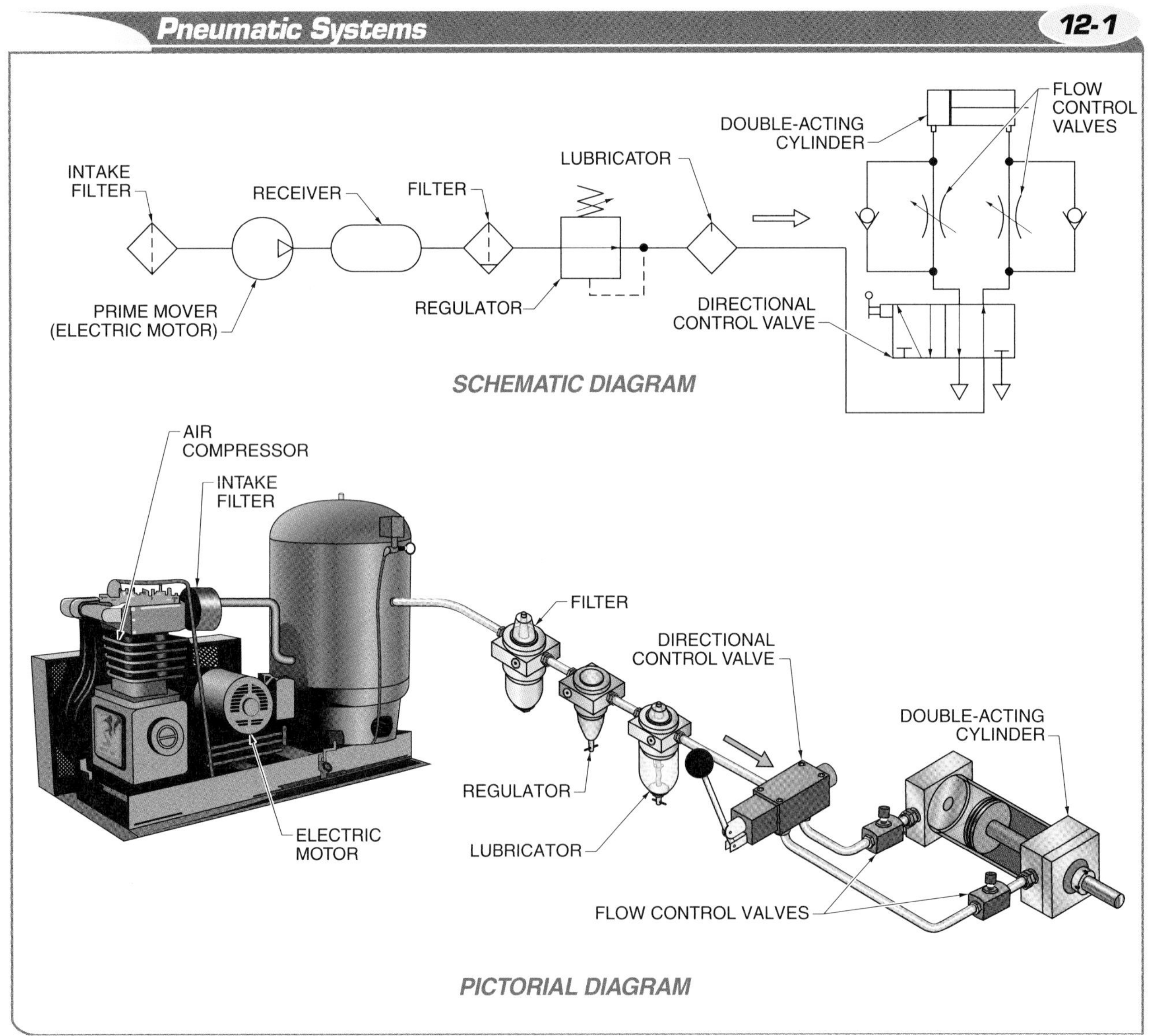

Figure 12-1. Pneumatic systems contain components that compress and store the air that is used to increase system pressure to accomplish work.

The advantage of having multiple compressors in smaller decentralized loops is that it is easier to control the air being sent to different parts of the facility. For example, when one section of the facility is not in operation, the compressor that serves that particular section can be shut down to reduce operating costs. A disadvantage of using multiple compressors is that there is a higher initial installation cost, although that cost is recouped over time. Another disadvantage is that the facility may have to perform more repairs.

Prime Movers

A *prime mover* is a device that supplies rotating mechanical energy to a fluid power system. Prime movers used with compressors are electric motors or internal combustion engines. However, internal combustion engines are used only for specific applications, such as when the location of the system makes it impossible to supply electricity for an electric motor. Prime movers are always attached to the air compressor.

The prime movers that are most often used with industrial and commercial compressors are electric motors. Alternating current (AC) motors are the most common type of electric motor. Pneumatic compressor systems that use AC voltage can vary in size. Small and/or portable pneumatic compressor systems (2.2 cfm to 19.0 cfm at 95 psi to 125 psi) use 1ϕ AC electric motors as prime movers. For example, small and/or portable pneumatic compressor systems are used to supply power (1 HP to 3 HP) to air-powered hand tools for light work applications.

Industrial facilities with pneumatic systems that require large amounts of compressed air typically use large compressors. These compressors include large air compressors (12,000 cfm to 500,000 cfm at 100 psi to 140 psi) and 3ϕ AC electric motors as prime movers. When a compressor uses a 3ϕ AC electric motor as its prime mover, the compressor will supply more airflow than a compressor with a 1ϕ motor.

Although internal combustion engines are not as commonly used as electric motors, they can be prime movers for certain types of pneumatic equipment. Internal combustion engines are used as the prime mover in applications where electricity is not readily available. For example, they are used in mobile construction equipment for road construction projects, such as jackhammers, rivet breakers, and different types of drills.

TERMS

A **prime mover** is a device that supplies rotating mechanical energy to a fluid power system.

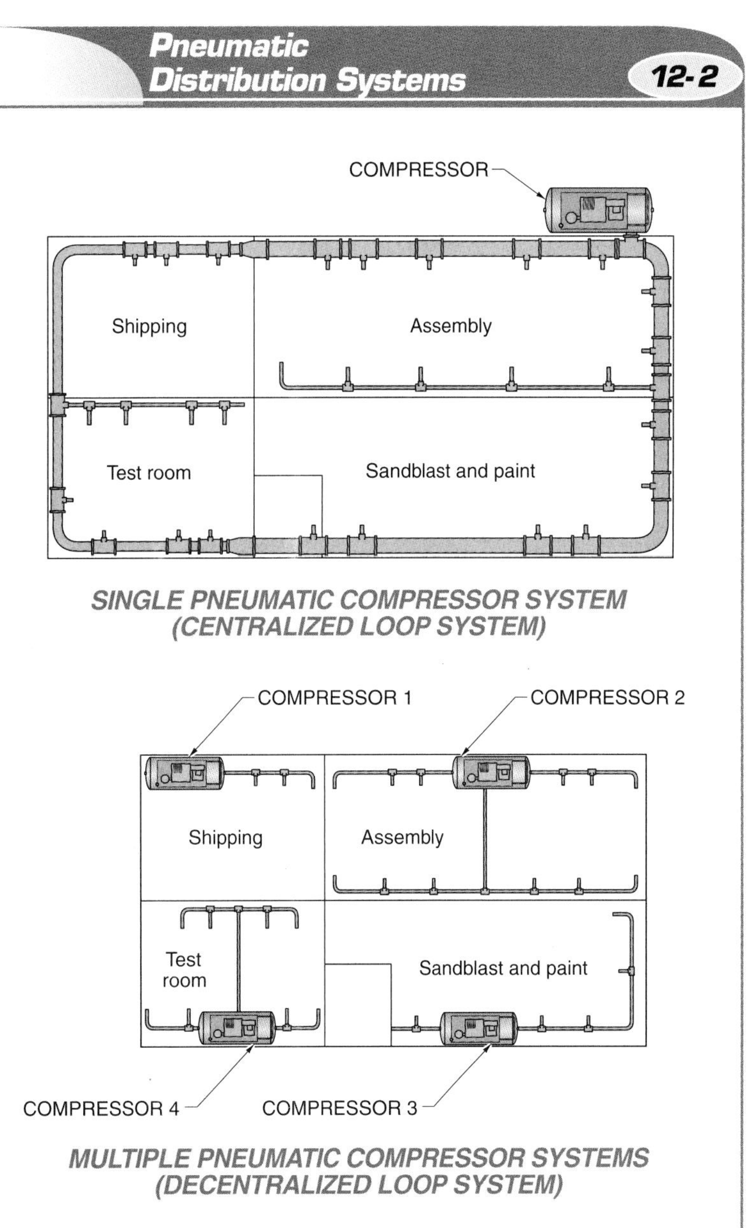

Figure 12-2. A single large compressor or multiple small compressors can be installed depending on the needs of a facility.

TERMS

A **nonpositive-displacement air compressor** is a type of dynamic air compressor that moves large volumes of air using high rotational speeds.

A **centrifugal air compressor** is a type of dynamic air compressor that uses an impeller rotating at high speed to compress air.

Air Compressor Types

Air compressors are either nonpositive displacement or positive displacement. **See Figure 12-3.** A *nonpositive-displacement air compressor* is a type of dynamic air compressor that moves large volumes of air using high rotational speeds. The most common type of nonpositive-displacement air compressor is a centrifugal air compressor. A *centrifugal air compressor* is a type of dynamic air compressor that uses an impeller rotating at high speed to compress air. The rotation of the impeller increases air speed, and the diffuser of the compressor changes the speed of the air into pressure and flow.

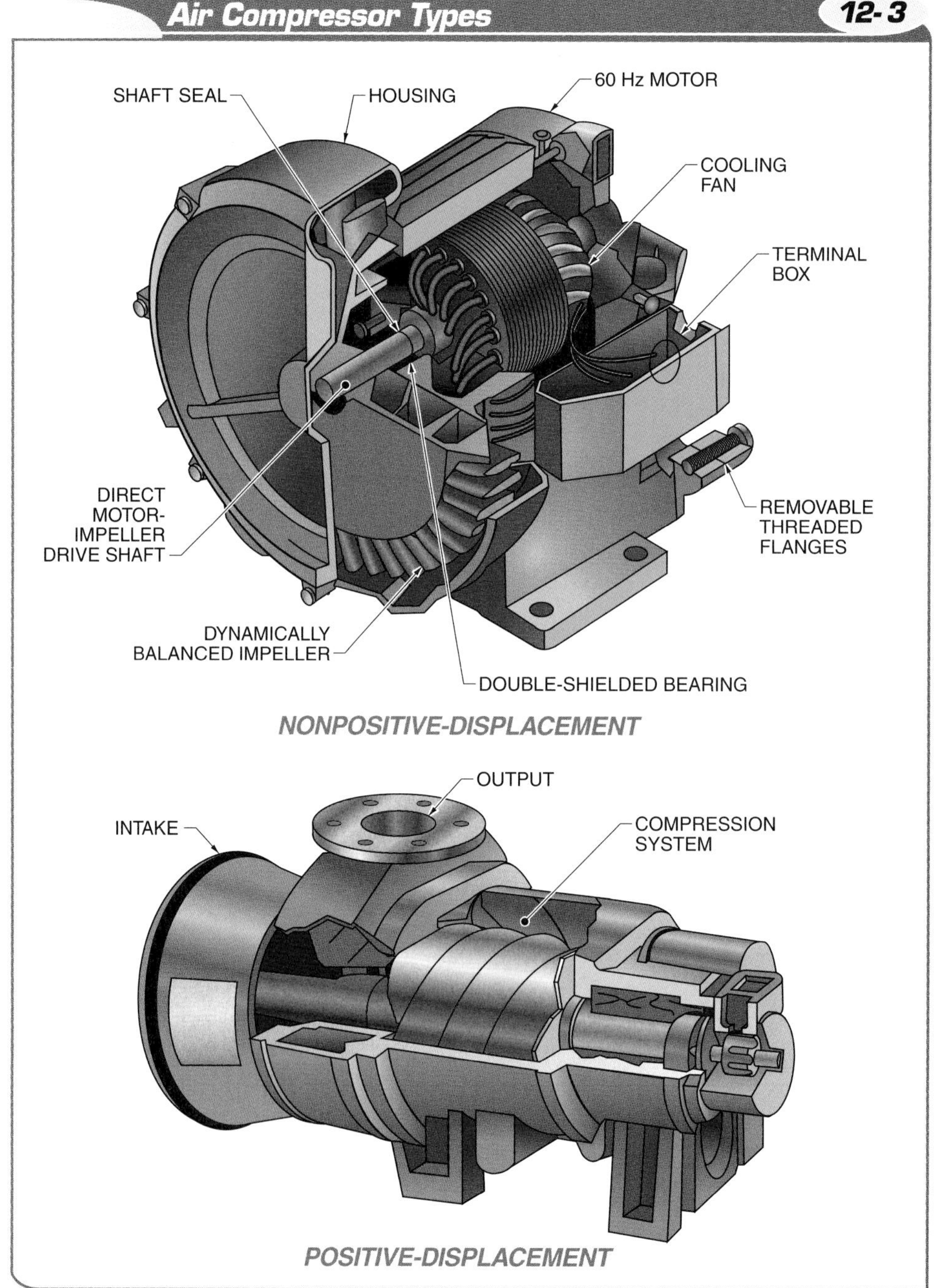

Figure 12-3. Air compressors are either nonpositive displacement or positive displacement.

Centrifugal air compressors are low-pressure, high-flow air compressors. They are capable of producing up to 500,000 cfm of airflow. The most common method used to control the capacity of a centrifugal air compressor is to adjust the inlet guide vanes. By closing the inlet guide vanes, the volume of airflow and air compressor capacity is reduced. A centrifugal air compressor also has shaft seals and atmospheric vents to separate the air from any oil-lubricated components.

A *positive-displacement air compressor* is a type of air compressor that discharges a fixed quantity of compressed air with each cycle. A positive-displacement air compressor also sends the air into the system piping after it has been compressed. The main types of positive-displacement compressors are reciprocating-piston, rotary-vane, and rotary-screw air compressors.

Reciprocating-Piston Air Compressors. A *reciprocating-piston air compressor* is a type of positive-displacement air compressor that compresses air by extending and retracting a piston inside a cylinder. A prime mover provides the mechanical force that moves the piston. A reciprocating-piston air compressor can be single-stage or multistage.

A reciprocating-piston air compressor consists of a crankshaft, connecting rod, piston, piston rings, cylinder, inlet valves, output valves, and unloading valves. **See Figure 12-4.** A reciprocating-piston air compressor operates through the following procedure:

1. The prime mover rotates the crankshaft. As the crankshaft rotates, the connecting rod moves downward and retracts the piston.
2. As the piston retracts, a vacuum is created inside the cylinder. A pressure differential is created because the pressure inside the cylinder is less than atmospheric pressure.
3. Due to the pressure differential, air flows into the cylinder through an inlet valve. The spring-loaded inlet valve traps the air inside the cylinder.
4. When the piston is completely retracted, the cylinder contains as much air as possible. The connecting rod then begins to move upward, extending the piston and compressing the air that has been trapped inside the cylinder.
5. The air cannot escape through the suction line because the compressed air closes the inlet valve. The compressed air is forced through the outlet valve instead. It flows through the discharge line either to a receiver or to piping to accomplish work.
6. When the crankshaft and connecting rod have completed a 360° rotation and the piston has both retracted and extended, the process repeats. When the piston retracts, air is pushed into the cylinder through the inlet valve. The output valve closes to prevent air from the system from traveling back into the compressor cylinder.

TERMS

A **positive-displacement air compressor** is a type of air compressor that discharges a fixed quantity of compressed air with each cycle.

A **reciprocating-piston air compressor** is a type of positive-displacement air compressor that compresses air by extending and retracting a piston inside a cylinder.

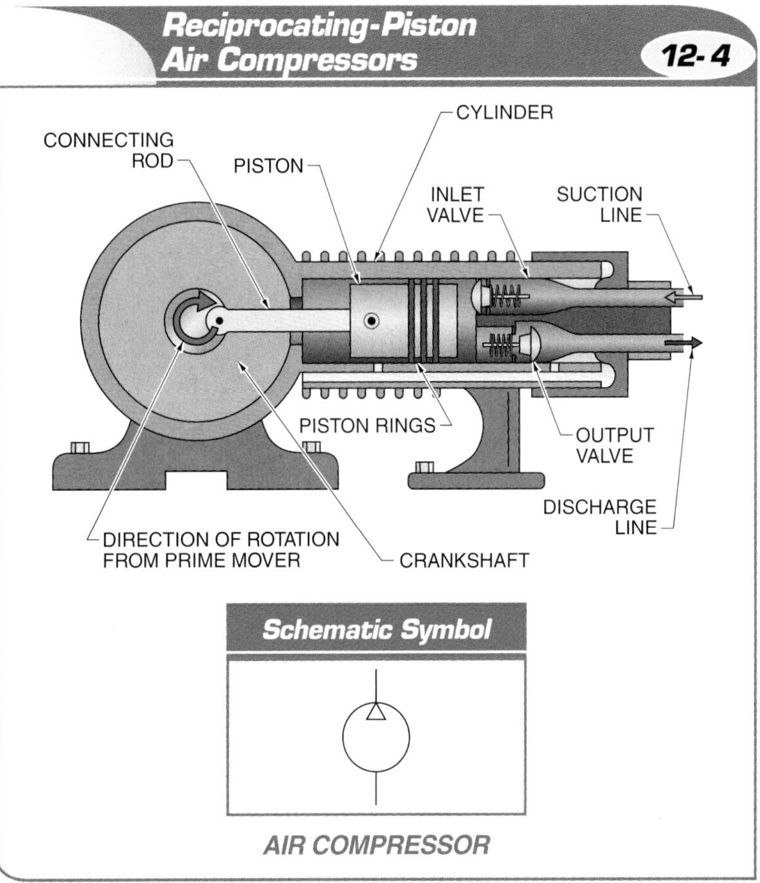

Figure 12-4. A reciprocating-piston air compressor compresses air by extending and retracting a piston inside a cylinder.

TERMS

A **multistage air compressor** is a type of air compressor where the compressor allows the airflow from the first stage to be sent to the second stage and so forth to allow for a more energy-efficient compression of air.

An **intercooler** is a type of pipe that is used to connect the different stages of an air compressor to cool the air that travels between the stages.

A **rotary-vane air compressor** a type of positive-displacement air compressor that uses vanes that slide in and out of a rotating rotor.

When a reciprocating-piston air compressor is operating, its piston extends and retracts hundreds of times per minute. The resulting friction causes heat to build up. To control and minimize heat, the compressor cylinders typically have water jackets to help reduce the buildup of heat in the compressor. Compressors that are air-cooled often have cooling fins added to dissipate some of the heat of compression. **See Figure 12-5.**

Multistage Air Compressors. When a single-stage air compressor is not capable of efficiently compressing air to high enough pressures, a multistage air compressor is used. A *multistage air compressor* is a type of air compressor where the compressor allows the airflow from the first stage to be sent to the second stage and so forth to allow for a more energy-efficient compression of air. The first stage of compression in a multistage air compressor takes place in the larger of the two cylinders, while the second stage takes place in the smaller cylinder.

During the first stage of compression, air from the atmosphere flows into the larger cylinder in the same manner as with a single-stage air compressor. Rather than releasing the compressed air to be stored or to do work, it is piped to the smaller cylinder for the second stage of compression. The smaller cylinder compresses the air again and sends it either to be stored or to do work. Because the air is compressed two or more times, a multistage air compressor is more efficient than a single-stage air compressor. Also, the compressed air from a multistage air compressor is cooler than the air from a single-stage air compressor.

TECH FACT

The term "multistage" usually refers to a two-stage air compressor. There are multistage air compressors that have more than two stages, such as three-, four-, and five-stage compressors.

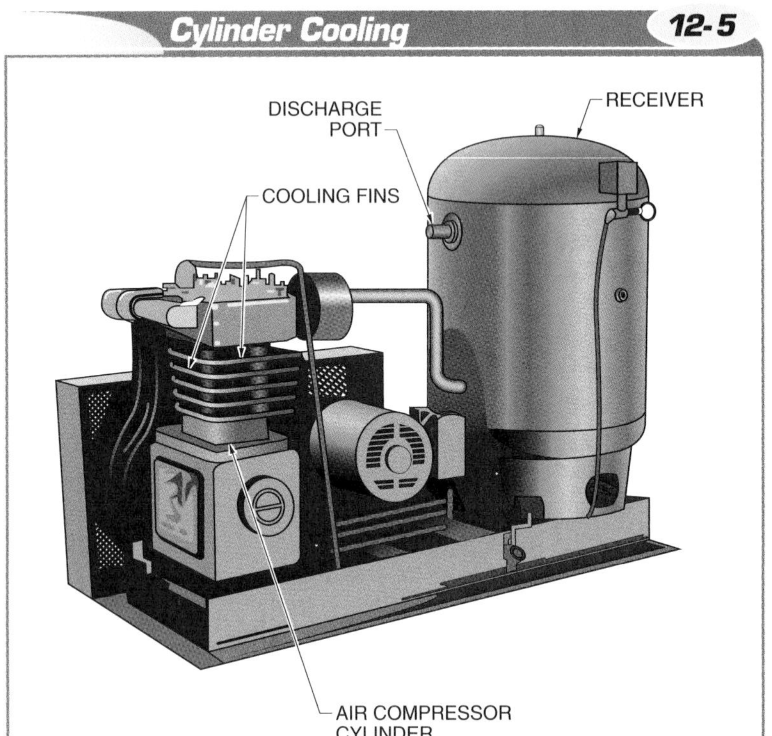

Figure 12-5. Air compressor cylinders often have cooling fins to dissipate some of the heat created by friction and air compression.

An *intercooler* is a type of pipe that is used to connect the different stages of an air compressor to cool the air that travels between the stages. An intercooler allows the heat from the first stage of compression to be dissipated before the air enters the second stage of compression. Also, both cylinders typically have cooling fins that are similar to the heat sinks on a single-stage compressor to help dissipate heat. **See Figure 12-6.**

A multistage compressor operates through the following procedure:

1. The prime mover rotates the crankshaft and the two connecting rods to retract and extend the two pistons.
2. As the piston for the first stage of compression retracts, air flows into the large cylinder.
3. As the piston for the first stage of compression extends to compress the air, the piston for the second stage of compression retracts, which allows in compressed air from the first stage cylinder.

4. The air flows through the intercooler to lower its temperature as it travels from the first stage cylinder to the second stage cylinder.
5. When the piston for the second stage of compression is fully retracted, its cylinder is filled with the air from the first stage cylinder.
6. The piston for the second stage of compression then extends, compressing the air more and sending it into the system piping or a receiver.

The prime mover for a multistage air compressor is typically a 3ϕ AC electric motor. In small commercial units, the prime mover is usually connected to the air compressor with V-belts pulleys. **See Figure 12-7.** As the prime mover rotates, it moves the V-belts that are attached to the air compressor pulley. The V-belts must be routinely inspected for cracks and tension to ensure that they are in good operating condition. In large industrial units, the prime mover is usually connected to the air compressor by a rigid coupling or drive shaft. When using V-belt drives, the prime mover and air compressor are positioned adjacent to each other.

Rotary Air Compressors. The two most common types of rotary air compressors are rotary-vane air compressors and rotary-screw air compressors. A *rotary-vane air compressor* is a type of positive-displacement air compressor that uses vanes that slide in and out of a rotating rotor. Rotary-vane air compressors use an offset rotor in a cam ring to create the increasing and decreasing volumes needed for airflow. The increasing volume allows air into the compressor, while the decreasing volume forces air out of the compressor.

TECH FACT

Standard cubic feet per minute (scfm) and pounds per square inch (psi) are used instead of horsepower (HP) to determine the operational capability of an air compressor. Ratings of cfm and psi are used to determine the types of equipment an air compressor can operate.

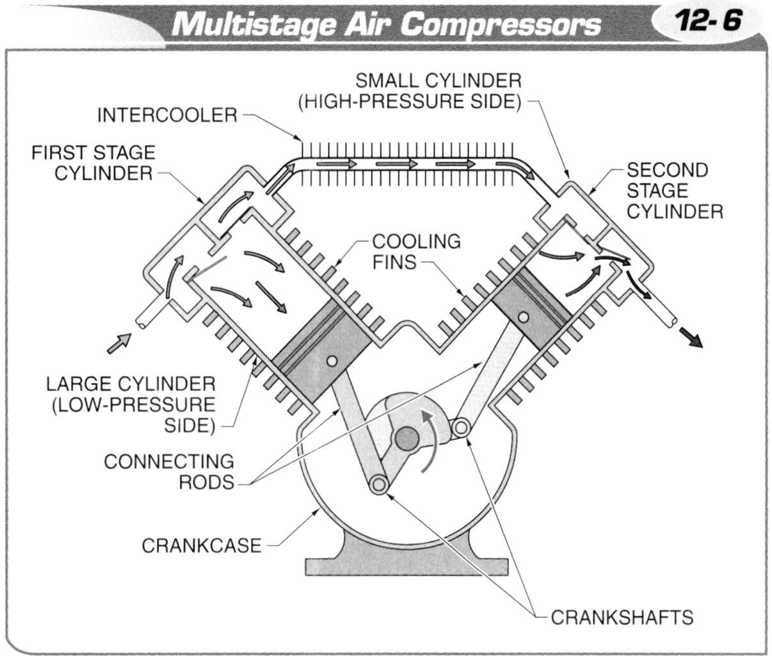

Figure 12-6. A multistage air compressor can be used to efficiently compress air to high levels.

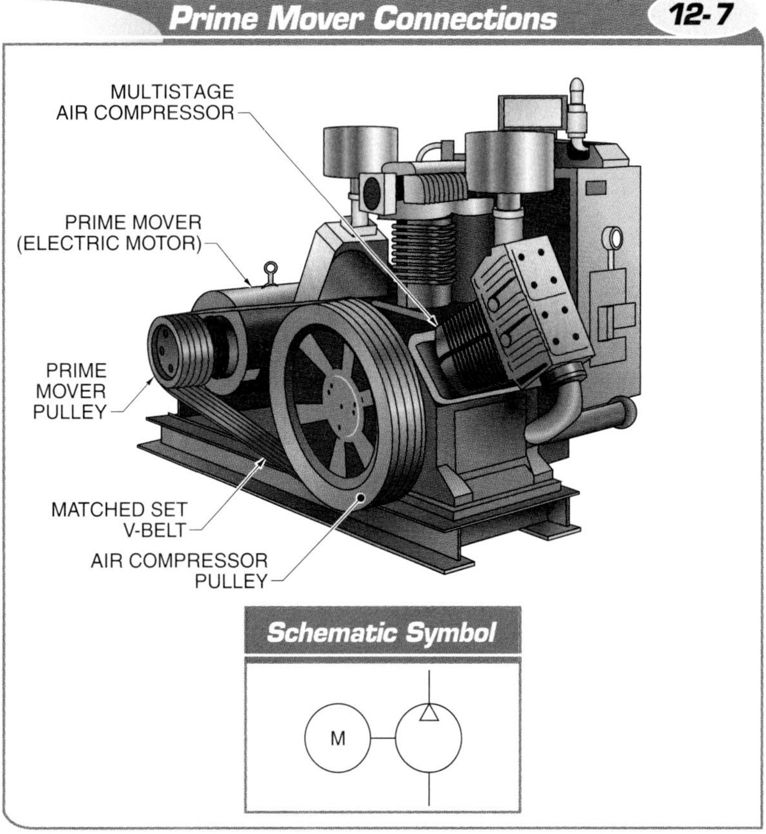

Figure 12-7. In small commercial units, the prime mover is usually connected to the air compressor with V-belts and pulleys.

TERMS

A **rotary-screw air compressor** is a type of positive-displacement air compressor that uses two intermeshing screw shafts or a screw shaft with two rotors to create airflow.

A rotary-vane air compressor consists of an offset rotor, an inlet, an outlet, sliding vanes, a cam ring, and the air compressor housing. **See Figure 12-8.** The airflow from a rotary-vane air compressor can be adjusted by changing the offset of the rotor and cam ring with a pressure adjustment screw. Rotary-vane air compressors can be single-stage (up to 50 psi) or multistage (50 psi to 125 psi). A single-stage rotary-vane air compressor operates through the following procedure:

1. The prime mover rotates the rotor shaft that is attached to the rotor.
2. As the rotor rotates, the sliding vanes are driven outward against the inside surface of the cam ring by centrifugal force, springs, or a combination of the two. The volume at the inlet port increases because of the offset between the rotor and the cam ring.
3. As the volume at the inlet port increases, a vacuum is created and atmospheric pressure forces air into the inlet side of the air compressor housing.
4. Air is trapped between the sliding vanes and is forced towards the outlet side of the air compressor housing.
5. As the air is forced towards the outlet side of the air compressor housing, the volume begins to decrease because of the offset between the rotor and the cam ring.
6. As the volume decreases, the air between the sliding vanes is compressed.
7. The compressed air then flows through the outlet port either to a receiver or to do work.

A *rotary-screw air compressor* is a type of positive-displacement air compressor that uses two intermeshing screw shafts or a screw shaft with two rotors to create airflow. Rotary-screw air compressors with two intermeshing screw (male and female) shafts are the more common design. The air is forced to flow along the screw shafts and is compressed as the volume between the two shafts decreases. Airflow is positive and continuous because air is constantly pushed and forced axially along the screw shafts. **See Figure 12-9.**

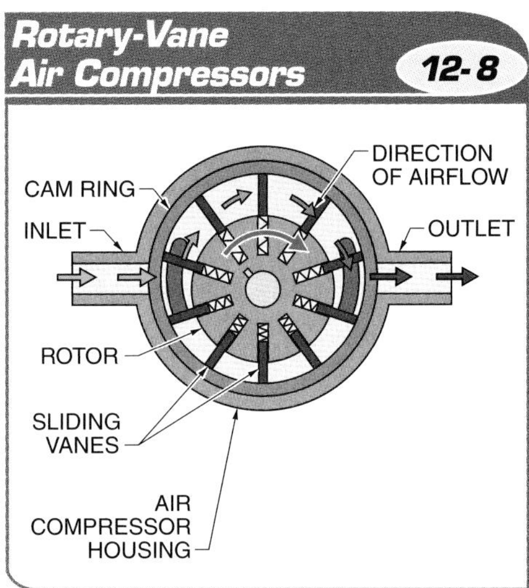

Figure 12-8. A rotary-vane air compressor uses sliding vanes to decrease volume as the rotor rotates.

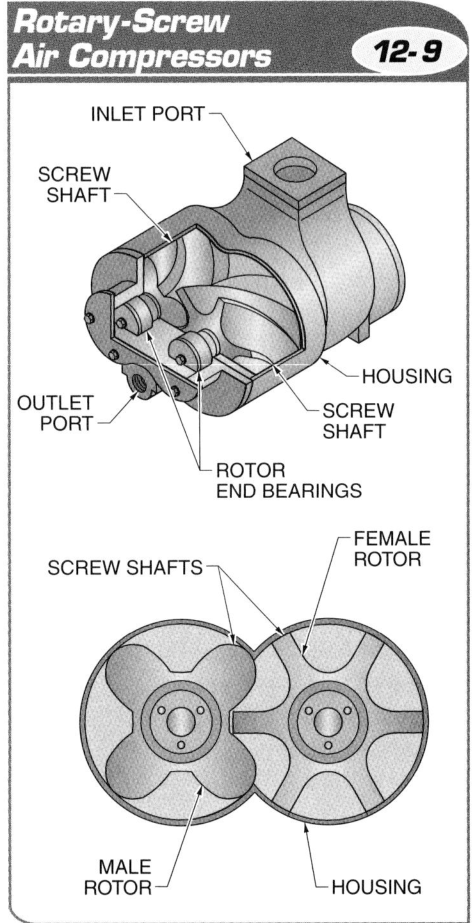

Figure 12-9. A rotary-screw air compressor traps air between the screw shafts and compresses air as the volume between the screw shafts is reduced.

Rotary-screw air compressors can contain two intermeshing screw shafts or a screw shaft and two rotors. Rotary-screw air compressors with two intermeshing screw shafts are the most common. In the operation of a rotary-screw air compressor with two intermeshing screw shafts, air entering the housing is compressed as the screw shafts mesh together. The screw shafts trap the air along the rotors and compress it. The air is progressively compressed until the screw shafts pass the outlet port and the compressed air is discharged.

Rotary-screw air compressors may have dry or oil-flooded compressing mechanisms. Dry compressing mechanisms compress air without the lobes contacting each other. This constant close separation of the lobes is accomplished by a set of timing gears that are driven by the prime mover. The prime mover normally rotates the rotors at speeds between 3000 rpm to 12,000 rpm. At these relatively high speeds, the rotors rotate freely with a carefully controlled clearance between the lobes and the housing. Dry compressing mechanisms are used when oil-free air is necessary, such as for instrumentation, paint spraying, food and beverage production, pharmaceutical production, and laboratory work.

Oil-flooded compressing mechanisms compress air with the lobes contacting each other. The male rotor is driven by the prime mover. It then meshes with and drives the female rotor. This driving motion causes surface contact between the rotors, which must be well-lubricated.

Oil-flooded compressing mechanisms lubricate the rotors and rotor end bearings by injecting a synthetic oil bath into their internal passages. The synthetic oil bath lubricates the rotors, seals the rotor clearances, and cools air temperature by absorbing the heat from compression. The synthetic oil is then separated from the compressed air, filtered, cooled, and returned to the air compressor for reuse. **See Figure 12-10.** Synthetic oil used in pneumatic systems must be in compliance with ISO Standard 32/46, *Synthetic Compressor Oil.*

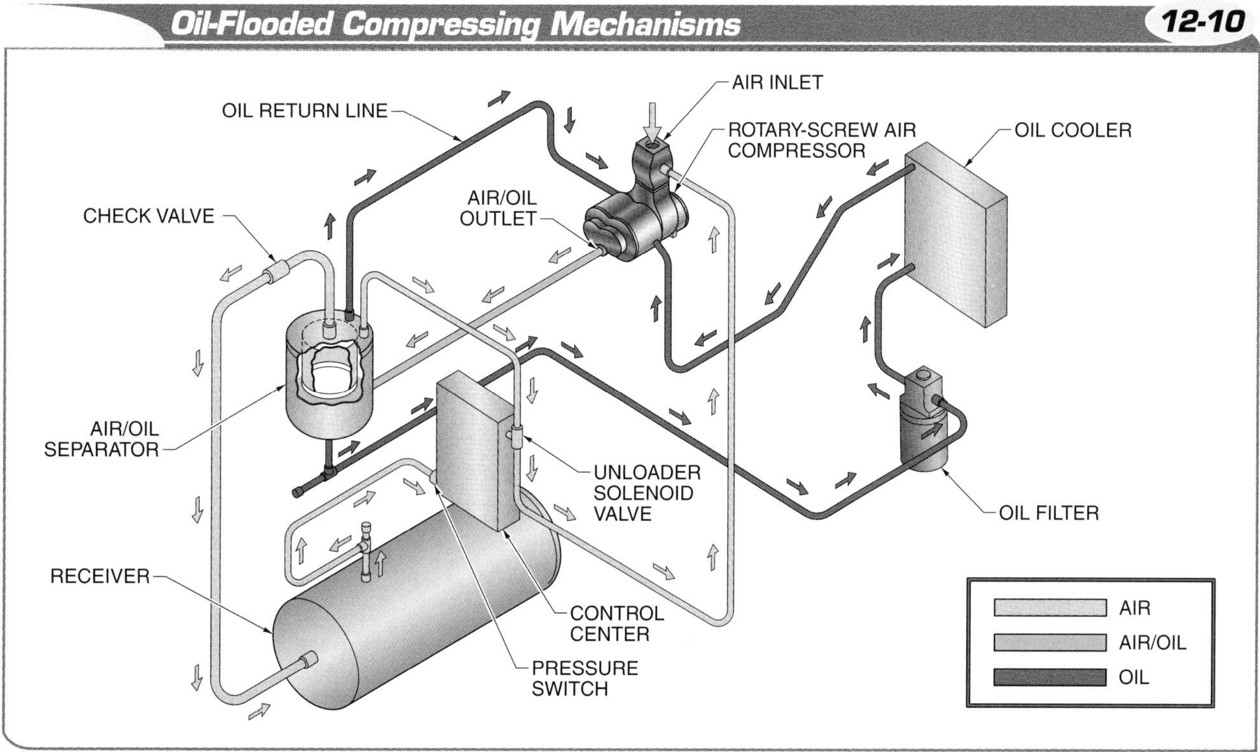

Figure 12-10. Oil-flooded compressing mechanisms lubricate the rotors and rotor end bearings by injecting a synthetic oil bath into their internal passages.

Receivers

A receiver is a container that holds gas in a pneumatic system. Many receivers are cylindrical steel tanks. **See Figure 12-11.** After air enters an air compressor and is compressed, it can flow into a system and accomplish work, or it can flow into a receiver and be stored until needed.

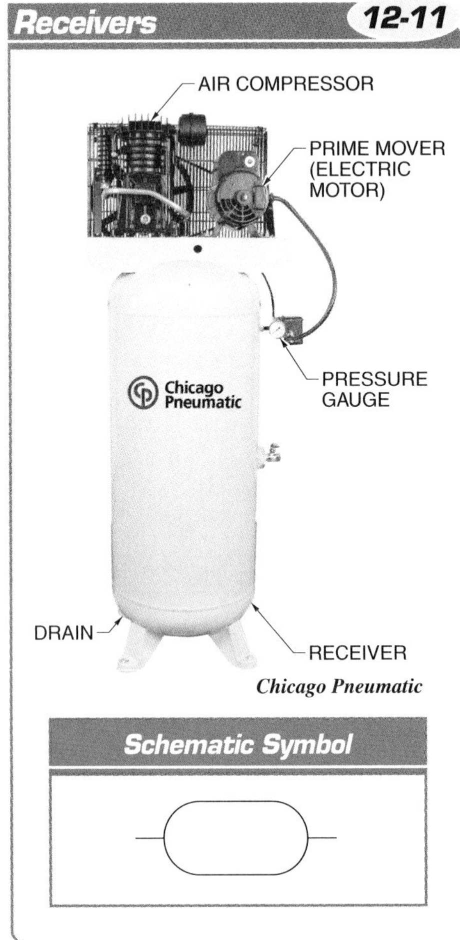

Figure 12-11. Many receivers are cylindrical steel tanks that are used to store compressed air until it is needed to accomplish work.

In all pneumatic systems, the compressor is stationary and constantly uses a high volume of air to supply the system. By law, receivers must have an operating safety valve (preset relief valve) connected to them. Depending on the demand of the pneumatic system, receiver safety valves can be found with preset pressures of 90 psi to 120 psi. Most facilities have the receiver safety valves set at 110 psi or 120 psi. A receiver stores fluid energy and allows airflow into a system when pressure is needed. The advantages provided by a receiver include the following:

- Smoother airflow—Receivers smooth out the pulsations of a reciprocating-piston air compressor so they are not felt in the system.
- Settling out contaminants—As the air rests and cools in the receiver, the contaminants that were not already filtered out settle to the bottom of the receiver.
- Settling out water—Water molecules that are in the air as humidity are brought into the compressor inlet and compressed. The molecules then settle to the bottom of the receiver as a liquid, which must be drained periodically. Most receivers have an automatic drain.
- Allowing air to cool—The air in the receiver cools as heat is transferred through the receiver walls and into the atmosphere.

CONTROL DEVICES

Compressors and the air in pneumatic systems must be under control at all times. If they are not controlled, pressure in the system can continue to build and either damage system components or create a hazardous condition for personnel working on or near the equipment. There are several different devices used to control pressure in a compressor system, including safety valves and pressure control valves.

TECH FACT

Flow control valves are commonly used in pneumatic systems to decrease, rather than increase, actuator speed by restricting the airflow exiting the actuator. The most common type of flow control valve used in pneumatic systems are needle valves. Needle valves allow the best control of airflow that is possible.

Compressor Control

Precise control of a compressor is necessary for energy efficiency. A compressor that runs continuously at full power has a duty cycle of 100%. Most industrial facilities attempt a maximum duty cycle of 100% for their compressors. *Duty cycle* is the percentage of time that a machine operates compared to not operating. For example, a compressor with a duty cycle of 30% operates 18 min/hr.

The duty cycle for most compressors is based on the size of the receiver. If the receiver is too large for the system, the duty cycle is lower. If the receiver is too small, the duty cycle is higher. To help keep the duty cycle low, a difference of 25 psi to 35 psi is usually maintained between the minimum pressure and maximum pressure in the receiver.

Duty cycle can be used to calculate the average operating cost of certain components. Average operating cost is calculated by applying the following procedure:

1. Calculate power. Power is calculated by applying the following formula:

$$P = \frac{I \times V}{1000}$$

where
P = power (in kW)
I = current (in A)
V = voltage (in V)
1000 = conversion factor

2. Calculate compressor run time. Compressor run time is calculated by applying the following formula:

$T_r = H_T \times D$
where
T_r = compressor run time (in hr)
H_T = total hours (in hr)
D = duty cycle (in %)

3. Calculate total daily energy consumed. Energy is calculated by applying the following formula:

$E_T = T_r \times P$
where
E_T = energy consumed (in kWh)
T_r = compressor run time (in hr)
P = power (in kW)

4. Calculate average daily cost. Average daily cost is calculated by applying the following formula:

$C_A = E_T \times C_E$
where
C_A = average daily cost (in $)
E_T = energy consumed (in kWh)
C_E = cost per kW (in $)

Example: An air compressor draws current of 11 A and operates on 120 V. What is the average daily cost to operate the air compressor?

1. Calculate power.

$$P = \frac{I \times V}{1000}$$

$$P = \frac{11 \times 120}{1000}$$

$$P = \frac{1320}{1000}$$

$$P = 1.32 \text{ kW}$$

2. Calculate compressor run time.

$T_r = H_T \times D$
$T_r = 24 \times 0.50$
$T_r = 12$ hr

3. Calculate total daily energy consumed.

$E_T = T_r \times P$
$E_T = 12 \times 1.32$
$E_T = 15.84$ kWh

4. Calculate average daily cost.

$C_A = E_T \times C_E$
$C_A = 15.84 \times 0.10$
$C_A =$ **$1.58**

For a 25 HP or less electric motor on a reciprocating-piston air compressor, a pressure switch is used for automatic start-stop control and the cycling of the unit. A *pressure switch* is a switch that electrically controls the energizing or de-energizing of a prime mover when a set pressure has been reached. A pressure switch is either located between the air compressor and the receiver or is attached directly to the receiver. A pressure switch consists of an inlet, adjustable biasing spring, electrical limit switch, and a diaphragm or piston. **See Figure 12-12.**

TERMS

Duty cycle is the percentage of time that a machine operates compared to not operating.

A **pressure switch** is a switch that electrically controls the energizing or de-energizing of a prime mover when a set pressure has been reached.

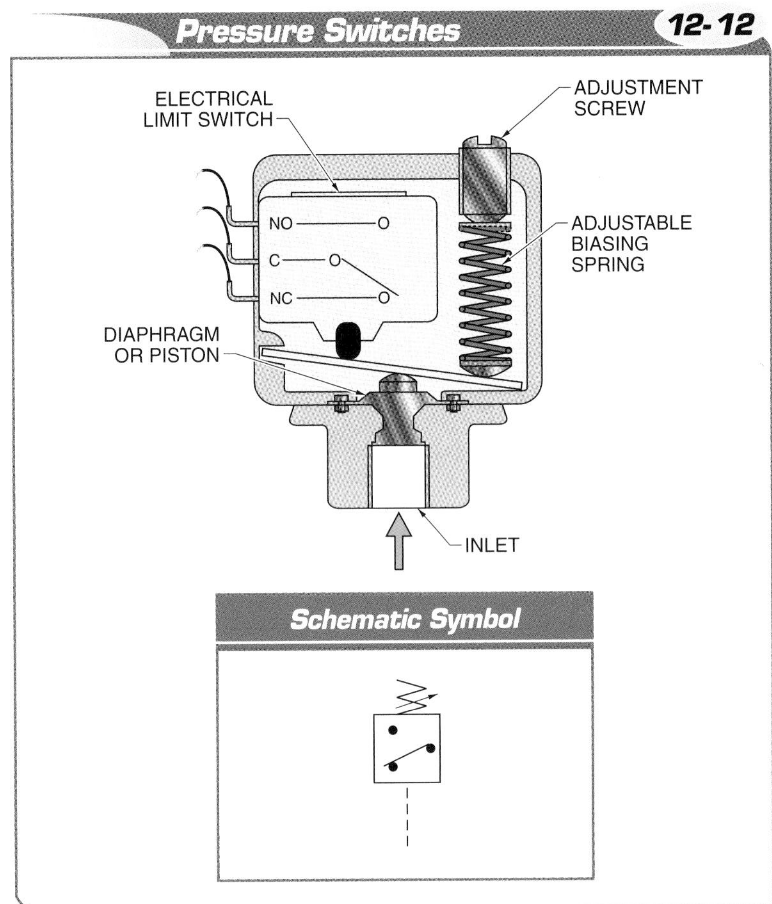

Figure 12-12. A pressure switch electrically controls the energizing or de-energizing of a prime mover when a set pressure has been reached.

Curtis-Toledo, Inc.

A pressure switch can be attached directly to the receiver in a pneumatic system.

A pressure switch operates through the following procedure:

1. The biasing spring is adjusted to the maximum system operating pressure.
2. When maximum system operating pressure in the receiver is reached, the air pressure overcomes the pressure of the biasing spring, deactivating the electrical limit switch and turning OFF the prime mover.
3. When system pressure reaches minimum pressure, the biasing spring overcomes air pressure and activates the electrical limit switch to turn ON the prime mover.

When automatic start-stop control is used, the compressor must be unloaded each time prior to startup. *Unloading* is a process that allows a compressor to run against no load. Unloading prevents high start-up power and premature wear on the prime mover.

One method of unloading a compressor with a duty cycle of less than 50% is accomplished with internal centrifugal control. This method releases air from the piping between the air compressor and the receiver until the air compressor is at the correct speed through a valving system. When the correct speed is reached, air can no longer be released and is forced to the receiver instead.

Another method of unloading allows the prime mover of a pneumatic compressor system with a duty cycle of higher than 50% to run continuously and does not use a pressure switch. Despite running continuously, the prime mover does not always engage the air compressor.

A compressor can be unloaded by holding the intake valve open (exhausting compressed air to the atmosphere), holding the discharge valve open (allowing air to move back and forth from the cylinder), or holding the intake valve closed (stopping air from entering the cylinder).

Safety Valves. A *safety valve* is a normally closed pressure control valve that is not adjustable and is used as overpressure protection on pneumatic components. The manufacturer of the valve presets the setting of the safety valve at a pressure that protects the pneumatic component it is attached to from overpressurization. Once system pressure is high enough to overcome the biasing spring, the safety valve opens and air from the receiver or any other pneumatic component is exhausted to the atmosphere. **See Figure 12-13.** Some safety valves include warning systems, that sound an audible alarm such as a whistle, bell, buzzer, or siren.

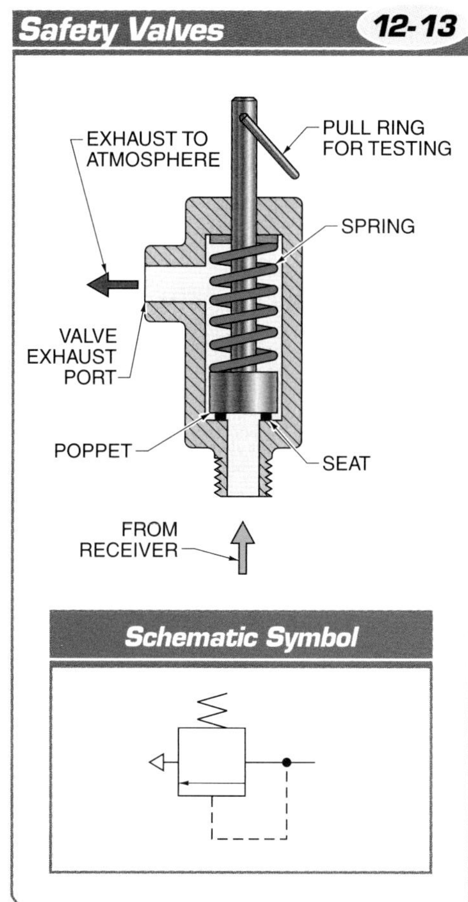

Figure 12-13. Once system pressure is high enough to overcome the biasing spring, a poppet in the safety valve opens and air from the receiver is exhausted into the atmosphere.

Safety valves are designed strictly for safety and are not intended for frequent operation. However, safety valves do require periodic maintenance. Safety valve maintenance consists of verifying whether the valve can move freely or not. To verify the operation of most safety valves, a test lever is moved or a ring is pulled to unseat the poppet. If the valve is regularly maintained, the vent port should remain clean due to the velocity of the exhausted air. However, if a safety valve is allowed to accumulate contaminants, it may leak even after being tested.

Relief Valves. A relief valve is a valve that protects a fluid power system from overpressure by setting a maximum operating pressure. Relief valves are used to prevent excessive pressure from building up in a pneumatic system by exhausting air to the atmosphere. The more pressure that is present, the more a relief valve opens.

If a pressure switch fails, the relief valve opens to exhaust air into the atmosphere. If pressure continues to build, the valve opens more to exhaust more air. The maximum release volume of a relief valve is higher than the maximum intake volume of the air compressor. Relief valves are either weight- or spring-loaded to hold them closed until the set pressure is reached. The pressure of the spring or the load of the weights can be adjusted with an adjustment screw.

The different types of relief valves used in pneumatic systems are poppet, ball, and diaphragm. **See Figure 12-14.** Poppet relief valves and ball relief valves are installed in small pneumatic systems that are 25 HP or less. These two types of relief valves are used because they exhaust air according to how much pressure is in the receiver. Diaphragm relief valves are used in large pneumatic systems that are more than 25 HP because they can exhaust a larger volume of air than similarly sized poppet or ball relief valves. Diaphragm relief valves are typically installed in systems where there is a large volume of airflow.

TECH FACT

Many individual facilities with large industrial compressors use custom-designed computerized control systems. These control systems are designed to maximize energy efficiency and safety.

TERMS

Unloading is a process that allows a compressor to run against no load.

A **safety valve** is a normally closed pressure control valve that is not adjustable and is used as overpressure protection on pneumatic components.

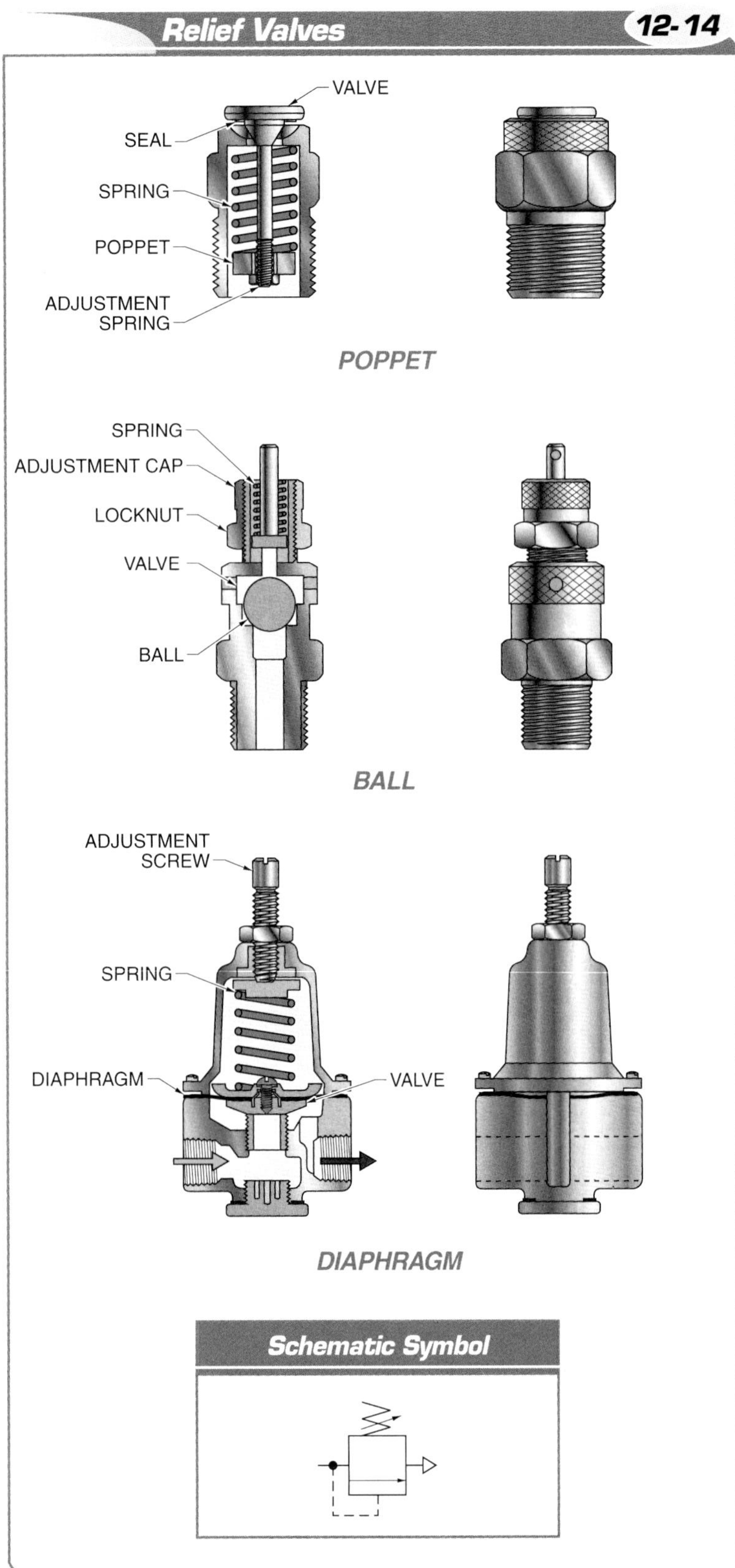

Figure 12-14. The different types of relief valves used in pneumatic systems are poppet, ball, and diaphragm.

Directional Control Valves

Directional control valves are used to control the movement of air through a system and to control the movement of actuators, which is similar to their function in hydraulic systems. Pneumatic directional control valves also have similar configurations and symbols, as well as similar actuation mechanisms, as hydraulic directional control valves. When a directional control valve is actuated, its spool changes positions to direct airflow.

However, there are differences between pneumatic directional control valves and hydraulic directional control valves. For example, pneumatic directional control valves can exhaust air into the atmosphere and do not need to pipe it back to the receiver. For this reason, the port that pipes air away from the cylinder is identified as the exhaust port. Since pneumatic systems do not have operating pressures as high as those in hydraulic systems, pneumatic directional control valves can be constructed of lighter metals than hydraulic directional control valves.

Pneumatic directional control valves typically have either two or three positions. **See Figure 12-15.** Two-position, three-way directional control valves and three-position, four-way directional control valves are commonly used in pneumatic systems. A two-position, three-way directional control valve has a pressure port (P), a cylinder port (A), and an exhaust port (EX). A two-position, three-way directional control valve is either normally open or normally closed.

Many three-position, four-way, five-ported directional control valves used in pneumatic systems actually have five ports. They have a pressure port (P), two cylinder ports (A and B), and two exhaust ports (EX). Because the two exhaust ports perform the same function, these directional control valves are often referred to as three-position, four-way directional control valves. At other times, they may be referred to as three-position, four-way, five-ported directional control valves. They operate similarly to hydraulic three-position, four-way directional control valves and are commonly used to control double-acting cylinders and air motors.

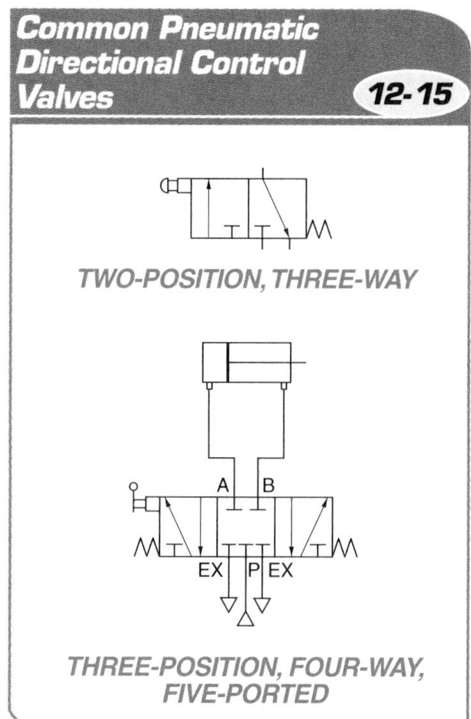

Figure 12-15. Two-position, three-way directional control valves and three-position, four-way directional control valves are commonly used in pneumatic systems.

Three-position, four-way directional control valves are available with different center positions to increase operational ability. Closed-center and float-center positions are the two most common center positions used with pneumatic directional control valves. Closed-center directional control valves are used to control double-acting cylinders because they allow the piston to be stopped anywhere during its travel in the cylinder. This allows for more motion control.

Float-center position directional control valves are typically used to control air motors. An *air motor* is a pneumatic device that uses airflow to create rotating mechanical energy. An operating air motor should not be stopped abruptly. A float-center position directional control valve allows an air motor to gradually reduce its speed.

In some facilities, it is necessary to control where the directional control valve exhausts the compressed air. In these facilities, compressed air must be exhausted away from the work that the pneumatic system is performing. For example, in a food-processing facility, the compressed air cannot be exhausted near the food product because any contaminants in the air would contaminate it. The air must be exhausted away from the production areas to make the pneumatic system compliant with any Food and Drug Administration (FDA) standards.

Mufflers. A *muffler* is a device that is attached to an exhaust port of a pneumatic directional control valve to deaden the noise of air as it exhausts. All exhaust ports in a pneumatic system must have mufflers attached to them. **See Figure 12-16.** Many mufflers also include a method of flow control.

TERMS

An **air motor** is a pneumatic device that uses airflow to create rotating mechanical energy.

A **muffler** is a device that is attached to an exhaust port of a pneumatic directional control valve to deaden the noise of air as it exhausts.

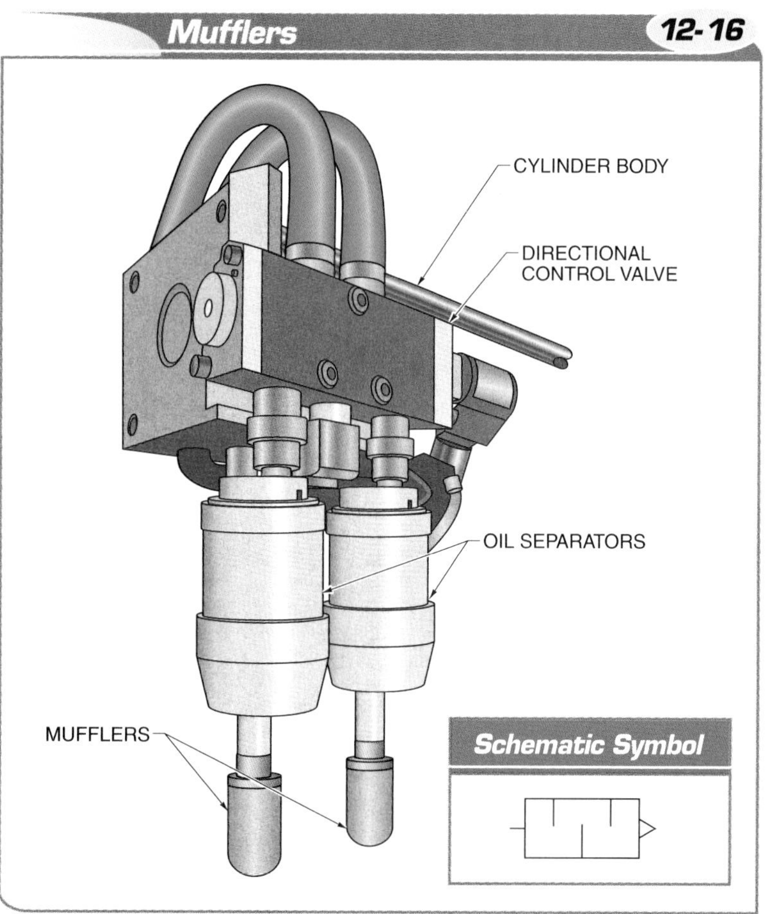

Figure 12-16. A muffler is attached to an exhaust port of a directional control valve to deaden the noise of air as it exhausts.

TERMS

A **metering valve** is a pneumatic flow control valve that controls the amount of airflow through a specific line at a given time.

Different muffler designs are available, depending on the application. Some mufflers have a plastic or metal body with a paper or foam muffler element to absorb the initial impact of the high-velocity air. The most commonly used types of mufflers are sintered bronze and brass screw-in type mufflers. They are the most commonly used because of their low cost, ability to minimize exhaust noise, low flow restriction, and reliable performance. These types of muffler designs usually allow for a dirty or worn muffler element to be replaced. Another type of muffler directs airflow through a series of baffles within an expansion chamber and exhausts the air at a reduced velocity.

Flow Control Valves

Flow control valves are used in pneumatic systems to control the rate of airflow out of cylinders. They can also be used to control the speed of a cylinder in one direction or both directions. The most common type of pneumatic flow control valve is a metering valve with an internal check valve.

A *metering valve* is a pneumatic flow control valve that controls the amount of airflow through a specific line at a given time. **See Figure 12-17.** A needle valve is the most common type of metering valve and is available with or without a check valve. A needle valve can be used to control the amount of airflow in one direction. It is common to use a needle valve without a check valve to control the speed of small cylinders at the exhaust port of a directional control valve.

When a flow control valve is installed between the cylinder and the directional control valve, a flow control valve with a check valve is used. This allows the cylinder speed to be controlled in one direction only. The needle valve is always used in a meter-out position to control airflow out of a cylinder.

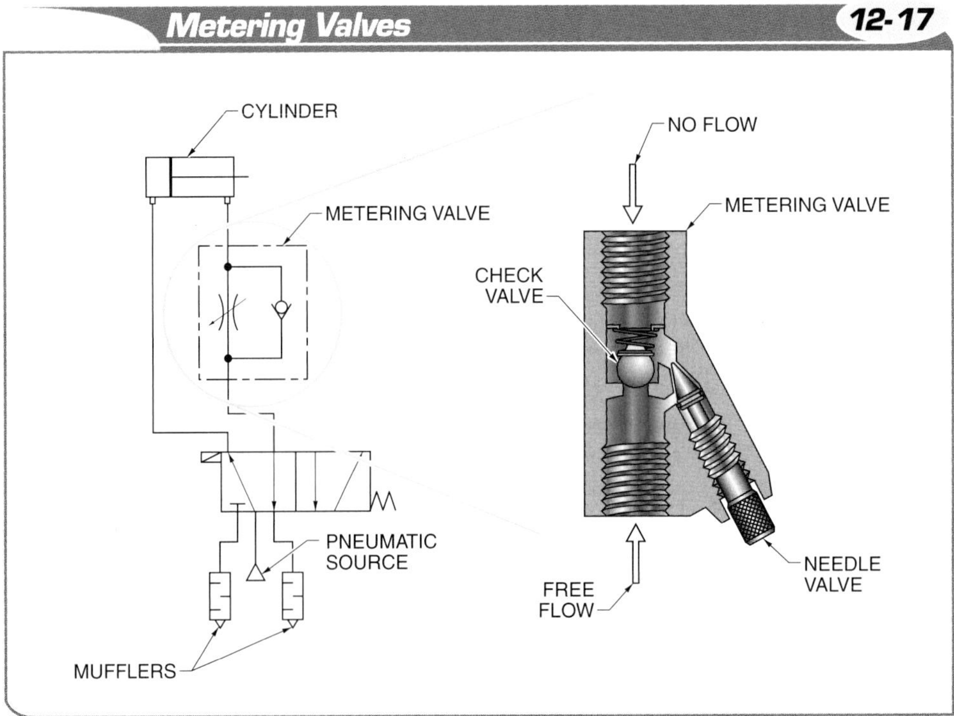

Figure 12-17. A needle valve is the most common type of pneumatic metering valve and is available with or without a check valve.

WORK DEVICES

The work that is accomplished by pneumatic systems is similar to hydraulic systems in that they both create linear and rotational movement. However, pneumatic systems differ from hydraulic systems in that they can also allow compressed air to flow through vacuum cups. Compressed air can be used to fill a container to a specific shape, such as an automobile tire, so that it can perform specific work. Pneumatic systems can also be used to operate equipment in environments such as medical, instrumentation, and food-processing facilities where contamination from oil is a concern.

Pneumatic Cylinders

A *pneumatic cylinder* is an actuator that moves in a straight line using compressed air. Pneumatic cylinders have similar parts as hydraulic cylinders, which include a cylinder barrel, a piston and rod, cap-end and rod-end ports, piston rings, cylinder seals, and end caps. However, pneumatic cylinders and hydraulic cylinders have differences in their pressure, speed, construction, and force. **See Figure 12-18.**

A pneumatic cylinder is used to produce linear force by extending and retracting a rod. Pneumatic cylinders can be single-acting, double-acting, double-rod, rodless, diaphragm, or rotary. Pneumatic cylinders are commonly used to move, position, clamp, sort, or drill product. The speed of a pneumatic cylinder is commonly measured in inches per second (ips) and its force is measured in pounds. The speed at which a pneumatic cylinder extends and retracts is fast and difficult to control compared to a hydraulic cylinder.

When the rod and piston of a pneumatic cylinder extend or retract, they are stopped at the end of their stroke by one of the end caps. The rod-end end cap stops the rod during extension and the cap-end end cap stops it during retraction. Abruptly stopping the rod during high-speed operation can damage the end caps. To prevent this damage, a cushion is often installed. A common cushion design is a tapered plug attached to the piston or rod that fills an exit hole on either end of the cylinder. A cushion sometimes includes a needle valve with a check valve in the end cap at the rod end, cap end, or both. **See Figure 12-19.**

TECH FACT

Pneumatic cylinders that require cushioning must be purchased with the cushions on one or both ends, depending on the application. Cushions cannot be added to cylinders that are in operation.

TERMS

A **pneumatic cylinder** is an actuator that moves in a straight line using compressed air.

Clippard Instrument Laboratory, Inc.

Pneumatic cylinders are available in a variety of sizes and designs, but they are typically smaller than hydraulic cylinders.

Pneumatic vs. Hydraulic Cylinders 12-18

Variable	Pneumatic Cylinders	Hydraulic Cylinders
Pressure	Low Pressure (20 psi to 120 psi)	High pressure (greater than 600 psi)
Speed	Fast (ips)	Slow (fpm)
Construction	Lightweight aluminum pressed together	Heavy duty steel with tie rods
Force	Small to medium— 0lb–300 lb	Medium to extremely large— 100 lb–100,000 lb

Figure 12-18. Although there are many similarities between pneumatic and hydraulic cylinders, there are also several differences.

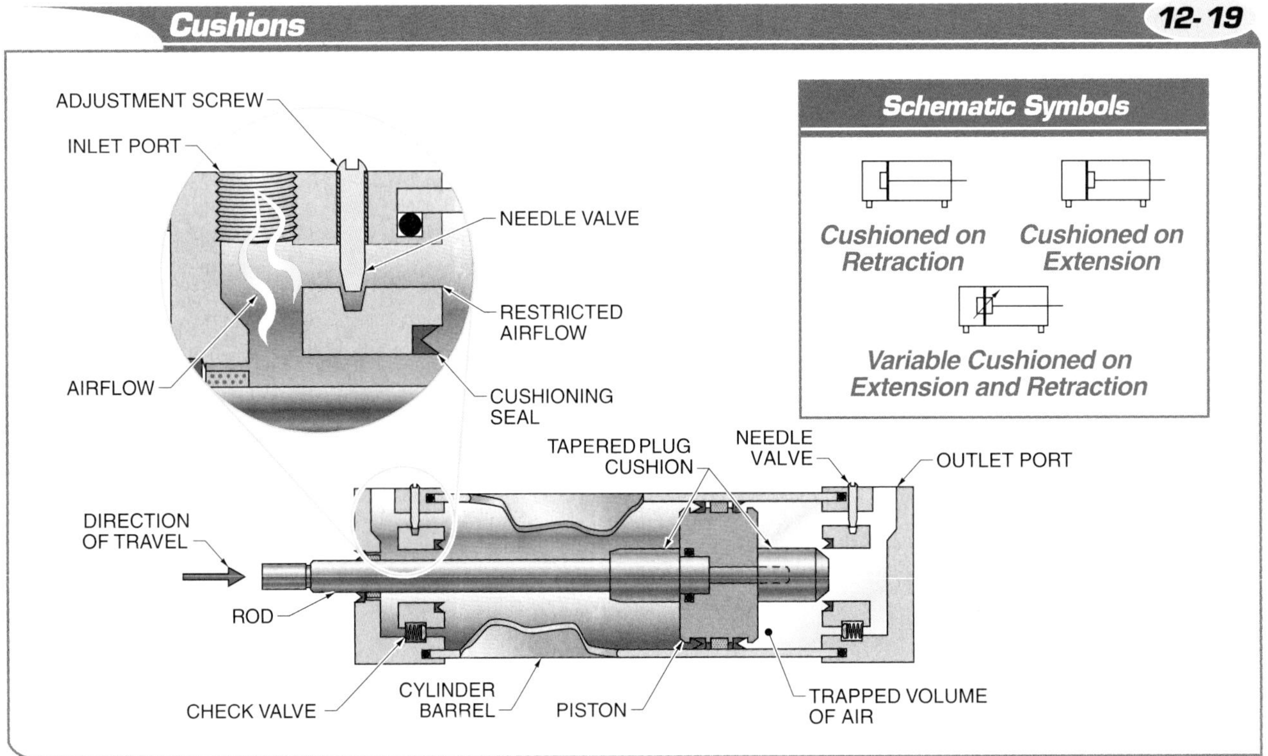

Figure 12-19. A cushion sometimes includes a needle valve with a check valve in the end cap at the rod end, cap end, or both.

Single-Acting Cylinders. Single-acting cylinders are pneumatic actuators that use compressed air for the linear movement of a piston in one direction and a spring to return it to its original position once the air is removed. The two most common types of single-acting cylinders are spring-return and spring-extend cylinders. A single-acting, spring-return cylinder holds the rod in a retracted position with spring pressure when no airflow is applied. When enough airflow is applied, it overcomes spring pressure and extends the rod. When air is allowed to exhaust, the spring returns the rod to its retracted position. **See Figure 12-20.**

For example, a step in an automated production line requires that a flap on a cardboard box be folded against a glued surface. A single-acting cylinder can be used to extend a rod to fold the flap into place against the glued surface. Once this step has been completed, the airflow to the cylinder is exhausted and the spring retracts the rod to its original position by exhausting air through the directional control valve.

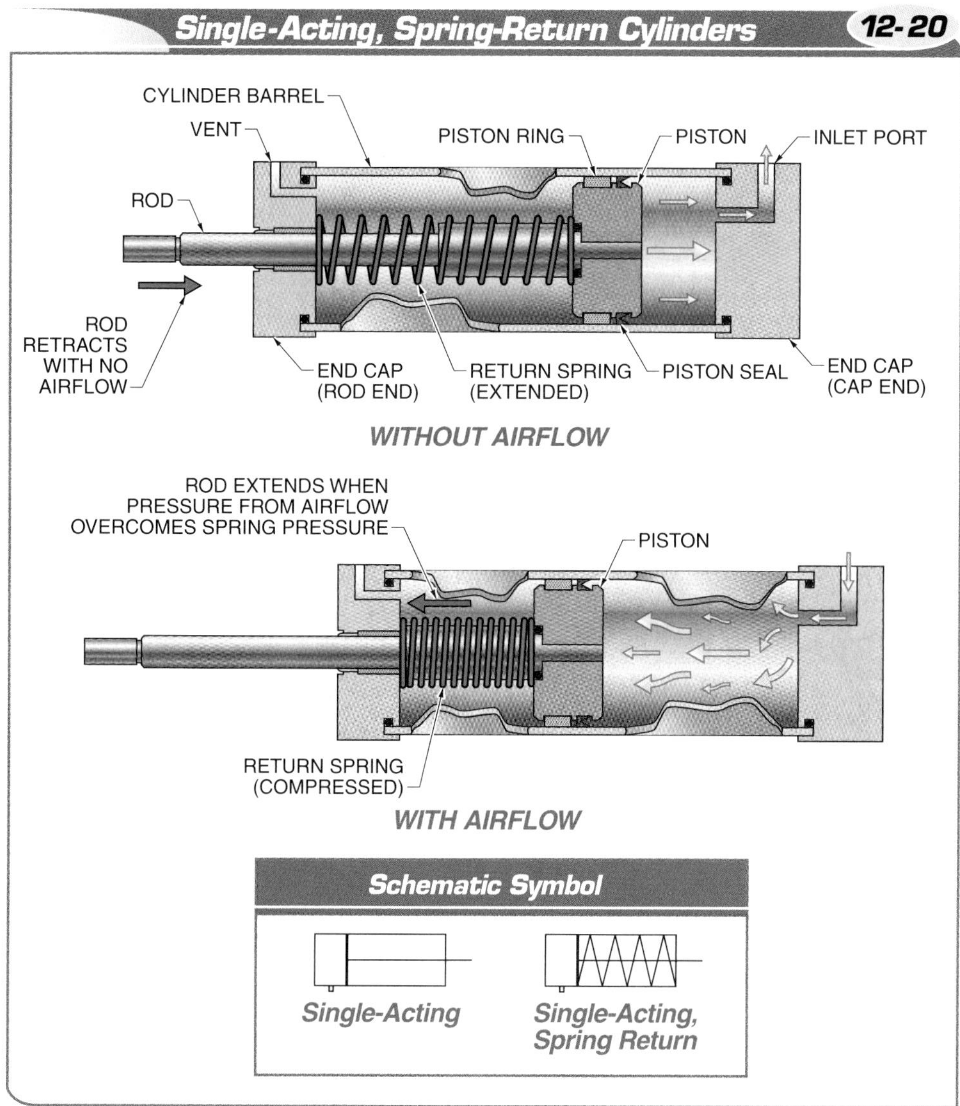

Figure 12-20. A single-acting, spring-return cylinder returns the rod to its retracted position when airflow is exhausted.

Because a single-acting, spring-extend cylinder holds the rod in an extended position when no airflow is applied, when enough airflow is applied, it overcomes spring pressure and retracts the rod. When airflow stops, the spring extends the rod to its original position.

For example, a step in an automated production line requires that each unit of product be inspected before it enters a moving conveyor line. A single-acting, spring-extend cylinder can be used to stop the product until it is inspected. When the product arrives, the extended rod holds it in place. Once the product has been inspected, airflow into the cylinder retracts the rod. After the product has passed by, airflow to the cylinder stops, and spring pressure extends the rod to its original position.

Double-Acting Cylinders. A double-acting cylinder is a pneumatic cylinder that uses compressed air to move the piston in both directions. Double-acting cylinders are able to produce force in both directions by applying pressure to either side of the piston. Double-acting cylinders are available with a variety of lengths, diameters, and mounting methods. **See Figure 12-21.**

TERMS

A **double-rod cylinder** is a pneumatic cylinder that has one piston and a rod that protrudes from both ends of the cylinder barrel.

A **rodless cylinder** is a pneumatic cylinder that has a cartridge attached to a piston that slides back and forth within the cylinder barrel.

A **diaphragm cylinder** is a pneumatic cylinder that uses a plastic or metal, or rubber diaphragm to extend the rod.

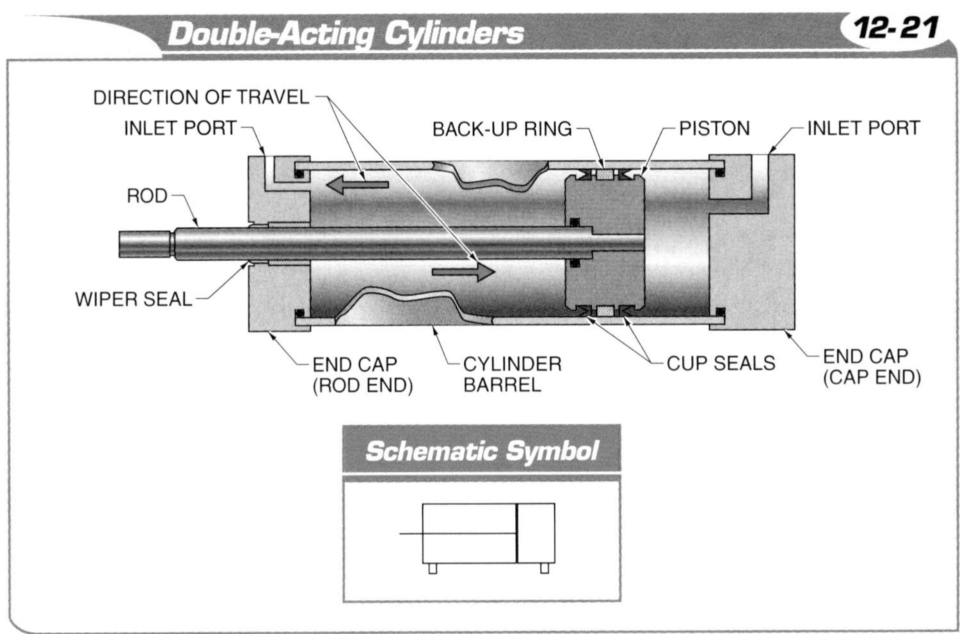

Figure 12-21. A double-acting cylinder uses compressed air to move a piston in both directions.

Double-acting cylinders can be used to extend a rod in an automatic drill press to drill holes in a workpiece. When the drilling cycle is complete, the rod is retracted. A double-acting cylinder is controlled by a directional control valve with two actuator ports.

Double-Rod Cylinders. A *double-rod cylinder* is a pneumatic cylinder that has one piston and a rod that protrudes from both ends of the cylinder barrel. The control of a double-rod cylinder is the same as with a double-acting cylinder. However, a double-rod cylinder can provide equal piston speed and force in both directions, allowing it to operate differently than a double-acting cylinder. **See Figure 12-22.**

Double-rod cylinders are used when there is a load on each end of the rod. They are commonly used in facilities that have labeling equipment, clamping/assembly lines, or pick-and-place machines. For example, if two conveyor belts are used to feed product into a single conveyor belt at the same time for efficient operation, a double-rod cylinder is used to feed product at separate intervals.

Rodless Cylinders. A *rodless cylinder* is a pneumatic cylinder that has a cartridge attached to a piston that slides back and forth within the cylinder barrel. The piston usually has a magnetic band around its diameter to allow for proper placement. **See Figure 12-23.** Rodless cylinders are typically used where space is limited, such as in sliding doors that are in confined spaces or in production manufacturing lines that include paint spraying, food placement, or parts-washing operations.

Diaphragm Cylinders. A *diaphragm cylinder* is a pneumatic cylinder that uses a plastic, metal, or rubber diaphragm to extend the rod. Most diaphragm cylinders apply compressed air to the diaphragm to extend the rod and spring pressure to retract it. The advantage of a diaphragm cylinder is that there is minimal friction, and no lubrication is required. **See Figure 12-24.**

Diaphragm cylinders work well in temperatures ranging from –45°F to 225°F. Custom-made diaphragm cylinders can work in temperatures ranging from –75°F to 400°F. Diaphragm cylinders are typically used in applications where a short, powerful piston stroke is required. Diaphragm cylinders may be installed in pneumatic devices such as impact absorbers, clamping mechanisms, valve actuation mechanisms, and positioning devices.

Rotary Cylinders. A *rotary cylinder* is a pneumatic cylinder that uses linear movement to create rotational movement. Rotary cylinders can create rotational movements in specific degrees and are used in applications where a twisting, turning, or back-and-forth motion is desired. A motor is typically used to create mechanical movement.

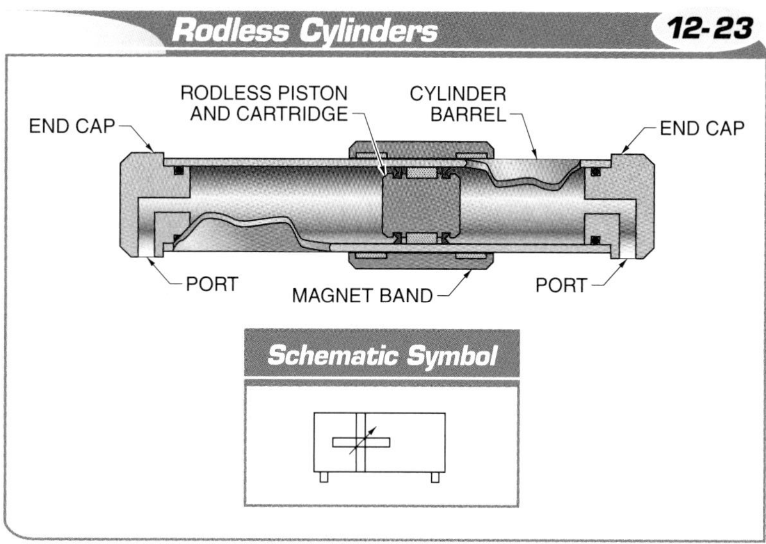

Figure 12-23. A rodless cylinder has a cartridge attached to the piston.

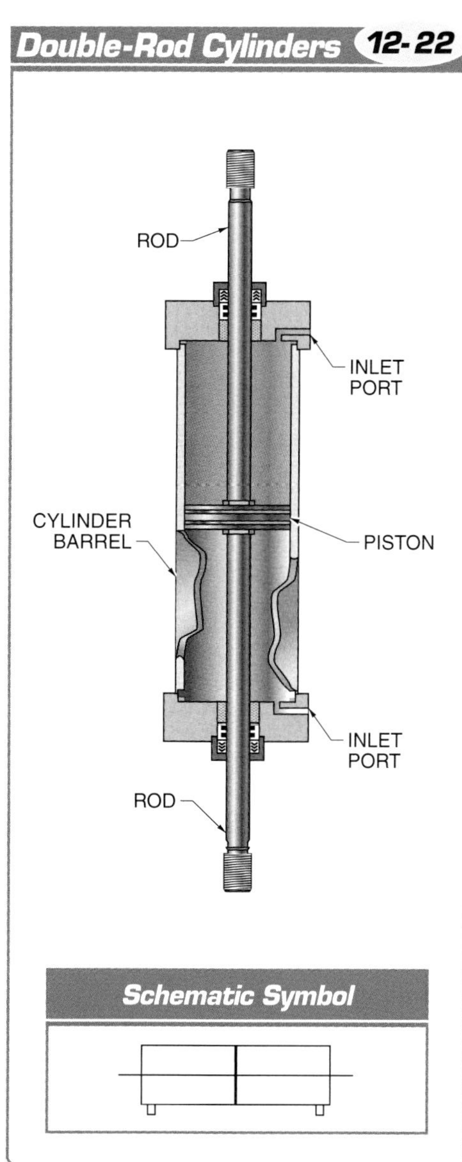

Figure 12-22. Double-rod cylinders are used when there is a load on each end of the rod.

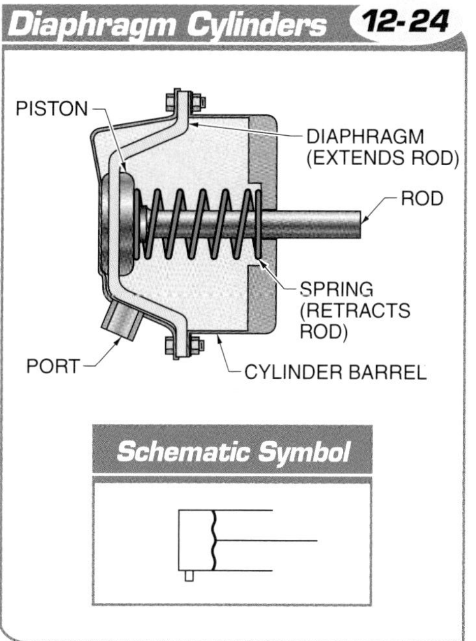

Figure 12-24. A diaphragm cylinder uses a plastic or metal diaphragm to extend or retract the rod.

TERMS

A **rotary cylinder** a pneumatic cylinder that uses linear movement to create rotational movement.

Rotary cylinders are often used for moving parts along an assembly line. The most commonly used rotary cylinders in pneumatic applications have rack-and-pinion mechanisms. **See Figure 12-25.** Rotary cylinders that rotate a specific amount of degrees for each movement or step are commonly used in equipment such as automated machinery.

TERMS

An **air motor** is a pneumatic device that uses airflow to create rotating mechanical energy.

A **vane air motor** is a type of motor that allows airflow to work against vanes to cause rotation of the shaft.

An **axial piston air motor** is a type of air motor that operates at low speed and high torque with its pistons horizontal to the motor's shaft.

A **radial piston air motor** is a type of air motor that operates at low speed and high torque with its pistons vertical to the shaft.

A **vacuum cup** is a device made of rubberized plastic that creates a vacuum when in contact with flat surfaces.

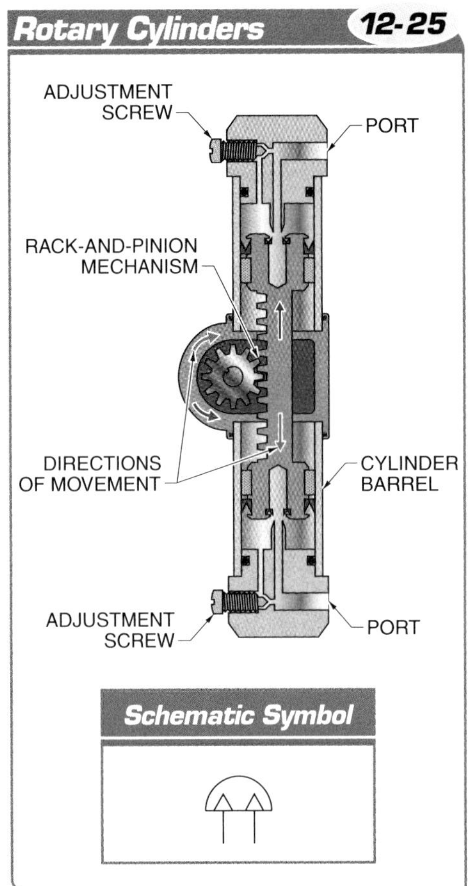

Figure 12-25. The most common type of rotary cylinder has a rack-and-pinion mechanism.

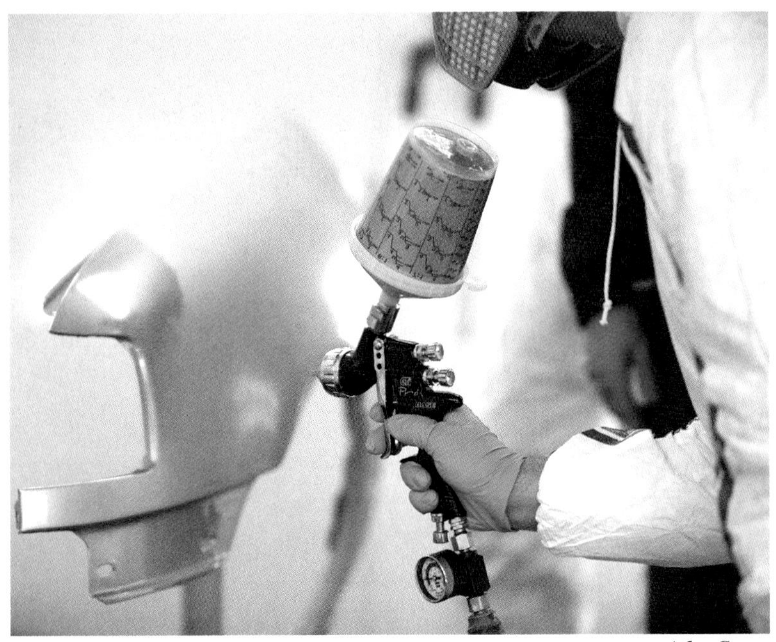

Atlas Copco

Pneumatic systems are often used with industrial coating production lines.

Air Motors

An *air motor* is a pneumatic device that uses airflow to create rotating mechanical energy. Air motors produce high operating speeds up to 15,000 rpm. The speed of an air motor depends on airflow and the load attached to the motor. An air motor can produce more speed than a hydraulic motor but not as much torque. Air motors can be vane, radial piston, or axial piston. **See Figure 12-26.**

There are also applications where an air motor is preferred over an electric motor. The advantages of an air motor over an electric motor include the following:

- Air motors can be used in environments where it is hazardous to use electric motors.
- Air motors produce less heat than electric motors.
- Air motors have better speed control.
- An air motor has a higher torque output than an electric motor of the same size.

A *vane air motor* is a type of motor that allows airflow to work against vanes to cause rotation of the shaft. Vane air motors are the most commonly used type of air motors. They are simple in design, available in many sizes, and easily maintained. Vane air motors are capable of speeds of up to 15,000 rpm at operating pressures of 100 psi.

Vane air motors create torque by allowing air pressure to act on the exposed surfaces of the vanes. As the rotor rotates, the vanes follow the interior surface of the housing due to springs and centrifugal force. The vanes slide in and out of the rotor, which is connected to the drive shaft. Portable hand tools, mixing motors, and air-operated hoists typically use vane air motors because they can produce different speeds and torque from the same input pressure.

An *axial piston air motor* is a type of air motor that operates at low speed and high torque with its pistons horizontal to the motor's shaft. Axial piston air motors are typically used in start-stop operations. They can also be installed for low-speed, high-torque applications, such as winches or conveyor drive systems.

A *radial piston air motor* is a type of air motor that operates at low speed and high torque with its pistons vertical to the shaft. Radial piston air motors are more commonly used than axial piston air motors because they have lower production costs. Radial piston air motors are typically used for continuous operations, such as pneumatic mixers. Both axial and radial piston air motors can produce higher torque than vane motors, but cannot operate at as high of speed as vane motors.

Vacuum Cups

A *vacuum cup* is a device made of rubberized plastic that creates a vacuum when in contact with flat surfaces. It operates by being connected to a vacuum that lowers pressure inside the vacuum cup until the pressure in the cup is less than atmospheric pressure, creating a pressure differential between the inside and outside of the cup. When the vacuum cup is in contact with the surface of an object, atmospheric pressure pushes on all sides of the vacuum cup and the object to create a vacuum. This allows the vacuum cup to lift and move the object through suction.

There are many different designs, sizes, materials, and manufacturers of vacuum cups. The two most common types of vacuum cups used with pneumatic systems are flat and bellow vacuum cups. **See Figure 12-27.** A flat vacuum cup is used to move objects vertically or horizontally. It can also be attached to a rotary cylinder to move or lift heavier or fragile objects a specific number of degrees. For example, if a packaging system needs to place a piece of cardboard between the two layers of an object, a flat vacuum cup is used to lift the piece of cardboard from its stack and place it between the layers of the object.

Bellow vacuum cups have a concave surface and are used to move lightweight objects. They are also used when the approach to lift the object is not perfectly aligned. A bellow vacuum cup compresses as it makes contact with the object it is attempting to lift.

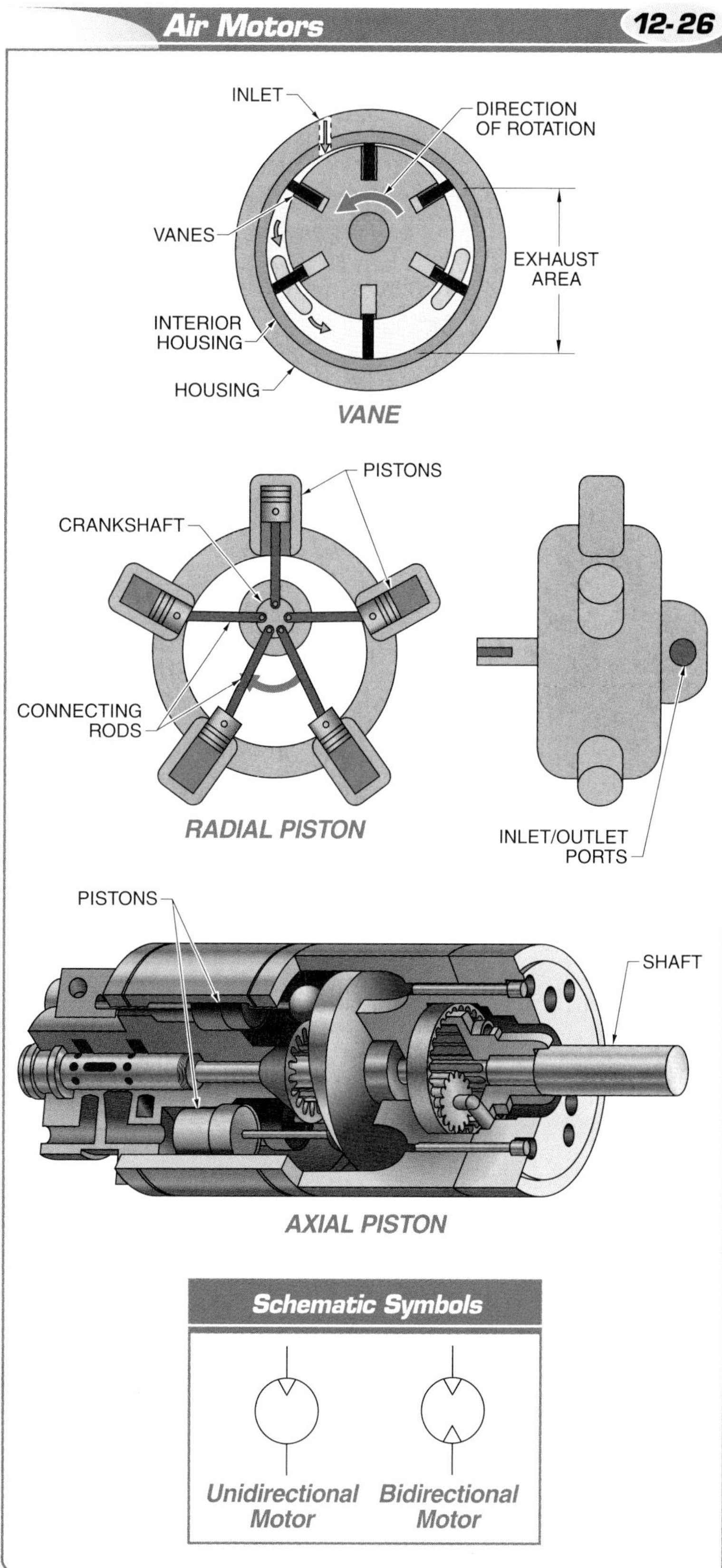

Figure 12-26. Air motors can be vane, radial piston, or axial piston.

TERMS

A **venturi valve** is a valve that creates the vacuum supplied to vacuum cups for use in pneumatic equipment applications.

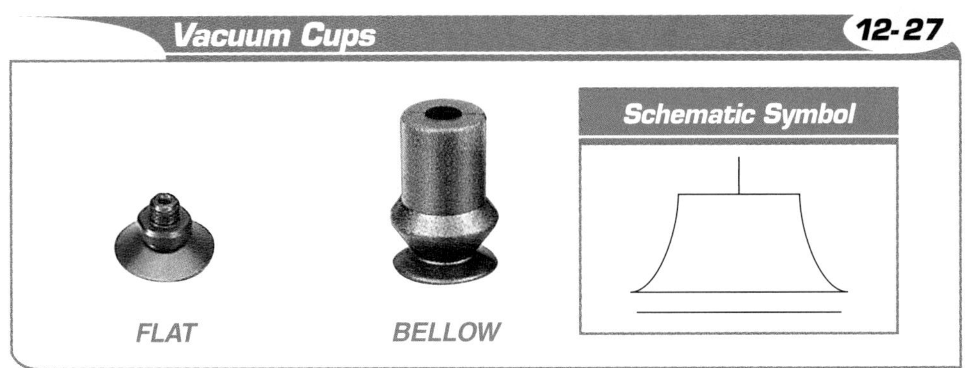

Figure 12-27. The two most common types of vacuum cups are flat and bellow vacuum cups.

A bellow vacuum cup can have up to four layers. The more layers it has, the more flexibility it has when lifting an object. For example, if a single sheet of paper must be lifted off a stack of 100 sheets of paper, a bellow vacuum cup is used. It can lift the single sheet of paper at an angle, and ensure that only one sheet of paper is lifted off the stack.

The strength of a vacuum cup is determined by the area of the vacuum and the pressure differential between the inside and outside of the vacuum cup. The larger the area, or the greater the pressure differential, the stronger the vacuum created. Vacuum cups can be used individually or with multiple cups to increase lifting capability. Vacuum is measured in either inches of mercury (in. Hg) or psia. Vacuum strength can be measured either with a Bourdon tube absolute pressure gauge or a digital vacuum meter.

A *venturi valve* is a valve that creates the vacuum supplied to vacuum cups for use in pneumatic equipment applications. **See Figure 12-28.** A venturi valve is sometimes referred to as a vacuum generator and creates a vacuum by taking compressed air that is supplied through the inlet and routes the airflow through an orifice. As the air flows through the orifice, its speed increases and flows to exhaust. As air speed increases, air pressure decreases. Because air pressure decreases below atmospheric pressure, the atmospheric pressure attempts to flow to the valve through the vacuum port. The vacuum port is attached to the vacuum cup. The atmospheric pressure is forced against the vacuum cup and holds in place whatever the vacuum cup is attached to.

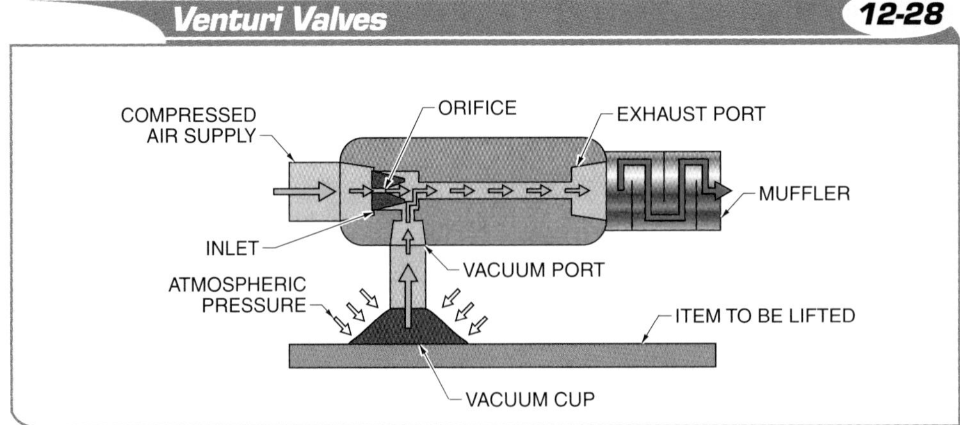

Figure 12-28. A venturi valve creates the vacuum supplied to vacuum cups in pneumatic equipment.

Digital Resources

Chapter Review

Pneumatic System Compression and Control

Chapter 12

Name: ______________________________ Date: ________________

MULTIPLE CHOICE

_______________ **1.** ___ vacuum cups have a concave surface and are used to lift objects that are not perfectly aligned.
A. Bellow
B. Diaphragm
C. Flat
D. Suction

_______________ **2.** Centrifugal air compressors are capable of producing up to ___ cfm of airflow.
A. 10,000
B. 50,000
C. 100,000
D. 500,000

_______________ **3.** A ___ cylinder has a cartridge attached to a piston that slides back and forth within the cylinder barrel.
A. diaphragm
B. double-rod
C. rodless
D. tandem

_______________ **4.** A ___ air compressor uses an offset rotor in a cam ring to create the increasing and decreasing volumes needed for airflow.
A. rotary-screw
B. rotary-vane
C. sliding-screw
D. sliding-vane

_______________ **5.** A ___ valve is normally closed pressure control valve that is not adjustable and is used as over pressure protection on pneumatic components.
A. directional control
B. pressure
C. relief
D. safety

_______________ **6.** A ___ cylinder uses compressed air to move the piston in both directions.
A. diaphragm
B. double-acting
C. rodless
D. single-acting

_______________ **7.** Depending on the demand of the pneumatic system, receiver safety valves can be found with preset pressures of ___ psi or ___ psi.
A. 90; 100
B. 90; 120
C. 100; 130
D. 110; 120

_______ **8.** ___ relief valves are typically installed in systems where there is a large volume of airflow.

A. Ball
B. Diaphragm
C. Gate
D. Poppet

_______ **9.** The duty cycle for most compressors is based on the size of the ___.

A. cylinders
B. piping
C. prime mover
D. receiver

_______ **10.** The advantage of having multiple compressors in smaller decentralized loops is that it is easier ___.

A. for maintenance technicians to move them, as required
B. to connect components such as valves and pipes
C. to control the air being sent to different parts of the facility
D. to keep them clean

_______ **11.** The prime mover in a pneumatic system normally rotates the rotors of a rotary-screw air compressor at speeds between ___ rpm to ___ rpm.

A. 1000; 5000
B. 3000; 10,000
C. 3000; 12,000
D. 3000; 15,000

_______ **12.** A ___ is a pneumatic flow control valve that controls the amount of airflow through a specific line at a given time.

A. metering valve
B. pneumatic muffler
C. safety valve
D. vacuum cup

_______ **13.** Most industrial facilities with pneumatic systems attempt a maximum duty cycle of ___% for their compressors.

A. 15
B. 30
C. 50
D. 100

_______ **14.** A ___ air compressor is a type of dynamic air compressor that uses an impeller rotating at high speed to compress air.

A. centrifugal
B. reciprocating-piston
C. rotary-screw
D. rotary-vane

COMPLETION

______________ **1.** A single-acting, ___ cylinder holds the rod in a retracted position with spring pressure when no airflow is applied.

______________ **2.** A(n) ___ air motor operates at low speed and high torque with its pistons vertical to the shaft.

______________ **3.** Synthetic oil used in pneumatic systems must be in compliance with ISO Standard ___.

______________ **4.** A(n) ___ is a device that is attached to an exhaust port of a pneumatic directional control valve to deaden the noise of air as it exhausts.

______________ **5.** A(n) ___ switch electrically controls the energizing or de-energizing of a prime mover when a set pressure has been reached.

______________ **6.** ___ valves protect systems from overpressure by setting a maximum operating pressure.

______________ **7.** A(n) ___ air compressor compresses air by extending and retracting a piston inside a cylinder.

______________ **8.** ___ cylinders can create rotational movements in specific degrees and are used in applications where a twisting, turning, or back-and-forth motion is desired.

______________ **9.** A(n) ___ is an actuator that moves in a straight line using compressed air.

______________ **10.** ___ is a process that allows a compressor to run against no load.

______________ **11.** A(n) ___ is a pneumatic device that uses pressurized air to create rotating mechanical energy.

______________ **12.** A(n) ___ is a pipe in a multistage air compressor that connects the different stages of an air compressor.

______________ **13.** A(n) ___ supplies rotating mechanical energy in a fluid power system.

______________ **14.** A(n) ___ air compressor uses vanes that slide in and out of a rotating rotor.

______________ **15.** A(n) ___ system uses gas under pressure to accomplish work.

______________ **16.** ___ is the percentage of time that a machine operates compared to not operating.

______________ **17.** A(n) ___ air compressor allows flow to be sent from the first stage to the second stage for more energy efficiency.

______________ **18.** A(n) ___ air compressor uses intermeshing screw shafts or rotors to create flow.

______________ **19.** A(n) ___ air compressor discharges a fixed quantity of compressed air with each cycle.

______________ **20.** A(n) ___ air compressor is a type of dynamic air compressor that moves a large volume of air using high speeds.

TRUE/FALSE

T F **1.** An axial piston air motor operates at low speed and high torque with its pistons horizontal to the motor's shaft.

T F **2.** The schematic symbols for hydraulic and pneumatic systems are similar.

T F **3.** A pressure switch is either located between the air compressor and the receiver or is attached directly to the receiver.

T F **4.** The prime mover for a multistage air compressor is typically a 1ϕ AC electric motor.

T F **5.** The oil-flooded compressing mechanisms in a rotary-screw air compressor compress air with their lobes contacting each other.

T F **6.** Diaphragm relief valves are used in small pneumatic systems that are less than 25 HP.

T F **7.** Cylinder cushions can be added to pneumatic cylinders while they are in operation.

T F **8.** A double-acting cylinder is a pneumatic cylinder that has one piston and a rod that protrudes from both ends of the cylinder barrel.

T F **9.** When a compressor uses a 3ϕ AC electric motor as its prime mover, it will apply more airflow than a compressor with a 1ϕ motor.

T F **10.** Vane air motors are the most commonly used type of air motor.

T F **11.** The speed of a pneumatic cylinder is commonly measured in feet per second (fps).

T F **12.** Many receivers are cylindrical steel tanks.

T F **13.** The most common type of nonpositive-displacement air compressor is a rotary-vane air compressor.

T F **14.** Poppet relief valves and ball relief valves are installed in small pneumatic systems that are 25 HP or less.

T F **15.** Safety valves are designed strictly for safety and are not intended for frequent operation.

T F **16.** Internal combustion engines are never used as prime movers for pneumatic equipment.

T F **17.** Rotary-screw air compressors only have oil-flooded compressing mechanisms.

T F **18.** All exhaust ports in a pneumatic system must have mufflers attached to them.

T F **19.** Many three-position, four-way, five-ported directional control valves used in pneumatic systems actually have four ports.

T F **20.** Centrifugal air compressors are high-pressure, low-flow air compressors.

T F **21.** Rotary-vane air compressors are only multistage and operate at 50 psi to 125 psi.

T F **22.** A compressor that runs continuously at full power has a duty cycle of 100%.

T F **23.** Prime movers are always attached to the air compressor.

T F **24.** Rotary-screw air compressors with two intermeshing screw shafts are the most common type of rotary-screw air compressor.

T F **25.** Because the air is compressed two or more times, a multistage air compressor is less efficient than a single-stage air compressor.

SHORT ANSWER

1. List four advantages provided by a receiver in a pneumatic system.

2. Explain why unloading is important in a pneumatic system.

3. List the three main types of positive-displacement air compressors.

4. List an advantage and disadvantage of using a single large compressor to supply compressed air in a facility.

5. List four advantages of using an air motor over an electric motor.

MATCHING

Schematic Symbols

_______ 1. Bidirectional air motor

_______ 2. Cylinder with cushion on extension

_______ 3. Cylinder with cushion on extension and retraction

_______ 4. Cylinder with cushion on retraction

_______ 5. Diaphragm cylinder

_______ 6. Double-acting cylinder

_______ 7. Double-rod cylinder

_______ 8. Electric motor

_______ 9. Metering valve

_______ 10. Muffler

_______ 11. Muffler with flow control

_______ 12. Pressure switch

_______ 13. Receiver

_______ 14. Reciprocating-piston air compressor (fixed displacement)

_______ 15. Relief valve

_______ 16. Rodless cylinder

_______ 17. Rotary cylinder

_______ 18. Safety valve

_______ 19. Single-acting cylinder

_______ 20. Single-acting, spring-return cylinder

_______ 21. Unidirectional air motor

_______ 22. Vacuum cup

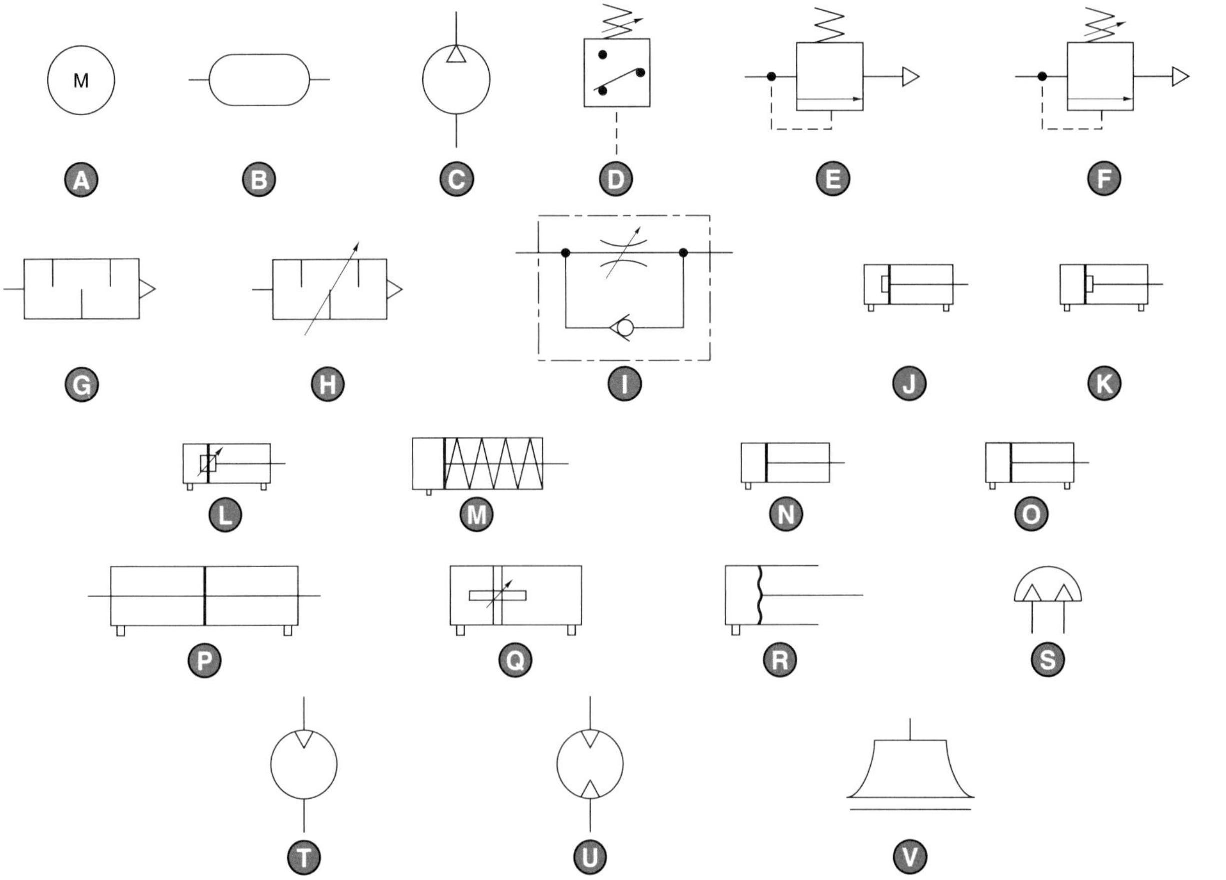

M

A B C D E F

G H I J K

L M N O

P Q R S

T U V

SCHEMATIC SYMBOLS

Activity 12-1: Calculating Duty Cycle

A production manufacturing facility must monitor the duty cycle of the air compressor in their quality control laboratory at three different time periods during the workday (at two hour cycles) to ensure that they are within the duty cycle rating for that particular air compressor. Duty Cycle Record indicates the duty cycle for the compressor at 8:00 AM, 4:00 PM, and 12:00 AM. The duty cycle rating for this compressor is 75%. Use Duty Cycle Record to answer questions 1 through 5.

Note: Duty cycle is calculated by dividing total minutes of air compressor operation by total minutes monitored and multiplying by 100.

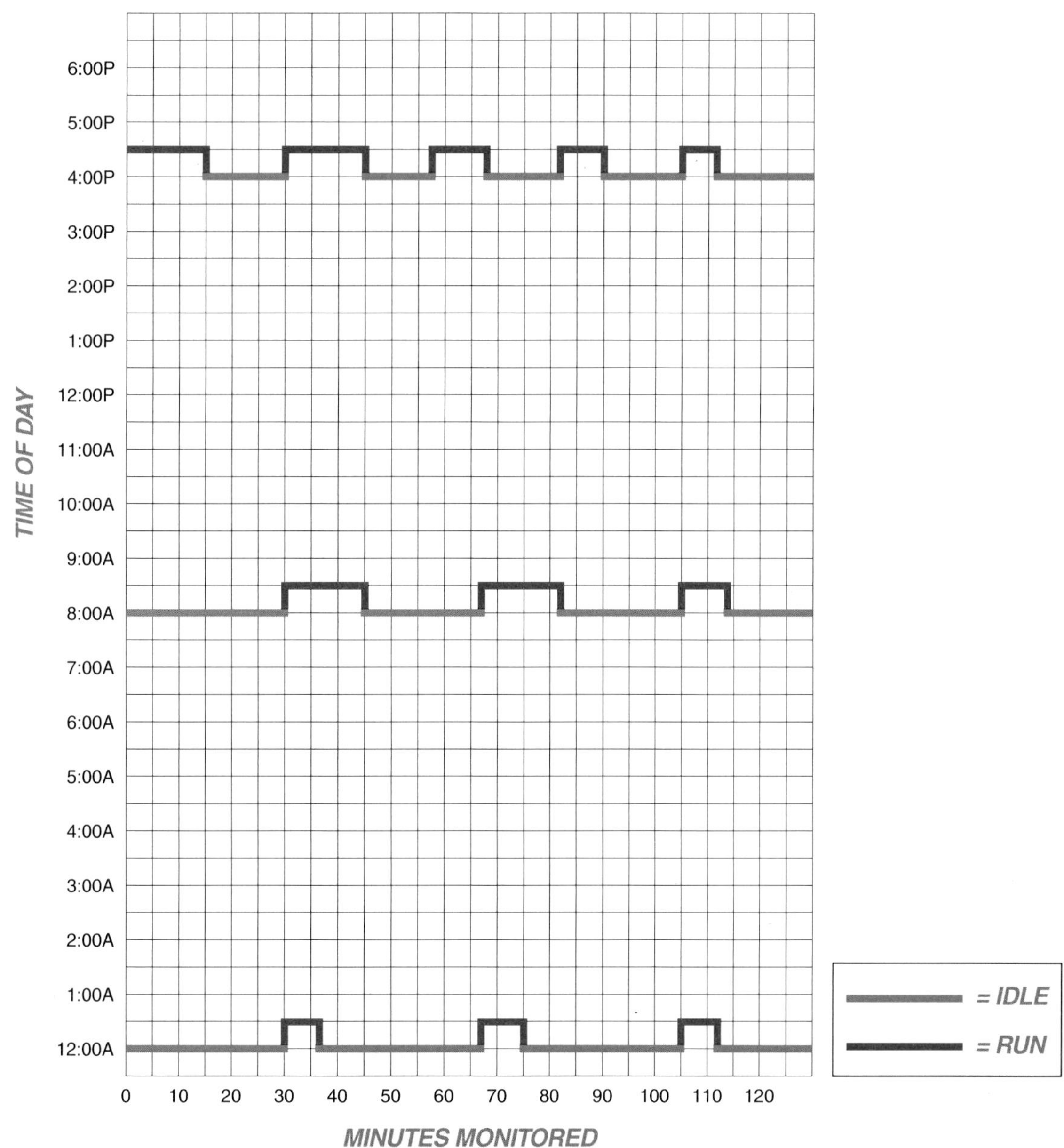

_____________________ **1.** What is the duty cycle at 8:00 AM?

_____________________ **2.** What is the duty cycle at 4:00 PM?

_____________________ **3.** What is the duty cycle at 12:00 AM?

4. Explain whether the quality control laboratory's compressor is or is not within the duty cycle rating.

5. What does the duty cycle at 12:00 AM indicate about the operation of the quality control laboratory at that time?

Activity 12-2: Compressor Run Time – Calculating Average Daily Cost

A packaging facility has decided to replace their air compressor. They need to decide between either a rotary-screw or rotary-vane air compressor. One of the selection factors is the daily cost to run the air compressor. Using the following information and the example below, calculate the daily cost to run each air compressor and answer the questions. *Note:* Assume that the factory requires compressed air 24 hours a day, has a duty cycle of 50%, and that the cost of electricity is $0.10 per kWh.

Air Compressor A is a rotary-screw air compressor and draws current of 8.5 A and operates on 220 V.

1. Calculate power.

2. Calculate compressor run time.

3. Calculate total daily power.

4. Calculate average daily cost.

Air Compressor B is a rotary-vane air compressor and draws current of 13 A and operates on 120 V.

5. Calculate power.

6. Calculate compressor run time.

7. Calculate total daily power.

8. Calculate average daily cost.

9. Which air compressor will cost the least to run each day?

10. Does the answer to question 9 change if the duty cycle changes?

11. What would be the cost to run Air Compressor A and B for a full calendar year?

Activity 12-3: Schematic Symbols – Conveyor Line

The shipping department for a distribution center wants to add a second conveyor line located in one of their shipping areas in order to separate differently sized products. The department has decided to use a pneumatic cylinder to direct the packages to the second conveyor when necessary. Either of the pushbutton switches can be used to extend the pneumatic cylinder.

1. Connect Schematic Symbols to indicate a schematic diagram that best indicates a second conveyor line.

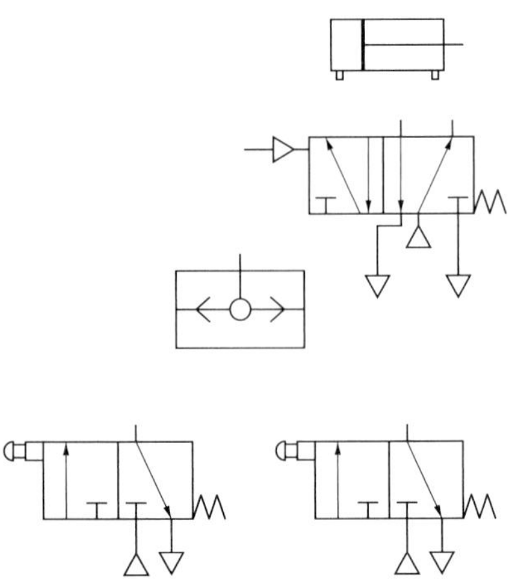

SCHEMATIC SYMBOLS

Activity 12-4: Schematic Symbols – Pneumatic Press Operation

A production manufacturing facility operates a pneumatic press that inserts ball bearings into the shaft end of electric motors.

1. Using Schematic Symbols, draw a schematic diagram that performs the following operations:
   - The cylinder automatically extends when the motor is in place by activating the pneumatic limit switch.
   - When the cylinder is fully extended, it activates another pneumatic limit switch and retracts the cylinder.

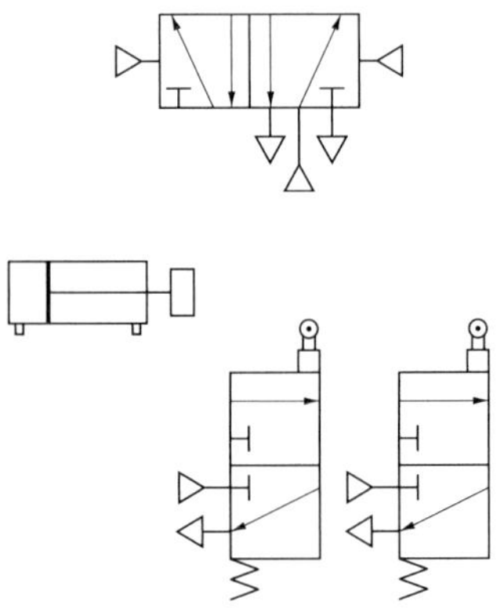

SCHEMATIC SYMBOLS

Pneumatic System Conditioning

Chapter 13

OBJECTIVES

- Describe the different types and effects of contaminants in a pneumatic system.
- List the different types of pneumatic filters.
- Describe how the different types of pneumatic filters operate.
- Describe the different types of components in pneumatic conditioning devices.
- Describe the sequence for conditioning and preparing compressed air.

INTRODUCTION

Compressed air in a pneumatic system serves the same purpose as hydraulic fluid in a hydraulic system. Compressed air must be cleaned of contaminants, cooled to the proper temperature, regulated to the required pressure, and lubricated for the proper lubrication of pneumatic system devices and components. Compressed air is cleaned of contaminants by filters and cooled by conditioning devices. When a pneumatic system is properly filtered and conditioned, it can operate problem-free for an extended time period.

TERMS

Atmospheric air is air that is located outside a facility.

Ambient air is the air that is located inside a facility.

PNEUMATIC SYSTEM CONTAMINANTS

Contaminants in a pneumatic system are similar to those present in a hydraulic system. A contaminant is any foreign object in the air of a pneumatic system. Solid contaminants, such as dirt, dust, and pipe scale, can be introduced through pipes or other pneumatic components. Liquid contaminants, such as moisture and crankcase oil from the air compressor, can also be entrained into the air. **See Figure 13-1.**

However, most contaminants enter a system through the air intake of a system. Pneumatic systems can use either atmospheric or ambient air. *Atmospheric air* is air that is located outside a facility. *Ambient air* is the air that is located inside a facility.

Contaminants negatively affect a pneumatic system. Even the smallest contaminants cause problems, such as disrupting airflow, in the tight clearances of valves. Since contaminants can be introduced through both external and internal sources, the compressed air must be filtered and conditioned as much as possible before it is used to accomplish work.

Solid Contaminants

Solid contaminants are particles of material that can damage a pneumatic system. They can include dirt, dust, metal chips, pipe scale, rubber particles, water droplet residue, rust particles, or any other solid substance that is not meant to be part of the system. Solid contaminants enter a pneumatic system through the air intake of the air compressor, from the natural wear of system components, and from the environment. For example, solid contaminants from the environment come from dirt on cylinder rods and motor shafts. Solid contaminants in pneumatic systems are rated in micrometers, or microns (μm).

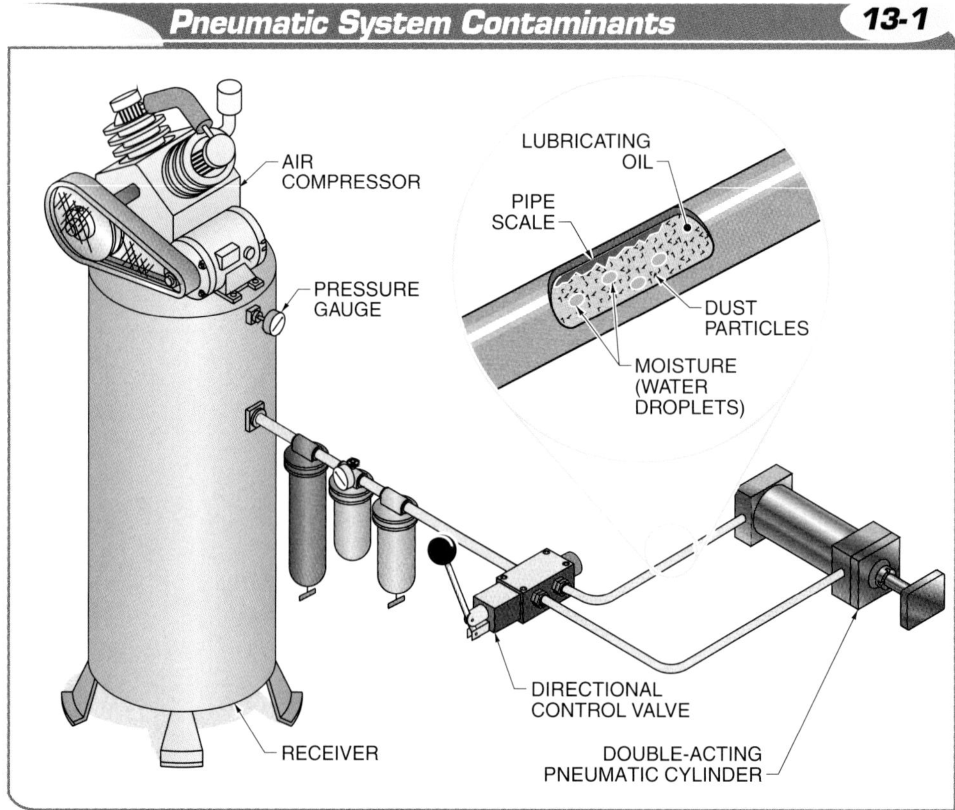

Figure 13-1. Common forms of pneumatic contaminants include dust particles, moisture (water droplets), and lubricating oil from the air compressor.

The inside air of a facility typically contains more airborne contaminants than outside air. Some exceptions include laboratories and clean rooms. Airborne contaminants are different inside every facility. For example, facilities that package product in cardboard boxes have solid contaminants from the cardboard material. A recycling facility that processes cardboard along with scrap paper and plastic has solid contaminants from these other materials and from the cardboard. In these cases, outside air is used and conditioned by the pneumatic system to avoid using ambient (inside) air.

For example, small dirt particles can lodge between a valve spool and its housing. This can cause the spool to seize or air to leak. Solid contaminants can also prevent airflow by blocking piping or the orifices of small devices, rendering equipment ineffective until the contaminant is removed. Contaminated pneumatic equipment must work harder to maintain air pressure, which leads to premature wear and wasted energy. Solid contaminants can also cause premature wear on seals, creating leaks within the pneumatic system and wasting energy.

Liquid Contaminants

Liquid contaminants can be as destructive as solid contaminants, causing problems such as breakdowns in lubrication. Air that is compressed by the air compressor can carry water vapor with it as it travels into the system. It can also accumulate oil vapor as it travels through the air compressor. Water and oil vapor must be removed from the air, otherwise they can affect the operation of the pneumatic system and contaminate the product. Each pneumatic system treats the removal of water and oil vapor differently. *Note:* In certain air compressors, accumulation of oil vapor is a sign that the air compressor requires maintenance.

Moisture. *Moisture* is small quantities of liquid condensation. *Condensation* is the change of a substance from a vapor state to a denser liquid state. This change in state is usually caused by a reduction in temperature. Moisture in a pneumatic system is typically in the form of water droplets, which are converted from water vapor (humidity) in the atmosphere after compression.

The temperature of the air is one of the determining factors for the percentage of water vapor in the air. The higher the temperature of the air, the higher the percentage of water vapor it can hold. The lower the temperature of the air, the lower the percentage of water vapor it can hold.

Water droplets that have evaporated into water vapor make up most of the moisture in atmospheric air. *Humidity* is the amount of moisture in the air. Water is a natural occurrence in a pneumatic system because the atmospheric air that is pushed into the compressor by the atmosphere contains significant amounts of water vapor. It takes 7.8 cu ft of atmospheric air to produce 1.0 cu ft of compressed air at 100 psi, which means that there is 7.8 times more moisture in compressed air than there is in atmospheric air.

Air generally contains less moisture than

TERMS

Moisture is small quantities of liquid condensation.

Condensation is the change of a substance from a vapor state to a denser liquid state.

Humidity is the amount of moisture in the air.

Atlas Copco

Because many air compressors are used in areas that have a high amount of dirt and dust in the air, incoming air must be filtered to prevent contaminating the pneumatic system.

TERMS

Saturated air is air that holds as much moisture as it is capable of holding.

Standard air is air that is at 68°F, has 36% humidity, and is at sea level (14.7 psia).

Dew point (saturation temperature) is the temperature at which the moisture in the air begins to condense.

it is capable of holding. However, as the moisture level of air rises to its maximum, the air becomes saturated. *Saturated air* is air that holds as much moisture as it is capable of holding. The total amount of moisture that air is capable of holding varies based on the temperature of the air.

Standard air is air that is at 68°F, has 36% humidity, and is at sea level (14.7 psia). Humidity and temperature determine the amount of water vapor that may be present in air. Standard air is typically used to rate the performance of system components. Standard air creates standard conditions for evaluating and comparing the operation of system components.

At standard pressure and temperature, the weight of 1 cu ft of air equals 0.076 lb. An average room (12 × 13 × 8 = 1248 cu ft) contains 94.848 lb of air (1248 × 0.076 = 94.848 lb). Saturated air at a temperature of 80°F is capable of holding about 0.022 lb of moisture/lb of dry air. The air in the average room (1248 cu ft) at saturation is capable of holding 2.086 lb of moisture (94.848 lb × 0.022 lb = 2.086 lb). Moisture in the air is more commonly expressed in grains (gr). There are 150 gr of moisture/lb of saturated air at 80°F. Therefore, an average room (1248 cu ft) contains 14,227.2 gr of moisture (150 gr × 94.848 lb = 14,227.2 gr).

Cooling the air at the air compressor reduces the moisture content in the receiver. For example, air that is at 100°F is saturated when it contains 19.766 gr of moisture/cu ft, but air that is at 60°F is saturated when it contains only 5.745 gr of moisture/cu ft. Cooler air cannot hold as much moisture as warmer air.

Water vapor condenses once the air sufficiently cools. *Dew point* (saturation temperature) is the temperature at which the moisture in the air begins to condense. **See Figure 13-2.** Typically, the atmospheric air that is used by an air compressor is not saturated because it does not contain visible moisture. However, since humidity changes from day to day, atmospheric air should still be tested before a conditioning plan is specified. Any form of water that is trapped in a pneumatic system can destroy equipment, contaminate product, or lower equipment performance.

TECH FACT

Even in well-maintained pneumatic systems, water may leak from the connections between the piping. This can cause a hazardous work environment by making the floor slippery or causing electrical short circuits. Water leaks can also contaminate or destroy product.

Oil. Oil contamination in a pneumatic system may occur from overlubricating the air compressor. An air compressor must be lubricated with crankcase oil in order to reduce the friction between internal moving parts. When air is compressed, it collects some of the crankcase oil and can carry it in the form of oil vapor into the system. **See Figure 13-3.**

Oil vapor tends to collect solid contaminants that are in the pneumatic system. Oil vapor and solid contaminants join to form large contaminant particles, which can easily clog small orifices. Flow control valves, filters, and cylinder cushions can also become clogged, causing them to improperly function and prematurely wear.

Oil vapor is collected in a pneumatic filter as compressed air flows through it. However, the more oil vapor that collects in the filter, the more difficult it is for compressed air to flow through the filter. This reduces the effectiveness of the filter, which can require frequent replacement.

PNEUMATIC FILTERS

To prevent contamination, the air that is used by a pneumatic system must be cleaned before it reaches any work devices. Ideally, pneumatic filters should be installed at every location where work is performed. However, this is not practical because the initial cost to install and maintain a large number of filters can be too high. Instead, many facilities install filters near work devices to filter out as many contaminants as possible.

Pressure drop is the pressure differential between upstream and downstream airflow caused by resistance. The resistance that creates pressure drop can be caused by friction between airflow and piping, by an intentionally restricted orifice, or by contaminants that are blocking airflow. Pressure drops on pneumatic filters should not exceed 2 psi.

In some pneumatic systems, multiple filters connected in series are used. The filter having the coarsest element is installed first, followed by one or more filters with increasingly finer elements. The success of the first filter elements at removing solid contaminants determines the operational lifetime of any downstream filter elements. The two basic types of pneumatic filters used are intake and inline filters.

Intake Filters

An *intake filter* is a pneumatic filter that is used to remove contaminants from air (typically atmospheric air) before it travels into the inlet of the air compressor. **See Figure 13-4.** There are two locations for intake filters that are typically used. One location for the intake filter is to directly attach it to the air compressor intake. The advantage to this location for the intake filter is that little or no piping is required. The disadvantage is that the environment where the air compressor is located can contain solid and chemical contaminants, and the air is typically much warmer than the outside air.

The other type of intake filter is a remote intake filter. A *remote intake filter* is a pneumatic filter located away from the air compressor and attached to pneumatic piping. The advantages to using a remote intake filter are that the air compressor can be located inside a building, the air compressor can be closer to the work devices, and the air can be pulled from a cleaner environment. The disadvantage is that extra piping is needed because the air pushed in is from a distant location. Remote intake filters are typically installed in facilities that have contaminated air but need to keep the air compressors near the work devices.

Saturated Air Moisture-Holding Properties 13-2

Dew Point*	Pounds of Moisture	Grains of Moisture†	Grains of Moisture‡
100	0.044	300	19.766
90	0.03	210	14.790
80	0.022	150	10.934
70	0.015	110	7.980
60	0.011	76	5.745
50	0.0075	53	4.076
40	0.005	36	2.749
30	0.0033	24	1.935
20	0.002	15	1.235
10	0.0014	9	0.776
0	0.0008	5.5	0.481

* in °F
† per lb of dry air
‡ per cu ft of dry air

Figure 13-2. The total amount of moisture that air is capable of holding varies based on the temperature of the air.

Oil Contamination 13-3

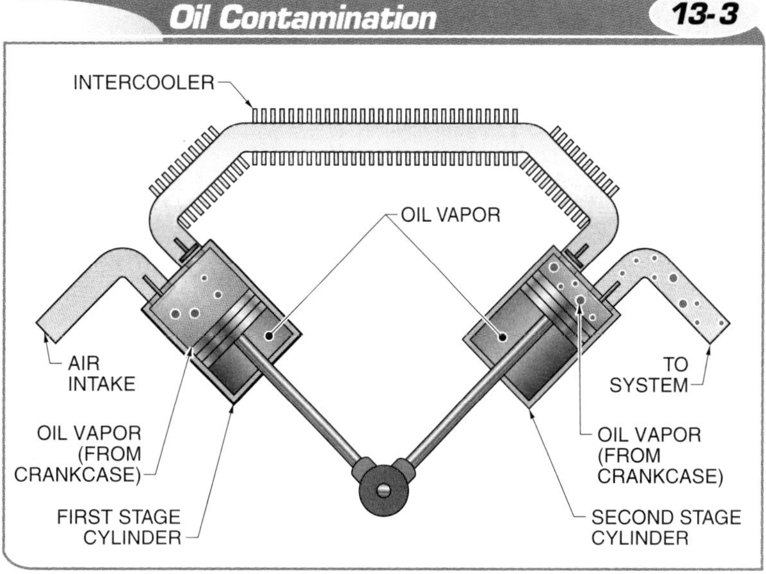

Figure 13-3. When air is compressed in an old and worn air compressor, oil can be discharged into the system.

TECH FACT

The coarsest grade of filter element that satisfactorily protects the system should be used because it provides the longest operational lifetime without causing sizable pressure drop.

TERMS

Pressure drop is the pressure differential between upstream and downstream airflow caused by resistance.

An **intake filter** is a pneumatic filter that is used to remove contaminants from air (typically atmospheric air) before it travels into the inlet of the air compressor.

A **remote intake filter** is a pneumatic filter located away from the air compressor and attached to pneumatic piping.

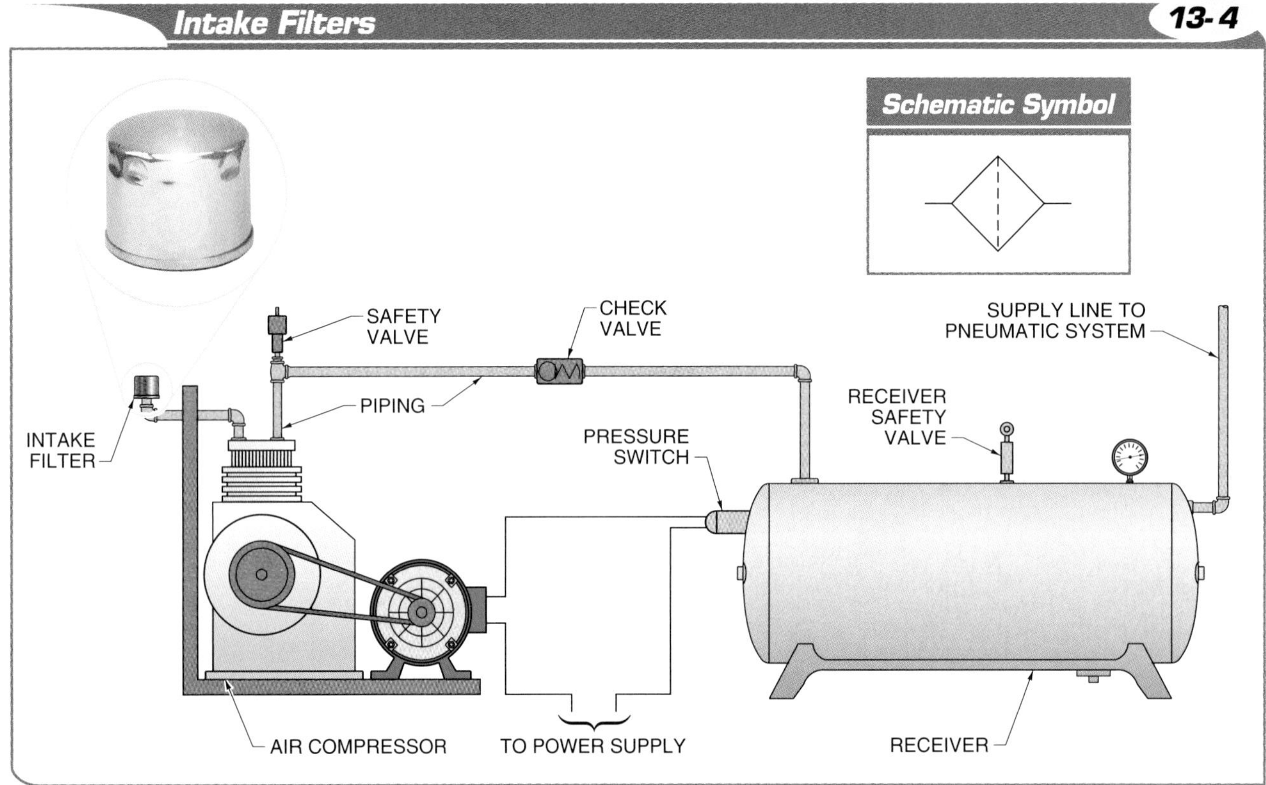

Flow Ezy Filters, Inc.

Figure 13-4. An intake filter removes contaminants from the air before it is pulled into the air compressor.

TERMS

An **inline filter** is a pneumatic filter that removes contaminants from within a pneumatic system.

Large industrial pneumatic systems use large intake filter units or specially designed rooms that work in three to four stages. The first stage prefilters contaminants sized 10 μm to 15 μm to protect the filters in the next stages. The second stage filters out specific contaminants that can damage pneumatic devices. The second stage filtration unit can be customized to facility specifications. The third stage is either the last stage or an additional prefilter for the fourth stage. It filters the air to 1 μm. If there is a fourth stage, it filters contaminants sized 0.3 μm and larger.

Note: Intake filters must be replaced per manufacturer recommendations. When an intake filter becomes clogged, the air compressor must work harder (create a deeper vacuum) to acquire its air, which can cause the filter element to collapse. Once an intake filter element collapses, airborne contaminants can enter the air compressor and contaminate the pneumatic system.

Inline Filters

Contaminants can be collected in the compressed air as it travels throughout a pneumatic system. An inline filter or separator is used to filter out the collected contaminants. An *inline filter* is a pneumatic filter that removes contaminants from within a pneumatic system. Inline filters include an inlet port, an outlet port, filter housing, a baffle, a filter element, and a filter screen. **See Figure 13-5.** An inline filter operates through the following procedure:

1. Air enters the inline filter and is forced downward by the baffle. This forces the air to rotate within the filter housing.
2. As the air rotates, large solid contaminants and liquid contaminants are forced against the inside wall of the filter housing.
3. All contaminants then fall past the baffle to the bottom section of the filter housing. *Note:* The bottom section of

an inline filter is referred to as the quiet zone because there is no air rotation beneath the baffle.

4. All compressed air is then forced through the filter element to filter out any remaining contaminants. Most inline filter elements remove contaminants sized 3 μm to 5 μm. However, custom filters can be installed to filter out smaller contaminants.
5. Compressed air then exits the inline filter through the outlet and continues into the pneumatic system.

Inline filters can have depth-type or surface-type elements. A *depth-type element* is a pneumatic filter element that has a deep, vertical filter. Contaminants are trapped in a depth-type element because the compressed air is forced to take an indirect path. Depth-type elements are typically constructed from a bronze alloy.

A *surface-type element* is a pneumatic filter element that allows compressed air to flow through the filter in a straight line to collect contaminants on its surface. **See Figure 13-6.** The surface that collects the contaminants must always face the airflow. Surface-type elements are usually constructed of treated paper.

Filter elements must be replaced at manufacturer-recommended intervals. The filter element can either be removed and manually cleaned, or cleaned by activating an automatic clean-out mechanism. Automatic clean-out mechanisms should never be placed near an area that has energized electrical equipment or near areas that need to be contaminant free.

TERMS

A **depth-type element** is a pneumatic filter element that has a deep, vertical filter.

A **surface-type element** is a pneumatic filter element that allows compressed air to flow through the filter in a straight line to collect contaminants on its surface.

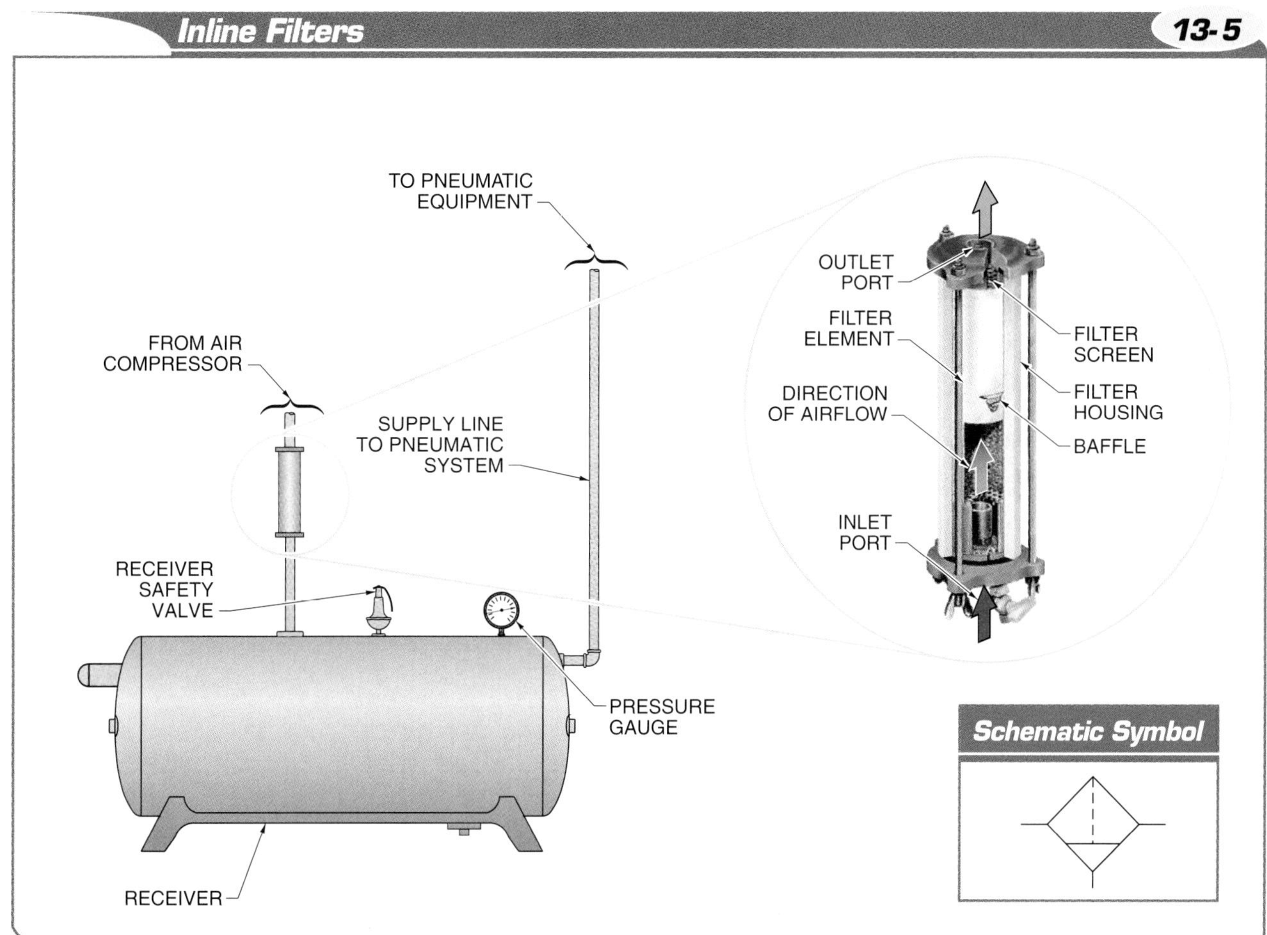

Figure 13-5. Inline filters are used to filter contaminants that have collected in a pneumatic system.

Surface-Type Filter Elements 13-6

PRESSURE CONTROL
NEEDLE HOUSING
DIFFUSER
DIAPHRAGM SPRING ADJUSTMENT SCREW
FILTER ELEMENT
DIAPHRAGM SPRING
DROPLET SHIELD
DIAPHRAGM
BODY
FILTER BOWL
NEEDLE VALVE
FILTER HOUSING
VALVE SPRING
NUT

Figure 13-6. A surface-type filter element allows compressed air to flow in a straight line through the filter and collects contaminants on its edge.

Coalescing Filters. A *coalescing filter* is a device that removes submicron solids, water vapor, and oil vapor by combining small droplets into larger droplets. Coalescing filters are sometimes referred to as inline liquid filters. Coalescing filters should be installed as close as possible to the point of work. **See Figure 13-7.**

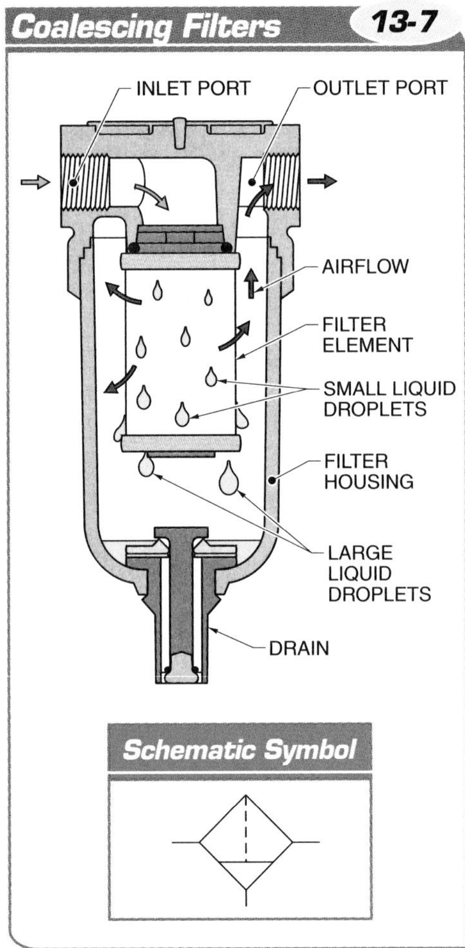

Figure 13-7. Coalescing filters remove submicron solids, water vapor, and oil vapor by combining small droplets into larger droplets.

TECH FACT

If a filter element becomes restricted by contaminant particles, the pressure drop through the filter will increase. An increase in the pressure drop through the filter causes pneumatic equipment to operate at slower speeds and with less force.

In a coalescing filter, fine liquid droplets are continuously trapped in the filter element. The liquid contaminant droplets grow in size and emerge on the exterior surface of the element. From there, they flow to the filter drain. Even when they are saturated, coalescing filters function at their original efficiency. They continue to function well until clogged by large amounts of contaminants. When coalescing filters become clogged, they must be discarded and replaced. When they collect too many contaminants, the pressure drop of a coalescing filter can be greater than 10 psi.

Some solid contaminants can enter a coalescing filter even after prefiltering, shortening its operational lifetime. A properly maintained coalescing filter should have approximately 2000 operational hours. Oil vapor not removed by a coalescing filter can enter the filter and destroy it. Coalescing filters are most effective when placed in the coolest locations in the system. If some types of coalescing filters are used in environments where the temperature is above 90°F, the water-absorbing material in the filter may transform into an acidic material that is pushed downstream towards the pneumatic components, degrading them.

PNEUMATIC CONDITIONING DEVICES

Compressed air must be properly conditioned before it can be sent throughout a pneumatic system. Compressed air must also be at the proper temperature to help control water vapor. Moisture is removed by using intercoolers and aftercoolers. Depending on the type of application using the compressed air, a dryer or a filter-regulator-lubricator (FRL) can also be used. FRLs provide filtering, regulate the pressure of compressed air to meet system requirements, and provide lubrication for work devices.

Intercoolers

Heat generated in a cylinder of an air compressor is removed by conduction through the cylinder walls and intercooling. *Intercooling* is a process that removes a portion of the heat of compression while air is piped from one stage of compression to another. An intercooler is a pipe in a multistage air compressor that connects the outlet port of a large cylinder with the inlet port of a small cylinder. **See Figure 13-8.** A tube-and-shell heat exchanger is a type of intercooler that commonly use water to cool compressed air in industrial compressors.

Intercooling reduces the temperature of the compressed air, as well as its volume, before it reaches the next stage. For every 5°F absorbed at the intercooler, approximately 1% of horsepower is saved. Without intercooling, there cannot be an overall reduction in the temperature of the compressed air discharged from a multistage air compressor.

An *aftercooler* is a heat exchanger used for cooling the discharged air from an air compressor. Aftercoolers control the amount of water vapor in a pneumatic system by lowering the temperature of the air, which condenses the water vapor into a liquid. Aftercoolers can be air-cooled or water-cooled.

TERMS

A **coalescing filter** is a device that removes submicron solids, water vapor, and oil vapor by combining small droplets into larger droplets.

Intercooling is a process that removes a portion of the heat of compression while air is piped from one stage of compression to another.

An **aftercooler** is a heat exchanger used for cooling the discharged air from an air compressor.

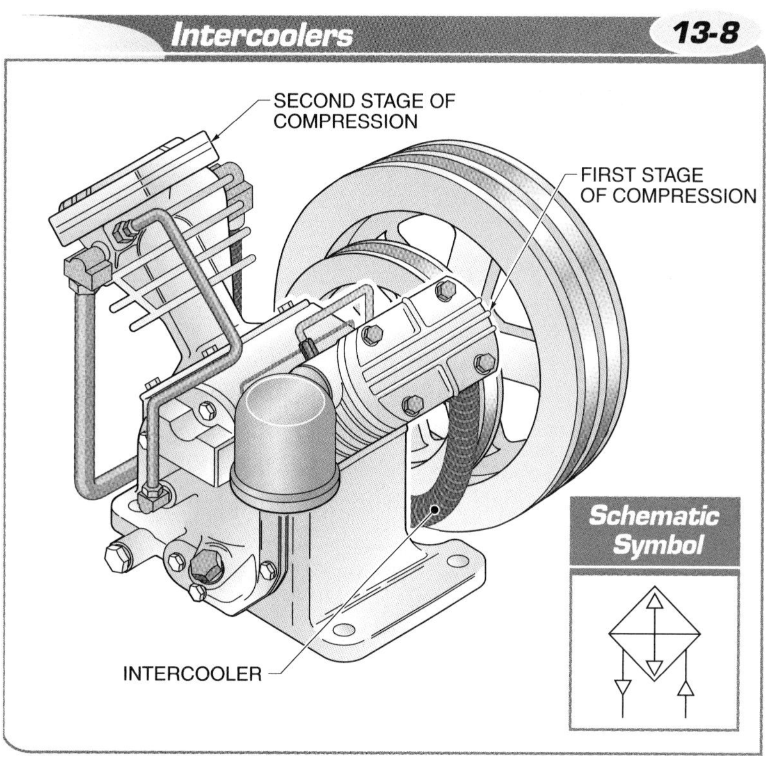

Figure 13-8. Intercoolers are used to remove a portion of the heat of compression as the air is piped from one stage of compression to another.

TERMS

A **moisture separator** is a pneumatic device that separates moisture from compressed air by forcing it to flow against baffles.

An **air dryer** is a pneumatic device that dries compressed air through cooling, condensing, or absorbing.

Dry air is air free of water and oil vapor.

A **refrigerated air dryer** is a device designed to lower the temperature of compressed air to 35°F to separate the water out.

A **desiccant air dryer** is a device that removes water vapor from compressed air through adsorption.

Adsorption is the adhesion of a gas or liquid to the surface of a porous material.

An air-cooled aftercooler blows atmospheric air against the pipes containing the compressed air. The pipes are typically bent and shaped to allow for more time and surface area to lower the temperature of the air. Air-cooled aftercoolers are not as efficient as water-cooled aftercoolers and tend to be used only in non-industrial applications.

In a water-cooled aftercooler, the pipes containing the compressed air are submerged in water. Compressed air flows through the pipes in one direction while cool water is sent through the aftercooler housing in the opposite direction. This transfers the heat from the compressed air to the water. The warm water then flows out of the aftercooler and the cool compressed air flows to a moisture separator. A water-cooled aftercooler can remove 80% or more of the moisture in the compressed air through condensation.

Moisture Separators

A moisture separator installed at the discharge port of an aftercooler can remove the largest amount of moisture from the compressed air. A *moisture separator* is a pneumatic device that separates moisture from compressed air by forcing it to flow against baffles. **See Figure 13-9.**

The baffles in a moisture separator are shaped and placed so that they cause the compressed air to rotate. This rotation causes air to flow in a centrifugal pattern, which forces large contaminants (mostly water particles 10 μm and larger) to strike against the interior wall of the moisture separator and collect at the bottom. The water at the bottom can then be drained.

Air Dryers

An *air dryer* is a pneumatic device that dries compressed air through cooling, condensing, or absorbing. Air dryers use either refrigerated air or a desiccant material. **See Figure 13-10.** *Dry air* is air free of water and oil vapor. Water and oil vapor is not always visible, most water vapor droplets are 0.5 μm to 2 μm in size, and the smallest droplets are visible at about 15 μm. Air dryers are used to dry air for sensitive equipment in instrumentation, medical applications, or food processing.

A *refrigerated air dryer* is a device designed to lower the temperature of compressed air to 35°F. The cooling provided by a refrigeration system causes the water in the air to condense by lowering the relative humidity and dew point of the air. Sometimes, cool dry air flows through a heat exchanger to precool the incoming air. Refrigerated air dryers are the most common type of air dryers used in industrial pneumatic systems. They are typically preassembled and only require a power supply once they are installed.

A *desiccant air dryer* is a device that removes water vapor from compressed air through adsorption. *Adsorption* is the adhesion of a gas or liquid to the surface of a porous material. A desiccant air dryer removes water vapor at the point of work using desiccant material such as silica gel or alumina. A desiccant air dryer is used to remove invisible water vapor when maximum drying is required.

Desiccant air dryers are normally placed in the coolest downstream location. To prevent contamination, desiccant dryers must be used for water removal only. Oil that is not removed before it reaches the desiccant air dryer builds up on the dryer shell walls. Desiccant air dryers are most effective when placed after coalescing filters and closest to the point of work. The air entering a desiccant air dryer can have a maximum temperature of 94°F.

Desiccant material becomes saturated because adsorption can remove 99.9% of the water vapor in the air. For this reason, two tanks are interconnected to allow for the regeneration of saturated desiccant. While one side of the desiccant dryer is drying compressed air, the other is reactivated by dry air. The desiccant used to dry compressed air can also adsorb any oil present in the system, which destroys the desiccant. The desiccant in the dryer must then be replaced. *Note:* The desiccant material that builds up on the dryer shell's walls is a hazardous material and must be disposed of, according to government regulations.

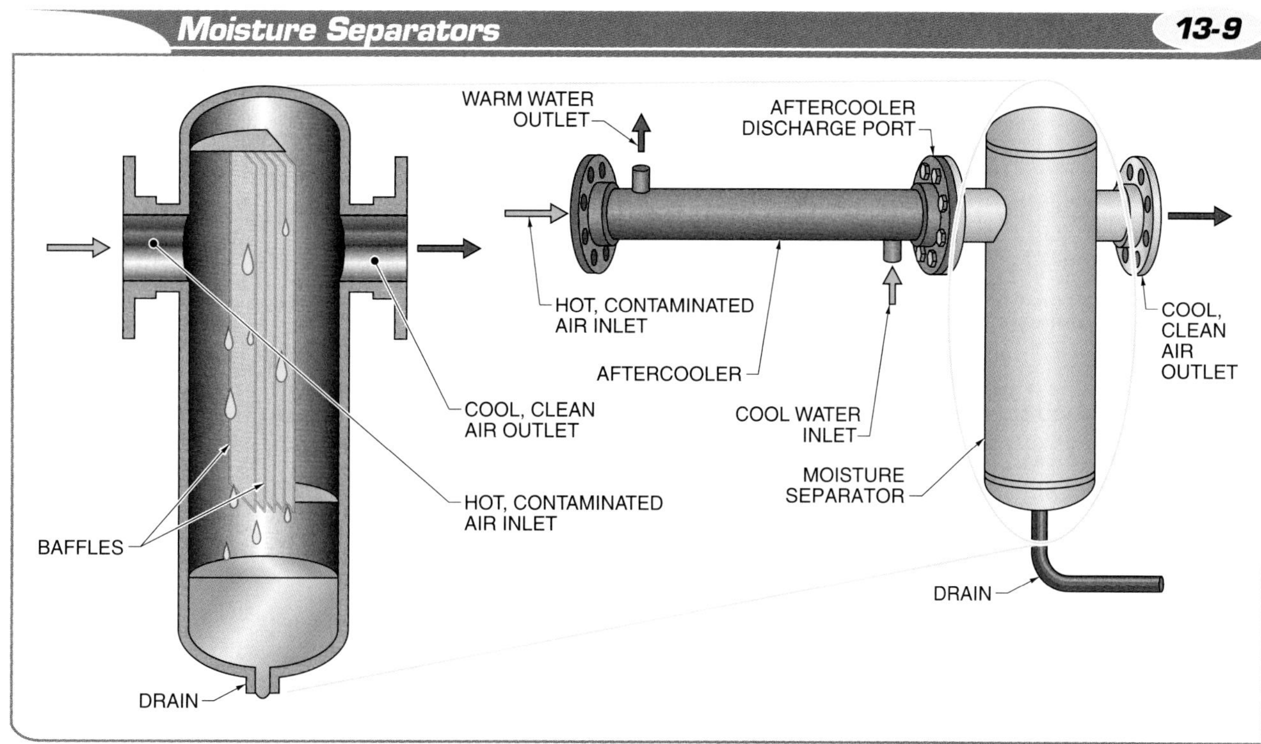

Figure 13-9. A moisture separator separates moisture from compressed air by forcing it to flow against baffles.

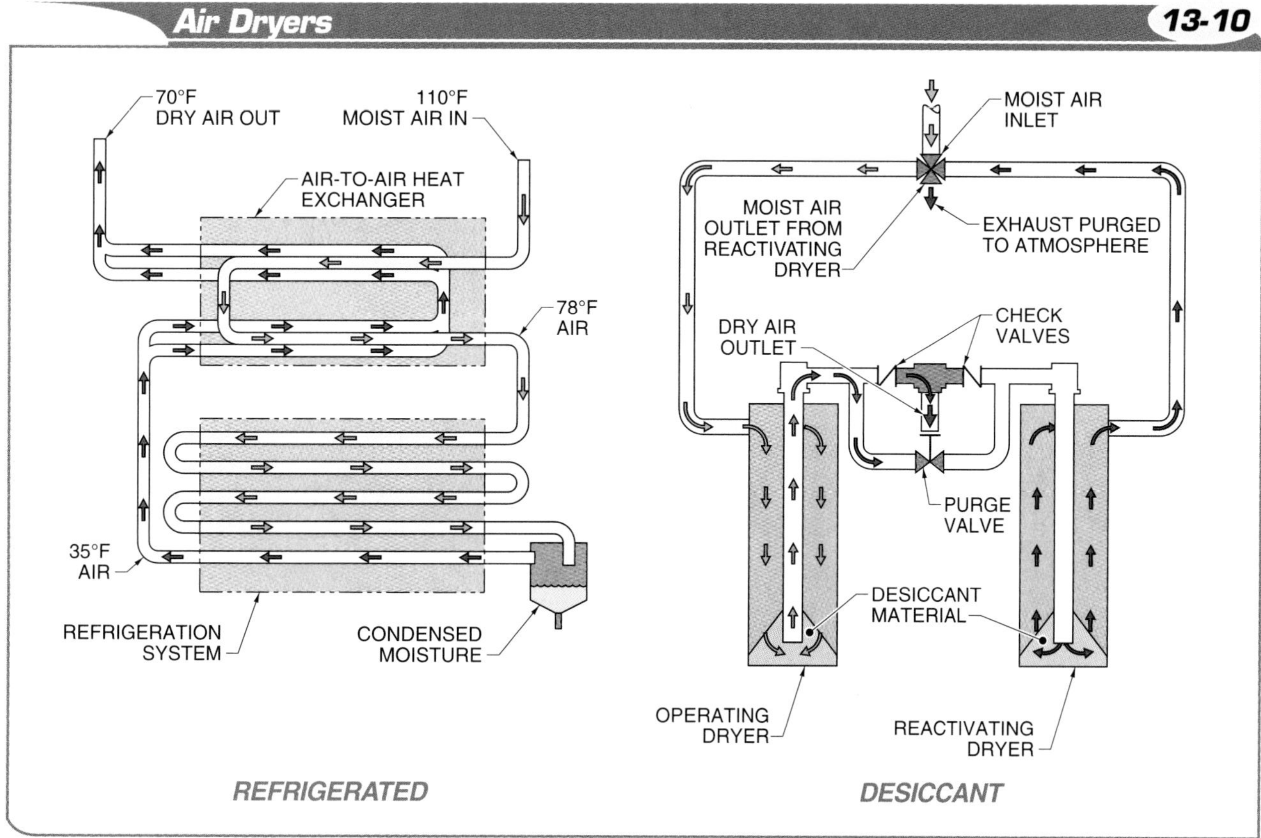

Figure 13-10. Compressed air in a pneumatic system is dried by either refrigeration or desiccants.

TERMS

A **deliquescent dryer** is a pneumatic device that removes moisture through absorption.

Absorption is the drawing of gases or vapors into permeable pores of porous material that results in physical and/or chemical changes in the material.

A **filter-regulator-lubricator (FRL)** is a pneumatic device that is used to filter, regulate, and lubricate compressed air before it reaches the point of work.

A **regulator** is a normally open pressure valve that is used to control downstream pressure.

A *deliquescent dryer* is a pneumatic device that removes moisture through absorption. *Absorption* is the drawing of gases or vapors into permeable pores of porous material that results in physical and/or chemical changes in the material. The moisture is removed by forcing the air through a chemical desiccant that removes moisture in vapor or condensed form by absorbing it. This type of dryer drops the dew point of the air about 25°F below the inlet temperature. The air temperature coming into a deliquescent dryer should never exceed the manufacture's recommendation because it will consume too much of the chemical desiccant and the desiccant will need to be replaced more often. One disadvantage to using deliquescent dryers is that the desiccant chemical must be removed from the dryer after several cycles by a manual release valve and disposed of according to OSHA regulations.

Filter-Regulator-Lubricators (FRLs)

A *filter-regulator-lubricator (FRL)* is a pneumatic device that is used to filter, regulate, and lubricate compressed air before it reaches the point of work. **See Figure 13-11.** The filters used with most FRLs are similar to inline filters. While the regulator lowers the pressure going to the pneumatic system, the lubricator injects lubricant into the airflow. FRLs must be installed as close as possible to the point of work to maximize pressure control and provide proper lubrication for components.

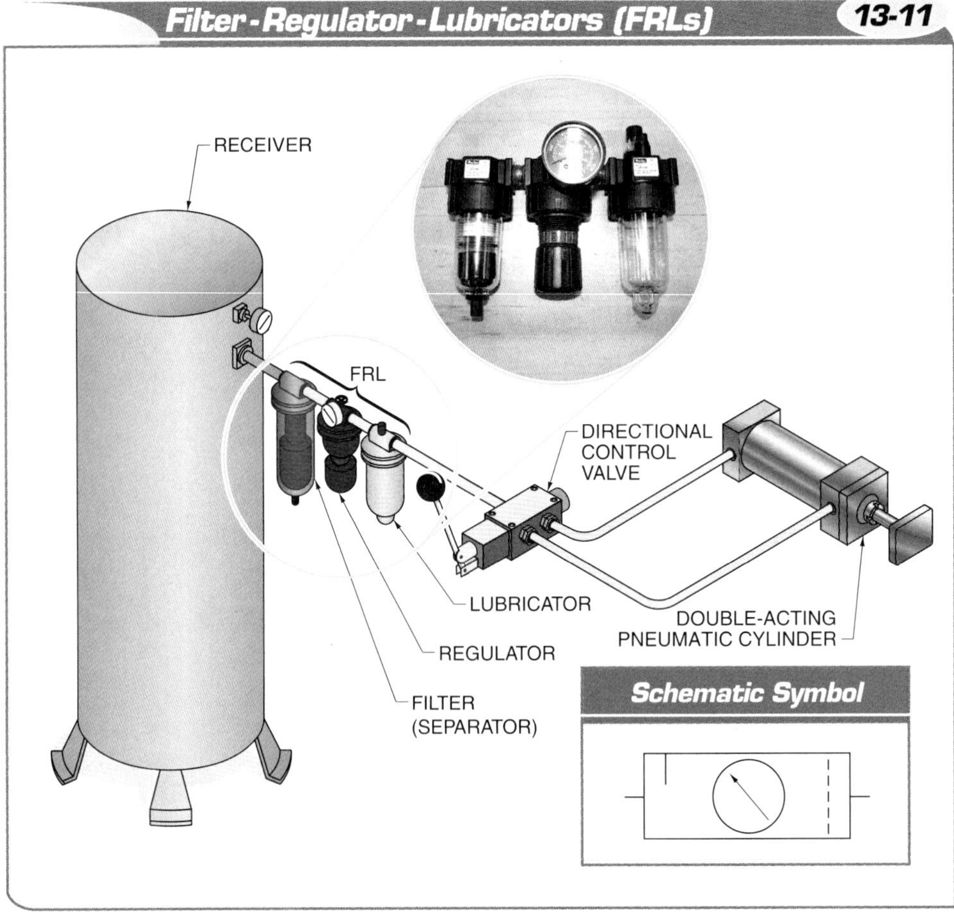

Figure 13-11. FRLs must be installed as close as possible to the point of work to maximize pressure control and provide proper lubrication for components.

TECH FACT

Desiccant material used in pneumatic systems is temperature sensitive and must be handled properly. Desiccant material becomes highly acidic when exposed to air temperatures of 94°F or higher, and therefore must be kept at a temperature below 94°F to avoid a hazardous condition.

A *regulator* is a normally open pressure valve that is used to control downstream pressure. A regulator includes an inlet port, an outlet port, a main biasing spring, a small biasing spring, a poppet, a diaphragm or piston, a pilot port, and an adjustment screw or knob. Regulators can have a diaphragm or piston design. **See Figure 13-12.** A regulator operates in the following manner:

1. Compressed air from the piping enters the inlet of the regulator. Compressed air then flows through the outlet and pilot ports.
2. If system pressure is higher than the pressure of the main biasing spring, the piston moves up.
3. As the piston moves up, the small biasing spring under the poppet forces the poppet to rise.
4. As the poppet rises, the size of the orifice that the compressed air is flowing through decreases, reducing the airflow, thereby after the regulator.

A regulator is adjusted by setting the adjustment screw or knob of the main biasing spring. A regulator can be used alone or with a filter and/or a lubricator. It should be set to the highest pressure needed by the pneumatic system.

Piping in a pneumatic system is usually designed to branch off into several different smaller subsystems that all receive compressed air from the one main header pipe or loop. For this reason, the system relief valve is set to the highest pressure needed by the system. This also saves energy and cost. After the system relief valve, each subsystem has at least a regulator to drop the pressure to the level that is needed for the subsystem.

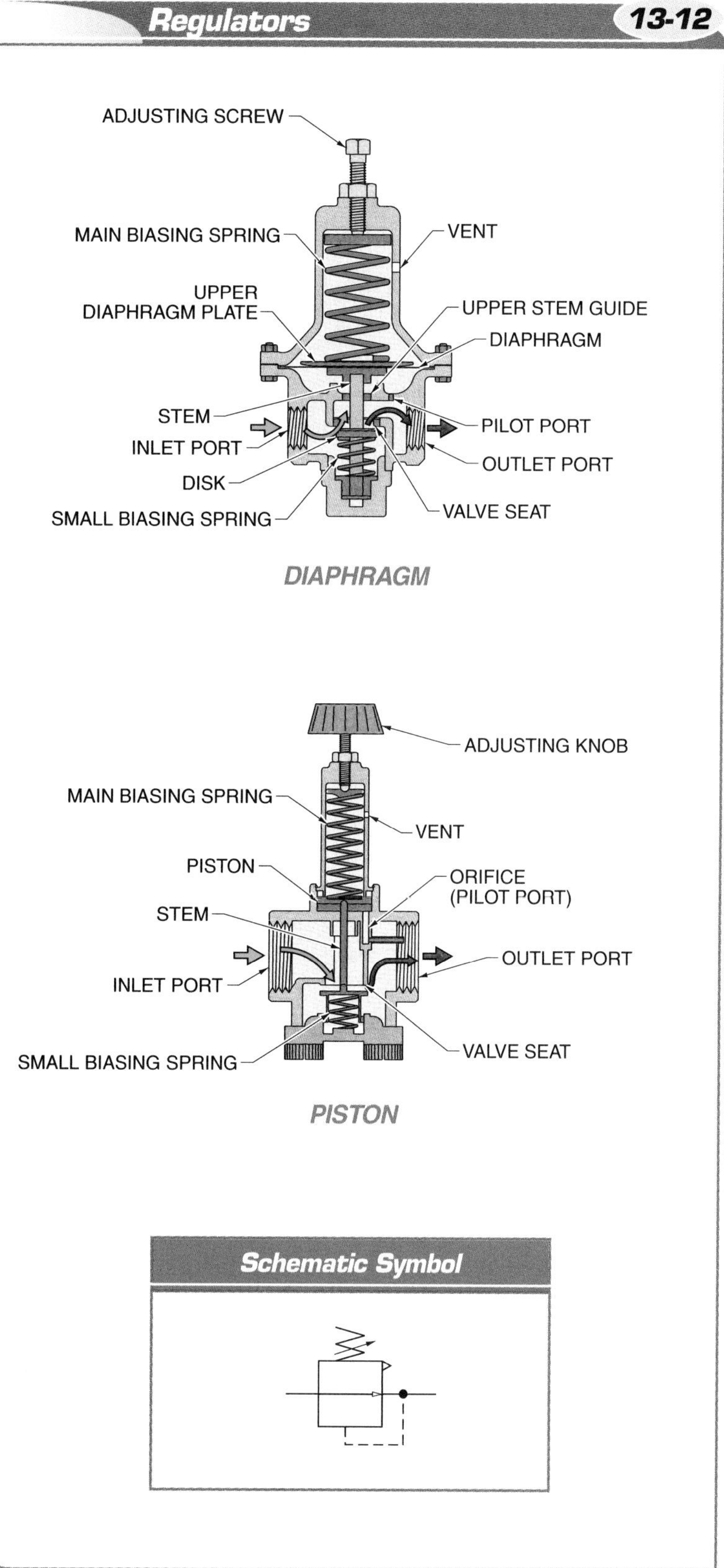

Figure 13-12. A regulator is a pressure-reducing valve used to control downstream pressure.

TERMS

A **lubricator** is a device that injects a lubricant into the airflow.

A *lubricator* is a device that injects a lubricant into the airflow. A lubricator is used to provide lubrication to component parts that move in a pneumatic system. The injected lubricant prevents the premature deterioration of seals and decreases friction between moving parts. A lubricator consists of an inlet port, an oil adjustment screw, a drip tube (on certain designs), a fill port, an outlet port, a mist generator, an oil filter, an oil reservoir, a manual drain, a metal bowl guard, a feeder tube, and an orifice. Lubricator designs can be mist or drip-feed. **See Figure 13-13.**

A mist lubricator operates through the following procedure:

1. Compressed air enters the inlet port of the lubricator.
2. The majority of the compressed air continues to flow through to an orifice that creates a slight pressure differential between the inlet and the outlet.
3. The compressed air that flowed through the pilot line applies pressure to the top of the lubricating oil stored in the reservoir.
4. Pressure forces the lubricating oil up the feeder tube.
5. Lubricating oil flows up the feeder tube through a needle valve, which regulates the amount of lubricating oil that is released into the airflow.
6. As drops of lubricating oil from the drip tube enter the airflow, they are atomized into an air mist and combined with the compressed air.

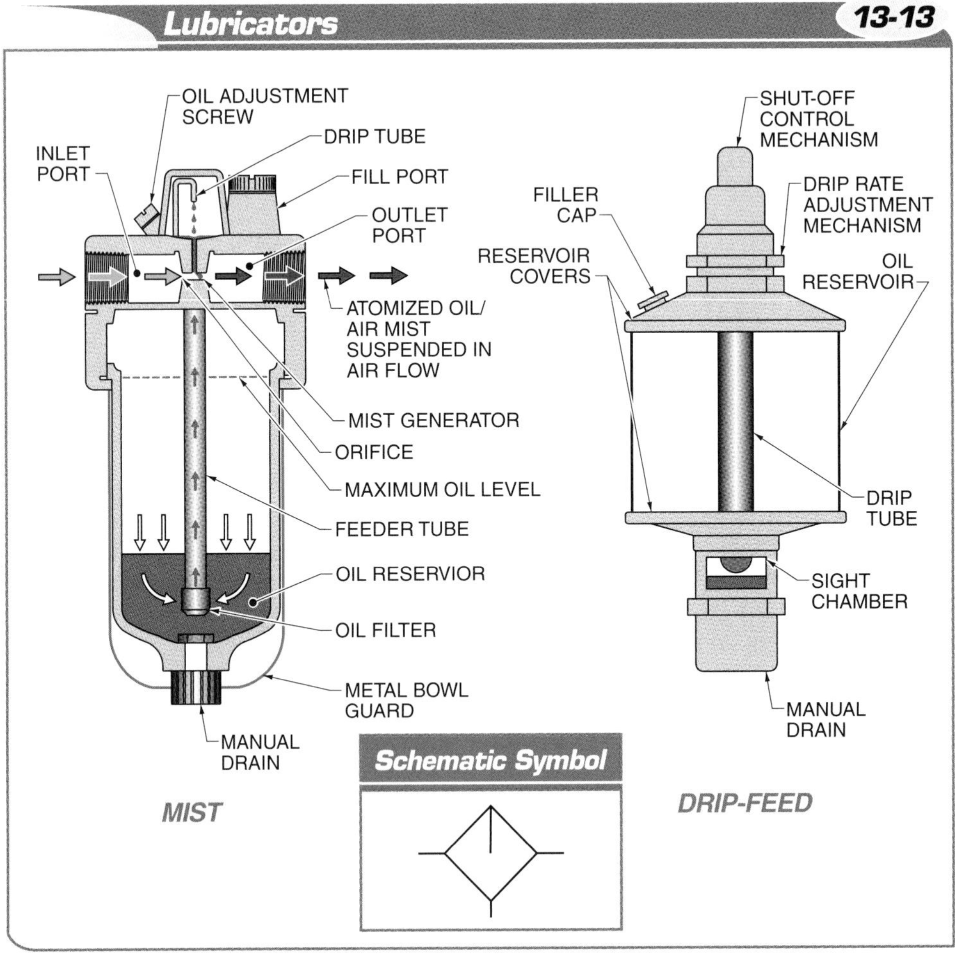

Figure 13-13. Lubricators provide lubrication for parts in a pneumatic system that actuate.

Mist lubricators are used to supply lubrication to one or two different pneumatic systems or components. A mist lubricator operates by sending the compressed air through a venturi (a tube that creates a vortex), which pulls oil from a reservoir through a tube. The compressed air creates a mist of oil particles 5 μm or smaller, while larger oil particles return to the oil reservoir. The smaller oil particles can travel up to 1000′ until gravity pulls them out of the airflow.

Drip lubricators must be installed no more than 10′ away from the components that require lubrication. Otherwise, gravity will pull the oil out of the airflow before it can provide lubrication. A drip lubricator should be the last component that air flows through before the point of work. Also, it must never be installed before a filter. A drip lubricator installed before a filter causes the lubricating oil to clog the filter, blocking airflow.

Certain pneumatic components are available prelubricated. The advantages of prelubricated pneumatic components are that they reduce the use of oil and eliminate the hazard of it entering the breathable air in the facility. Breathable air contaminated with oil can cause health hazards. When a lubricator is not required, an air filter-regulator is used. An *air filter-regulator* is a pneumatic device that filters and regulates airflow through prelubricated components. With the exception that it does not use a lubricator, an air filter-regulator is designed and operated similar to a FRL. **See Figure 13-14.**

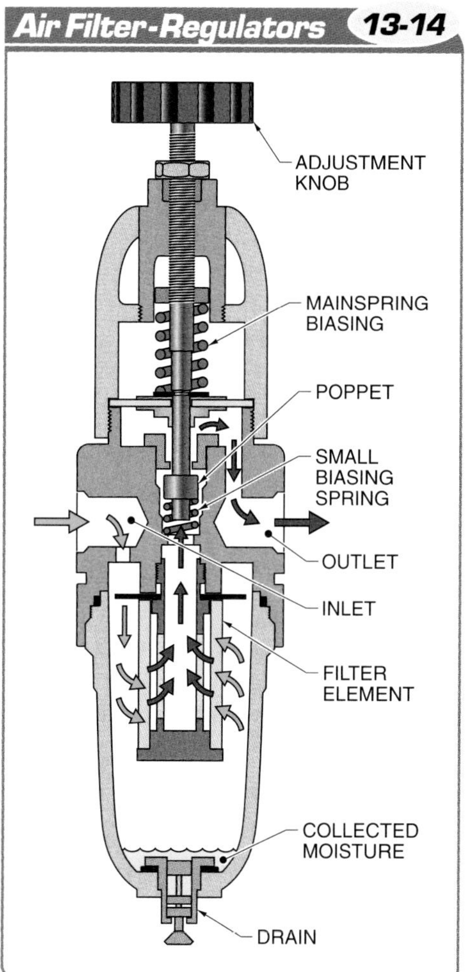

Figure 13-14. An air filter-regulator filters and regulates airflow through prelubricated components.

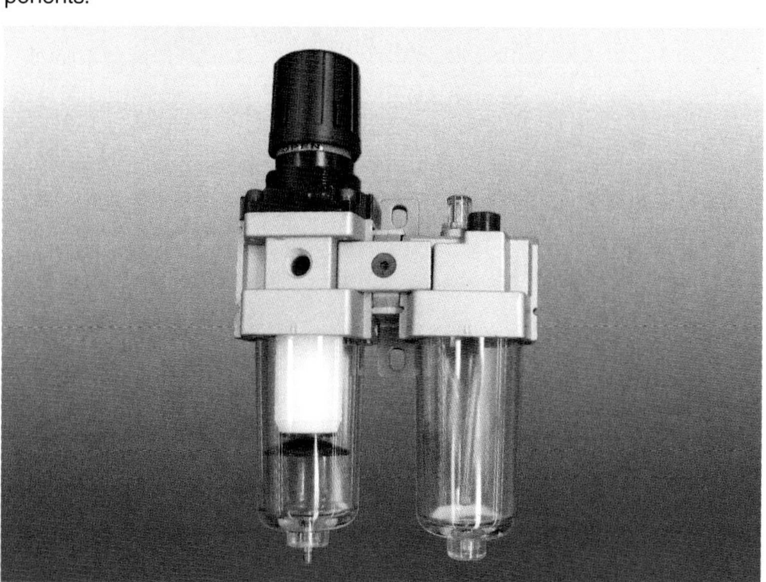

Pneumatic filters and regulators are typically combined into a single unit.

TERMS

An **air filter-regulator** is a pneumatic device that filters and regulates airflow through prelubricated components.

TECH FACT

Leaks in a regulator must be repaired immediately because a complete rupture of the diaphragm could allow full-pressure air to flow to downstream valves and cylinders. In addition, some directional control valves can shift themselves without being energized, which can cause a hazard such as premature cylinder extension.

COMPRESSED AIR PREPARATION PROCEDURE

After compression, air must be properly prepared to efficiently accomplish work without damaging the pneumatic system or contaminating the environment where the system is installed. To prepare compressed air and remove contaminants, components must be installed in the proper sequence. **See Figure 13-15.** Compressed air is prepared through the following procedure:

1. Atmospheric air is pushed into the air compressor. Before it reaches the air compressor, the air flows through an intake filter to remove the greatest amount of atmospheric contaminants as possible. In some systems, the air is run through a separator before it gets to the air compressor.
2. When the air is compressed, its temperature rises and it can sometimes collect oil vapor from the pneumatic compressor system. Note: With two-stage air compressors, the air runs through an intercooler.
3. Compressed air flows through an aftercooler, which lowers its temperature to the dew point of the facility. As the compressed air reaches the dew point temperature, the water vapor liquefies. The compressed air then flows through a moisture separator, which extracts up to 80% of the oil and water in the air.
4. The air is then stored in a receiver where water and containments settle at its bottom.
5. In many industrial pneumatic systems, the compressed air flows through a dryer. If compressed air flows through a dryer, it must also flow through an inline filter and coalescing filter.
6. The compressed air is regulated down to the lowest level that the pneumatic system needs. This is typically accomplished in locations where a line supplies compressed air to the point of work.
7. Compressed air flows through an air filter-regulator for additional filtering and regulating.

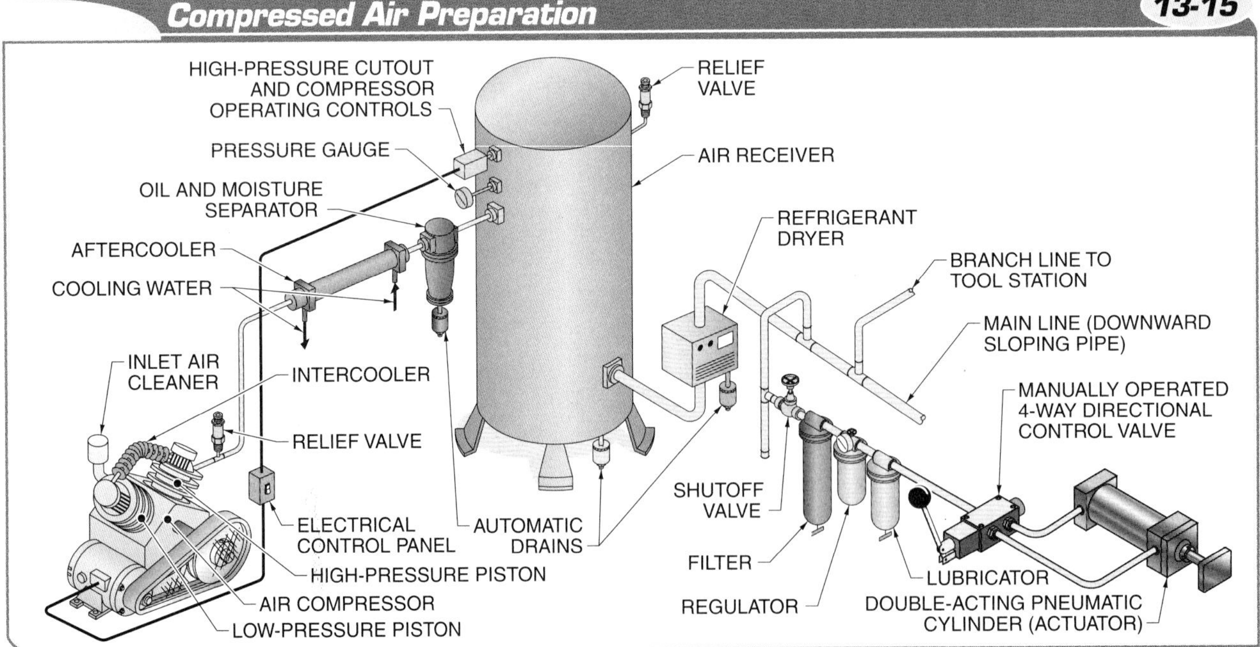

Figure 13-15. When preparing compressed air for use in a pneumatic system, the correct type and sequence of filtering components must be used.

Chapter Review

Pneumatic System Conditioning — Chapter 13

Name: ______________________ Date: ____________

MULTIPLE CHOICE

_______________ **1.** The air entering a desiccant air dryer can have a maximum temperature of ___°F.

A. 88
B. 94
C. 100
D. 105

_______________ **2.** Drip lubricators must be installed no more than ___′ away from the components that require lubrication.

A. 3
B. 8
C. 10
D. 12

_______________ **3.** A(n) ___ is a device designed to lower the temperature of compressed air to 35°F.

A. aftercooler
B. intercooler
C. moisture separator
D. refrigerated air dryer

_______________ **4.** It takes ___ cu ft of atmospheric air to produce 1.0 cu ft of compressed air at 100 psi.

A. 5.8
B. 6.7
C. 7.8
D. 8.3

_______________ **5.** Oil contamination in a pneumatic system can possibly come from ___.

A. a failed filter
B. oil vapor in the atmosphere
C. overlubrication of the air compressor
D. protective lubricants on components

_______________ **6.** At standard pressure and temperature, the weight of 1 cu ft of air equals ___ lb.

A. 0.066
B. 0.067
C. 0.076
D. 0.084

_______________ **7.** Pressure drops on pneumatic filters should not exceed ___ psi.

A. 2
B. 3
C. 10
D. 12.5

______________ **8.** A properly maintained coalescing filter should have approximately ___ operational hours.

A. 1500
B. 2000
C. 3000
D. 5000

______________ **9.** ___ is small quantities of liquid condensation.

A. Condensation
B. Moisture
C. Oil vapor
D. Water vapor

______________ **10.** Most contaminants enter a pneumatic system through the ___.

A. air intake
B. motor
C. pressure lines
D. receiver

COMPLETION

______________ **1.** A(n) ___ is a device that removes water vapor from compressed air through adsorption.

______________ **2.** A(n) ___ is used to remove contaminants from the air before it travels to the inlet of the compressor.

______________ **3.** A(n) ___ is a heat exchanger used for cooling the discharged air from an air compressor.

______________ **4.** Dew point is also known as ___.

______________ **5.** ___ is the pressure differential between upstream and downstream airflow caused by resistance.

______________ **6.** A venturi is a tube that creates a(n) ___.

______________ **7.** Surface-type elements are usually constructed of ___.

______________ **8.** ___ is the conversion of a substance from a vapor state to a denser solid state, usually initiated by a reduction in temperature during its vapor state.

______________ **9.** A(n) ___ is a pneumatic device that is used to filter, regulate, and lubricate compressed air before it reaches the point of work.

______________ **10.** The higher the temperature of the air, the higher percentage of ___ it can hold.

______________ **11.** Coalescing filters are sometimes referred to as ___ filters.

______________ **12.** Solid contaminants in pneumatic systems are rated in ___.

______________ **13.** ___ is the amount of moisture in the air.

______________ **14.** Depth-type elements are typically constructed from a(n) ___.

______________ **15.** A(n) ___ is a pneumatic device that separates moisture from compressed air by forcing it to flow against baffles.

______________________ **16.** A(n) ___ is a device that injects a lubricant into the airflow.

______________________ **17.** ___ air is air that holds as much moisture as it is capable of holding.

______________________ **18.** A(n) ___ is a normally-open pressure valve that is used to control downstream pressure.

______________________ **19.** A(n) ___ filter is a filter that removes contaminants from within a pneumatic system.

______________________ **20.** ___ air is air that is located inside a facility.

TRUE/FALSE

T F **1.** Breathable air contaminated with oil can cause health hazards.

T F **2.** Intercooling reduces the temperature of the compressed air, as well as its volume, before it reaches the next stage.

T F **3.** Pneumatic components are never available prelubricated.

T F **4.** Contaminants in a pneumatic system are different than those present in a hydraulic system.

T F **5.** Desiccant material becomes saturated because adsorption can remove 99.9% of the water vapor in the air.

T F **6.** The temperature is a determining factor for the percentage of water vapor in the air.

T F **7.** Liquid contaminants can be as destructive as solid contaminants.

T F **8.** Air is always pushed into a pneumatic system, rather than pulled.

T F **9.** The filters used with most FRLs are different than inline filters.

T F **10.** Ambient air is air that is located outside a facility.

T F **11.** A regulator can only be used alone.

T F **12.** The total amount of moisture that air is capable of holding varies based on the temperature of the air.

T F **13.** The inside air of a facility typically contains more airborne contaminants than outside air.

T F **14.** The higher the temperature of the air, the lower percentage of water vapor it can hold.

T F **15.** Coalescing filters are most effective when placed in the coolest locations in the system.

T F **16.** When an intake filter becomes clogged, the air compressor must work harder to acquire its air, which can cause the filter element to collapse.

T F **17.** Coalescing filters should be installed as far as possible from the point of work.

T F **18.** Lubricators must never be installed before a filter.

T F 19. Remote intake filters are never installed in facilities that have contaminated air but need to keep the air compressors near the work devices.

T F 20. Desiccant air dryers are normally placed in the warmest downstream location.

SHORT ANSWER

1. Explain how a regulator operates.

2. Explain how solid contaminants can damage a pneumatic system.

3. List five common solid contaminants found in pneumatic systems.

MATCHING

Schematic Symbols

Note: Some symbols may be used more than once.

_______________ 1. Coalescing filter

_______________ 2. Filter-regulator-lubricator

_______________ 3. Inline filter

_______________ 4. Intake filter

_______________ 5. Intercooler

_______________ 6. Lubricator

_______________ 7. Regulator

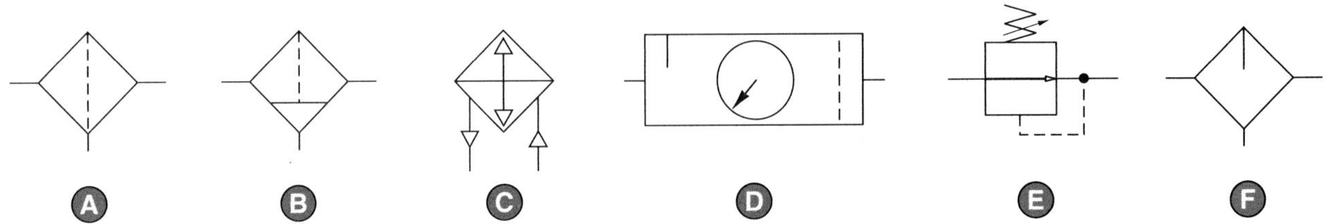

Activity 13-1: Schematic Diagrams

1. Redraw Pneumatic Power Distribution System by adding the schematic diagram of the equipment that would be used to condition the air.

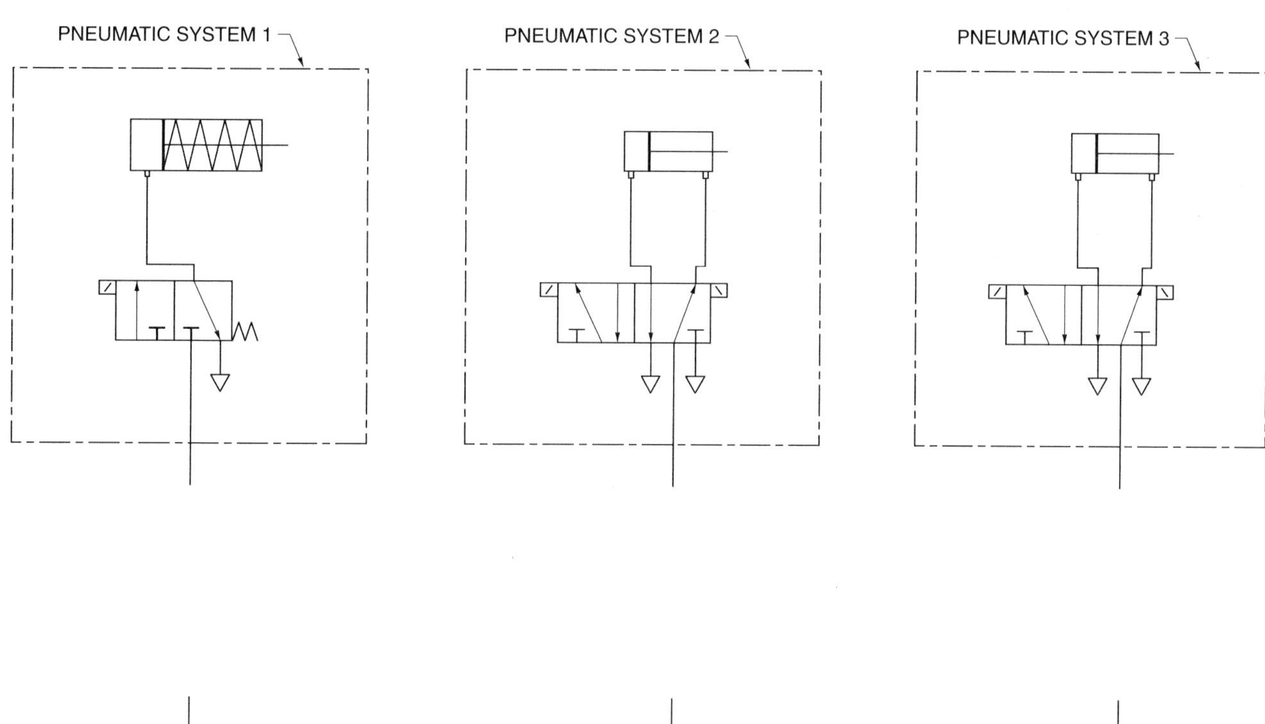

PNEUMATIC POWER DISTRIBUTION SYSTEM

Activity 13-2: Pneumatic System Pressure Level

1. Redraw Pneumatic Power Distribution System by adding the schematic diagrams of the valves that will regulate pressure to each system close to where each pipeline splits. *Note:* Cylinder speed controls should be as meter out.
2. Add an FRL to each system.
3. Add a control valve to System 1 that can control the speed of the cylinder on retraction.
4. Add a control valve to System 2 that can control the speed of the cylinder on retraction.
5. Add a control valve to System 3 that can control the speed of the cylinder on extension.

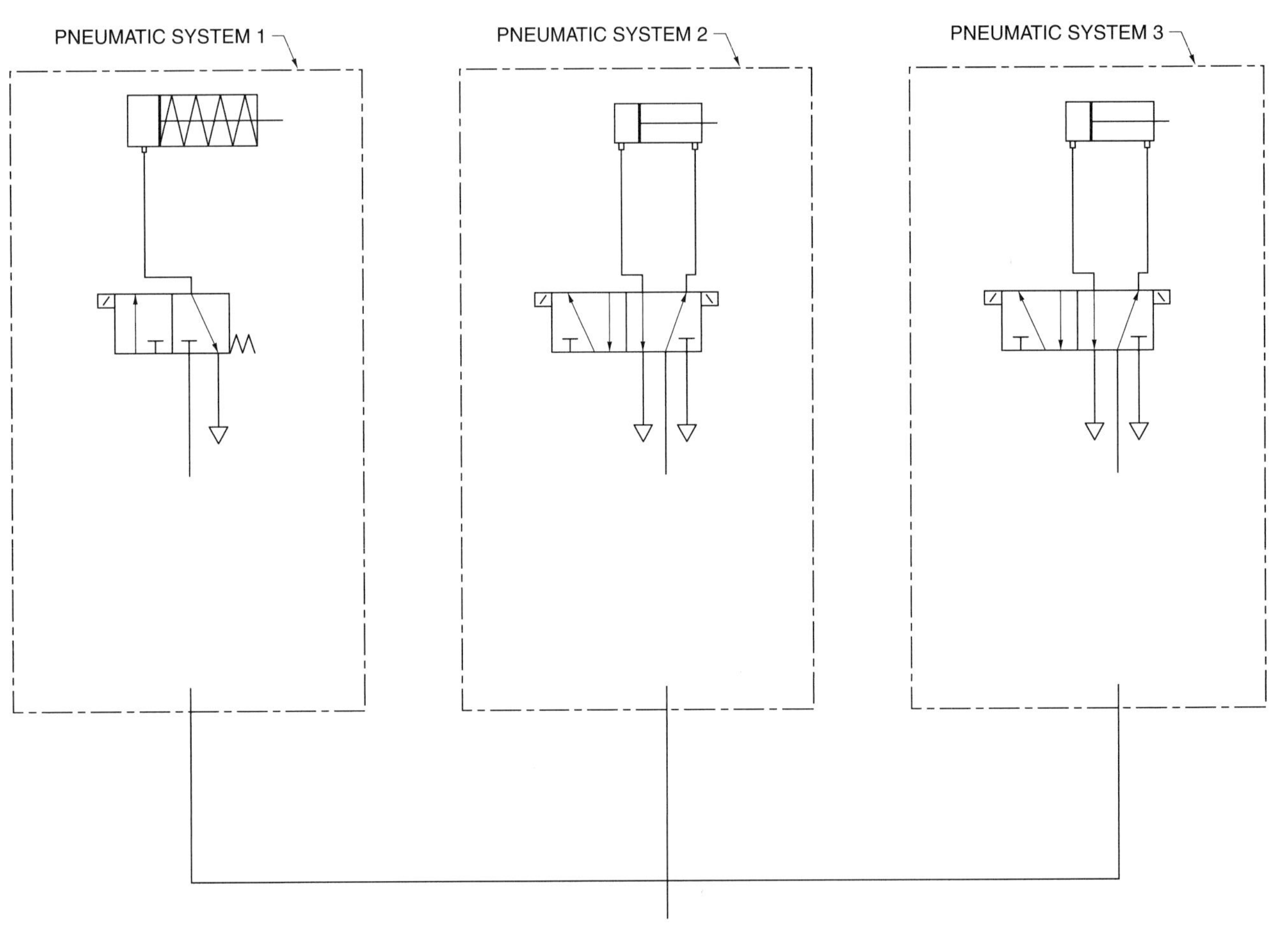

PNEUMATIC POWER DISTRIBUTION SYSTEM

Activity 13-3: Pressure Regulator Repair

A regulator that is part of an FRL in a pneumatic system has failed. Rather than replacing the entire FRL, it is determined that the regulator can be repaired by replacing the diaphragm assembly.

1. List the steps required to disassemble, repair, and reassemble the regulator.

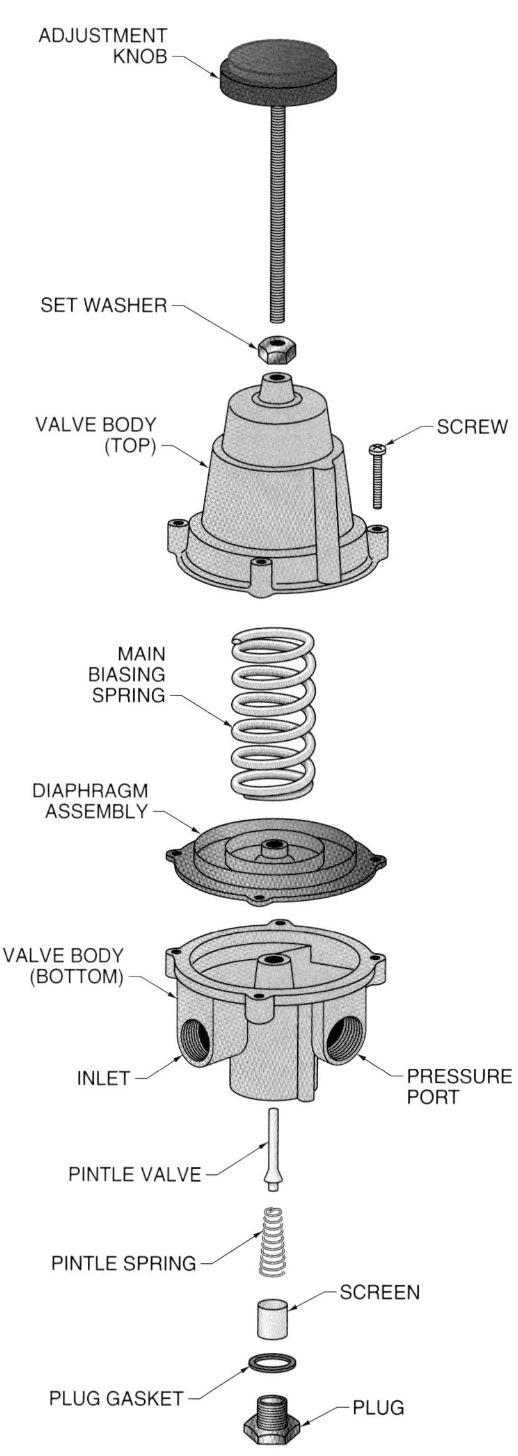

Chapter 14

Fluid Power System Electrical Control

OBJECTIVES

- Identify the basic electrical quantities.
- Identify parallel, series, and series-parallel circuits.
- Describe the basic electrical components used to control fluid power systems.
- Describe the different types of control equipment used to control fluid power systems.
- Identify the different schematic diagrams used for fluid power systems.
- Describe the different electrical control circuits used in fluid power systems.

INTRODUCTION

Fluid power systems can be controlled by mechanical or electrical means. Electrical control is the most common method of controlling commercial and industrial fluid power systems. It allows a greater degree of flexibility than mechanical control by operating the fluid power system more efficiently, using fewer components, increasing reliability, and reducing operational costs.

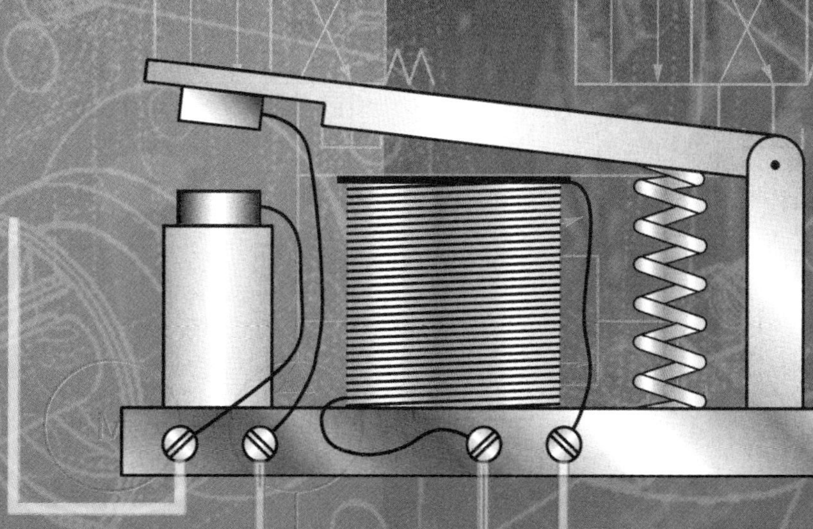

TERMS

Electricity is the energy released by the flow of electrons in a conductor.

Current (I) is the amount of electrons flowing through an electrical circuit.

Direct current (DC) is current that flows in one direction only.

Alternating current (AC) is current that reverses its direction of flow at regular intervals.

An **ampere** (A) is the amount of current that flows when 1 volt (V) is applied across 1 ohm (Ω).

Voltage (V) is the electrical pressure, or electromotive force, that causes electrons to move in an electrical circuit.

Polarity is the positive (+) or negative (–) state of an object.

ELECTRICAL QUANTITIES

Electricity is the energy released by the flow of electrons in a conductor. The flow of electrons in a conductor produces energy and controls fluid power systems through the basic electrical quantities of current, voltage, and resistance. A technician needs to understand these basic electrical quantities to operate and troubleshoot the controls of a fluid power system.

Current

Current flows through an electrical circuit when a power source is connected to a component that uses electricity. *Current (I)* is the amount of electrons flowing through an electrical circuit. *Direct current (DC)* is current that flows in one direction only. *Alternating current (AC)* is current that reverses its direction of flow at regular intervals. **See Figure 14-1.**

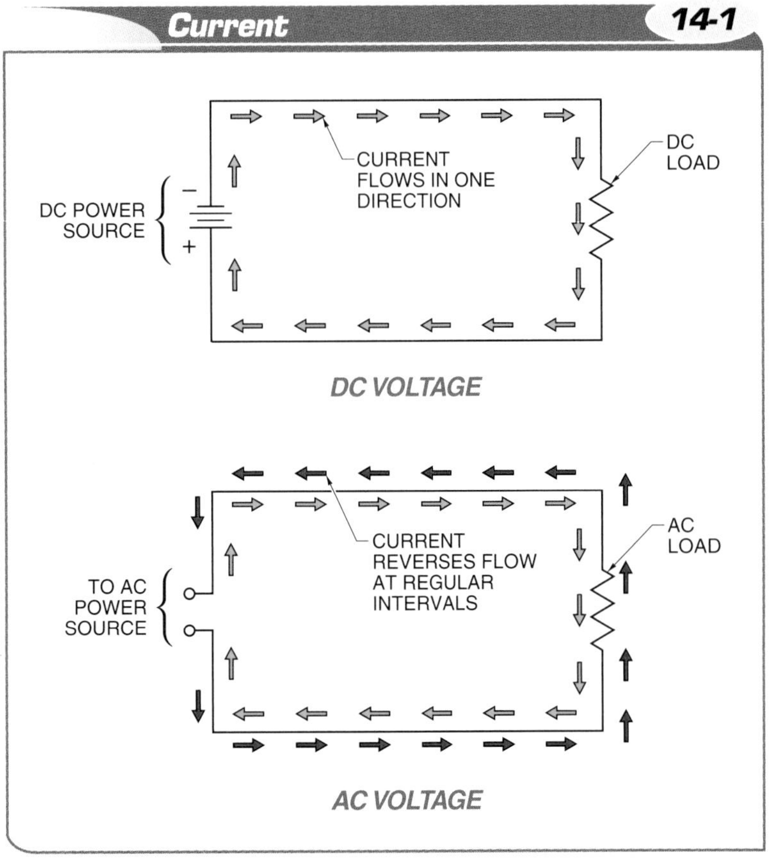

Current 14-1

Figure 14-1. Direct current flows in one direction only, while alternating current reverses its direction of flow at regular intervals.

Current is measured in amperes. An *ampere (A)* is the amount of current that flows when 1 volt (V) is applied across 1 ohm (Ω) of resistance. The more power required by an electrical device (load), the larger the amount of current flow. For example, a 10 HP motor draws approximately 28 A when wired for 230 V, and a 20 HP motor draws approximately 54 A when wired for 230 V. The current in a wire in an electrical circuit is similar to the flow of hydraulic fluid through a pipe in a hydraulic system.

Voltage

Voltage (V) is the electrical pressure, or electromotive force, that causes electrons to move in an electrical circuit. Voltage is measured in volts. A difference in electrical polarity is required for electricity to flow in a circuit. *Polarity* is the positive (+) or negative (–) state of an object. The voltage needed to move electrons in an electrical circuit is comparable to the fluid pressure needed to cause hydraulic fluid to flow through a pipe in a hydraulic system. **See Figure 14-2.**

AC voltage reverses its polarity and alternates its flow at regular intervals. AC voltage is the voltage that is provided from an electrical service provider. DC voltage does not change its polarity, except in some specialized electrical circuits such as electric motor speed controllers. DC voltage is used in most portable equipment such as mobile hydraulic cranes and automobiles. DC voltage can be obtained directly from batteries or rectified from an AC voltage supply. The greater the difference in the number of electrons between the negative (–) and positive (+) terminals in an electrical circuit, the greater the voltage.

TECH FACT

Electrical current can flow from positive to negative and negative to positive terminals. Electrical current that flows from positive to negative is known as conventional current flow and electrical current that flows from negative to positive is known as electron current flow.

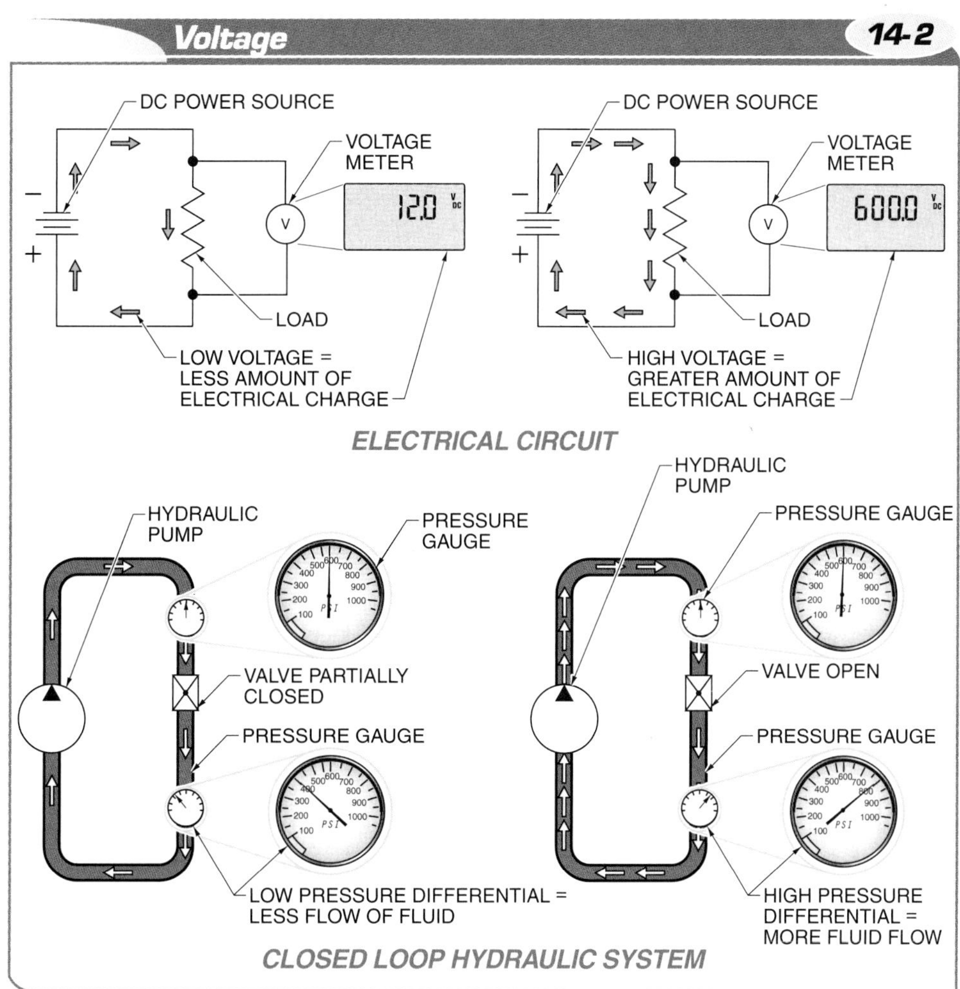

Figure 14-2. The voltage needed to move electrons in an electrical circuit is comparable to the fluid pressure needed to cause hydraulic fluid to flow through a pipe in a hydraulic system.

TERMS

Resistance (R) is the opposition to the flow of electrons.

A **conductor** is a material that has low resistance and permits electrons to move easily through it.

Resistance

Resistance (R) is the opposition to the flow of electrons. Resistance is measured in ohms (Ω). A resistance of 100 ohms is written as $R = 100\ \Omega$. Resistance limits the flow of current in an electrical circuit. The higher the resistance, the lower the current, and the lower the resistance, the higher the current. **See Figure 14-3.**

Every material has resistance to the flow of electrons. The amount of resistance determines the characteristic of the material. If a material has a low amount of resistance, it is considered a conductor. A *conductor* is a material that has low resistance and permits electrons to move easily through it. Wires are typically the conductors used for hydraulic electrical control systems.

Curtis-Toledo, Inc.

Electric motors are rated by the amount of voltage and current required for operation.

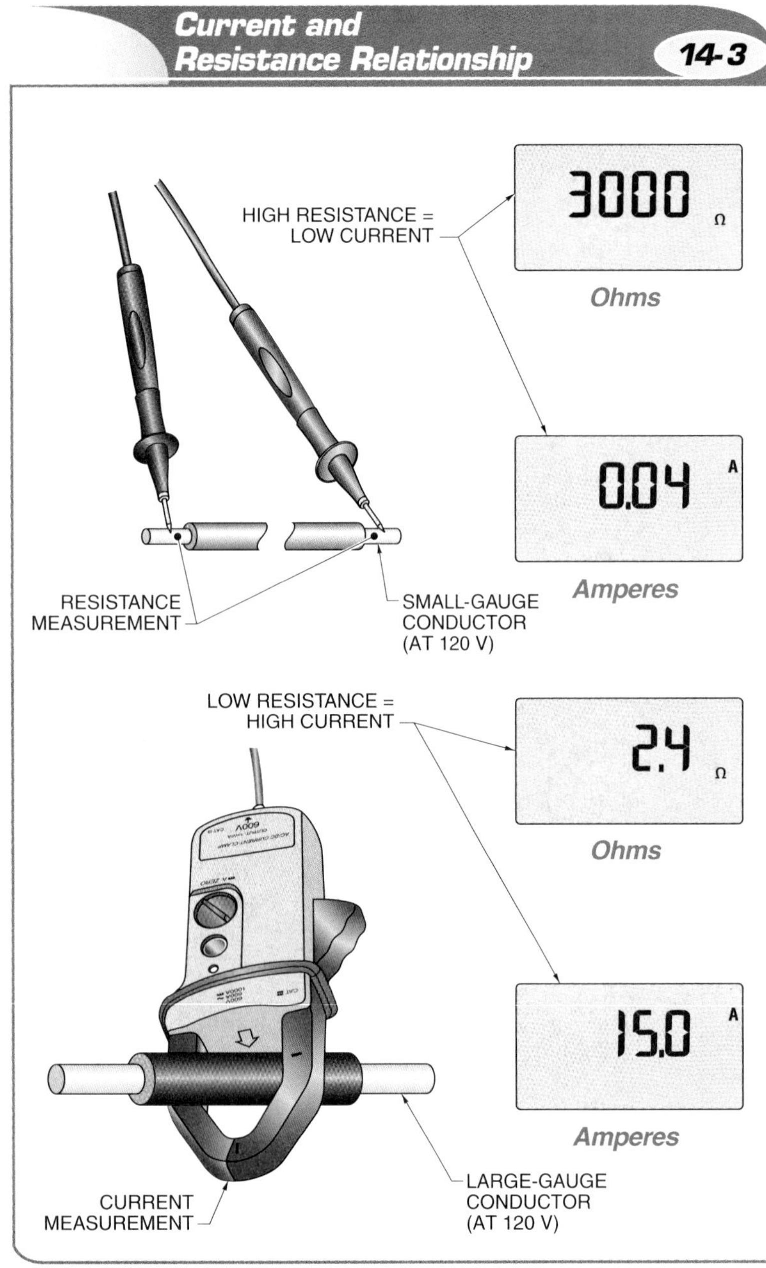

Figure 14-3. The higher the resistance, the lower the current, and the lower the resistance, the higher the current.

If a material has a high amount of resistance, it is considered an insulator. An *insulator* is a material that has a high resistance and resists the flow of electrons. Insulation commonly used to cover fluid power electrical wiring is typically made of rubber, vinyl, or plastic. **See Figure 14-4.**

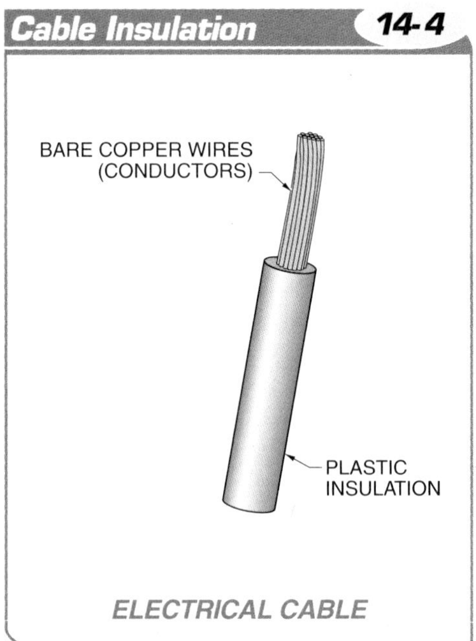

Figure 14-4. Insulation commonly used in fluid power systems are plastic coverings over electrical wires, cables, and other conductors.

TERMS

An **insulator** is a material that has a high resistance and resists the flow of electrons.

A **contact** is a conducting part of a switch that operates with another conducting part of the switch to make or break a circuit.

TECH FACT

Most wiring used in fluid power systems have copper conductors. Copper is typically used because it offers the least resistance to the flow of electrons, has good flexibility in small sizes, and has stranded construction. Typical stranded constructions are 7-strand, 19-strand, and 37-strand.

BASIC ELECTRICAL CIRCUITS

The three different types of electrical circuits are parallel circuits, series circuits, and series-parallel circuits. In the electrical control of a fluid power system, the configuration of switches and contacts determines when and how system components activate. A *contact* is a conducting part of a switch that operates with another conducting part of the switch to make or break a circuit.

Electrical devices (loads) in fluid power systems can be controlled by switches connected in different configurations. To change electrical control, the configuration of the switches must be changed. Electrical control can be changed by rewiring the circuit, adding additional switches into the circuit, or by removing switches from the circuit.

Parallel Circuits

A *parallel circuit* is an electrical circuit with two or more paths for current to flow. In a parallel circuit, current passes through all terminals simultaneously. When two or more switches are connected in parallel, current is able to flow through any one of the switches if they are closed. There is no limit to how many switches can be connected in parallel with each other. For example, if a hydraulic cylinder needs to be activated from three different places, three different switches are connected in parallel to actuate the directional control valve solenoids that control the cylinder.

Electrical devices (loads) used in fluid power systems are typically connected in parallel so that they can receive the same amount of voltage. However, they are connected in parallel when they can be controlled by the same formation of switches. For example, if three cylinders with three separate directional control valves need to be extended at the same time, their solenoids are connected in parallel. When the switches are closed, current flows to all three solenoids and actuates them. **See Figure 14-5.**

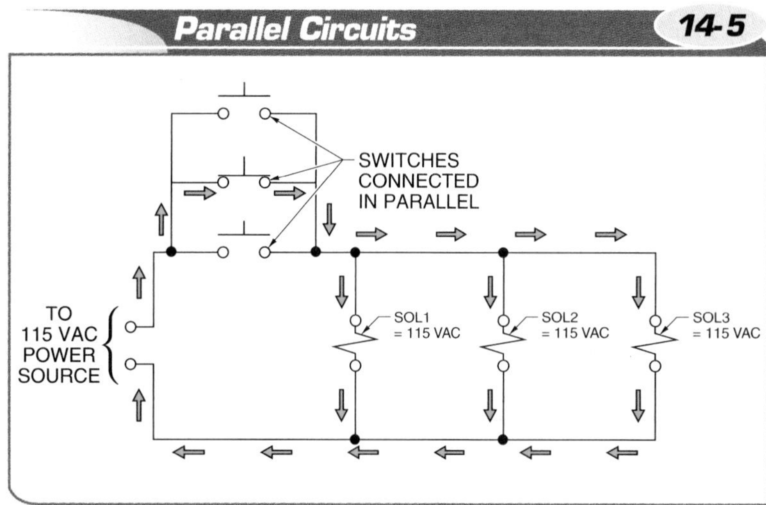

Figure 14-5. Electrical devices used in fluid power systems are typically connected in parallel so that they can have the same amount of voltage.

A *limit switch* is an electrical device that uses a mechanical actuator to control its electrical contacts. For example, if a limit switch and a pushbutton need to be closed for a solenoid to actuate, they are connected in series. In electrical circuits, it is common to have many types of switches and contacts in series. Switches and their contacts can be normally open (NO) or normally closed (NC).

TERMS

A **parallel circuit** is an electrical circuit with two or more paths for current to flow.

A **series circuit** is an electrical circuit with two or more components connected so that there is only one path for current to flow.

A **limit switch** is an electrical device that uses a mechanical actuator to control its electrical contacts.

TECH FACT

Electrical devices in fluid power systems are never connected in series because each one would receive only part of the power supply, which would not be enough to move the actuator.

Series Circuits

A *series circuit* is an electrical circuit with two or more switches connected so that there is only one path for current to flow. When switches are connected in series, the current enters the first switch and then flows to the next switch if the first switch is closed. Once the current has flowed through all of the switches connected in series, it reaches the load. When two or more switches are connected in series, all of the switches must be closed for current to flow to the load. **See Figure 14-6.**

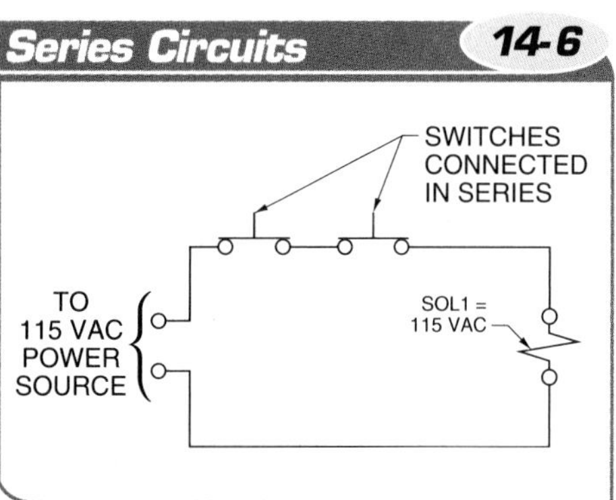

Figure 14-6. A series circuit has two or more components connected in a manner so that there is only one path for current to flow.

TERMS

A **series-parallel circuit** is a circuit with a combination of series- and parallel-connected components.

A **holding circuit** is an electrical circuit that allows a load to remain on after the switch that controls it is released.

An **electrical control circuit** is an electrical circuit that determines when the output component is energized or de-energized.

A **power circuit** is a circuit controlled by the electrical control circuit to accomplish work in a fluid power system.

Series-Parallel Circuits

A *series-parallel circuit* is a circuit with a combination of series- and parallel-connected components. A series-parallel circuit has at least one switch connected in series and two or more switches connected in parallel. A series-parallel circuit can have many combinations, which are always created with at least three switches or relay contacts. A series-parallel circuit is typically used in industrial applications that use a lamp to indicate when a solenoid is on. **See Figure 14-7.**

Series-Parallel Circuits 14-7

SWITCHES CONNECTED IN SERIES-PARALLEL

TO 115 VAC POWER SOURCE

Figure 14-7. A series-parallel circuit has at least one load connected in series with two or more loads connected in parallel.

Pushbutton override switches are typically used with hydraulic systems in mobile equipment.

Series-parallel circuits are common in fluid power systems because they offer a wide range of options for control. The most common type of series-parallel circuit is a holding circuit. A *holding circuit* is an electrical circuit that allows a load to remain on after the switch that controls it is released. **See Figure 14-8.** A holding circuit consists of an NC momentary pushbutton connected in series with an NO momentary pushbutton that is in parallel with a set of NO relay contacts.

When the NO momentary pushbutton of a holding circuit is pushed, current flows to the control relay and actuates it. The contacts of the control relay change states as the current creates a magnetic field around the coil. This actuates the contacts. The NO contact of the control relay connected in parallel with the NO momentary pushbutton closes and current flows to the control relay. When the NO momentary pushbutton is released, current continues to flow to the coil through the NO contact. When the NC momentary pushbutton is actuated, the control relay loses power because current flow stops. This also causes the NO contacts to return to their normal state.

A holding circuit is sometimes referred to as a start-stop station or a three-wire control circuit. For example, in an electrical control circuit, the solenoid is connected through the control relay, which allows the solenoid to be powered whenever the relay has power.

ELECTRICAL CONTROL CIRCUITS

Specific combinations of electrical components are used in electrical control circuits to operate a fluid power system. An *electrical control circuit* is an electrical circuit that determines when the output component is energized or de-energized. A *power circuit* is a circuit controlled by the electrical control circuit to accomplish work in a fluid power system. **See Figure 14-9.**

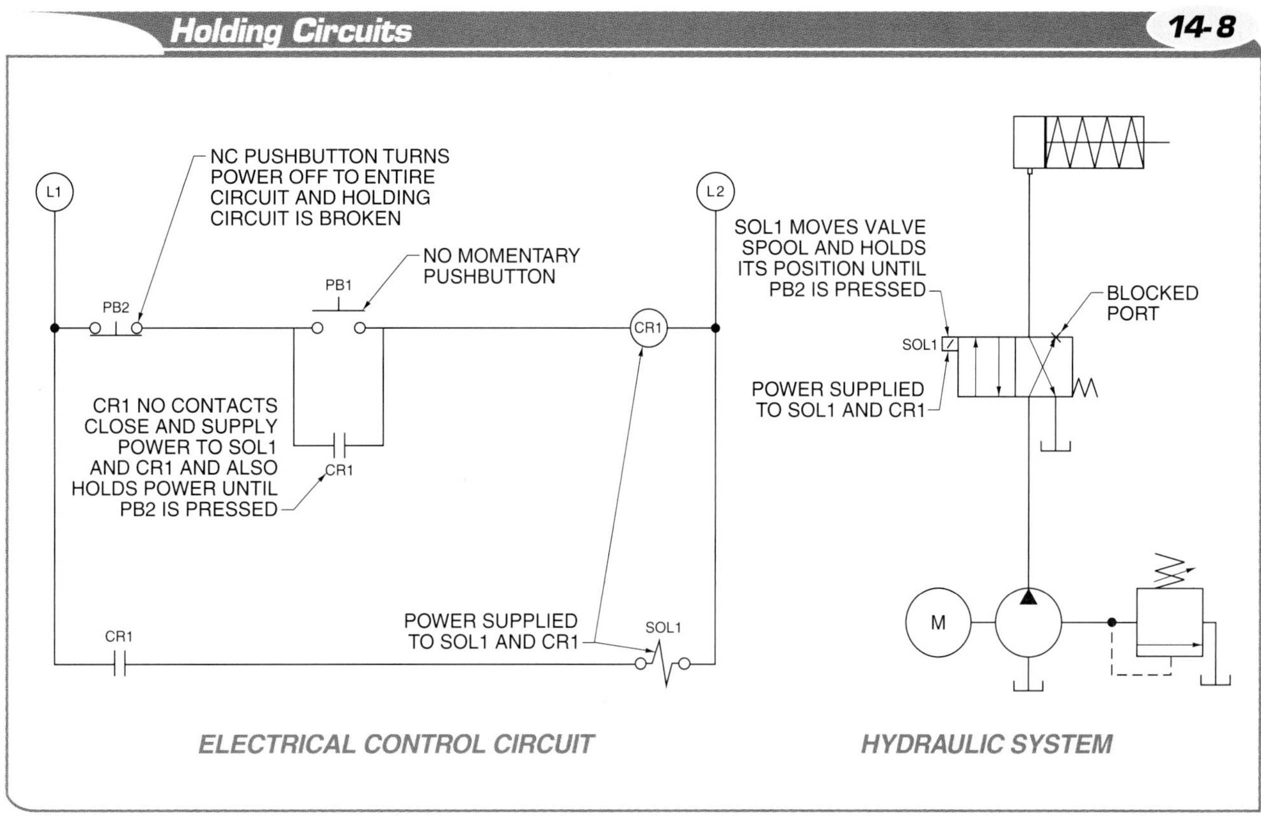

Figure 14-8. A holding circuit allows a load to remain on after the switch that controls it is released.

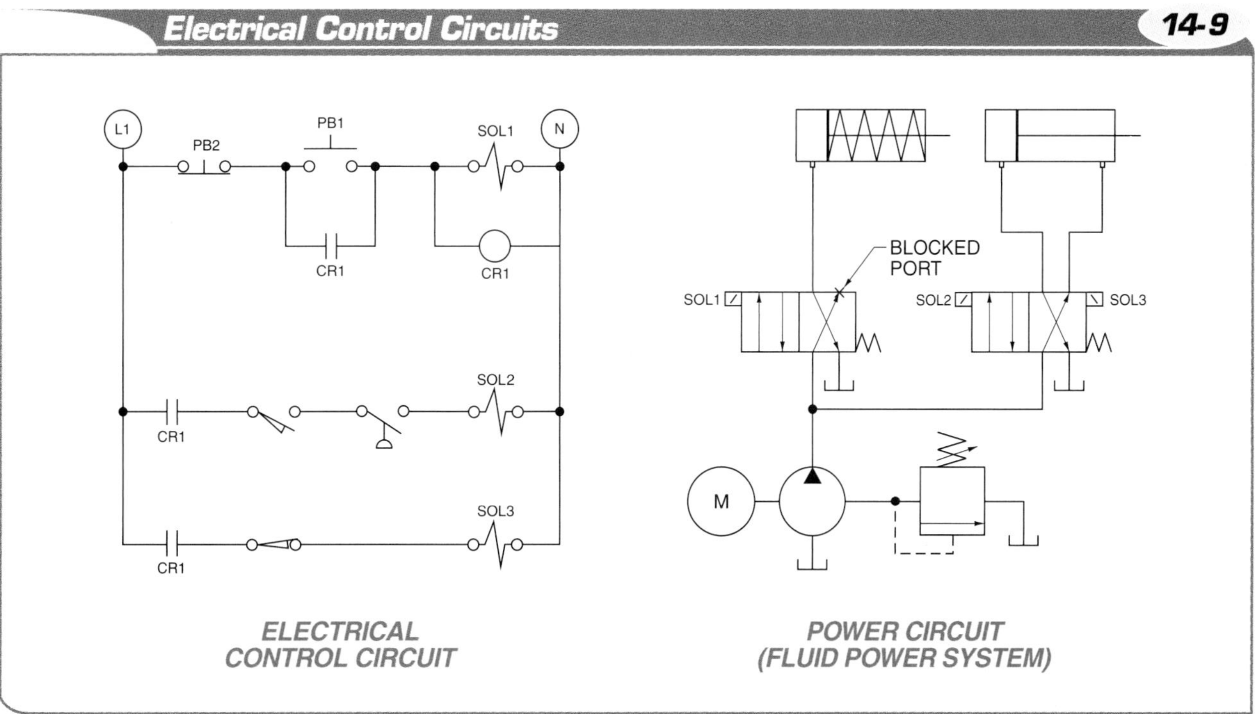

Figure 14-9. An electrical control circuit is an electrical circuit that determines when the output component is energized or de-energized, while a power circuit accomplishes work in a fluid power system.

TERMS

An **electrical power source** is a system that produces, transmits, distributes, and delivers electrical power to satisfactorily operate electrical loads designed to connect to the system.

The **American wire gauge (AWG)** is a standardized wire sizing system used in the United States for the diameters of round, solid, nonferrous, electrically conducting wire.

Electrical Control Circuit Components

All electrical control circuits are comprised of an electrical power source, conductors, and a load. Without these three basic components, an electrical control circuit cannot operate. Along with these three basic components, most electrical control circuits also have a control switch. **See Figure 14-10.** The components that comprise a common electrical control circuit come in many different forms and accomplish different types of work.

Electrical Power Sources. An *electrical power source* is a system that produces, transmits, distributes, and delivers electrical power to satisfactorily operate electrical loads designed to connect to the system. An electrical power source can be DC or AC, and it can be small and simple or large and complex. For example, a utility company emergency crew often uses a portable generator, which is a small, self-contained AC electrical power source. A utility company uses large AC generators to supply power to metropolitan areas.

Regardless of the size of the electrical power source, the power supplied must allow loads to operate satisfactorily. Damage to electrical equipment can occur if power is not supplied at the proper voltage, current, type (single-phase and three-phase AC or DC), and/or condition. For example, electricity used to power an electric motor that is coupled to a hydraulic pump must be at the correct voltage or it can cause the motor to prematurely wear. If the power received by the motor is so poor that it affects motor speed, the hydraulic pump or air compressor can also prematurely wear.

Conductors. A conductor is a material that has low resistance and permits electrons to easily move through it. Wire conductors are used to transmit current in an electrical control circuit. The most common type of conductor material is copper. The size of the wire used is determined by the amount of current the load needs in order to operate.

Wire conductor size is indicated by a wire gauge number and is standardized by the American wire gauge (AWG) identification system. The *American wire gauge (AWG)* is a standardized wire sizing system used in the United States for the diameters of round, solid, nonferrous, electrically conducting wire. The diameter of the cross-sectional area of a wire is the factor used for determining its capacity for carrying current. When using AWG, the larger the number, the smaller the diameter of the wire. **See Figure 14-11.** Common AWG sizes used with electrical control circuits are 14 AWG, 16 AWG and 18 AWG. Size 14 AWG is the largest, while size 18 AWG is the smallest.

Electrical Control Components 14-10

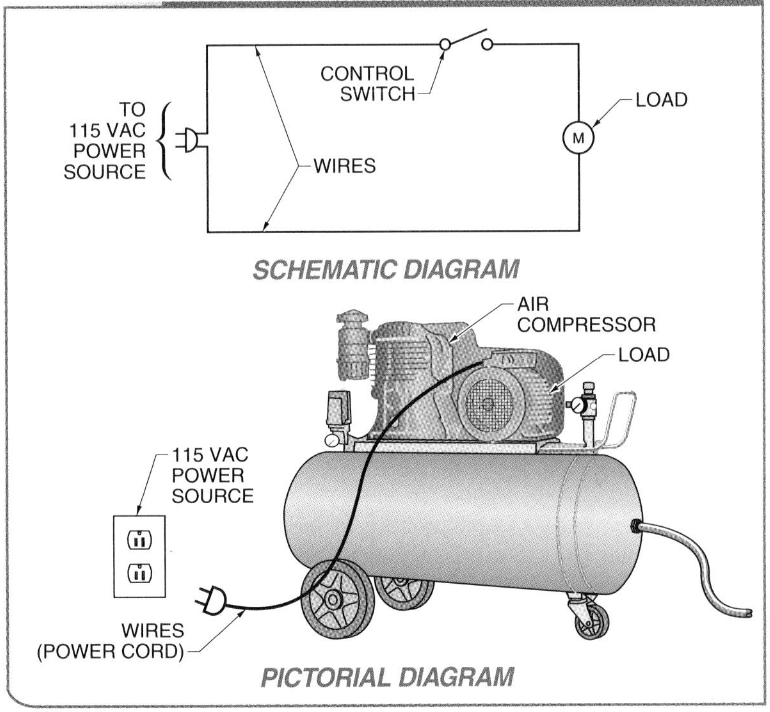

Figure 14-10. All electrical circuits are comprised of an electrical power source, conductors, load, and a control switch or a relay.

TECH FACT

As the diameter of a wire increases and the AWG numerical value decreases, the maximum allowable current-carrying capacity of the wire increases. Conductors normally have insulation and are referred to as insulated wires. Although the term "wire" refers just to the metal, the term generally also includes the insulation.

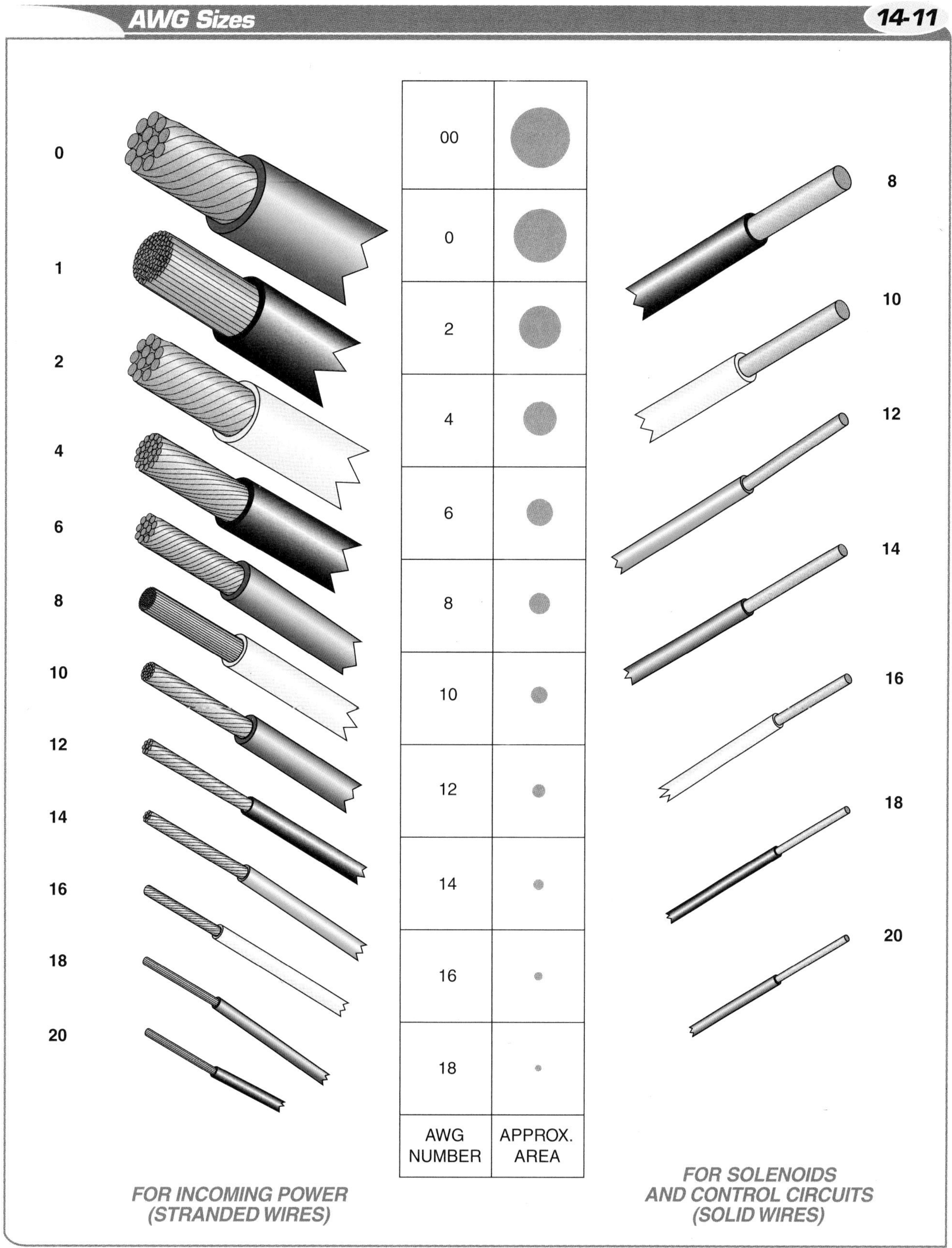

Figure 14-11. When using the AWG identification system, the larger the number, the smaller the diameter of the wire.

Loads. A *load* is an electrical device with a specific amount of resistance that allows current to flow through it to accomplish work. The load that is typically used for the electrical control of a fluid power system is a solenoid. A solenoid controls the spool in a directional control valve. For example, when a pushbutton is actuated, current flows through the conductor to the solenoid. The solenoid uses the current to create a magnetic field, which moves the directional control valve spool. **See Figure 14-12.**

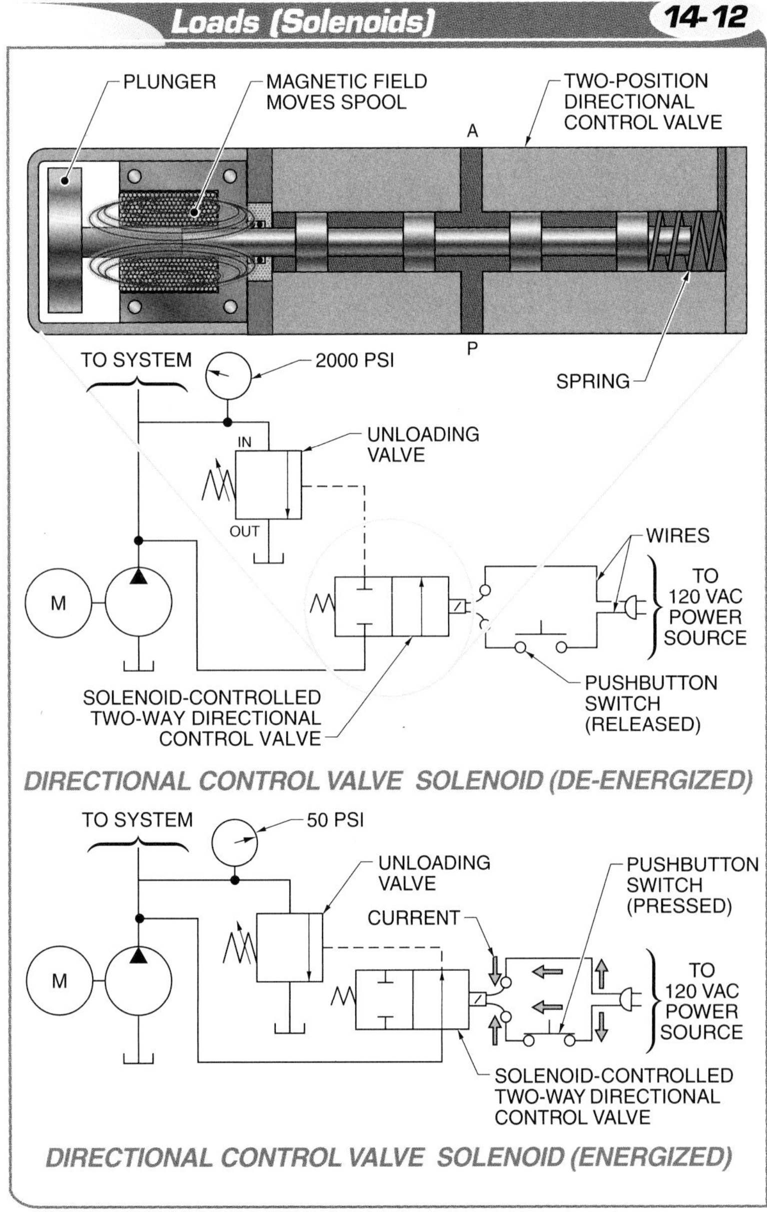

Figure 14-12. When a switch is actuated, current flows through conductors and actuates a load (solenoid).

Switches. A *switch* is an electrical control mechanism that connects and disconnects the power source from the load when it is actuated. A switch can be either normally open or normally closed. A *normally open (NO) switch* is an electrical control device that when actuated, allows current to flow through it. A *normally closed (NC) switch* is an electrical control device that when actuated, does not allow current to flow through it.

Switches are available in many shapes and with manual or automatic operation. Once actuated, a switch changes the position of contacts. The contacts are used to start and stop the flow of current in an electrical circuit. **See Figure 14-13.** The type of switch that is used for the electrical control of a fluid power system depends on the application. For example, a pushbutton is a switch that is used to extend the rod of a cylinder. When the rod is completely extended, it actuates a limit switch, which shifts a directional control valve and causes the rod to retract.

Manually operated switches are the most common type of switch used for electrical control of fluid power systems. They include pushbutton switches, selector switches, and foot switches. They are commonly used to turn directional control valve solenoids on and off. They are also used to override automatic switches.

The ability to override automatic switches is necessary when troubleshooting and in an emergency. The symbol for an NO manually operated switch shows the operator above the terminals. The symbol for an NC manually operated switch shows the operator below the terminals.

All mechanically operated switches are limit switches. Limit switches detect the physical presence of an object and are normally used as safety devices. For example, they are used to determine if a door is closed, if safety guards are in place, or if workpieces are placed properly. The symbol for an NO mechanically operated switch shows the operator below the terminals. The symbol for an NC mechanically operated switch shows the operator above the terminals.

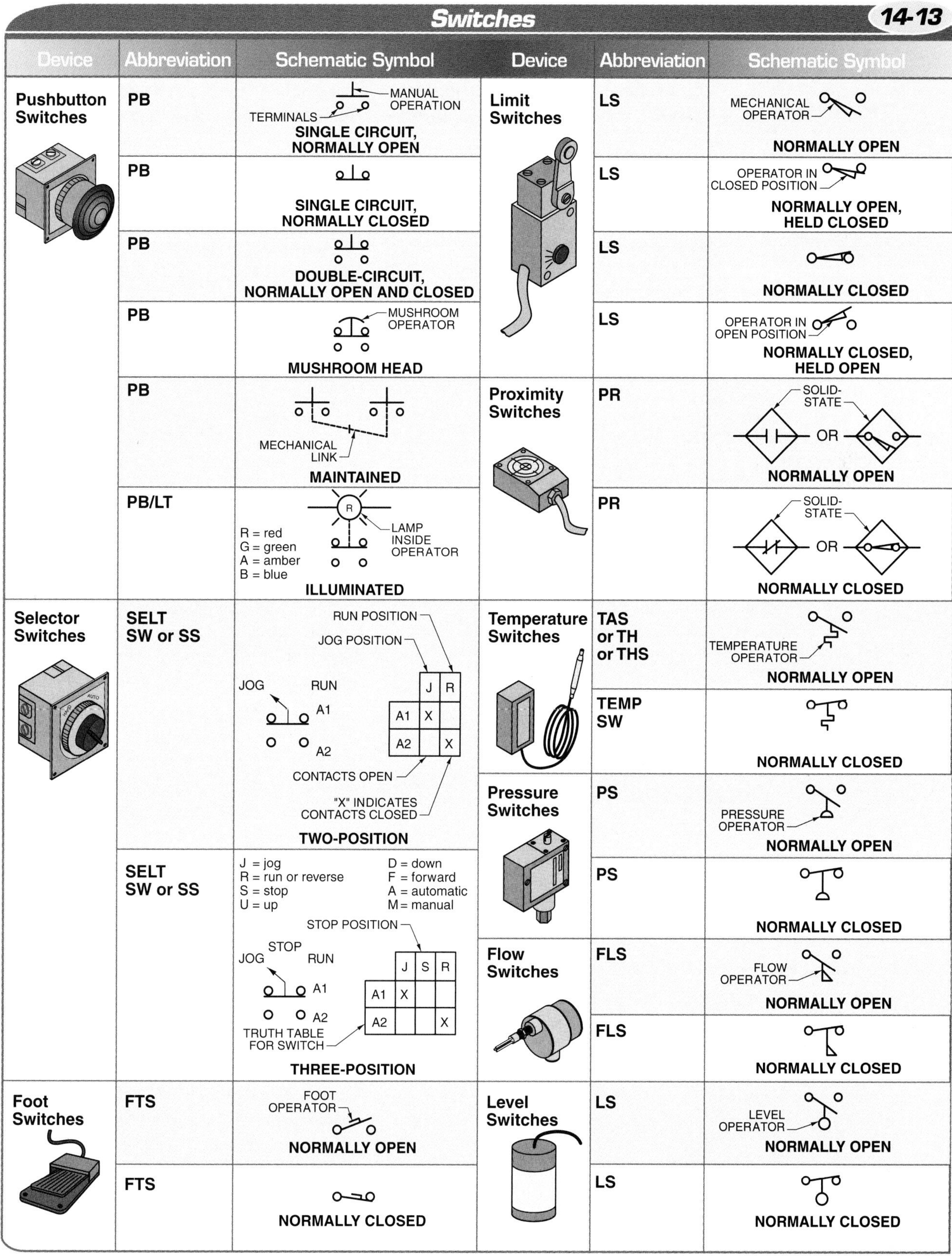

Switches 14-13

Device	Abbreviation	Schematic Symbol
Pushbutton Switches	PB	TERMINALS; MANUAL OPERATION SINGLE CIRCUIT, NORMALLY OPEN
	PB	SINGLE CIRCUIT, NORMALLY CLOSED
	PB	DOUBLE-CIRCUIT, NORMALLY OPEN AND CLOSED
	PB	MUSHROOM OPERATOR MUSHROOM HEAD
	PB	MECHANICAL LINK MAINTAINED
	PB/LT	R = red G = green A = amber B = blue LAMP INSIDE OPERATOR ILLUMINATED
Selector Switches	SELT SW or SS	RUN POSITION; JOG POSITION; JOG; RUN; A1; A2 Truth table (J, R): A1: X, –; A2: –, X CONTACTS OPEN "X" INDICATES CONTACTS CLOSED TWO-POSITION
	SELT SW or SS	J = jog; R = run or reverse; S = stop; U = up D = down; F = forward; A = automatic; M = manual STOP POSITION; STOP; JOG; RUN; A1; A2 Truth table (J, S, R): A1: X, –, –; A2: –, –, X TRUTH TABLE FOR SWITCH THREE-POSITION
Foot Switches	FTS	FOOT OPERATOR NORMALLY OPEN
	FTS	NORMALLY CLOSED

Device	Abbreviation	Schematic Symbol
Limit Switches	LS	MECHANICAL OPERATOR NORMALLY OPEN
	LS	OPERATOR IN CLOSED POSITION NORMALLY OPEN, HELD CLOSED
	LS	NORMALLY CLOSED
	LS	OPERATOR IN OPEN POSITION NORMALLY CLOSED, HELD OPEN
Proximity Switches	PR	SOLID-STATE; OR NORMALLY OPEN
	PR	SOLID-STATE; OR NORMALLY CLOSED
Temperature Switches	TAS or TH or THS	TEMPERATURE OPERATOR NORMALLY OPEN
	TEMP SW	NORMALLY CLOSED
Pressure Switches	PS	PRESSURE OPERATOR NORMALLY OPEN
	PS	NORMALLY CLOSED
Flow Switches	FLS	FLOW OPERATOR NORMALLY OPEN
	FLS	NORMALLY CLOSED
Level Switches	LS	LEVEL OPERATOR NORMALLY OPEN
	LS	NORMALLY CLOSED

Figure 14-13. There are many types of switches used for the electric control of fluid power systems.

TERMS

A **load** is an electrical device with a specific amount of resistance that allows current to flow through it to accomplish work.

A **switch** is an electrical control mechanism that connects and disconnects the power source from the load when it is actuated.

A **normally open (NO) switch** is an electrical control device that when actuated, allows current to flow through it.

A **normally closed (NC) switch** is an electrical control device that when actuated, does not allow current to flow through it.

A **general-purpose relay** is an electrical switch operated by a magnetic coil.

A **ladder diagram** is a diagram that contains the logic for an application and is typically used to test and troubleshoot hardwired applications.

Automatically operated switches operate electrical circuits with little or no manual input. They include pressure, flow, and level switches. They are used to keep hydraulic systems running at a set temperature, to keep hydraulic fluid at the proper level in a reservoir, to maintain pressure in a hydraulic or pneumatic system, and to keep actuators in the correct position. The symbol for an NO automatically operated switch shows the operator below the terminals, while an NC automatically operated switch shows the operator above the terminals.

General-Purpose Relays. A *general-purpose relay* is an electrical switch operated by a magnetic coil. General-purpose relays typically include two, three, or four sets of nonreplaceable NO and NC contacts. **See Figure 14-14.** The contacts are normally rated at 5 A to 15 A. Special attention must be given to the contact current rating when using general-purpose relays because the rating for switching DC is lower than the rating for switching AC. For example, a 15 A AC contact is normally rated for only 8 A to 10 A DC.

General-Purpose Relays 14-14

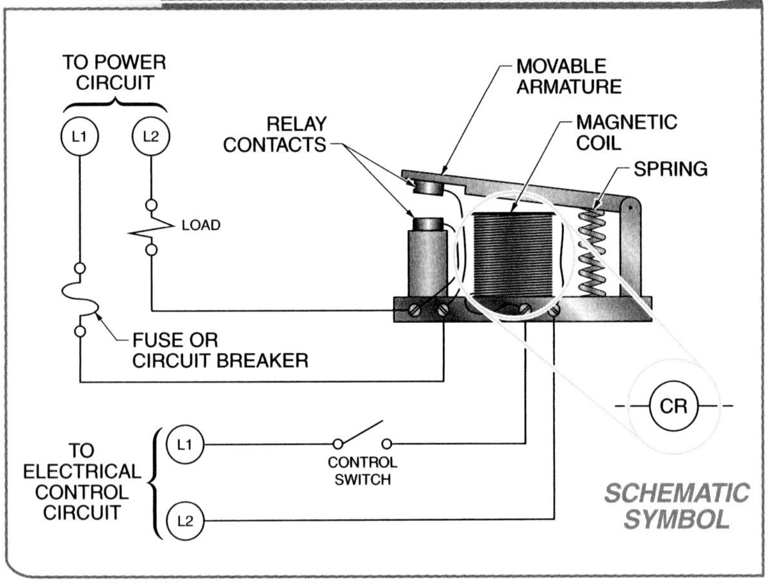

Figure 14-14. A general-purpose relay is a mechanical switch operated by a magnetic coil.

Several different styles of general-purpose relays are available. A general-purpose relay is often used with applications that have a disposable plug-in feature to simplify troubleshooting and reduce costs. These relays are designed for commercial and industrial applications where economy and fast replacement are high priorities.

Programmable Logic Controllers. A programmable logic controller (PLC) is a solid-state control device that is programmed and reprogrammed to automatically control electrical systems in residential, commercial, and industrial facilities. PLCs use a computer program to control simple or complex electrical circuits. An advantage of a PLC is that it can increase the reliability of a fluid power system by decreasing the number of control devices needed to operate the system. For example, control relays or timer relays are not required when using some types of PLCs because both can be replaced by programmable instructions.

Another advantage of a PLC is that it can reduce the total amount of wiring required in an electrical control circuit. Electrical devices such as switches and solenoids are connected directly into the PLC, which reduces the number of conductors needed, the overall cost, and the amount of time for installation. PLCs also have faster response times than direct, hard wired systems because they are controlled by solid-state electronics that change states at a faster rate. PLCs also allow real-time monitoring of electrical system switching, which allows for easier troubleshooting of the hydraulic system.

It is also easy to change the electrical control circuit of a hydraulic system when using a PLC. When a circuit using direct-wired components, also called hardwired logic, must be changed, it must be rewired. This is not the case when a PLC is used to control the logic. If the control circuit has to be changed, no hardwiring changes are required because only the PLC program must be changed.

The programming logic most often used to control PLCs is a ladder diagram. A *ladder diagram* is a diagram that contains the logic for an application and is typically used to test and troubleshoot hardwired applications. A ladder diagram has vertical rails with individual horizontal rungs that have at least one load each. When a switch controls a load, the switch will be in the same rung as the load in the ladder diagram. Using a ladder diagram allows the user to input control functions in a ladder format and makes the logic of the electrical control circuit easy to follow. It is also easy for technicians that are familiar with control relay diagrams to interpret PLC programming ladder diagrams.

The difference between a PLC programming diagram and a ladder diagram is that a ladder diagram uses different symbols for each switch, while a programming diagram uses almost the same symbol for all inputs and almost the same symbol for all outputs. **See Figure 14-15.** Each input and output terminal on a PLC is labeled with a number and connected to a device with the same number. That number is used to identify the device in the program.

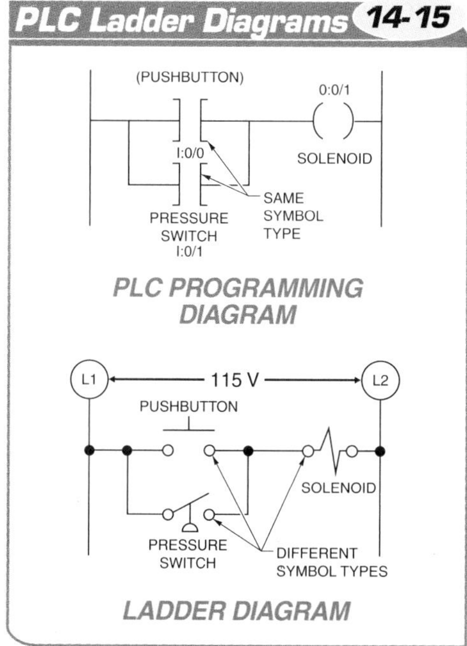

Figure 14-15. The difference between a PLC programming diagram and a ladder diagram is that the inputs and outputs have the same symbol in a PLC programming diagram.

ELECTRIC DIAGRAMS AND APPLICATIONS

Electric diagrams for fluid power systems are in schematic form. The two types of circuits that are usually represented in schematic diagrams are the power circuit (fluid power system) and the electrical control circuit. The power circuit (fluid power system) schematic diagram represents the fluid power system components, while the electrical control circuit schematic diagram represents the electrical components. The control of actuators, such as single-acting and double-acting cylinders, is different from the control of electrical devices, such as lights. When controlling a fluid power system with electric controls, the type of directional control valve that is used determines the electrical control circuit configuration.

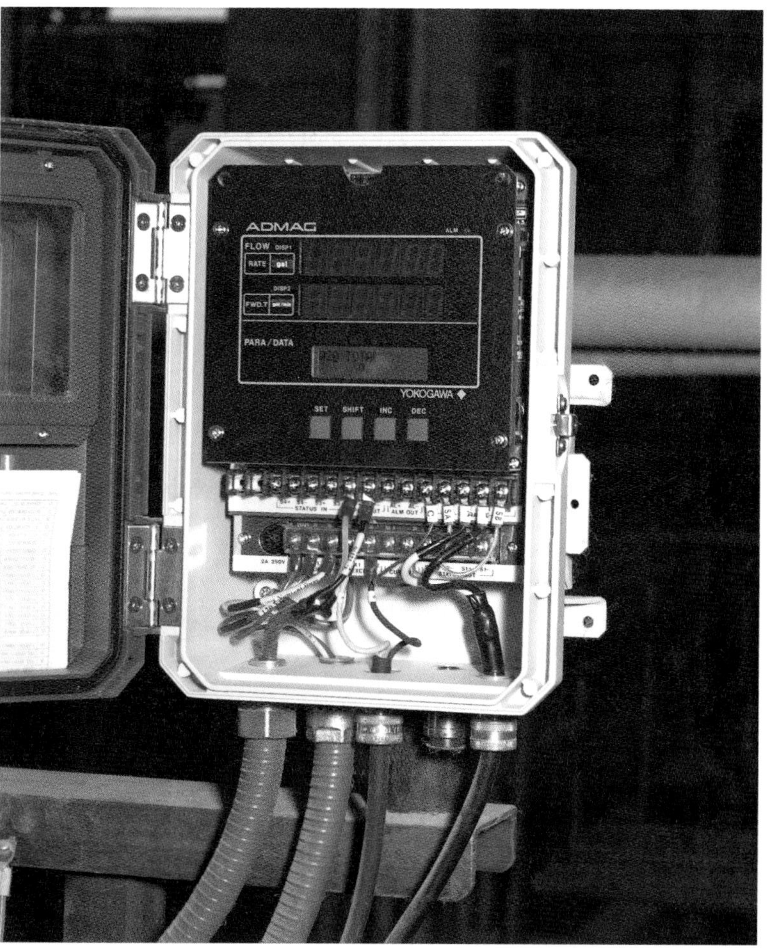

Programmable logic controllers (PLCs) are solid-state devices that can be used for the electrical control of fluid power systems.

TERMS

An **emergency stop pushbutton** is an NC switch that shuts off power to the electrical control circuit.

Single-Acting Cylinder Control Applications

The electrical control of a single-acting cylinder is usually accomplished with a two-position, three-way, spring-offset, solenoid-actuated directional control valve for a pneumatic system and a two-position, four-way, spring-offset, solenoid-actuated directional control valve with actuator port B plugged for a hydraulic system. Electrical control is typically used for single-acting cylinders in hydraulic systems, such as those operating industrial cardboard compactors. The single-acting cylinder in an industrial cardboard compactor has a crush plate attached to its rod that extends into a hopper when actuated. **See Figure 14-16.**

An industrial cardboard compactor operates through the following procedure:

1. An NO momentary pushbutton is actuated and current flows to the solenoid.
2. Current flowing through the coil of the solenoid creates a magnetic field around the solenoid.
3. The linear force created by the solenoid shifts the directional control valve spool, directing fluid flow to the cylinder.
4. The rod extends from the cylinder.
5. When the NO momentary pushbutton to extend the rod is released, the magnetic field of the coil collapses and the spring of the directional control valve moves the valve's spool back to its normal position.

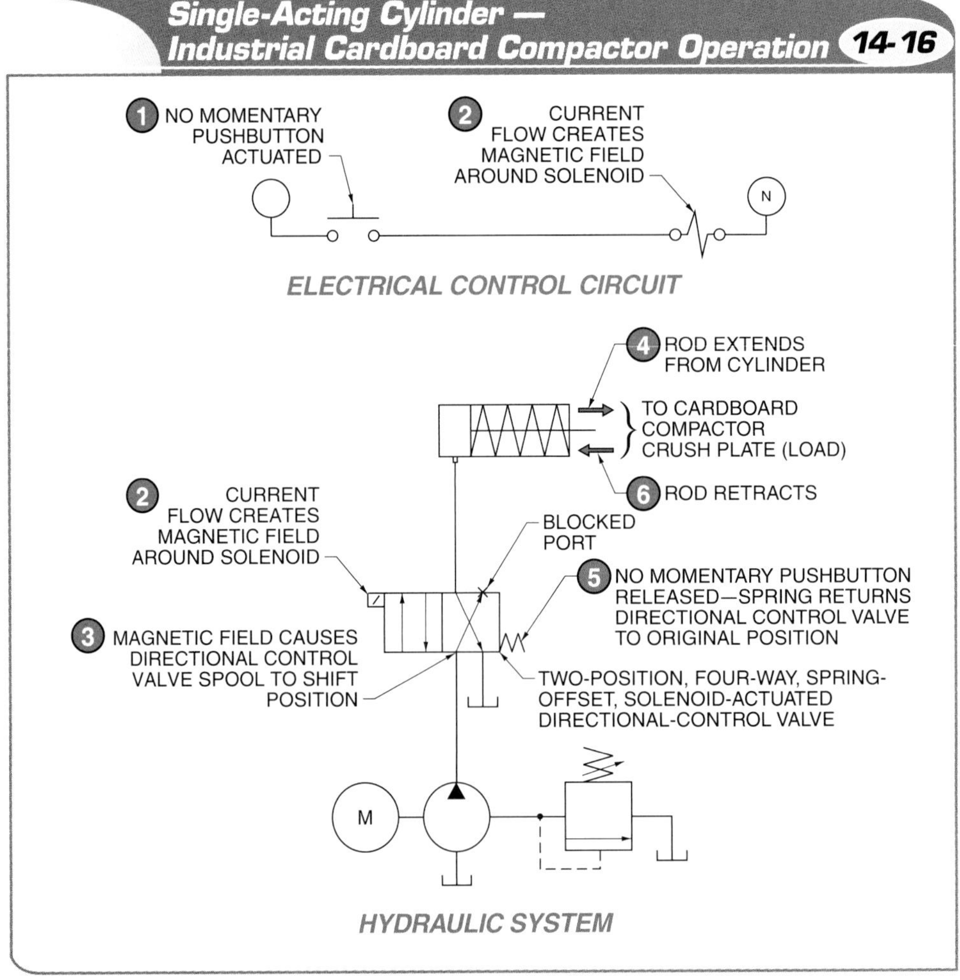

Figure 14-16. The electrical control of a single-acting cylinder is usually accomplished with a two-position, four-way, spring-offset, solenoid-actuated directional control valve.

6. When the spool changes position, fluid flow to the cylinder stops, and the spring in the cylinder retracts the rod.

The control of the single-acting cylinder in the industrial cardboard compactor can be improved by connecting an NC pressure switch in series with the pushbutton and solenoid. The NC pressure switch senses pressure at the inlet of the cylinder. If the pressure reaches a point that is too high because there is either too much material or incompressible material in the hopper, the cylinder cannot extend any further and retracts. **See Figure 14-17.**

The hydraulic system of an industrial cardboard compactor with an NC pressure switch connected in series with the pushbutton and solenoid operates through the following procedure:

1. The rod extends from the cylinder against incompressible material in the hopper.
2. The NC pressure switch opens once it senses that cylinder inlet pressure is too high.
3. The rod retracts.

The disadvantage of both of these electrical control circuits is that the operator has to hold the pushbutton down for the cylinder to operate. To correct this, a holding circuit is used with an NO start pushbutton and an emergency stop pushbutton so that the start pushbutton only needs to be pushed for a moment to extend the rod. An *emergency stop pushbutton* is an NC switch that shuts off power to the electrical control circuit.

The holding circuit keeps the solenoid engaged after the start pushbutton is actuated and released. The contacts of the control relay change their state, and current flows through the NO contacts to the control relay, keeping it energized. To disengage the control relay, two input devices are connected in series with the start pushbutton and NO relay contacts. The two input devices are an emergency stop pushbutton and an NC limit switch. If an emergency occurs, the emergency stop pushbutton can be actuated to stop current from flowing to the control relay. This causes the NO relay contacts to open, which de-energizes the solenoid and allows the spring of the directional control valve to shift the spool. The cylinder rod is then also allowed to retract.

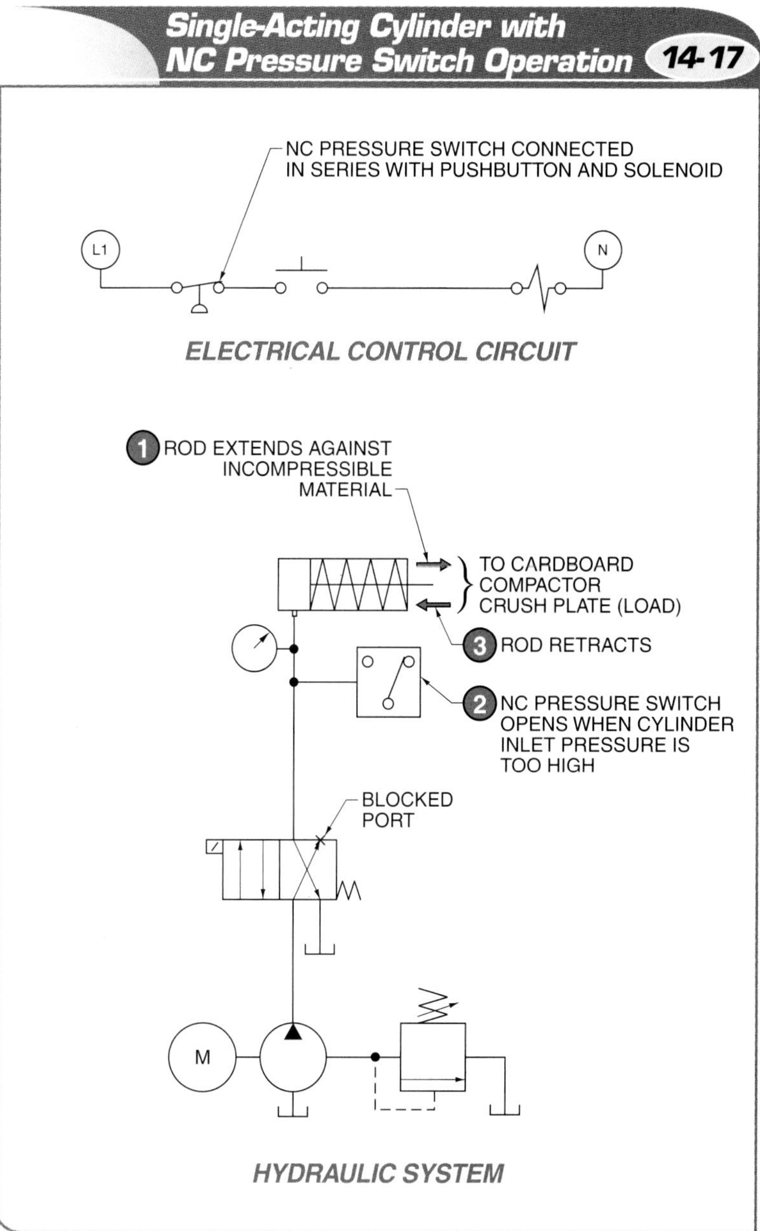

Figure 14-17. The control of the single-acting cylinder in the industrial cardboard compactor can be improved by connecting an NC pressure switch in series with the pushbutton and solenoid.

To retract the rod when it is fully extended, an NC limit switch is installed. When the rod has fully extended, the NC limit switch has the same effect on the electrical control circuit as an emergency stop pushbutton. **See Figure 14-18.** An industrial cardboard compactor with a start-stop station and an NC limit switch operates through the following procedure:

1. When the start pushbutton is actuated, current flows to the solenoid and the control relay.
2. When current flows through the solenoid and control relay, magnetic fields are created around both.
3. When the coil of the control relay is energized, its NO contacts, which are connected in parallel with the start pushbutton, close.
4. The magnetic field around the solenoid causes the directional control valve spool to shift positions and hydraulic fluid flows to the cylinder.
5. When the start pushbutton is released, current flows through the contacts of the control relay, keeping the solenoid and the control relay coil energized.
6. The rod continues to extend after the start pushbutton is released, until the NC limit switch is actuated.
7. When the NC limit switch is actuated, current flow stops. This causes the magnetic fields around the solenoid and control relay to collapse, and the spring returns the directional control valve to its normal position, retracting the rod.

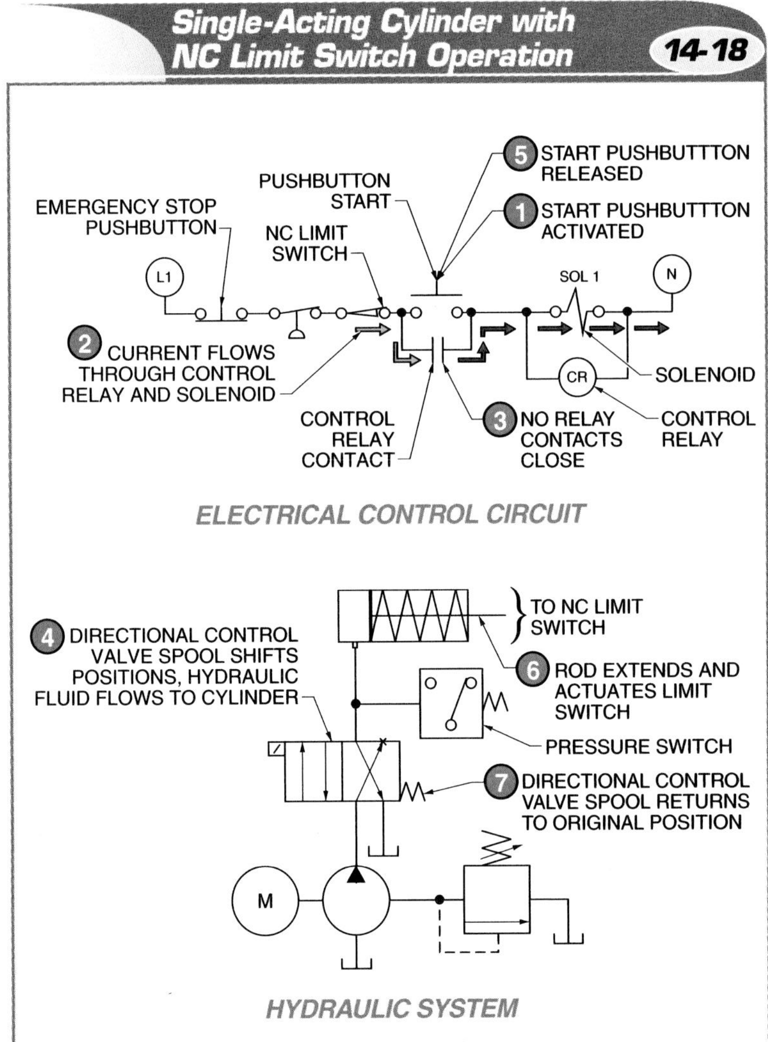

Figure 14-18. When the rod has fully extended, an NC limit switch has the same effect on the electrical control circuit as an emergency stop pushbutton.

If pressure is too high before the rod reaches the NC limit switch, the NC pressure switch opens to stop current flow to the solenoid, which collapses the magnetic fields, and allows the spring to return the spool to its normal position. The direction of fluid flow changes when the spool returns to its normal position, retracting the rod.

Double-Acting Cylinder Control Applications

The control of a double-acting pneumatic cylinder is commonly performed by a two-position, four-way, five-ported, spring-centered, NC, solenoid-actuated directional control valve. This often requires electrical control with separate pushbuttons to allow the directional control valve to control the extension and retraction of the rod. **See Figure 14-19.**

For example, electrical control can be used in a pneumatic stamping press. The rod must automatically extend when the workpiece is in position and automatically retract after the workpiece is stamped. To accomplish automatic extension and retraction, an interlocking circuit is incorporated into the controls of the pneumatic cylinder.

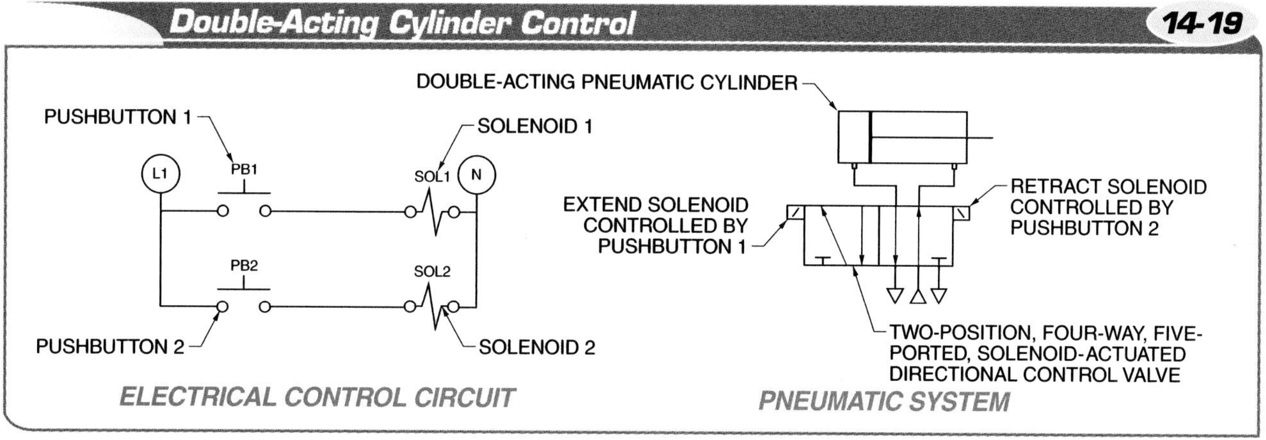

Figure 14-19. The electrical control of a double-acting pneumatic cylinder is commonly performed a two-position, four-way, five-ported, spring-centered, NC, solenoid-actuated directional control valve.

An *interlocking circuit* is a circuit in which one branch can be energized at a time with a limit switch that has both NO and NC contacts. The NC switching capability can be used to disengage a load that must be turned off before another load can be turned on. **See Figure 14-20.**

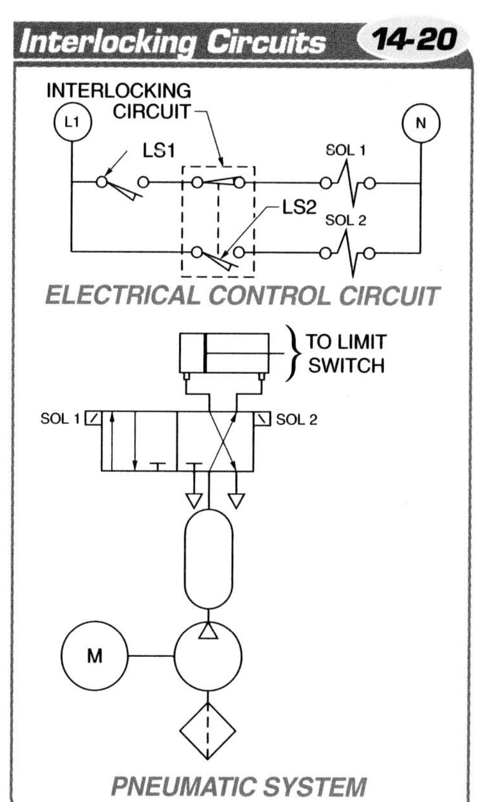

Figure 14-20. In an interlocking circuit, one branch can be energized at a time with a limit switch that has both NO and NC contacts.

A double-acting cylinder in a pneumatic stamping press operates through the following procedure:

1. A conveyor moving the workpiece stops when it is in the proper position to be stamped.
2. Switch LS1 closes when the workpiece is in position and current flows to the first solenoid.
3. The solenoid is actuated and air flows to the cap end of the cylinder. Air on the opposite side of the piston exits through the rod end of the cylinder and exhausts through the directional control valve.
4. The rod extends to stamp the workpiece and activate the limit switch.
5. The limit switch changes states and switch LS2 contact opens, turning off the first solenoid. At the same time, the NO contact on LS2 closes and current flows to the second solenoid. The directional control valve spool moves back to its normal position and the piston retracts.

Electrical control of a double-acting cylinder can also be used in a pneumatic thermoforming press. The pneumatic thermoforming press requires a specific amount of force and heat to be applied to a plastic workpiece for six seconds. The operator actuates an NO momentary pushbutton and the rod extends. Upon

TERMS

An **interlocking circuit** is a circuit in which one branch can be energized at a time with a limit switch that has both NO and NC contacts.

reaching the workpiece, the rod forms it and holds it in position for six seconds while heat is applied. After six seconds, the rod retracts. An on-delay timer properly controls the time needed for heating and forming the workpiece. **See Figure 14-21.**

A pneumatic thermoforming press operates through the following procedure:

1. An operator actuates an NO momentary pushbutton, actuating solenoid 1.
2. Air flows into the cap end of the cylinder.
3. The rod extends to the workpiece.
4. When the rod reaches the workpiece, a limit switch closes and current flows to the on-delay timer.
5. After six seconds, the on-delay timer changes the state of its contacts and the NO contact closes.
6. Current flows to solenoid 2, returning the directional control valve spool to its normal position.
7. Air flows to the rod end of the cylinder, causing the rod to retract and opening the limit switch.

TECH FACT

Solenoid burnout is more common on valves with AC coils than valves with DC coils because of high inrush current when the coil is energized.

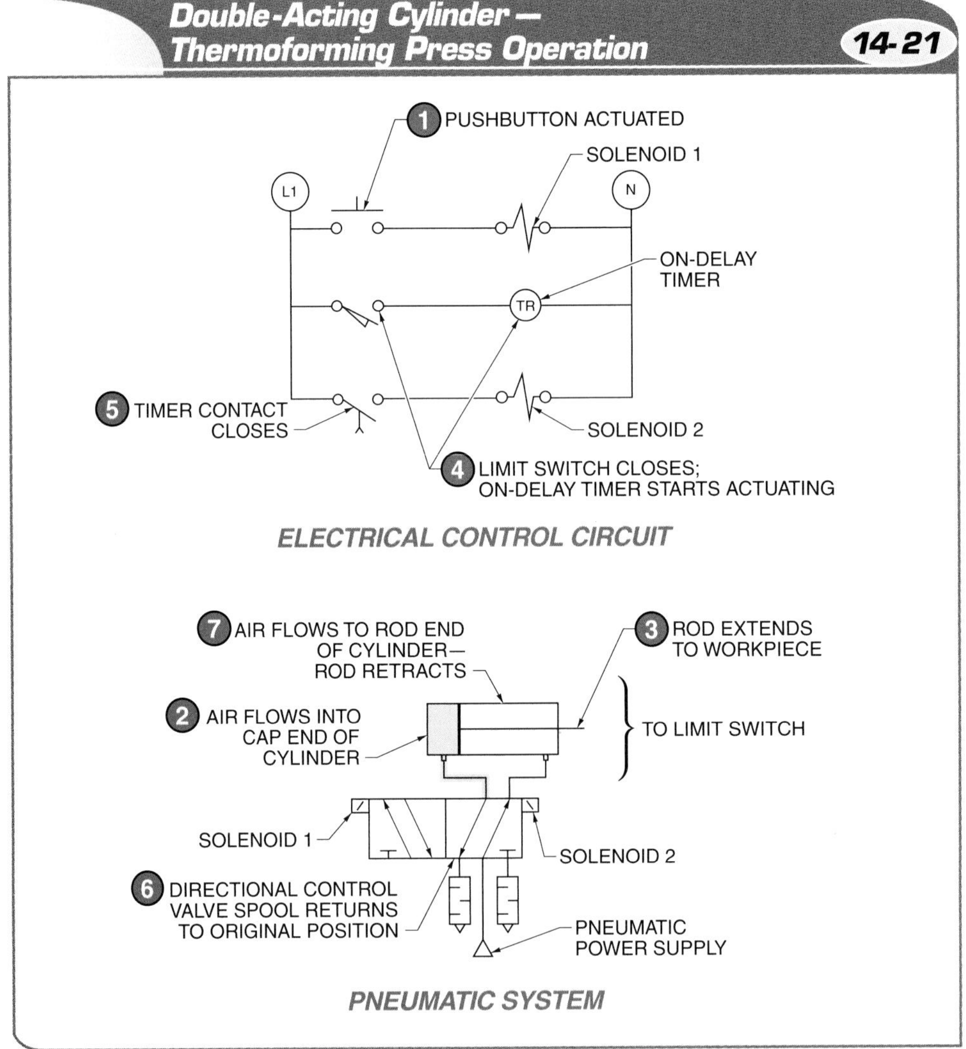

Figure 14-21. The electrical control of a double-acting cylinder can be used in a thermoforming press that requires a specific amount of force and heat to be applied to a plastic workpiece for a specific amount of time.

The electrical control of a double-acting cylinder that is controlled by a solenoid-controlled, spring-return directional control valve may need more control functions to ensure proper operation. For example, a double-acting cylinder that is controlled by a two-position, four-way, spring-offset, NO, solenoid-actuated directional control valve used in a metal shearing process needs a relay to ensure that the spool remains in the proper position for the appropriate amount of time.

In a metal shearing press, the electrical control circuit must use a holding circuit to extend the rod and lock the directional control valve spool into position. An NC limit switch is used to disengage the relay and allow the spring to return the spool to its normal position, retracting the rod. **See Figure 14-22.** A metal shearing press operates through the following procedure:

1. An NO pushbutton is actuated and current flows to the solenoid and control relay.
2. The NO contact of the control relay, which is connected in parallel with the pushbutton, closes.
3. At the same time, the solenoid actuates the directional control valve and hydraulic fluid flows to the cap end of the cylinder.
4. The rod extends and actuates the limit switch, stopping current flow. The magnetic field around the solenoid collapses, causing the spring to return the directional control valve spool to its normal position.
5. Hydraulic fluid flows through the directional control valve, retracting the rod. Hydraulic fluid that exits from the cylinder is returned to the reservoir.

It is sometimes necessary to sequence two cylinders for the operation of hydraulic equipment. Electrical control circuits can sequence two cylinders without the use of a sequencing valve. For example, the cylinders in a hydraulic press must be sequenced so that the workpiece can be clamped before it is pressed. Once the workpiece has been pressed, the cylinders must retract so the workpiece can be removed.

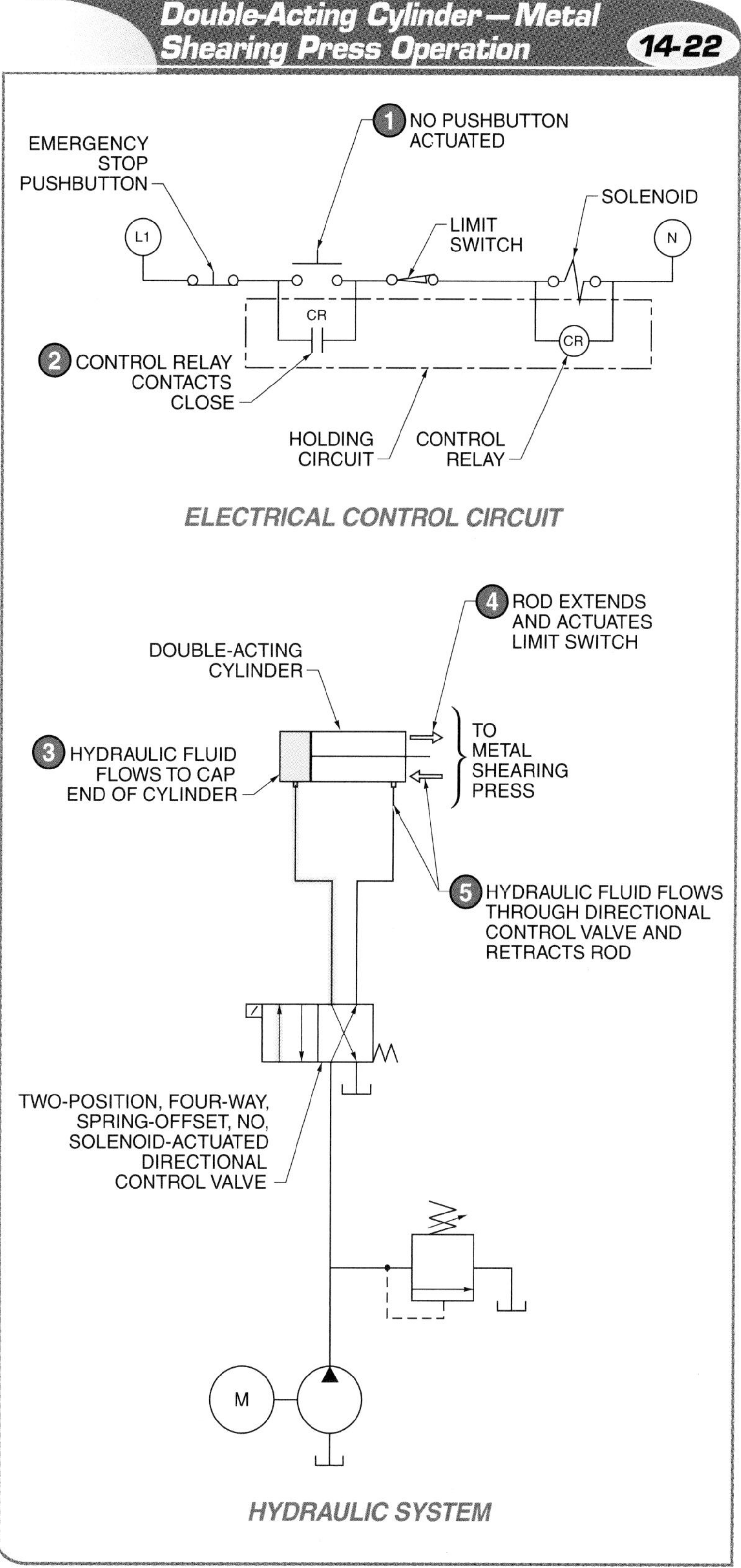

Figure 14-22. In a metal shearing press, a holding circuit must be used to extend the rod and place the directional control valve spool into position.

The sequencing of clamping and pressing cylinders can be performed electrically with two two-position, four-way, spring-offset, NO, solenoid-actuated, directional control valves, three control relays, a limit switch, a proximity sensor, a magnetic reed switch to sense the position of the cylinder piston, and a control circuit that allows the proper logic of the hydraulic system to be followed. **See Figure 14-23.** A sequenced hydraulic press operates through the following procedure:

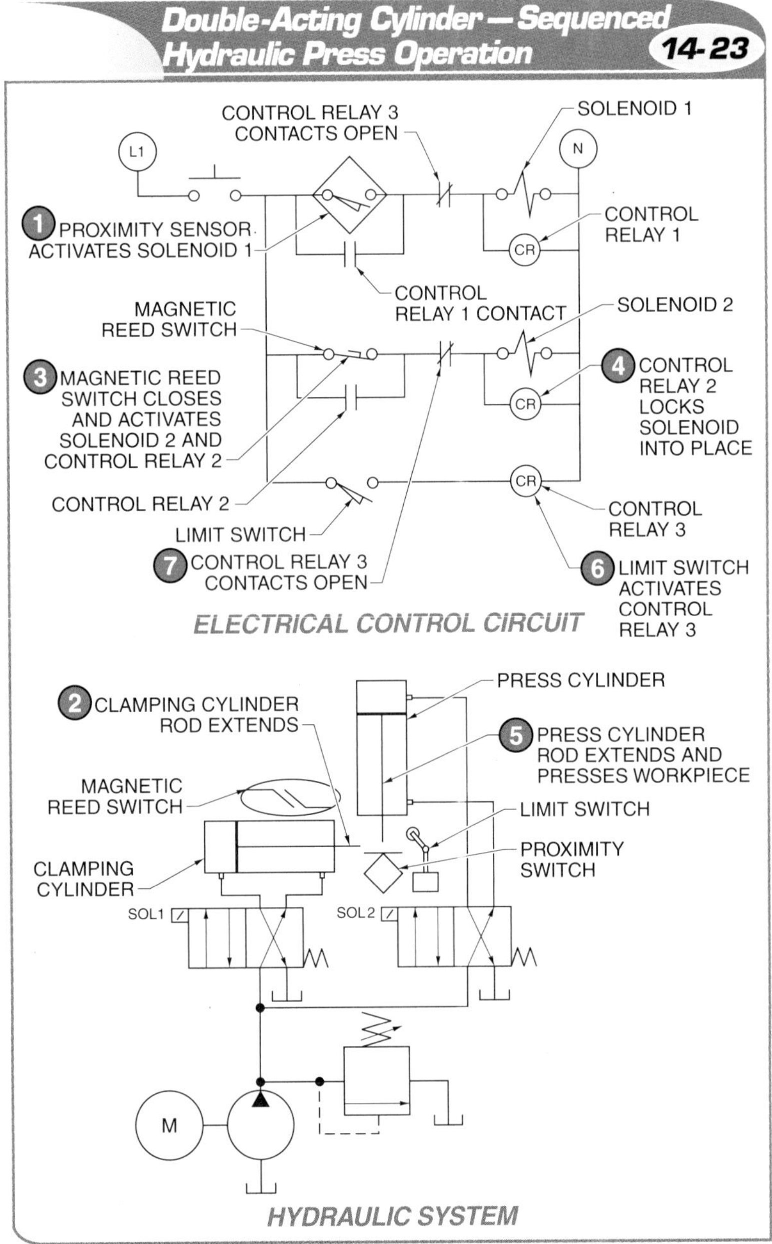

Figure 14-23. Electrical control circuits can sequence two cylinders without the use of a sequencing valve.

1. When the workpiece is in place, a proximity sensor activates solenoid 1, and control relay 1 locks solenoid 1 into place.
2. The clamping cylinder rod extends, clamping the workpiece and activating the magnetic reed switch.
3. As the magnetic reed switch closes, it activates solenoid 2 and control relay 2.
4. Control relay 2 locks solenoid 2 into place.
5. The press cylinder extends its rod, pressing the workpiece and activating the limit switch.
6. The limit switch activates control relay 3.
7. The NC contacts of control relay 3 are connected in series with control relays 1 and 2. When control relay 3 is activated, its contacts open, and power to solenoids 1 and 2 is lost. The directional control valve spools return to their normal positions and both rods retract.

In some applications, such as a metal bending machine, it is necessary to control three parallel-connected double-acting cylinders in sequence. When one workpiece must be bent at two different angles, three double-acting cylinders can be used to complete the process. One cylinder clamps the workpiece, another cylinder makes the first bend and holds it in place, and the third cylinder makes the second bend. Once all of the bends are made, the rods retract and release the workpiece. Because the three cylinders are usually located in a tight space, magnetic reed switches are used to sense the position of the three cylinders.

A master start-stop station is used to start the operation or stop it in an emergency. There are usually other safety features as well that do not allow the operation to start unless all three rods are fully retracted. For example, the metal bending machine will not start until the magnetic reed switches sense the correct position of the rods, the proximity sensor activates, and a second start pushbutton is actuated. **See Figure 14-24.**

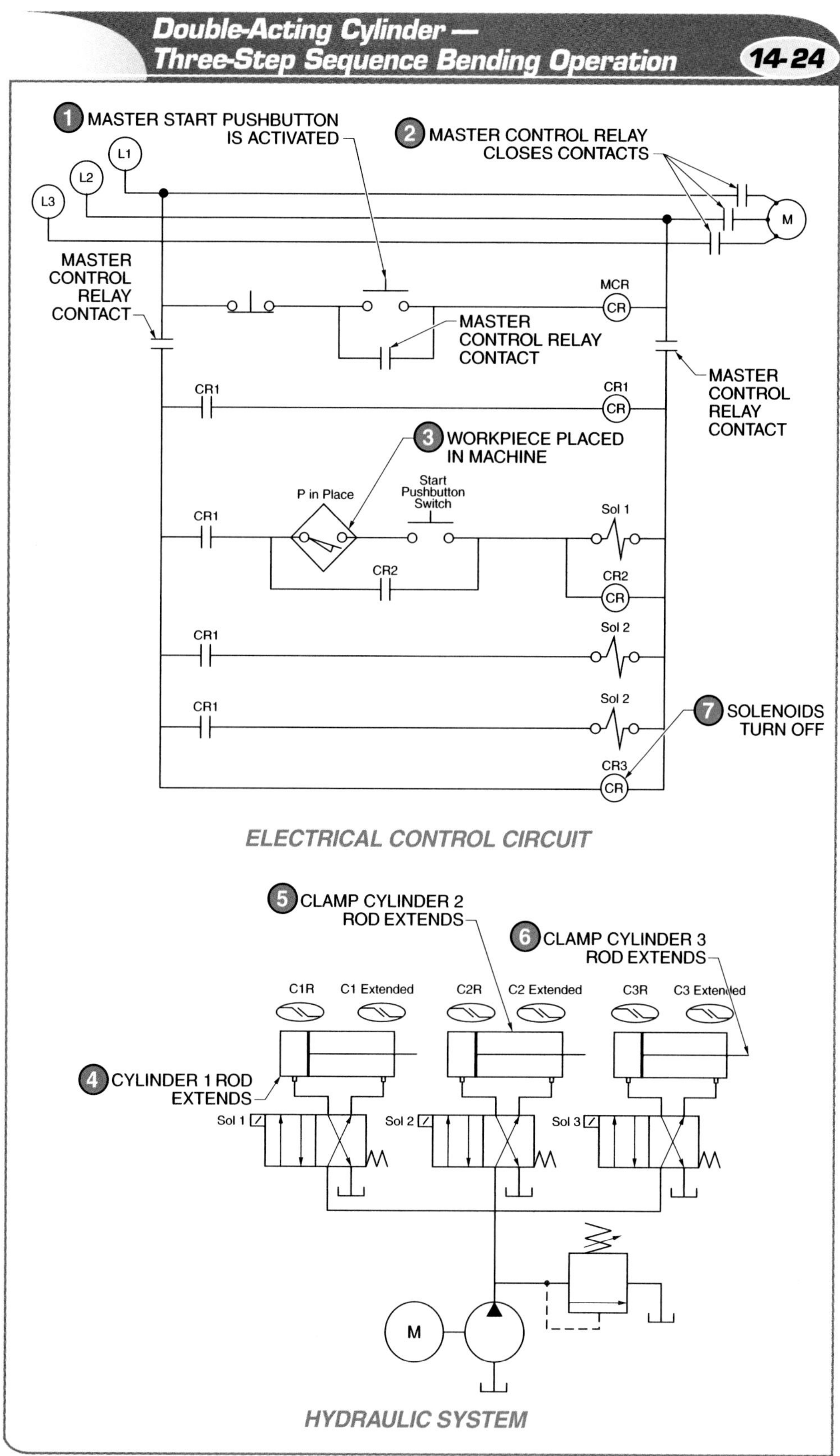

Figure 14-24. Three double-acting cylinders connected in series can operate in sequence.

To start the metal bending machine, all three rods must be retracted. When it is started, it activates control relay 1. Control relay 1 closes its NO contact and allows power to all of the solenoids. A three-step sequenced bending operation operates through the following procedure:

1. The master start pushbutton is actuated to start the machine, which activates the master control relay.
2. The NO contact of the master control relay closes, turning power ON for the control circuit and the electric motor. If the stop pushbutton is actuated, the motor and the rods stop.
3. The workpiece is placed in the machine, which activates the proximity sensor.
4. The start pushbutton is actuated and solenoid 1 and control relay 2 actuate. Control relay 2 locks in the power to solenoid 1 to extend the rod from cylinder 1.
5. When the rod from cylinder 1 has completely extended, the magnetic reed switch at the end of the cylinder actuates and sends power to solenoid 2. Solenoid 2 extends the rod from cylinder 2.
6. When the rod from cylinder 2 has completely extended and completed its bend, the magnetic reed switch at the end of the cylinder actuates. Power is sent to solenoid 3 and solenoid 3 extends the rod from cylinder 3 to make the final bend. When the rod from cylinder 3 has extended, the magnetic reed switch at the end of the cylinder actuates and sends power to control relay 3.
7. When control relay 3 is actuated, its NC contact opens and turns the power OFF for control relay 1 and all of the solenoids. The springs return the directional control valve spools to their normal positions and retracts all three rods.

Digital Resources

Chapter Review

Fluid Power System Electrical Control

Chapter 14

Name: ______________________________ Date: ______________

MULTIPLE CHOICE

______________ **1.** A(n) ___ circuit is a circuit in which one branch can be energized at a time with a limit switch that has both NO and NC contacts.
- A. holding
- B. interlocking
- C. safety
- D. all of the above

______________ **2.** A(n) ___ circuit is an electrical circuit with two or more paths for current to flow.
- A. low-voltage
- B. overhead power line
- C. parallel
- D. series

______________ **3.** A(n) ___ is a solid-state control device that is programmed and reprogrammed to automatically control electrical systems in residential, commercial, and industrial facilities.
- A. automatic switch
- B. general-purpose relay
- C. programmable logic controller (PLC)
- D. thermostat

______________ **4.** The ___ is a standardized wire sizing system used in the United States for the diameters of round, solid, nonferrous, electrically conducting wire.
- A. American wire gauge (AWG)
- B. North American wire gauge (NAWG)
- C. standard wire gauge (SWG)
- D. steel wire gauge (SWG)

______________ **5.** A ___ circuit allows a load to remain on after the switch that controls it is released.
- A. control
- B. holding
- C. normally closed
- D. normally open

______________ **6.** ___ is the electrical pressure, or electromotive force, that causes electrons to move in an electrical circuit.
- A. Current (A)
- B. Power (P)
- C. Resistance (R)
- D. Voltage (V)

______________ **7.** ___ is the opposition to the flow of electrons.
- A. Current (I)
- B. Power (P)
- C. Resistance (R)
- D. Voltage (V)

_______________ 8. The most common type of wire conductor material is ___.
A. aluminum
B. copper
C. copper-clad aluminum
D. steel

_______________ 9. The programming logic most often used to control programmable logic controllers (PLCs) is a ___ diagram.
A. binary
B. ladder
C. pictorial
D. schematic

_______________ 10. ___ is the positive (+) or negative (–) state of an object.
A. Amperage
B. Polarity
C. Resistance
D. Voltage

_______________ 11. A series-parallel circuit can have many combinations, which are always created with at least ___ switches or relay contacts.
A. two
B. three
C. four
D. six

_______________ 12. Current is measured in ___.
A. amperes (A)
B. electrons (E)
C. ohms (Ω)
D. volts (V)

_______________ 13. ___ is the amount of electrons flowing through an electrical circuit.
A. Current (I)
B. Power (P)
C. Resistance (R)
D. Voltage (V)

_______________ 14. A ___ is an electrical control mechanism that connects and disconnects the power source from the load when it is actuated.
A. motor
B. multimeter
C. pump
D. switch

_______________ 15. A(n) ___ circuit is an electrical circuit with two or more switches connected so that there is only one path for current to flow.
A. low-voltage
B. overhead power
C. parallel
D. series

COMPLETION

______ 1. A(n) ___ circuit is sometimes referred to as a start-stop station or a three-wire control circuit.

______ 2. A(n) ___ contains the logic for an application and is typically used to test and troubleshoot hardwired applications.

______ 3. Resistance is measured in ___.

______ 4. ___ is current that flows in one direction only.

______ 5. A(n) ___ is an NC switch that shuts off power to the electrical control circuit.

______ 6. A(n) ___ is often used to control the time needed for heating and forming a workpiece in an industrial thermoforming operation.

______ 7. A(n) ___ is a system that produces, transmits, distributes, and delivers electrical power to satisfactorily operate electrical loads designed to connect to the system.

______ 8. ___ switches detect the physical presence of an object and are normally used as safety devices.

______ 9. ___ is current that reverses its direction of flow at regular intervals.

______ 10. A(n) ___ circuit is a circuit with a combination of series- and parallel-connected components.

______ 11. A(n) ___ is an electrical control device that when actuated allows current to flow through it.

______ 12. A(n) ___ is a material that has a high resistance and resists the flow of electrons.

______ 13. A(n) ___ is the amount of current that flows when 1 volt (V) is applied across 1 ohm (Ω).

______ 14. A(n) ___ is an electrical switch operated by a magnetic coil.

______ 15. A(n) ___ is an electrical device that uses a mechanical actuator to control its electrical contacts.

______ 16. A(n) ___ is an electrical device with a specific amount of resistance that allows current to flow through it to accomplish work.

______ 17. A(n) ___ is a material that has low resistance and permits electrons to move easily through it.

______ 18. ___ is the energy released by the flow of electrons in a conductor.

______ 19. A(n) ___ circuit is an electrical circuit that determines when the output component is energized or de-energized.

______ 20. A(n) ___ is a conducting part of a switch that operates with another conducting part of the switch to make or break a circuit.

TRUE/FALSE

T F 1. Only certain materials have resistance to the flow of electrons.

T F 2. Series-parallel circuits are not commonly used in fluid power systems.

T F 3. In the electrical control of a fluid power system, the configuration of switches and contacts determines when and how system components activate.

T F 4. A switch can only be normally open.

T F 5. When using AWG, the larger the number, the larger the diameter of the wire.

T F 6. AC voltage is the voltage that is provided from an electrical service provider.

T F 7. When controlling a fluid power system with electric controls, the type of pressure control valve that is used determines the electrical control circuit configuration.

T F 8. If a material has a high amount of resistance, it is considered an insulator.

T F 9. When two or more switches are connected in parallel, current is able to flow through any one of the switches if they are closed.

T F 10. The greater the difference in the number of electrons between the negative (–) and positive (+) terminals in an electrical circuit, the lesser the voltage.

T F 11. Insulation commonly used to cover fluid power electrical wiring is composed of cloth.

T F 12. DC voltage can be obtained directly from batteries or rectified from an AC voltage supply.

T F 13. In an electrical circuit, there is a limit to how many switches can be connected in parallel with each other.

T F 14. A normally open (NO) switch is an electrical control device that when actuated does not allow current to flow through it.

T F 15. The higher the resistance, the lower the current, and the lower the resistance, the higher the current.

T F 16. Damage to electrical equipment can occur if power is not supplied at the proper voltage, current, type (single-phase and three-phase AC or DC), and/or condition.

T F 17. Programmable logic controllers (PLCs) use a switch to control simple or complex electrical circuits.

T F 18. A material with a high amount of resistance is considered a conductor.

T F 19. Electric diagrams for fluid power systems are in schematic form.

T F 20. The voltage needed to move electrons in an electrical circuit is comparable to the fluid pressure needed to cause hydraulic fluid to flow through a pipe in a hydraulic system.

T F 21. Automatically operated switches are the most common type of switch used for electrical control of fluid power systems.

T F 22. A power circuit is a circuit controlled by the electrical control circuit to accomplish work in a fluid power system.

T F 23. Automatically operated switches include pressure, flow, and level switches.

T F 24. The load that is typically used for the electrical control of a fluid power system is a pushbutton.

T F 25. An electrical power source can only be AC.

SHORT ANSWER

1. Explain how electrical control of a circuit can be changed.

2. Explain the difference between a PLC programming diagram and a ladder diagram.

3. Explain what type of equipment would most likely use DC voltage.

4. Explain when override capability is necessary for automatic switches.

5. List the three types of electrical circuits.

MATCHING

Schematic Symbols

______ 1. control relay

______ 2. flow switch, NC

______ 3. flow switch, NO

______ 4. foot switch, NC

______ 5. foot switch, NO

______ 6. level switch, NC

______ 7. level switch, NO

______ 8. limit switch, NC

______ 9. limit switch, NC, held open

______ 10. limit switch, NO

______ 11. limit switch, NO, held closed

______ 12. pressure switch, NC

______ 13. pressure switch, NO

______ 14. proximity switch, NC

______ 15. proximity switch, NO

______ 16. pushbutton switch, double circuit, NO and NC

______ 17. pushbutton switch, illuminated

______ 18. pushbutton switch, maintained

______ 19. pushbutton switch, mushroom head

______ 20. pushbutton switch, single circuit, NC

______ 21. pushbutton switch, single circuit, NO

______ 22. selector switch, two-position

______ 23. selector switch, three-position

______ 24. solenoid

______ 25. temperature switch, NC

______ 26. temperature switch, NO

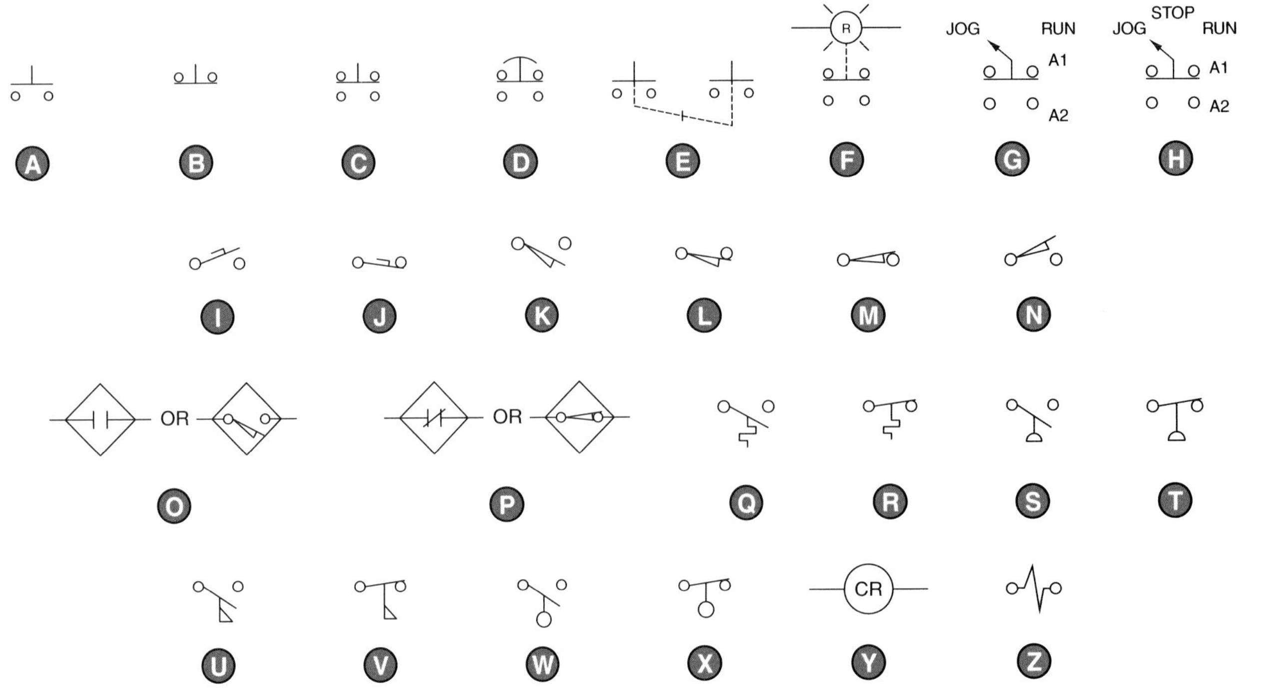

SCHEMATIC SYMBOLS

Activity 14-1: Hydraulic vs. Electrohydraulic Systems

Refer to Two-Cylinder Hydraulic System to answer the following question. *Note:* This activity can also be performed with the FluidSIM® software included on the CD-ROM.

1. Draw a schematic diagram for an electrohydraulic system that operates the same as the hydraulic system.

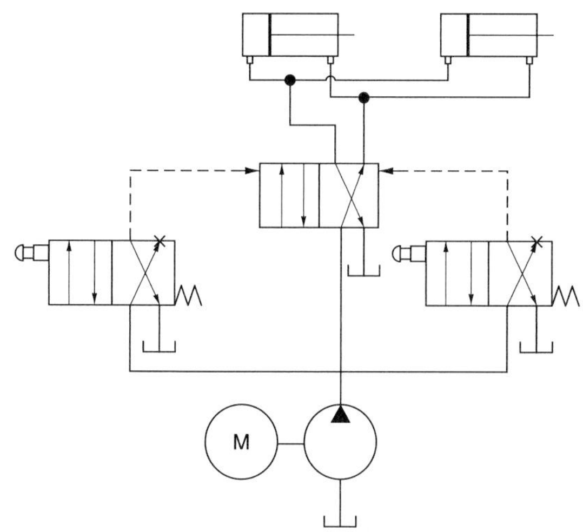

TWO-CYLINDER HYDRAULIC SYSTEM

Activity 14-2: Schematic Diagrams—Product Positioning

A production manufacturing facility has two differently sized boxes that are initially moved along one conveyor line until the larger boxes are separated and moved onto a second conveyor line.

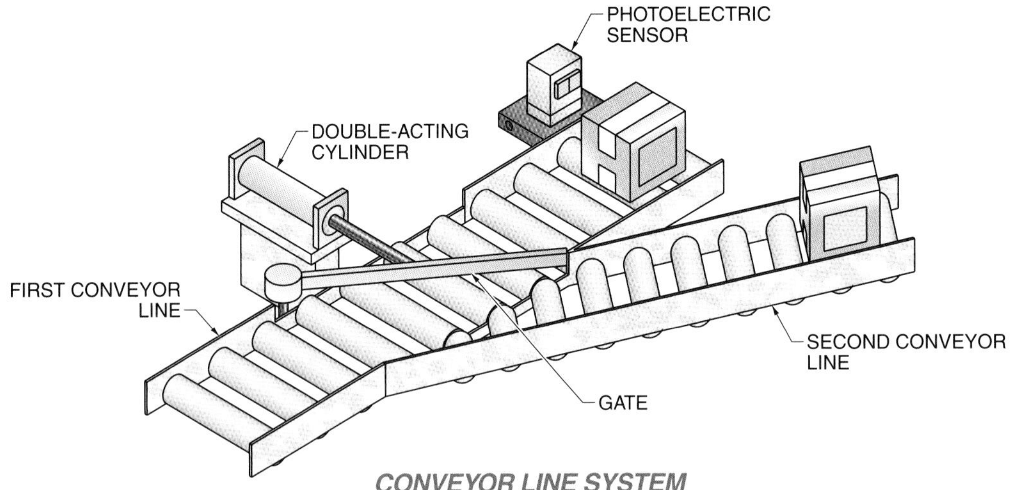

CONVEYOR LINE SYSTEM

1. Draw a fluid power schematic diagram that illustrates the following operations (*Note:* This activity can also be performed with the FluidSIM® software included on the CD-ROM):
   - When the larger boxes move along the initial conveyor line, a photoelectric sensor reads them.
   - When the photoelectric sensor is activated, a double-acting pneumatic cylinder extends and moves a gate that directs the larger boxes onto the second conveyor line.
   - When the photoelectric sensor is not activated, the double-acting pneumatic cylinder must retract.
   - To minimize cost, this operation must be accomplished using only one photoelectric sensor and a two-position, four-way, spring-return, solenoid-actuated directional control valve.

2. Draw a ladder diagram that illustrates the solenoid operation of the two-position, four-way, spring-return, solenoid-actuated directional control valve.

Activity 14-3: Schematic Diagrams—Pneumatic Stamping Press

In a pneumatic stamping operation, the press operator must depress and hold the pushbutton switch until the stamping is completed. The operator can then release the pushbutton switch, which causes the cylinder to retract. *Note:* This activity can also be performed with the FluidSIM® software included on the CD-ROM.

1. Connect symbols in Schematic Diagram to illustrate an operation in which the press operator only has to momentarily depress the pushbutton switch to allow the cylinder to extend. Also, the cylinder must retract automatically when the cylinder activates a proximity switch indicating that the cylinder has fully extended.

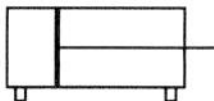

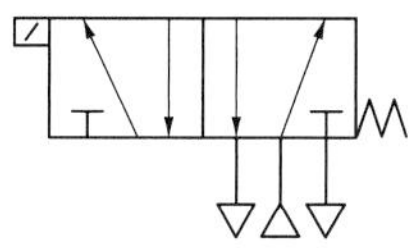

SCHEMATIC DIAGRAM

2. Use Electrical Schematic Symbols to draw a ladder diagram that illustrates the operation of the pneumatic stamping press in problem 1.

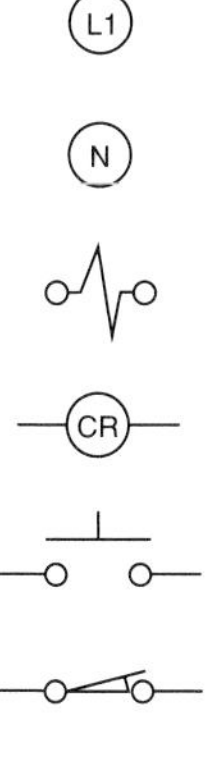

ELECTRICAL SCHEMATIC SYMBOLS

Activity 14-4: Schematic Diagram—Hydraulic Lift Application

A box is moved along one conveyor line to a hydraulic lift platform, which is raised with a double-acting cylinder to lift the box to a second conveyor line. Once the hydraulic lift platform is fully raised, a single-acting cylinder extends and pushes the box onto the second conveyor line. Refer to Hydraulic Lift—Pictorial Diagram and Hydraulic Lift—Schematic Diagram to answer the following question. *Note:* This activity can also be performed with the FluidSIM® software included on the CD-ROM.

1. Draw a ladder diagram and that illustrates the following operations:

- When the box is placed on the hydraulic lift platform, a limit switch is activated that raises the hydraulic lift platform.
- When the double-acting cylinder is fully extended, another limit switch is activated that extends the single-acting cylinder to push the box onto the second conveyor line.
- When the single-acting cylinder has pushed the box onto the second conveyor line, a limit switch is activated that returns both the double-acting and single-acting cylinders to their starting positions.

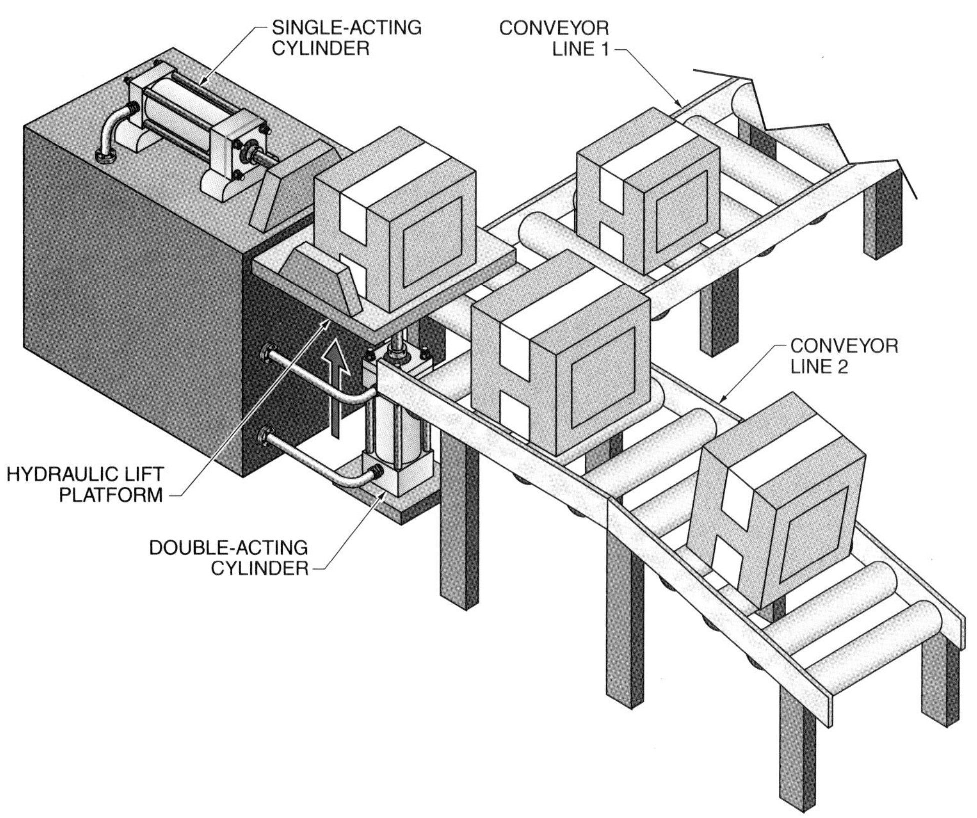

HYDRAULIC LIFT — PICTORIAL DIAGRAM

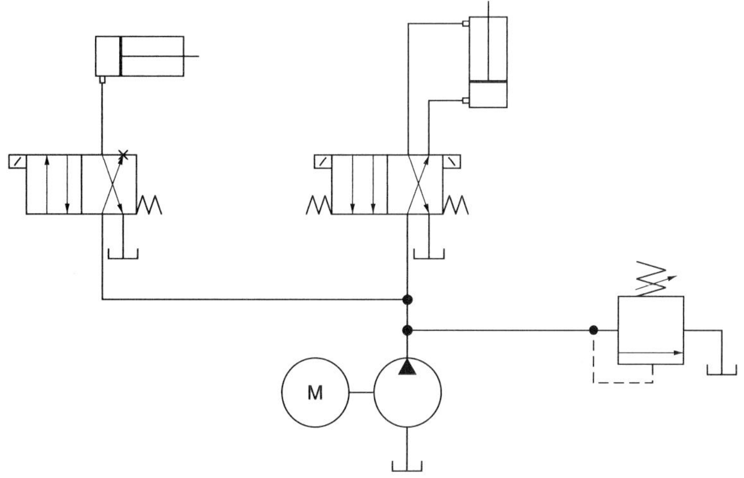

HYDRAULIC LIFT — SCHEMATIC DIAGRAM

Activity 14-5: Hydraulic Car Lift Application

A three-position, four-way, closed-center, lever-detent-actuated directional control valve is used as a brake for a hydraulic car lift. It is decided to replace the lever-detent-actuated directional control valve with a solenoid-actuated, spring-centered directional control valve.

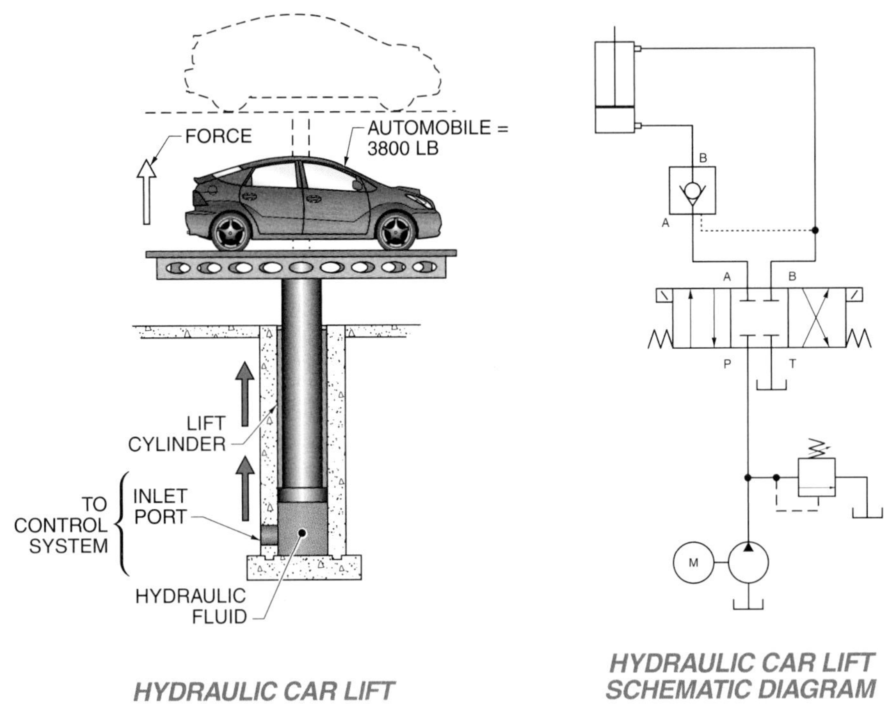

HYDRAULIC CAR LIFT

HYDRAULIC CAR LIFT SCHEMATIC DIAGRAM

1. Sketch a schematic diagram with a solenoid-actuated, spring-centered directional control valve that accomplishes all of the following:
   - When an NO pushbutton is momentarily actuated, the cylinder extends until it reaches the correct height.
   - When the cylinder reaches the correct height, the operator momentarily presses a stop pushbutton switch, which allows the directional control valve to shift to the center position.
   - When the lift needs to retract, another NO pushbutton switch is actuated and the lift moves down.
   - When the lift reaches the bottom, an NC limit switch opens, allowing the directional control valve to shift to the center position.
   - An interlocking system must ensure that both solenoids do not activate at the same time.

Activity 14-6: Normally Closed and Normally Open Pushbutton Switches

Use the FluidSIM® software included on the CD-ROM to build the hydraulic system in activity 14-5. *Note:* Use the ladder symbols to create the electrical control circuit. To link the solenoid in the electrical control circuit and the solenoid in the power circuit, they must be labeled with the same names. To label symbols in FluidSIM®, the symbol must be double clicked. For the contacts off of the control relay, use the "make" and "break" switches. (When they are labeled the same as the control relay, they will change to the symbols in the diagram). There is no limit switch included in FluidSIM®, so a normally closed pushbutton can be used to simulate it and labeled as "LS."

1. After building the ladder diagram with FluidSIM®, describe the fluid flow in the hydraulic system when the NC pushbutton on first rung of ladder diagram is actuated.

2. Describe the fluid flow in the hydraulic system when the NC pushbutton on second rung of ladder diagram is actuated.

3. How fast does the cylinder move when either one of the NC pushbuttons is actuated?

4. What happens with the cylinder when either one of the NO pushbuttons is pressed?

Fluid Power System Maintenance and Troubleshooting

Chapter 15

OBJECTIVES

- Identify common maintenance issues and problems that arise for hydraulic systems and pneumatic systems.
- Describe the procedure for successfully troubleshooting fluid power systems.
- Describe common hydraulic and pneumatic system troubleshooting applications.
- Describe common fluid power system troubleshooting methods.

INTRODUCTION

Fluid power technicians must have training and work experience to maintain or troubleshoot fluid power systems. Fluid power systems are often so integrated into a mechanical system that if they stop operating, the entire system must be shutdown. Equipment that is not in operation can cost an end user thousands of dollars for every hour it is not in use. For this reason, a fluid power technician must be able to troubleshoot quickly, efficiently, and in a cost-effective manner, while following all safety procedures. Various procedures and methods can be used for troubleshooting fluid power systems, depending on the type of system that is installed.

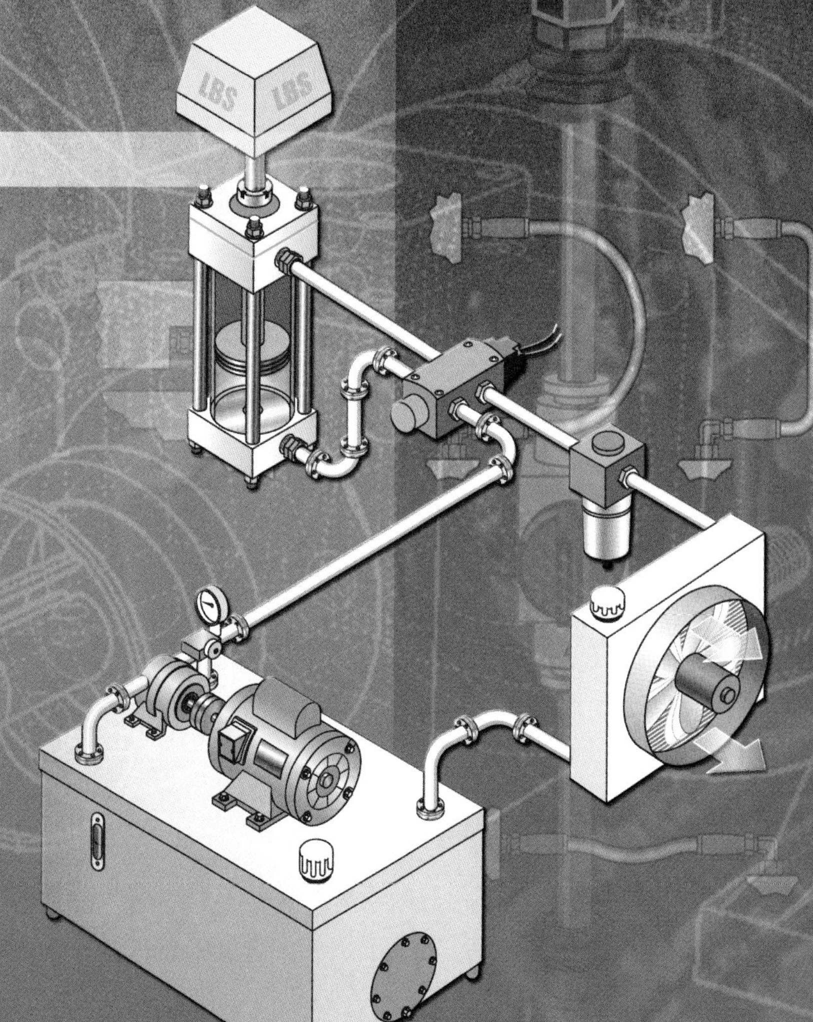

TERMS

Maintenance is the practice of periodically inspecting and repairing equipment.

FLUID POWER SYSTEM MAINTENANCE

The most productive and cost-effective way to maintain a fluid power system is to routinely inspect and test it, as well as follow the best practices for maintenance. *Maintenance* is the practice of periodically inspecting and repairing equipment. Most commercial and industrial facilities have routine maintenance programs. For example, both hydraulic and pneumatic filters are changed regularly according to a routine maintenance program.

The advantage of a routine maintenance program is that it reduces the amount of troubleshooting that may need to be performed. Many future problems can be avoided. Separate maintenance programs are applied for hydraulic and pneumatic systems.

Flo-Ezy Filters, Inc.

Many maintenance problems with fluid power equipment can be avoided by replacing system filters at regularly scheduled intervals.

Hydraulic System Maintenance

Hydraulic system maintenance is performed based on the results of routine inspections of equipment and systems within a specific period of time. The frequency of inspection varies with the manufacturer's recommendations. If those are not available, mobile hydraulic equipment typically should be inspected every 400 hr to 600 hr, or every three months. Industrial hydraulic equipment should also be inspected every three months if the manufacturer's recommendations are not available. Inspection intervals are based on conditions that include operating temperature, pressure, accessibility, operating time, and environmental factors such as shock, vibration, operating time, and cleanliness. **See Figure 15-1.**

Most hydraulic system maintenance problems are caused by the contamination and breakdown of hydraulic fluid. Air, water, dirt particles, and other contaminants can accumulate in hydraulic fluid to create sludge and corrosive acids. Contaminants can clog small openings and cause severe operating problems. Acids corrode metal components, plastic seals, and hoses. Contaminated hydraulic fluid also has diminished lubricating capabilities. This can cause a component with mating surfaces, such as a cam ring in a vane pump, to rapidly wear. **See Figure 15-2.**
The most important routine maintenance is replacing hydraulic filters according to the manufacturer's recommendations or when indicated by devices such as pressure gauges, flow meters, or filter indicators. The correct filter for the system must be installed. Contaminants must not enter the system when adding hydraulic fluid to the reservoir. Also, hydraulic fluid sampling must be performed on a routine schedule to ensure the contaminant level is low in the fluid.

The tubes and funnels for adding hydraulic fluid must be cleaned with lint-free rags before and after use. The breather/filler cap should fit snugly. The breather/filler cap filter must keep contaminants out of the hydraulic system and allow air to flow freely in and out of the reservoir.

TECH FACT

Many new hose failures in fluid power equipment are attributed to installing the wrong type of hose, hose fitting, or twisted layline. Manufacturer's procedures should always be followed when installing new hoses.

Hydraulic System Maintenance Intervals 15-1

Hydraulic System Component: Tasks Required	Frequency					
	Continuously or Daily	First Time After Operation*	Weekly	Monthly	Biannually	Annually
Accumulator:						
Check gas charge pressure		10 to 15				
Reservoir:						
Check hydraulic fluid level	x					
Monitor fluid temperature	x					
Check for leaks	x					
Take fluid test sample						
Change fluid		50				
Filter:						
Monitor level of contamination indicators	x					
Clean or replace filter element				x	x	
Clean or replace breather filler cap		10 to 15		x	x	
Drive:						
Check couplings between drive motor and pump (running noises)					x	x
Valves:						
Check settings of pressure and flow control valves		10 to 15		x	x	x
Signal Elements:						
Check settings of pressure switches and positions of limit switches		10 to 15		x	x	x
Cylinder(s):						
Visually check rod and wiper seals						
Clean and grease suspension points				x	x	

* in hr

NOTE:
Hydraulic components not listed here are not usually subject to inspection and servicing regulations, per manufacturer's instructions.

Figure 15-1. Hydraulic system maintenance is performed based on the results of routine inspections of equipment and systems within a specified period of time.

Hydraulic System Maintenance Problems 15-2

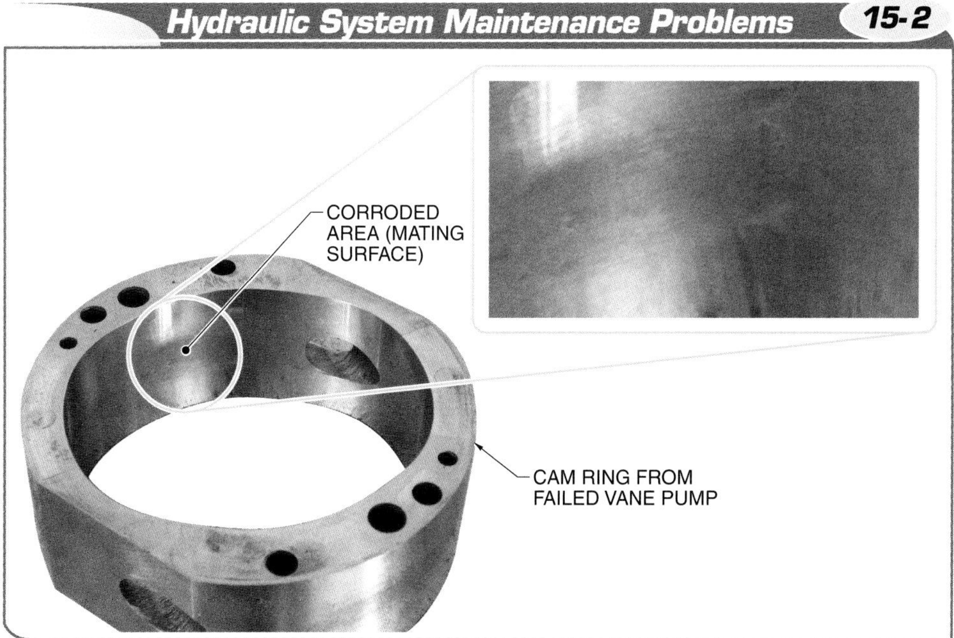

Figure 15-2. Mating surfaces in a hydraulic system can often have maintenance problems due to contamination and breakdown of the hydraulic fluid.

Hydraulic cylinders that are not in service must have their rods retracted to prevent contaminants from settling on any exposed metal and entering the system when the rod is moved. **See Figure 15-3.** Rods that must be stored exposed should be coated with grease that can be easily removed before startup.

Hydraulic Cylinder Storage 15-3

FRONT-END LOADER (NOT IN SERVICE)

DOUBLE-ACTING CYLINDER WITH ROD RETRACTED TO PREVENT CONTAMINATION

Figure 15-3. Hydraulic cylinders that are not in service must have their rods retracted to prevent contaminants from settling on any exposed metal and entering the system when the rod is moved.

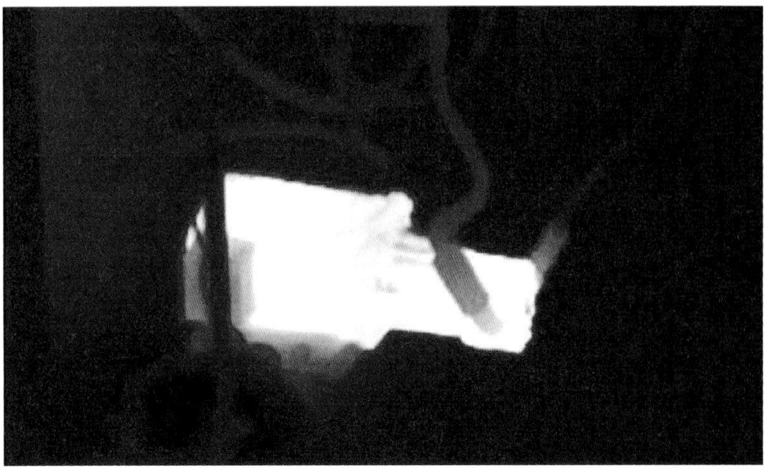

The Snell Group

Thermal images from specialized test instruments can be used to safely troubleshoot hydraulic equipment such as overheated pumps.

The correct hydraulic fluid must be used for each application. Hydraulic fluids contain additives such as foaming and rust inhibitors. Some hydraulic fluid additives can cause certain hose and valve materials to deteriorate.

Hydraulic fluid must maintain the required viscosity as the system warms up. Improper viscosity can cause serious operating problems. Hydraulic fluid that has higher viscosity than that recommended by the manufacturer can cause sluggish operation. Hydraulic fluid that has lower viscosity than that recommended by the manufacturer can cause internal leakage and poor lubrication. Viscosity problems cause extra power consumption that wastes energy. Hydraulic fluid must be drained and replaced according to the manufacturer's recommendations. Used hydraulic fluid must be disposed of according to state and local municipal codes.

Overheating can damage hydraulic fluid and system components. Operating temperatures should be kept within the range suggested by the manufacturer. The reservoir and system components should be clean so heat can dissipate easily. Kinked lines can also cause excessive heat buildup. The relief valves should be set at the recommended level because excessive pressures generate additional heat. To prevent heating hydraulic fluid, overspeeding or overloading a system needs to be avoided. Also, hydraulic fluid coolers must be kept clean and operating efficiently.

Hoses should be regularly inspected for signs of wear, cracking, abrasion, fluid seepage, blisters in the outer covering, deterioration, uneven twists in the hose line, coupling corrosion, or hydraulic fluid leaks. Hoses with any of these types of problems need to be replaced immediately. Fluid stains or puddles under hydraulic equipment and hoses indicate the presence of a leak in the line. **See Figure 15-4.**

Warning: A body part should never touch or pass within a few feet of a pressurized hydraulic hose assembly. Small, pinhole-size hydraulic fluid leaks are sometimes difficult to see and are extremely dangerous because high-pressure hydraulic fluid can pierce the skin and enter the body, causing serious injury or death.

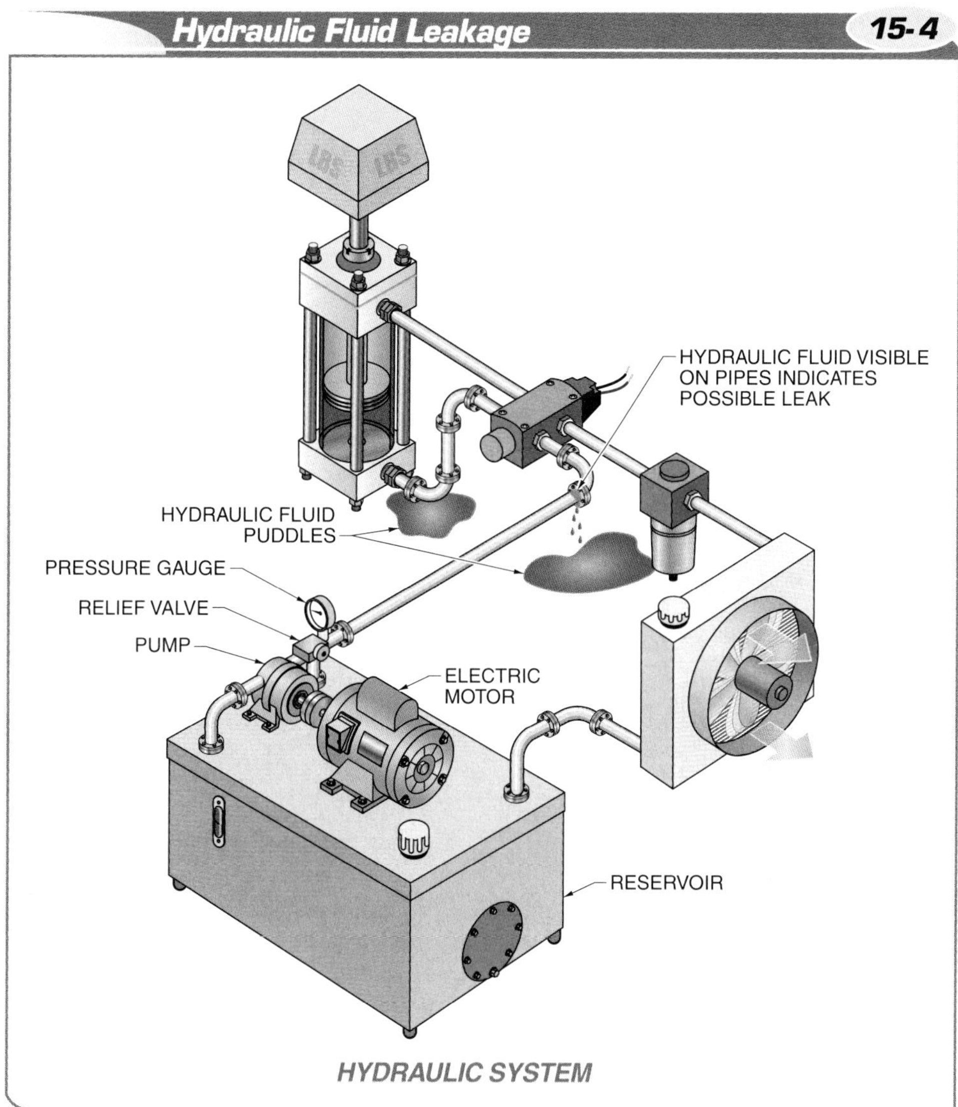

Figure 15-4. Fluid stains or puddles under hydraulic equipment and hoses indicate the presence of a leak in the line.

TERMS

A **minimum hose bend radius** is a standard given by the manufacturer that indicates the least amount of bend that can be used for a specific type of hose.

Hydraulic hoses can be damaged if improperly sized, constructed, or installed. Elbows and adapters are used to relieve strain on hoses. Hoses should not be allowed to rub against anything that could remove their outer layer. They should be properly installed for maximum operating life. **See Figure 15-5.** Clamps are often needed to support long hose lengths or to keep the hose away from moving parts.

The manufacturer's recommendation for a minimum hose bend radius should be followed to avoid hose collapse and flow restriction. A *minimum hose bend radius* is a standard given by the manufacturer that indicates the least amount of bend that can be used for a specific type of hose. Typically, the more layers in a hose, the higher the minimum bend radius.

TECH FACT

Typically, the higher the operating pressure used in a hydraulic system, the more layers a hose will have. While standard hoses usually have three layers, high-pressure hoses can have five to ten layers.

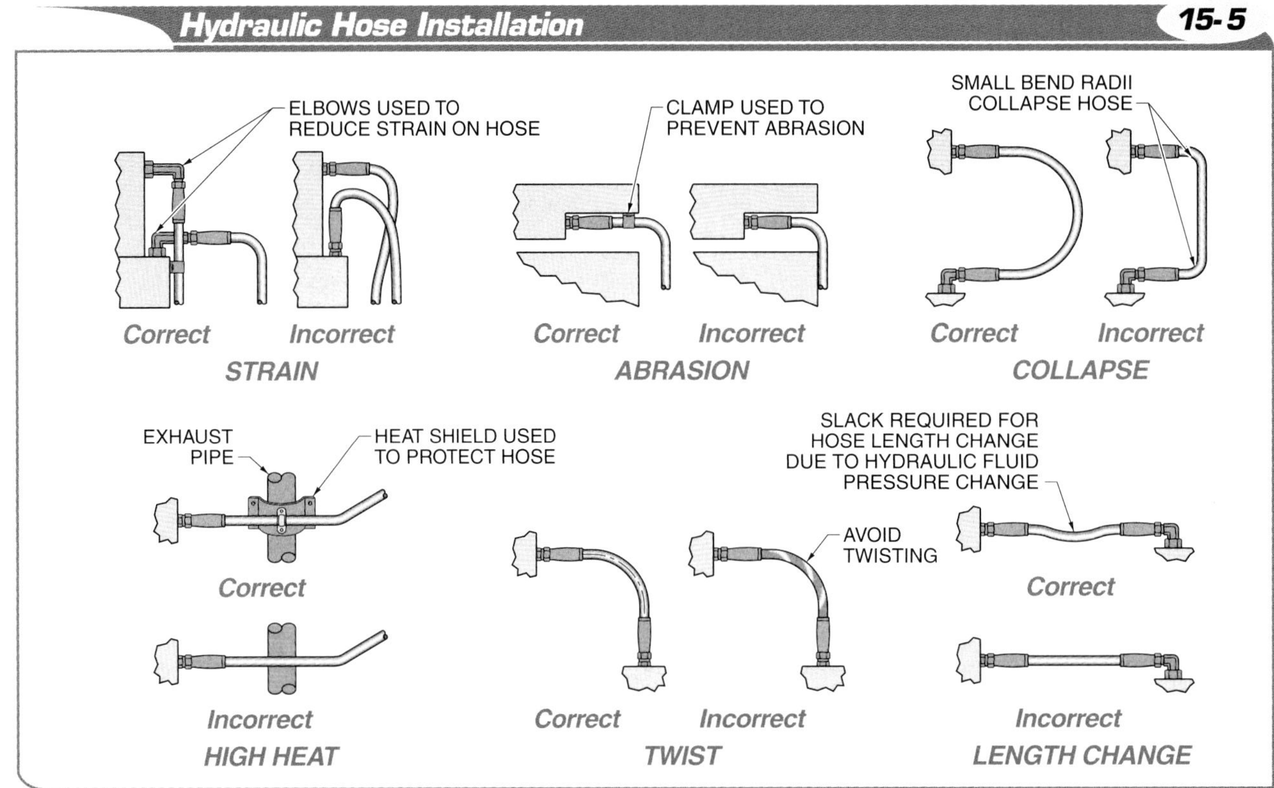

Figure 15-5. Hydraulic hoses must be properly installed for maximum operating life.

Hoses exposed to excessive heat deteriorate at the outer layer, which heats the hydraulic fluid. The inner layer of synthetic rubber can be damaged from overheated hydraulic fluid, rough treatment, or incompatible hydraulic fluid. Breakdown of the inner layer can cause internal blockage that results in noisy pump operation or erratic actuator operation.

Hose ends should always be installed properly. If a hose end is over- or under-crimped, it could leak or come loose at the end and cause premature failure. Hoses should not be stretched, twisted, or kinked. When a hose is installed in a straight line, enough slack should be allowed to provide for length changes, which occur when pressure is applied.

Care is necessary when removing packaging material from a hose. A hose that is nicked by the sharp edge of a blade can be weakened and fail. Hoses should be stored in cool, dry areas away from direct sunlight. Hoses should never be hung on nails or hooks. Stored hoses should be evenly coiled to avoid kinking.

Hoses that are designed for one application should never be used in another. A hose should never be kinked or folded to stop flow. Also, the number of shutoff valves necessary for stopping fluid flow should be verified. Hose spring guards should be used to avoid flexing. When reinforcing wire in the hose is deformed, it should be replaced because it cannot be bent back to its original shape. Hoses stored in temperatures below 32°F should be gradually warmed to room temperature before being placed in service.

Hydraulic valves, pumps, and actuators are precision devices and must be kept free from contaminants. Hands, tools, and the work area should always be kept clean when work is performed on hydraulic components. Sheets of lint-free paper should be used to cover the work area, while lint-free rags should be used to clean components. The work area should be prepared to contain any hydraulic fluid discharges during the repair. The specific recommendations of the manufacturer must always be followed.

Before they are dismantled, hydraulic components need to have their exteriors cleaned. Openings should be covered with clean plastic plugs. An approved cleaning solvent is used to clean separate components as soon as they are removed. Only recommended cleaning solvents should be used on plastic components. Components are dried with low-pressure, clean, dry compressed air. The dried parts are coated with hydraulic fluid and stored in clean plastic bags. When a component is dismantled, its parts should be placed in the order that they were removed. This will help ensure correct reassembly.

Only compliant flushing fluids should be used to clean and flush a hydraulic system as recommended by the manufacturer. The use of noncompliant flushing fluids can leave residual chemicals on the interior of the system that can break down hydraulic fluid or hoses. When flushing a hydraulic system, the system should be operated to circulate the flushing fluid. However, this should only be done if recommended by the manufacturer. The flushing fluid is then drained and the system is refilled with the correct hydraulic fluid. The system should be operated slowly to remove any trapped air.

Air can be kept out of a hydraulic system by maintaining the correct hydraulic fluid level in the reservoir. Also, the system should be cycled under no-load conditions several times to purge air after refilling or working on the system and then given time for the air to settle out. Before operation, the level and condition of the hydraulic fluid should always be checked. In cold weather, a system should be started and operated slowly. Cold hydraulic fluid, due to high fluid viscosity, can cause problems such as sluggish operation and poor lubrication.

The correct replacement parts must always be used. Springs, seals, and O-rings should be replaced any time a component is serviced. The O-ring material should be verified for resistance to chemical additives in the hydraulic fluid and for its capability to handle expected system pressures. The shaft or working area needs to be cleaned before installing new O-rings. Nicks or sharp edges should be removed from the O-ring grooves or from along the shaft. The O-ring is then lubricated with the same hydraulic fluid used in the system and installed carefully to avoid contact with sharp edges and overstretching. The O-ring size should be verified as the correct size for the application.

The correct positions for the installed packing and seals should be verified. Seals normally expand to make a tighter seal when hydraulic system pressure is applied. Seals should be installed using the same precautions as with O-rings. Hardness, brittleness, or spongy consistency of the seal material indicates damage from incompatible hydraulic fluid or excessive temperatures.

Metal surfaces can be damaged due to excessive force. When preparing metal surfaces for use in fluid power systems, they should be polished with soft, nonabrasive, manufacturer-approved cleaning pads. Abrasive grit can remove surface metal and cause internal leakage. Only approved cleaning products and materials supplied or recommended by the manufacturer should be used. **See Figure 15-6.**

Figure 15-6. When preparing metal surfaces for use in fluid power systems, they should be polished with soft, nonabrasive, manufacturer-approved cleaning pads.

Schroeder Industries

Portable hydraulic test kits are used to help technicians test, troubleshoot, and arrange preventive maintenance data on hydraulic equipment.

It is always necessary to check for leaks after service. External leaks usually leave stains and accumulate dirt. Internal leaks caused by worn or damaged seals or metal components are more difficult to detect. They cause slippage of pistons and poor operation of valves and motors. They also generate heat at the site of the leak. Areas where work was performed should be cleaned thoroughly.

Pneumatic System Maintenance

Pneumatic systems should be inspected routinely for leaks and clogged filters. Also, lubricator oil levels should be checked routinely. A hose rupture in a pneumatic system can cause serious injury and damage to system components. Pneumatic system components require regular maintenance to prevent any problems that may lead to hose ruptures.

Pneumatic components must be kept clean and damage free to provide efficient heat dissipation. To prevent corrosion damage and efficiently dissipate heat, a pneumatic component manufacturer coats their components with protective paint. Improper or excessive coats of paint applied to a component can increase insulation and prevent proper heat transfer. When a component must be recoated, the manufacturer's recommendations for coatings procedures should be followed.

Ensuring that the supply of air is clean and dry is the most important task when performing pneumatic system maintenance. Depending on how critical the operation is, the intake filter may need to be inspected weekly or daily. Drains and dryers must be inspected routinely for proper operation. Pneumatic equipment such as filters are maintained according to the manufacturer's recommendations.

The lubrication, cooling, and filtration systems of large air compressors, such as those used in industrial facilities, must be monitored carefully. The correct type and amount of compressor oil is critical to proper air compressor operation. The alignment of the prime mover and the air compressor must regularly be inspected. Compressed air filters must be inspected and replaced on a regular schedule. **See Figure 15-7.** If a belt-driven air compressor system is used, the condition of the belt must be inspected. Equipment maintenance requirements are compiled and scheduled into the overall pneumatic system maintenance procedures.

The amount and type of lubricant supplied to actuators is critical. Too much lubricant can clog an actuator, cause environmental hazards, and mask other problems. Too little lubricant can result in premature wear of components and heat build-up. The lubricant used must meet the requirements from the manufacturer. The wrong types of lubricant can also cause premature wear on moving parts. The mounting and alignment of all actuators, especially heavily loaded pistons and motors, should be checked regularly. Actuators that are used intermittently or held in one position for a long time should be operated occasionally to verify free movement.

FLUID POWER SYSTEM TROUBLESHOOTING

Troubleshooting fluid power systems should only be performed by trained fluid power technicians and should always be done with safety as the first priority. In addition to fluid power system troubleshooting charts, one of the most useful tools to troubleshoot fluid power equipment is a schematic diagram.

There are many similarities between troubleshooting a hydraulic system and a pneumatic system, such as using pressure gauges to help isolate the problem. However, there are also many differences. For example, troubleshooting moving parts is different because a pneumatic system uses lubricating oil for lubrication, while a hydraulic system uses hydraulic fluid.

There are troubleshooting procedures and methods that a fluid power technician can use to help troubleshoot broken or malfunctioning equipment. Following troubleshooting methods and procedures can provide a fluid power technician with the means necessary to quickly and efficiently determine the solution to a problem.

Troubleshooting Hydraulic Systems

Hydraulic system problems are usually related to a malfunction in the force or speed of the actuators. If the system is not producing enough force, the problem is likely related to a lack of pressure. The pressure settings and pressure control valve operation should be checked. When equipment speed is incorrect, the pump and flow control valve operation should be checked.

Internal and external leaks can affect pressure and fluid flow in a hydraulic system. External leaks are usually easy to locate if the exterior of the equipment is kept clean. For external leaks that are difficult to locate, dye and a "black" light can be used to help determine the leak location. Internal leaks are harder to locate and are commonly caused by the wear on couplings or the deterioration of seals or O-rings. Hydraulic fluid leaks reduce the capacity of the equipment to perform work. Serious leaks may cause hydraulic systems to stall and fail when heavily loaded.

Internal leaks from hydraulic components, such as pumps, relief valves, and directional control valves, can be isolated using hydraulic testers. A *hydraulic tester* is a device that measures the pressure, flow rate, and temperature of hydraulic fluid in a system. Flow rate and system pressure are tested at various points in the hydraulic system. **See Figure 15-8.**

Figure 15-7. Pneumatic system maintenance requires regular inspection and replacement of compressed air filters.

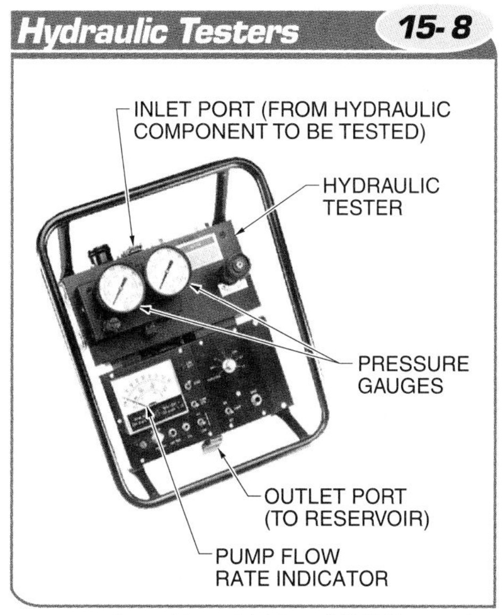

Figure 15-8. A hydraulic tester is used to measure the pressure, flow rate, and temperature of hydraulic fluid in a system.

TERMS

A **hydraulic tester** is a device that measures the pressure, flow rate, and temperature of hydraulic fluid in a system.

Flow control problems are identified using a flow meter connected into the system. In addition, actuator operation speed can be timed and compared to the specifications of the hydraulic system. Tachometers are used to measure electric or hydraulic motor speed. Excessive heating of a system or hydraulic fluid can also cause changes in pressure and fluid flow. Temperatures should not exceed the recommendations of the equipment and hydraulic fluid manufacturers.

When working on a hydraulic system, troubleshooting charts from the manufacturer should always be consulted because complex hydraulic systems are simplified into specific parts and component functions. Hydraulic testers can then be used to determine the proper function of specific parts and to test component functions. Flow and pressure can be measured at various points in a hydraulic system. The operation of pumps, relief valves, and directional control valves can also be tested.

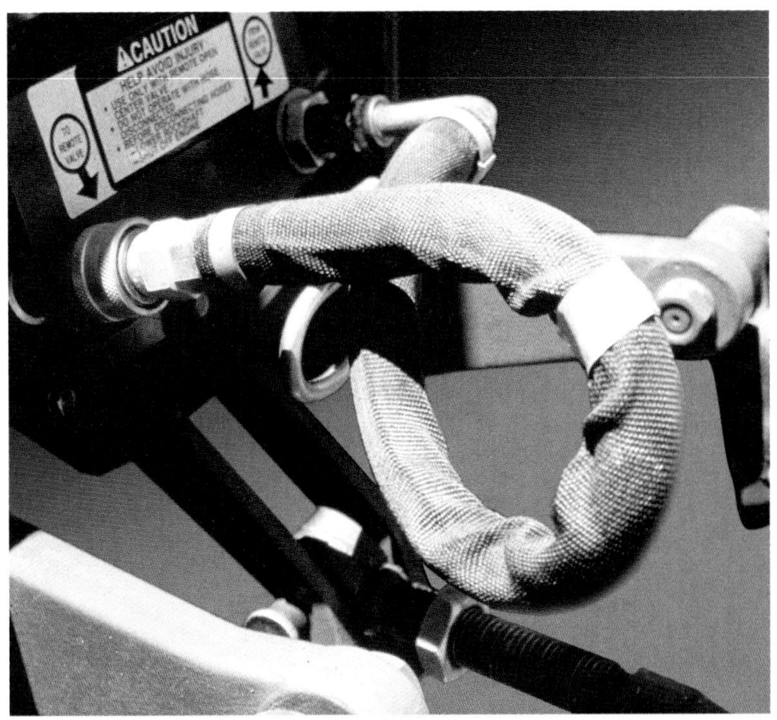

The Gates Rubber Company

Hoses used in fluid power systems should typically be in lengths of 3′ or less.

Hydraulic systems that are critical to production must operate without interruption, and their downtime must be kept to a minimum. Troubleshooting procedures are followed to isolate and repair problems as quickly as possible. Troubleshooting procedures and results should always be documented for future reference. **See Figure 15-9.**

Troubleshooting Pneumatic Systems

The most common problem in pneumatic systems is insufficient air pressure or airflow. This problem is usually caused by air leaks from worn, disconnected, or kinked hoses or tubing lines. Air leaks in a pneumatic system commonly occur at connection points of hoses, tubing, and fittings. Hoses can also have air leaks at worn areas on their outer layers. High-pressure pneumatic hoses have linear reinforcement cords along the inner layer of the hose. After repeated use, the outer layer of the hose can become cracked and worn, exposing the reinforcement cords and causing air leaks.

Depending on the manufacturer, some pneumatic hoses have reinforcement layers rather than cords. In these hoses, the inner reinforcement layers can collapse and are not seen on the surface. Many modern pneumatic systems use clear or semiclear polyurethane (plastic) tubing without reinforcement layers or cords. Damaged inner linings are visible in these types of plastic tubing. Hoses and tubing can also become kinked, blocking airflow. **See Figure 15-10.**

TECH FACT

Because hoses and tubing used in fluid power systems can become easily damaged and are susceptible to other problems such as kinks or premature wear, it is recommended that they only be used with moving components and in lengths of 3′ or less.

Hydraulic System Troubleshooting 15-9

Problem

A crane winch used to load and unload cargo from coastal freighters is powered by a hydraulic motor. The winch is rated for 15 t. The hydraulic pump is driven by an electric motor. During the second shift, the winch begins to malfunction. The winch operates erratically; sometimes sluggishly, sometimes not lifting over 1 t. The maintenance technician is notified of the problem by phone from the dock supervisor.

Troubleshooting Step 1 - Investigate

Check operation of winch - operates properly for short period then starts to work sluggishly and does not lift heavy load. After rest, winch operates properly but malfunctions again after short period.

Troubleshooting Step 2 - Isolate

Likely Causes	Action/Result
Air in system.	Check oil in reservoir sightglass for air bubbles - oil appears milky (air present).
	Pump makes rumbling noises from air in oil.
	Inspect intake line from reservoir to pump. Apply oil to connections with pump operating - no change in pump noise. No leak at intake piping connections.
	Remove top inspection plate on reservoir. Oil level low. Whirlpool formed around pump inlet pipe when operating - low oil level allows air to enter system at pump inlet pipe. Oil added to reservoir - whirlpool disappears.
	Operate winch to force air though system. Bleed air through air vents on lines at top of crane.
Low oil in reservoir/oil leak in system.	Inspect lines mounted on crane - oily residue followed to leaking coupling. Coupling leaks when hose flexed in certain position.

Troubleshooting Step 3 - Remedy

Replace hose to stop leak.	Replace defective hose with new longer hose. Secure hose to minimize stress on coupling in all positions.

Troubleshooting Step 4 - Documentation

File reports.	Complete troubleshooting report. File work orders for hose inspection on cranes during scheduled maintenance.

PROCEDURE

Troubleshooting Report

Maintenance Technician Identification Number: 6704

Department: Crane Maintenance

Equipment Identification Number: Crane #1

Problem: Erratic operation, speed, and force.

Symptoms: Not lifting rated weight. Not moving quickly.

Cause(s): Air in the oil caused by low oil level in reservoir. Oil leak found at hose connection.

Repair Procedures: Replaced section of hose with longer hose. Secured hose to reduce stress from flexing.

Preventive Maintenance Action: Check all hoses for leaks, proper length, and mounting.

DOCUMENTATION

Figure 15-9. Troubleshooting procedures are followed and documented to isolate and correct hydraulic system problems.

Air leaks in a pneumatic system are costly and must be repaired quickly. In many applications, air leakage can waste up to 25% or higher of the pneumatic system energy. **See Figure 15-11.** Air leaks can be located by listening for the noise of discharging air when the work area is quiet or by applying soapy water to the suspected location of the air leak. The soapy water will produce bubbles when applied to an air leak.

When checking for pressure or airflow problems, troubleshooting should begin at the air compressor and move towards the

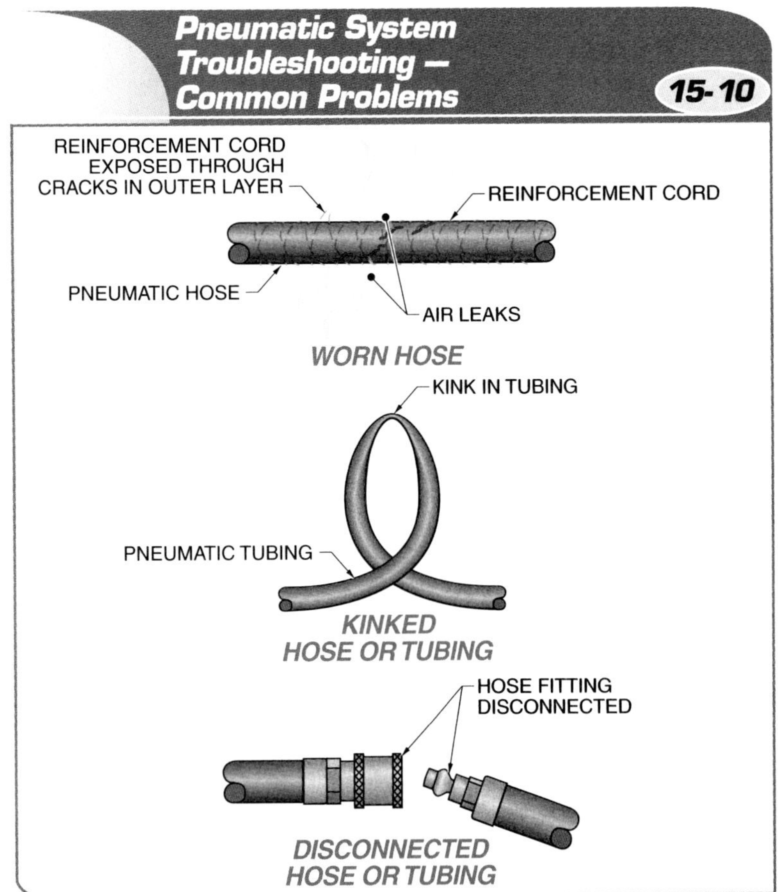

Figure 15-10. The most common problem in pneumatic systems is insufficient airflow caused by leaks from worn, kinked, or disconnected hoses or tubing.

TERMS

A **packing nut** is a mechanical device on a pneumatic actuator that applies tension to the seals or packing to prevent air leaks around the circumference of a rod or shaft.

A **plug-in pressure gauge** is a pneumatic device that can be added to or removed from a pressurized pneumatic system without removing pressure from the system.

actuators, or vice versa. Couplings, valve operation, and pressure and airflow readings are checked until the air leak, disconnection, or blockage is located. A common cause of pneumatic system problems is air compressor controls that are at incorrect pressure settings. The settings of pressure controls or pressure gauges on the air compressor or receiver should be checked for proper levels.

If the air compressor is working properly, each component should be checked in sequence starting with the air compressor in the system piping. For example, a blocked filter at an FRL unit restricts airflow at that portion of the system. If there are no other problems in the system piping, the actuators should be tested because they are the only moving parts that are visible.

When testing an actuator, it should be deenergized and disconnected from the load. Once the load is disconnected, the actuator is reenergized and operated without a load attached. If it functions normally without a load attached, the load could be too large, the actuator may have internal leaks, or there may be insufficient pressure to operate the load. If the actuator does not function normally without a load attached, the problem is usually with the actuator or air supply components. The air supply components should be tested for proper operation by starting at the actuator and moving toward the air compressor.

Actuator packing nuts should not be overtightened. A *packing nut* is a mechanical device on a pneumatic actuator that applies tension to the seals or packing to prevent air leaks around the circumference of a rod or shaft. Depending on the application, plug-in pressure gauges can be used to trace pressure levels through a pneumatic system. A *plug-in pressure gauge* is a pneumatic device that can be added to or removed from a pressurized pneumatic system without removing pressure from the system. For example, plug-in pressure gauges are typically used to verify that the pressure in a pneumatic hand tool is at the proper level. **See Figure 15-12.**

Fluid Power System Troubleshooting Methods

A technician can use several different troubleshooting methods to identify which component is causing the fluid power system to malfunction, but will commonly become comfortable with one method. Typically, a technician should become proficient with the method that works best after experiencing all methods. The three most commonly used methods for troubleshooting fluid power systems are part replacement, beginning-to-end, and end-to-beginning.

TECH FACT

Flexible components, such as hoses and tubing, used in fluid power systems can have additional protection from cuts and abrasion through the use of polyethylene spiral wrapping. Polyethylene spiral wrapping is an inexpensive method of covering hoses and tubing that can extend their operational lifetimes.

Part Replacement Troubleshooting Method. The *part replacement troubleshooting method* is a fluid power system troubleshooting method that replaces parts without the use of equipment diagnosis tools. This is the type of least accurate, troubleshooting method. A technician using this method replaces parts until the problem is corrected, spending less time diagnosing an unknown fluid power problem. However, it is only beneficial if the first replacement corrects the problem. This is an efficient method for fluid power technicians who have a lot of experience troubleshooting the same equipment and who are familiar with common fluid power problems.

Beginning-to-End Troubleshooting Method. The *beginning-to-end troubleshooting method* is a fluid power troubleshooting method that traces problems from the beginning of the fluid power system (prime mover, hydraulic pump, or air compressor) to the end (actuator). This method is preferred for fluid power systems that have completely stopped operating. For example, the beginning-to-end method is used if a rod is not extending and all gauges are reading zero. Because there is no pressure in the system, the problem is most likely near the beginning of the system.

End-to-Beginning Troubleshooting Method. The *end-to-beginning troubleshooting method* is a fluid power troubleshooting method that traces problems from the end of the fluid power system (actuator) to the beginning (prime mover, hydraulic pump, or air compressor). This method is preferred when actuators are not operating, but the gauges at the beginning of the system indicate normal levels. For example, the end-to-beginning method is used if a rod is not extending and the gauges after the directional control valve indicate normal levels. The gauge readings indicate that the problem is most likely at the end of the system.

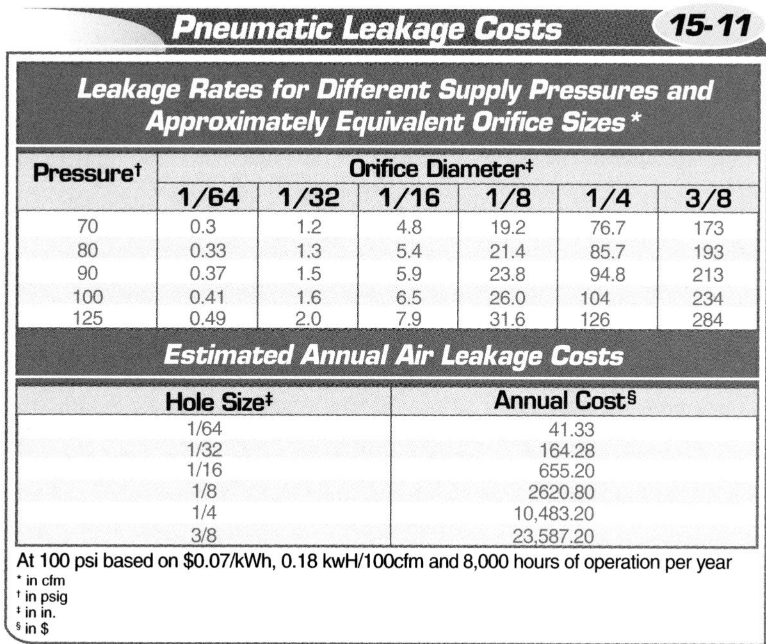
Pneumatic Leakage Costs 15-11

Leakage Rates for Different Supply Pressures and Approximately Equivalent Orifice Sizes *

Pressure†	Orifice Diameter‡					
	1/64	1/32	1/16	1/8	1/4	3/8
70	0.3	1.2	4.8	19.2	76.7	173
80	0.33	1.3	5.4	21.4	85.7	193
90	0.37	1.5	5.9	23.8	94.8	213
100	0.41	1.6	6.5	26.0	104	234
125	0.49	2.0	7.9	31.6	126	284

Estimated Annual Air Leakage Costs

Hole Size‡	Annual Cost§
1/64	41.33
1/32	164.28
1/16	655.20
1/8	2620.80
1/4	10,483.20
3/8	23,587.20

At 100 psi based on $0.07/kWh, 0.18 kwH/100cfm and 8,000 hours of operation per year
* in cfm
† in psig
‡ in in.
§ in $

Figure 15-11. Air leaks can waste up to 25% of a pneumatic system's energy.

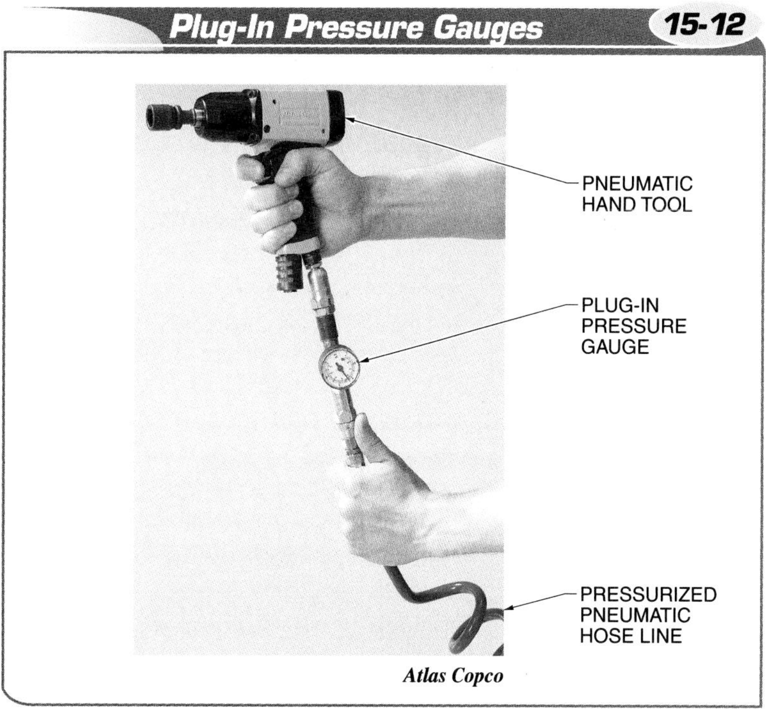

Figure 15-12. Plug-in pressure gauges can be used to trace pressure levels through a pneumatic system.

Half-Back Troubleshooting Method. The *half-back troubleshooting method* is a method for testing for problems in the middle of a hydraulic circuit and working backward or forward on the system. This is the most

TERMS

The **part replacement troubleshooting method** is a fluid power system troubleshooting method that replaces parts without the use of equipment diagnosis tools.

The **beginning-to-end troubleshooting method** is a fluid power troubleshooting method that traces problems from the beginning of the fluid power system (prime mover, hydraulic pump, or air compressor) to the end (actuator).

The **end-to-beginning troubleshooting method** is a fluid power troubleshooting method that traces problems from the end of the fluid power system (actuator) to the beginning (prime mover, hydraulic pump, or air compressor).

The **half-back troubleshooting method** is a method for testing for problems in the middle of a hydraulic circuit and working backward or forward on the system.

common and practical approach when a problem occurs that could have multiple causes. The advantage of this method is that if the cause is an unusual problem or could logically be a number of components, this will help to isolate the problem.

For example, the cause for a hydraulic cylinder operating too slowly could be from a number of components that are restricting flow. If a technician starts to troubleshoot the flow problem in the middle of the circuit, it will indicate whether the problem is in the front or back half of the circuit. If in the middle of the circuit the flow is at its normal rate, then the problem exists in the back half of the circuit. If the flow is restricted to the middle, then the problem is in the front half of the circuit. This method can quickly help to isolate complex problems and save time and costs when troubleshooting. This method can be used to troubleshoot electrical problems as also.

Fluid Power System Troubleshooting Procedures

There are procedures that a technician can apply when troubleshooting a fluid power system for problems. A technician can troubleshoot a specific problem more efficiently if a facility maintains accessible and accurate records of equipment problems. This way, the technician may not have to perform the entire troubleshooting procedure. As a technician becomes more adept at troubleshooting a fluid power system, the technician may add or subtract steps to or from the troubleshooting method. Troubleshooting a fluid power system is performed by applying the following procedure:

1. Follow all safety regulations that are in place to protect the personnel and equipment. Failure to follow all facility safety regulations and the safety regulations of the equipment manufacturer can result in serious injury or death to those working on or near the equipment being serviced.
2. Understand normal equipment operation. This will help narrow down the list of potential problems. A good practice is to read all literature that is associated with the equipment.
3. Gather information concerning the specific equipment failure. Typically, the best source for specific equipment information is the equipment operator. This can help reduce the amount of time required for the troubleshooting. **See Figure 15-13.**
4. Operate the equipment. When the equipment is operating erratically, or only one part of the operating sequence is experiencing a malfunction, watching the equipment operate may reveal the malfunction.
5. Perform a sensory inspection of the system. Read pressure gauges, temperature gauges, and flow meters. Also, listen for unusual sounds, or smell the area for unusual odors coming from the equipment. *Note:* Many types of injuries can occur from direct contact with fluid power equipment, such as burns, pinched body parts, loss of body parts, or death. Never use hands to inspect a pressurized line for leaks because pressurized fluid or air can penetrate the skin and contaminate the body.
6. List the possible causes of equipment failure. Sometimes, a problem will not be visible through the operation and inspection of equipment. Multiple problems may even be the source of the failure. When this is the case, compile a list of the components that could be causing the malfunction. List them in order from the most probable to the least probable. From that list, form another list of the probable causes in order of which malfunctions are the easiest and quickest to troubleshoot and repair.
7. Verify the conclusion for probable malfunction. For example, if the likely cause of a fluid power malfunction is a directional control valve with a manual override, activate the manual override to determine if the directional control valve is the problem. Continue to the next step if the directional control valve operates properly.

Equipment Operation Checklists 15-13

Equipment Operating Condition	Operator Response
1. Describe equipment operation immediately prior to failure (slow, fast, normal).	________
2. Did equipment seize at failure?	________
3. Did equipment gradually slow down prior to failure?	________
4. Describe any loud sounds prior to failure (metal grinding, vibrations, etc.).	________
5. Describe any unusual odors when equipment failed (smoke, burning materials, oil, etc.).	________
6. List any mechanical or electrical work performed on equipment within last maintenance period.	________
7. Describe any indications of system leakage (air or oil).	________

Machine No. ________

Date: ________

Time: ________

Shift: ________

Operator: ________

Figure 15-13. An itemized equipment operation checklist can help an equipment operator provide information that can reduce the amount of time needed for the troubleshooting process.

8. Verify that all power sources are locked and tagged out before repairing the cause of the malfunction. In a hydraulic system, verify that there is no pressure from the two possible sources, the accumulator or the load attached to the rod. In a pneumatic system, verify that all pressure in the system was safely released after the power was locked out. Also, even if the equipment that needs to be repaired can be isolated with a shutoff valve, residual pressure in the pneumatic lines must be released.

   Note: All facility safety procedures must be followed when locking out any power supplies. If at any time a technician feels uncertain about the safety procedure or feels that safety is compromised, the technician should cease working on the equipment and contact a supervisor or other qualified person immediately.
9. Repair or replace the device that is causing the equipment to malfunction. Carefully reconnect all pipes and hoses to prevent leaks and properly tighten the device to its subplate. After all connections have been securely tightened and all lockouts have been removed, start the equipment to verify that replacing the malfunctioning component has corrected the problem. If the problem has been successfully corrected, continue to step 10. If the problem persists, repeat steps 7 through 9.
10. Clean the work area after the problem is corrected. Follow facility procedures when removing and disposing hydraulic fluid that has spilled onto the floor or equipment. Remove all tools from the

TERMS

A **digital multimeter (DMM)** is a portable meter that can measure more than one electrical property and display the measured properties as numerical values.

equipment area and secure the safety doors or entryways. Once the cleanup is done, complete any required written reports as soon as possible.

Note: Troubleshooting procedures should be used as a guide or a starting point. Consult the equipment manufacturer's instructions for specific troubleshooting procedures.

Troubleshooting Fluid Power Electrical Control Circuits. When a problem occurs with an electrically controlled fluid power system, the most important factor to determine is whether the problem is with the electrical control circuit or the fluid power system. To help determine this, most fluid power components that are electrically controlled have a light emitting diode (LED) to indicate if electrical power is present at the component's solenoid.

If a component's solenoid is not actuating when it should, a technician can locate the LED to check for electrical power. If the LED lights up, it indicates that the solenoid is receiving electrical power and that the problem is in the fluid power system. If the LED does not light up, the solenoid may not be receiving power and there could be a problem in the electrical control circuit. Another possibility is that the LED may have been damaged, so a technician should always perform a voltage check across the solenoid, even if the LED does not light up.

Electrical readings are typically taken with a digital multimeter. A *digital multimeter (DMM)* is a portable meter that can measure more than one electrical property and display the measured properties as numerical values. If the problem is determined to be electrical, checking for voltage and continuity is the most common method of locating electrical problems. Checking for voltage and continuity is performed by applying the following procedure:

1. Determine if there is voltage to the electrical circuit with a DMM.
2. Refer to the ladder diagram to determine which switches must be closed to activate the component's solenoid.
3. Verify that there is voltage to the solenoid with a DMM.
4. Verify that each switch is in proper working order by doing a voltage measurement or a continuity test across each switch with a DMM. For example, if one switch controls the component's solenoid, a voltage measurement must be performed to determine if the switch is faulty.

If the switch is open, the voltage measurement should be the same as the source voltage. If the switch is closed, the voltage should read close to 0 V. If neither of the measurements is indicated by the DMM, the switch could be faulty. **See Figure 15-14.**

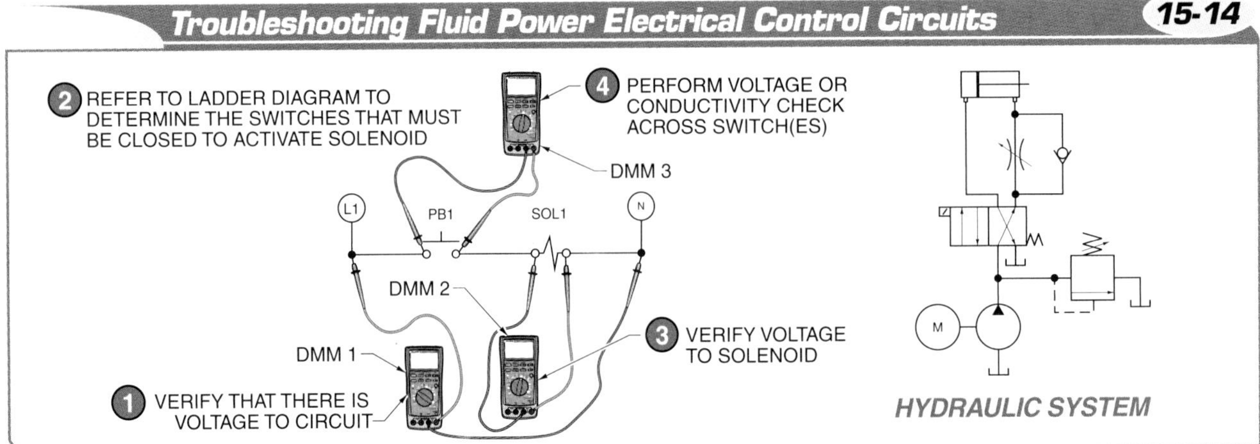

Figure 15-14. Troubleshooting electrical problems with an electrically controlled hydraulic system is performed by taking voltage or continuity readings.

The Five "Whys" Troubleshooting Method

The five "whys" troubleshooting method is a troubleshooting method with a series of "why" questions that leads to the root cause of a hydraulic circuit failure. This is a good method to use on hard-to-solve or recurring problems. The basic method is to ask the question "why?" five times. This will typically allow a technician to find the root cause of a problem. The five whys troubleshooting method is not always required to solve a particular problem. For example, if a cylinder stops working, and the problem is due to normal wear and operation, there may not be any reason for a more complex system inspection. The cylinder simply needs to be replaced to solve the problem.

If a problem keeps reoccurring, however, then the five whys troubleshooting method can lead to the root cause of the recurring failure. For example, a cylinder does not retract, and the problem is discovered to be a bad counterbalance valve. The counterbalance valve is replaced and the same problem occurs in six months. The five whys troubleshooting method can then be applied to help determine the root cause.

First "why" question: Why is the counterbalance malfunctioning again?
Answer: *The internal check valve is locked.*

Second "why" question: Why is the internal check valve locked?
Answer: *Contaminants have destroyed the seal.*

Third "why" question: Why have contaminants entered the system?
Answer: *The filter has not been replaced as specified by the OEM and is allowing particles through the filter and into the system.*

Fourth "why" question: Why has the filter not been replaced?
Answer: *The filter indicator on the equipment is malfunctioning.*

Fifth "why" question: Why was the malfunctioning filter indicator not discovered during previous inspections?
Answer: *It was not a part of the routine maintenance plan to check the filter indicator.*

Solution to problem: Replace the counterbalance valve, change the filter, replace the filter indicator, and change the routine maintenance plan to now include inspecting the filter indicator.

The five "whys" troubleshooting method is not always necessary. Sometimes it will only take three questions to find the root cause, and sometimes six or seven questions may be required. Although, typically with fluid power systems, if a technician understands the basics of fluid power and the system to be worked on, five questions are enough to solve a problem.

Digital Resources

Chapter Review Fluid Power System Maintenance and Troubleshooting

Chapter 15

Name: ______________________________ Date: ______________

MULTIPLE CHOICE

______________ **1.** Components that have been dismantled from a hydraulic system and cleaned are dried with ___.

A. ambient air
B. high-pressure, clean, dry compressed air
C. low-pressure, clean, dry compressed air
D. nitrogen

______________ **2.** ___ are used to measure electric or hydraulic motor speed.

A. Flow meters
B. Multimeters
C. Tachometers
D. Thermal imagers

______________ **3.** A ___ is a standard given by the manufacturer that indicates the least amount of bend that can be used for a specific type of hose.

A. layer count
B. material flexibility number
C. minimum hose bend radius
D. wall thickness

______________ **4.** The ___ troubleshooting method is a fluid power troubleshooting method that traces problems from the beginning of the fluid power system (prime mover, hydraulic pump, or air compressor) to the end (actuator).

A. beginning-to-end
B. end-to-beginning
C. hydraulic tester
D. part replacement

______________ **5.** A pneumatic system uses ___ for lubrication.

A. compressor oil
B. lubricating oil
C. motor oil
D. water

______________ **6.** If manufacturer's recommendations are not available, mobile hydraulic equipment typically should be inspected every 400 hr to 600 hr, or every ___ month(s).

A. one
B. two
C. three
D. six

_______________ **7.** A(n) ___ is a mechanical device on a pneumatic actuator that applies tension to the seals or packing to prevent air leaks around the circumference of a rod or shaft.
A. collar
B. O-ring
C. packing nut
D. rod wiper

_______________ **8.** Because hoses and tubing used in fluid power systems can become easily damaged and are susceptible to other problems such as kinks or premature wear, it is recommended that they only be used with moving components and in lengths of ___′ or less.
A. 2
B. 3
C. 4
D. 6

_______________ **9.** The ___ troubleshooting method is preferred when actuators are not operating, but the gauges at the beginning of the system indicate normal levels.
A. beginning-to-end
B. end-to-beginning
C. hydraulic tester
D. part replacement

_______________ **10.** Air leaks in a pneumatic system can waste up to ___% or higher of the pneumatic system energy.
A. 20
B. 25
C. 35
D. 50

COMPLETION

_______________ **1.** The most common problem in pneumatic systems is insufficient air pressure or ___.

_______________ **2.** Hoses stored in temperatures below ___°F should be gradually warmed to room temperature before being placed in service.

_______________ **3.** A(n) ___ is a pneumatic device that can be added to or removed from a pressurized pneumatic system without removing pressure from the system.

_______________ **4.** In addition to the prevention of hose ruptures, pneumatic components must be kept clean and damage free to provide efficient ___.

_______________ **5.** The ___ troubleshooting method is a fluid power troubleshooting method that traces problems from the end of the fluid power system (actuator) to the beginning (prime mover, hydraulic pump, or air compressor).

_______________ **6.** The ___ troubleshooting method is a fluid power system troubleshooting method that replaces parts without the use of equipment diagnosis tools.

_______________ **7.** The most important routine maintenance for hydraulic equipment is replacing ___ according to the manufacturer's recommendations or when indicated by devices such as pressure gauges, flow meters, or filter indicators.

______________ **8.** A(n) ___ is a device that measures the pressure, flow rate, and temperature of hydraulic fluid in a system.

______________ **9.** Rods that must be stored exposed should be coated with ___ that can be easily removed before startup.

______________ **10.** A common cause of pneumatic system problems is ___ that are at incorrect pressure settings.

______________ **11.** Many modern pneumatic systems use clear or semiclear ___ without reinforcement layers or cords.

______________ **12.** ___ is the practice of periodically inspecting and repairing equipment.

______________ **13.** The ___ filter is used to keep contaminants out of the hydraulic system and allow air to flow freely in and out of the reservoir.

______________ **14.** Fluid stains or puddles under hydraulic equipment and hoses indicate the presence of a(n) ___ in the line.

______________ **15.** The advantage of a(n) ___ program is that it reduces the amount of troubleshooting that may need to be performed.

TRUE/FALSE

T F **1.** In addition to the noise, escaping pressurized air from a pneumatic system can create a clean area and wear grooves in surrounding surfaces.

T F **2.** Improper viscosity of hydraulic fluid can cause serious operating problems with hydraulic equipment.

T F **3.** Springs, seals, and O-rings do not need to be replaced every time a component is serviced.

T F **4.** When testing an actuator, it must be deenergized and disconnected from the load.

T F **5.** Hydraulic cylinders that are not in service must have their rods retracted to prevent contaminants from settling on any exposed metal and entering the system when the rod is moved.

T F **6.** Excessive coats of paint applied to a component can increase insulation and provide proper heat transfer.

T F **7.** Typically, the less layers in a hose, the higher the minimum bend radius.

T F **8.** Air can be kept out of a hydraulic system by maintaining the correct hydraulic fluid level in the reservoir.

T F **9.** Very few hydraulic system maintenance problems are caused by the contamination and breakdown of hydraulic fluid.

T F **10.** Hoses that are designed for one application should never be used in another.

T F **11.** The frequency of hydraulic equipment maintenance inspection varies with the manufacturer's recommendations.

T F **12.** A hydraulic system uses lubricating oil for lubrication.

T F 13. The least common problem in pneumatic systems is insufficient airflow caused by leaks from worn, disconnected, or kinked hoses or tubing.

T F 14. Flow control problems are identified using a flow meter connected into the system.

T F 15. Hydraulic hoses that show any type of wear or damage should be replaced immediately, rather than repaired.

T F 16. The part replacement troubleshooting method is the quickest and most accurate fluid power system troubleshooting method.

T F 17. Troubleshooting procedures should be used as a guide or a starting point, while equipment manufacturer's instructions must be consulted for specific troubleshooting procedures.

T F 18. The same maintenance programs are applied to hydraulic and pneumatic systems.

T F 19. Many new hose failures in fluid power equipment are attributed to using the wrong type of hose, hose fitting, or both.

T F 20. The beginning-to-end troubleshooting method is preferred for fluid power systems that have completely stopped operating, without any readings registering on the system gauges.

SHORT ANSWER

1. Explain why troubleshooting charts from the manufacturer should be used when working on a hydraulic system.

2. Explain why a body part should never touch or pass near a pressurized hydraulic line.

3. Explain why proper viscosity of hydraulic fluid is important for equipment operation.

Activity 15-1: Determining Hydraulic System Inefficiencies

A hydraulic system's electrical motor blows its fuses after the system operates for less than five minutes due to overcurrent. The maintenance department believes the problem is a combination of inefficiencies in the system components. The components that the maintenance department believes may have the most inefficiencies include the cylinder, pump, pressure filter with bypass, pump and motor coupling, strainer, filler/breather cap, electric motor, and return line filter. *Note:* The cylinder is rated to move a maximum load of 9500 lb.

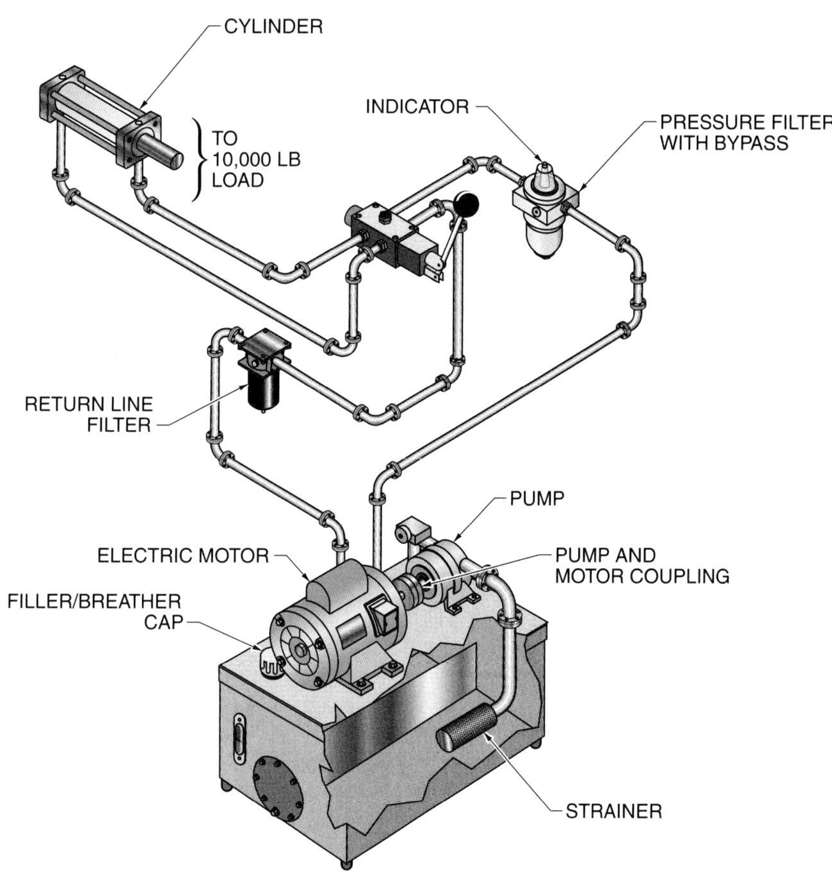

HYDRAULIC SYSTEM

Refer to Hydraulic System and list the most common reason why inefficiencies could lead to an overcurrent in each of the listed component conditions.

1. The cylinder is struggling to move.

2. The pressure filter with bypass indicator is red.

3. There is a loud or unusual pump noise.

4. The pump and motor coupling is operating erratically.

5. Particles are in hydraulic fluid after it passes through the strainer.

6. Particles are in hydraulic fluid after it passes through the filler/breather cap.

7. The electric motor is running hot.

8. Particles are in hydraulic fluid after it passes through the return line filter.

Activity 15-2: Pneumatic Leakage Costs

A manufacturing facility is planning to perform an unofficial self-audit on a portion of their compressed air system. They have decided to use an ultrasonic acoustic detector to inspect 25% of the pneumatic system. After using the ultrasonic detector, they located the following amount of air leaks on various pieces of equipment:

Hole Size*	Number of Leaks
1/64	17
1/32	4
1/16	2
1/8	1
1/4	0
3/8	0

* in in.

Pneumatic Leakage Rates and Costs

Leakage Rates for Different Supply Pressures and Approximately Equivalent Orifice Sizes *

Pressure†	Orifice Diameter‡					
	1/64	1/32	1/16	1/8	1/4	3/8
70	0.3	1.2	4.8	19.2	76.7	173
80	0.33	1.3	5.4	21.4	85.7	193
90	0.37	1.5	5.9	23.8	94.8	213
100	0.41	1.6	6.5	26.0	104	234
125	0.49	2.0	7.9	31.6	126	284

Estimated Annual Air Leakage Costs

Hole Size‡	Annual Cost§
1/64	41.33
1/32	164.28
1/16	655.20
1/8	2,620.80
1/4	10,483.20
3/8	23,587.20

At 100 psi based on $0.07/kWh, 0.18 kWh/100 cfm and 8000 hours of operation per year
* in cfm
† in psig
‡ in in.
§ in $

Refer to Pneumatic Leakage Rates and Costs chart to answer the following questions:

1. How much money is wasted annually in the portion of the system that was inspected?

2. If the 25% of the system that was inspected is a true representation of the remaining 75% of the factory, how much money could be saved annually if all the leaks are fixed?

3. If the cost for the inspection and repair to the equipment was $10,000, how much time (in months) would it require recovering this cost after repairing all leaks?

4. Which leaks should be fixed first?

Activity 15-3: Troubleshooting with Schematic Diagrams

Hydraulic System 1 and Hydraulic System 2 are schematic diagrams of the same hydraulic system. In both systems, the rod will not extend. Refer to each schematic diagram to answer the following questions.

900 PSI
900 PSI
900 PSI
RELIEF VALVE SET TO 900 PSI
M

HYDRAULIC SYSTEM 1

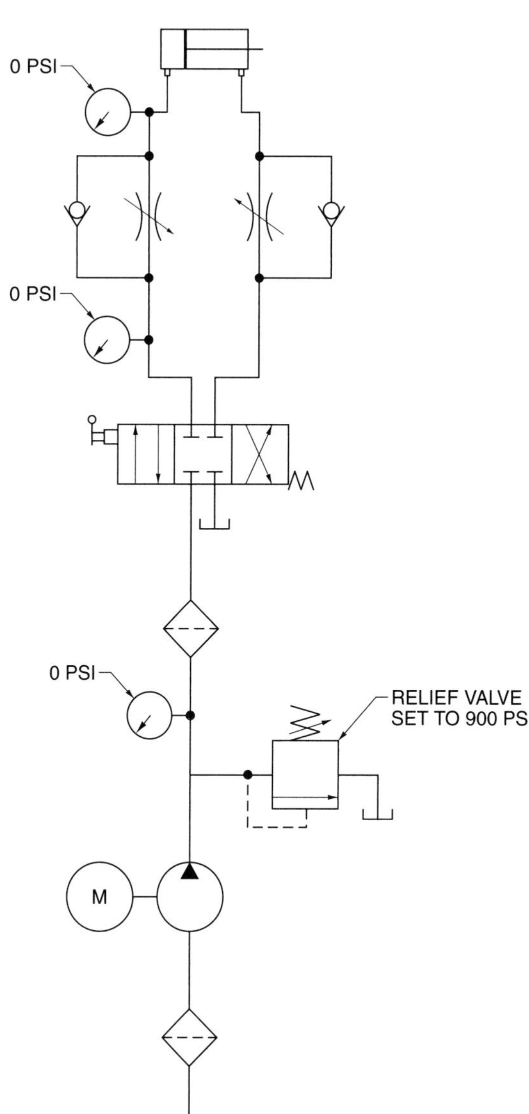

HYDRAULIC SYSTEM 2

1. By reading the pressure gauges, state which method of troubleshooting would work best to troubleshoot Hydraulic System 1.

2. By reading the pressure gauges, state which method of troubleshooting would work best to troubleshoot Hydraulic System 2.

3. List at least two problems in Hydraulic System 1.

4. List at least two problems in Hydraulic System 2.

Activity 15-4: Troubleshooting Hydraulic Door System

A hydraulic door on an airplane hangar has malfunctioned and a maintenance technician troubleshoots the hydraulic system. After troubleshooting the hydraulic system, the maintenance technician determines that the NC directional control valve is inoperable. After replacing the NC directional control valve, the hydraulic system is tested by turning on the system and opening and closing the door. For normal operation, an operator must manually activate the system to open or close the door. After replacing the NC directional control valve, the door opens automatically without the operator actuating the directional control valve.

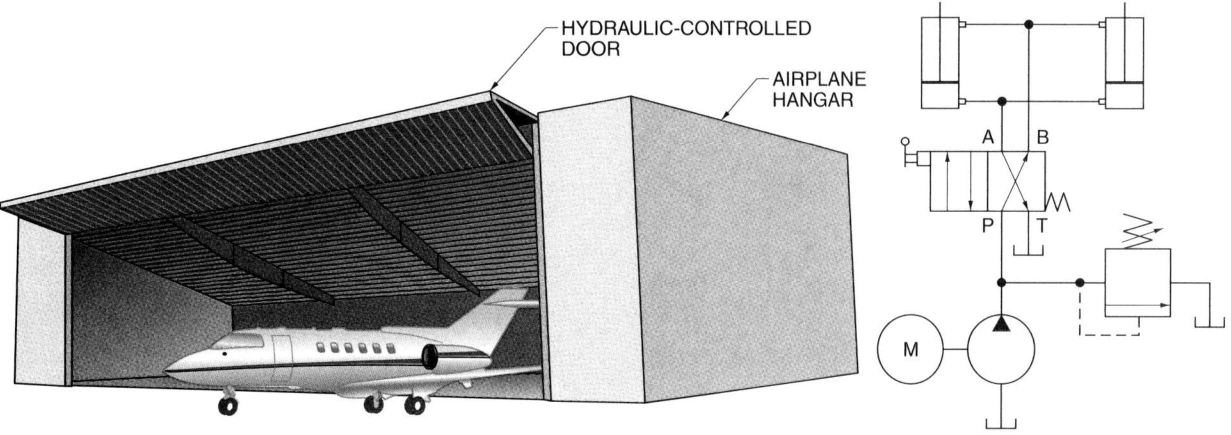

Use the FluidSIM® software included on the CD-ROM to build and operate the hydraulic door system.

1. List at least two reasons why the door does not operate properly after the technician replaced the NC directional control valve.

Appendix

Hydraulic Pump Selection Considerations . . . 484
Air Compressor Types . . . 484
Fluid Power Graphic Symbols . . . 485
Industrial Electrical Symbols . . . 488
Logic Symbols . . . 492
Filter Pressure Loss Constants . . . 493
Pipe Fitting Pressure Loss Constants . . . 493
Cylinder Size Selection . . . 494
Safety Procedure for Shutting Down Machines . . . 495
Fluid Weights/Temperature Standards . . . 495
Volume and Capacity Conversion Factors . . . 495
Fluid Power Abbreviations/Acronyms . . . 496
Pounds Per Square Inch—Kilopascals Conversion Table . . . 497
Ambient Pressure and Engine Coolant Boiling Point . . . 497
Ohm's Law . . . 497
Hazardous Material Container Labeling . . . 498
Hazardous Materials . . . 499
Hazardous Locations . . . 500
Recommended Hydraulic System Flushing Guidelines . . . 500
Hose Installation Recommendations . . . 501
Hydraulic Cylinder Speed . . . 503
Motor Horsepower . . . 504
Horsepower Formulas . . . 504
Pipe . . . 505
AC/DC Formulas . . . 505
Subplate Designs . . . 506
Voltage Drop Formulas—1φ, 3φ . . . 506
ISO Hydraulic Fluid Viscosity Ratings . . . 507
SAE Hydraulic Fluid Viscosity Ratings . . . 507
Cleanliness Levels of Hydraulic Fluids (SAEJ1165 OCT80) . . . 507
Fluid Power Troubleshooting . . . 508
Industry Organizations . . . 512
Abbreviations . . . 513
Printreading Symbols . . . 515
Prefixes . . . 516
Conversion Table . . . 516

HYDRAULIC PUMP SELECTION CONSIDERATIONS

All hydraulic pumps create fluid flow by using variations of the four basic operational steps of a positive-displacement pump. Each type of pump has applications where it functions better. In addition to a facility's replacement parts inventory, there are a number of variables to consider:

- **Equipment cost** – Typically, the least expensive pump that accomplishes the facility requirements is installed.
- **Operating environment** – The type of operating environment that the fluid power system is installed in can determine which type of pump to use. Facilities that operate in clean, dirty, or excessively loud environments can require certain types of pumps.
- **Pressure operating range** – A pump with a maximum pressure rating that is higher than that of the application requirements is typically installed.
- **Efficiency** – Pump efficiency must be taken into consideration. Certain applications may require high efficiency pumps.
- **Pump design** – Pumps can be fixed or variable. If fluid flow needs to be adjusted, a variable pump must be used.
- **Design complexity** – Design complexity is associated with equipment cost. A less complex pump, such as a gear pump, has lower cost than that of a more complex pump, such as a piston pump.
- **Loudness** – Although not as expensive as piston or vane pumps, gear pumps have the highest level of loudness when in operation.

A pump comparison chart can help show the differences between different types of pumps.

Pump Comparison

	Gear	Vane	Piston
Equipment cost	Least expensive	Intermediate	Most expensive
Operating environment	Least clean	Clean	Most clean
Pressure operating range	1500 to 3000*	200 to 3000*	3000 to 12,000*
Efficiency	Least	Intermediate	Most
Pump design	Fixed only	Fixed or variable	Fixed or variable
Design complexity	Least	Intermediate	Most
Loudness	100†	70†	80†

* in psi
† in dB

Air Compressor Types

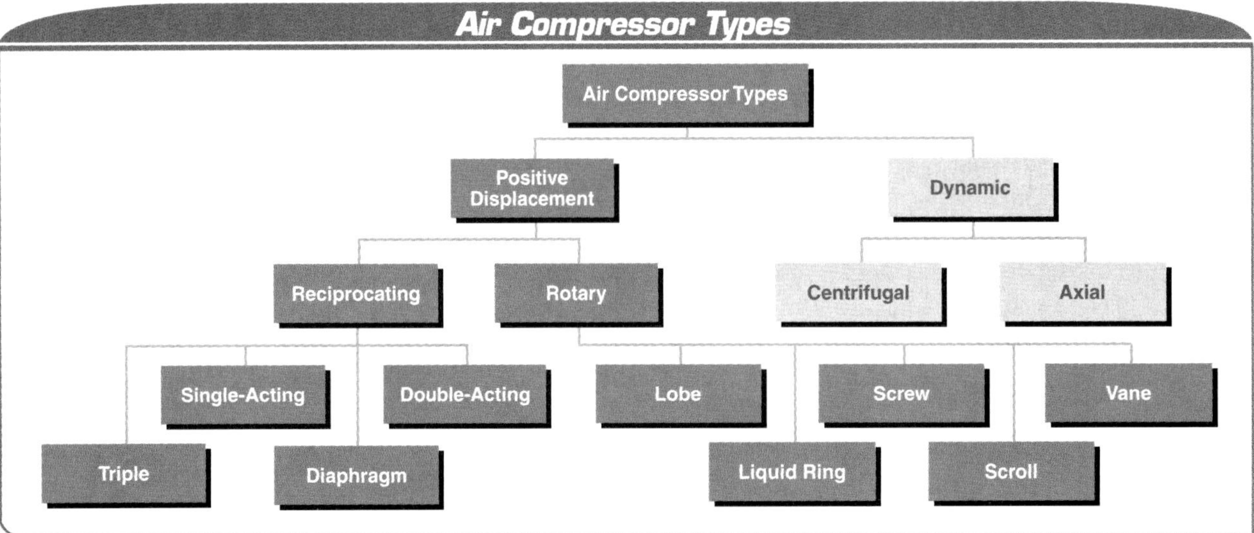

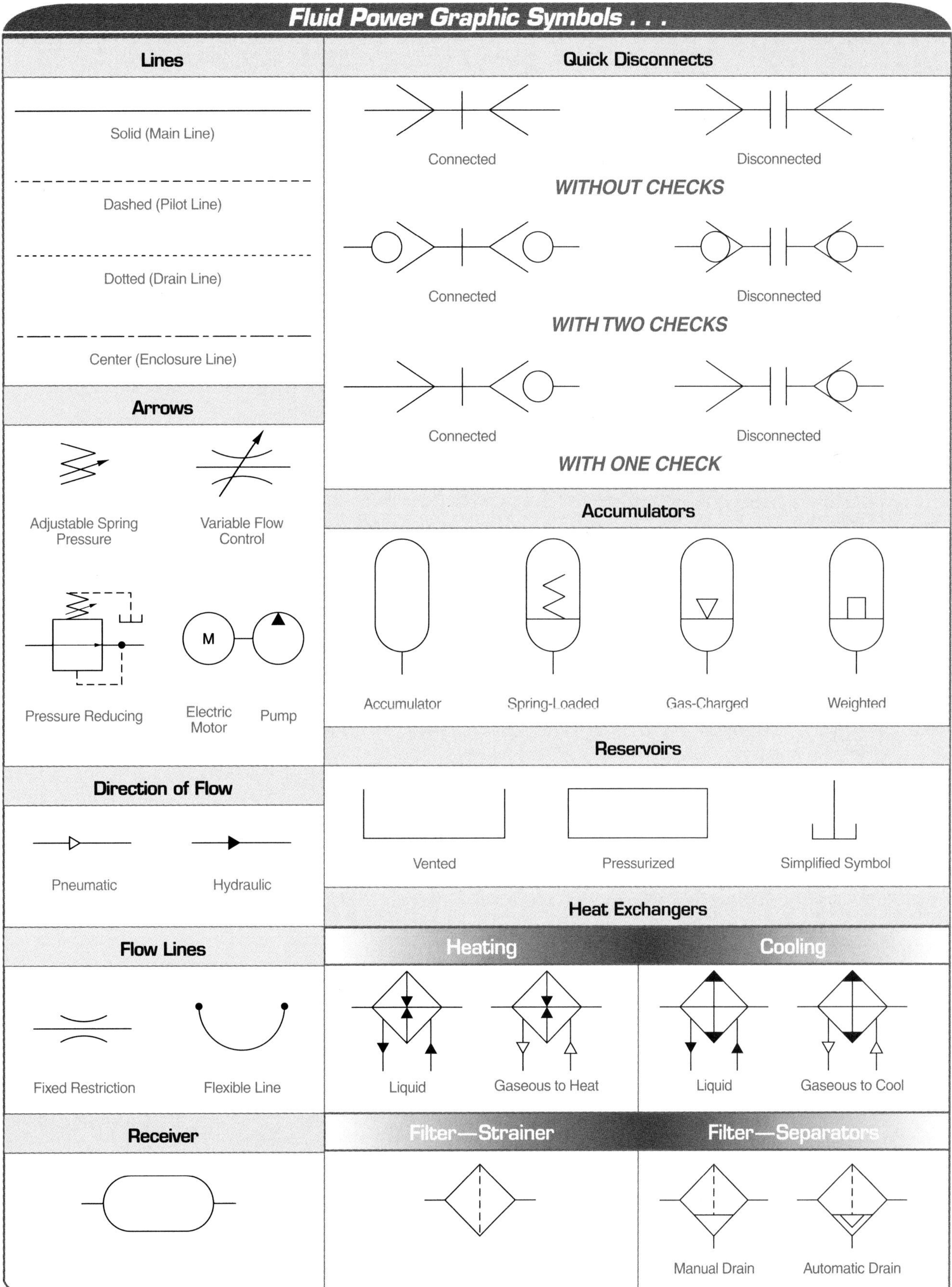
Fluid Power Graphic Symbols . . .
Lines
Solid (Main Line)
Dashed (Pilot Line)
Dotted (Drain Line)
Center (Enclosure Line)
Arrows
Adjustable Spring Pressure
Variable Flow Control
Pressure Reducing
M
Electric Motor
Pump
Direction of Flow
Pneumatic
Hydraulic
Flow Lines
Fixed Restriction
Flexible Line
Receiver
Quick Disconnects
Connected
Disconnected
WITHOUT CHECKS
Connected
Disconnected
WITH TWO CHECKS
Connected
Disconnected
WITH ONE CHECK
Accumulators
Accumulator
Spring-Loaded
Gas-Charged
Weighted
Reservoirs
Vented
Pressurized
Simplified Symbol
Heat Exchangers
Heating
Cooling
Liquid
Gaseous to Heat
Liquid
Gaseous to Cool
Filter—Strainer
Filter—Separators
Manual Drain
Automatic Drain

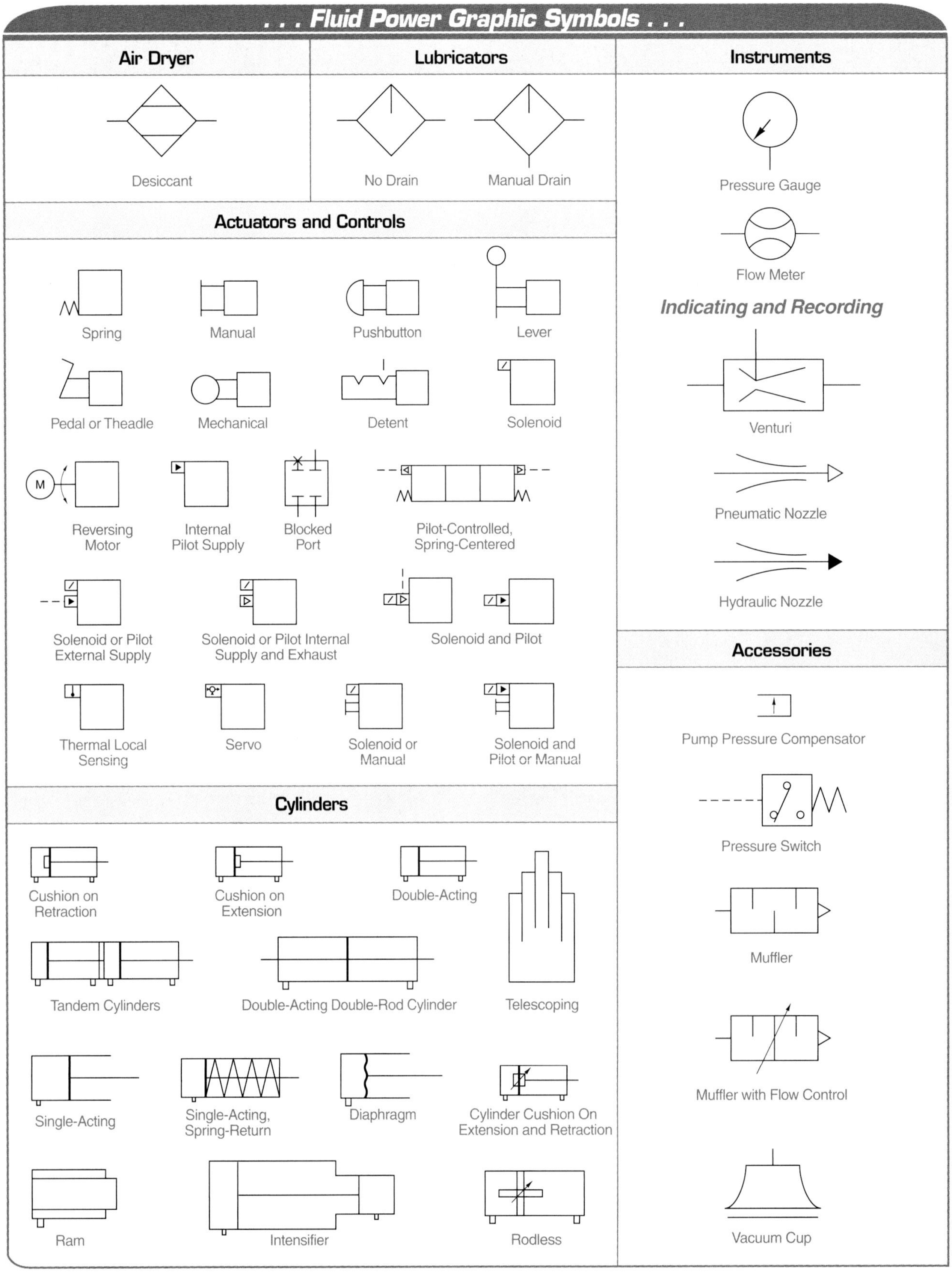

. . . Fluid Power Graphic Symbols . . .

Air Dryer

Desiccant

Lubricators

No Drain

Manual Drain

Instruments

Pressure Gauge

Flow Meter

Indicating and Recording

Venturi

Pneumatic Nozzle

Hydraulic Nozzle

Actuators and Controls

Spring

Manual

Pushbutton

Lever

Pedal or Theadle

Mechanical

Detent

Solenoid

Reversing Motor

Internal Pilot Supply

Blocked Port

Pilot-Controlled, Spring-Centered

Solenoid or Pilot External Supply

Solenoid or Pilot Internal Supply and Exhaust

Solenoid and Pilot

Thermal Local Sensing

Servo

Solenoid or Manual

Solenoid and Pilot or Manual

Accessories

Pump Pressure Compensator

Pressure Switch

Muffler

Muffler with Flow Control

Vacuum Cup

Cylinders

Cushion on Retraction

Cushion on Extension

Double-Acting

Tandem Cylinders

Double-Acting Double-Rod Cylinder

Telescoping

Single-Acting

Single-Acting, Spring-Return

Diaphragm

Cylinder Cushion On Extension and Retraction

Ram

Intensifier

Rodless

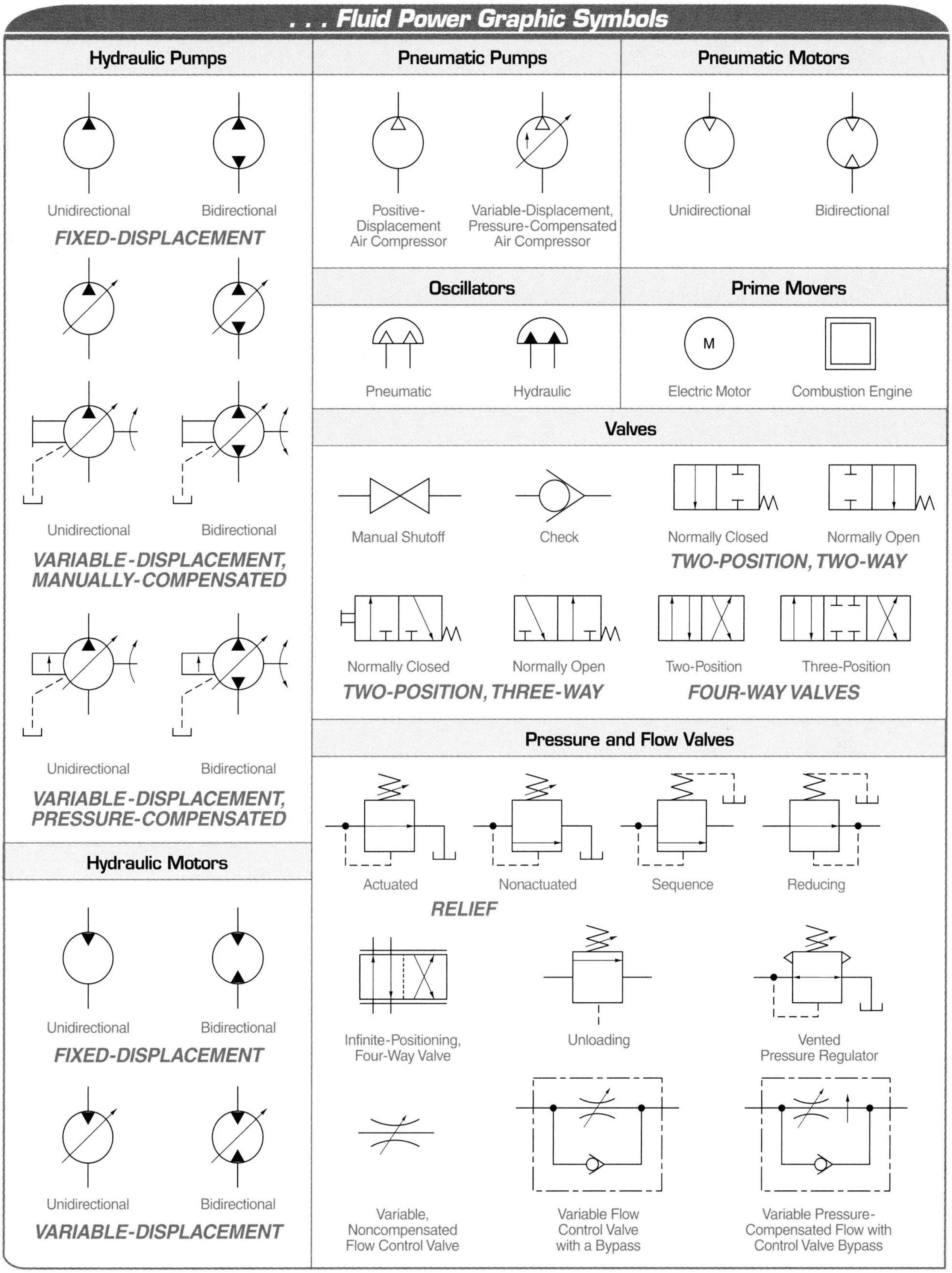

. . . Fluid Power Graphic Symbols

Hydraulic Pumps

Unidirectional
Bidirectional

FIXED-DISPLACEMENT

Unidirectional
Bidirectional

VARIABLE-DISPLACEMENT, MANUALLY-COMPENSATED

Unidirectional
Bidirectional

VARIABLE-DISPLACEMENT, PRESSURE-COMPENSATED

Hydraulic Motors

Unidirectional
Bidirectional

FIXED-DISPLACEMENT

Unidirectional
Bidirectional

VARIABLE-DISPLACEMENT

Pneumatic Pumps

Positive-Displacement Air Compressor
Variable-Displacement, Pressure-Compensated Air Compressor

Pneumatic Motors

Unidirectional
Bidirectional

Oscillators

Pneumatic
Hydraulic

Prime Movers

M

Electric Motor
Combustion Engine

Valves

Manual Shutoff
Check
Normally Closed
Normally Open

TWO-POSITION, TWO-WAY

Normally Closed
Normally Open

TWO-POSITION, THREE-WAY

Two-Position
Three-Position

FOUR-WAY VALVES

Pressure and Flow Valves

Actuated
Nonactuated

RELIEF

Sequence
Reducing

Infinite-Positioning, Four-Way Valve
Unloading
Vented Pressure Regulator

Variable, Noncompensated Flow Control Valve
Variable Flow Control Valve with a Bypass
Variable Pressure-Compensated Flow with Control Valve Bypass

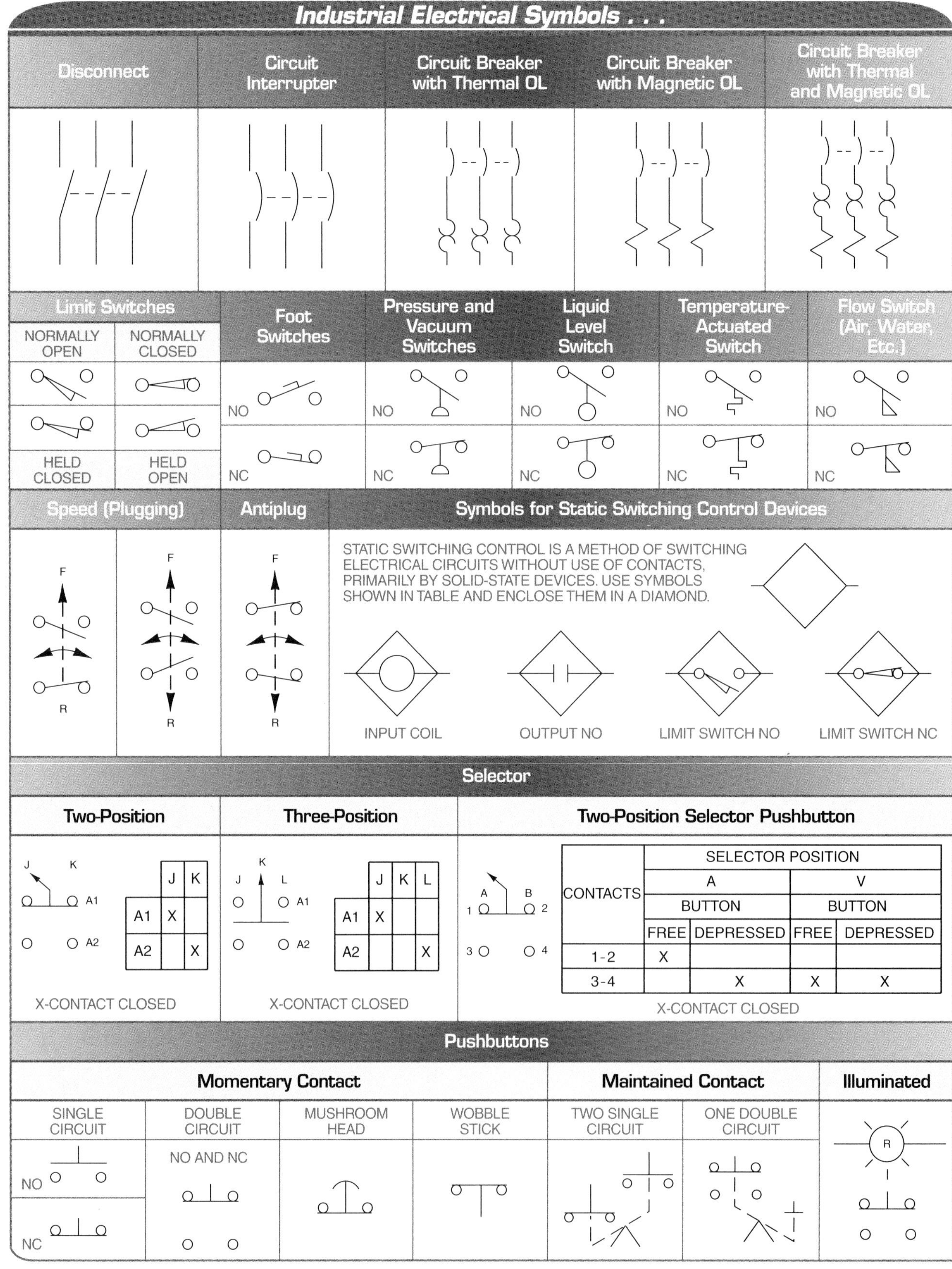

	J	K
A1	X	
A2		X

	J	K	L
A1	X		
A2			X

CONTACTS	SELECTOR POSITION			
	A		V	
	BUTTON		BUTTON	
	FREE	DEPRESSED	FREE	DEPRESSED
1-2	X			
3-4		X	X	X

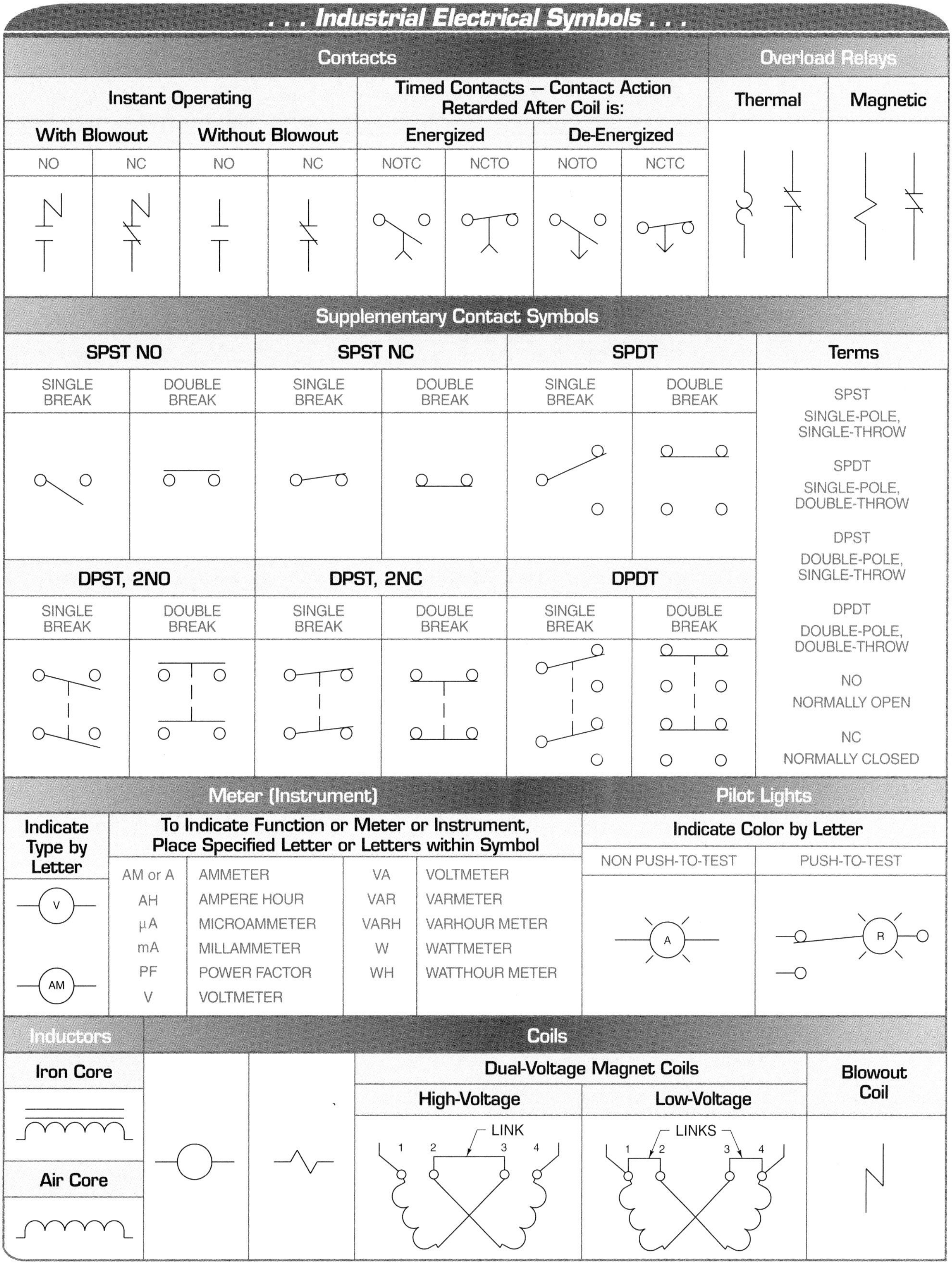

. . . Industrial Electrical Symbols . . .

Contacts

Instant Operating				Timed Contacts — Contact Action Retarded After Coil is:			
With Blowout		Without Blowout		Energized		De-Energized	
NO	NC	NO	NC	NOTC	NCTO	NOTO	NCTC

Overload Relays

Thermal	Magnetic

Supplementary Contact Symbols

SPST NO		SPST NC		SPDT	
SINGLE BREAK	DOUBLE BREAK	SINGLE BREAK	DOUBLE BREAK	SINGLE BREAK	DOUBLE BREAK

DPST, 2NO		DPST, 2NC		DPDT	
SINGLE BREAK	DOUBLE BREAK	SINGLE BREAK	DOUBLE BREAK	SINGLE BREAK	DOUBLE BREAK

Terms

SPST
SINGLE-POLE, SINGLE-THROW

SPDT
SINGLE-POLE, DOUBLE-THROW

DPST
DOUBLE-POLE, SINGLE-THROW

DPDT
DOUBLE-POLE, DOUBLE-THROW

NO
NORMALLY OPEN

NC
NORMALLY CLOSED

Meter (Instrument)

Indicate Type by Letter

V

AM

To Indicate Function or Meter or Instrument, Place Specified Letter or Letters within Symbol

AM or A	AMMETER	VA	VOLTMETER
AH	AMPERE HOUR	VAR	VARMETER
μA	MICROAMMETER	VARH	VARHOUR METER
mA	MILLAMMETER	W	WATTMETER
PF	POWER FACTOR	WH	WATTHOUR METER
V	VOLTMETER		

Pilot Lights

Indicate Color by Letter

NON PUSH-TO-TEST	PUSH-TO-TEST
A	R

Inductors

Iron Core

Air Core

Coils

Dual-Voltage Magnet Coils

High-Voltage	Low-Voltage
LINK 1 2 3 4	LINKS 1 2 3 4

Blowout Coil

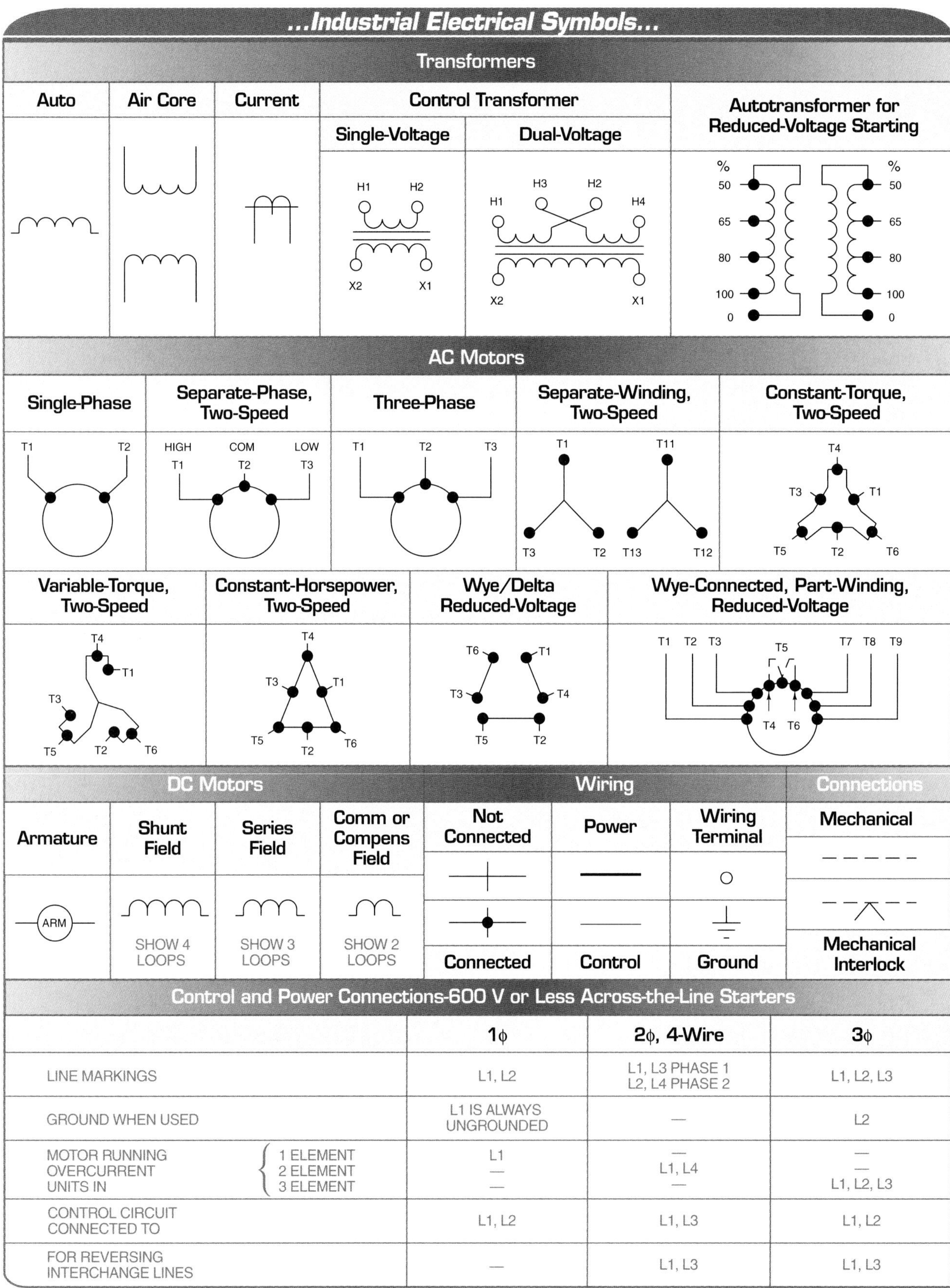

Control and Power Connections-600 V or Less Across-the-Line Starters

	1ϕ	2ϕ, 4-Wire	3ϕ
LINE MARKINGS	L1, L2	L1, L3 PHASE 1 L2, L4 PHASE 2	L1, L2, L3
GROUND WHEN USED	L1 IS ALWAYS UNGROUNDED	—	L2
MOTOR RUNNING OVERCURRENT UNITS IN — 1 ELEMENT	L1	—	—
MOTOR RUNNING OVERCURRENT UNITS IN — 2 ELEMENT	—	L1, L4	—
MOTOR RUNNING OVERCURRENT UNITS IN — 3 ELEMENT	—	—	L1, L2, L3
CONTROL CIRCUIT CONNECTED TO	L1, L2	L1, L3	L1, L2
FOR REVERSING INTERCHANGE LINES	—	L1, L3	L1, L3

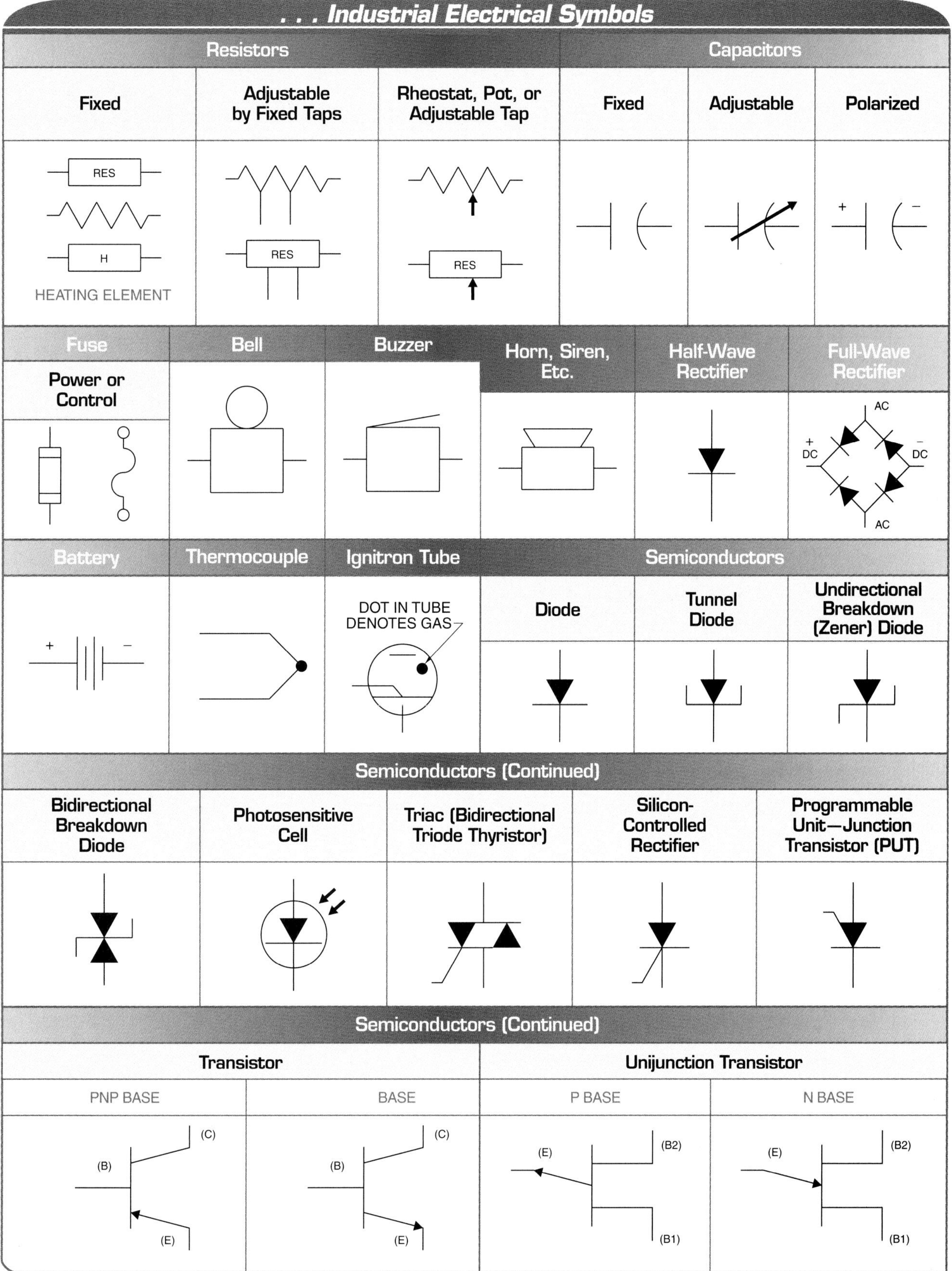
. . . Industrial Electrical Symbols
Resistors
Capacitors
Fixed
Adjustable by Fixed Taps
Rheostat, Pot, or Adjustable Tap
Fixed
Adjustable
Polarized
RES
H
HEATING ELEMENT
RES
RES
+
−
Fuse
Bell
Buzzer
Horn, Siren, Etc.
Half-Wave Rectifier
Full-Wave Rectifier
Power or Control
AC
+ DC
− DC
AC
Battery
Thermocouple
Ignitron Tube
Semiconductors
+
−
DOT IN TUBE DENOTES GAS
Diode
Tunnel Diode
Undirectional Breakdown (Zener) Diode
Semiconductors (Continued)
Bidirectional Breakdown Diode
Photosensitive Cell
Triac (Bidirectional Triode Thyristor)
Silicon-Controlled Rectifier
Programmable Unit—Junction Transistor (PUT)
Semiconductors (Continued)
Transistor
Unijunction Transistor
PNP BASE
BASE
P BASE
N BASE
(C)
(B)
(E)
(C)
(B)
(E)
(E)
(B2)
(B1)
(E)
(B2)
(B1)

Logic Symbols

Logic Element	AND	OR	NOT	NAND	NOR
LOGIC ELEMENT FUNCTION	OUTPUT IF ALL CONTROL INPUT SIGNALS ARE ON	OUTPUT IF ANY ONE OF THE CONTROL INPUTS IS ON	OUTPUT IF SINGLE CONTROL INPUT SIGNAL IS OFF	OUTPUT IF ALL CONTROL INPUT SIGNALS ARE ON	OUTPUT IF ANY OF THE CONTROL INPUTS ARE ON
MIL-STD-806B AND ELECTRONIC LOGIC SYMBOL					
ELECTRICAL RELAY LOGIC SYMBOL	1 2 CR CR	1 2 CR CR	1 CR CR	1 2 CR CR	1 2 CR CR
ELECTRICAL SWITCH LOGIC SYMBOL					
ASA (JIC) VALVING SYMBOL	P	P	P	P	P
ARO PNEUMATIC LOGIC SYMBOL			N 1	N 1	N 1
NFPA STANDARD			N SUPPLY	N SUPPLY	N SUPPLY
BOOLEAN ALGEBRA SYMBOL	$(\ ) \cdot (\ )$	$(\ ) + (\ )$	$\overline{(\ )}$	$\overline{(\ ) \cdot (\ )}$	$\overline{(\ ) + (\ )}$
FLUIDIC DEVICE TURBULENCE AMPLIFIER					

Filter Pressure Loss Constants

Filter Length*	Micron Size	Pipe Size*						
		1/8	1/4	3/4	1	1 1/4	1 1/2	2
3.5	5	115.0	55.0	—	—	—	—	—
	25	112.0	49.0	—	—	—	—	—
	100	92.0	41.0	—	—	—	—	—
14	5	—	—	0.47	0.34	0.34	0.34	—
	25	—	—	0.34	0.23	0.20	0.20	—
	50	—	—	0.32	0.20	0.19	0.19	—
	75	—	—	0.32	0.20	0.19	0.19	—
17	25	—	—	—	—	—	0.05	0.028
	50	—	—	—	—	—	0.036	0.020
	75	—	—	—	—	—	0.032	0.018

* in in.

Pipe Fitting Pressure Loss Constants

Pipe Fitting	Pipe Size*								
	1/8	1/4	3/8	1/2	3/4	1	1 1/4	1 1/2	2
45 Elbow	8.30	2.20	0.53	0.2	0.05	0.02	0.00	0.00	0.001
90 Elbow	15.40	4.09	1.09	0.4	0.1	0.04	0.01	0.00	0.002
Gate Valve	6.7	1.76	0.47	0.1	0.0	0.01	0.00	0.00	0.001
Globe Valve	175.3	46.40	12.70	4.7	1.3	0.4	0.1	0.0	0.03
Tee-Side Flow	31.0	8.14	2.37	0.8	0.24	0.0	0.0	0.01	0.005
Tee-Run Flow	10.4	2.74	0.80	0.2	0.0	0.0	0.00	0.00	0.002

* in in.

Cylinder Size Selection

Cyl. Bore Dia.*	Piston Rod Dia.*	Work Area†	Hydraulic Working Pressure‡						Fluid Req'd per Stroke*		Port Size*	Fluid Velocity§	
			500	750	1000	1500	2000	3000	Gal.	Cu In.		Flow GPM	Piston Vel.‖
1½	—	1.767	883	13,325	1767	2651	3534	5301	0.00765	1.767	½	11.0	24.0
	⅝	1.460	730	1095	1460	2190	2920	4380	0.00632	1.460			29.0
	1	0.982	491	736	982	1473	1964	2946	0.00425	0.982			43.1
2	—	3.141	1571	2356	3141	4711	6283	9423	0.01360	3.141	½	11.0	13.5
	1	2.356	1178	1767	2356	3534	4712	7068	0.01020	2.356			18.0
	1⅜	1.656	828	1242	1656	2484	3312	4968	0.00717	1.656			25.6
2½	—	4.909	2454	3682	4909	7363	9818	14,727	0.02125	4.909	½	11.0	8.6
	1	4.124	2062	3093	4124	6186	8248	12,372	0.01785	4.124			10.3
	1⅜	3.424	1712	2568	3424	5136	6848	10,272	0.01482	3.424			12.5
	1¾	2.504	1252	1878	2504	3756	5008	7512	0.01084	2.504			16.9
3¼	—	8.296	4148	6222	8296	12,444	16,592	24,888	0.0359	8.296	¾	20.3	9.4
	1⅜	6.811	3405	5108	6811	10,216	13,622	20,433	0.0295	6.811			11.5
	1¾	5.891	2945	4418	5891	8836	11,782	17,673	0.0255	5.891			13.3
	2	5.154	2577	3865	5154	7731	10,308	15,462	0.0223	5.154			15.2
4	—	12.566	6283	9425	12,566	18,849	25,132	37,698	0.0544	12.566	¾	20.3	6.2
	1¾	10.161	5080	7621	10,161	15,241	20,300	30,483	0.0440	10.161			7.7
	2	9.424	4712	7068	9424	14,136	18,848	28,272	0.0408	9.424			8.3
	2½	7.657	3828	5743	7657	11,485	15,314	22,971	0.0331	7.657			10.2
5	—	19.635	9818	14,726	19,635	29,453	39,270	58,905	0.0850	19.635	¾	20.3	4.0
	2	16.492	8246	12,369	16,492	24,738	32,984	49,476	0.0714	16.492			4.7
	2½	14.726	7363	11,044	14,726	22,089	29,542	44,178	0.0637	14.726			5.3
	3	12.566	6283	9424	12,566	18,849	25,132	37,698	0.0544	12.566			6.2
	3½	10.014	5007	7510	10,014	15,021	20,028	30,042	0.0433	10.014			7.8
6	—	28.274	14,137	21,205	28,274	42,411	56,548	84,822	0.1224	28.274	1	33.8	4.6
	2½	23.365	11,682	17,524	23,365	35,047	46,730	70,095	0.1011	23.365			5.6
	3	21.205	10,602	15,904	21,205	31,807	42,410	63,615	0.0918	21.205			6.1
	4	15.708	7854	11,781	15,708	23,562	31,416	47,124	0.0680	15.708			8.3
7	—	38.485	19,242	28,864	38,485	57,728	76,970	115,455	0.1666	38.485	1¼	60.2	6.0
	3	31.416	15,708	23,562	31,416	47,124	62,832	94,248	0.1360	31.416			7.4
	4	25.919	12,960	19,439	25,919	38,878	51,838	77,757	0.1122	25.919			8.9
	5	18.850	9425	14,137	18,850	28,275	37,700	56,550	0.0816	18.850			12.3
8	—	50.265	25,133	37,699	50,256	75,398	100,530	150,795	0.2176	50.265	1½	83.0	6.4
	3½	40.644	20,322	30,483	40,644	60,966	81,288	121,932	0.1759	40.644			7.9
	4	37.699	18,850	28,274	37,699	56,548	75,398	113,097	0.1632	37.699			8.5
	5½	26.507	13,253	19,880	26,507	39,760	53,014	79,521	0.1147	26.507			12.0
10	—	78.540	39,270	58,905	78,540	117,810	157,080	235,620	0.3400	78.540	2	139	6.8
	4½	62.636	31,318	46,977	62,636	93,954	125,272	187,908	0.2711	62.636			8.5
	5½	54.782	27,391	41,086	54,782	82,173	109,564	164,346	0.2371	54.782			9.8
	7	40.055	20,027	30,041	40,055	60,082	80,110	120,165	0.1734	40.055			13.4
12	—	113.10	56,550	84,825	113,100	169,650	226,200	339,300	0.4896	113.10	2½	199	6.8
	5½	89.34	44,670	67,005	89,340	134,010	178,680	268,020	0.3867	89.34			8.6
	7	74.62	37,310	55,965	74,620	111,930	149,240	223,860	0.3230	74.62			10.3
	8	62.84	31,420	47,130	62,840	94,260	125,680	188,520	0.2720	62.84			12.2
14	—	153.94	76,970	115,455	153,940	230,910	307,880	461,820	0.6664	153.94	2½	199	5.0
	7	115.46	57,730	86,595	115,460	173,190	230,920	346,380	0.4998	115.46			6.6
	8	103.68	51,840	77,760	103,680	155,520	207,360	311,040	0.4488	103.68			7.4
	10	75.40	37,700	56,550	75,400	113,100	150,800	226,200	0.3264	75.40			10.2

* in ips
† in sq in.
‡ in psi
§ 15 fps
‖ in ips

Safety Procedure for Shutting Down Machines

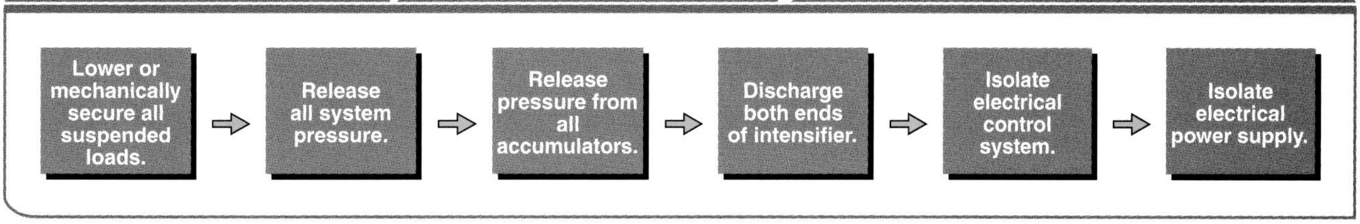

Lower or mechanically secure all suspended loads. → Release all system pressure. → Release pressure from all accumulators. → Discharge both ends of intensifier. → Isolate electrical control system. → Isolate electrical power supply.

Fluid Weights/Temperature Standards

Fluid	Weight*	Temperature†
Air	4.33×10^{-5}	20°C/68°F @ 29.92 in. Hg
Gasoline	0.0237 – 0.0249	20°C/68°F
Kerosene	0.0296	20°C/68°F
Mercury	0.49116	0°C/32°F
Oil, lubricating	0.0307 – 0.0318	15°C/59°F
Oil, fuel	0.0336 – 0.0353	15°C/59°F
Water	0.0361	4°C/39°F
Sea water	0.0370	15°C/59°F

* in lb/cu in.
† laboratory temperature under which numerical values are defined

Volume and Capacity Conversion Factors

	Cubic Inches	Cubic Feet	Cubic Centimeters	Liters	U.S. Gallons	Imperial Gallons	Water at Max Density	
							Pounds of Water	Kilograms of Water
cu in.	1	0.0005787	16.384	0.016384	0.004329	0.0036065	0.361275	0.0163872
cu ft.	1728	1	0.037037	28.317	7.48052	6.23210	62.4283	28.3170
cu cm	0.0610	0.0000353	1	0.001	0.000264	0.000220	0.002205	0.0001
L	61.0234	0.0353145	0.001308	1	0.264170	0.220083	2.20462	1
gal.	231	0.133681	0.004951	3.78543	1	0.833111	8.34545	3.78543
gal. (imperial)	277.274	0.160459	0.0059429	4.54374	1.20032	1	10.0172	4.54373
water lb.	27.6798	0.0160184	0.0005929	0.453592	0.119825	0.0998281	1	0.453593

Fluid Power Abbreviations/Acronyms

Abbr/Acronym	Meaning	Abbr/Acronym	Meaning
BTU	British Thermal Unit	IN.-LB	Inch-Pound
C	Degrees Celsius	INT	Internal
CC	Closed Center (valves)	I/O	Input/Output
CCW	Counterclockwise	IPM	Inches Per Minute
CFM	Cubic Feet Per Minute	IPS	Inches Per Second
CFS	Cubic Feet Per Second	LB	Pound
CIM	Cubic Inches Per Minute	MAX	Maximum
COM	Common	MIN	Minimum
CPM	Cycles Per Minute	NC	Normally Closed
CPS	Cycles Per Second	NO	Normally Open
CW	Clockwise	NPT	National Pipe Thread
CYL	Cylinder	OC	Open Center (valves)
D	Drain	OZ	Ounce
DIA	Diameter	PO	Pilot Operated
EXT	External	PRES or P	Pressure
F	Degrees Fahrenheit	PSI	Pounds Per Square Inch
FL	Fluid	PSIA	PSI Absolute
FPM	Feet Per Minute	PSIG	PSI Gauge
FS	Full Scale	PT	Pint
FT	Foot	QT	Quart
FT-LB	Foot-Pound	R	Radius
GA	Gauge	RMS	Root Mean Square
GAL.	Gallon	RPM	Revolutions Per Minute
GPM	Gallons Per Minute	RPS	Revolutions Per Second
HP	Horsepower	SOL	Solenoid
Hz	Hertz	T	Torque; Thrust; Tank
ID	Inside Diameter	VAC	Vacuum
IN.	Inches	VI	Viscosity Index
		VISC	Viscosity

Pounds Per Square Inch – Kilopascals Conversion Table*

Pounds Per Square Inch to Kilopascals (1 lb/in.² = 6.894757 kPa)

lb/in.²	0	1	2	3	4	5	6	7	8	9
0	—	6.895	13.790	20.864	27.579	34.474	41.369	48.263	55.158	62.053
10	68.948	75.842	82.737	89.632	96.527	103.421	110.316	117.211	124.106	131.000
20	137.895	144.790	151.685	158.579	165.474	172.369	179.264	186.158	193.053	199.948
30	206.843	213.737	220.632	227.527	234.422	241.316	248.211	255.106	262.001	268.896
40	275.790	282.685	289.580	296.475	303.369	310.264	317.159	324.054	330.948	337.843
50	344.738	351.633	358.527	365.422	372.317	379.212	386.106	393.001	399.896	406.791
60	413.685	420.580	427.475	434.370	441.264	448.159	455.054	461.949	468.843	475.738
70	482.633	489.528	496.423	503.317	510.212	517.107	524.002	530.896	537.791	544.686
80	551.581	558.475	565.370	572.265	579.160	586.054	592.949	599.844	606.739	613.633
90	620.528	627.423	634.318	641.212	648.107	655.002	661.897	668.791	675.686	682.581
100	689.476	696.370	703.265	710.160	717.055	723.949	730.844	737.739	744.634	751.529

Kilopascals to Pounds Per Square Inch (1 kPa = 0.1450377 lb/in.²)

kPa	0	1	2	3	4	5	6	7	8	9
0	—	0.145	0.290	0.435	0.580	0.725	0.870	1.015	1.160	1.305
10	1.450	1.595	1.740	1.885	2.031	2.176	2.321	2.466	2.611	2.756
20	2.901	3.046	3.191	3.336	3.481	3.626	3.771	3.916	4.061	4.206
30	4.351	4.496	4.641	4.786	4.931	5.076	5.221	5.366	5.511	5.656
40	5.802	5.947	6.092	6.237	6.382	6.527	6.672	6.817	6.962	7.107
50	7.252	7.397	7.542	7.687	7.832	7.977	8.122	8.267	8.412	8.557
60	8.702	8.847	8.992	9.137	9.282	9.427	9.572	9.718	9.863	10.008
70	10.153	10.298	10.443	10.588	10.733	10.878	11.023	11.168	11.313	11.458
80	11.603	11.748	11.893	12.038	12.183	12.328	12.473	12.618	12.763	12.908
90	13.053	13.198	13.343	13.489	13.634	13.779	13.924	14.069	14.214	14.359
100	14.504	14.649	14.794	14.939	15.084	15.229	15.374	15.519	15.664	15.809

* 1 kPa = 1 kilonewton/m²

Ambient Pressure and Engine Coolant* Boiling Point

PSI Above Atmospheric Pressure (14.7) PSI	Boiling Point†
1	229
2	232
3	235
4	238
5	241
6	244
7	247
8	249
9	251
10	253
11	255
12	257
13	259
14	261
15	264
16	266
17	268
18	271
19	274
20	278

* 50/50 mixture of antifreeze and water
† in °F

Ohm's Law

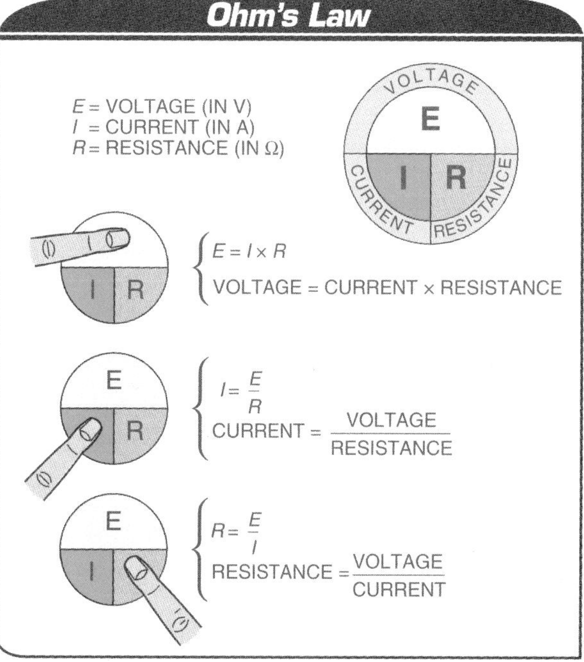

Hazardous Material Container Labeling . . .

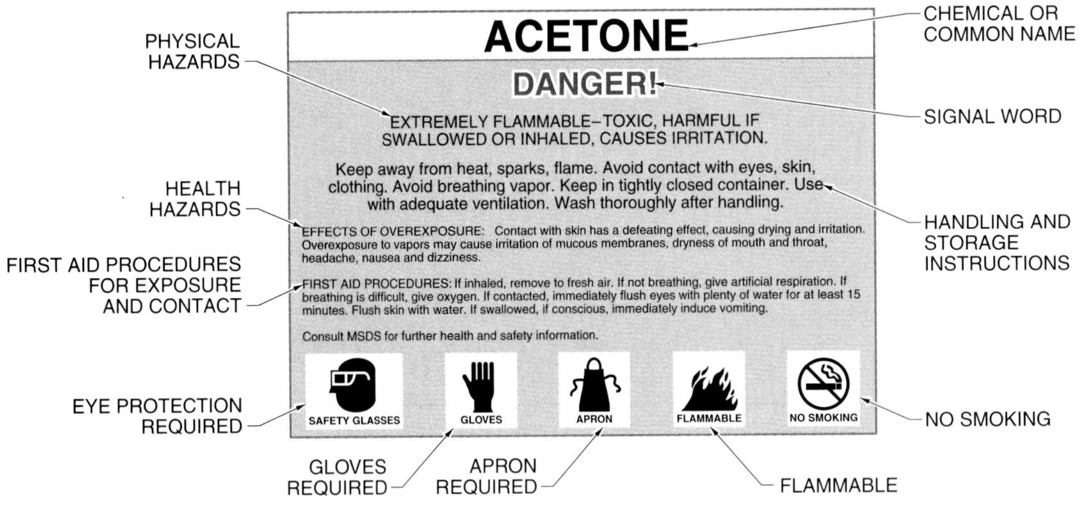

ACETONE

DANGER!

EXTREMELY FLAMMABLE–TOXIC, HARMFUL IF SWALLOWED OR INHALED, CAUSES IRRITATION.

Keep away from heat, sparks, flame. Avoid contact with eyes, skin, clothing. Avoid breathing vapor. Keep in tightly closed container. Use with adequate ventilation. Wash thoroughly after handling.

EFFECTS OF OVEREXPOSURE: Contact with skin has a defeating effect, causing drying and irritation. Overexposure to vapors may cause irritation of mucous membranes, dryness of mouth and throat, headache, nausea and dizziness.

FIRST AID PROCEDURES: If inhaled, remove to fresh air. If not breathing, give artificial respiration. If breathing is difficult, give oxygen. If contacted, immediately flush eyes with plenty of water for at least 15 minutes. Flush skin with water. If swallowed, if conscious, immediately induce vomiting.

Consult MSDS for further health and safety information.

SAFETY GLASSES | GLOVES | APRON | FLAMMABLE | NO SMOKING

PHYSICAL HAZARDS
HEALTH HAZARDS
FIRST AID PROCEDURES FOR EXPOSURE AND CONTACT
EYE PROTECTION REQUIRED
GLOVES REQUIRED
APRON REQUIRED
FLAMMABLE
NO SMOKING
HANDLING AND STORAGE INSTRUCTIONS
SIGNAL WORD
CHEMICAL OR COMMON NAME

RTK LABEL

HEALTH HAZARD
4 DEADLY
3 EXTREME DANGER
2 HAZARDOUS
1 SLIGHTLY HAZARDOUS
0 NORMAL MATERIAL

SPECIFIC HAZARD
OX OXIDIZER
ACID ACID
ALK ALKALI
COR CORROSIVE
W̶ USE **NO WATER**
☢ RADIATION HAZARD

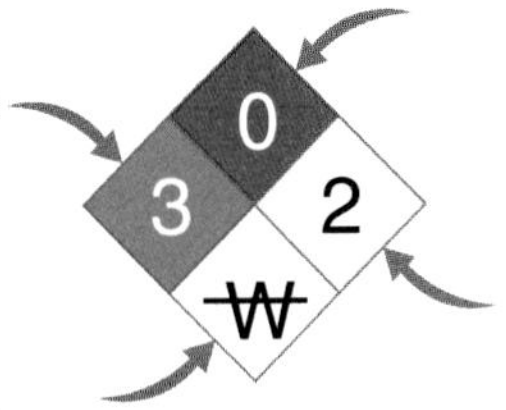

0
3 2
W̶

FIRE HAZARD
FLASH POINTS
4 BELOW 73°F
3 BELOW 100°F
2 BELOW 200°F
1 ABOVE 200°F
0 WILL NOT BURN

REACTIVITY
4 MAY DETONATE
3 SHOCK AND HEAT MAY DETONATE
2 VIOLENT CHEMICAL CHANGE
1 UNSTABLE IF HEATED
0 STABLE

Identification of Health Hazard Color Code: BLUE		*Identification of Flammability Color Code: RED*		*Identification of Reactivity (Stability) Color Code: YELLOW*	
Signal	**Type of Possible Injury**	**Signal**	**Susceptibility of Materials to Burning**	**Signal**	**Susceptibility to Release of Energy**
4	Materials that on very short exposure could cause death or major residual injury	4	Materials that will rapidly or completely vaporize at atmospheric pressure and normal ambient temperature, or that are readily dispersed in air and that will burn readily	4	Materials that in themselves are readily capable of detonation or reaction at normal temperatures and pressures
3	Materials that on short exposure could cause serious, temporary, or residual injury	3	Liquids and solids that can be ignited under almost all ambient temperature conditions	3	Materials that in themselves are capable of detonation or explosive decomposition or reaction, but require a strong initiating source, or which must be heated under confinement before initiation, or which react explosively with water
2	Materials that on intense or continued, but not chronic, exposure could cause temporary incapacitation or possible residual injury	2	Materials that must be moderately heated or exposed to relatively high ambient temperatures before ignition can occur	2	Materials that readily undergo violent chemical change at elevated temperatures and pressures, or which react violently with water, or which may form explosive mixtures with water
1	Materials that on exposure would cause irritation, but only minor residual injury	1	Materials that must be preheated before ignition can occur	1	Materials that in themselves are normally stable, but which can become unstable at elevated temperatures and pressures
0	Materials that on exposure under fire conditions would offer no hazard beyond that of ordinary combustible material	0	Materials that will not burn	0	Materials that in themselves are normally stable, even under fire exposure conditions, and which are not reactive with water

NFPA HAZARD SIGNAL SYSTEM

. . . Hazardous Material Container Labeling

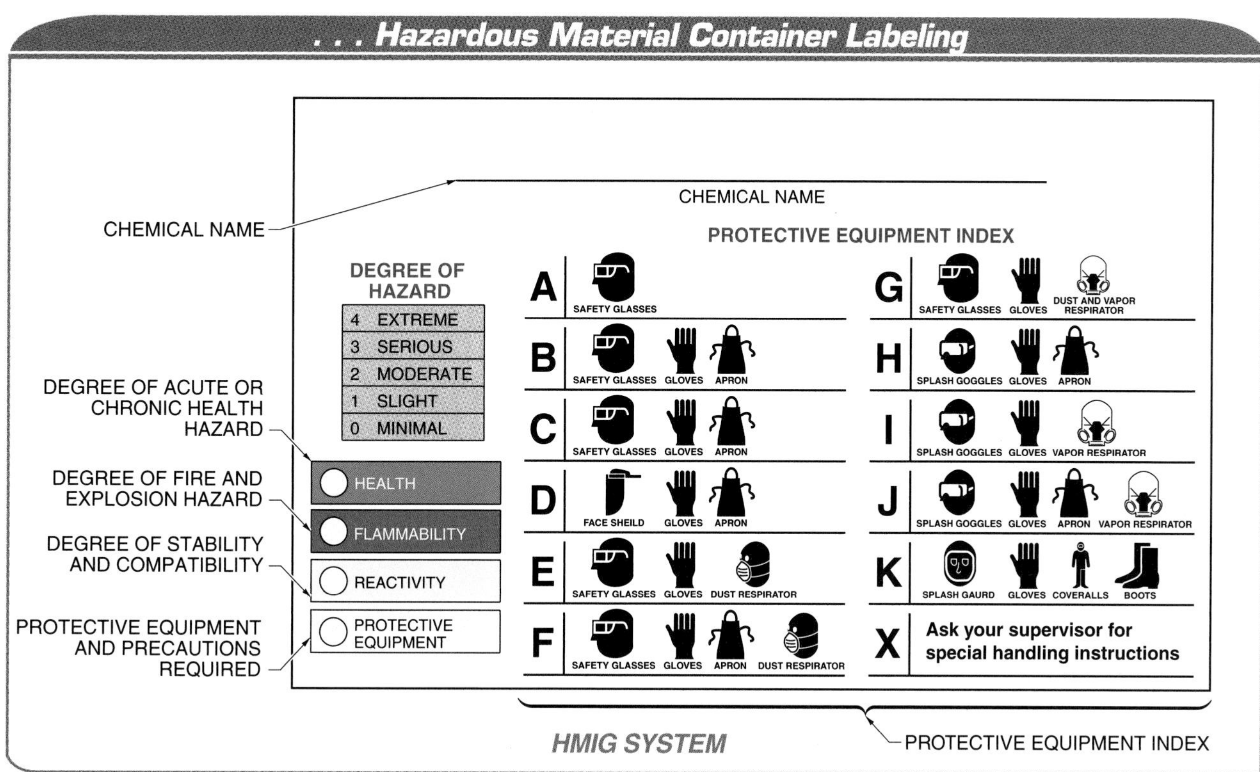

Hazardous Materials

Chemical	Hazard Rating				Chemical Abstract Service Number	Chemical	Hazard Rating				Chemical Abstract Service Number
	H	F	R	S/H			H	F	R	S/H	
Acetic acid	2	2	0	—	64-19-7	Isopropyl ether	2	3	1	—	108-20-3
Acetone	1	3	0	—	67-64-1	Methanol	1	3	0	—	67-56-1
Acetonitrile	3	3	0	—	75-05-8	Methyl acetate	1	3	0	—	79-20-9
Acrolein	3	3	3	—	107-02-8	Methyl bromide	3	1	0	—	74-83-9
Allyl alcohol	3	3	1	—	107-16-6	Methyl isobutyl ketone	2	3	0	—	108-10-1
Ammonia anhydrous	3	1	0	—	7664-41-7	Methylamine	3	4	0	—	74-89-5
Aniline	3	2	0	—	65-53-3	Morpholine	2	3	0	—	110-91-8
Bromine	3	0	0	OX	7726-95-6	Naphtha	1	4	0	—	8030-30-6
1 – 3 Butadiene	2	4	2	—	106-99-0	Naphthalene	2	2	0	—	91-20-3
Butyl acetate	1	3	0	—	123-86-4	Nitric acid	3	0	0	OX	7697-37-2
tert-Butyl alcohol	1	3	0	—	75-65-0	Nitrobenzene	3	2	1	—	98-95-3
Caustic soda	3	0	1	—	1310-73-2	p-Nitrochlorobenzene	2	1	3	—	100-00-5
Chlorine	3	0	0	OX	7782-50-5	Octane	0	3	0	—	111-65-9
Chloroform	2	0	0	—	67-66-3	Oxalic acid	2	1	0	—	144-62-7
o-Cresol	3	2	0	—	1319-77-3	Pentane	1	4	0	—	109-66-0
Cumene	2	3	1	—	98-82-8	Petroleum distillates	1	4	0	—	8002-05-9
Cyclohexane	1	3	0	—	110-82-7	Phenol	3	2	0	—	108-95-2
Cyclohexanol	1	2	0	—	108-93-0	Propane gas	1	4	0	—	74-98-6
Cyclohexanone	1	2	0	—	108-94-1	1-Propanol	1	3	0	—	71-23-8
Diborane	3	4	3	₩	19287-45-7	Propyl acetate	1	3	0	—	109-60-7
Dimethylamine	3	4	0	—	124-40-3	n-Propyl alcohol	1	3	0	—	71-23-8
p-Dioxane	2	3	1	—	123-91-1	Propylene oxide	4	2	2	—	75-56-9
Ethyl acetate	1	3	0	—	141-78-6	Pyridine	2	3	0	—	110-86-1
Ethyl ether	2	4	1	—	60-29-7	Sodium cyanide	3	0	0	—	143-33-9
Formic acid	3	2	0	—	64-18-6	Sodium hydroxide	3	0	1	—	1310-73-2
n-Heptane	1	3	0	—	142-82-5	Stoddard solvent	0	2	0	—	8052-41-3
n-Hexane	1	3	0	—	110-54-3	Sulfur dioxide	3	0	0	—	7446-09-5
Hydrazine	3	3	3	—	302-01-2	Sulfuric acid	3	0	2	₩	7664-93-9
Hydrochloric acid	3	0	0	—	7647-01-0	Tetrahydrofuran	2	3	1	—	109-99-9
Hydrogen peroxide	2	0	1	OX	7722-84-1	1-1-1 Trichloroethane	2	1	0	—	71-55-6
Iodine	—	—	—	—	7553-56-2	Triethylamine	2	3	0	—	121-44-8
Isobutyl alcohol	1	3	0	—	78-83-1	Xylene	2	3	0	—	1330-20-7
Isopropyl alcohol	1	3	0	—	67-63-0						

Lab Safety Supply, Inc.

Hazardous Locations

Class	Group	Material
I	A	Acetylene
	B	Hydrogen, butadiene, ethylene oxide, propylene oxide
	C	Carbon monoxide, ether, ethylene, hydrogen sulfide, morpholine, cyclopropane
	D	Gasoline, benzene, butane, propane, alcohol, acetone, ammonia, vinyl chloride
II	E	Metal dusts
	F	Carbon black, coke dust, coal
	G	Grain dust, flour, starch, sugar, plastics
III	No Groups	Wood chips, cotton, flax, nylon

Recommended Hydraulic System Flushing Guidelines

- Remove precision system components before flushing and install spool pieces, flushing plates, or replacement pieces in their place.
- Remove filter elements from main lines being flushed.
- Flushing velocity should be two to two and one-half times the system flow rate.
- Use a low-viscosity fluid at 185°F (85°C).
- Be sure the flushing fluid is compatible with hydraulic fluids.
- Always flush in only one direction of flow.
- Flush each system leg off the main leg one at a time, starting with the one closest to the flushing pump and proceeding downstream.
- In blind systems, provide vertical dirt traps by including short standpipes below the level of branch piping.
- Do not use the system pump as the flushing pump. A pump such as a centrifugal pump provides adequate head and greater flow rates, operates more economically, and has a better tolerance for contaminants that circulate during flushing.
- Use a cleanup filter in the flushing system with a capacity that matches the flow rates used. Filter rating should be as fine as practical, but not coarser than the proposed system filter rating.
- If practical, use a separate flushing fluid reservoir to avoid trapping contaminants in the system's reservoir.
- To determine when to complete the flushing procedure, take fluid samples to check contamination levels.
- Avoid introducing contaminants while reinstalling working components.

Hose Installation Recommendations . . .

Bend Radius Measurement

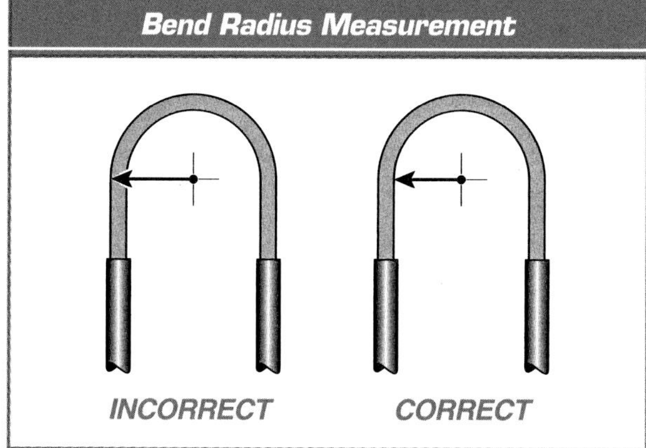

Tight Bends

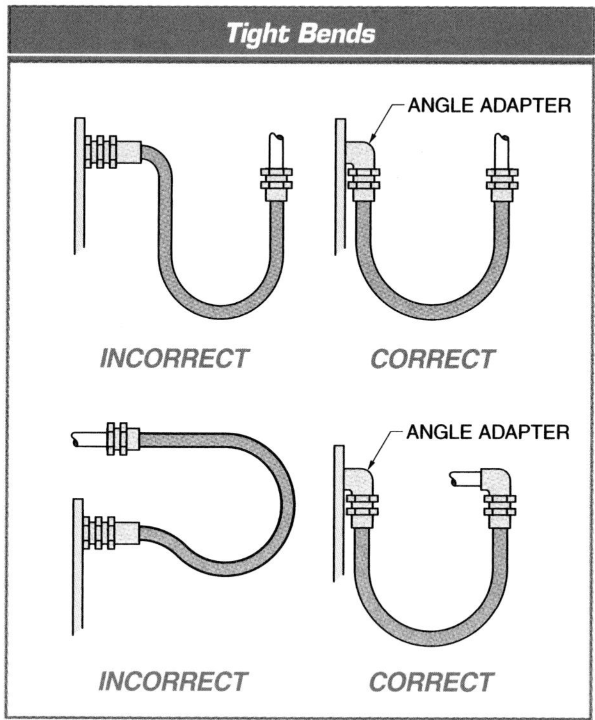

Movement Flexing

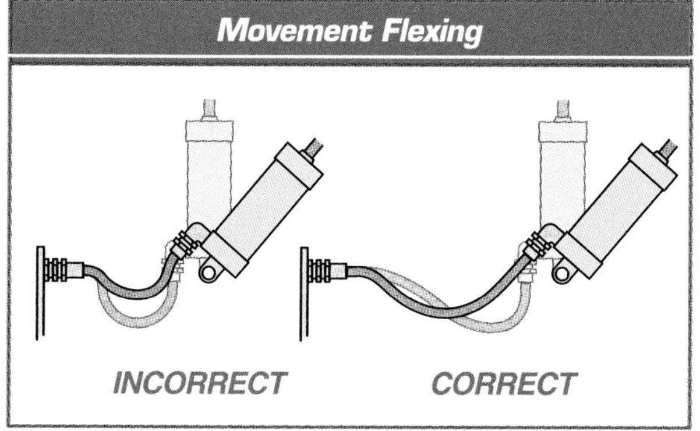

Length Changes

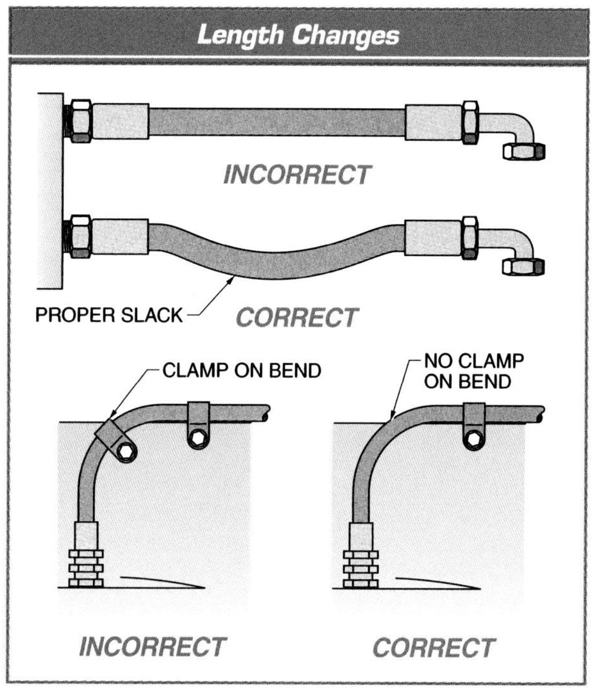

Strain

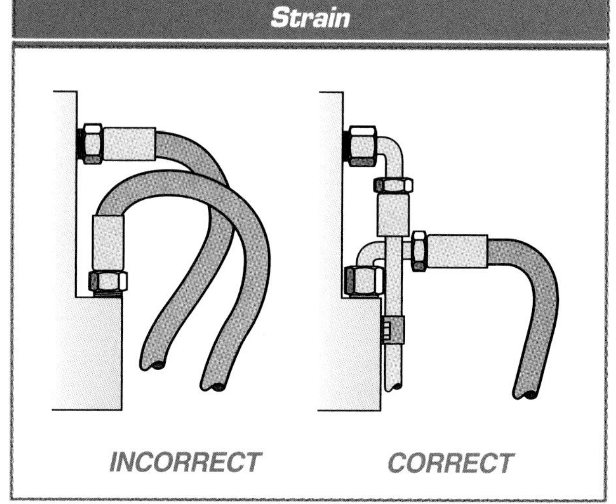

Reduced Connections

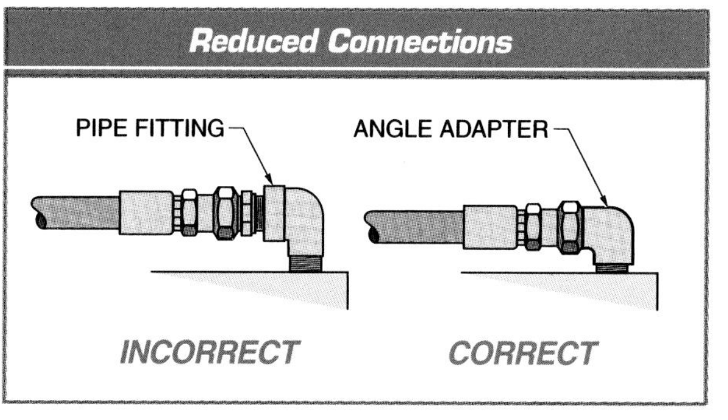

. . . Hose Installation Recommendations

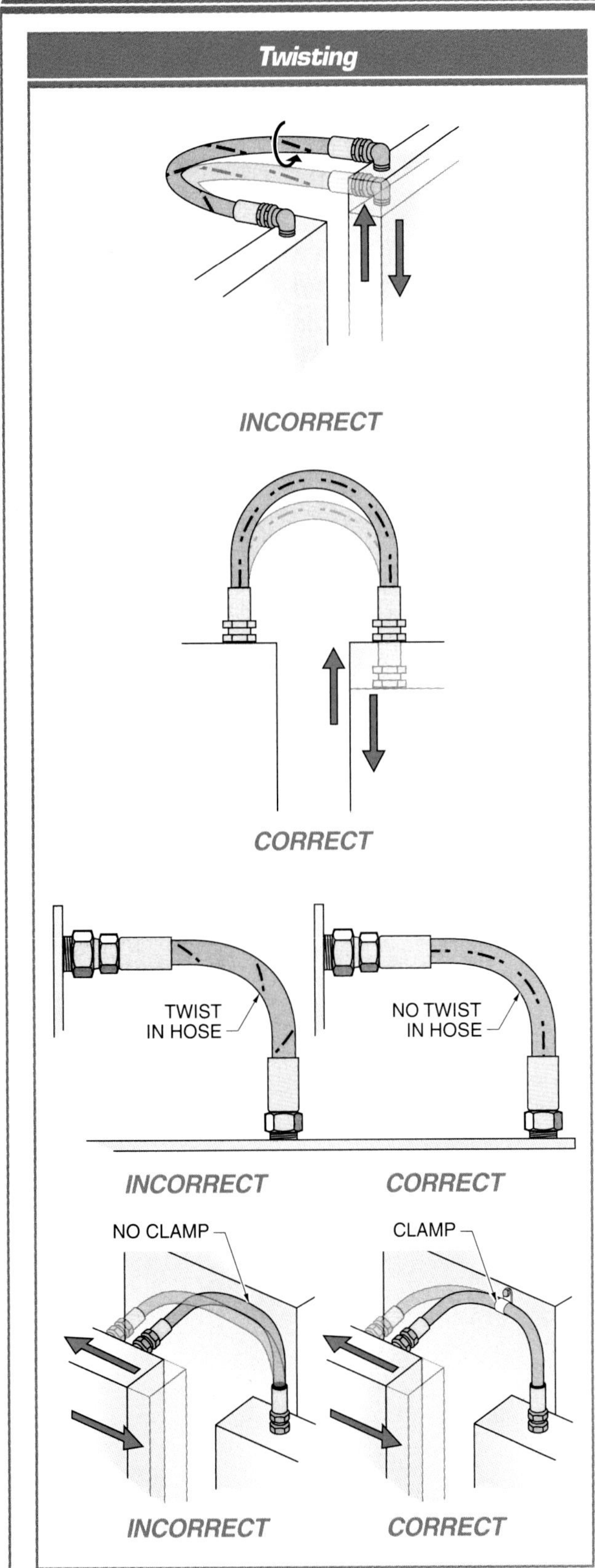
Twisting
INCORRECT
CORRECT
TWIST
IN HOSE
NO TWIST
IN HOSE
INCORRECT
CORRECT
NO CLAMP
CLAMP
INCORRECT
CORRECT

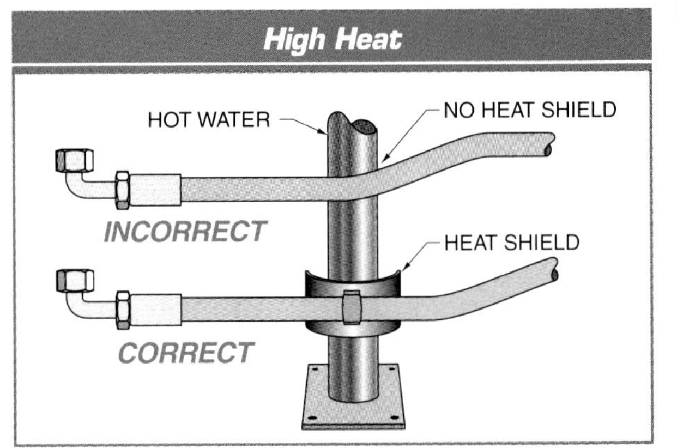
High Heat
HOT WATER
NO HEAT SHIELD
INCORRECT
HEAT SHIELD
CORRECT

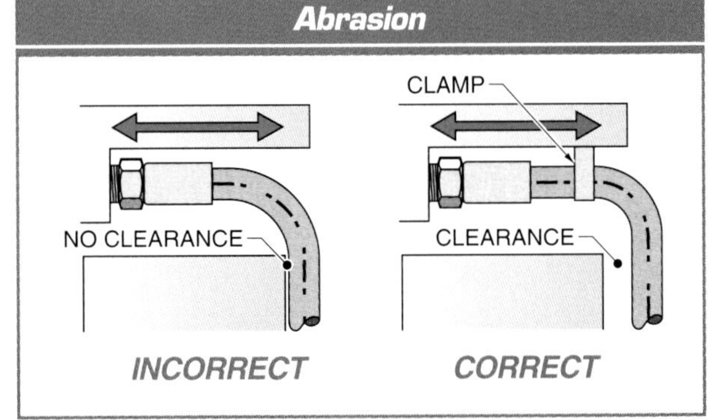
Abrasion
CLAMP
NO CLEARANCE
CLEARANCE
INCORRECT
CORRECT

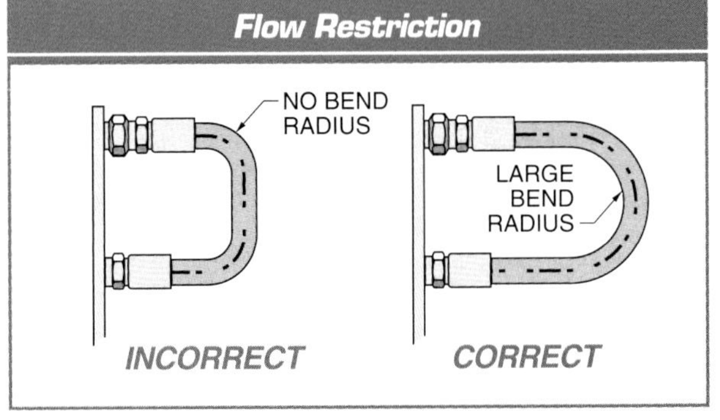
Flow Restriction
NO BEND
RADIUS
LARGE
BEND
RADIUS
INCORRECT
CORRECT

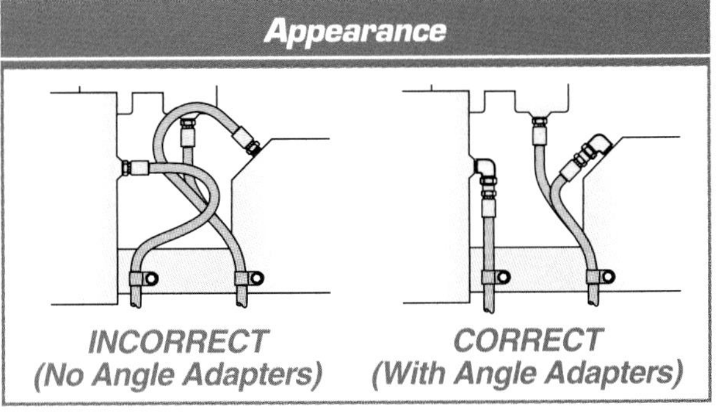
Appearance
INCORRECT
(No Angle Adapters)
CORRECT
(With Angle Adapters)

Hydraulic Cylinder Speed *

Piston Diameter‡	Rod Diameter‡	Flow† 1	2	3	5	10	12	15	20	25	50	75
1	-	298	596	849	149							
	½	392	784	1176	196							
1½	-	130	260	392	654	1308						
	⅝	158	316	476	792	1584						
	1	235	470	706	1176	2352						
2	-	73	146	221	368	736	883	1120				
	¾	85	170	257	428	956	1025	1283				
	1	97	184	294	490	980	1175	1465				
	1⅜	139	278	418	697	1394	1673	2090				
2½	-	47	94	141	235	470	565	675	940	1175		
	1	56	112	168	280	560	672	840	1120	1400		
	1⅜	67	134	203	339	678	813	1015	1355	1695		
	1¾	92	184	277	463	926	1110	1385	1850	2310		
3	-	32	64	98	163	326	392	490	653	817		
	1	36	72	110	184	368	440	551	735	920		
	1½	43	86	131	218	436	523	655	872	1090		
	2	58	116	176	294	588	705	882	1175	1470		
3½	-	24	48	72	120	240	288	360	480	600	1200	
	1¼	27	54	82	137	274	330	411	548	685	1370	
	1¾	32	64	96	160	320	384	480	640	800	1600	
	2	35	70	107	178	356	428	534	712	890	1780	
4	-	18	36	55	92	184	220	276	368	460	920	
	1¼	20	40	61	102	240	244	306	408	510	1020	
	1¾	22	44	68	113	226	273	339	452	565	1130	
	2	24	48	73	122	244	294	366	488	610	1220	
	2½	30	60	90	150	300	362	450	600	750	1500	
5	-	12	24	35	58	116	141	174	232	290	580	870
	1½	13	26	39	64	128	155	193	258	320	640	960
	2	14	28	42	70	140	168	210	280	350	700	1050
	2½	16	32	47	78	156	188	235	315	390	780	1170
	3	18	36	55	92	184	220	375	365	460	920	1380
	3½	22	44	66	111	222	266	333	444	555	1110	1665
6	-	8	16	24	41	82	98	123	162	202	404	606
	1¾	9	18	27	45	90	107	135	180	225	450	675
	2½	10	20	30	50	100	118	150	200	250	500	750
	3	11	22	33	54	108	130	165	206	270	540	810
	3½	12	24	37	62	124	148	185	245	310	620	930
	4	15	30	44	73	146	176	220	295	365	730	1095
8	-	4	8	14	23	46	55	69	92	115	230	345
	3½	5½	11	17	28	56	68	85	115	140	280	420
	4	6	12	18	30	60	73	90	122	150	300	450
	5	7½	15	22	38	76	90	114	150	185	375	555
	5½	8½	17	26	43	86	104	129	172	215	430	645
10	-	3	6	9	15	30	35	44	60	73	146	220
	4½	3½	7	11	18	36	44	55	75	92	184	275
	5	4	8	12	20	40	47	60	80	100	200	300
	5½	4½	9	13	21	42	50	84	84	105	210	315
	7	5½	11	17	29	58	69	115	115	145	290	435

* in ipm
† in gpm
‡ in in.
§ in sq in.

NOTE: Calculations based on:

$$V = \frac{231 \times F^{\dagger}}{\textit{cylinder area}^{\S}}$$

Motor Horsepower

Pump Flow*	Pump Pressure†										
	100	250	500	750	1000	1250	1500	2000	3000	4000	5000
1	0.07	0.18	0.36	0.54	0.72	0.91	1.09	1.45	2.18	2.91	3.64
2	0.14	0.38	0.72	1.09	1.45	1.82	2.18	2.91	4.37	5.83	7.29
3	0.21	0.54	1.09	1.64	2.18	2.73	3.28	4.37	6.56	8.75	10.93
4	0.29	0.72	1.45	2.18	2.91	3.64	4.37	5.83	8.75	11.66	14.58
5	0.36	0.91	1.82	2.73	3.64	4.55	5.46	7.29	10.93	14.58	18.23
8	0.58	1.45	2.91	4.37	5.83	7.29	8.75	11.66	17.50	23.33	29.17
10	0.72	1.82	3.64	5.46	7.29	9.11	10.93	14.58	21.87	29.17	36.46
12	0.87	2.18	4.37	6.56	8.75	10.93	13.12	17.50	26.25	35.00	43.75
15	1.09	2.73	5.46	8.20	10.93	13.67	16.40	21.87	32.81	43.75	54.69
20	1.45	3.64	7.29	10.93	14.58	18.23	21.87	29.17	43.75	58.34	72.92
25	1.82	4.55	9.11	13.67	18.23	22.79	27.34	36.46	54.69	72.92	91.16
30	2.18	5.46	10.93	16.40	21.87	27.34	32.81	43.75	65.63	87.51	109.39
35	2.55	6.38	12.76	19.14	25.52	31.90	38.28	51.05	76.57	102.10	127.62
40	2.91	7.29	14.58	21.87	29.17	36.46	43.75	58.34	87.51	116.68	145.85
45	3.28	8.20	16.40	24.61	32.81	41.02	49.22	65.63	98.45	131.27	164.08
50	3.64	9.11	18.23	27.34	36.46	45.58	54.69	72.92	109.39	145.85	182.32
55	4.01	10.20	20.05	30.08	40.11	50.13	60.16	80.22	120.33	160.44	200.55
60	4.37	10.93	21.87	32.81	43.75	54.69	65.63	87.51	131.27	175.02	218.78
65	4.74	11.85	23.70	35.55	47.40	59.25	71.10	94.80	142.21	189.61	237.01
70	5.10	12.76	25.52	38.28	51.05	63.81	76.57	102.10	153.13	204.20	255.25
75	5.46	13.67	27.36	41.02	54.69	68.37	82.04	109.39	164.08	218.78	273.48
80	5.83	14.58	29.17	43.75	58.34	72.92	87.51	116.68	175.02	233.37	291.71
90	6.56	16.40	32.81	49.22	65.63	82.04	98.45	131.27	196.90	262.54	328.17
100	7.29	18.23	36.46	54.69	72.92	91.16	109.39	145.85	218.78	291.71	364.64

* in gpm
† pump pressure in psi (efficiency assumed to be 80%)

Horsepower Formulas

To Find	Use Formula	Example		
		Given	Find	Solution
HP	$HP = \frac{I \times E \times E_{ff}}{746}$	240 V, 20 A, 85% E_{ff}	HP	$HP = \frac{I \times E \times E_{ff}}{746}$ $HP = \frac{20\text{ A} \times 240\text{ V} \times 85\%}{746}$ $HP = 5.5$
I	$I = \frac{HP \times 746}{E \times E_{ff} \times PF}$	*10 HP*, 240 V, 90% E_{ff}, 88% PF	I	$I = \frac{HP \times 746}{E \times E_{ff} \times PF}$ $I = \frac{10\text{ HP} \times 746}{240\text{ V} \times 90\% \times 88\%}$ $I = 39\text{ A}$

Pipe

Nominal ID*	OD (BW Gauge)	Inside Diameter (BW Gauge)			Nominal Wall Thickness		
		Standard	Extra-Heavy	Double Extra-Heavy	Schedule 40	Schedule 60	Schedule 80
1/8	0.405	0.269	0.215	—	0.068	0.095	—
1/4	0.540	0.364	0.302	—	0.088	0.119	—
3/8	0.675	0.493	0.423	—	0.091	0.126	—
1/2	0.840	0.622	0.546	0.252	0.109	0.147	0.294
3/4	1.050	0.824	0.742	0.434	0.113	0.154	0.308
1	1.315	1.049	0.957	0.599	0.133	0.179	0.358
1 1/4	1.660	1.380	1.278	0.896	0.140	0.191	0.382
1 1/2	1.900	1.610	1.500	1.100	0.145	0.200	0.400
2	2.375	2.067	1.939	1.503	0.154	0.218	0.436
2 1/2	2.875	2.469	2.323	1.771	0.203	0.276	0.552
3	3.500	3.068	2.900	2.300	0.216	0.300	0.600
3 1/2	4.000	3.548	3.364	2.728	0.226	0.318	—
4	4.500	4.026	3.826	3.152	0.237	0.337	0.674
5	5.563	5.047	4.813	4.063	0.258	0.375	0.750
6	6.625	6.065	5.761	4.897	0.280	0.432	0.864
8	8.625	7.981	7.625	6.875	0.322	0.500	0.875
10	10.750	10.020	9.750	8.750	0.365	0.500	—
12	12.750	12.000	11.750	10.750	0.406	0.500	—

* in in.

AC/DC Formulas

To Find	DC	AC		
		1ϕ, 115 or 220 V	1ϕ, 208, 230, or 240 V	3ϕ – All Voltages
I, HP known	$\frac{HP \times 746}{HP \times E_{ff}}$	$\frac{HP \times 746}{E \times E_{ff} \times PF}$	$\frac{HP \times 746}{E \times E_{ff} \times PF}$	$\frac{HP \times 746}{HP \times E_{ff}}$
I, kW known	$\frac{kW \times 1000}{E}$	$\frac{kW \times 1000}{E \times PF}$	$\frac{kW \times 1000}{E \times PF}$	$\frac{kW \times 1000}{1.73 \times E \times PF}$
I, kVA known		$\frac{kVA \times 1000}{E}$	$\frac{kVA \times 1000}{E}$	$\frac{kVA \times 1000}{1.763 \times E}$
kW	$\frac{I \times E}{1000}$	$\frac{I \times E \times PF}{1000}$	$\frac{I \times E \times PF}{1000}$	$\frac{I \times E \times 1.73 \times PF}{1000}$
kVA		$\frac{I \times E}{1000}$	$\frac{I \times E}{1000}$	$\frac{I \times E \times 1.73}{1000}$
HP (output)	$\frac{I \times E \times E_{ff}}{746}$	$\frac{I \times E \times E_{ff} \times PF}{746}$	$\frac{I \times E \times E_{ff} \times PF}{746}$	$\frac{I \times E \times 1.73 \times E_{ff} \times PF}{746}$

E_{ff} = efficiency

Subplate Designs

12.00″ 7.75″ T P Y L X A B D10

8.00″ 4.50″ T P Y L X A B D08

3.75″ 2.25″ T B A P D03

3.50″ 2.36″ FLOW CONTROL

3.75″ 4.20″ PRESSURE RELIEF

4.00″ 4.50″ P B T A D05

Voltage Drop Formulas – 1ϕ, 3ϕ

Phase	To Find	Use Formula	Example		
			Given	Find	Solution
1ϕ	VD	$VD = \frac{2 \times R \times L \times I}{1000}$	240 V, 40 A, 60 L, 0.764 R	VD	$VD = \frac{2 \times R \times L \times I}{1000}$ $VD = \frac{2 \times 0.764 \times 60 \times 40}{1000}$ $VD = 3.67\text{ V}$
3ϕ	VD	$VD = \frac{2 \times R \times L \times I}{1000} 0.866$	208 V, 110 A, 75 L, 0.194 R, 0.866 Multiplier	VD	$VD = \frac{2 \times R \times L \times I}{1000} \times 0.866$ $VD = \frac{2 \times 0.194 \times 75 \times 110}{1000} \times 0.866$ $VD = 2.77\text{ V}$

$^{*}\frac{\sqrt{3}}{2} = 0.866$

ISO Hydraulic Fluid Viscosity Ratings

ISO Grade‡	Viscosity Range†
32	28.8-35.2
46	41.4-50.6
68	61.2-74.8
100	90.0-110
150	135-165
220	198-242
320	288-352
460	414-506
680	612-748
1000	900-1100
1500	1350-1650

† in cST
‡ @40°C

SAE Hydraulic Fluid Viscosity Ratings

SAE Grade*	Viscosity Range†
5	3.80
10	4.10
20	5.60-9.29
30	9.30-12.49
40	12.50-16.29
50	16.30-21.89
60	21.90-26.09
80	7.0-11.00
90	13.50-23.99
140	24.00-40.99
250	41.00-UP

* @100°C
† in cST

Cleanliness Levels of Hydraulic Fluids (SAEJ1165 OCT80)

ISO Code	Particles per Millimeter <10 μm
26/23	140,000
2/23	85,000
23/20	14,000
21/18	4500
20/18	2400
20/17	2300
20/16	1400
19/16	1200
18/15	580
17/14	280
16/13	140
15/12	70
14/12	40
14/11	35
13/10	14
12/9	9
11/8	5
10/8	3
10/7	2.3
10/6	1.4
9/6	1.2
8/5	0.6
7/5	0.3
6/3	0.14
5/2	0.04
2/0.8	0.01

Fluid Power Troubleshooting . . .

System Inoperative

Possible Causes:	Solution
No hydraulic fluid in system	Fill to full mark. Check system for leaks.
Hydraulic fluid low in reservoir	Check level and fill to full mark. Check system for leaks.
Hydraulic fluid of wrong viscosity	Refer to specifications for proper viscosity.
Filter dirty or plugged	Drain hydraulic fluid and replace filters. Locate source of contamination.
Restriction in system	Hydraulic fluid lines could be dirty or have inner walls that are collapsing, cutting off hydraulic fluid supply. Clean or replace lines. Clean orifices.
Air leaks in suction line	Repair or replace lines.
Dirt in pump	Clean and repair pump. If necessary, drain and flush hydraulic system. Locate source of contamination.
Badly worn pump	Repair or replace pump. Check for problems causing pump wear such as misalignment or contaminated hydraulic fluid.
Badly worn components	Examine and test valves, motors, cylinders, etc. for external and internal leaks. If wear is abnormal, try to locate the cause.
Hydraulic fluid leak in pressure lines	Tighten fittings or replace defective lines. Examine mating surfaces on couplers for irregularities.
Components not properly adjusted	Refer to machine technical manual for proper adjustment of components.
Relief valve defective	Test relief valves to make sure they are opening at their rated pressure. Examine seals for damage that could cause leaks. Clean relief valves and check for broken springs, etc.
Pump rotating in wrong direction	Reverse to prevent damage.
Excessive load on system	Check specification of unit for load limits.
Hoses attached improperly	Attach properly and tighten securely.
Slipping or broken pump drive	Replace couplers or belts if necessary. Align them and adjust tension.
Pump not operating	Check for shut off device on pump or pump drive.

Erratic System Operation

Possible Causes:	Solution
Air in system	Examine suction side of system for leaks. Make sure hydraulic fluid level is correct. Hydraulic fluid leaks on the pressure side of system could account for loss of hydraulic fluid.
Cold hydraulic fluid	Viscosity of hydraulic fluid may be too high at start of warm-up period. Allow hydraulic fluid to warm up to operating temperature before using hydraulic functions.
Components sticking or binding	Check for dirt or deposits. If contaminated, locate the source of contamination. Check for worn or damaged parts.
Pump damaged	Check for worn or damaged parts. Determine cause of pump damage.
Dirt in relief valves	Clean or replace relief valves.
Restriction in filter or suction line	Suction line could be dirty or have inner walls that are collapsing, cutting off hydraulic fluid supply. Clean or replace suction line. Also, check filter line for restrictions.

. . . Fluid Power Troubleshooting . . .

Hydraulic Fluid Overheating

Possible Causes:	Solution
Operator holds control valves in operating position too long, causing relief valve to open	Return control lever to neutral position when not in use.
Using incorrect hydraulic fluid	Use hydraulic fluid recommended by manufacturer. Be sure hydraulic fluid viscosity is correct.
Low hydraulic fluid level	Fill reservoir. Check for leaks.
Dirty hydraulic fluid	Drain and refill with clean hydraulic fluid. Check for source of contamination and replace filters.
Engine running too fast	Reset governor or reduce throttle.
Incorrect relief valve pressure	Check pressure and clean or replace relief valves.
Internal component hydraulic fluid leakage	Examine and test valves, cylinders, motors, etc. for external and internal leaks. If wear is abnormal, locate cause.
Restriction in pump suction line	Clean or replace.
Dented, obstructed, or undersized hydraulic fluid lines	Replace defective or undersized hydraulic fluid lines. Remove obstructions.
Oil cooler malfunctioning	Clean or repair.
Control valve stuck open	Free all spools so that they return to neutral position.
Heat not radiating properly	Clean dirt and mud from reservoir, hydraulic fluid lines, coolers, and other components.
Automatic unloading control inoperative	Repair valve.

System Operation Too Slow

Possible Causes:	Solution
Cold hydraulic fluid	Allow hydraulic fluid to warm up before operating machine.
Hydraulic fluid viscosity too heavy	Use hydraulic fluid recommended by the manufacturer.
Insufficient engine speed	Refer to operator's manual for recommended speed. If machine has a governor, it may need adjustment.
Low hydraulic fluid supply	Check reservoir and add hydraulic fluid if necessary. Check system for leaks that could cause loss of hydraulic fluid.
Adjustable orifice restricted too much	Back out orifice and adjust it. Check machine specifications for proper setting.
Air in system	Check suction side of the system for leaks.
Badly worn pump	Repair or replace pump. Check for problems causing pump wear such as misalignment or contaminated hydraulic fluid.
Restriction in suction line or filter	Suction line could be dirty or have inner walls that are collapsing to cut of hydraulic fluid supply. Clean or replace suction line. Examine filter for plugging.
Relief valves not properly set or leaking	Test relief valves to make sure they are opening at their rated pressure. Examine valves for damaged seats that could leak.
Badly worn components	Examine and test valves, motors, cylinders, etc. for external and internal leaks. If wear is abnormal, try to locate the cause.
Valve or regulators plugged	Clean dirt from components. Clean orifices. Check for source of dirt and correct.
Hydraulic fluid leak in pressure lines	Tighten fittings or replace defective lines. Examine mating surfaces on couplers for irregularities.
Components not properly adjusted	Refer to machine manufacturer's specifications for proper adjustment of components.

. . . Fluid Power Troubleshooting . . .

System Operation Too Fast

Possible Causes:	Solution
Adjustable orifice installed backward or not installed	Install orifice parts correctly and adjust.
Obstruction of foreign material under seat of orifice	Remove foreign material. Readjust orifice.
Overspeeding of engine	Refer to manufacturer's specifications for recommended speed. If machine has a governor, it may need adjustment.

Foaming of Hydraulic Fluid in System

Possible Causes:	Solution
Low hydraulic fluid level	Fill reservoir. Look for leaks. Drain and replace hydraulic fluid.
Water in hydraulic fluid	Check breather/filler cap assembly on reservoir. Heat exchanger may be cracked.
Wrong kind of hydraulic fluid being used	Use hydraulic fluid recommended by the manufacturer.
Air leak in line from reservoir to pump	Tighten or replace suction line.
Kink or dent in hydraulic fluid lines	Replace hydraulic fluid lines.
Worn pump shaft seal	Clean sealing area and replace seal. Check hydraulic fluid for contamination or pump for misalignment.

Noisy Pump

Possible Causes:	Solution
Low hydraulic fluid level	Fill reservoir. Check system for leaks.
Hydraulic fluid viscosity too high	Change to lighter hydraulic fluid.
Pump speed too fast	Operate pump at recommended speed.
Suction line plugged or pinched	Clean or replace line between reservoir and pump.
Sludge and dirt in pump	Disassemble and inspect pump and lines. Clean hydraulic system. Determine cause of dirt.
Reservoir air vent plugged	Remove breather/filler cap, flush, and clean air vent.
Air in hydraulic fluid	Tighten or replace suction line. Check system for leaks. Replace pump shaft seal.
Worn or scored pump bearings or shafts	Replace worn parts or complete pump if parts are badly worn or scored. Determine cause of scoring.
Inlet screen plugged	Clean screen.
Broken or damaged pump parts	Repair pump. Look for cause of damage such as contamination or too much pressure.
Sticking or binding parts	Repair binding parts. Clean parts and change oil if necessary.

Pump Leaks Hydraulic Fluid

Possible Causes:	Solution
Damaged seal around drive shaft	Tighten packing or replace seal. Trouble may be caused by contaminated hydraulic fluid. Check hydraulic fluid for abrasives and clean entire hydraulic system. Try to locate source of contamination. Check the pump drive shaft. Misalignment could cause the seal to wear. If shaft is not aligned, check the pump for other damage.
Loose or broken pump parts	Make sure all bolts and fittings are tight. Check gaskets. Examine pump castings for cracks. If pump is cracked, look for a cause like too much pressure or hoses that are attached incorrectly.

. . . Fluid Power Troubleshooting

Load Drops with Control Valve in Neutral Position

Possible Causes:	Solution
Leaking or broken hydraulic fluid lines from control valve to cylinder	Check for leaks. Tighten or replace lines. Examine mating surfaces on couplers for irregularities.
Hydraulic fluid leaking past cylinder packings or O-rings	Replace worn parts. If wear is caused by contamination, clean hydraulic system and determine the contamination source.
Hydraulic fluid leaking past control valves or relief valves	Clean or replace valves. Wear may be caused by contamination. Clean hydraulic system and determine the contamination source.
Hydraulic fluid leaking past load holding valve	Check for proper adjustment. Remove and replace cartridge with spare. (Support boom before removing cartridge.) Do not attempt to repair.
Control lever not centering when released	Check linkage for binding. Make sure valve is properly adjusted and has no broken or binding parts.

Control Valve Sticks

Possible Causes:	Solution
Misalignment or seizing of control linkage	Correct misalignment. Lubricate linkage joints.
Tie bolts too tight on stacking valves	Use manufacturer's recommendation to adjust tie bolt torque.
Valve broken or worn internally	Repair broken or worn parts. Locate source of contamination that caused scoring.

Control Valve Leaks Hydraulic Fluid

Possible Causes:	Solution
Tie bolts too loose (on valve stacks)	Use manufacturer's recommendation to adjust tie bolt torque.
Worn or damaged O-rings	Replace O-rings, especially between valve stacks. If contamination has caused O-rings to wear, clean system and locate source of contamination.
Cracked valve parts	Check pressure for pipe fittings that are overtightened.

Cylinders Leak Hydraulic Fluid

Possible Causes:	Solution
Damaged cylinder barrel	Replace cylinder barrel. Correct cause of barrel damage.
Rod seal leaking	Replace seal. If contamination caused seal to wear, determine source of contamination. Wear may be caused by external as well as internal contaminants. Check rod for scratches or misalignment.
Loose parts	Tighten parts until leakage has stopped.
Rod damaged	Check rod for nicks or scratches that could cause seal damage or allow hydraulic fluid leakage. Replace defective rods.

Cylinders Lower When Control Valve Is in Raise Position

Possible Causes:	Solution
Damaged check valve in lift circuit	Repair or replace check valve.
Leaking cylinder packing	Replace packing. Check hydraulic fluid for contamination that could cause wear. Check alignment of cylinder.
Leaking lines or fittings to cylinder	Check and tighten. Examine mating surfaces on couplers for irregularities.

Industry Organizations

American National Standards Institute (ANSI)
1899 L Street, NW, 11th Floor
Washington, DC 20036
202-293-8020
www.ansi.org

American Petroleum Institute (API)
1220 L St. NW
Washington, DC 20005-4070
202-682-8000
www.api.org

ASTM International
100 Barr Harbor Drive
PO Box C700
West Conshohocken, PA 19428-2959
610-832-9500
www.astm.org

American Society of Agricultural and Biological Engineers
2950 Niles Road
St. Joseph, MI 49085-9659
269-429-0300
www.asabe.org

ASME International
Two Park Ave.
New York, NY 10016-5990
800-843-2763
www.asme.org

Association of Equipment Manufacturers (AEM)
6737 W. Washington St., Ste. 2400
Milwaukee, WI 53214-5647
414-272-0943
www.aem.org

British Fluid Power Association (BFPA)
Cheriton House, Cromwell Business Park
Chipping Norton, Oxon
OX7 5SR
United Kingdom
44-1608-647900
www.bfpa.co.uk

Canadian Fluid Power Association (CFPA)
1250 Marlborough Court, Unit 25
Oakville, ON
L6H 2W7
905-844-6822
www.cfpa.ca

Compressed Air and Gas Institute (CAGI)
1300 Summer Ave.
Cleveland, OH 44115-2851
216-241-7333
www.cagi.org

The FPDA Motion and Control Network
105 Eastern Ave, Suite 104
Annapolis, MD 21403-3300
410-940-6347
www.fpda.org

Fluid Power Educational Foundation (FPEF)
1930 E. Marlton Pike, Suite A2
Cherry Hill, NJ 08003
856-424-8998
www.fpef.org

Fluid Power Institute (FPI)
Milwaukee School of Engineering
1025 N. Broadway
Milwaukee, WI 53202-3109
414-277-7191
www.msoe.edu/fpi

Fluid Power Safety Institute™ (FPSI)
2170 South 3140 West, Suite B
West Valley City, UT 84119
801-908-5456
www.fluidpowersafety.org

Institute of Electrical and Electronics Engineers (IEEE)
2001 L Street, NW, Suite 700
Washington, DC, 20036-4910
202-785-0017
www.ieee.org

Instrument Society of America (ISA)
67 Alexander Dr.
Research Triangle Park, NC 27709
919-549-8411
www.isa.org

International Fluid Power Society (IFPS)
1930 E. Marlton Pike Suite A-2
Cherry Hill, NJ 08003
800-308-6005
www.ifps.org

International Organization for Standardization (ISO)
1, rue de Varembe, Case postale 56
CH-1211
Geneva 20, Switzerland
41-22-749-0111
www.iso.ch

International Safety Equipment Association (ISEA)
1901 North Moore St.
Arlington, VA 22209-1762
703-525-1695
www.safetyequipment.org

National Electrical Manufacturers Association (NEMA)
1300 N. 17th St., Ste. 1847
Rosslyn, VA 22209
703-841-3200
www.nema.org

National Fluid Power Association (NFPA)
3333 N. Mayfair Rd.
Milwaukee, WI 53222-3219
414-778-3344
www.nfpa.com

Society of Automotive Engineers (SAE)
400 Commonwealth Dr.
Warrendale, PA 15096-0001
724-776-4841
www.sae.org

Society of Manufacturing Engineers (SME)
One SME Dr.
PO Box 930
Dearborn, MI 48121-0930
313-271-1500
www.sme.org

Abbreviations . . .

A
Absolute	ABS
Actual	ACT
Adapter	ADPT
Addendum	ADD
Adjust	ADJ
Advance	ADV
Allowance	ALLOW
Alloy	ALY
Altitude	ALT
Aluminum	AL
American Standard	AMER STD
American Wire Gauge	AWG
Amount	AMT
Anneal	ANL
Apparatus	APPAR
Approved	APP
Approximate	APPROX
Arc Weld	ARC/W
Area	A
Arrangement	ARR
Assemble	ASSEM
Assembly	ASSY
Authorized	AUTH
Auxiliary	AUX

B
Babbitt	BAB
Back-feed	BF
Back Pressure	BP
Ball Bearing	BB
Base Line	BL
Base Plate	BP
Bearing	BRG
Benchmark	BM
Bending Moment	M
Between	BET
Bevel	BEV
Bill of Material	B/M
Bolt Circle	BC
Both Faces	BF
Both Sides	BS
Both Ways	BW
Bottom	BOT
Bottom Chord	BC
Bracket	BRKT
Brass	BRS
Brazing	BRZG
Break	BRK
Brinell Hardness	BH
British Standard	BR STD
Broach	BRO
Bronze	BRZ
Bushing	BUSH

C
Cabinet	CAB
Cadmium Plate	CD PL
Capacity CAP	
Cap Screw	CAP SCR
Carbon	C
Carburize	CARB
Carriage	CRG
Case Harden	CH
Cast	C
Cast Iron	CI
Cast Iron Pipe	CIP
Cast Steel	CS
Casting	CSTG
Castle Nut	CAS NUT
Center	CTR
Centerline	CL
Center of Gravity	CG
Center Punch	CP
Ceramic	CER
Chamfer	CH or CHAM
Channel	CHAN
Chrome Molybdenum	CR MOLY
Chromium Plate	CR PL
Chrome Vanadium	CR VAN
Circle	CIR
Circular Pitch	CP
Circumference	CIRC
Clearance	CL
Clockwise	CW
Coated	CTD
Cold Drawn	CD
Cold Drawn Steel	CDS
Cold Finish	CF
Cold Punched	CP
Cold Rolled Steel	CRS
Concentric	CONC
Copper Plate	COP PL
Corrosion Resistant	CRE
Corrosion Resistant Steel	CRES
Cotter	COT
Counterclockwise	CCW
Counterbore	CB or CBORE
Counterdrill	CD or CDRILL
Countersink	CSK or CSINK
Countersink Other Side	CSKO
Coupling	CPLG
Cross Section	XSECT
Cubic	CU
Cubic Foot	CU FT
Cubic Inch	CU IN
Cylinder	CYL

D
Decimal	DEC
Dedendum	DED
Depth	DP or DEEP
Deep Drawn	DD
Degree	DEG
Density	D
Design	DSGN
Detail	DET
Diagonal	DIAG
Diagram	DIAG
Diameter	DIA
Diametral Pitch	DP
Dimension	DIM
Dovetail	DVTL
Dowel	DWL
Drafting	DFTG
Draftsman	DFTSMN
Drawing	DWG
Drill	DR
Drive	DR
Drop Forge	DF
Duplicate	DUP

E
Each	EA
Eccentric	ECC
Electric	ELEC
Elongation	ELONG
Enclose	ENCL
Engineer	ENGR
Envelope	ENV
Equipment	EQUIP
Equivalent	EQUIV
Existing	EXIST
Extension	EXT
Extrude	EXTR

F
Fabricate	FAB
Far Side	FS
Feet	FT
Feet Per Minute	FPM
Feet Per Second	FPS
Figure	FIG
Fillet	FIL
Finish	FIN
Finish All Over	FAO
Fitting	FTG
Fixture	FIX
Flange	FLG
Flashing	FL
Flat	F
Flat Head	FH
Flexible	FLEX
Forged Steel	FST
Forging	FORG
Forward	FWD
Foundry	FDRY
Fractional	FRAC
Furnish	FURN

G
Gauge	GA
Galvanize	GALV
Galvanized	Iron GI
Galvanized	Steel GS
Gasket	GSKT
General	GEN
Glass	GL
Grade	GR
Grind	GRD or GND
Groove	GRV

H
Half-Round	½RD
Hard	H
Hard Drawn	HD
Harden	HDN
Hardware	HDW
Head	HD
Headless	HDLS

. . . Abbreviations . . .

Term	Abbreviation
Heat	HT
Heat Treat	HT TR
Heavy	HVY
Height	HGT
Hexagon	HEX
High-Speed	HS
High-Speed Steel	HSS
High-Tensile Cast Iron	HTCI
High-Tensile Steel	HTS
Horizontal	HORIZ
Hot Rolled	HR
Hot Rolled Steel	HRS
I	
Impregnate	IMPG
Inch	IN.
Inches Per Minute	IPM
Indicate	IND
Inside Diameter	ID
Install	INSTL
Internal	INT
International Pipe Standard	IPS
Intersect	INT
Iron	I
Irregular	IRREG
J	
Joint	JT
Junction	JCT
K	
Key	K
Keyseat	KST
Keyway	KWY
Knockout	KO
L	
Laboratory	LAB
Laminate	LAM
Lateral	LAT
Left Hand	LH
Length	LG
Limit	LIM
Linear	LIN
Locate	LOC
Low-Speed	LS
Lubricate	LUB
M	
Machine	MACH
Machine Steel	MS
Malleable	MALL
Malleable	Iron MI
Manual	MAN
Manufacture	MFR
Manufactured	MFD
Manufacturing	MFG
Material	MAT or MATL
Material List	ML
Maximum	MAX
Mechanical	MECH
Metal	MET
Micrometer	MIC
Military	MIL
Millimeter	MM
Minimum	MIN
Minute	MIN
Miscellaneous	MISC
Mold Line	ML
Molded	MLD
Molding	MLDG
Morse Taper	MOR T
Mounted	MTD
Mounting	MTG
Multiple	MULT
N	
National	NATL
Near Face	NF
Near Side	NS
New British Standard (Imperial Wire Gauge)	NBS
Nipple	NIP
Nominal	NOM
Normal	NORM
Not To Scale	NTS
Number	NO.
O	
Octagon	OCT
On Center	OC
Opening	OPNG
Opposite	OPP
Original	ORIG
Outside Diameter	OD
Overall	OA
P	
Pair	PR
Parallel	PAR
Part	PT
Patent	PAT
Pattern	PATT
Permanent	PERM
Perpendicular	PERP
Phenolic	PHEN
Pitch	P
Pitch Circle	PC
Pitch Diameter	PD
Plate	PL
Point	PT
Position	POS
Pound	LB
Pounds Per Square Inch	PSI
Precast	PRCST
Prefabricated	PREFAB
Preferred	PFD
Primary	PRIM
Production	PROD
Profile	PF
Project	PROJ
Proposed	PROP
Punch	PCH
Q	
Quadrant	QUAD
Quality	QUAL
Quantity	QTY
Quarter Round	¼RD
R	
Radial	RAD
Radians	RAD
Radius	R
Ream	RM
Reassemble	REASM
Received	RECD
Rectangle	RECT
Reference	REF
Reference Line	REF L
Reinforce	REINF
Relief	REL
Remove	REM
Require	REQ
Required	REQD
Return	RET
Reverse	REV
Revolutions Per Minute	RPM
Right Hand	RH
Rivet	RIV
Rockwell Hardness	RH
Roller Bearing	RB
Root Diameter	RD
Root Mean Square	RMS
Round	RD
S	
Schedule	SCH
Schematic	SCHEM
Screw	SCR
Secondary	SEC
Section	SECT
Semi-Finished	SF
Set Screw	SS
Shaft	SFT
Sheet	SH
Shop Order	SO
Shoulder	SHLD
Side	S
Similar	SIM
Sketch	SK
Sleeve	SLV
Sleeve Bearing	SB
Slotted	SLOT
Socket	SOC
Space	SP
Special Treatment Steel	STS
Specification	SPEC
Speed	SP
Spherical	SPHER
Spotface	SF or SFACE
Spring	SPG
Square	SQ
Stainless	STN
Stainless Steel	SST or SS
Standard	STD

. . . Abbreviations

Steel	STL
Stock	STK
Straight	STR
Stress Anneal	SA
Structural	STR
Supplement	SUPP
Supply	SUP
Surface	SURF
Symbol	SYM
Symmetrical	SYM
Synthetic	SYN
T	
Tachometer	TACH
Tangent	TAN
Taper	TPR
Technical	TECH
Tee	T
Teeth	T
Teeth Per Inch	TPI
Temperature	TEMP
Template	TEMP
Tensile Strength	TS
Tension	TENS
Thick	THK
Thread	THD
Threads Per Inch	TPI
Through	THRU
Tolerance	TOL
Tool Steel	TS
Total	TOT
Total Indicator Reading	TIR
Tubing	TUB
Typical	TYP
U	
United States Gauge	USG
United States Standard	USS
V	
Velocity	V
Vertical	VERT
Vibrate	VIB
Volume	VOL
W	
Washer	WASH
Weight	WT
Wheel Base	WB
Width	W
Wire	W
With	W/
Without	W/O
Woodruff	WDF
Wrought	WRT
Wrought Iron	WI

Printreading Symbols

Meaning	Symbol	Meaning	Symbol
Straightness	—	Projected Tolerance Zone	Ⓟ
Flatness	▱	Diameter	Ø
Circularity	○	Basic dimension	[50]
Cylindricity	⌭	Reference dimension	(50)
Profile of a line	⌒	Conical taper	⌲
Profile of a surface	⌓	Taper	⌳
All around	↙○	Counterbore/Spotface	⌴
Angularity	∠	Countersink	⌵
Perpendicularity	⊥	Depth/deep	↧
Parallelism	//	Square	□
Position	⌖	Dimension not to scale	<u>15</u>
Concentricity	◎	Arc length	⌒105
Symmetry	⌯	Radius	R
Circular runout	*↗	Spherical radius	SR
Total runout	*⌰	Spherical diameter	SØ
Maximum Material Condition	Ⓜ	Between	*←→
Least Material Condition	Ⓛ	Statistical tolerance	⟨ST⟩

* may be filled or not filled

Prefixes

Multiples and Submultiples	Prefixes	Symbols	Meaning
$1,000,000,000,000 = 10^{12}$	tera	T	trillion
$1,000,000,000 = 10^{9}$	giga	G	billion
$1,000,000 = 10^{6}$	mega	M	million
$1000 = 10^{3}$	kilo	k	thousand
$100 = 10^{2}$	hecto	h	hundred
$10 = 10^{1}$	deka	d	ten
Unit $1 = 10^{0}$			
$.1 = 10^{-1}$	deci	d	tenth
$.01 = 10^{-2}$	centi	c	hundredth
$.001 = 10^{-3}$	milli	m	thousandth
$.000001 = 10^{-6}$	micro	µ	millionth
$.000000001 = 10^{-9}$	nano	n	billionth
$.000000000001 = 10^{-12}$	pico	p	trillionth

Conversion Table

Initial Units	Final Units											
	giga	mega	kilo	hecto	deka	base	deci	centi	milli	micro	nano	pico
giga		3R	6R	7R	8R	9R	10R	11R	12R	15R	18R	21R
mega	3L		3R	4R	5R	6R	7R	8R	9R	12R	15R	18R
kilo	6L	3L		1R	2R	3R	4R	5R	6R	9R	12R	15R
hecto	7L	4L	1L		1R	2R	3R	4R	5R	8R	11R	14R
deka	8L	5L	2L	1L		1R	2R	3R	4R	7R	10R	13R
base	9L	6L	3L	2L	1L		1R	2R	3R	6R	9R	12R
deci	10L	7L	4L	3L	2L	1L		1R	2R	5R	8R	11R
centi	11L	8L	5L	4L	3L	2L	1L		1R	4R	7R	10R
milli	12L	9L	6L	5L	4L	3L	2L	1L		3R	6R	9R
micro	15L	12L	9L	8L	7L	6L	5L	4L	3L		3R	6R
nano	18L	15L	12L	11L	10L	9L	8L	7L	6L	3L		3R
pico	21L	18L	15L	14L	13L	12L	11L	10L	9L	6L	3L	

R = move the decimal point to the right
L = move the decimal point to the left

Glossary

A

abrasion: The wearing of a material due to grinding or friction.

absolute pressure: The sum of gauge pressure and atmospheric pressure.

absolute rating: A value that indicates the diameter of the largest hard, sphere-shaped particle that can pass through a filter under specific test conditions.

absolute temperature: Any temperature on a scale that begins with absolute zero.

absolute zero: The hypothetical temperature at which molecular motion ceases, precisely –273.15°C (–459.67°F).

absorption: The drawing of gases or vapors into permeable pores of porous material that results in physical and/or chemical changes in the material.

acceleration: An increase in speed and is measured in feet per second squared (ft/sec^2).

accumulator: A vessel in which fluid is stored under pressure for future release of energy.

actuator: A mechanical device used for moving or controlling movement of a load.

adsorption: The adhesion of a gas or liquid to the surface of a porous material.

aftercooler: A heat exchanger used for cooling the discharged air from an air compressor.

air: A gas primarily composed of nitrogen and oxygen, with traces of other gases.

air compressor: A device that increases the pressure of a gas by mechanically decreasing its volume.

air dryer: A pneumatic device that dries compressed air through cooling, condensing, or absorbing.

air filter-regulator: A pneumatic device that filters and regulates airflow through prelubricated components.

airflow: The movement of air in a pneumatic system through pipes, tubing, and hoses that transmit energy produced by the air compressor to the location where it is needed to accomplish work.

air motor: A pneumatic device that uses airflow to create rotating mechanical energy.

alternating current (AC): Current that reverses its direction of flow at regular intervals.

ambient air: The air that is located inside a facility.

American wire gauge (AWG): A standardized wire sizing system used in the United States for the diameters of round, solid, nonferrous, electrically conducting wire.

ampere (A): The amount of current that flows when 1 volt (V) is applied across 1 ohm (Ω) of resistance.

AND logic valve: A type of logic valve that allows fluid flow out of the valve when pressure is present on input 1 and input 2.

area: The number of unit squares equal to the surface of an object.

atmospheric air: Air that is located outside a facility.

atmospheric pressure: The pressure created by the weight of the atmosphere at sea level under standard air conditions.

atom: The smallest particle that an element can be reduced to while keeping the properties of that element.

automated actuator: A type of directional control valve actuator that can be actuated without a person being involved.

automatic detent: A detent that releases a spool to its original position when force at the detent reaches a certain point.

axial piston air motor: A type of air motor that operates at low speed and high torque with its pistons horizontal to the motor's shaft.

B

back pressure: The level of pressure on the return line of a device or hydraulic system.

back-up ring: A ring that supports the O-ring and is installed on the side receiving the least amount of pressure.

baffle: An artificial surface, usually a plate or wall, for deflecting, retarding, or regulating the flow of hydraulic fluid.

balanced vane pump: A pump that consists of a cam ring, rotor, vanes, and a port plate with opposing inlet and outlet ports.

ball check valve: A check valve that uses a ball located between the biasing spring and the seat to block fluid flow.

ball valve: An infinite-position flow control valve with a ball that has an orifice through the center to allow fluid flow.

beginning-to-end troubleshooting method: A fluid power troubleshooting method that traces problems from the beginning of the fluid power system (prime mover, hydraulic pump, or air compressor) to the end (actuator).

bellows: A device that expands to draw air in through a flapper valve and contracts to expel the air through a nozzle.

bent-axis piston pump: A piston pump in which the pistons and cylinders are at an angle to the drive shaft and thrust plate.

beta ratio: A number that represents the number of particles at a given size or larger that can pass through a filter.

bias spring: A spring that has an adjustable force.

bias spring adjustment screw: A screw that is used to adjust the amount of force in a bias spring.

bio-hydraulic fluid: An environmentally nonhazardous hydraulic fluid that is composed of synthetic chemicals and vegetable-based oil to lower the hazardous effects from leaks and spills.

bladder gas-charged accumulator: An accumulator consisting of a seamless steel shell, rubber bladder with a gas valve, and poppet valve.

bleed-off: A type of flow control where a needle valve (no check) controls the fluid flow to the reservoir in a parallel leg of the system to control the speed of an actuator.

Bourdon tube: A hollow metal tube made of brass or similar material that is bent in the shape of the letter C.

Bourdon tube pressure gauge: A measurement device that is used to register and measure pressure in fluid power systems.

Boyle's law: A gas law that states that when temperature remains constant, there is an inverse relationship between pressure and the volume of a gas.

brake valve: A normally closed pressure control valve that prevents a hydraulic motor from running away, or speeding out of control.

breakaway torque: The torque required to get a nonmoving load to begin moving.

breather/filler cap assembly: A device used to allow air in and out of a reservoir as the pump delivers hydraulic fluid into the system.

bypass filter: A hydraulic filter equipped with a bypass check valve to allow hydraulic fluid to bypass the filter element as the element becomes clogged.

C

capacity: The ability to hold or contain something.

cartridge assembly: A cartridge located in a vane pump that houses the vanes, rotor, and cam ring, which are all placed between two end plates.

case drain: A line or passage from the cam ring area in a hydraulic device that carries hydraulic fluid leakage.

cavitation: A localized gaseous condition within a stream of fluid, which occurs when pressure is reduced to vapor pressure.

centrifugal air compressor: A type of dynamic air compressor that uses an impeller rotating at high speed to compress air.

Charles's law: A gas law that states that when the pressure of a gas is constant, the volume of the gas has a direct relationship with its absolute temperature.

check valve: A valve that allows fluid flow in one direction, but stops it in the opposite direction.

clean-out cover: A device used to access a reservoir when it requires cleaning from the build-up of solid materials.

clevis: A steel or iron U-shaped device with holes in the ends that receive a pin.

coalescing filter: A device that removes submicron solids, water vapor, and oil vapor by combining small droplets into larger droplets.

compression fitting: A type of fitting where the seal is created by a fitting component that is deformed to fit the tubing without threads.

condensation: The change of a substance from a vapor state to a denser liquid state.

conductor: 1. A material that has low resistance and permits electrons to move easily through it. 2. A device used to transmit hydraulic fluid between various components in a hydraulic system.

contact: A conducting part of a switch that operates with another conducting part of the switch to make or break a circuit.

contaminant: A foreign substance in the hydraulic fluid that is not meant to be a part of fluid.

counterbalance valve: A normally closed pressure control valve that stops a vertically mounted cylinder from running away because of the load weight.

cracking pressure: The amount of pressure required to slightly unseat a ball or a poppet and start to allow fluid flow through the check valve.

crescent seal: A crescent-moon-shaped seal between the gears and between the inlet and outlet sides of an internal gear pump.

cup seal: A piston seal that has a sealing edge formed into a lip.

current (I): The amount of electrons flowing through an electrical circuit.

custom-designed power unit: A hydraulic power unit that is designed for a specific use.

cutaway diagram: A diagram showing the internal details of components and the path of fluid flow.

cylinder bellows: An expandable and retractable protective cover that is attached to the rod.

cylinder cushion: A tapered plug attached to the piston or rod that fills an exit hole on either end of the cylinder and is used to prevent the piston from colliding with the inside surfaces of the end caps.

D

deliquescent dryer: A pneumatic device that removes moisture through absorption.

depth-type element: 1. A filter element that is composed of multiple layers of fabric or threaded material. 2. A pneumatic filter element that has a deep, vertical filter.

desiccant air dryer: A device that removes water vapor from compressed air through adsorption.

detent: An actuator that is used to hold a directional control valve spool in selected positions.

dew point (saturation temperature): The temperature at which the moisture in the air begins to condense.

diaphragm cylinder: A pneumatic cylinder that uses a plastic, metal, or rubber diaphragm to extend the rod.

diaphragm gas-charged accumulator: An accumulator with a flexible diaphragm separating the gas and the hydraulic fluid.

digital flow meter: A flow meter that uses electrical signals to indicate the flow rate of a fluid on a digital display.

digital multimeter (DMM): A portable meter that can measure more than one electrical property and display the measured properties as numerical values.

digital pressure gauge: A pressure gauge that converts fluid pressure into an electrical signal.

direct-acting check valve: A check valve that is directly activated or moved by fluid flow from the primary port.

direct-acting pressure control valve: A valve that senses pressure through an internal pilot line connected to the inlet or the outlet of the valve in order to direct fluid flow to the bottom of its spool where it works against a biasing spring.

direct-acting relief valve: A normally closed hydraulic system safety valve that sets the maximum system pressure and must be capable of allowing full pump flow through to the reservoir.

direct current (DC): Current that flows in one direction only.

directional control valve: A valve that allows or prevents fluid flow to specific piping or system actuators.

directional control valve actuator: A mechanism that is used to move the position of the spool position in a directional control valve.

displacement: The volume of hydraulic fluid moved during each revolution of a pump's shaft.

distance: The extent of advance from one point to another.

double-acting cylinder: A cylinder that has fluid pressure flow alternately to both sides of the cylinder to extend and retract the piston.

double-rod cylinder: 1. A hydraulic cylinder that has a single piston and rod that protrudes from both end caps of the cylinder. 2. A pneumatic cylinder that has one piston and a rod that protrudes from both ends of the cylinder barrel.

double-vane oscillator: An oscillator that has two diametrically opposed vanes and stops.

drain plug: A threaded device installed at the lowest point of a reservoir to allow for the removal of hydraulic fluid.

dry air: Air free of water and oil vapor.

duplex cylinder: A hydraulic cylinder that consists of two or more in-line cylinders that do not have their rods connected to form two or more cylinders in one housing.

duty cycle: The percentage of time that a machine operates compared to not operating.

dynamic seal: A device that creates a positive seal between parts that move relative to one another.

E

earmuff: An ear protection device worn over the ears to reduce the level of noise reaching the eardrum.

earplug: A moldable device inserted into the ear canal to reduce the level of noise reaching the eardrum.

electrical control circuit: An electrical circuit that determines when the output component is energized or de-energized.

electrical energy: The flow of electrons (subatomic elements) from atom to atom.

electrical power source: A system that produces, transmits, distributes, and delivers electrical power to satisfactorily operate electrical loads designed to connect to the system.

electricity: The energy released by the flow of electrons in a conductor.

electron: A negatively charged particle that orbits the nucleus of an atom.

emergency stop pushbutton: An NC switch that shuts off power to the electrical control circuit.

end-to-beginning troubleshooting method: A fluid power troubleshooting method that traces problems from the end of the fluid power system (actuator) to the beginning (prime mover, hydraulic pump, or air compressor).

energy: The capacity to do work.

exerted force: The amount of weight in pounds a fluid power system must produce to move an object.

exhaust flow: The fluid flow from the cylinder back through the directional control valve to the reservoir.

external drain line: A line that is externally connected to a pressure control valve and reservoir and used to drain fluid from the top of the spool where the secondary port of the valve is pressurized.

external gear pump: A gear pump that consists of two externally toothed gears that form a seal within the pump housing.

F

face shield: An eye and face protection device that covers the entire face with a plastic shield and is used for protection from flying objects.

fastener: A mechanical device used to attach two or more members in position, or join two or more members.

ferrule: A metal sleeve used for joining one piece of tubing to another.

filter: A porous device used to separate solid material from a liquid or gas.

filter-regulator-lubricator (FRL): A pneumatic device that is used to filter, regulate, and lubricate compressed air before it reaches the point of work.

fixed cushion: A tapered plug attached to the piston or rod that fills an exit hole on either end of the cylinder.

fixed cylinder mount: A mounting method that holds a hydraulic cylinder rigidly in place, only allowing rod movement.

fixed-displacement pump: A positive-displacement pump where the fluid flow rate (gpm) cannot be changed.

fixed orifice: An orifice that cannot be adjusted.

flange mount: A fixed rectangular-shaped plate that is attached to either end cap in the cylinder and is used to bolt the cylinder into position.

flared fitting: A fitting that is connected to tubing in which the end is spread outward.

flow: The movement of fluid through piping in gallons per minute (gpm) for hydraulic systems and in standard cubic feet per minute (scfm) in pneumatic systems.

flow control valve: A valve whose primary function is to control the rate (gpm) of fluid flow.

flow divider: A valve that divides flow from a single source equally to two or more sections or components of a hydraulic system.

flow meter: A meter that measures the flow of hydraulic fluid within a system.

flow rate: The amount of fluid that passes a given point in one minute.

fluid: A liquid or a gas that takes the shape of its container.

fluid flow: The movement of fluid caused by a difference in pressure between two points.

fluid power: The technology of using a fluid to transmit power from one location to another.

fluid power system efficiency: The useful amount of output energy from a fluid power system compared to the amount of input energy.

fluorocarbon (Viton®): A dense elastomeric compound commonly used as a sealing material in hydraulic systems.

flush-side mount: A fixed cylinder mounting method where both end caps in the cylinder have predrilled holes through them for bolting the cylinder into position.

force: Anything that changes or tends to change the state of rest or motion of a body.

force multiplication system: A system that allows the force applied on one cylinder to be increased on another cylinder.

friction: The resistance to movement between two mating surfaces.

G

gallons per minute (gpm): A measure of fluid flow that is used to measure small volumes of intermittently flowing fluids such as pump discharges.

gas-charged accumulator: An accumulator that applies force to a hydraulic fluid by using compressed gas.

gasket: A seal used between machined parts or around the circumference of flange pipe joints to prevent hydraulic fluid leaks.

gate valve: A two-position valve with an internal gate between the two seats of the orifice.

gauge pressure: The amount of pressure above the existing atmospheric pressure and is used to measure pressure inside a closed fluid power system.

Gay-Lussac's law: A gas law that states that when the volume of a gas is held constant, the pressure exerted by the gas has a direct relationship to its absolute temperature.

gear motor: A hydraulic motor that has meshing gears, which are rotated by hydraulic fluid flow.

gear pump: A hydraulic pump that consists of gears that mesh together in various manners to create fluid flow.

general-purpose relay: An electrical switch operated by a magnetic coil.

gerotor pump: A gear pump that has an inner rotor that meshes with the gear teeth of an outer rotor.

globe valve: A valve with a disk that is raised or lowered over an orifice.

H

half-back troubleshooting method: A method for testing for problems in the middle of a hydraulic circuit and working backward or forward on the system.

heat exchanger: A mechanical device used to transfer heat from one fluid to another fluid.

holding circuit: An electrical circuit that allows a load to remain on after the switch that controls it is released.

horsepower (HP): A mechanical unit of measure equal to the force required to move 550 lb, 1 ft in 1 sec.

hose: A flexible tube used for carrying fluids under pressure in both hydraulic and pneumatic systems and allows for movement of components.

humidity: The amount of moisture in the air.

hydraulic cylinder: An actuator that moves its shaft in a straight line called a stroke.

hydraulic fluid: The liquid used in hydraulic systems to transfer energy.

hydraulic motor: A device that converts hydraulic energy into rotating mechanical energy.

hydraulic power unit: A self-contained unit that includes all the components required to create flow, regulate pressure, and filter hydraulic fluid.

hydraulic pump: A mechanical device that changes mechanical energy into hydraulic energy (fluid flow).

hydraulic system: A system that uses liquid under pressure to create movement.

hydraulic tester: A device that measures the pressure, flow rate, and temperature of hydraulic fluid in a system.

I

implosion: An inward bursting.

inline axial piston pump: A piston pump that consists of pistons in a rotating piston block parallel to the drive shaft.

inline filter: A pneumatic filter that removes contaminants from within a pneumatic system.

in-line mounting: A mounting method that uses tubing or threaded pipes that connect directly to the directional control valve.

inside diameter: The diameter of a pipe measuring between the inside walls.

insulator: A material that has a high resistance and resists the flow of electrons.

intake filter: A pneumatic filter that is used to remove contaminants from air (typically atmospheric air) before it travels into the inlet of the air compressor.

intake flow: The fluid flow from the reservoir through the filters to the pump.

intensifier: A device that converts low-pressure, high-flow-rate fluid to high-pressure, low-flow-rate fluid.

intercooler: A type of pipe that is used to connect the different stages of an air compressor to cool the air that travels between the stages.

intercooling: A process that removes a portion of the heat of compression while air is piped from one stage of compression to another.

interlocking circuit: A circuit in which one branch can be energized at a time with a limit switch that has both NO and NC contacts.

internal drain line: A line that is internally machined into the body of a pressure control valve and used to drain fluid from the top of the spool to the secondary port where the valve is not pressurized.

internal gear pump: A gear pump that consists of a small external drive gear mounted inside a large internal spur gear (ring gear).

K

kinetic energy: The energy of motion.

L

ladder diagram: A diagram that contains the logic for an application and is typically used to test and troubleshoot hardwired applications.

laminar flow: Fluid flow in a hydraulic system passage that is characterized by a slow, smooth movement of fluid in a straight path along the centerline of the passage.

law of conservation of energy: A law that states that energy can neither be created nor destroyed.

layline: A line mark or print on a hydraulic hose that indicates if an installed hose has been improperly twisted.

limit switch: An electrical device that uses a mechanical actuator to control its electrical contacts.

liquid: A relatively noncompressible fluid that can readily flow and assume the shape of a confined space.

load: An electrical device with a specific amount of resistance that allows current to flow through it to accomplish work.

lobe pump: A positive-displacement pump that has two external-driven, intermeshing, lobe-shaped gears.

lockout: The use of locks, chains, or other lockout devices to prevent the startup and operation of specific equipment.

logic valve: A valve that uses pressure signals at its input ports to determine when fluid flow will occur from its output ports.

lubricator: A device that injects a lubricant into the airflow.

lug mount: A fixed cylinder mounting method that has lugs attached to the side of each of the end caps.

M

maintenance: The practice of periodically inspecting and repairing equipment.

manifold subplate: A subplate that can be used for mounting multiple directional control valves of the same configuration.

manual actuator: A type of directional control valve actuator that requires a person for actuation.

mechanical energy: Machine energy.

mercury barometer: An instrument used to measure atmospheric pressure using a column of mercury (Hg).

meter-in: A type of flow control where a needle valve controls the fluid flow into an actuator to control its speed.

metering: The controlling of the rate of fluid flow and how the fluid flow is being accomplished.

metering valve: A pneumatic flow control valve that controls the amount of airflow through a specific line at a given time.

meter-out: A type of flow control where a needle valve controls the fluid flow out of an actuator to control its speed.

micron (μm): One-millionth of a meter and 39-millionths of an inch.

minimum hose bend radius: A standard given by the manufacturer that indicates the least amount of bend that can be used for a specific type of hose.

moisture: Small quantities of liquid condensation.

moisture separator: A pneumatic device that separates moisture from compressed air by forcing it to flow against baffles.

molecule: Matter that is composed of atoms and is the smallest particle that a compound can be reduced to while still possessing the chemical properties of that compound.

muffler: A device that is attached to an exhaust port of a pneumatic directional control valve to deaden the noise of air as it exhausts.

multistage air compressor: A type of air compressor where the compressor allows the airflow from the first stage to be sent to the second stage and so forth to allow for a more energy-efficient compression of air.

N

needle valve: An infinite-position flow control valve that has a narrowly tapered stem or needle positioned in-line with an orifice that is the same size as the stem or needle.

nitrile (NBR): An elastomeric compound commonly used as a sealing material in hydraulic systems because it has elasticity, tensile strength, and hardness.

noise reduction rating (NRR) number: A number that indicates how many decibels the noise level is reduced by.

nominal rating: A value that indicates the average size or larger contaminant that a filter can trap.

nomographic chart: A chart used to determine the ID of a conductor, flow velocity in fps, and fluid flow in gpm when two variables are found.

nonpositive-displacement air compressor: A type of dynamic air compressor that moves large volumes of air using high rotational speeds.

nonpositive seal: A seal that allows a small amount of hydraulic fluid to pass to provide lubrication between mating surfaces.

normally closed (NC) switch: An electrical control device that when actuated, does not allow current to flow through it.

normally closed (NC) valve: A valve that does not allow fluid flow from the pump port to an actuator port in the spring-actuated (normal) position.

normally open (NO) switch: An electrical control device that when actuated, allows current to flow through it.

normally open (NO) valve: A valve that allows fluid flow from the pump port to an actuator port in the spring-actuated (normal) position.

O

orifice: A restricted passage in a fluid power line or component and is used to control fluid flow or to create a pressure differential.

O-ring: A molded synthetic rubber seal with a circular cross section.

oscillator: A motor that moves back and forth over a fixed arc that is less than 360°.

outside diameter: The diameter of a pipe measuring across the outside walls.

P

packing: A bulk deformable material reshaped by compression.

packing nut: A mechanical device on a pneumatic actuator that applies tension to the seals or packing to prevent air leaks around the circumference of a rod or shaft.

parallel circuit: An electrical circuit with two or more paths for current to flow.

parallel flow path: A flow path configuration in which there are two or more paths for hydraulic fluid to flow.

part replacement troubleshooting method: A fluid power system troubleshooting method that replaces parts without the use of equipment diagnosis tools.

Pascal's law: A fluid power law that states that when a force is applied to a confined fluid, the force is felt throughout the fluid undiminished.

permanent chip magnet: A magnetic device placed in a reservoir to attract and hold ferrous metal particles that have contaminated the system but not been recovered by system filters or strainers.

personal protective equipment (PPE): Clothing and/or equipment worn by a worker to reduce the possibility of an injury.

pictorial diagram: A graphic representation that shows how devices interconnect in a fluid power system.

pickling: A method of removing scale and rust from metal through chemical treatment.

piggyback valve: A combination valve that consists of a small solenoid-actuated directional control valve that is used to control the main pilot-operated valve.

pilot choke: An assembly that is mounted between a main valve and a pilot valve that is used to slow or briefly delay reversals in spool movement.

pilot line: A line used to transmit pressure in a hydraulic system for control purposes.

pilot piston: A device that allows a directional control valve spool to shift quickly by placing the pilot piston in a pilot-pressure chamber near the spool in the main valve.

pilot pressure: A valve control method that uses fluid pressure from somewhere in the hydraulic system to control or shift a valve.

pilot pressure actuator: A mechanism that uses fluid pressure to shift a directional control valve spool.

pilot-operated check valve: A check valve that operates with a pilot line that allows or stops fluid flow in both directions when activated.

pilot-operated pressure control valve: A type of pressure control valve that is remote controlled and, instead of a spring, uses pilot pressure set at a level specified by a pilot valve to bias the main spool.

pilot-operated pressure-reducing valve: A type of pilot-operated relief valve that is used to control pressures in one leg of a system.

pilot-operated relief valve: A type of pilot-operated pressure control valve that is used to control the maximum pressure in a hydraulic system.

pipe fitting: A piece of pipe that is used to interconnect pipes and allows them to change direction.

pipe threader: A tool used to cut threads in steel pipe.

piping: A passage or series of passages in a fluid power system that is constructed of metal, plastic, or plasticized rubber and conforms to ANSI standards.

piston gas-charged accumulator: An accumulator with a floating piston acting as a barrier between the gas and the hydraulic fluid.

piston motor: A hydraulic motor that has internal pistons, which are extended by fluid flow to produce rotational movement.

piston pump: A hydraulic pump in which fluid flow is produced by reciprocating pistons.

piston stroke: The distance the piston and rod of the cylinder travel.

pivot cylinder mount: A cylinder mounting method that allows the entire hydraulic cylinder to pivot while the rod moves.

plug-in pressure gauge: A pneumatic device that can be added to or removed from a pressurized pneumatic system without removing pressure from the system.

pneumatic cylinder: An actuator that moves in a straight line using compressed air.

pneumatic system: A system that uses gas under pressure to create movement.

polarity: The positive (+) or negative (–) state of an object.

polytetrafluoroethylene (PTFE): Commonly referred to as Teflon®, a low-friction plastic material used as sealing in low-pressure hydraulic systems.

polyurethane: A hard, chemical-resistant plastic used as a sealing material in hydraulic systems.

poppet: A movable part within the valve body that separates the hydraulic fluid and the bias spring.

poppet check valve: A check valve that uses a poppet located between the biasing spring and the seat to block fluid flow.

port: The section of a directional control valve that connects a pipe, hose, or tube to the internal passages of the directional control valve.

portable filtration unit: A nonfixed filtration machine that filters hydraulic fluid as it is being transferred from a storage vessel to the reservoir.

position: A schematic representation of the direction in which a spool forces fluid to flow.

position box: A symbol that represents of the number of different positions that the spool of a directional control valve is capable of moving into.

positive-displacement air compressor: A type of air compressor that discharges a fixed quantity of compressed air with each cycle.

positive-displacement pump: A pump that has a positive seal between its inlet and outlet and moves a specific volume of hydraulic oil with each revolution of the shaft.

positive seal: A seal that does not allow any hydraulic fluid to pass.

power: The amount of work accomplished over a specific period of time.

power circuit: A circuit controlled by the electrical control circuit to accomplish work in a fluid power system.

precharge pressure: The pressure of the compressed gas in an accumulator prior to the admission of hydraulic fluid.

pressure: The resistance to flow.

pressure-compensated flow control valve: A flow control valve that changes flow due to changes in pressure before or after the value to keep flow constant from the orifice.

pressure drop: 1. The pressure differential between any two points in a hydraulic system or component. 2. The pressure differential between upstream and downstream airflow caused by resistance.

pressure gauge: An instrument used to measure pounds per square inch (psi) in a closed system.

pressure line filter: A type of filter that is designed to operate in high pressures and is placed before one or more hydraulic system components to protect them from contaminants in the hydraulic fluid.

pressure rating: The highest amount of pressure at which a pump can continually create flow without premature wear.

pressure-reducing valve: A normally open pressure control valve that controls pressure in one leg (one small part) of a system.

pressure switch: A switch that electrically controls the energizing or de-energizing of a prime mover when a set pressure has been reached.

pressure compensated variable-displacement inline axial piston pump: A piston pump in which the angle of the swash plate can be varied.

prime mover: A device that supplies rotating mechanical energy to a fluid power system.

programmable logic controller (PLC): A solid-state control device that can be programmed and reprogrammed to automatically control electrical systems in commercial and industrial facilities.

proportional actuator: A type of actuator that uses analog signaling to actuate a proportional directional control valve with a spool that can be placed in an infinite number of positions.

pseudocavitation: Artificial cavitation caused by air being allowed into the pump suction line.

Q

Quad-ring®: A molded synthetic rubber seal with a rounded-off, X-shaped cross section.

R

radial piston air motor: A type of air motor that operates at low speed and high torque with its pistons vertical to the shaft.

radial piston pump: A piston pump that consists of a cylinder barrel, pistons with shoes, a ring, and a valve block located perpendicular to the pump shaft.

ram cylinder: A single-acting cylinder that has a large piston and a rod with the same diameter as the piston, and is capable of producing large amounts of linear force during extension.

receiver: A specialized vessel that holds gas for future use in a pneumatic system.

reciprocating-piston air compressor: A type of positive-displacement air compressor that compresses air by extending and retracting a piston inside a cylinder.

refrigerated air dryer: A device designed to lower the temperature of compressed air to 35°F.

regulator: A normally open pressure valve that is used to control downstream pressure.

relief valve: A valve that protects a fluid power system from overpressure by setting a maximum operating pressure.

remote-controlled, pilot-operated relief valve: A type of pilot-operated relief valve that uses one or more external pilot valves to set the pressure(s) at which the relief valve will open.

remote-controlled counterbalance valve: A normally closed pressure control valve that uses an external pilot line to sense pressure from a remote section of the hydraulic system and counteract the pressure from a hanging cylinder load.

remote-controlled sequence valve: A pressure control valve that receives a pilot signal from a remote section of a hydraulic system.

remote intake filter: A pneumatic filter located away from the air compressor and attached to pneumatic piping.

reservoir: A tank that stores hydraulic fluid that is not in use by the hydraulic system.

resilience: The capability of a material to regain its original shape after being bent, stretched, or compressed.

resistance: The force that stops, slows, or restricts the movement of fluid or devices in a fluid power system.

resistance (R): The opposition to the flow of electrons.

restriction (orifice) check valve: A type of poppet check valve with an orifice placed in the center of the poppet to permit restricted fluid flow through the valve in the NC position.

return line filter: A type of filter that is designed to operate in low pressures and is connected to a main return line of the hydraulic system before the line enters the reservoir.

right angle check valve: A check valve that has its inlet and outlet ports set at a right angle to each other.

rodless cylinder: A pneumatic cylinder that has a cartridge attached to a piston that slides back and forth within the cylinder barrel.

rod wiper: The part of a cylinder that keeps foreign materials that have attached themselves to the rod from entering the cylinder and contaminating the hydraulic fluid.

rotary cylinder: A pneumatic cylinder that uses linear movement to create rotational movement.

rotary-screw air compressor: A type of positive-displacement air compressor that uses two intermeshing screw shafts or a screw shaft with two rotors to create airflow.

rotary-vane air compressor: A type of positive-displacement air compressor that uses vanes that slide in and out of a rotating rotor.

running torque: The torque produced by a rotating motor.

S

safety glasses: An eye protection device with special impact-resistant glass or plastic lenses, reinforced frames, and side shields.

safety valve: A normally closed pressure control valve that is not adjustable and is used as overpressure protection on pneumatic components.

saturated air: Air that holds as much moisture as it is capable of holding.

Saybolt viscometer: A test instrument used to measure fluid viscosity.

schematic diagram: A diagram that uses standardized lines, shapes, and symbols, also called a graphic diagram. Interconnecting lines represent the function of each component in a system.

Schrader gauge: A pressure gauge that uses fluid pressure to push a piston against a compression spring that is attached to a pointer.

seal: An airproof and/or fluidproof joint between two members.

sequence valve: A normally closed pressure control valve that makes various operations occur in an orderly manner, or one after another.

series circuit: An electrical circuit with two or more switches connected so that there is only one path for current to flow.

series flow path: A flow path configuration in which there is only one path for hydraulic fluid to flow.

series-parallel circuit: A circuit with a combination of series- and parallel-connected components.

shuttle valve (OR logic valve): A logic valve that permits fluid flow from the highest pressure of two different input signals.

sight gauge: A device used to allow a visual inspection of the hydraulic fluid level in a reservoir at any given time.

single-acting, spring-return cylinder: A hydraulic cylinder that uses fluid flow for extension and a spring for retraction.

single-acting cylinder: A cylinder that uses fluid flow for extension of movement and spring, gravity, or other mechanical means for retraction.

single-vane oscillator: An oscillator that has a cylindrical chamber in which a vane connected to a drive shaft rotates in a 280° arc.

solenoid: An electrical control device that converts electrical energy into linear mechanical energy when a current passes through a magnetic coil in the solenoid.

solenoid actuator: An actuator that uses electricity to actuate a directional control valve spool.

sound energy: Transmitted vibration that can be sensed by the human ear.

spool: An internal component of a directional control valve that is used to control fluid flow and to connect internal passages and ports.

spring actuator: A mechanism that uses one or more springs to move a directional control valve spool to normal position.

spring-loaded accumulator: An accumulator that applies force to a fluid by means of a spring.

stand-alone power unit: A self-contained hydraulic power unit.

standard air: Air that is at 68°F, has 36% humidity, and is at sea level (14.7 psia).

standard cubic feet per minute (scfm): A unit of measure that assumes a constant atmospheric environment for air before it is pushed into the system.

starting torque: The torque produced when a motor is started under a load.

static energy (potential energy): Stored energy ready to be used.

static seal: A device that creates a positive seal between parts that do not move relative to one another.

stop: A stationary barrier used to prevent the continuous rotation of an oscillator.

stop tube: A short hollow metal tube attached to the rod end of a hydraulic cylinder that changes the piston stroke.

strainer: A screen, mesh, or perforated obstruction used to separate a solid from a liquid.

subplate mounting: A mounting method where a directional control valve is mounted to a plate that attaches to system piping.

suction filter: A type of filter that is designed to operate in a vacuum and is connected to the hydraulic system before the pump and outside of the reservoir to protect the pump from contaminants in the reservoir.

surface-type element: 1. A filter element that is composed of a single closely woven fabric or cellulose layer. 2. A pneumatic filter element that allows compressed air to flow through the filter in a straight line to collect contaminants on its surface.

swash plate: An angled plate in contact with the piston heads that causes the pistons in the cylinders of a pump to extend and retract.

switch: An electrical control mechanism that connects and disconnects the power source from the load when it is actuated.

T

tagout: The process of attaching a danger tag to the source of power to indicate that the equipment may not be operated until the tag is removed.

tandem cylinder: A hydraulic cylinder that consists of two or more in-line cylinders with their rods connected to form a common rod.

telescoping ram cylinder: A ram cylinder that extends its rod to a length longer than the housing of the cylinder by extending it in stages.

temperature- and pressure-compensated flow control valve: A flow control valve that compensates for changes in hydraulic fluid temperature and pressure.

thermal energy: The addition of heat to make a particle or molecule move faster or the dissipation of heat that makes a particle or molecule move slower.

thread sealant: A material applied to the male pipe threads to ensure an airtight connection with the female pipe threads.

thread screw gauge: A hand tool used to determine the dimensions and pitch of round and cut threads.

three-position, four-way directional control valve: A directional control valve with three positions and four ports.

tie-rod mount: A fixed cylinder mounting method that has tie rods extending out of the cap end, the rod end, or both.

torque: A turning or twisting force that causes a shaft, or other object, to rotate.

total energy: The combined forces of different forms of energy.

trunnion: A pivot mechanism consisting of two cylindrical objects that protrude from an end cap.

tubing: An extruded tubular material used to convey fluids.

turbulent flow: Fluid flow in a hydraulic system passage that is characterized by a rapid movement of fluid in an erratic, nonlayered, and random pattern.

two-position, four-way directional control valve: A directional control valve with two positions and four ports.

two-position, three-way directional control valve: A directional control valve that has two positions and three ports.

two-position, two-way directional control valve: A directional control valve with two positions and two ports.

U

unbalanced vane pump: A fixed- or variable-displacement hydraulic pump in which the pumping action occurs in the chambers on one side of the rotor and shaft.

union: A fitting used to connect or disconnect two tubes or two pipes that would otherwise require the removal of other pipes for repair.

unloading: A process that allows a compressor to run against no load.

unloading valve: A normally closed pressure control valve that can set maximum system pressure and unload the pump, which operates under 100 psi, at the same time.

U-ring seal: A cup seal with a U-shaped cross section.

V

vacuum: Any pressure less than atmospheric pressure.

vacuum cup: A device made of rubberized plastic that creates a vacuum when in contact with flat surfaces.

valve: A device that controls the pressure, direction, and/or rate of fluid flow.

valve body: A housing for valve components.

vane air motor: A type of motor that allows airflow to work against vanes to cause rotation of the shaft.

vane motor: A hydraulic motor that contains a rotor with vanes, which are rotated by fluid flow.

vane oscillator: A type of oscillator that has a fixed rotation, typically either 100° or 280°.

vane pump: A hydraulic pump that creates a vacuum by rotating a rotor inside a cam ring while trapping fluid between vanes that expand and retract from the rotor while moving the fluid toward the output.

variable cushion: A needle flow control valve with a check valve in the rod-end end cap, cap-end end cap, or both and a tapered plug attached to the piston or rod to slow the piston at the end of its stroke.

variable-displacement, pressure-compensated vane pump: A pump that automatically adjusts the amount of volume it displaces per rotation by centering the rotor when the pressure in the system starts to build.

variable-displacement pump: A positive-displacement pump that can have its flow rate (gpm) changed.

variable orifice: An orifice that allows an adjustable amount of fluid flow.

varnish: A coating of a sludge-like residue that results from the breakdown of hydraulic fluid from the heat and pressure within a hydraulic system.

velocity: The speed of fluid flow through a hydraulic line.

venturi valve: A valve that creates the vacuum supplied to vacuum cups for use in pneumatic equipment applications.

viscosity: A measurement of the resistance to flow against an established standard.

viscosity index (VI): A number that represents how much the viscosity of a fluid changes in respect to a specific change in temperature.

voltage (V): The electrical pressure, or electromotive force, that causes electrons to move in an electrical circuit.

volume: The size of a space or chamber measured in cubic units.

volumetric efficiency: The relationship between actual and theoretical fluid flow, or pump gpm.

V-ring seal: A cup seal with a V-shaped cross section.

W

wall thickness: The difference between the inside diameter and outside diameter divided by two.

way: The number of piping ports in a directional control valve.

weight-loaded accumulator: An accumulator that applies force to a fluid by means of heavy weights.

witness mark: A mark that is used as a guide to join two parts together.

work: The movement of an object (in lb) through a distance (in ft).

working pressure: The measure of the force applied to a given area.

Index

Page numbers in italics refer to figures.

A

abrasion, 303
absolute pressure, *39*, 39–40, 337
absolute ratings, 304, *305*
absolute temperature, 337
absolute zero, *337*, 337
absorption, 404
AC, 418
acceleration, 94–95
accumulators, 71–74, *73*
  actuators, 28–29, *30*. *See also* directional control valve actuators
  hydraulic. *See* hydraulic cylinders; hydraulic motors
  valve actuators, *250*, 250
adsorption, 402
aftercoolers, 401–402
air, *344*, 344
air compressor control, 369–371
air compressors, *345*, 345–346, 360–361, *361*
  control, 369–371
  cylinder cooling, *364*, 364
  types, *362*, 362–367
air-cooled heat exchangers, 312, *313*
air dryers, 402–404, *403*
air filter-regulators, *407*, 407
airflow, 345–346, *347*
airflow restrictions, 348
air leaks, 463–464, *465*
air motors, 373, 380–381, *381*
air speed, 347
alternating current (AC), 418
ambient air, 394
American wire gauge (AWG), 424, *425*
ampere, 418
AND logic valves, *180*, 180–181
area, *34*, 34, 36–37, *63*
arm protection, 8–9
associations, *13*, 13–14
atmospheric air, 394
atmospheric pressure, *38*, 38
atoms, 28
automated actuators, 174
automatic detents, 178
AWG, 424, *425*
axial piston air motors, 380, *381*

B

back pressure, 310
back-up rings, *239*, 239
baffles, 317
balanced vane pumps, *132*, 132
ball check valves, 156
ball valves, *202*, 202
beginning-to-end troubleshooting method, 465
bellows, 344
bent-axis piston pumps, 136, *137*
beta ratios, *306*, 306
bias spring adjustment screws, 69
bias springs, 69
bio-hydraulic fluids, 3
bladder gas-charged accumulators, 73
bleed-off, 208
boosters, 75
Bourdon tube pressure gauges, *41*, 41–42
Bourdon tubes, 41
Boyle's law, 337–338, *339*
brake valves, 278, *279*
breakaway torque, 246
breather/filler cap assemblies, 302, *317*, 317
bypass filters, 308, *309*

C

capacity, *63*, 64–65
cardboard compactor operation, *430*, 430–432, *431*, *432*
cartridge assemblies, *132*, 132
case drains, 248
cavitation, 138, *139*
center position selection, 172
centrifugal air compressors, 362–363
certifications, 13–14, *14*
cfm ratings, 347–348
Charles's law, 338, *339*
check valves, 107, 156, *158*
  direct-acting, *156*, 156–158
  pilot-operated, 158–159, *159*, *160*
chemical analyses, *311*
circuits. *See* electrical circuits; electrical control circuits
cleaning hydraulic fluid, *317*, *318*
clean-out covers, *319*, 319
clevises, *243*, 244
closed-center positions, *169*, 169–170, *170*
closed fluid power systems, *2*, 2
clothing, 9
coalescing filters, *400*, 400–401
combined gas law, 340–342, *342*
compressed air, *408*, 408
compressing air, 344–345, *345*
compression fittings, *104*, 104

compression seals, *241*, 241
compressors. *See* air compressors
condensation, 395
conductors, 94, 97, 110, 419
  hoses, 105–108, *106*, *108*
  piping, *97*, 97, 99–102, *101*
  tubing, 102–105, *105*
connectors, 108–109
contacts, 420
contaminants, 298–303, *311*, 394–396
  analysis reports, *311*
  effects of, 303
  particle sizes, 298, *300*, *303*
  pneumatic systems, *394*, 394–396, *397*
  sources, 298–303
control devices, 368, 372–375
  directional control valves, 372–374, *373*
  flow control valves, 374–375
cooling hydraulic fluid, 317, *318*
counterbalance valves, *270*, 270–271, 276–278, *277*
cracking pressure, 157
crescent seals, 127
cubic feet per minute (cfm) ratings, 347–348
cup seals, *240*, 240–241
current, *418*, 418, *420*
custom-designed power units, *78*, 78
cutaway diagrams, 78
cylinder bellows, *242*, 242
cylinder cooling, *364*, 364
cylinder cushions, 235, 375, *376*
cylinders, 375–379, *376*. *See also* hydraulic cylinders

D

DC, 418
deliquescent dryers, 404
depth-type elements, *305*, 305, 399
desiccant air dryers, 402
detents, 178, *179*
dew point, 396
diagrams, 78–80, *79*, *80*
diaphragm cylinders, 378–379, *379*
diaphragm gas-charged accumulators, 73
digital flow meters, 66
digital multimeters (DMMs), *468*, 468
digital pressure gauges, *43*, 43–44
direct-acting check valves, *156*, 156–158
direct-acting pressure control valves, 266
  brake valves, 278, *279*
  counterbalance valves, *270*, 270–271, 276–278, *277*
  direct-acting relief valves, 266, *267*
  pressure-reducing valves, *271*, 271–272
  remote-controlled counterbalance valves, 276–278, *277*
  remote-controlled sequence valves, *276*, 276
  sequence valves, 267–268, *268*, *269*, *276*, 276
  unloading valves, *272*, 272–275, *274*, *275*
direct-acting relief valves, 266, *267*
direct current (DC), 418
directional control valve actuators, 174
  detents, 178, *179*
  pilot pressure, *175*, 175–176
  proportional, 179–180
  solenoid, *176*, 176–178, *177*
  spring, *174*, 174–175
directional control valves, *161*, 161
  mounting methods, *173*, 173–174
  naming standards, *165*, 165
  pneumatic, 372–374, *373*
  schematic diagrams for, 161–165, *162*, *163*, *164*, *165*
  three-position, four-way, 168–172, *169*
  two-position, four-way, 167, *168*
  two-position, three-way, *167*, 167
  two-position, two-way, 165, *166*
displacement, 119
distance, 31
DMMs, *468*, 468
double-acting cylinders, *233*, 233–238
  control applications, 432–438, *433*, *434*, *435*, *436*, *437*
  hydraulic systems, 209–212, *210*, *211*
  for pneumatic systems, 377–378, *378*
double pumps, *140*, 140
double-rod cylinders, 234, *235*, 378, *379*
double-vane oscillators, 250
drain lines, *266*, 266
drain plugs, *319*, 319
dry air, 402
duplex cylinders, *234*, 234, 236
duty cycles, 369
dynamic seals, 238

E

ear protection, *9*, 9
efficiency, 29–30
electrical circuits, 420–422, *421*, *422*
electrical control circuit components, *424*, 424–429
electrical control circuits, 422, *423*, *468*, 468–469
electrical energy, 28
electrical power sources, 424
electric applications, 429
  double-acting cylinders, 432–438, *433*, *434*, *435*, *436*, *437*
  single-acting cylinders, *430*, 430–432, *431*, *432*
electric diagrams, 429
  double-acting cylinders, *433*, *434*, *435*, *436*, *437*
  single-acting cylinders, *430*, *431*, *432*
electricity, 418
electronic displacement, 138, *139*
electrons, 28
emergency showers, *8*, 8
emergency stop pushbuttons, 431
end-to-beginning troubleshooting method, 465
energy, 26–28, *27*, *28*, *61*, 61
energy loss, 29
energy transmission, 28–29

exerted force, 31
exhaust flow, 79
external drain lines, *266*, 266
external gear pumps, *124*, 125, *126*
eye and face protection, *7*, 7–8
eyewash stations, *8*, 8

F

face shields, *7*, 8
Fahrenheit (°F), *337*, 337
fasteners, 36
ferrules, 104
filter ratings, 304–306, *305*, *306*
filter-regulator-lubricators (FRLs), *404*, 404–407
filters, *304*, 304, 306–310, 396–401
  hydraulic systems, 306–310, *307*, *308*, *309*
  pneumatic systems, 396–401, *399*, *400*
fixed cushions, 236–238, *237*
fixed cylinder mounts, *243*, 243
fixed-displacement pumps, *121*, 121
fixed orifices, 198
flange mounts, *243*, 244
flared fittings, *103*, 103–105, *104*
flareless tube fittings, *105*
float-center positions, *169*, 171, *172*
flow, *62*, 62
flow control, 196
flow control valves, 198–199, *199*, 374–375
  ball, *202*, 202
  gate, 200–201, *201*
  globe, *200*, 200
  needle, 202, *203*
  pressure-compensated, 203–204, *204*
  temperature- and pressure-compensated, 204–205, *205*
flow meters, 66, *67*
flow paths, 62–63, 66, *68*, 68
flow rates, 66–68
fluid. *See* hydraulic fluid
fluid flow, *62*, 62–63, 94–95
  acceleration, 94–95
  resistance, 66–68, *67*, *68*
  velocity of, *94*, 94–95
fluid power, 2
fluid power circle, *35*, 35
fluid power formulas, *35*, 35–37
fluid power system efficiency, 29–30
fluid power systems, 4–6, 454–468
  advantages of, 6
  applications, 4–6, *5*
  disadvantages of, 6
  maintenance, 454–460, *455*, *461*
  troubleshooting, 460–468
    hydraulic systems, 461–462, *463*
    methods, 464–466
    pneumatic systems, 462–464, *464*
    procedures, 466–468, *467*, *468*
fluid power variables, 33
fluids, 2
fluid storage, 315, *316*
fluorocarbon, 242
flush-side mounts, *243*, 244
foot-pounds (ft-lb), 31
foot protection, *10*, 10
force, 30–31, 33
force multiplication systems, *74*, 74
formulas, *35*, 35–37
friction, *30*, 30, 66, *67*
FRLs, *404*, 404–407
future of fluid power, 3–4

G

gallons per minute (gpm), *119*, 119
gas-charged accumulators, 72–74
gaskets, *241*, 241
gas physics, 336–342
  absolute pressure, 337
  absolute temperature, 337
  Boyle's law, 337–338, *339*
  Charles's law, 338, *339*
  combined gas law, 340–342, *342*
  Gay-Lussac's law, 340, *341*
  volume, *336*, 336
gate valves, 200–201, *201*
gauge pressure, *39*, 39
Gay-Lussac's law, 340, *341*
gear motors, 246–247, *247*
gear pumps, *124*, 124–129, *129*
general-purpose relays, *428*, 428
gerotor pumps, *124*, 124, 128
globe valves, *200*, 200
gloves, 8–9, *9*
goggles, *7*, 8
graphic diagrams. *See* schematic diagrams

H

hand protection, 8–9, *9*
hard hats, 6–7, *7*
hazards, 12
head protection, 6–7, *7*
hearing protection, *9*, 9–10
heat exchangers, 312–314, *313*, *314*
helmets, 6–7, *7*
history of fluid power, 2
holding circuits, 422, *423*
horsepower control, 138
horsepower (HP), *32*, 32
hoses, 105–108, *106*, *108*
humidity, 395
hydraulic actuators. *See* hydraulic cylinders; hydraulic motors

hydraulic cylinders, 230–238
  double acting, *233*, 233–238
  mounting methods, *243*, 243–244
  ram, *230*, 230, *231*
  single acting, *230*, 230–232, *231*, *232*
hydraulic cylinders, *230*, 230, 456
hydraulic flow control, 196
hydraulic fluids, *60*, 60, 300–302, 306–310, 314–318
  cleaning, 317, *318*
  cleanliness ratings, 310, *311*
  cooling, 317, *318*
  filtration, 300–302, *301*, 306–310, *307*
  heating, 314
  maintenance, 456, *457*
  receiving, 318
  sending, 318
  storage, 315, *316*
hydraulic hoses, 457–458, *458*
hydraulic motors, 244–250, *245*
  torque produced, 244–246, *245*
  types of, 246–250
hydraulic power units, *76*, 76–78, *77*
hydraulic principles, 60
hydraulic pump ratings, 118–120
hydraulic pumps, *118*, 118–123
hydraulic pump schematic symbols, 140, *141*
hydraulic pump types, 124–140, *142*
  external gear pumps, *124*, 125, *126*
  gear pumps, *124*, 124–129, *129*
  piston pumps, 133–138
  vane pumps, *129*, 129–133, *130*, *132*, *133*
hydraulic systems, *2*, 2, 26–29, 33, 454–468
  characteristics, *29*
  energy transmission, 26, *27*
  heat generation, 312
  maintenance, 454–460, *455*
  pressure supplements, 71
  safety, 12
  troubleshooting, 461–462, *463*, 464–468, *467*, *468*
hydraulic testers, *461*, 461–462
hydraulic vs. pneumatic cylinders, *376*

I

implosions, 138, *139*
industry associations, *13*, 13–14
inline axial piston pumps, *134*, 134
inline filters, 398–401, *399*
in-line mounting, 173–174
inside diameter (ID), 99
insulators, *420*, 420
intake filters, 397–398, *398*
intake flow, 79
intensifiers, *74*, 74–76, *75*
intercoolers, 364
intercooling, *401*, 401–402
interlocking circuits, *433*, 433
internal drain lines, *266*, 266
internal gear pumps, *124*, *127*, 127

K

kilograms, 33
kilopascals (kPa), 38
kinetic energy, 26, *27*

L

ladder diagrams, 428–429, *429*
laminar flow, *62*, 62
law of conversation of energy, 26
laylines, *108*
limit switches, 421, *432*, 432
lines, *81*, 81
liquids, 33, *60*, 60–61, *61*
loads, *426*, 426
load sensing, *138*, 138
lobe pumps, *124*, 125, *126*
lockout, 10, *11*
logic valves, *180*, 180–181, *181*
lubricators, *406*, 406–407
lug mounts, *243*, 244

M

maintenance, 454–460, *455*, *461*
manifold subplates, 173
manual actuators, 174
maximum resistance, 70, *71*
mechanical accumulators, 72–73
mechanical energy, *28*, 28
mercury barometers, *40*, 40–41
metal bending machine operation, 436–438, *437*
metal shearing press operation, *435*, 435
metal surface preparation, *459*, 459
meter-in, 207, 210, *212*
metering, 198, *207*, 207–212, *212*
  bleed-off systems, *213*, 213
  double-acting cylinders, 209–212, *210*, *211*
  single-acting cylinders, *208*, 208–209, *209*
metering valves, *374*, 374
meter-out, 207, 210–212, *212*
microns (μm), 298, *300*
minimum hose bend radius, 457
moisture, 395
moisture separators, 402, *403*
molecules, 60, *336*, 336
motor torque, 244–246, *245*
mounting hydraulic cylinders, *243*, 243–244
mufflers, *373*, 373–374
multistage air compressors, 364–365, *365*

N

NBR, 242
NC limit switches, *432*, 432
NC pressure switches, *431*, 431
NC switches, *431*, 431, 432, 426–428, *427*
NC valves, 165
needle valves, 202, *203*
neoprene gloves, *9*, 9
nitrile (NBR), 242
noise reduction rating (NRR) number, 10
nominal ratings, 304, *305*
nomographic charts, 97, *98*
nonpositive-displacement air compressors, *362*, 362
nonpositive seals, 104, 238
normally closed (NC) switches, 426–428, *427*, *431*, 431, 432
normally closed (NC) valves, 165
normally open (NO) switches, 426–428, *427*
normally open (NO) valves, 165
NRR number, 10

O

oil contamination, 396, *397*
oil-flooded compressing mechanisms, *367*, 367
open-center positions, *169*, 169, *170*
orifice check valves, *157*, 157
orifices, 157, 198, *199*
O-rings, *239*, 239
OR logic valves, *181*, 181
oscillators, *249*, 249–250
outside diameter (OD), 99

P

packing, 241
packing nuts, 464
parallel circuits, *421*, 421
parallel flow paths, 63
particles, 298, *300*, *303*
part replacement troubleshooting method, 465
Pascal's law, *33*, 33
pascals (Pa), 38
permanent chip magnets, *319*, 319
personal protective equipment (PPE), 6–10, *7*, *9*, *10*
pickling, 110
pictorial diagrams, *78*, 78
piggyback valves, 176–177, *177*
pilot chokes, 177–178, *178*
pilot lines, 158
pilot-operated check valves, 158–159, *159*, *160*
pilot-operated pressure control valves, 280–283
pilot-operated pressure-reducing valves, 283
pilot-operated relief valves, *280*, 280–282, *281*, *283*
pilot pistons, 178, *179*
pilot pressure, 175
pilot pressure actuators, *175*, 175–176
pilot-to-close check valves, 159, *160*
pilot-to-open check valves, 158–159, *159*
pipe fittings, 100
pipe threaders, *100*, 100
piping, 97, *99*, 99–102, *101*
piston gas-charged accumulators, 72
piston motors, 247–249, *248*
piston pumps, 133–138
piston speed, *206*, 206–207
piston strokes, 238
pivot cylinder mounts, *243*, 244
PLC programming ladder diagrams, *429*, 429
PLCs, 179–180, 428–429
plug-in pressure gauges, 464, *465*
pneumatic conditioning devices, 401–407
  air dryers, 402–404, *403*
  filter-regulator-lubricators (FRLs), *404*, 404–407
  intercoolers, *401*, 401–402
  moisture separators, 402, *403*
pneumatic cylinders, 375–379
pneumatic filters, 396–401, *398*, *399*, *400*
pneumatic stamping press operation, 433
pneumatic systems, 2, 33, 342–344, *343*
  characteristics, *29*
  compression and control, *360*, 360–361, *361*
  contaminants, *394*, 394–396
  maintenance, 460, *461*
  safety, 12
  troubleshooting methods, 462–466, *464*
  troubleshooting procedures, 466–468, *467*, *468*
pneumatic thermoforming press operation, *434*, 434–435
pneumatic vs. hydraulic cylinders, *376*
polarity, 418
polytetrafluoroethylene (PTFE), 241
polyurethane, 241
poppet check valves, 157
poppets, 69
portable filtration units, *301*, 301–302
ports, 162
position boxes, 162, *164*
positions, 165, *169*, 169–172, *170*, *171*, *172*
positive-displacement air compressors, *362*, 363
positive-displacement pumps, 120–123, *123*
positive seals, 104, 238
pounds, 33
pounds per square inch gauge (psig), *39*, 39
pounds per square inch (psi), 34
power, 31–32, *32*
power circuits, 422
PPE, 6–10
precharge pressure, 74
pressure, 34–40, *35*, *36*, 69, 340–342
  and resistance, 69
  example calculations, 36–37
  in fluid power circle, *35*, 35
  in Gay-Lussac's law, 340, *341*

pressure (*continued*)
in the combined gas law, 340–342
types of, 38–40
pressure-compensated flow control valves, 203–204, *204*
pressure-compensated variable-displacement inline axial piston pumps, *135*, 135–136
pressure control, 69
pressure differential, 196–198, *197*
pressure drop, 66, 396
pressure gauges, 38
pressure line filters, 307–308, *308*
pressure measurement, *40*, 40–44
pressure ratings, 120
pressure-reducing valves, *271*, 271–272
pressure supplements, 71
pressure switches, 369–370, *370*, *431*, 431
prime movers, *119*, 119, 361, 365
programmable logic controllers (PLCs), 179–180, 428–429
proportional actuators, 179–180
pseudocavitation, 138
psi, 34
psig, *39*, 39
PTFE, 241

Q

Quad-rings®, 239–240, *240*

R

rack and pinion valve actuators, *250*, 250
radial piston air motors, *381*, 381
radial piston pumps, 136, *137*
ram cylinders, *230*, 230, *231*
Rankine (°R), *337*, 337
receivers, 337, *368*, 368
reciprocating intensifiers, 76
reciprocating-piston air compressors, *363*, 363–364
refrigerated air dryers, 402
regulators, *405*, 405
relief valves, 69, *70*, 371, *372*
remote-controlled counterbalance valves, 276–278, *277*
remote-controlled pilot-operated relief valves, 281–282, *282*
remote-controlled sequence valves, *276*, 276
remote intake filters, 397
reservoirs, 315–319, *316*, *318*, *319*
resilience, 240
resistance, 30
electrical, 419–420, *420*
and pressure, 69–70, *71*
resistance to fluid flow, 66–68, *67*, *68*
restriction check valves, *157*, 157
restrictions to airflow, 348
return line filters, 308–310, *309*
right angle check valves, 157
rodless cylinders, 378, *379*
rod wipers, *242*, 242
rotary cylinders, 379, *380*
rotary-screw air compressors, *366*, 366–367
rotary-vane air compressors, 365–366, *366*
running torque, 246

S

safety, 6, 12
hazards, 12
lockout devices, 10, *11*
personal protective equipment (PPE), 6–10, *7*, *9*, *10*
rules, 12
tagout devices, 10, *11*
training programs, 12
safety glasses, *7*, 8
safety shoes, *10*, 10
safety valves, 370–371, *371*
saturated air, 396, *397*
Saybolt viscometers, *65*, 65
scfm, 347
schematic diagrams, *80*, 80, 429–437
double-acting cylinders, *433*, *434*, *435*, *436*, *437*
single-acting cylinders, *430*, *431*, *432*
schematic symbols, 161–165, *163*, *164*, *165*
Schrader gauges, 42–43, *43*
seals, 123, 238–242, *239*
sequenced hydraulic press operation, *436*, 436
sequence valves, 267–268, *268*, *269*, *276*, 276
series circuits, *421*, 421
series flow paths, 62
series-parallel circuits, *422*, 422
showers, emergency, *8*, 8
shuttle valves, *181*, 181
sight gauges, 318, *319*
single-acting cylinders, *230*, 230–232, *231*, *232*, 430–432
applications, *430*, 430–432, *431*, *432*
for pneumatic systems, 376–377, *377*
hydraulic systems, *208*, 208–209, *209*
single-acting spring-return cylinders, *232*, 232
single stroke intensifiers, 75
single-vane oscillators, 249
solenoid actuators, *176*, 176–178, *177*
solenoids, 176, *426*, 426
sound energy, 28
spools, *161*, 161
spring actuators, *174*, 174–175
spring-loaded accumulators, 72
spring-loaded piston (Schrader) gauges, 42–43, *43*
STAMPED, 106
stamping press operation, 433
stand-alone power units, 77
standard air, 396
standard cubic feet per minute (scfm), 347
starting torque, 246
static energy, 26, 27
static seals, 238
stops, 249
stop tubes, *238*, 238
strainer ratings, 305–306

strainers, *304*, 304
subplate mounting, *173*, 173
suction line filters, 306–307, *307*
surface-type elements, *305*, 305, 399, *400*
swash plates, 134
switches, 426–428, *427*, *431*, 431, 432
symbols, *81*, 81
synthetic oil baths, 367

T

tagout, 10, *11*
tandem-center positions, *169*, *171*, 171
tandem cylinders, *234*, 234, *235*
tanks. *See* receivers
Teflon®, 241
telescoping ram cylinders, 232, *233*
temperature- and pressure-compensated flow control valves, 204–206, *205*
thermal energy, 26
thermoforming press operation, *434*, 434–435
threading pipes, 100–102
thread pitch, 109, *110*
threads, 109–110, *110*
thread screw gauges, *109*
thread sealants, 100–101, *102*
three-position, four-way directional control valves, 168–172, *169*
tie-rod mounts, *243*, 244
torque, 244–246, *245*
torque control, 138
total energy, 26, *27*
trade associations, *13*, 13–14
transmission
  hydraulic means of
    hoses, 105–110
    pipes, 97–100
    tubes, 102–105
  of energy, *61*, 61
transmission of energy, 28–29
triple pumps, *140*, 140
troubleshooting, 460–468
  hydraulic systems, 461–462, *463*
  methods, 464–466
  pneumatic systems, 462–464, *464*
  procedures, 466–468, *467*, *468*
trunnions, *243*, 244
tubing, 102–105, *105*
turbulent flow, *62*, 62
two-position, four-way directional control valves, 167, *168*
two-position, three-way directional control valves, *167*, 167
two-position, two-way directional control valves, 165, *166*

U

unions, 102
unloading, 370
unloading valves, *272*, 272–275, *274*, *275*
U-ring seals, *240*, 241

V

vacuum, 38, *44*, 44
vacuum cups, 381–382, *382*
valve actuators, *250*, 250
valve bodies, 69
valves, 69, *70*
vane air motors, 380, *381*
vane motors, 247, *248*
vane oscillators, 249
vane pumps, *129*, 129–133, *130*, *131*, *132*, *133*
variable cushions, 236, *237*
variable-displacement pressure-compensated vane pumps, *131*, 131
variable-displacement pumps, *122*, 122
variable orifices, 198, *199*
variables, 33
varnish, 303
velocity, *94*, 94, 95
venting remote-controlled, pilot-operated relief valves, 282, *283*
VI numbers, 66
viscometers, *65*, 65
viscosity, *65*, 65
viscosity index (VI), 66
Viton®, 242
voltage, 418, *419*
volume, *63*, 63, 336
  Boyle's law, 337–338, *339*
  Charles's law, 338, *339*
  combined gas law, 340–342, *342*
volumetric efficiency, *120*, 120
V-ring seals, *240*, 240–241

W

wall thicknesses, 99
water-cooled heat exchangers, 313–314, *314*
ways, 162, *163*
weight-loaded accumulators, 72–73
witness marks, *104*, 104
work, 30–31, *31*
work devices, 375
  air motors, 373, 380–381, *381*
  pneumatic cylinders, 375–379
  vacuum cups, 381–382, *382*
working pressure, 34

USING THE *FLUID POWER SYSTEMS* INTERACTIVE DVD

Before removing the Interactive DVD from the protective sleeve, please note that the book cannot be returned for refund or credit if the DVD sleeve seal is broken.

System Requirements

To use this Windows®-compatible DVD, your computer must meet the following minimum system requirements:

- Microsoft® Windows® 7, Windows Vista®, or Windows® XP operating system
- Intel® 1.3 GHz processor (or equivalent)
- 128 MB of available RAM (256 MB recommended)
- 335 MB of available hard disk space
- 1024 × 768 monitor resolution
- DVD drive (or equivalent optical drive)
- Sound output capability and speakers
- Microsoft® Internet Explorer® 6.0 or Firefox® 2.0 web browser
- Active Internet connection required for Internet links

Opening Files

Insert the Interactive DVD into the computer DVD drive. Within a few seconds, the home screen will be displayed allowing access to all features of the DVD. Information about the usage of the DVD can be accessed by clicking on Using This Interactive DVD. The Quick Quizzes®, Illustrated Glossary, Flash Cards, Interactive Schematics, Chapter Reviews, Media Library, and ATPeResources.com can be accessed by clicking on the appropriate button on the home screen. Clicking on the ATP logo (www.atplearning.com) accesses information on related educational products. Unauthorized reproduction of the material on this DVD is strictly prohibited.